AF477994

COATINGS TECHNOLOGY HANDBOOK

COATINGS TECHNOLOGY HANDBOOK

Edited by

D. SATAS

Satas & Associates
Warwick, Rhode Island

MARCEL DEKKER, INC.   New York · Basel · Hong Kong

Library of Congress Cataloging-in-Publication Data

Coatings technology handbook / edited by D. Satas
      p.   cm.
    Includes bibliographical references and index.
    ISBN 0-8247-8410-3
    1. Coating processes—Handbooks, manuals, etc. I. Satas,
Donatas.
    TP 156.C57C62 1991
    667'.9—dc20
                                     91-13981
                                         CIP

This book is printed on acid-free paper

MARCEL DEKKER, INC.
270 Madison Avenue, New York, New York 10016

Current printing (last digit):
10 9 8 7 6 5 4 3 2

PRINTED IN THE UNITED STATES OF AMERICA

Preface

Coatings are widely used in industry for many different purposes. They have permeated practically all areas of manufacturing, and it would be difficult to find a finished product that has escaped the application of a coating. Coatings are applied to protect bulk materials from corrosion and other detrimental effects of the ambient atmosphere. They are used to change surface properties and color, gloss, and general appearance. Adhesive coatings are used in laminating and in the preparation of composites. Coatings are used as barriers for gases and liquids, and for many other reasons. Application techniques, coating types, and their purposes make coating technology an extremely diverse field.

Since coating technology is transferable from one application to another, it was thought that a comprehensive handbook would be of interest and value to those involved in coating. A person practicing one aspect of coating technology may not be at all familiar with techniques or materials used for another application. Awareness and understanding of these techniques and materials will help solve problems and improve levels of technology.

This handbook provides short discussions of many of the various techniques, processes, and materials used for coatings. Its main purpose is to expose those practicing coating technology to techniques and materials used in related technologies, and perhaps offer some useful ideas.

We gratefully acknowledge the numerous contributors and the staff of Marcel Dekker, Inc., without whose work this book would not have been possible.

D. Satas

Contents

PART 2: COATING AND PROCESSING TECHNIQUES

CONTENTS

PART 4: SURFACE COATINGS

CONTENTS

Contributors

N. J. Abbott Albany International Research Company, Dedham, Massachusetts

Walter Alina General Magnaplate Corporation, Linden, New Jersey

Robert D. Athey, Jr. Athey Technologies, El Cerrito, California

Brian E. Aufderheide W. H. Brady Company, Milwaukee, Wisconsin

William F. Beach Bridgeport, New Jersey

Edward A. Bernheim Exxene Corporation, Corpus Christi, Texas

Deepak G. Bhat GTE Valenite Corporation, Troy, Michigan

Thomas P. Blomstrom Monsanto Chemical Company, Springfield, Massachusetts

Claire Bluestein Epolin, Inc., Newark, New Jersey

Kenneth Bourlier Union Carbide Corporation, Bound Brook, New Jersey

J. David Bower Hoechst Celanese Corporation, Somerville, New Jersey

Donald L. Brebner E. I. du Pont de Nemours & Company, Wilmington, Delaware

Patrick Brennan The Q-Panel Company, Cleveland, Ohio

George E. F. Brewer George E. F. Brewer Coating Consultants, Birmingham, Michigan

Lisa A. Burmeister Aqualon Company, Wilmington, Delaware

Peter A. Callais Pennwalt Corporation, Buffalo, New York

Naomi Luft Cameron Datek Information Services, Newtonville, Massachusetts

Robert W. Carpenter Windsor Plastics, Inc., Evansville, Indiana

Chi-Ming Chan Raychem Corporation, Menlo Park, California

Gary W. Cleary Cygnus Research Corporation, Redwood City, California

Murray S. Cohen Epolin, Inc., Newark, New Jersey

Carl A. Dahlquist 3M Company, St. Paul, Minnesota

B. Davis ABM Chemicals, Limited, Stockport, Cheshire, England

Richey M. Davis Hercules Incorporated, Wilmington, Delaware

David R. Day Micromet Instruments, Inc., Cambridge, Massachusetts

Marcel Dery Chemical Fabrics Corporation, Merrimack, New Hampshire

Arnold H. Deutchman BeamAlloy Corporation, Dublin, Ohio

Richard P. Eckberg General Electric Company, Schenectady, New York

Jesse Edenbaum Consultant, Cranston, Rhode Island

Carol Fedor The Q-Panel Company, Cleveland, Ohio

William C. Feist U.S. Department of Agriculture Forest Service, Madison, Wisconsin

R. H. Foster Eval Company of America, Lisle, Illinois

James D. Gasper ICI Resins US, Wilmington, Massachusetts

Sam Gilbert Sun Chemical Corporation, Northlake, Illinois

K. B. Gilleo Sheldahl, Inc., Northfield, Minnesota

F. A. Goossens Stork Brabant, Boxmeer, The Netherlands

Joseph Green, FMC Corporation, Princeton, New Jersey

William F. Harrington Jr. Uniroyal Adhesives and Sealant Company, Inc., Mishawaka, Indiana

J. Rufford Harrison E. I. du Pont de Nemours & Company, Wilmington, Delaware

Krister Holmberg Berol Kemi AB, Stenungsund, Sweden

Albert G. Hoyle Hoyle Associates, Lowell, Massachusetts

H. F. Huber Hüls Troisdorf AG, Troisdorf/Marl, Germany

Joseph L. Johnson Aqualon Company, Wilmington, Delaware

Stephen L. Kaplan Plasma Science, Inc., Belmont, California

Lisa C. Klein Rutgers University, New Brunswick, New Jersey

Artur Koch Ahlbrandt System GmbH, Lauterbach, Germany

Joseph V. Koleske Charleston, West Virginia

Alan Lambuth Boise Cascade, Boise, Idaho

Kenneth Lawson DeSoto, Inc., Des Plaines, Illinois

B. H. Lee Ciba-Geigy Corporation, Ardsley, New York

Peter A. Lewis Sun Chemical Corporation, Cincinnati, Ohio

Raimond Liepins Los Alamos National Laboratory, Los Alamos, New Mexico

H. Thomas Lindland Flynn Burner Corporation, New Rochelle, New York

Harry G. Lippert Extrusion Dies, Inc., Chippewa Falls, Wisconsin

Ronald A. Lombardi ICI Resins US, Wilmington, Massachusetts

Donald M. MacLeod Industry Tech, Oldsmar, Florida

Algirdas Matukonis Kaunas Technical University, Kaunas, Lithuania

John A. McClenathan IMD Corporation, Birmingham, Alabama

Christopher W. McGlinchey The Metropolitan Museum of Art, New York, New York

Frederic S. McIntyre Acumeter Laboratories, Inc., Marlborough, Massachusetts

Timothy B. McSweeney Screen Printing Association International, Fairfax, Virginia

R. Milker Lohmann GmbH, Neuwied, Germany

Wayne E. Mozer Oxford Analytical, Inc., Andover, Massachusetts

Helmut W. J. Müller BASF AG, Ludwigshafen/Rhein, Germany

Richard Neumann Windmöller & Hölscher, Lengerich/Westfalen, Germany

Robert E. Norland Norland Products, Inc., North Brunswick, New Jersey

Michael O'Mary The Armoloy Corporation, DeKalb, Illinois

Robert J. Partyka BeamAlloy Corporation, Dublin, Ohio

John A. Pasquale III Liberty Machine Company, Paterson, New Jersey

Kim S. Percell Witco Corporation, Memphis, Tennessee

Edwin P. Plueddemann Dow Corning Corporation, Midland, Michigan

Charles P. Rader Monsanto Chemical Company, Akron, Ohio

Valentinas Rajeckas Kaunas Polytechnic University, Kaunas, Lithuania

H. Randhawa Vac-Tec Systems Inc., Boulder, Colorado

Richard Rathmell Londonderry, New Hampshire

Donald A. Reinke Oliver Products Company, Grand Rapids, Michigan

Peter W. Rose Plasma Science, Inc., Belmont, California 94002

D. Satas Satas & Associates, Warwick, Rhode Island

Milton C. Schmit Plymouth Printing Company, Cranford, New Jersey

Jaykumar (Jay) J. Shah Decora, Fort Edward, New York

D. Stoye Hüls Troisdorf AG, Troisdorf/Marl, Germany

Larry S. Timm Findley Adhesives Inc., Wauwatosa, Wisconsin

Harry H. Tomlinson Witco Corporation, Memphis, Tennessee

A. Vaškelis Lithuanian Academy of Sciences, Vilnius, Lithuania

Subbu Venkatraman Raychem Corporation, Menlo Park, California

Theodore G. Vernardakis Sun Chemical Corporation, Cincinnati, Ohio

Leonard E. Walp Witco Corporation, Memphis, Tennessee

Daniel M. Zavisza Hercules Incorporated, Wilmington, Delaware

Randall W. Zempel Dow Chemical Company, Midland, Michigan

Alexander Zettl BASF AG, Ludwigshafen/Rhein, Germany

Ulrich Zorll Forschungsinstitut für Pigmente und Lacke, Stuttgart, Germany

COATINGS TECHNOLOGY HANDBOOK

Part 1
Fundamentals and Testing

1

Rheology and Surface Chemistry

K. B. Gilleo

Sheldahl, Inc., Northfield, Minnesota

1.0 INTRODUCTION

A basic understanding of rheology and surface chemistry, two primary sciences of liquid flow and solid–liquid interaction is necessary for understanding coating and printing processes and materials. A generally qualitative treatment of these subjects will suffice to provide the insight needed to use and apply coatings and inks and to help solve the problems associated with their use.

Rheology, in the broadest sense, is the study of the physical behavior of all materials when placed under stress. Four general categories are recognized: elasticity, plasticity, rigidity, and viscosity. Our concern here is with liquids and pastes. The scope of rheology of fluids encompasses the changes in the shape of a liquid as physical force is applied and removed. Viscosity is a key rheological property of coatings and inks. Viscosity is simply the resistance of the ink to flow—the ratio of shear stress to shear rate.

Throughout coating and printing processes, mechanical forces of various types and quantities are exerted. The amount of shear force directly affects the viscosity value for non-Newtonian fluids. Most coatings undergo some degree of "shear thinning" phenomenon when worked by mixing or running on a coater. Heavy inks are especially prone to shear thinning. As shear rate is increased, the viscosity drops, in some cases, dramatically.

This seems simple enough except for two other effects. One is called the yield point. This is the shear rate required to cause flow. Ketchup often refuses to flow until a little extra shear force is applied. Then it often flows too freely. Once the yield point has been exceeded the solidlike behavior vanishes. The loose network structure is broken up. Inks also display this yield point property, but to a lesser degree. Yield point is one of the most important ink properties.

Yield value, an important, but often ignored attribute of liquids, will also be discussed. We must examine rheology as a dynamic variable and explore how it changes throughout the coating process. The mutual interaction, in which the coating process alters viscosity

3

and rheology affects the process, will be a key concept in our discussions of coating technology.

The second factor is time dependency. Some inks change viscosity over time even though a constant shear rate is being applied. This means that viscosity can be dependent on the amount of mechanical force applied and on the length of time. When shearing forces are removed, the ink will return to the initial viscosity. That rate of return is another important ink property. It can vary from seconds to hours.

Rheology goes far beyond the familiar snapshot view of viscosity at a single shear rate which is often reported by ink vendors. It deals with the changes in viscosity as different levels of force are applied, as temperature is varied, and as solvents and additives come into play. Brookfield viscometer readings, although valuable, do not show the full picture for non-Newtonian liquids.

Surface chemistry describes wetting (and dewetting) phenomena resulting from mutual attractions between ink molecules, as well as intramolecular attractions between ink and the substrate surface. The relative strengths of these molecular interactions determine a number of ink performance parameters. Good print definition, adhesion, and a smooth ink surface all require the right surface chemistry. Bubble formation and related film formation defects have their basis in surface chemistry also.

Surface chemistry, for our purposes, deals with the attractive forces liquid molecules exhibit for each other and for the substrate. We will focus on the wetting phenomenon and relate it to coating processes and problems. It will be seen that an understanding of wetting and dewetting will help elucidate many of the anomalies seen in coating and printing.

The two sciences of rheology and surface tension, taken together, provide the tools required for handling the increasingly complex technology of coating. It is necessary to combine rheology and surface chemistry into a unified topic to better understand inks and the screen printing process. We will cover this unification in a straightforward and semiqualitative manner. One benefit will be the discovery that printing and coating problems often blamed on rheology have their basis in surface chemistry. We will further find that coating leveling is influenced by both rheology and surface chemistry.

2.0 RHEOLOGY

Rheology, the science of flow and deformation, is critical to the understanding of coating use, application, and quality control. Viscosity, the resistance to flow, is the most important rheological characteristic of liquids and therefore of coatings and inks. Even more significant is the way in which viscosity changes during coating and printing. Newtonian fluids, like solvents, have an absolute viscosity that is unaltered by the application of mechanical shear. However, virtually all coatings show a significant change in viscosity as different forces are applied. We will look at the apparent viscosity of coatings and inks and discover how these force-induced changes during processing are a necessary part of the application process.

Viscosity, the resistance of a liquid to flow, is a key property describing the behavior of liquids subjected to forces such as mixing. Other important forces are gravity, surface tension, and shear associated with the method of applying the material. Viscosity is simply the ratio of shear stress to shear rate (Eq. 3). A high viscosity liquid requires considerable force (work) to produce a change in shape. For example, high viscosity coatings are not as easily pumped as are the low viscosity counterparts. High viscosity coatings also take longer to flow out when applied.

Table 1 Viscosities of Common Industrial Liquids

Liquid	Viscosity (cP)
Acetone	0.32
Chloroform	0.58
Toulene	0.59
Water, standard 2(20°C)	1.0000
Cyclohexane	1.0
Ethyl alcohol	1.2
Turpentine	1.5
Mercury, metal	1.6
Creosote	12.0
Sulfuric acid	25.4
Lindseed oil	33.1
Olive oil	84.0
Castor oil	986.0
Glycerine	1490.0
Venice turpentine	130,000.0

Values are for approximately 20°C.
Source: *Handbook of Chemistry and Physics*, 64th Ed., 1984,
CRC Press, Boca Raton, FL.

$$\text{Shear rate, } D = \frac{\text{velocity}}{\text{thickness}} \quad (\text{sec}^{-1}) \tag{1}$$

$$\text{Shear stress, } \tau = \frac{\text{force}}{\text{area}} \quad (\text{dynes/cm}^2) \tag{2}$$

$$\text{Viscosity, } \eta = \frac{\text{shear stress}}{\text{shear rate}} = \frac{\tau}{D} \quad (\text{dynes} \cdot \text{sec}/\text{cm}^2) \tag{3}$$

As indicated above, shear stress, the force per unit area applied to a liquid, is typically in dynes per square centimeter, the force per unit area. Shear rate is in reciprocal seconds (sec^{-1}), the amount of mechanical energy applied to the liquid. Applying Equation 3, the viscosity unit becomes dyne-seconds per square centimeter or poise (P). For low viscosity fluids like water ($\approx$0.01 P), the poise unit is rather small, and the more common centipoise (0.01 P) is used. Since 100 centipoise = 1 poise, water has a viscosity of about 1 centipoise (cP). Screen inks are much more viscous and range from 1000 to 10,000 cP for graphics and as high as 50,000 cP for some highly loaded polymer thick film (PTF) inks and adhesives. Viscosity is expressed in pascal-seconds (Pa·sec) in the international system of units (SI: 1 Pa·sec = 1000 cP). Viscosity values of common industrial liquids are provided in Table 1.

Viscosity is rather a simple concept. Thin, or low viscosity liquids flow easily, while high viscosity ones move with much resistance. The ideal, or Newtonian, case has been assumed. With Newtonian fluids, viscosity is constant over any region of shear. Very few

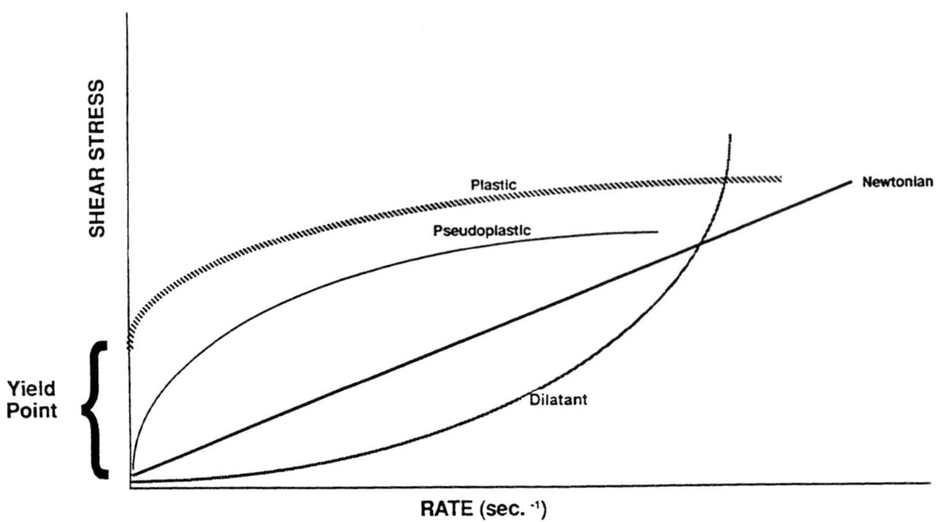

Figure 1 Shear stress shear rate curves.

liquids are truly Newtonian. More typically, liquids drop in viscosity as shear or work is applied. The phenomenon was identified above as shear thinning. It is, therefore, necessary to specify exactly the conditions under which a viscosity value is measured. Time must also be considered in addition to shear stress. A liquid can be affected by the amount of time that force is applied. A shear-thinned liquid will tend to return to its initial viscosity over time. Therefore, time under shearing action and time at rest are necessary quantifiers if viscosity is to be accurately reported.

It should be apparent that we are really dealing with a viscosity curve, not a fixed point. The necessity of dealing with viscosity curves is even more pronounced in plastic decorating. A particular material will experience a variety of different shear stresses. For example, a coating may be mixed at relatively low shear stress of 10–20 cP, pumped through a spray gun line at 1000 cP, sprayed through an airless gun orifice at extreme pressure exceeding 10^6 cP, and finally allowed to flow out on the substrate under mild forces of gravity (minor) and surface tension. It is very likely that the material will have a different viscosity at each stage. In fact, a good product should change in viscosity under applications processing.

2.1 Types of Viscosity Behavior

2.1.1 Plasticity

Rheologically speaking, plastic fluids behave more like plastic solids until a specific minimum force is applied to overcome the yield point. Gels, sols, and ketchup are extreme examples. Once the yield point has been reached, the liquid begins to approach Newtonian behavior as shear rate is increased. Figure 1 shows the shear stress–shear rate curve and the yield point. Although plastic behavior is of questionable value to ketchup, it has some benefit in inks and paints. Actually it is the yield point phenomenon that is of practical value. No-drip paints are an excellent example of the usefulness of yield point. After the brush stroke force has been removed, the paint's viscosity builds quickly until flow stops. Dripping is prevented because the yield point exceeds the force of gravity.

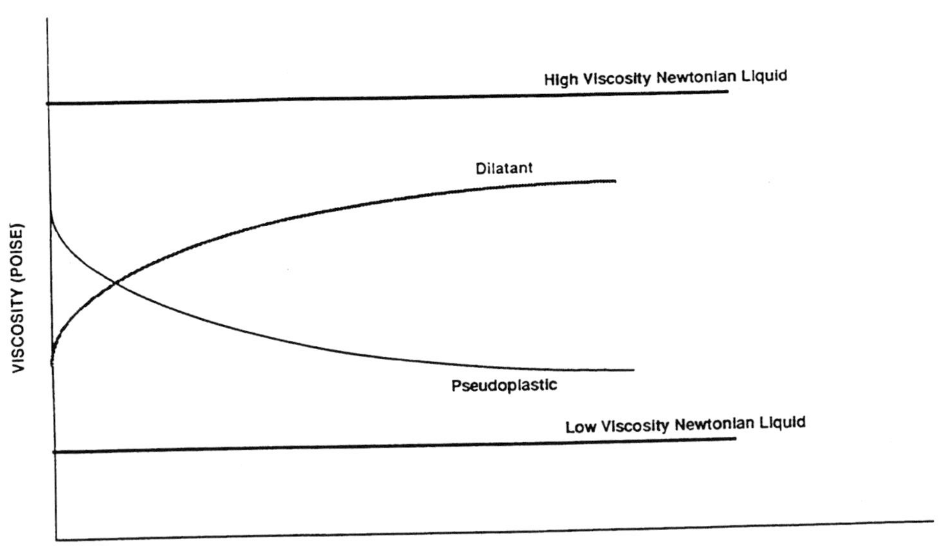

Figure 2   Viscosity shear rate curves.

Ink bleed in a printing ink, the tendency to flow beyond the printed boundaries, is controlled by yield point. Inks with a high yield point will not bleed, but their flow out may be poor. A very low yield point will provide excellent flow out, but bleed may be excessive. Just the right yield point provides the needed flow out and leveling without excessive bleed. Both polymer binders and fillers can account for the yield point phenomenon. At rest, polymer chains are randomly oriented and offer more resistance to flow. Application of shear force straightens the chains in the direction of flow, reducing resistance. Solid fillers can form loose molecular attraction structures, which break down quickly under shear.

2.1.2   Pseudoplasticity

Like plastic-behaving materials, pseudoplastic liquids drop in viscosity as force is applied. There is no yield point, however. The more energy applied, the greater the thinning. When shear rate is reduced, the viscosity increases at the same rate by which the force is diminished. There is no hysteresis; the shear stress–shear rate curve is the same in both directions as was seen in Figure 1. Figure 2 compares pseudoplastic behavior using viscosity–shear rate curves.

Many coatings exhibit this kind of behavior, but with time dependency. There is a pronounced delay in viscosity increase after force has been removed. This form of pseudoplasticity with a hysteresis loop is called thixotropy. Pseudoplasticity is generally a useful property for coatings and inks. However, thixotropy is even more useful.

Thixotropy.   Thixotropy is a special case of pseudoplasticity. The material undergoes "shear thinning"; but as shear forces are reduced, viscosity increases at a lesser rate to produce a hysteresis loop. Thixotropy is very common and very useful. Dripless house paints owe their driplessness to thixotropy. The paint begins as a moderately viscous material that stays on the brush. It quickly drops in viscosity under the shear stress of brushing for easy,

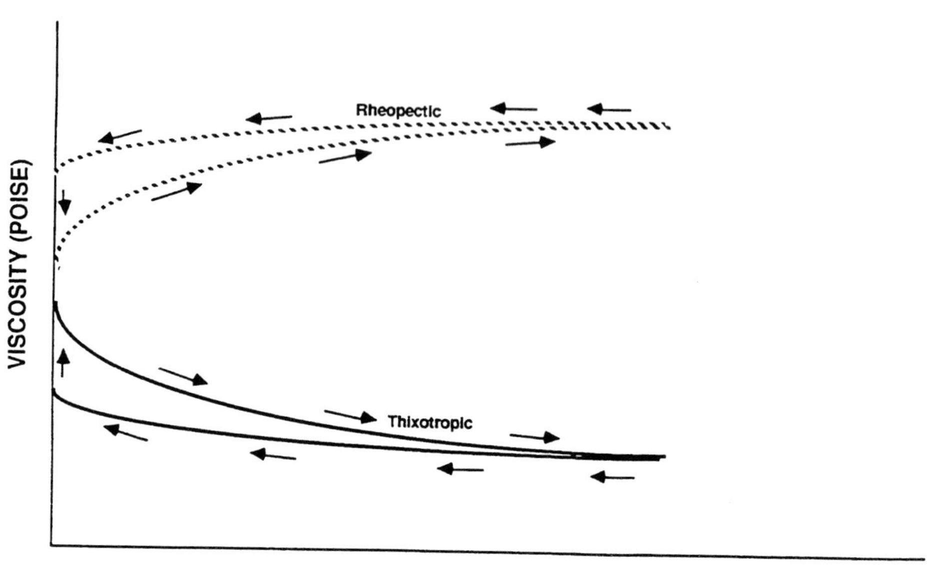

Figure 3　Shear stress shear rate curves: hysteresis loop.

smooth application. A return to higher viscosity, when shearing action stops, prevents dripping and sagging.

Screen printing inks also benefit from thixotropy. The relatively high viscosity screen ink drops abruptly in viscosity under the high shear stress associated with being forced through a fine mesh screen. The momentary low viscosity permits the printed ink dots to merge together into a solid, continuous film. Viscosity returns to a higher range before the ink can "bleed" beyond the intended boundaries.

Thixotropic materials yield individual hysteresis loops. Shear stress lowers viscosity to a point at which higher force produces no further change. As energy input to the liquid is reduced, viscosity begins to build again, but more slowly than it initially dropped. It is not necessary to know the shape of the viscosity loop, but merely to realize that such a response is common in decorating inks, paints, and coatings.

The presence in decorating vehicles of pigments, flatting agents, and other solid fillers usually produces or increases thixotropic behavior. More highly loaded materials, such as inks, are often highly thixotropic. Thixotropic agents, consisting of flat, platelet structures, can be added to liquids to adjust thixotropy. A loose, interconnecting network forms between the platelets to produce the viscosity increase. Shearing breaks down the network, resulting in the viscosity drop.

Mixing and other high shear forces rapidly reduce viscosity. However, thixotropic inks continue to thin down while undergoing shearing even is the shear stress is constant. This can be seen with a Brookfield viscometer, where measured viscosity continues to drop while the spindle turns at constant rpm's. When the ink is left motionless, viscosity builds back to the initial value. This can occur slowly or rapidly. Curves of various shapes are possible, but they will all display a hysteresis loop. In fact this hysteresis curve is used to detect thixotropy (see Fig. 3). The rate of viscosity change is an important characteristic of an ink which is examined later as we take an ink step by step through screen printing.

Thixotropy is very important to proper ink behavior, and the changing viscosity attribute makes screen printing possible.

Dilatancy.   Liquids that show an increase in viscosity as shear is applied are called dilatants. Very few liquids possess this property. Dilatant behavior should not be confused with the common viscosity build, which occurs when inks and coatings lose solvent. For example, a solvent-borne coating applied by a roll coater will show a viscosity increase as the run progresses. The rotating roller serves as a solvent evaporator, increasing the coating's solids content and, therefore, the viscosity. True dilatancy occurs independently of solvent loss.

Rheoplexy.   Sounding more like a disease than a property, rheoplexy is the exact opposite of thixotropy. It is the time-dependent form of dilatancy where mixing causes shear thickening. Figure 3 showed the hysteresis loop. Rheoplexy is fortunately rare, since it is totally useless as a characteristic for screen print inks.

2.2 Temperature Effects

Viscosity is strongly affected by temperature. Measurements should be taken at the same temperature (typically 23°C). A viscosity value is incomplete without a temperature notation.

Although each liquid is affected differently by a temperature change, the change per degree is usually a constant for a particular material. The subject of temperature effects has been covered thoroughly elsewhere.[2] It will suffice to say that a coating's viscosity may be reduced by heating, a principle used in many coating application systems.

Viscosity reduction by heating may also be used after a material has been applied. Preheating of UV-curable coatings just prior to UV exposure is often advantageous for leveling out these sometimes viscous materials.

2.3 Solvent Effects

Higher resin solids produce higher solution viscosity, while solvent addition reduces viscosity. It is important to note that viscosity changes are much more pronounced in the case of soluble resins (polymers) than for insoluble pigments or plastic particles. For example, although a coating may be highly viscous at 50% solids, a plastisol suspension (plastic particles in liquid plasticizer) may have medium viscosity at 80% solids. Different solvents will produce various degrees of viscosity reduction depending on whether they are true solvents, latent solvents, or nonsolvents. This subject has been treated extensively elsewhere.[2,3]

2.4 Viscosity Measurement

Many instruments are available. A rheometer is capable of accurately measuring viscosities through a wide range of shear stress. Much simpler equipment is typically used in the plastic decorating industry. As indicated previously, perhaps the most common device is the Brookfield viscometer, in which an electric motor is coupled to an immersion spindle through a tensiometer. The spindle is rotated in the liquid to be measured. The higher the viscosity (resistance to flow), the larger the reading on the tensiometer. Several spindle diameters are available, and a number of rotational speeds may be selected. Viscosity must be reported along with spindle size and rotational speed and temperature.

The Brookfield instrument is a good tool for incoming quality control. Although certainly not a replacement for the rheometer, the viscometer may be used to estimate viscosity change with shear. Viscosity readings are taken at different rpm's and then compared. A highly thixotropic material will be easily identified.

An even simpler viscosity device is the flow cup, a simple container with an opening at the bottom. The Ford cup and the Zahn cup are very common in the plastic painting and coating field. The Ford cup, the more accurate of the two, is supported on a stand. Once filled, the bottom orifice is unstoppered and the time for the liquid to flow out is recorded. Unlike the Brookfield, which yields a value in centipoise, the cup gives only a flow time. Relative flow times reflect different relative viscosities. Interconversion charts permit Ford and other cup values to be converted to centipoise (Table 2).

The Zahn cup is dipped in a liquid sample by means of its handle and quickly withdrawn, whereupon time to empty is recorded. The Zahn type of device is commonly used on line, primarily as a checking device for familiar materials.

2.5 Yield Value

The yield value is the shear stress in a viscosity measurement, but one taken at very low shear. The yield value is the minimum shear stress, applied to a liquid, that produces flow. As force is gradually applied, a liquid undergoes deformation without flowing. In essence, the liquid is behaving as if it were an elastic solid. Below the yield value, viscosity approaches infinity. At a critical force input (the yield value) flow commences.

The yield value is important in understanding the behavior of decorating liquids after they have been deposited onto the substrate. Shear stress, acting on a deposited coating or ink, is very low. Although gravity exerts force on the liquid, surface tension is considerably more important.

If the yield value is greater than shear stress, flow will not occur. The liquid will behave as if it were a solid. In this situation, what you deposit is what you get. Coatings that refuse to level, even though the apparent viscosity is low, probably have a relatively high yield value. As we will see in the next section, surface tension forces, although alterable, cannot be changed enough to overcome a high yield value. Unfortunately, a high yield value may be an intrinsic property of the decorative material. Under these circumstances, changing the material application method may be the only remedy.

Although a high yield value can make a coating unusable, the property can be desirable for printing inks. Once an ink has been deposited, it should remain where placed. Too low a yield value can allow an ink to flow out, producing poor, irregular edge definition. An ink with too high a value may flow out poorly. Since pigments tend to increase yield value, color inks are not a problem. Clear protective inks can be a problem, especially when a thick film is deposited, as in screen printing. When it is not practical to increase yield value, wettability can sometimes be favorably altered through surface tension modification. Increasing surface tension will inhibit flow and therefore ink or coating bleed.

3.0 SURFACE CHEMISTRY

Surface chemistry is the science that deals with the interface of two materials. The interface may exist between any forms of matter, including a gas phase. For the purpose of understanding the interfacial interaction of decorative liquid materials, we need only analyze the liquid–solid interaction. Although there is a surface interaction between a liquid coating and the air surrounding it, the effect is small and may be ignored.

Table 2 Viscosity Conversions[a]

Viscosity device	Consistency								
	Watery			Medium			Heavy		
Poise:	0.1	0.5	1.0	2.5	5.0	10	50	100	150
Centipoise:	10	50	100	250	500	1000	5000	10,000	15,000
Fisher #1	20								
Fisher #2		24	50						
Ford #4 cup	5	22	34	67					
Parlin #10	11	17	25	55					
Parlin #15				12	25	47	232	465	697
Saybolt	60	260	530	1240	2480	4600	23,500	46,500	69,500
Zahn #1	30	60							
Zahn #2	16	24	37	85					
Zahn #3			12	29	57				
Zahn #4			10	21	37				

[a]Liquids are at 25°C. Values are in seconds for liquids with specific gravity of approximately 1.0.
Source: Binks Inc.

Table 3 Surface Tension of Liquids

Liquid	Surface tension (dynes/cm)
SF$_6$	5.6
Trifluoroacetic acid	15.6
Heptane	22.1
Methanol	24.0
Acetone	26.3
Dimethylformamide (DMF)	36.8
Dimethyl sulfoxide (DMSO)	43.5
Ethylene glycol	48.4
Formamide	59.1
Glycerol	63.1
Diiodomethane	70.2
Water	72.8
Mercury, metal	490.6

Source: *Lang's Handbook*, 13th Ed., McGraw Hill, New York, 1985.

3.1 Surface Tension

All liquids are made up of submicroscopic combinations of atoms called molecules (a very few liquids are made up of uncombined atoms). All molecules that are close to one another exert attractive forces. It is these mutual attractions that produce the universal property called surface tension. The units are force per unit length: dynes per centimeter.

A drop of liquid suspended in space quickly assumes a spherical shape. As surface molecules are pulled toward those directly beneath them, a minimum surface area (sphere) results. The spherical form is the result of an uneven distribution of force; molecules within the droplet are attracted from all directions, while those at the surface are pulled only toward molecules below them. All liquids attempt to form a minimum surface sphere. A number of counterforces come into play, however. A liquid placed on a solid provides a liquid–solid interface. This type of interface is critically important to the plastic decorator. liquid molecules are attracted not only to each other (intramolecular attraction) but also to any solid surface (intermolecular attraction) with which they come in contact. We need only concern ourselves with these two interactions; intra- and intermolecular. A fundamental understanding of this interfacial interaction will permit the decorator to optimize materials and processes.

3.2 Measuring Surface Tension

Every liquid has a specific surface tension value. Liquids with high surface tensions, such as water (73 dynes/cm), demonstrate a high intramolecular attraction and a strong tendency to bead up (form spheres). Liquids with low values have a weak tendency toward sphere formation that is easily overcome by countering forces.

A variety of methods are available for measuring liquid surface tension. Table 3 gives values for common solvents. Methods are also available for determining the surface tension of solids, which is usually referred to as surface energy. Table 4 gives surface energy values

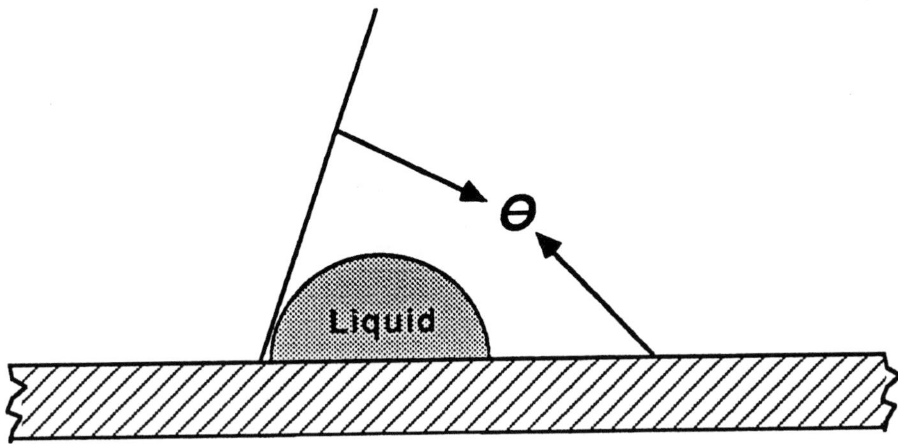

Figure 4   Contact angle.

for plastics. We need be concerned only with ways of estimating surface tension and with techniques for determining relative differences.

3.3 Wetting

A liquid placed on a flat, horizontal solid surface either will wet and flow out, or it will dewet to form a semispherical drop. An in-between state may also occur in which the liquid neither recedes or advances but remains stationary. The angle that the droplet or edge of the liquid makes with the solid plane is called the *contact angle* (Fig. 4).

A nonwetting condition exists when the contact angle exceeds 0°—that is, when the angle is measurable. The liquid's intramolecular attraction is greater than its attraction for the solid surface. The liquid surface tension value is higher than the solid's surface energy. A wetting condition occurs when the contact angle is 0°. The liquid's edge continues to advance, even though the rate may be slow for high viscosity materials. The intermolecular

Table 4   Surface Tension of Polymers

Polymer	Surface tension (dynes/cm)
Polyperfluoropropylene	16
Polytetrafluoroethylene (Teflon)	18.5
Polydimethyliloxane	24
Polyethylene	31
Polystyrene	34
Polymethylmethacrylate (acrylic)	39
Polyvinyl chloride (PVC)	40
Polyethylene terephthalate (polyester)	43
Polyhexamethylene adipate (nylon)	46

Source: Ref. 5.

Table 5 Surface Tension Test Kit

Surface Tension	Castor oil	Toluene	Heptane	FC48/FC77
15 dynes/cm				0/100
17				100/100
19				100/0
22			100	
22.4	12.0	49.2	38.8	
24.5	55.2	25.0	19.8	
27	74.2	14.4	11.4	
30	0	100.0	0	
32.5	88.0	4.5	3.5	
35	100.0			
63	(100 glycerol)			
72.8	(100 water)			

Mixtures are in weight percent.
Source: Various sources and tests by author.

(solid–liquid) attraction is greater in this case. The surface energy of the solid is higher than the liquid's surface tension.

Measuring the contact angle is a simple technique for determining the relative difference between the two surface tensions. A high contact angle signifies a large departure, while a small angle suggests that the two values are close, but not equal.

One can estimate liquid surface tension by applying drops of the liquid onto smooth surfaces of known values until a wetting just occurs, signifying that the two surface tensions are equal. Conversely, the surface energy of a solid may be estimated by applying drops of standard surface tension liquids until wetting is achieved. A surface tension kit can be made up from simple mixtures for testing surfaces. Table 5 provides formulas.

Low energy surfaces are difficult to wet and can give poor results for coating, painting, and printing. The standard surface tension kit may be used to estimate the surface energy of a plastic to be decorated. If the particular plastic shows a much lower value than that reported in Table 4, contamination is suspected. Mold release agents, unless specially made compatible for decorating materials, can greatly lower surface energy of a plastic part, making it uncoatable.

3.4 Surfactants

Agents that alter interfacial interactions are called surfactants. The surfactant possesses two different chemical groups, one compatible with the liquid to be modified, and the other having a lower surface tension. For example, the surface tension of an epoxy may be reduced by adding a surfactant with an alcohol group (epoxy-compatible) at one end and a fluorochemical group at the other. The alcohol group will associate with the epoxy resin, presenting the incompatible fluorochemical "tail" to the surface. The epoxy coating will behave as if it were a low surface tension fluorochemical. The addition of a small amount of surfactant will permit the epoxy coating to wet difficult, low energy surfaces, even oil-contaminated plastic.

Surfactants efficiently lower the surface tension of inks, coatings, and paints. Typically, 1% or less is sufficient. When dewetting occurs because of intrinsically low surface energy of the substrate, use of surfactants, also called wetting agents, is indicated. These materials are not a substitute for good housekeeping and proper parts preparation. Contamination can cause adhesion failure later.

Fluorochemicals, silicones, and hydrocarbons are common categories of surfactants. Fluorochemicals have the lowest surface tension of any material and are the most efficient wetting agents. Silicones are next in efficacy and are lower in cost. Certain types of silicone, however, can become airborne, causing contamination of the substrate.

Although it may be desirable to lower the surface tension of a coating, the opposite is true for the substrate. The very agent that helps the decorating material renders the substrate useless. Silicone contamination will produce the notorious dewetting defect called "fish-eyes."

Coatings, paints, and inks, once modified with surfactants, are usually permanently changed, even after curing. Their low surface energy will make them difficult to wet over if, for example, it is necessary to apply a top coat. There are several options for overcoming this problem. The best practice is to use the smallest amount of the least potent surfactant that will do the job. Start with the hydrocarbon class. Also make sure that the substrate is clean to begin with.

Another possibility is to use reactive surfactants. Agents possessing a functional group that can react with coating or binder are rendered less active after curing. Once the surfactant has completed the role of wetting agent, it is no longer needed. One other approach is to add surfactant to the second material to be applied. Often the same surfactant will work, especially at a slightly higher loading.

3.5 Leveling

Leveling depends on both rheology and surface chemistry. It is a more complex phenomenon and a more difficult one to control. Coatings applied by spraying, dipping, roll coating, and most other methods are often not smooth enough for aesthetic appeal. Splatters, runs, ridges, and other topological defects require that the liquid material level out. It is therefore important to understand the dynamics of leveling.

We will first assume that proper wetting has been achieved, by wetting agents if necessary. Important parameters affecting leveling are viscosity, surface tension, yield value, coating thickness, and the degree of wet coating irregularity. Several workers have developed empirical relationships to describe leveling. The leveling equation (Equation 4) is quite useful.[6]

$$a_t = a_0 \frac{\exp(\text{const } \sigma h^3 t)}{3\lambda^4 \eta} \tag{4}$$

where

a_t = amplitude (height) of coating ridge
σ = the surface tension of the coating
η = coating viscosity
h = coating thickness or height
t = the time for leveling
λ = wavelength or distance between ridges

Equation 4 shows that leveling is improved by one or more of the following:

1. Longer time (t)
2. Higher surface tension of coating (σ)
3. Lower viscosity (η)
4. Greater coating thickness (h)
5. Small repeating distance between ridges (λ)

Note that h, the coating thickness, is raised to the third power. Doubling the thickness provides an eightfold (2^3) improvement in leveling. Also note that λ, wavelength between ridges, is raised to the fourth power. This means that ridges that are very far apart create a very difficult leveling situation.

Earlier, it was pointed out that a high yield value could prevent leveling. The shear stress on a wet coating, must be greater than the yield value for leveling to take place. Equation 5 shows the relationship between various parameters and shear stress.[7]

$$T_{max} = \frac{4\pi^3 \sigma\, ah}{\lambda^3} \quad \text{or} \quad D(\text{coating ridge depth}) = \frac{\tau\lambda^3}{4\pi^3 \sigma h} \tag{5}$$

where

σ = surface tension of coating
a = amplitude of coating ridge
h = coating height
λ = coating ridge wavelength

Since Equation 5 deals with force, the time factor and the viscosity value drop out. It is seen that increasing surface tension and coating thickness produce the maximum shear stress. Coating defect height (a) increases shear, while wavelength (λ), strongly reduces it. If coating ridges cannot be avoided, higher, more closely packed ones, are preferable.

When the yield value is higher than the maximum shear (T_{max}), leveling will not occur. Extending leveling time and reducing viscosity will not help to overcome the yield value barrier, since these terms are not in the shear equation. Increasing surface tension and coating thickness are options, but there are practical limits.

Since yield value is usually affected by shear (thixotropy), coating application rate and premixing conditions may be important. Higher roller speed (for roll coaters) and higher spray pressure (for spray guns) can drop the yield value temporarily. It should be apparent that best leveling is not achieved by lowest surface tension. Although good wetting may require a reduction in surface tension, higher surface tension promotes leveling. This is one more reason to use the minimum effective level of surfactant.

4.0 SUMMARY

A comprehension of the basic principles that describe and predict liquid flow and interfacial interactions is important for the effective formulation and the efficient application of coatings and related materials. The theoretical tools for managing the technology of coatings are rheology, the science of flow and deformation, considered with surface chemistry, the science of wetting and dewetting phenomena. Viewing such rheological properties as viscosity in terms of their time dependency adds the necessary dimension for practical application of theory to practice. Such important coating attributes as leveling are affected by both viscosity and surface tension. Knowing the interrelationships allows the coating specialist to make adjustments and take corrective actions with confidence.

REFERENCES

1. Handbook of Chemistry and Physics, CRC Press, Boca Raton, FL, 1984, 64th Ed.
2. Temple C. Patton, *Paint Flow and Pigment Dispersion*, 2nd Ed. New York: Wiley, 1979.
3. Charles R. Martens, *Technology of paint, Varnish and Lacquers*, New York: Krieger Pub. Co., 1974.
4. *Lang's Handbook*, 13th Ed. New York: McGraw Hill, 1985.
5. Norbert M. Bikales, *Adhesion and Bonding*, New York: Wiley-Interscience, 1971
6. S. Orchard, *Appl. Sci. Res.*, *A11*, 451 (1962).
7. N. D. P. Smith, S. E. Orchard, and A. J. Rhind-Tutt, "The Physics of Brush Marks," *JOCCA*, *44*, 618–633, Sept. (1961).

BIBLIOGRAPHY

Bikales, N. M., *Adhesion and Bonding*. New York: Wiley-Interscience, 1971.
Martens, C. R., *Technology of Paint, Varnish and Lacquers*. New York: Krieger Pub. Co., 1974.
Nylen, P. and S. Sunderland, *Modern Surface Coatings*. New York: Wiley, 1965.
Patton, T. C., *Paint Flow and Pigment Dispersion*, 2nd Ed. New York: Wiley-Interscience, 1979.

2
Coating Rheology

Chi-Ming Chan and Subbu Venkatraman

Raychem Corporation, Menlo Park, California

1 INTRODUCTION

Depending on the nature of the starting material, coatings can be broadly classified into solvent-borne and powder coatings. The solvent-borne coatings include both solutions (high and low solid contents) and suspensions or dispersions. Methods of application and the markets for these coatings are listed in Table 1.

2.0 DEFINITIONS AND MEASUREMENT TECHNIQUES

2.1 Surface Tension

Surface tension is defined as the excess force per unit length at the surface; it is reckoned as positive if it acts in such a direction as to contract the surface.[1] The tendency for a system to decrease its surface area is the result of the excess surface energy, because the surface atoms are subjected to a different environment as compared to those in the bulk. Surface tension of liquids and polymer melts can be measured by methods such as capillary tube,[1] Du Nuoy ring[2] Wilhelmy plate,[3] and pendent drop.[4] We shall focus our discussion on two methods: the capillary-height and pendant-drop methods.

The capillary-height method is the most suitable for low viscosity liquids because the system takes a long time to reach equilibrium for high viscosity liquids. It is reported that as much as 4 days is needed to attain equilibrium for a polystyrene melt at 200°C.[5] Figure 1 illustrates the capillary-height method. At equilibrium, the force exerted on the meniscus periphery due to the surface tension must be balanced by the weight of the liquid column. Neglecting the weight of the liquid above the meniscus, an approximate equation can be written

$$\Delta \varrho g h = 2\gamma \frac{\cos \theta}{r} \tag{1}$$

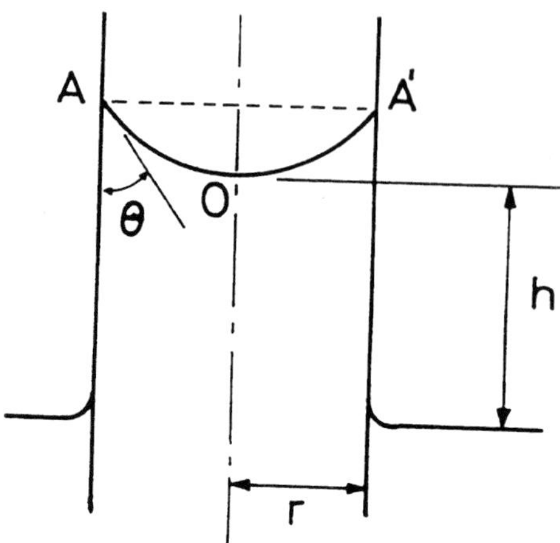

Figure 1   The capillary-method.

where $\Delta\rho$ is the density difference between the liquid and air, g is the gravitational constant, h is the height of the liquid column, γ is the surface tension, θ is the contact angle, and r is the radius of the capillary. In practice, it is difficult to measure the contact angle accurately vertical, and of known and uniform radius. For a more accurate determination of the surface tension various methods are available to calculate the weight of the liquid above the meniscus.

The pendant-drop method is a very versatile technique to measure the surface tension of liquids and also the interfacial tension between two liquids. Andreas et al.[9] used this method to measure the surface tension of various organic liquids. Wu[10] and Roe[11] have applied this method extensively to measure the surface and interfacial tensions of many polymer liquids and melts.

The experimental set up shown in Figure 2 consists of a light source, a pendant-drop cell, and a syringe assembly in a constant-temperature chamber, as well as a photomicrographic arrangement. A typical shape of a pendant drop is shown in Figure 3. The surface tension of the liquid is given by[9]

$$\gamma = \Delta\varrho g \frac{d_e}{H} \tag{2}$$

Table 1   Application Methods and Markets for Solvent-borne and Powder Coatings

Coating type	Method of application	Market
Solvent-borne	Brushing, rolling	Consumer paints
	Spraying	Automotive, Industrial
	Spin-coating	Microelectronics
	Electrodeposition	Automotive, Industrial
Powder	Electrostatic	Automotive, Industrial

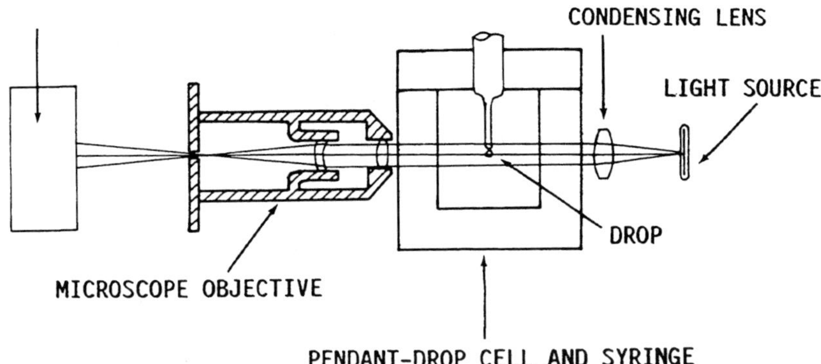

Figure 2 Experimental setup for the pendant-drop method.

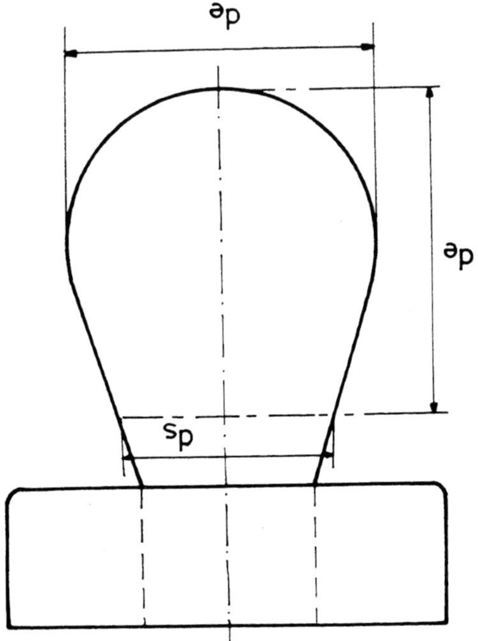

Figure 3 Typical pendant-drop profile.

where d_e is the maximum (equatorial) diameter of the pendant drop and H is a correction factor that depends on the shape of the drop; H is related to a measurable shape-dependent factor S, which is defined by

$$S = \frac{d_s}{d_e} \tag{3}$$

where d_s is the diameter of the pendant drop in a selected plant at a distance d_e from the apex of the drop (see Fig. 3). Tables showing the values of $1/H$ as a function of S are available.[12–14]

Recently there have been a number of significant improvements in both data acquisition and analysis of the pendant-drop profiles.[15–17] The photographic recording and measurement of the pendant drop are replaced by direct digitization of a video image. The ability to measure the entire drop profile has led to the development of new algorithms for the drop-profile analysis.[16,17]

2.2 Viscosity

The shear viscosity is defined as the ratio of the shear stress to the shear strain rate, at the strain rate of interest. Although the viscosity is usually quoted as a number without reference to the strain rate, it is really a function of strain rate. The strain rate dependence and, in certain situations, the time dependence, of the viscosity need to be determined if a meaningful correlation is the be made with coating phenomena. In the case of coatings, the shear strain rate range of interest extends from about a few thousand reciprocal seconds (during spraying, for instance) down to a hundredth of a reciprocal second (following application).

A variety of techniques is available to measure viscosity of coating formulations. Some of them are listed in Table 2.[18] Instruments with a single or undefined strain rate should be avoided in the study of coating rheology, if meaningful correlations are to be made with coating phenomena, the viscosity must be measured over a wide range of strain rates.

The most acceptable technique for determining the strain-rate dependence of the viscosity is the use of the constant rate-of-strain experiment in torsion. This can be done in either a cone-and-plate (for low rates) or a concentric cylinder geometry (for higher rates). However, the oscillatory, or dynamic measurement is also commonly employed for the same purpose. It is assumed that the shear strain rate and the frequency are equivalent quantities and the complex viscosity is equal to the steady state constant rate viscosity (i.e., the Cox–Merz rule is valid). The applicability of the Cox–Merz rule, however, is by no means universal, and its validity must be demonstrated before the dynamic measurements can be substituted for the steady-state ones. The capillary technique, as employed in several commercial instruments, is not suitable for coating studies in general, because it is more suitable for measuring viscosity at higher strain rates.

2.3 Thixotropy

Thixotropy is a much abused term in the coatings industry. In this review, we shall define the phenomenon of thixotropy as the particular case of the time dependence of the viscosity, that is its decrease during a constant rate-of-strain experiment. This time dependence manifests itself in hysteresis in experiments involving increasing and decreasing rates of strain. The area under the hysteresis loop has been used as a quantitative estimate of thixotropy, although its validity is still a matter of debate.[18,19] Another attempt at quantifying thixotr-

TABLE 2 Some Commercially Available Rheological Instrumentation

Name of instrument	Geometries available	Shear-rate range	Modes available
Weissenberg Rheogoniometer	Couette, cone and plate, parallel plate	Broad	Steady shear, oscillatory
Rheometrics Mechanical Spectrometer	Couette, plate and cone, parallel plate	Broad	Steady shear, oscillatory
Carri-Med Controlled Stress Rheometer (CSR)	Couette, parallel plate	Fixed stress	Creep and recovery, oscillatory
Rheo-Tech Viscoelastic Rheometer (VER)	Cone and plate	Fixed stress	Oscillatory , creep and recovery
Contraves Rheomat 115	Cone and plate, Couette	Broad	Steady shear
Rheometrics Stress Rheometer	Cone and plate	Fixed stress	Oscillatory, creep and recovery
Haake Rotovisco	Couette, cone and plate	Broad	Steady state
Shirley-Ferranti	Cone and plate	Broad	Steady shear
ICI Rotothinner	Couette	Single high rate	Steady shear
Brookfield Cone and Plate	Cone and plate	Medium to high	Steady
Brookfield Spindle	Undefined	Undefined	Steady shear
Gardner-Holdt	Rising buble	Undefined	
Cannon-Ubbelohde	Poiseuille	Limited range, high end	Shear
Brushometer	Couette	High end only; single	Steady shear

opy[20] involves the measurement of a peak stress (σ_p) and a stress at a long time (σ_∞) in a constant rate-of-strain experiment. In this instance, the thixotropy index β is defined as:

$$\beta = \sigma_p - \sigma \frac{\infty}{\sigma_p} \qquad (4)$$

The utility of these different definitions is still unclear, and their correlation to coating phenomena is even less certain.

In a purely phenomenological sense, thixotropy can be studied by monitoring the time-dependence of the viscosity, at constant rates of strain. Quantification of the property is, however, rather arbitrary. The coefficient of thixotropy, β, appears to be the most reasonable, and is measurable in torsional rheometers such as those mentioned in Table 2. It should be noted that this index, as defined above, increases with increase in the rate of strain. In addition, the thixotropic behavior is influenced considerably by the shear history of the material. In comparative measurements, care should be taken to ensure a similar or identical history for all samples. The phenomenon of thixotropy is also responsible for the increase in viscosity after the cessation of shear. If after a constant rate-of strain experiment, the material viscosity is monitored using a sinusoidal technique, it will be found to increase to a value characteristic of a low shear rate-of-strain measurement.

2.4 Dilatancy

The original definition of dilatancy,[21] an increase in viscosity with increasing rate of strain, is still the most widely accepted one today.[22–24] The term has been used, however, to mean the opposite of thixotropy.[25] The constant rate-of-strain experiment, outlined above for viscosity measurements, can obviously be employed to determine shear thickening, or dilatancy

2.5 Yield Stress

In the case of fluids, the yield stress is defined as the minimum shear stress required to initiate flow. It is also commonly referred to as the "Bingham" stress, and a material that exhibits a yield stress is commonly known as a "Bingham plastic" or viscoplastic.[26] Though easily defined, this quantity is not as easily measured. Its importance in coating phenomena is, however, quite widely accepted.

The most direct method of measuring this stress is by creep experiments in shear. This can be accomplished in the so-called stress-controlled rheometers (see Table 2). The minimum stress that can be imposed on a sample varies with the type of instrument, but by the judicious use of geometry, stress (in shear) in the range of 1 to 5 dynes/cm^2 can be applied. This is the range of yield stresses exhibited by most paints with a low level of solids. However, the detection of flow is not straightforward. In the conventional sense, the measured strain in the sample must attain linearity in time when permanent flow occurs. This may necessitate the measurement over a long period of time.

An estimate of the yield stress may be obtained from constant rate-of-strain measurements of stress and viscosity. When the viscosity is plotted against stress, its magnitude appears to approach infinity at low stresses. The asymptote on the stress axis gives an estimate of the yield stress.

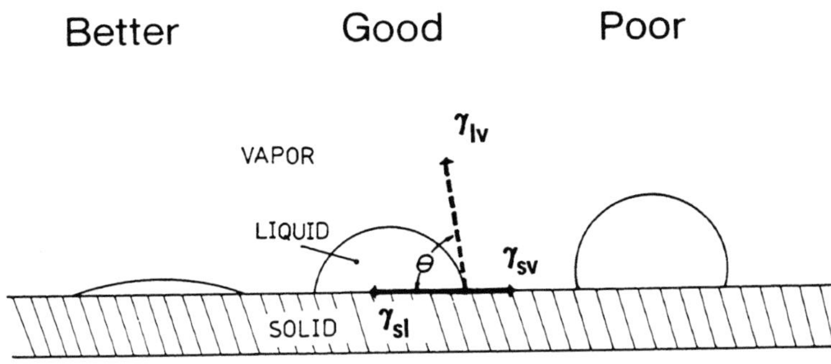

Figure 4   Schematic illustration of good and poor wetting.

Another method used is the stress relaxation measurement after the imposition of a step strain. For materials exhibiting viscoplasticity, the stress decays to a nonzero value which is taken as the estimate of the yield stress.

2.6 Elasticity

Elasticity of coating materials is frequently mentioned in the literature[18,19] as being very important in determining the coating quality, particularly of leveling. However, most of the reported measurements of elasticity are indirect, either through the first normal stress difference or through the stress relaxation measurement. Correlations are shown to exist, in paints, between high values of the first normal stress difference and the leveling ability.[18] However, no satisfactory rationalization has been put forward for a cause-and-effect relationship. Also, direct measurement of the elasticity of a coating through the creep-and-recovery experiment is virtually nonexistent. We shall not discuss the role of elasticity in this chapter.

3.0 RHEOLOGICAL PHENOMENA IN COATING

Coalescence, wetting, leveling, cratering, sagging, and slumping are the processes that are strongly influenced by surface tension and viscoelasticity. These, in turn, are the two important parameters that control the quality and appearance of coatings, hence their effects on the coating process are discussed in detail.

3.1 Wetting

Surface tension is an important factor that determines the ability of a coating to wet and adhere to a substrate. The ability of a paint to wet a substrate has been shown to be improved by using solvents with lower surface tensions.[27] Wetting may be quantitatively defined by reference to a liquid drop resting in equilibrium on a solid surface (Fig. 4). The smaller the contact angle, the better the wetting. When θ is greater than zero, the liquid wets the solid completely over the surface at a rate depending on the liquid viscosity and the solid surface roughness. The equilibrium contact angle for a liquid drop sitting an ideally smooth, homogeneous, flat, and nondeformable surface is related to various interfacial tensions by Young's equation

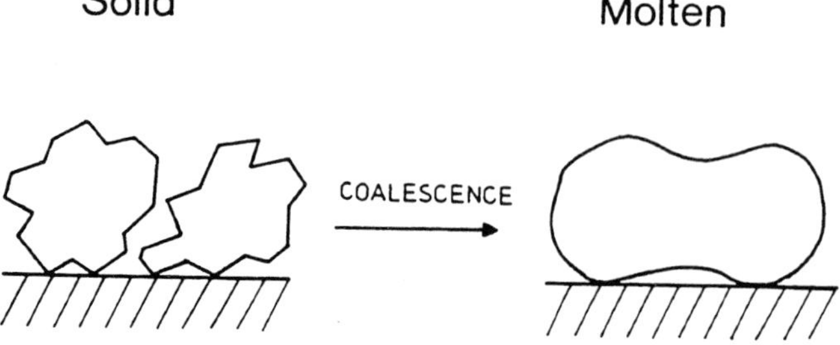

Figure 5   Schematization of the coalescence of molten powders.

$$\gamma_{lv}\cos\theta = \gamma_{sv} - \gamma_{sl} \tag{5}$$

where γ_{lv} is the surface tension of the liquid in equilibrium with its won saturated vapor, γ_{sv} is the surface tension of the solid in equilibrium with the saturated vapor of the liquid, and γ_{sl} is the interfacial tension between the solid and liquid. When θ is zero and assuming γ_{sv} to be approximately equal to γ_s (which is usually a reasonable approximation), then from Equation 5 it can be concluded that for spontaneous wetting to occur, the surface tension of the liquid must be greater than the surface tension of the solid. It is also possible for the liquid to spread and wet a solid surface when θ is greater than zero, but this required the application of a force to the liquid.

3.2   Coalescence

Coalescence is the fusing of molten particles to form a continuous film. It is the first step in powder coating. The factors that control coalescence are surface tension, radius of curvature, and viscosity of the molten powder. Figure 5 shows a schematic diagram of the coalescence of molten powder. Nix and Dodge[28] related the time of coalescence to those factors by the equation,

$$t_c = f\left(\eta\,\frac{R_c}{\gamma}\right) \tag{6}$$

where t_c is the coalescence time and R_c is the radius of the curvature (the mean particle radius). To minimize the coalescence time such that more time is available for the leveling-out stage, low viscosity, small particles, and low surface tension are desirable.

3.3   Sagging and Slumping

Sagging and slumping are phenomena that occur in coatings applied to inclined surfaces, in particular to vertical surfaces. Under the influence of gravity, downward flow occurs and leads to sagging or slumping, depending on the nature of the coating fluid. In the case of purely Newtonian or shear thinning fluids, sagging (shear flow) occurs; Figure 6 represents "gravity-induced" flow on a vertical surface. On the other hand, a material with a yield stress exhibits slumping (plug flow and shear flow).

For the case of Newtonian fluids, the physics of the phenomenon has been treated.[29,30] The extension to other types of fluid, including shear thinning and viscoplastic fluids, has

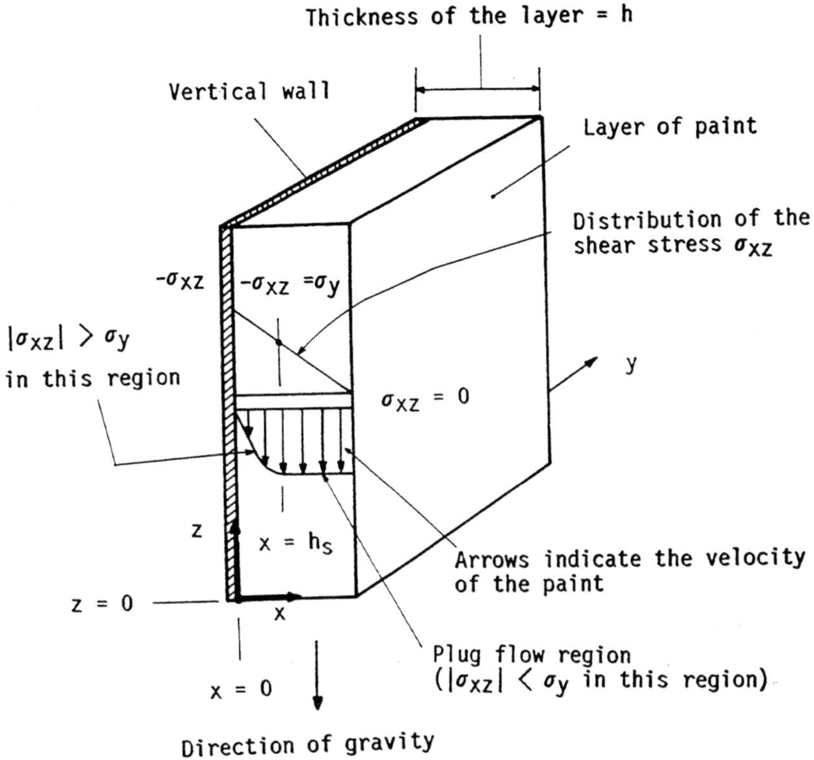

Figure 6 Gravity-induced flow on a vertical surface.

been done as well.[31] The treatment that follows is based largely on these three sources (i.e., Refs. 29–31. The parameters of interest in the analysis are the velocity V_0 of the material in flow at the fluid–air interface and the resulting sag or slump length, S. For the general case of a power-law fluid of index n,[31] these above quantities can be calculated:

$$V_0 = \left(\varrho \frac{g}{\eta_0} \right)^{(1/n)} \frac{n}{n+1} h^{(n+1)/n} \tag{7}$$

and

$$S = V_0 t \tag{8}$$

where η_0 is the zero-shear viscosity and h is the film thickness. The special case of Newtonian fluids is obtained by putting $n = 1$ in Equation 8. The final sag or slump length S is determined by the velocity as well as a time factor t, which is really a time interval for which the material remains fluid (or the time the material takes to solidify). The velocity v_0 depends inversely on the zero-shear viscosity. When all other things are equal, a shear thinning fluid ($n < 1$) will exhibit lower sag/slump velocities. In general, therefore, a Newtonian or a shear-thinning fluid will sag or slump under its own weight until its viscosity increased to the point at which V_0 is negligible. However, sagging might not occur at all, provided certain conditions are met. One of these is the existence of the yield stress. No sagging occurs if the yield stress (σ_y) is larger than the force due to gravity, ρgh. However, if the coating is thick enough (large h), this condition may no longer be satisfied, and both sagging and slumping can occur if the film thickness is larger than h_s, which is given by

$$h_s = \frac{\sigma_y}{\varrho\, g} \tag{9}$$

Between $h = 0$ and $h = h_s$, sagging occurs. The velocity can be obtained by substituting ($h - h_s$) for h in Equation 7:

$$V_0 = \left(\frac{\varrho\, g}{\eta_0}\right)^{1/n} \frac{n}{n+1} (h - h_s)^{(n+1)/n} \tag{10}$$

For $h > h_s$, plug flow occurs (see Fig. 6).

Wu[31] also finds that the tendency to sag, in general, increases in the order: shear-thinning fluids < viscoplastic fluids < Newtonian fluids < shear-thickening fluids, provided that all these materials have the same zero- shear viscosity, η_0. The significance of η_0 for viscoplastic fluids is unclear, although it is used in the equations derived by Wu.[31]

For the particular case of sprayable coatings, Wu finds that a shear thinning fluid with $n = 0.6$, without a yield stress, can exhibit good sag control while retaining adequate sprayability.

3.4　Leveling

Leveling is the critical step to achieve a smooth and uniform coating. During the application of coatings, imperfections such as waves or furrows usually appear on the surface. For the coating to be acceptable, these imperfections must disappear before the wet coating (fluid) solidifies.

Surface tension has been generally recognized as the major driving force for the flow-out in coating, and the resistance to flow is the viscosity of the coating. The result of leveling is the reduction of the surface tension of the film. Figure 7 illustrates the leveling out of a newly formed sinusoidal surface of a continuous fused film. For a thin film with an idealized sinusoidal surface, as shown in Figure 7, an equation that relates leveling speed t_v with viscosity and surface tension has given by Rhodes and Orchard[32]

$$t_v = \frac{16\pi^4 h^3 \gamma}{3\lambda^3 \eta} \ln\left(\frac{a_t}{a_0}\right) \tag{11}$$

where a_t and a_0 are the final and initial amplitudes, λ is the wavelength, and h is the averaged thickness of the film. This equation is valid only when λ is greater than h. From Equation 11 it is clear that leveling is favored by large film thickness, small wavelength, high surface tension, and low melt viscosity.

However, the question of the relevant viscosity to be used in Equation 11 is not quite settled. Lin[18] suggests computing the stress generated by surface tension with one of several available methods.[33,34] Then, from a predetermined flow curve, obtain the viscosity at that shear stress; this may necessitate the measurement of viscosity at a very low strain rate. On

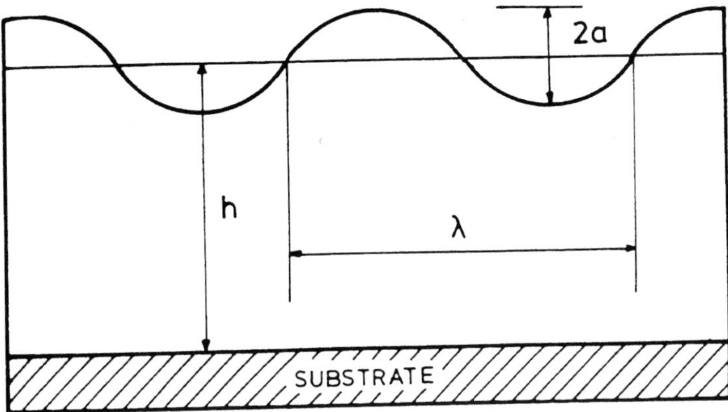

Figure 7   An ideal sinusoidal surface.

the other hand, Wu proposes[31] the use of the zero-shear value for the viscosity in Equation 11. These two approaches will yield similar results except when the material is highly sensitive to strain rate ($n \ll 1$).

When the material possesses a yield stress, the surface tension force must overcome the yield stress to initiate the flow or leveling. Thus we replace λ in Equation 11 by λ':

$$\gamma' = \gamma - \frac{\sigma_y \lambda}{8\pi^3 a_t h} \tag{12}$$

This equation implies that a coating fluid with low yield stress should level out quickly. This requirement for leveling is in conflict with that for low sag or slump (high yield stress). Wu[35] claims that a shear thinning fluid of index 0.6 exhibits the lowest sag, provided the viscosity is 50 poise at 1 reciprocal second. Since such a fluid does not have a yield stress, it should level out well. This kind of rheological behavior may be attainable in an oligomeric powder coating at temperatures close to its melting point, or in a solution coating with a high solid content. It is difficult to see how this behavior could be realized in all situations, in particular for latex dispersions that possess yield stresses.

3.5   Viscosity Changes After Application

After a wet or fluid coating has been applied to a substrate, its viscosity starts to increase. This increase is due to several factors; some of the more important ones are depicted in Figure 8. The magnitudes of the viscosity increases due to the different factors shown in Figure 8 are typical of a solution coating with a low solid content. The relative magnitudes will, of course, differ for solution coatings with a high solid level, as well as for powder coatings. In powder coatings, the principal increase will be due to freezing, as the temperature approaches the melting point.

The measurement of the viscosity increase is important because it gives us an idea of how much time is available for the various phenomena to occur before solidification. The leveling and sagging phenomena discussed above can occur only as long as the material remains fluid; as the viscosity increases, these processes become less and less significant because of the decrease in the sagging velocity and leveling speed in accordance with Equations 7 and

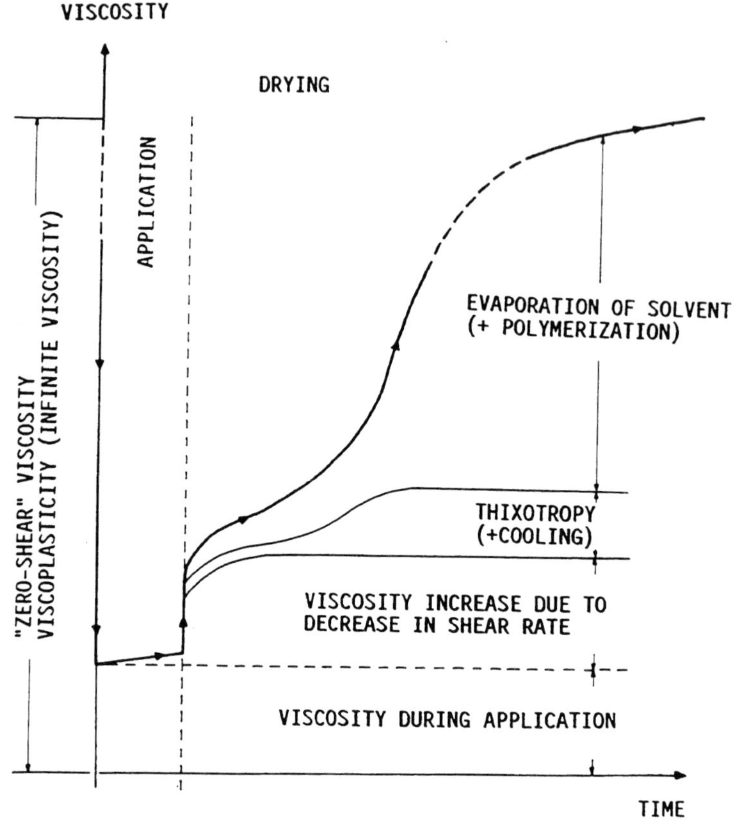

Figure 8　Schematic plot of coating viscosity during application and film formation.

11. In fact, using the measured time dependence of the viscosity, one can estimate the time t (time taken to solidify) to be used in Equation 8, as well as the time of leveling, in Equation 11. In general, if the viscosity is higher than approximately 100,000 P, then leveling and sagging phenomena occur to a negligible extent.

Experimentally, one can monitor the viscosity increase using an oscillatory technique (see Section 2.2). This method is the preferred one, since measurements can be made under the condition of low shear amplitude, which approximates the condition after a coating application. Also, the solidification point can be estimated from the measurement of the elastic modulus. To mimic the condition immediately after coating, the oscillatory measurement should be preceded by shearing at a fairly high rate, corresponding to the method of application. [36] In such an experiment, the average amplitude of the torque/stress wave increases with time after the cessation of a ramp shear. Although it is not easy to compute the viscosity change from the amplitude change, an estimation is possible.[37] Alternatively, one can use just the amplitude of the stress for correlation purposes. Dodge[36] finds a correlation between the viscosity level after application and the extent of leveling as quantified by a special technique he developed. Another method that has been used[38] involves rolling a sphere down a coating applied to an inclined surface. The speed of the sphere can be taken as an indicator of the viscosity, after suitable calibration with Newtonian fluids. This method can be very

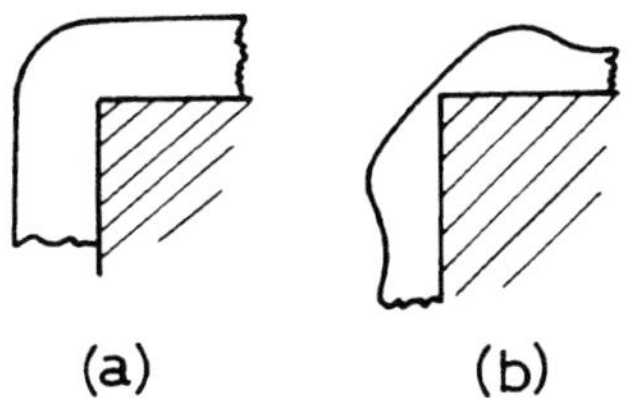

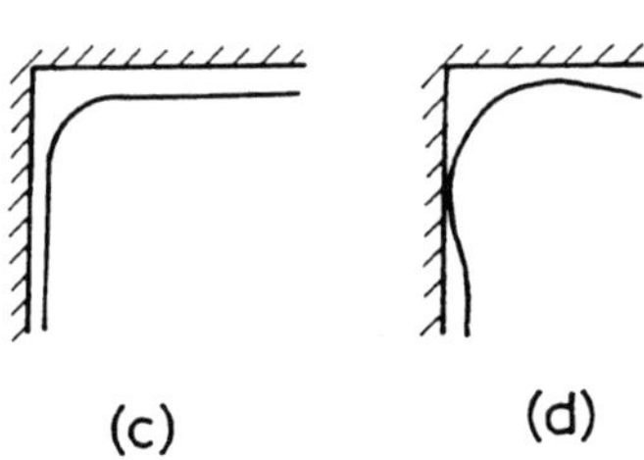

Figure 9 (a) Newly applied thick film at a corner. (b) Decrease in the film thickness at the corner due to surface tension. (c) Newly applied thin film at a corner. (d) Increase in the film thickness at the corner due to surface tension.

misleading because the flow is not viscometric, and it is not applicable to non-Newtonian fluids. A more acceptable technique is to use a simple shear, with a plate being drawn at constant velocity over a horizontal coating.[19]

3.6 Edge and Corner Effects

When a film is applied around a corner, surface tension, which tends to minimize the surface area of the film, may cause a decrease or increase in the film thickness at the corners as shown in Figures 9b and 9d respectively. In the case of edges of coated objects, an increase in the thickness has been observed. This phenomenon is related to surface tension variation with the solvent concentration.[40] In a newly formed film, a decrease in film thickness at the edge is caused by the surface tension of the film. Consequently, the solvent evaporation is much faster at the edge of the film because there is a larger surface area per unit volume of fluid near the edge (Fig. 10a). As more solvent (which usually has a lower surface tension than the polymer) evaporates, a higher surface tension exists at the edge, hence causes a material transport toward the edge from regions 2 to 1 (Fig. 10b). The newly formed surface in region 2 will have a lower surface tension due to the exposure of the underlying material, which has a higher solvent concentration. Consequently, more materials are transported from region 2 to the surrounding areas (regions 1 and 3) because of the surface tension gradient across the regions (Fig. 10c).

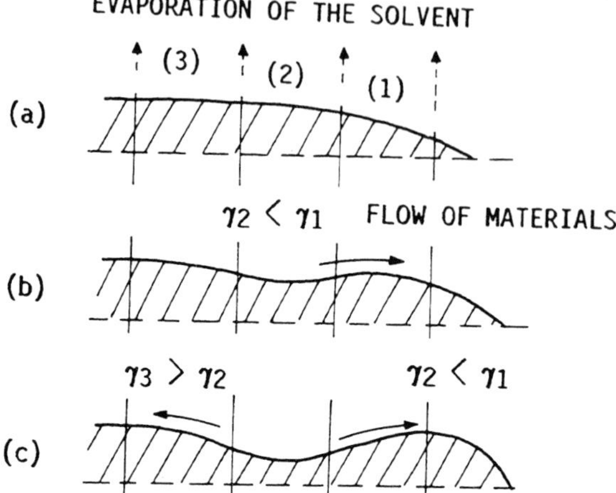

Figure 10   (a) Newly formed film near an edge. (b) Flow of materials from regions 2 to 1. (c) Further flow of materials from region 2 to the surroundings.

3.7 Depressions: Bernard Cells and Craters

Local distortions (depressions) in a coating can be caused by a surface tension gradient (due to composition variation or temperature variation). This phenomenon is known as the Maragoni effect.[41] The flow of a liquid from a region of lower to higher surface tension caused by the surface tension gradient results in the formation of depressions on the liquid surface.Such depressions come in two types: Bernard cells and craters.

Bernard cells usually appears as hexagonal cells with raised edges and depressed centers.[42–44] The increase in the polymer concentration and the cooling due to solvent evaporation cause the surface tension and surface density to exceed those of the bulk. This creates an unstable configuration, which tends to move into a more stable one in which the material at the surface has a lower surface tension and density. Theoretical analysis[45] has established two characteristic numbers: the Raleigh number R_a and the Marangoni number M_a, given by

$$R_a = \frac{\varrho \, g\alpha\tau h^4}{K\eta} \tag{13}$$

$$M_a = \frac{\tau h^2(-d\gamma/dT)}{K\eta} \tag{14}$$

where ρ is the liquid density, g is the gravitational constant, α is the thermal expansion coefficient, τ is the temperature gradient on the liquid surface, h is the film thickness, K is the thermal diffusivity, and T is the temperature. If the critical Marangoni number is exceeded, the cellular convective flow is formed by the surface tension gradient. As shown in Figure 11a, the flow is upward and downward beneath the center depression and the raised edge, respectively. But if the critical Raleigh number is exceeded, the cellular convective flow, which is caused by density gradient, is downward and upward beneath the depression and the raised edge, respectively (Fig. 11b). In general, the density-gradient-driven flow pre-

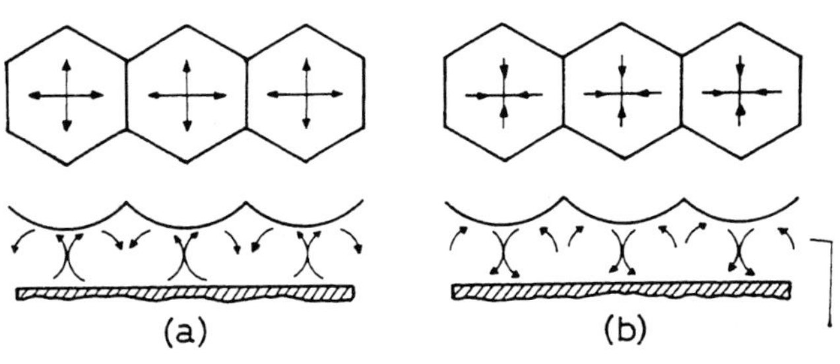

Figure 11 Schematic illustration of the formation of the Bernard cells due to (a) the surface tension gradient and (b) the density gradient.

dominates in thicker liquid layers (> 4 mm), while the surface tension gradient is the controlling force for thinner films.

Cratering is similar to the Bernard cell formation in many ways. Craters, which are circular depressions on a liquid surface, can be caused by the presence of a low surface tension component at the film surface. The spreading of this low surface tension component causes the bulk transfer of film materials, resulting in the formation of a crater. The flow q of material during crater formation is given by[46]

$$q = \frac{h^2 \Delta \gamma}{2\eta} \tag{15}$$

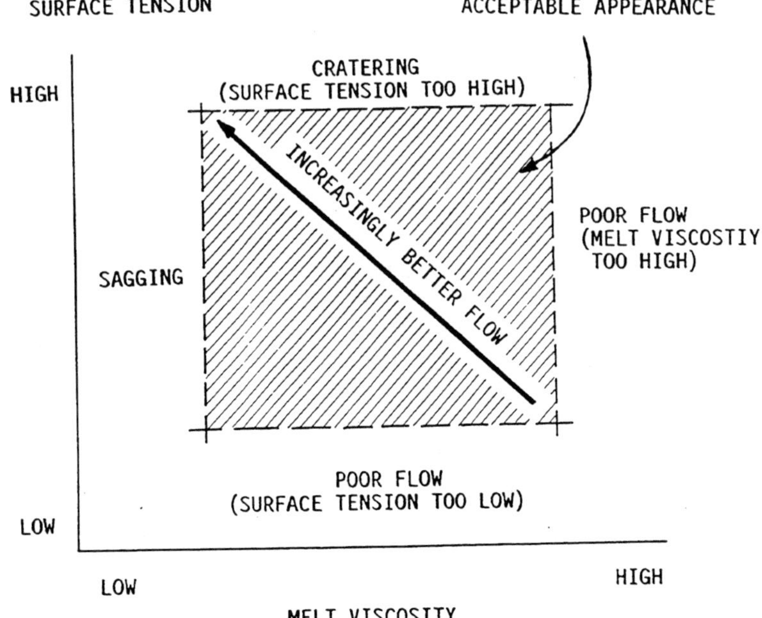

Figure 12 The effects of surface tension and melt viscosity on coating appearance.

where $\Delta\gamma$ is the surface tension difference between the regions of high and low surface tension. The crater depth d_c is given by[47]

$$d_c = \frac{3\Delta\gamma}{\varrho\, gh} \tag{16}$$

The relationship between the cratering tendency and the concentration of surfactant was investigated by Satoh and Takano.[48] Their results indicate that craters appear whenever paints contain silicon oils (a surfactant) in an amount exceeding their solubility limits.

In the discussion above, high surface tension and low viscosity are required for good flow-out and leveling. But high surface tension can cause cratering, and excessively low viscosity would result in sagging and poor edge coverage. To obtain an optimal coating, the balance between surface tension and viscosity is important. Figure 12 illustrates coating performance as a function of surface tension and melt viscosity. Coating is a fairly complex process; achieving an optimal result calls for the consideration of many factors.

ACKNOWLEDGMENTS

We are grateful to Steve Trigwell for preparing the figures.

REFERENCES

1. A. W. Adamson, *Physical Chemistry of Surfaces*, 4th ed. New York: Wiley, 1982.
2. L. Du Nouy, *J. Gen. Physiol.* 1, 521 (1919).
3. R. H. Dettre and R. E. Johnson, Jr., *J. Colloid Interface Sci.*, 21, 367 (1966).
4. D. S. Ambwani and T. Fort, Jr., *Surface Colloid Sci.*, 11, 93 (1979).
5. J. R. J. Harford and E. F. T. White, *Plast. Polym.*, 37, 53 (1969).
6. J. Twin, *Phil. Trans.*, 29–30, 739 (1718).
7. J. W. Strutt (Lord Rayleigh), *Proc. R. Soc. London*, A92, 184 (1915).
8. S. Sugden, *J. Chem. Soc.*, 1483 (1921).
9. J. M. Andreas, E. A. Hauser, and W. B. Tucker, *J. Phys. Chem.*, 42, 1001 (1938).
10. S. Wu, *J. Polym. Sci.*, C34, 19 (1971).
11. R. J. Roe, *J. Colloid Interface Sci.*, 31, 228 (1969).
12. S. Fordham, *Proc. R. Soc. London.*, A194, 1 (1948).
13. C. E. Stauffer, *J. Phys. Chem.*, 69, 1933 (1965).
14. J. F. Padday and A. R. Pitt, *Phil. Trans. R. Soc. London*, A275, 489 (1973).
15. H. H. Girault, D. J. Schiffrin, and B. D. V. Smith, *J. Colloid Interface Sci.*, 101, 257 (1984).
16. C. Huh and R. L. Reed, *J. Colloid Interface Sci.*, 91, 472 (1983).
17. Y. Rotenberg, L. Boruvka, and A. W. Neumann, *J. Colloid Interface Sci.*, 93, 169 (1983).
18. O. C. Lin, *Chemtech*, January 1975, p. 15.
19. L. Kornum, *Rheol. Acta.*, 18, 178 (1979).
20. O. C. Lin, *J. Apl. Polym. Sci.*, 19, 199 (1975).
21. *H. Freundlich and A. D. Jones*, J. Phys. Chem., Vol. 4, 40, 1217 (1936).
22. W. H. Bauer and E. A. Collins, in *Rheology* 4, F. Eirich, Ed. New York: Academic press, 1967, Chapter 8.
23. P. S. Roller, *J. Phys. Chem.*, 43, 457 (1939).
24. S. Reiner and G. W. Scott Blair, in *Rheology*, Vol 4, F. Eirich, Ed. New York: Academic Press, 1967, Chapter 9.
25. S. LeSota, *Paint Varnish. Prod.*, 47, 60 (1957).

26. R. B. Bird, R. C. Armstrong, and O. Hassager, *Dynamics of Polymeric Fluids*, Vol, 1. New York: Wiley-Interscience, 19787, p. 61.
27. S. J. Storfer, J. T. DiPiazza, and R. E. Moran, *J. Coating Technol.*, 60, 37 (1988).
28. V. G. Nix and J. S. Dodge, *J. Paint Technol.*, 45, 59 (1973).
29. T. C. Patton, *Paint Flow and Pigment Dispersion*, 2nd ed. New York: Wiley-Interscience, 1979.
30. A. G. Frederickson, *Principles and Applications of Rheology*, Prentice Hall, Englewood Cliffs, NJ, 1964.
31. S. Wu, *J. Appl. Polym. Sci.*, 22, 2769 (1978).
32. J. F. Rhodes and S. E. Orchard, *J. Appl. Sci. Res. A*, 11, 451 (1962).
33. R. K. Waring, *Rheology*, 2, 307 (1931).
34. N. O. P. Smith, S. E. Orchard, and A. J. Rhind-Tutt, *J. Oil Colour. Chem. Assoc.*, 44, 618 (1961).
35. S. Wu, *J. Appl. Polym. Sci.*, 22, 2783 (1978).
36. J. S. Dodge, *J. Paint Technol.*, 44, 72 (1972).
37. K. Walters and R. K. Kemp, in *Polymer Systems: Deformation and Flow*. R. E. Wetton and R. W. Wharlow, Eds. New York: Macmillan, 1967, p. 237.
38. A. Quach and C. M. Hansen, *J. Paint Technol.*, 46, 592 (1974).
39. L. O. Kornum and H. K. Raaschou Nielsen, *Progr. Org. Coatings*, 8, 275 (1980).
40. L. Weh, *Plaste Kautsch*, 20, 138 (1973).
41. C. G. M. Marangoni, *Nuovo Cimento*, 2, 239 (1971).
42. C. M. Hansen and P. E. Pierce, *Ind. Eng. Chem. Prod. Res. Dev.*, 12, 67 (1973).
43. C. M. Hansen and Pierce, *Ind. Eng. Chem. Prod. Res. Dev.*, 13, 218 (1974).
44. J. N. Anand and H. J. Karma, *J. Colloid Interface Sci.*, 31, 208 (1969).
45. J. R. A. Pearson, *J. Fluid Mech.*, 4, 489 (1958).
46. P. Fink-Jensen, *Farbe Lack*, 68, 155 (1962).
47. A. V. Hersey, *Phys. Ser.*, 2, 56, 204 (1939).
48. T. Satoh and N. Takano, *Colour Mater.*, 47, 402 (1974).

3
Leveling

D. Satas

Satas & Associates, Warwick, Rhode Island

1.0 INTRODUCTION

A coating is applied to a surface by a mechanical force: by a stroke of a brush, by transfer from a roll, by removing the excess with a knife's edge, or by other means. Most of these coating processes leave surface disturbances: a brush leaves brush marks; a reverse roll coater leaves longitudinal striations; knife coating leaves machine direction streaks; roll coating leaves a rough surface, when the coating splits between the roll and the substrate; and spraying may produce a surface resembling orange peel.

2.0 YIELD VALUE

These surface disturbances may disappear before the coating is dried, or they may remain, depending on the coating properties and time elapsed between the coating application and its solidification. The surface leveling process is driven by surface tension and resisted by viscosity. Some coatings, especially thickened aqueous emulsions, may exhibit pseudo-plastic flow characteristics and may have a yield value: minimum force required to cause the coating to flow (see Fig. 1). For such coating to level, the driving force (surface tension) must be higher than the yield value. Solution coatings are usually Newtonian (have no yield value) and level rather well. Hot melt coatings solidify fast and may not level adequately. Some typical yield values for various coatings are given in Table 1.

Viscosity measurements at very low shear rates are required to determine the yield value. Some of the shear rates experienced in various processes are shown in Table 2. A Brookfield viscometer, which operates at shear rates of 0.6–24 sec^{-1}, is not suitable for investigating the leveling effects that appear at much lower shear rates. Shear rates experienced during various coating processes are very high, and the viscosity measurements at low shear rates might not disclose coating behavior at these high shear rates.

A yield value of 0.5 dynes/cm^2 produces very fine brush marks, while a yield value of 20 dynes/cm^2 produces pronounced brush marks. The yield stress necessary to suppress sagging is estimated at 5 dynes/cm^2, which is not compatible with the best leveling properties.

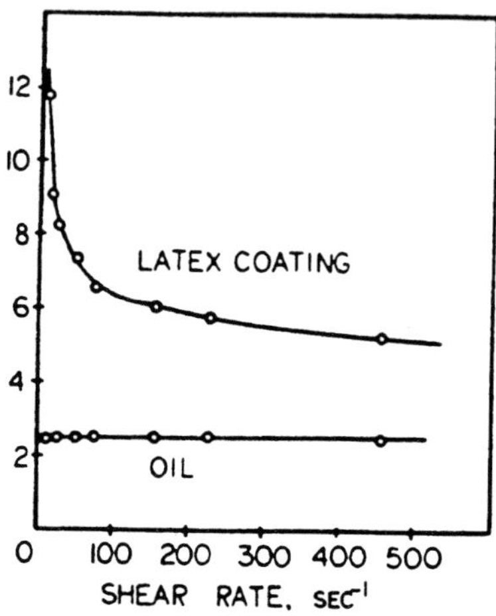

Figure 1   Viscosity of pseudoplastic emulsion coating as a function of shear rate compared to Newtonian oil.

Therefore, to produce both adequate leveling and nonsagging behavior, thixotropic properties must be introduced to a coating.

3.0 LEVELING AND VISCOSITY

The driving force of the leveling process is surface tension. The force that resists leveling is viscosity and to a lesser degree, the elasticity of the coating. To facilitate leveling, therefore, it is desirable to employ coating of low viscosity. Low viscosity coatings, however, cannot always be used. It is difficult to deposit heavy coatings if the viscosity is low. If the coating is applied on a vertical surface, a high viscosity is needed to prevent sagging.

Aqueous coatings are often pseudoplastic: they exhibit a rate-dependent viscosity. They may have a low viscosity (20–30 mPa·sec) at high shear rates (10^4 sec^{-1}), such as experienced in roll coating operations, and a much higher viscosity (1–3 Pa·sec) at low shear rates

Table 1   Yield Values and Thixotropy Ratings for Various Coatings

Coating type	Yield Value (dynes/cm^2)	Thixotropy rating
Enamels, glossy	0–30	None to slight
Enamels, semiglossy	50–120	Slight
Flat paints	20–100	Slight to marked
Aqueous wall coatings	10–100	Slight to marked
Metal primers	0–100	None to marked
Varnishes	0	None

TABLE 2 Shear Rates for Various Proesses

Process	Shear rate (sec^{-1})
Leveling	0.01–0.10
Pouring if solution	10–100
Mixing	50–1000
Reverse roll coating	100–10,000
Knife over roll coating	1000–10,000
Brushing	10,000–50,000
Spraying	10,000–70,000

(0.01–10 sec^{-1}, prevalent during leveling. Such coatings do not level well because of the high viscosity at low shear rates.

The rheology index is sometimes used as an indicator of the leveling capabilities of a coating. The rheology index is defined as the ratio of high shear rate viscosity to low shear rate viscosity. If the rheology index is one, the coating is Newtonian; if it is larger than one, the coating is dilatant; if it is smaller than one, the coating is pseudoplastic. A large rheology index favors good leveling; it should exceed 0.25 for aqueous systems of acceptable leveling properties.

3.1 Thixotropy

Thixotropic behavior of some coatings is utilized to circumvent the dilemma of having a coating of sufficiently low viscosity that levels well and a coating of sufficiently high viscosity that does not sag. Thixotropy is the dependence of viscosity on time. There are coatings that retain a low viscosity for a short period after shearing, thus allowing good leveling, but thicken fast enough to prevent sagging.

The thixotropic behavior of a coating is usually characterized by the thixotropy loop as shown in Figure 2. At the beginning, the coating is sheared at a continuously increasing shear rate, producing curve *a*. Then the coating is sheared at a constant shear rate until constant viscosity (curve *b*) is reached. The shear rate is then gradually decreased, producing curve *c*. The area enclosed by the thixotropy loop indicates the degree of thixotropy:: the larger the area, the more pronounced is the thixotropic behavior of the coating. A pseudoplastic coating that exhibits no thixotropy would form no loop and curve *a* would coincide with curve *c*. Curve *b* would not be formed, because the viscosity is not time dependent in nonthixotropic coatings. Thixotropic behavior is quite common in many aqueous coatings and high viscosity inks, and it is utilized to improve the coatability.

4.0 LEVELING AND SURFACE TENSION

If the coating contains ingredients of differing surface tension and volatility, a surface tension gradient may be formed during drying, which results in poor leveling.[1] This behavior was observed when a drop of alkyd resin was dried in heptane solution. A high surface tension area is created around the outer edges, because of a faster solvent evaporation rate in

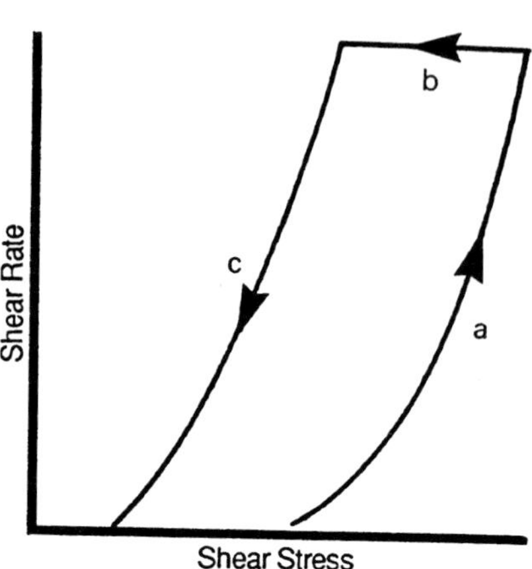

Figure 2   Thixotropic loop.

that region. This causes a flow of the solution from the center of the drop, resulting in formation of a donut-shaped resin deposit. When xylene is employed as the solvent, the result is the opposite. A region of lower surface tension is produced around the outer perimeter of the drop, resulting in a thicker center of the dried deposit. The addition of a solvent-soluble surface active agent eliminates the formation of the surface tension gradient, resulting in a dried deposit of uniform thickness.

5.0   LEVELING OF BRUSH AND STRIATION MARKS

Brush application produces brush marks dependent on the rush fineness,. Reverse roll and other roll coaters in which the coating splits between the rolls produce regular longitudinal striation marks in the coating. Knife coaters produce longitudinal streaks, which are caused by the restriction of the flow under the knife by gel and other particles. The geometry of such coating striations or brush marks is shown schematically in Figure 3.

According to Orchard's[2] mathematical model of the leveling process, the leveling half-time may be expressed in the following manner:
where γ is the surface tension and η_L the viscosity (at low shear rate).

The equation states that a low viscosity and a high surface tension favor the rate of leveling. The geometry of the coating (distance between striation marks and the coating thick-

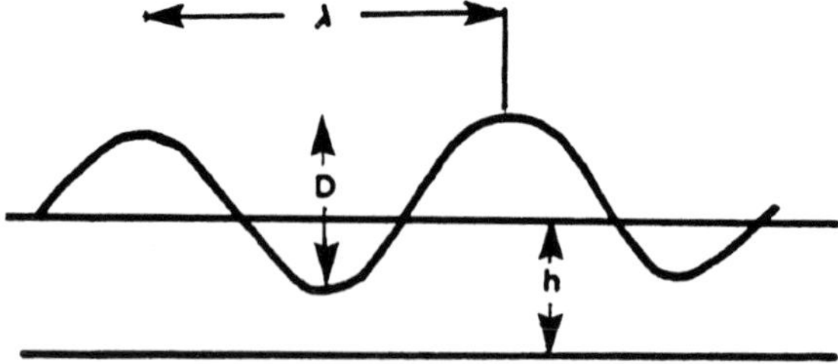

Figure 3   Profile of coating striation marks.

$$T_{1/2} \propto \frac{\eta_L}{\gamma} \frac{\lambda^4}{h^3}$$

ness) has a very large effect on leveling. Decreasing the distance between the striation or brush marks is very effective in accelerating leveling, as is an increase in coating thickness. It has also been stated by some authors[3] that coating elasticity has a retarding effect on leveling, although the effect is less important than that of viscosity.

REFERENCES

1. D. A. Bultman and M. T. Pike, *J. Chem. Spec. Manuf. Assoc.*, January 1981.
2. S. Orchard, *Appl. Sci. Res., A11, 451 (1962).*
3. M. Bierman, *Rheol. Acta*, 138 (1968).

BIBLIOGRAPHY

Beeferman, H. L., and D. A. Bregren, *J. Paint Technol.*, 38(492), 9–17 (1966).
Kooistra, M. F., "Film formation phenomena: Fundamentals and Applications," in *Third International Conference in Organic Coatings Science and Technology*, G. D. Parfitt and A. V. Patsis, Eds., Westport, CT: Technomic, 1979.
Matsuda, T., and W. H. Brendly, Jr., *J. Coating Technol.*, 51(658), 46–60 (1979).
Overdiep, W. S. "The leveling of paints," in *Proceedings of the Conference on Physical Chemistry and Hydrodynamics*, Oxford, July 1–13, 1977.
Patton, T. C., *Paint Flow and Pigment Dispersion*, 2nd ed. New York: Wiley, 1979.
Satas, D., "Surface appearance," in *Plastics Finishing and Decoration*, D. Satas, Ed. New York: Van Nostrand Reinhold, 1986, pp. 17–20.
Smith, N. D. P., S. E. Orchard, and A. J. Rhind–Tutt, *J. Oil Chem. Colour. Assoc.*, 44, 618–633 (September 1961).

4

Structure–Property Relationships in Polymers

Subbu Venkatraman

Raychem Corporation, Menlo Park, California

Most of the binders used in paints, varnishes, lacquer films, and photolithographic coatings are made up of macromolecules. The final dry coating consists predominately of a polymer, either cross-linked or un-cross-linked. The material may have been polymeric before application, or cured to become a polymer after application. In either case, a knowledge of the properties of polymers as related to structural features helps in obtaining coatings with desired performance characteristics.

1.0 STRUCTURAL PARAMETERS

We begin by defining some important structural parameters of polymers.

1.1 Molecular-Weight Averages

Since all polymers contain a distribution of molecules of differing masses, it is customary to define averages of the distribution

$$M_n \text{ (number } - \text{ average)} = \frac{\sum N_i M_i}{\sum N_i}$$

$$M_w \text{ (weight } - \text{ average)} = \frac{\sum w_i M_i}{\sum w_i}$$

$$M_v \text{ (viscosity } - \text{ average)} = \left[\sum w_i M_i^\alpha \right]^{1/\alpha}$$

where N_i = number of molecules of molar mass M_i, and w_i their weight, and α is the Mark–Houwink exponent defined by

$$[\eta] = KM_v^{\alpha} \tag{1}$$

where $[\eta]$ is the intrinsic viscosity.

The number M_n is usually measured by NMR spectrometry or osmometry; M_w can be obtained via light scattering techniques, while intrinsic viscosity measurements yield estimates of M_v. Size-exclusion chromatography or gel permeation chromatography (GPA) can in principle be used to obtain all the averages mentioned above; care must be taken to ensure proper calibration of the column with standards that have the same molecular structure as the polymer of interest.

Although there exist different definitions for the breadth of a distribution, we will use the most common one involving the averages defined above:

$$\text{MWD (molecular weight dispersity)} = \frac{M_w}{M_n}$$

A value of unity for this quantity defines a "narrow distribution" polymer; a value of 2 is obtained in condensation polymers, and higher values indicate considerable breadth of molecular weights. A measure of this quantity can be obtained via GPC, or a combination of NMR and light-scattering techniques.

1.2 Molecular Weight Between Cross-Links

This is defined as the average molar mass between successive cross-link sites in a network polymer and is denoted by the symbol M_c. It is a measure of the density of cross-linking and can be estimated from measurements of the equilibrium degree of swelling or of the modulus.

1.3 Particle Size and Particle Size Distribution

In the case of latexes, many properties of the wet and dry coatings are determined by the sizes of the latex particles. Estimations can be obtained directly through scanning electron micrography (SEM) if a film can be made. For the suspension, however, it is more customary to use light-scattering techniques (Coulter model N4, Brookhaven model DCP-1000) or optical sedimentation techniques (Horiba CAPA-700). In either case, it is possible to obtain a major portion of the particle size distribution.

2.0 PROPERTIES OF WET COATINGS

Described below are some of the more important properties of coatings that are relevant to their ease of application, either in solution or as suspensions. Most wet coatings are brushed on (as with paints) or sprayed (as with some epoxies used as insulation). The solution coatings are mostly polymer based, and thus a survey of the rheological properties of polymer solutions is given; in addition, some properties of suspensions are discussed.

Table 1   Critical Molecular Weight of Source Polymers

Polymer	M_c
Polyvinyl chloride	6,200
Polyethylene	3,500
Polyvinyl acetate	25,000
Polymethyl acrylate	24,000
Polystyrene	35,000

Source: Ref. 3.

2.1   Viscosity of Polymer Solutions

Although several theories of polymer solutions[1] examine the dependence of viscoelastic properties on molecular parameters, we shall not discuss these here. Instead, we shall focus on some generally accepted empirical relationships. Most of these are covered extensively by Ferry.[2]

2.1.1   Dependence on Molecular Weight

For pure polymers, the molecular weight dependence is usually expressed by the following type of relationship:

$$\eta_0 = KM^\beta \tag{2}$$

where η_0 is the "zero-shear" viscosity, and K is a solvent- and temperature-dependent constant. The value of the exponent β is determined by the molecular weight range under consideration:

$$\text{for } M < M_c, \ \beta = 1 \quad \text{and} \quad \text{for } M > M_c, \ \beta = 3.4 \tag{3}$$

where M_c is a critical molecular weight that expresses the onset of entanglements between molecules. The magnitude of M_c is characteristic of the polymer structure; Table 1 gives some representative numbers.

Although M_c signals the onset of topological effects on the viscosity, it is not identical to the molecular weight between entanglements, M_e. (The latter quantity is estimated from the magnitude of the rubbery plateau modulus.) Approximately, we have

$$M_c \sim 2M_e \tag{4}$$

Also, M_c is a function of polymer concentration. In the pure polymer (denoted by superscript zero), it attains its lowest value, M_c^0; in a solution of concentration C, its magnitude varies as discussed in Section 2.1.2.

The exponent β assumes the values quoted in Equation 3 only if the measured viscosity is in the so-called zero-shear-rate limit. At higher rates, β assumes values lower than unity and 3.4, in the two regimes.

2.1.2 Concentration Dependence of the Viscosity

As mentioned in Section 2.1.1, below a certain concentration, C^*, entanglement effects are not significant. This concentration is estimated from:

$$C^* = \varrho \frac{M_c^0}{M} \tag{4}$$

where M is the molecular weight of the polymer in the coating solution. The concentration C^* cannot be estimated from a plot of η_0 against concentration, however; the transition is not sharp, but gradual.[4]. No single expression for the concentration exists below C^*; however, in the entangled regime, the expression

$$\eta_0 = C^5 M^{3.4} \tag{5}$$

works well for some polymers.[5,6] This relation does not hold all the way to the pure polymer, where higher exponents are found.[6] Equation 5 also does not hold in the case of polymer solutions in which there are other specific attractive forces, such as in poly(n-alkyl acrylates).[7]

There are two reasons for a reduction in viscosity of a polymer upon dilution: (a) the dilution effect, which causes the solution viscosity to be between those of the two pure components, and (b) a decrease of viscosity due to a lowering of T_g upon dilution. The latter is solvent-specific, and is the main reason for the apparent difficulty in establishing a universal viscosity–concentration relationship for polymer solutions.

2.2 Viscosity of Suspensions

Many latex paints are suspensions in water or in an organic solvent. Their rheological properties differ from those of polymer solutions in several ways. The concentration dependence is of a different form, and in addition, there is a dependence on particle size. Also, at high concentrations, these suspensions tend to have structure, which usually refers to an aggregated network. The immediate consequences of the existence of a pseudonetwork are the phenomena of yield stress and thixotropy. We explore next the relationship of these quantities to the characteristics of the particles making up the suspension.

2.2.1 Concentration Dependence of the Viscosity

In dilute suspensions, the concentration dependence is expressed by an extension of the Einstein equation:

$$\eta_{rel} = \frac{\eta}{\eta_s} = 1 + 2.5\phi + 14.1\phi^2 \tag{6}$$

where η_s is the solvent viscosity and ϕ is the volume fraction of the suspension. Equation 6 is valid for spherical particles without any interparticle interaction. Inclusion of long-range interaction (such as volume exclusion) merely changes the coefficient of the ϕ^2 term.

Of greater interest are the rheological phenomena that occur in suspensions of particles that have short-range interactions, attractive or repulsive. In a comprehensive study, Matsumoto et al.[8] have established the conditions for the existence of yield stresses in suspension. Their conclusions are:

1. For particles with repulsive interactions, no yield stresses exist.
2. Suspensions of neutral particles, or particles with attractive forces do exhibit yield stresses.
3. The magnitude of the yield stress increases with the concentration of the particles and with increasing ratio of surface area to volume.

In this study, the existence of the yield stress was inferred from the presence of a plateau in the elastic modulus G', at very low frequencies; the magnitude of the yield stress was deduced from the height of the plateau modulus.

A detailed and critical survey of the literature is given by Meitz.[9]

The other important rheological consequences of a pseudonetwork is thixotropy, defined elsewhere in this volume. The phenomenon is attributed to a time-dependent but reversible breakdown of the network.

3.0 PROPERTIES OF DRIED FILMS

3.1 The Glass Transition Temperature

The T_g is defined in various ways, but in a broad sense, it signals the onset of small-scale motion in a polymer. It is heavily influenced by the chemical structure, in particular, by the bulkiness (steric hindrance) of pendant groups. (See Van Kreleven[10] for an excellent discussion.)

The molecular weight dependence of the glass transition is fairly straight forward and is given by:

$$T_g = T_g^\infty - \frac{K}{M_n} \tag{7}$$

where T_g^∞ is the limiting value of T_g at high molecular weights.

The effect of plasticizers on T_g is well documented.[10] According to a theoretical treatment by Bueche,[11] the T_g reduction by plasticizers can be calculated from:

$$T_g \text{ (solution)} = \frac{T_{g,p} + (KT_{g,s} - T_{g,p})\phi_s}{1 + (K-1)\phi_s} \tag{8}$$

where $T_{g,p}$ and $T_{g,s}$ are the glass transition temperatures of the polymer and solvent, respectively, and ϕ_s is the volume fraction of solvent. Then we write

$$K = \frac{\alpha_{1s} - \alpha_{gs}}{\alpha_{1p} - \alpha_{gp}} \, , \; 1 < K < 3$$

where
$\quad \alpha_1 = $ volume expansion coefficient above T_g
$\quad \alpha_g = $ volume expansion coefficient below T_g

To calculate the T_g of a solution, the T_g of the solvent should be known[12] or estimated as $2T_m/3$, K is usually taken to be 2.5.

3.2 Tensile and Shear Moduli

Both tensile and shear moduli are functions of temperature, and of time in the case of viscoelastic polymers. We shall restrict our discussion to the temperature dependence of the isochronal moduli. (Since the tensile and shear moduli are related to each other through the use of an equation involving the Poisson's ratio, the comments made here on the shear modulus G can be extended to the tensile modulus E, as well.)

For semicrystalline polymers below T_g, the modulus can be estimated from:[13]

$$G = G_g + X_c^2(G_c - G_g) \tag{9}$$

where G_g, G_c are the moduli for the fully amorphous and fully crystalline polymer, respectively, and X_c is the degree of crystallinity.

Above T_g, the same equation can be used, but G_c far exceeds G_g, and Equation 9 reduces to

$$G \simeq X_c^2(G_c) \tag{10}$$

For amorphous polymers above T_g, the modulus is given by the rubber elasticity expression:

$$G_e = \frac{\varrho RT}{M_e} \tag{11}$$

where M_e^o is the entanglement molecular weight (see Section 2.1.2). This rubbery plateau for un-cross-linked polymers is observed only at molecular weights higher than M_e^o.

For cross-linked amorphous polymers above T_g (elastomers), the modulus is given in analogous fashion:

$$G \simeq \frac{\varrho RT}{M_c} \tag{12}$$

where M_c is the molecular weight between permanent cross-linked junctions. When M_c exceeds M_e, trapped entanglements also play a part in determining modulus:[14]

$$G \simeq \frac{\varrho RT}{M_c} + f_e G_e \tag{13}$$

where G_e is given by Equation 11 and f_e is a probability factor for trapped entanglements.

In the case of network imperfections, Equation 12 is modified.[14,15] The quantity f_e can be calculated if the reaction parameters for network formation are known.[14,16,17]

3.3 Other Properties

Several other properties of dried films influence its performance characteristics. Examples are the coefficient of thermal expansion, ultimate mechanical properties, stress relaxation and creep, and dielectric properties. However, correlation of these properties with structure for polymeric films is not well established; some of the more successful attempts are treated in Refs. 2 and 3.

REFERENCES

1. R. B. Bird, R. C. Armstrong, and O. Hassager, *Dynamics of Polymeric Fluids*, Vol. 2. New York: Wiley-Interscience, 1987.
2. J. D. Ferry, *Viscoelastic Properties of Polymers*. New York: Wiley, 1980.
3. D. W. Van Krevelen, *Properties of Polymers*. New York: Elsevier, 1976.
4. Ref. 2, see discussion in Chapter 17.
5. J. W. Berge and J. D. Ferry, *J. Colloid Sci., 12*, 400 (1957).
6. G. Pezzin and N. Gligo, *J. Appl. Polym. Sci., 10*, 1 (1966).
7. Ref. 2, p. 510.
8. T. Matsumoto, O. Yamamoto, and S. Onogi, *J. Rheol., 24*, 279 (1980).
9. D. W. Meitz, Ph.D. thesis, Carnegie-Mellon University, December 1984.
10. Ref. 3, p. 383.
11. F. Bueche, *Physical Properties of Polymers*. New York: Wiley, 1962.
12. Ref. 3, p. 384.
13. Ref. 3, p. 266.
14. E. M. Valles and C. W. Macosko, *Macromolecules, 12*, 673 (1979).
15. P. J. Flory, *Principles of Polymer Chemistry*. Ithaca, NY: Cornell University Press, 1953, p. 458.
16. M. Gottlieb, C. W. Macosko, G. S. Benjamin, K. O. Meyers, and E. W. Merill, *Macromolecules, 14*, 1039 (1981).
17. D. S. Pearson and W. W. Graessley, *Macromolecules, 13*, 1001 (1980).

5
The Theory of Adhesion

Carl A. Dahlquist

3M Company, St. Paul, Minnesota

When pressure-sensitive adhesive is applied to a smooth surface, it sticks immediately. The application pressure can be very slight, not, more than the pressure due to the weight of the tape itself. The adhesive is said to "wet" the surface, and, indeed, if the tape is applied to clear glass and one views the attached area through the glass, it is found that in certain areas the adhesive–glass interface looks like a liquid–glass interface. From this one would infer that a pressure-sensitive adhesive, even though it is a soft, highly compliant solid, also has liquidlike characteristics. Some knowledge of the interaction between liquids and solids is beneficial to the understanding of adhesion.

1.0 CONTACT ANGLE EQUILIBRIUM

When a drop of liquid is placed on a surface of a solid that is smooth, planar, and level, the liquid either spreads out to a thin surface film, or it forms a sessile droplet on the surface. The droplet has a finite angle of contact (Fig. 1). The magnitude of the contact angle depends on the force of attraction between the solid and the liquid and the surface tension of the liquid. The contact angle equilibrium has received a great deal of attention, principally because it is perhaps the simplest direct experimental approach to the thermodynamic work of adhesion.

Many years ago Young[1] proposed that the contact angle represents the vectorial balance of three tensors, the surface tension of the solid in air (γ_{sa}), the surface tension of the liquid in equilibrium with the vapor (γ_{lv}), and the interfacial tension between the solid and the liquid (γ_{sl}). The force balance can be written

$$\gamma_{sa} = \gamma_{lv} \cos\theta + \gamma_{sl} \tag{1}$$

Young's equation has come under criticism on the grounds that the surface tension of a solid is ill defined, but most surface chemists find his equation acceptable on theoretical grounds.

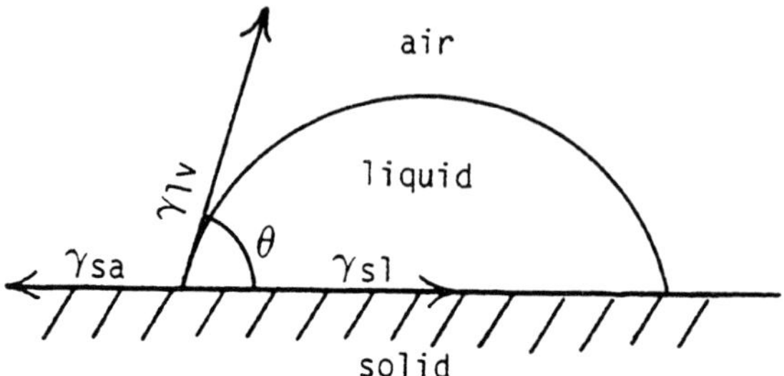

Figure 1 Contact angle equilibrium of surface and interfacial tensions.

The equation can be written as a force equilibrium or as an energy equilibrium, because the surface tension, expressed as a force per unit of length, will require an energy expenditure of the same numerical value when it acts to generate a unit area of new surface.

Harkins and Livingston[2] recognized that Young's equation must be corrected when the exposed surface of the solid carries an adsorbed film of the liquid's vapor. The solid–"air-plus-vapor" tensor, γ_{sv}, is less than the solid–air tensor, γ_{sa}. Harkins and Livingston introduced a term, πe, to indicate the reduction thus:

$$\gamma_{sv} = \gamma_{sa} - \pi e \tag{2}$$

and the modified Young equation becomes accordingly

$$\gamma_{sa} = \gamma_{lv} \cos\theta + \gamma_{sl} + \pi e \tag{3}$$

The work of adhesion (W_a) is the thermodynamic work necessary to separate the liquid from the solid without performing any additional work, such as viscous or elastic deformation of either the liquid or the solid. Dupré[3] is credited with the definition of the work of adhesion. When a liquid is separated from a solid, new solid–air and liquid–air interfaces are created, and solid–liquid interface is destroyed (Fig. 2).

Dupré equation becomes

$$W_a = \gamma_{lv} + \gamma_{sa} - \gamma_{sl} \tag{4}$$

Combining Equations 4 and 3 yields

$$W_a = \gamma_{lv}(1 + \cos\theta) + \pi e \tag{5}$$

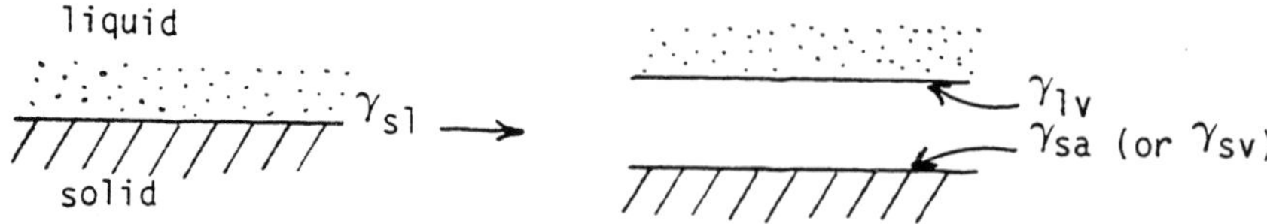

Figure 2 $W_a = \gamma_{lv} + \gamma_{sa} - \gamma_{sl}$.

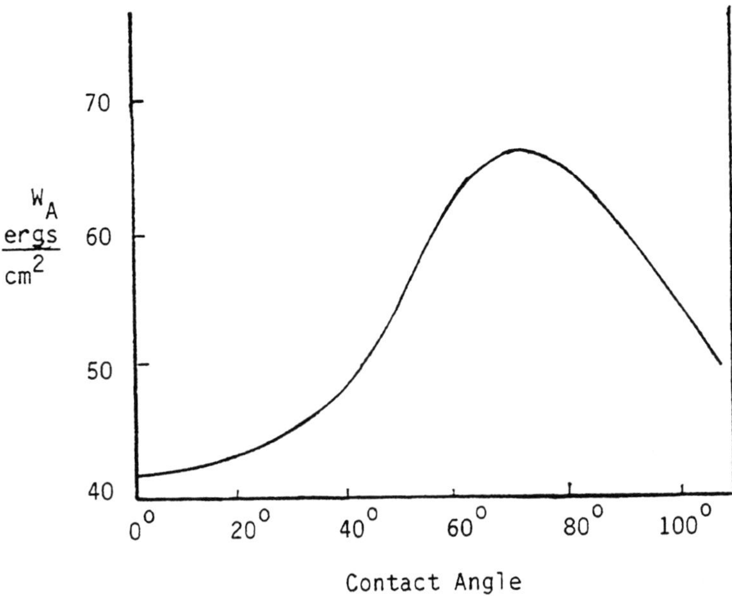

Figure 3 Work of adhesion of liquids on a surface having a γ_c of 21.5 ergs/cm².

Under some circumstances πe is negligible, and Equation 5 simply reduces to the original Young–Dupré equation

$$W_a = \gamma_{lv} \, (1 + \cos\theta) \tag{6}$$

The work of adhesion can then be obtained by measuring the surface tension of the liquid and the contact angle; γ_{sa} need not be known.

Zisman and his associates[4,5] at the Naval Research Laboratories have done extensive and careful investigation of the contact angles of liquids on solids. They found that when the surface tensions of a family of liquids were plotted against the cosines of their contact angles on a given solid, the data fell, with some scatter, along a relatively straight line. The surface tension value at which this line intersected the ordinate [i.e., where $\cos\theta = 1.00$ (or $\theta = 0°$)], was designated the critical surface tension, γ_c.

Some have equated γ_c of a solid to its surface tension or surface energy, but this was never done or condoned by Zisman. The critical surface tension of a solid is simply the highest surface tension among the surface tensions of liquids that will spread on the solid. It is, of course, a measure of the attractive force a liquid experiences when it comes in contact with the solid.

Some have postulated that a contact angle of 0° is essential for the formation of a strong adhesive bond. This is said to be requisite for wetting. Wetting, however, is a relative term. Surely the molecules of a liquid and a solid will be in close contact at the interface even though the contact angle is greater than 0°. In fact, if one measures the works of adhesion between a family of liquids and a solid, one finds, as illustrated in Figure 3, that the work of adhesion can maximize at a relatively high contact angle. Thus spreading is not requisite for good adhesion on thermodynamic grounds, but it may be highly desirable in real situations for maximizing contact area and minimizing interfacial flaws and defects.

2.0 FORCES OF ATTRACTION

All atoms or molecules, when brought into sufficiently close proximity, exhibit attraction for each other as a result of asymmetric charge distributions in their electron clouds. Under special circumstances, if there are fixed electrical charge distributions (e.g., in electrets) or fixed magnetic polarization, repulsion may occur, but these are special and rare cases. Repulsion also occurs when matter is compressed to the point that the outer electron orbits begin to interpenetrate each other; but when electrical charges and magnetic domains are randomly distributed, and the liquids and solids are not under extremely high compressive forces, the forces of attraction prevail.

The forces take several forms. The forces due to the fluctuation in the electron cloud density are universal. These forces were first recognized by Fritz London[6] and are sometimes called the "London forces," but more frequently they are called "dispersion forces." The fluctuation in electron density produces temporary electrical dipoles, which give rise to universal attraction in matter. These are the forces that account for cohesion and adhesion in aliphatic hydrocarbons. They are relatively weak atomic interactions, but since all the atoms are involved, the cumulative effect is great. The dispersion forces depend markedly on the distance of separation, decreasing as the seventh power of the separation between isolated atom pairs, and as the third or fourth power of the separation between planar surfaces.

The dispersion forces are common to all matter and can account for universal adhesion in condensed media, but other forces can contribute substantially to adhesion in specific cases. There may be attractive forces between permanent dipoles, where the potential energy of head-to-tail interaction of two dipoles having dipole moments μ_1 and μ_2 is given in the lowest energy configuration, as

$$U_{nt}^P = \frac{-2\mu_1\mu_2}{r^3}$$

where r is the center-to-center distance between the dipoles.

If the rotational energy is less than the thermal energy of the system, then

$$U_K = \text{Keesom potential} = \frac{-2\mu_1^2\mu_1^2}{3kTr^6}$$

where k is Boltzmann's constant (0.0821 l·atm/mol deg) and T is absolute temperature (K).

There may be dipole-induced dipoles, where the potential energy of interaction is given by

$$U^I = \frac{-\mu_1^2\alpha_2 + \mu_2^2\alpha_1}{r^6}$$

where α_2 and α_1 are the molecular polarizabilities.

There may be acid–base interactions[7,8] across the interface which can lead to strong bonding. Examples are hydrogen bonding, Lewis acid–base interactions, and Brênsted-type acid–base interactions.

Covalent bonding between adhesive and adherend, if achievable either by chemical reactions or by high energy radiation, can lead to very strong bonds.

Interdiffusion, usually not achievable except between selected polymers, can also lead to high adhesion.

The force of attraction between planar surfaces has been derived from quantum mechanical considerations by Casimir, Polder,[9] and Lifshitz.[10]

Lifshitz calculated the attractive forces between nonmetallic solids at distances of separation sufficiently large that the phase lag due to the finite velocity of electromagnetic waves becomes a factor. He obtained the following relationship between the attractive force and the known physical constants:

$$F = \frac{\pi^2 hc(e_o - 1)[\phi(e_o)]}{240d^4(e_o + 1)}$$

where F is the attractive force per unit of area, h is Planck's constant, C is the velocity of light, d is the distance of separation, e_o is the dielectric constant, and $\phi(e_o)$ is a multiplying factor that depends on the dielectric constant as follows:

$1/e_o$:	0	0.025	0.10	0.25	0.50	1
$\phi(e_o)$:	1	0.53	0.41	0.37	0.35	0.35

Strictly speaking, the dielectric constant in this expression should be measured at electron orbital frequency, about 10^{15} Hz. However, if we assume handbook values of the dielectric constant at 10^6 Hz, which for nylon, polyethylene, and polytetrafluoroethylene, are 3.5, 2.3, and 2.0, respectively, the corresponding $\phi(e_o)$ values are 0.37, 0.36, and 0.35. The force values then stand in the ratios 0.11 to 0.056 to 0.039. When normalized to F (nylon) = 1.0, they fall to the following ratios:

$$F(\text{nylon}), = 1.00; \quad F(\text{PE}), 0.51; \quad F(\text{PTFE}), 0.35$$

When the γ_c values (dynes/cm) of these three materials are similarly normalized to the γ_c values for nylon, the values fall in remarkedly similar ratios.

	Nylon	PE	PTFE
γ_c	56	31	18.5
γ_c(norm)	1.00	0.55	0.33

In the Lifshitz equation, the force of attraction is shown to decrease as the inverse fourth power of the distance of separation. However, when the separation becomes so small that the phase lag in the interaction no longer is significant (it is of the order of 6° at a separation of 50 Å), the attractive force varies as the inverse third power of the distance of separation. This has been verified experimentally,[11] although the direct measurement is extremely difficult (Fig. 4). The forces existing at separations greater than 50 Å contribute very little to adhesion.

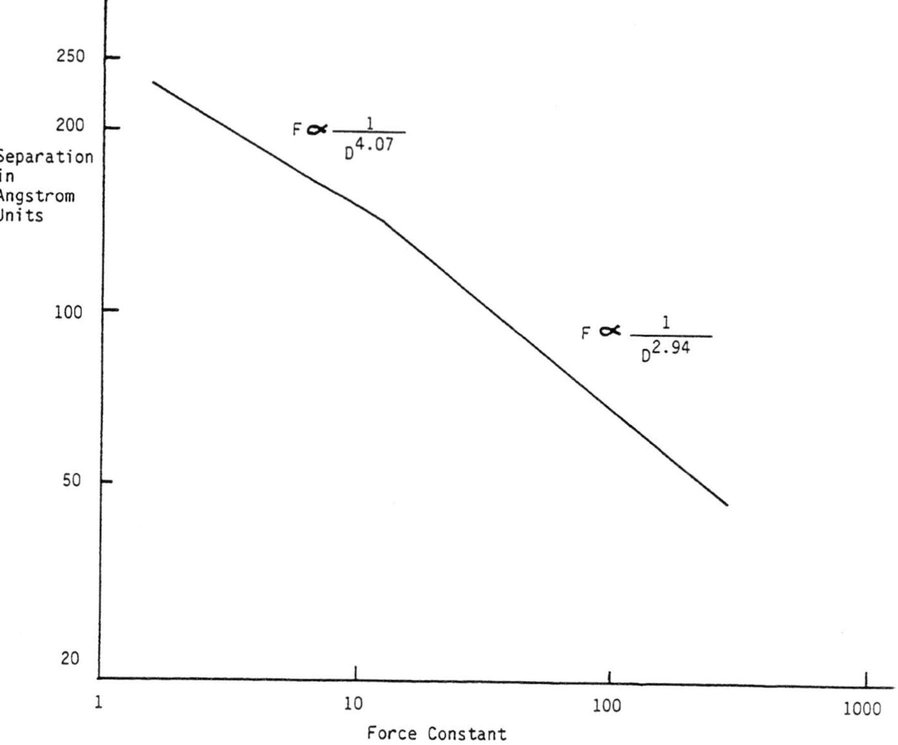

Figure 4 Attraction between ideally planar solids.

Some 30 years ago Good and Girifalco reexamined the interfacial tensions between dissimilar liquids and developed a theory of adhesion.[12] They found that the work of adhesion, given by

$$W_a = \gamma_{L_1} + \gamma_{L_2} - \gamma_{L_1 L_2}$$

could be approximated quite well by the geometric mean of the works of cohesion of the two liquids when the only attractive forces of cohesion are dispersion forces:

$$W_a = 2(\gamma_{L_1}\gamma_{L_2})^{1/2}$$

However, in some liquid pairs (e.g., water and hydrocarbons), this did not hold, and they coined an "interaction parameter," Φ, given by

$$\Phi = \frac{\gamma_{L_1} + \gamma_{L_2} - \gamma_{L_1 L_2}}{2(\gamma_{L_1}\gamma_{L_2})^{1/2}}$$

Thus

$$W_a = 2\Phi(\gamma_{L_1}\gamma_{L_2})^{1/2}$$

For water on a paraffinic hydrocarbon, where the contact angle is 108°, Φ would have a value of about 0.55. For hexadecane on polyethylene, Φ is very near unity. Good and his associates[11,12] have provided directions for calculating Φ, and they give experimental and calculated values for several combinations of water and organic liquids.

Fowkes[13] approached the problem from a different point of view. He reasoned that the only forces operable at the interface between water and an aliphatic hydrocarbon molecule contains no hydrogen bonding groups and no fixed dipoles.

Fowkes also assumed that the work of adhesion would be given by twice the geometric mean of the surface energies of the two liquids on either side of the interface, but now taking into consideration only the dispersion force components of the surface energies. For the work of adhesion between water (L_1) and n-octane (L_2), we have

$$W_a = 2\left(\gamma_{L_1}^{D}\gamma_{L_2}^{D}\right)^{1/2} = \gamma_{L_1} + \gamma_{L_2} - \gamma_{L_1 L_2}$$

where the superscript D stands for the dispersion energy component of the total surface energy. Accepted values for the surface energies and interfacial energies are:

$$\gamma_{L_1} = 72.8 \text{ ergs/cm}^2; \quad \gamma_{L_2} = \gamma_{L_2}^{D} = 21.8 \text{ ergs/cm}^2; \quad \gamma_{L_1 L_2} = 50.8 \text{ ergs/cm}^2$$

If these values are substituted into the equation above to solve for $\gamma_{H_2O}^{D}$, we get 22.0 ergs/cm². Fowkes evaluated several water–aliphatic hydrocarbon systems and found that they all yielded essentially the same value for the dispersion energy component of the surface energy of water, 21.8 ± 0.7 ergs/cm².

Turning now to the work of adhesion and the interfacial energy between mercury and aliphatic hydrocarbon, Fowkes calculated the dispersion energy component of the surface energy of mercury.

Using n-octane as the hydrocarbon liquid having a surface energy of 21.8 ergs/cm² (all of it attributed to dispersion forces), the surface energy of mercury, 484 ergs/cm², and the interfacial energy, 375 ergs/cm², we have

$$W_a = 2(\gamma_{Hg}^{D}\gamma_{n\text{-oct}}^{D})^{1/2} = \gamma_{Hg} + \gamma_{n\text{-oct}} - \gamma_{(Hg, n\text{-oct})}$$

$$W_a = 2(\gamma_{Hg}^{D} \times 21.8)^{1/2} = 484 + 21.8 - 375$$

$$\gamma_{Hg}^{D} = 196.2$$

The average γ_{Hg}^{D} for a series of mercury–aliphatic hydrocarbon systems yielded 200 ± 7 ergs/cm² for the dispersion energy component of the surface energy of mercury.

Since the remaining forces that contribute to the surface energy of mercury are metallic forces, the only interacting forces at the water–mercury interface are the dispersion forces, and the work of adhesion is given by:

$$W_a = 2(200 \times 21.8)^{1/2} = 484 + 72.8 - \gamma_{(Hg, H_2O)}$$

from which

$$\gamma_{(Hg, H_2O)} = 424.7 \text{ ergs/cm}^2$$

This compares very favorably with the measured value of 426 ergs/cm^2.

The work of adhesion due to dispersion forces is numerically small in work or energy units. For example, the work of adhesion of methylene iodide on polyethylene is 82 ergs/cm^2 ($\theta = 52°$). This small value is not, however, indicative of a small force of attraction across the interface. Keep in mind that the work is the product of force and displacement, and that the attractive force, at separation distances less than 50 Å (5×10^{-7} cm) increases as the inverse of displacement raised to the third power (Fig. 4).

The molecules at the interface are at an equilibrium distance of separation where attractive forces and repulsive forces balance. The variation in the repulsive forces with distance of separation has a dependence several orders of magnitude higher than the attractive forces (of the order of 10^{12} for atom pairs and 10^8 for repulsion forces across a hypothetical plane). We can calculate the maximum force of attraction by equating the work of adhesion to the work of separation.

Let F_a indicate the attractive force, F_r the repulsive force, x the distance separation, and d the equilibrium distance. We cannot measure d directly, but we can estimate it from calculations of the distance between molecular centers in a liquid of known specific gravity and molecular weight. In the case of methylene iodide (sp g 3.325, mol wt 267.9) we calculate the separation to be about 5×10^{-8} cm between the centers of adjacent molecules.

If we take 5×10^{-8} cm as a reasonable distance of separation across the interface between methylene iodide and polyethylene, and we accept the force versus distance relationships for attraction (a) and repulsion (r) we can write:

$$F_a = (F_a)_e \left(\frac{d}{x}\right)^3$$

$$F_r = (F_r)_e \left(\frac{d}{x}\right)^8$$

where the subscript e stands for "equilibrium." At equilibrium we have the condition that $(F_a)_e = (F_r)_e$. We can then express the work of adhesion as

$$W_a = \int_d^\infty (F_e)\left(\frac{d}{x}\right)^3 dx - \int_d^\infty (F_e)\left(\frac{d}{x}\right)^8 dx$$

The solution is

$$W_a = F_e \left[\frac{d}{2} - \frac{d}{7} \right]$$

For methylene iodide on polyethylene, W_a is 82 ergs/cm^2. Taking d as 5×10^{-8} cm, $F_e = F = F_r = 4.92 \times 10^9$ dynes/cm^2.

The maximum attractive force is encountered where the difference between the attractive forces and the repulsive forces maximizes as separation proceeds. This occurs where $(d/x)^3 - (d/x)^8$ maximizes, at about $x = 1.22d$.

At this displacement $F = 0.347F_e$, or, in the case of methylene iodide and polyethylene, at 1.71×10^9 dynes/cm^2 (about 25,000 psi). This would be the maximum attractive force experienced when separation of the materials is attempted; it far exceeds the average stresses that are typically observed when adhesive bonds are broken.

Others have calculated theoretical forces of adhesion by other approaches. All yield results that predict breaking strengths far exceeding the measured breaking strengths.

3.0 REAL AND IDEAL ADHESIVE BOND STRENGTHS

How, then, does one account for the fact that theoretical or ideal bond strengths do not seem to be attainable? The measured cohesive strengths of solids also fall far short of theoretical values. Only whisker crystals of silicon, graphite, iron, and the like have measured tensile strengths approaching their theoretical tensile strengths.

J. J. Bikerman contended that all adhesive bonds are flawed and contain weak boundary layers, hence always fall short of their theoretical strengths. Undoubtedly this is often the case; but flaws, at least gross flaws, can be minimized, and fracture in weak boundary layers, which can be detected, is not always observed.

All destructive tests, whether done in tension, shear, or peel, involve stress concentrations. Unless the stress is uniformly distributed over a very small area, as in the tensile strength tests on the "whiskers," there will be localized stresses that far exceed the average stress. It is well known that the measured average breaking strengths of adhesive bonds broken in tension decrease as the adhesive thickness increases (Fig. 5), and it can be shown that tensile stresses at the periphery of the bond are higher than interior stresses. If a flaw leads to highly localized stress, cleavage may be initiated and proceed catastrophically.

Bonds pulled in shear not only experience shearing stress concentrations but also tearing stresses.[14,15] Peeling is a deliberate application of stress concentration.

Deformation and flow also contribute to stress concentrations and failure. It would be advantageous to match the mechanical properties of the adhesive to the mechanical properties of the adherend, but this is rarely feasible. Instead, the adhesive designer resorts to the expedient of toughening the adhesive and incorporating materials that arrest crack propagation, thereby maximizing the work necessary to destroy the bond.

In many instances the adhesive–adherend interface is more accurately described as an "interphase." A case in point is the bonding of aluminum to aluminum with structural adhesive. The surface of aluminum is really aluminum oxide, which, depending on the manner in which it was formed, will vary in strength and porosity. The adhesive penetrates and locks into the oxide film, and the bond strength can be greatly enhanced by use of a surface treatment that produces a strong, well-bonded oxide layer.

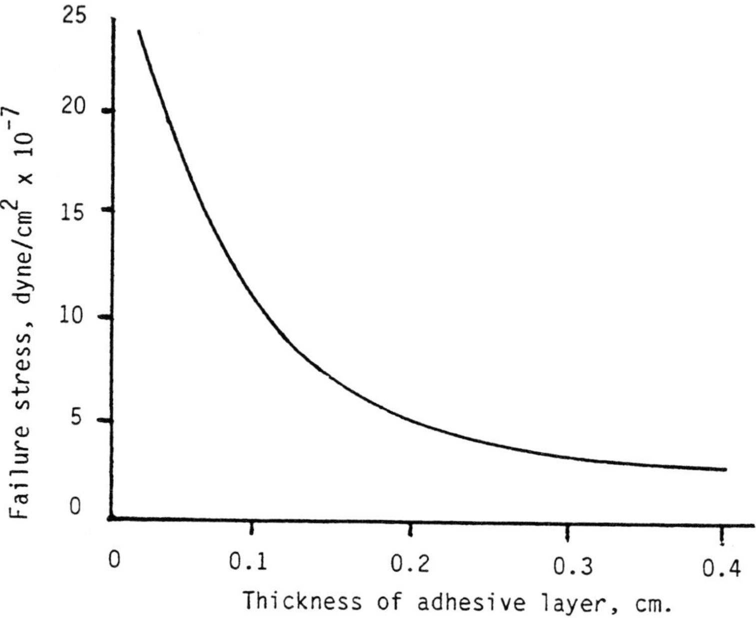

Figure 5   Adhesive layer thickness and strength of butt joints.

Adhesive–adherend bond strengths are often enhanced by priming. A classic example is the bonding of vulcanized rubber to steel, in which the steel is first electroplated with a thin coat of copper and the rubber compound is cured on the copper surface under heat and pressure. It is believed that the sulfur in the vulcanizate bridges to the copper by chemical bonding to give a strong bond, and the copper in turn is strongly bonded to the steel.

Priming is often used on bonding adhesives to plastic films. For example, the first transparent pressure-sensitive tape, which comprised a cellophane film and a natural rubber rosin adhesive, would undergo separation of the adhesive from the film under humid conditions. The problem was solved by first applying to the cellophane, a thin prime coat, a blend of natural rubber and casein, then coating the adhesive over the primer.

Surfaces notoriously difficult to bond to, such as polyethylene, polypropylene, and Teflon, are modified by treatments that make the surfaces more polar and possible chemically active, for example, by corona, plasma, or chemical treatments. These treatments may also remove weak boundary layers.

Though strong, durable adhesive bonds are usually the goal of adhesive technology, there is also a need for bonds that are deliberately made weak. This need arises in the pressure-sensitive tape industry, where it is desirable to have tape that unwinds easily from the roll, and especially where pressure-sensitive adhesives are to be transferred from a carrier film to another surface. The surfaces that provide easy release typically have low critical surface energies, almost totally dominated by the "dispersion energy" component. Also, for release coatings to function well, there must be no mutual solubility between them and the adhesives. Silicone release coatings, which consist mainly of polydimethyl siloxane, provide the easiest release. They have low critical surface energies, though not as low as certain fluorocarbon polymers. They also have a high degree of incompatibility with the pressure-sensitive adhesives that release well from them, but these criteria alone do not explain the low level of adhesion. In addition, they differ from other release coatings by being soft and elastic rather than hard. This feature may serve to enhance the stress concentration when the adhesive is separated from the release liner.

In conclusion, adhesive bond strengths measured by destructive tests will never approach theoretical values, but the intrinsic attractive forces can be manipulated by the choice of materials and surface treatments to produce a wide range of practical bond strengths.

REFERENCES

1. T. Young, *Phil. Trans. R. Soc. London, 95,* 65 (1805).
2. W. D. Harkins and H. K. Livingston, *J. Chem. Phys., 10,* 342 (1942).
3. A. Dupré, *Théorie Mécanique de la Chaleur,* Paris, 1869, p. 393.
4. W. A. Zisman, *Ind. Eng. Chem., 55,* 18 (1963).
5. E. G. Shafrin, in *Polymer Handbook,* J. Brandrup and E. M. Immergut, Eds. New York: Wiley-Interscience, 1966, pp. 111–113.
6. F. London, *Trans. Faraday Soc., 33,* 8 (1936).
7. R. S. Drago, L. B. Parr, and C. S. Chamberlain, *J. Am. Chem. Soc., 99,* 3203 (1977).
8. F. M. Fowkes, *J. Adhes. Sci. Technol., 1,* 7 (1987).
9. H. B. C. Casimir and D. Polder, *Phys. Rev., 73,* 360 (1948).
10. E. M. Lifshitz, *C.R. Acad. Sci. USSR, 97,* 643 (1954).
11. D. Tabor and R. N. S. Winterton, *Nature, 219,* 1120 (1968).
12. L. A. Girifalco and R. J. Good, *J. Phys. Chem., 61,* 904 (1957).
13. F. M. Fowkes, *J. Phys. Chem., 66,* 382 (1962).
14. O. Volkersen, *Luftfahrforschung, 15,* 41 (1938).
15. M. Goland and E. Reissner, *J. Appl. Mech., Trans. ASME, 66,* 17 (1944).

6
Adhesion Testing

Ulrich Zorll

Forschungsinstitut für Pigmente und Lacke,
Stuttgart, Germany

1.0 FUNDAMENTALS OF ADHESION

Without sufficient adhesion, a coating of otherwise excellent properties in terms of resistance to weather, chemicals, scratches, or impact would be rather worthless. It is therefore necessary to provide for good adhesion features when paint materials are formulated. There must also be adequate means for controlling the level of adhesion strength after the coating has been spread and cured on the substrate. Moreover, methods should be available that allow for the detection of any failure in the case of the dissolution of the bond between coating and substrate, under any circumstances whatsoever.

1.1 Components at the Interface

In chemical terms, there is a considerable similarity between paints on one side and adhesives or glues on the other (Fig. 1). Both materials appear in the form of organic coatings; thus it is appropriate in this chapter to concentrate on the behavior of paint materials. Adhesion is the property requested in either case, though perhaps with different emphasis on its intensity, according to the intended use.

Such a coating is, in essence, a polymer consisting of more or less cross–linked macromolecules, and a certain amount of pigments and fillers. Metals, woods, plastics, paper, leather, concrete, or masonry, to name only the most important materials can form the substrate for the coating.

It is, however, important to keep in mind that these substrate materials may inhibit a rigidity higher than that off the coating. Under such conditions, fracture will occur within the coating, if the system experiences external force of sufficient intensity. Cohesive failure will be the consequence, however, if the adhesion at the interface surpasses the cohesion of the paint layer. Otherwise, adhesive failure is obtained, indicating a definite separation between coating and substrate.

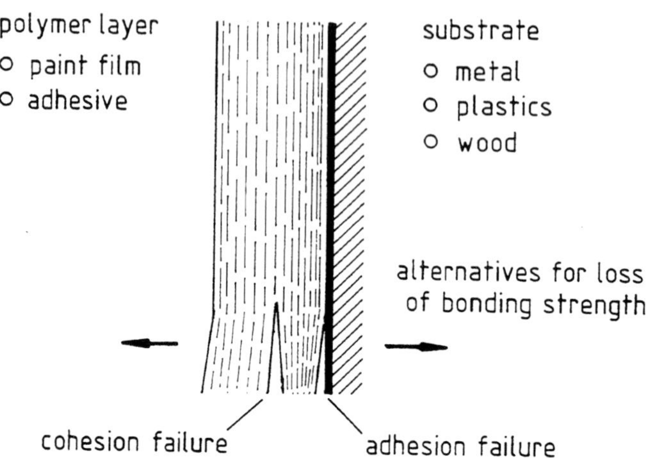

Figure 1 Bonding situation at the interface of polymer layer and substrate.

The energetic aspects at the interface and the effects of elasticity within coating and substrate have been taken into consideration for a theoretical approach, on which it is possible to base predictions of where failure will occur.[1]

Both types of adhesion damage are encountered in practice. The existence of cohesion would signal the attainment of an optimum adhesion strength. Any further improvement of the systems should then be sought in the direction of increasing the inherent strength of the coating material.

1.2 Causes of Failure

The bond between coating and substrate can be put under stress, and may thus finally fail, by means of several external factors, acting either alone or in combination (Fig.2). First, there may be regular mechanical stress, affecting not only the bulk of the materials but also the

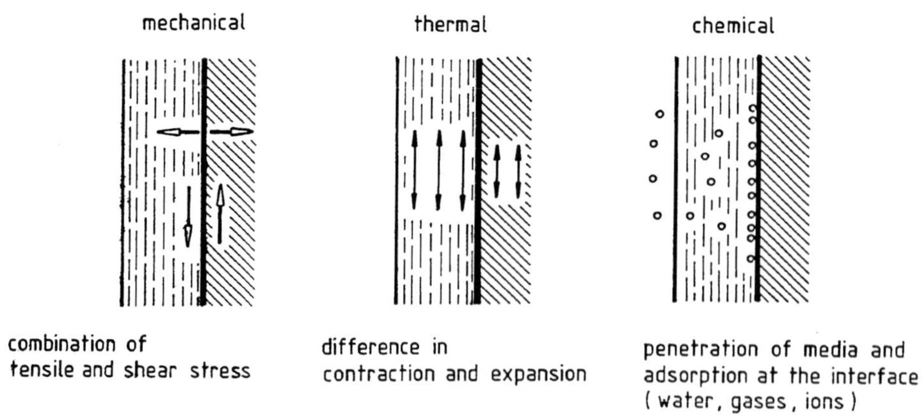

Figure 2 External situation at the interface of polymer layer and substrate.

bond strength at the interface. It is useful here to distinguish between the two most common types of stress: tensile stress, effective perpendicularly to the interface, and shear stress, appearing along the plane of contact.

Moreover, since coatings may undergo changes in temperature, sometimes even rather rapidly, any difference in the coefficient of expansion can cause at the interface stress conditions of such high intensity that the paint film comes detached from the substrate. This event may be especially disadvantageous because the temperature effects tend to be less obvious than the mechanical and the chemical factors.

There may be, of course, an effect of a chemical, which penetrates through the coating and becomes absorbed at the interface, causing loss of adhesion here.

It is always useful to take these effects into consideration when adhesion must be measured, because the method of testing the coating should reproduce the end–use conditions.

1.3   Measures of Adhesion

There are various possibilities for characterizing the results obtained in an adhesion test. If it is necessary to evaluate the bonding strength at the interface, the quantity to be measured is obviously the maximum mechanical stress that can be attained at the interface. This is adhesion strength in the strict sense, expressed as force per unit area, and specified as either tensile or shear stress. Consequently, in several test methods, the result is obtained in that form.

The energy that must be provided for breaking the bond at the interface can also be an informative quantity. It is expressed as work of adhesion and is formally equivalent to the product of adhesion strength, as defined earlier, and the distance between the separated surfaces of coating and substrate immediately after detachment. Thus, that quantity has the dimensions of force per unit length, and this is exactly the value obtained with some other test methods e.g., the peeling test.

2.0   STANDARDIZATION OF ADHESION TESTS

Since a specification for the degree of adhesion must be provided in nearly each paint formulation, it is not surprising that methods for routine measurement of that key quantity have been established in the field of quality testing., This is true for the cross–cut test, the paint technicians' first choice when adhesion must be estimated. However, the cross–cut test is nowadays more and more complemented by the pull–off methods.

Both methods have been the subject of national standardization. However, the differences in the documents of various countries are of minor importance, and it was relatively easy to formulate international standards, ensuring that these fundamental tests can be carried out in a uniform manner.

2.1   Cross–cut Test

Scope and procedure of practical and instructive method have been laid down by the International Organization for Standardization (ISO).[2] To obtain an idea of the adhesion of the coating, a lattice pattern is cut into it, penetrating through the film and into the substrate. Various cutting tools can be used either manually or mechanically for this purpose. A good choice is the multiple cutting tool, a set of six "knives," 1 or 2 mm apart, yielding a uniform pattern.

The test results are evaluated according to the scheme indicated in Figure 3. The classification is based on estimating the amount of paint flakes separated from the substrate. If in

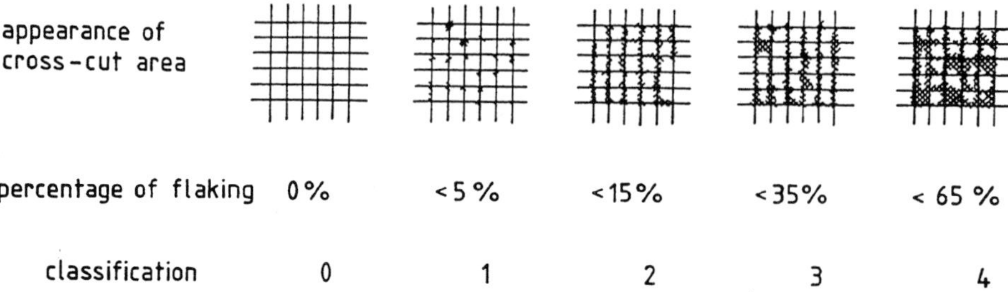

Figure 3 Principle of classifying paint film adhesion in the cross-cut test.

doubt about the real percentage of detachment, one may brush off the loose parts, or remove them by means of an adhesive tape.

It is not always necessary to base the judgment about the degree of adhesion on the whole six–step classification. The ISO recommends standard considering the test for "go-no go" statements. In such case, class "0" would indicate perfect adhesion, whereas class "2," or even class "1," should be interpreted as an objectionable result. All higher classes would then signal, although with different distinctness, that something must be done to improve the coating's adhesion.

2.2 Tensile Methods

The typical stress patterns at the interface, caused by loads acting predominantly either normal or parallel to the plane of contact, have been used as the basis for pertinent test methods (Fig. 4). The pull–off method is the most widely used procedure and has already been stan-

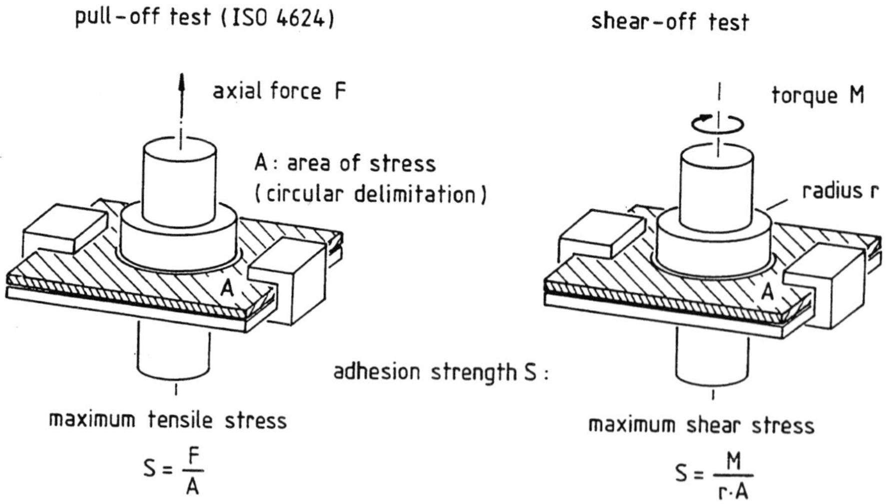

Figure 4 Methods for quantatative determination of adhesion strength.

dardized internationally.[3] As a preparation for the test, a stud, normally made of steel, is glued with the coating and is subjected to axial tension until detachment of the paint film occurs. The result i.e., the adhesion strength is the maximum tensile stress that is possible at the interface.

If, however, a torque is applied about the axis of the stud, the process of detachment reveals the maximum shear stress that can be attained at the interface, thus also leading to a characteristic measure of adhesion.

It has been shown[4] that the value of adhesion strength obtained from either method are of the same order of magnitude. However, there is a tendency to obtain results with the torque principle in the case of cohesive failure, but lower results for adhesive failure.

The accuracy, with which the tests can be carried out, and the existence of a well–defined mechanical principle for them, must not, however, lead to the idea that the values obtained in this way can be considered to be material constants for the bonding components. There is, instead, an additional influence of several parameters, such as temperature, speed of deformation, and even form and size of the stud.

Also of importance are the rigidity of the test piece and the possibility of securing it for measurement.[5] As shown in Figure 5, for coatings on undeformable substrates, using either clamps at the edge of the support in the vicinity of the attached stud is an adequate means of fastening the test piece. For flexible substrate such as plastics or leather, however, the sandwich principle is recommended.

Since an adhesive is used for fastening the stud firmly at the test area, it is necessary to discuss what type of adhesive would be suitable, especially with regard to avoiding any negative effects of its use. In general, solventless epoxy resin adhesives, cured with polyamines, or the fast–hardening cyanoacrylates, can be used for this purpose. To ensure reliability of the results, the constituents of the adhesive must not interact with the coating in a way that causes complete swelling. Penetration of the coating only to the uppermost layers, would be beneficial in terms of bonding strength between adhesive and coating. The wide variety of adhesives, available commercially, ought to make it possible to find the most suitable type.

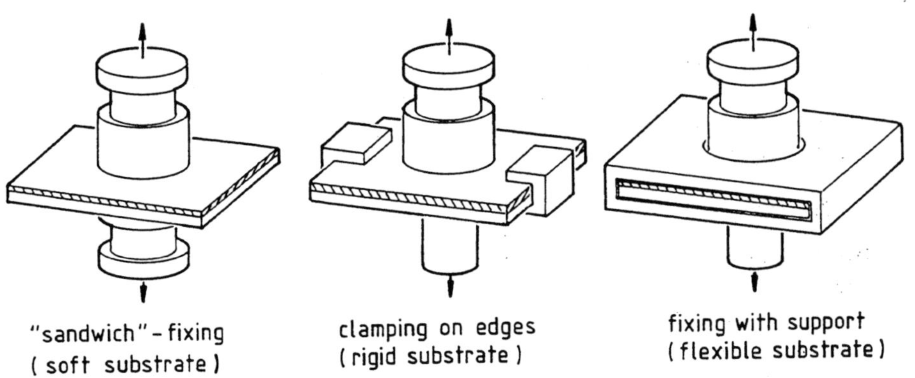

Figure 5 Formation of specimens for measuring adhesion strength in the pull-off test.

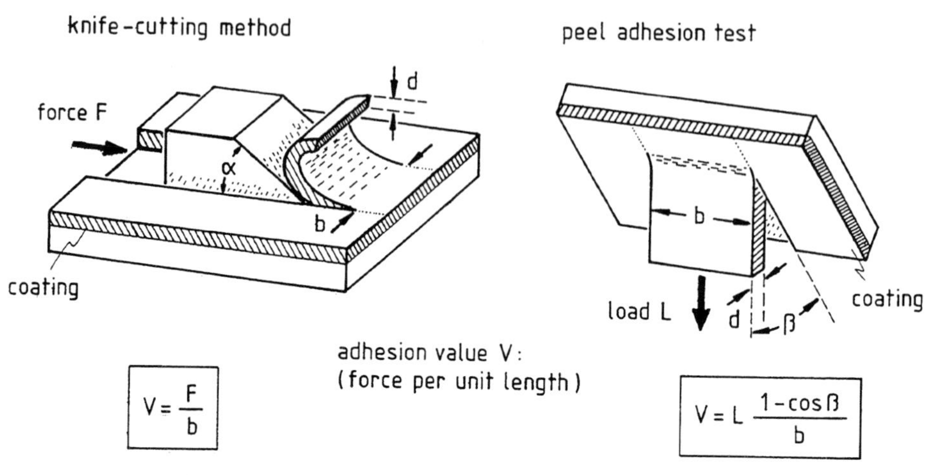

Figure 6   Devices for measuring adhesion on the basis of delamination procedures.

3.0 DELAMINATION PROCEDURES

There is a certain contrast between the tensile methods, in which stress conditions at the interface are of primary concern, and another group of test methods, in which the delamination effects that sometimes occur in the practical use of the coating are duplicated. In these methods, the test piece is, under idealized conditions, subjected to peeling forces. These forces attack the bond between paint film and substrate either at well-defined lines, so that the coating is detached in strips, or at one point only, thus causing delamination that progresses radially in the form of a blister. Both situations are encountered in practice. Thus, an appropriate method can always be selected according to the actual needs.

Such methods are equivalent to a model of the typical treatment that the coating may undergo. This is the most essential advantage of the methods in practical terms. But they do not as a rule lead to results so clearly defined as the maximum stresses at the interface. Only under certain conditions can the values of the stresses be calculated from results of these tests, and this is possible only if all geometrical and dynamical factors, on which the test procedure depends are available.

As a test result, the work of adhesion is regularly obtained. This quantity is in any case useful for comparative purposes (i.e., without reference to a particular absolute adhesion standard). But the work of adhesion also provides some insight into the mechanism of bonding in energetic terms.

3.1 Knife–cutting Method

There are two special methods for testing the degree of adhesion, and the procedures for loosening the bond between paint film and substrate are rather similar (Fig. 6). In one method, film separation is obtained by means of a sharp knife, pushed along the interface with an exactly measured force.[6] Although this seems to be a simple test method, the process of detachment is, in fact, complicated, comprising both shear and tensile stresses, which finally cause disbonding of the film. Moreover, a leverage effect plays a part in the separa-

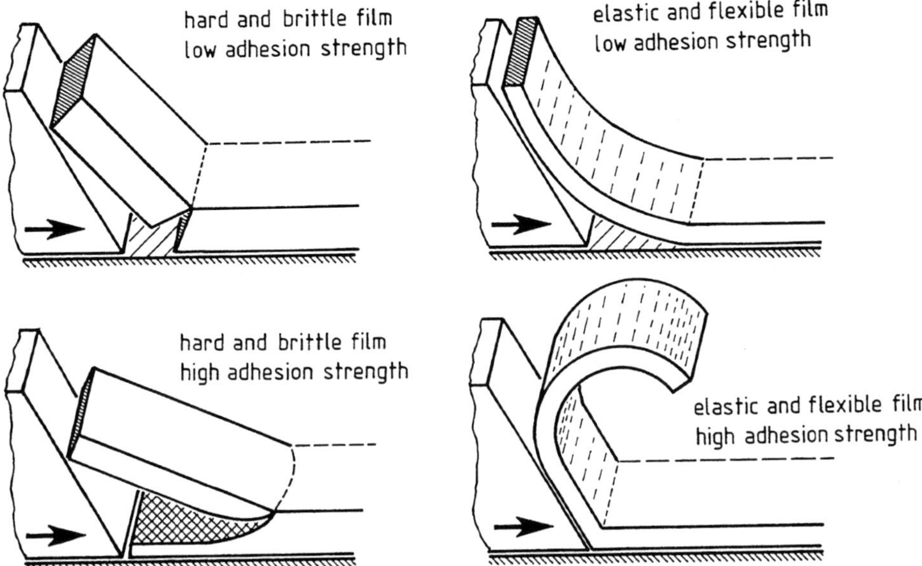

Figure 7 Influence of mechanical paint film properties on the results of the knife-cutting test.

tion process of the coating, and the intensity of that effect can be characterized by the angle under which the force is acting upon the film.

In the knife–cutting method, the principle of removing the paint film is analogous to the machining of metal in a lathe. A particular system of forces becomes effective here. It is determined by rake angle of the knife, coating thickness, friction between cutting tool and coating as well as substrate, amount of energy stored elastically in the film and energy losses caused by plastic deformation, fracture energy occurring during decomposition within the film, and other effects of minor importance.

To obtain meaningful results, all those parameters should be controlled strictly, or their influence at least estimated as exactly as possible. In many cases, however, the details of the separation process reveal, as indicated in Figure 7, how much relevance can be attributed to the various factors.

3.2 Peel Test

In practice, a coating often fails not by becoming simultaneously detached over an extended area, but by gradual peeling, for example, starting from a scarcely covered edge, or from a line–shaped damage zone. Thus, it appears sensible to duplicate these conditions in an appropriate test procedure.

The peel test was originally developed for adhesive tape, where its use appears natural and simple. For measuring the adhesion of a coating, however, a strip of adequate width must first be marked on the sample by two parallel cuts of sufficient length.

An analysis of the test conditions, as indicated in Figure 6, reveals the influence of the angle under which the load is applied.[7] Thus, in contrast to the first impression, that the test procedure ought to be simple, the details of the separation process turns out to be rather complicated.[8] Detachment of the film occurs under the combined effects of both positive and negative tensile stresses and shear stress.

Besides, the viscoelastic character of the coating material must be taken into account, and this feature is also affected by the amount of pigmentation in the film.[9] As a consequence of the viscoelasticity, the test results depend to a remarkable degree, on the velocity of the detachment process.

Thus, the general experience is that the peel test, when used for measuring adhesion of paint films, can certainly be considered to be a practical method, but it is also understandable that the results obtained cannot be interpreted in terms of the bonding mechanism.

3.3 Blister Method

Formation of blisters is generally a first sign of deterioration of any coating designed for protection against corrosion. It is therefore sensible to investigate the behavior of such a coating material under conditions that finally may lead to film detachment in the form of blisters.

This test (Fig. 8) is normally carried out in the following way. First, before the liquid paint is spread over the surface of the sample, a hole is bored into the substrate at the site at which the blister is to occur. The hole is plugged with a material such as Teflon, on which there will be no adhesion during painting at all, to permit easy removal of the plug after film formation.[10] It is also possible to generate the hole after painting, using spark erosion techniques.

The detachment of the film is initiated, and supported, by providing hydrostatic pressure in the hole, either with a fluid (oil, mercury etc.) or with pressurized air.[11] In any case, the pressure is a primary measure of the progress of the debonding process. To obtain adhesion values, either as maximum stress or as bonding energy (work of adhesion), both height and diameter of the blister must be monitored. An optical system with relatively low magnification can prove sufficient for that purpose.[12] From these geometrical data, together with the tensile modulus of the film and its thickness a critical pressure value can be calculated. It is pressure that causes the blister to grow, and thus pressure can serve as a basis for determining the adhesion strength.[13]

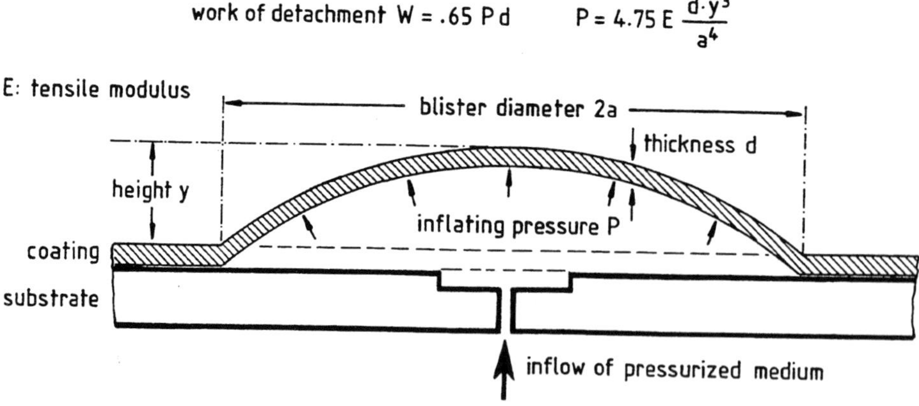

Figure 8 Blow-off test for measuring adhesion based on determination of blister dimensions and pressure.

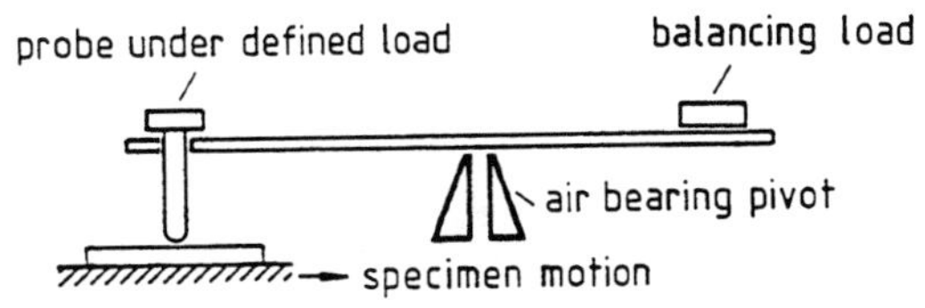

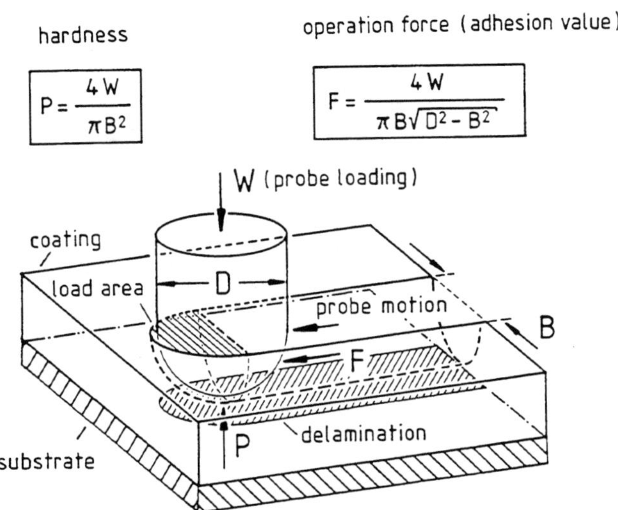

Figure 9 Scratch technique for determination the bonding of polymer films subjected to local surface forces.

Preparation of test pieces as well as procedures of measurement are comparatively complicated, but this disadvantage is virtually compensated for by the basic information the test results yield in terms of the mechanism of blister formation.

4.0 LOCAL DEBONDING SYSTEMS

In practice, mechanical damage on the coating often occurs locally, in form of scratches or impact deformation. As a consequence, not only are the appearance properties of the film deteriorated, but its adhesion may be lost as well, if the intensity of the external forces, transferred through the film, is high enough. Idealized conditions for such treatment have become the basis of special adhesion test methods. Prototypes of pertinent devices exist, and their scope of application has been studied in detail. Moreover, the deformation caused by such external loading processes has been analyzed, and theoretical relations are now available for assessing the extent of debonding at the interface.

4.1 Scratch Technique

Figure 9 illustrates the principle of a test method in which scratch resistance as well as adhesion of the coating can be measured. A loaded stylus is drawn across the film surface. There is obviously a simple relationship between the applied load, of which the intensity can be controlled easily in the balance system carrying the stylus, and the shearing force eventually causing detachment of the coating. A critical quantity is the radius at the tip of the tool, and it must be measured with the same accuracy as the width of the contact area generated during the scratching procedure.

It is recommended[14] that the information obtained directly from the test be supplemented by additional data, such as results of surface profilemeters or scanning electron microscopes, which provide insight into the details of the scratch topography. Under such provision, even the adhesion values in multilayer systems can be estimated. It can be seen what type of film failure occurs in those systems when they are subjected to scratch loading.

4.2 Indentation Debonding

If a needlelike indenter is perpendicularly pressed into the surface of a coating, which is bonded to a virtually undeformable substrate, most of the deformation will occur within the film, but there also will be a certain debonding effect at the interface.

According to that situation, indicated in Figure 10, a peeling moment can be calculated, which may serve as a measure of the film's capacity to withstand delamination in the vicinity of the indentation site.[15] It is convenient, especially for thin coatings, to monitor the

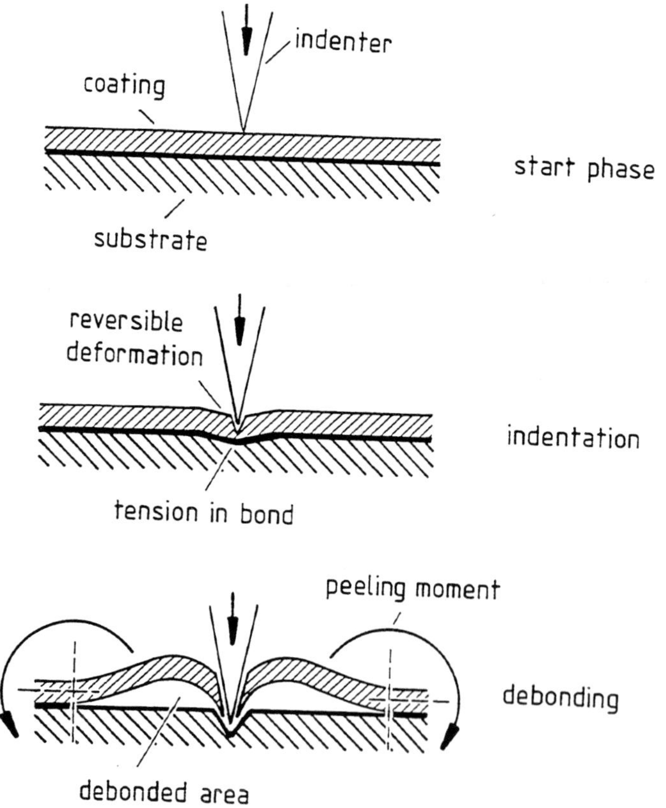

Figure 10 Principle of the indentation process for measuring adhesion at the polymer–substrate interface.

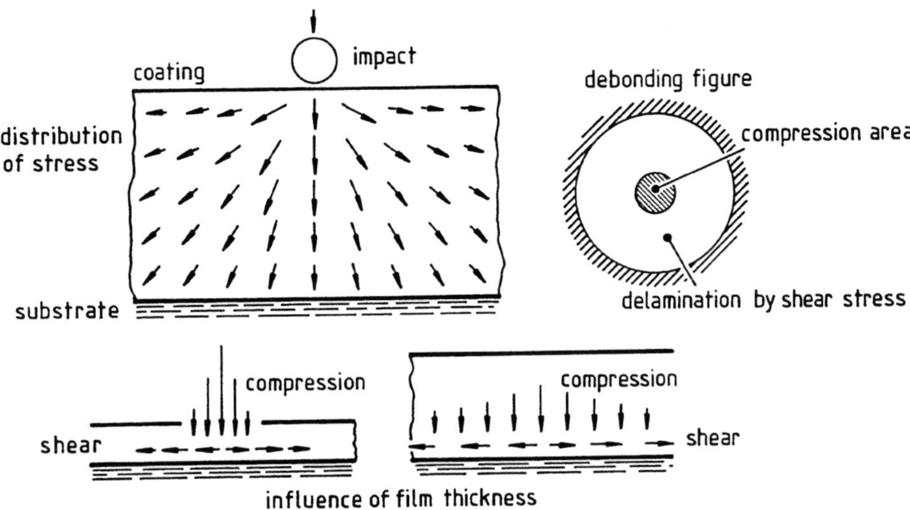

Figure 11 Information about adhesion on the basis of the circular debonding figure produced by impact loading of the paint film.

gradually increasing area of debonding with optical devices capable of leading to an evaluation on the basis of Newton's rings, for instance, with transparent coatings.

It is by no means necessary to restrict the scope of the indentation test to using a needle for penetration of the coating. Indenters of other typical shapes also are used successfully. It has been shown that the best approach for taking into account the boundary conditions at the interface, as a basis for the calculation of adhesion values, is to use a 60° angle cone as an indenter.[16] An essential advantage of the identification test can be seen in the fact that it yields values for the bond strength in absolute terms, as well as information about the durability of the connection between coating and substrate under those specific loading conditions.

4.3 Impact Tests

Especially for determining the stone–chip resistance of coatings that are supposed to provide efficient protection against corrosion of a metal substrate, the value of adhesion at the interface is of primary interest. The situation encountered in practice can be duplicated, under only slightly idealized conditions, by means of a steel ball impinging on the test piece (Fig. 11). The transfer of forces through the film is, in a first approximation, equal to the case of static loading; that is, it can be calculated in virtually the same way as for an indentation test.

Thus, in debonding area, two types of stress are active: negative tensile (i.e., compressive) stress in the center of the circular detachment site, and, still more essential, shear stress in the annular region. The dimensions of the debonding area, mainly the maximum diameter, can serve as a measure of adhesion at the interface. it is convenient to use either that diameter or, better still, the area of the debonding zone, both as (reciprocal) measures of adhesion. An extended site of separation is thus an indication of a low adhesion level.[17]

It is possible, in principle at least, to calculate the value of adhesion in absolute terms, but then some additional parameters must be taken into account. There is first the remarkable influence of film thickness. For a thin film, the detachment area is more concentrated in the region in the vicinity of the axis of impact, areas with thicker coatings, under the same test conditions, a larger area is affected by the stresses leading to delamination. The details of this relationship can easily be understood on the basis of the particular scheme that is valid for the stress distribution within the film as a whole.

The other parameters for the calculation of adhesion strength are related to the impinging steel ball. Its mass, diameter, and speed-before and after impact-must be known, as well as the (generally very short) time of impact, in which the energy is being transferred into the film.

5.0 FLAW DETECTION METHODS

In practice, it is often necessary to know as fast as possible about any deterioration of the bonding strength between film and substrate; quantitative details of the adhesion values however, are not required. The occurrence at the interface, of any flaws, even those of tiny dimensions, is an event belonging to this category. Thus tests allowing for identification of the first signs of adhesion failure are of great interest.

Some methods, based on acoustical or thermo–optical principles, have been proposed and, to a considerable degree, tested already for that purpose. The unique potential of these test methods, most of them introduced only recently, promises very interesting applications in the future.

5.1 Ultrasonic Pulse–Echo System

Monitoring of ultrasonic echoes, according to the time domain principle,[18] is widely used to identify any irregularities in materials that are supposed to be flawless. This method thus can be applied also to evaluate the quality of bonding between a polymer coating and the substrate, whatever its nature.

According to Figure 12, an incident of ultrasound will be partly reflected and transmitted at each interface at the test piece, including the backing. It is the pulse partly transmitted at

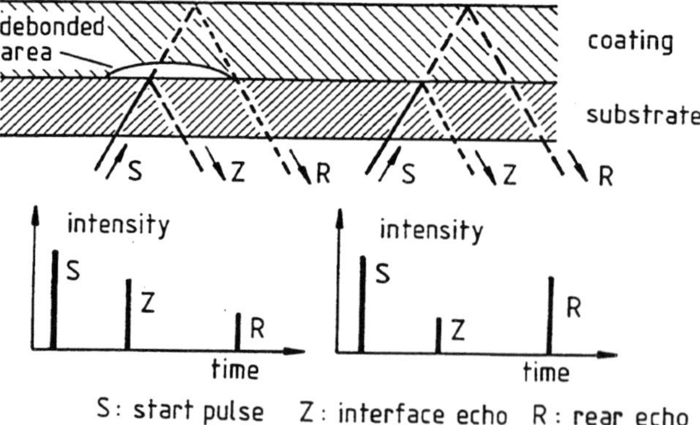

Figure 12 Ultrasonic pulse-echo technique for determining position and size of adhesion flaws.

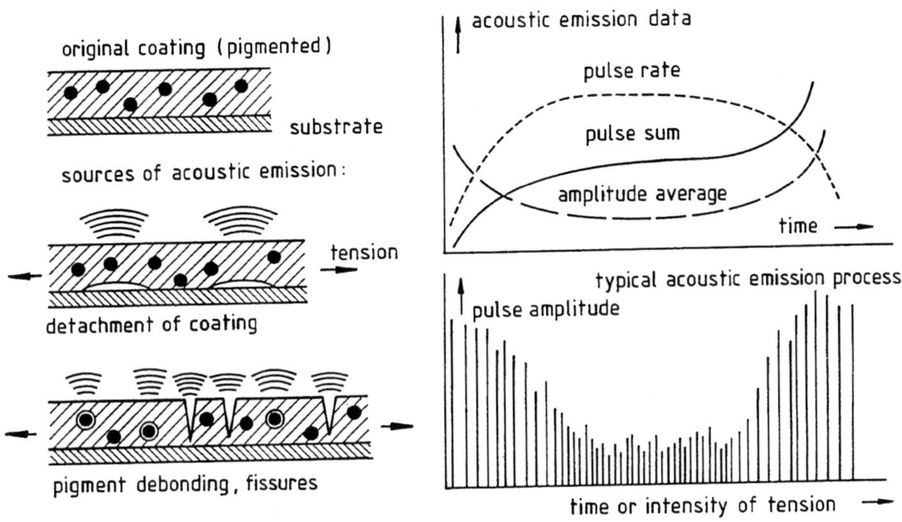

Figure 13 Application of acoustic emission analysis for monitoring the onset of coating detachment.

the interface that then undergoes more or less total reflection at this free surface. The bonding quality is expressed in the relative heights of the pulses.

With an intact bonding at the interface, the amplitude of the pulse reflected there will be fairly low, in contrast to the amplitude of the transmitted pulse that travels through the substrate and is reflected on its free boundary, appearing later at the sensor accordingly. If a defect is assumed at the interface, containing air or indicating in another way that the joint has been disbonded, the amplitude of the related ultrasonic pulse will increase considerably, owing to the very low acoustic impedance now prevailing at that site. There is thus less intensity available for the transmitted pulse, and its amplitude would then decrease considerable.

The region subjected to that ultrasonic treatment can be restricted to a relatively small size. Consequently, there is a possibility of virtually scanning the test piece, line after line, to identify any sites at which adhesion strength might have been lost. The extent, even the particular shape, of that defective area can be determined in the test procedure. [19]

5.2 Acoustic Emission Analysis

Since it has been observed, by means of very sensitive acoustical sensors, that any debonding effects within an originally uniform and coherent material are accompanied by a specific burst of (mostly ultrasonic) pulses, this principle has also been proposed for monitoring the behavior of adhesive joints under load.[20]

Thus, the method can also be applied with slight modifications (Fig.13), to examine the extent of debonding phenomena occurring in a coating system on a given substrate, if the whole test piece is subjected to gradual stretching, normally in one dimension.[21] The acoustical signals obtained are related to individual fracture events in the test system. As a rule, it will be the detachment process of the film that is indicated in this way. Also, however, any

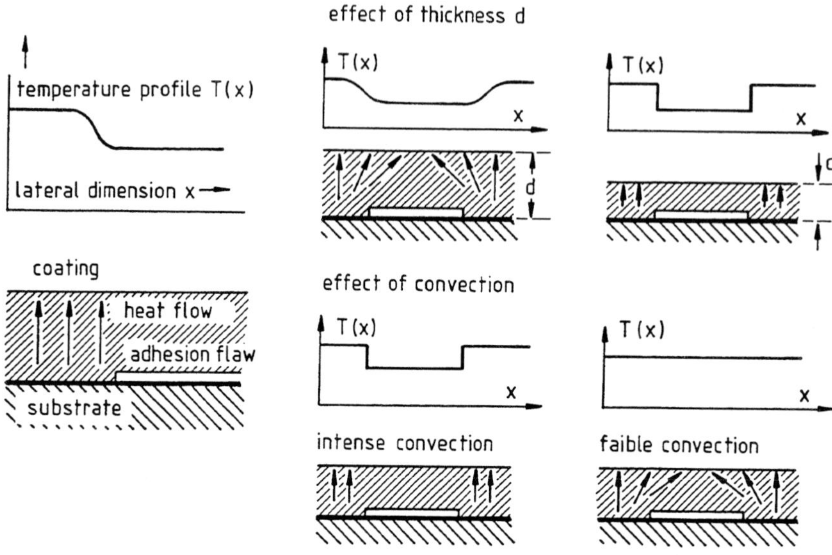

Figure 14 Thermographic detection of debonding areas by their impedance to heat flow through the substrate–film system.

separation within the film (e.g., between binder matrix and pigment), could cause acoustic signals, although most probably, of course, on a lower level of intensity.

This possibility, however, together with the knowledge that there may be a wide variety in form and distribution of the acoustic pulse spectrum, elucidates the present situation with this test. It provides for a promising potential in terms of flaw detection in coating systems; nevertheless, a great deal of information is required if the primary results are to be interpreted correctly.

5.3 Thermographic Detection of Defects

The main advantage of thermographic methods can be seen in the fact that they yield remote mapping of the distribution of temperature on a surface, whatever the cause may have been. This principle has been applied successfully for the detection of subsurface flaws, which may occur as a consequence of any deterioration in the bonding quality of coating and substrate.[22]

As indicated schematically in Figure 14, it is the heat flux in the test piece, proceeding toward the surface, that, if there is an obstacle in its way in the form of a flaw or a debonded area, produces more or less pronounced differences in the otherwise uniform temperature on the surface. The shape of the area in which the temperature decreases is made visible at the surface by means of an infrared sensor. It is a rather exact reproduction of the real defect, which would have remained invisible otherwise. The agreement between real shape of debonded area and its "image" on the surface will be the better, the lower the distance is between the position of the flaw and the surface. The situation is similar, on optical grounds, to the formation of a shadow from a distant object.

As a rule of thumb, the linear dimension of the debonded area should be at least twice the distance from the interface to the surface. Under these conditions, a distinct reproduction of the defect on the surface can be expected.

6.0 OUTLOOK

Although adhesion is in principle a well–defined quantity for a system consisting of paint film, or adhesives, and its pertinent substrate, there are so many possibilities for practical combinations that it would be unrealistic to look for universal adhesion test systems. On the contrary, the relatively great number of test methods proposed so far is the best proof of the diversity of the means available for tackling the problem of adhesion measurement in practical terms.

In many cases, the selected method will lead to satisfactory results, but it is nevertheless advisable to make it a practice to follow up the development activities in the field of adhesion testing. Many of the existing methods were proposed when there was a need for a particular reproduction of typical bonding situations. It is thus not unrealistic to expect that the successful methods of today will be supplemented, in the not so distant future, by even more sophisticated methods with accordingly high performance qualities.

REFERENCES

1. R. J. Good, "Theory of cohesive vs adhesive separation in an adhering system," *J. Adhes.*, 4, 133-154, (1972).
2. International Standard ISO 2409, Paint and Varnishes-Cross–cut Test,"
3. International Standard ISO 4624, Paint and Varnishes-Pull–off Test for Adhesion,"
4. J. Sickfeld, "Experimental aspects of adhesion testing-An IUPAC study," *J. Oil Color. Chem. Assoc.*, 61 292-298 (1978).
5. P. C. Hopman, "The direct pull–off test," *J. Oil Color. Chem. Assoc.*, 67, 179-184 (1984).
6. W. K. Asbeck, "The measurement of adhesion in absolute units by knife–cutting methods: The Hesiometer," presented at the Eleventh FATIPEC Congress, Brussels, 1972. *Congress Book*, pp. P78-P87.
7. A. N. Gent and G. R. Hamed, "Peel mechanics," *J. Adhes.*, 7, 97-95 (1975).
8. A. D. Crocombe and R. D. Adams, "Peel analysis using the finite element method," *J. Adhes.*, 12, 127-139 (1981).
9. P. M. Heertjes and J. de Jong, "The peeling–off of paint film," *J. Oil Color. Chem. Assoc.*, 55, 1096-1106 (1972).
10. A. Elbasir, J. D. Scantlebury, and L. M. Callow, "The adhesion of chlorinated rubber to mild steel using a blister technique," *J. Oil Color. Chem. Assoc.*, 68, 282-284 (1985).
11. H. Dannenberg, "Measurement of adhesion by a blister method," *J. Appl. Polym. Sci.*, 5, 125-134 (1961).
12. L. A. van der Meer–Lerk and P. M. Heertjes, "The influence of pressure on blister growth," *J. Oil Color. Chem. Assoc.*, 64, 30-38 (1981).
13. M. G. Allen and S. D. Senturia, "analysis of critical debonding pressures of stressed thin films in the blister test," *J. Adhes.*, 25, 303-315 (1988).
14. J. Ahn, K. L. Mittal, and R. H. McQueen, "Hardness and adhesion of filmed structures as determined by the scratch technique," ASTM STP 640. Philadelphia: American Society for Testing and Materials, 1978 pp. 134-157.
15. P. A. Engel and D. D. Roshon, "Indentation–debonding of an adhered surface layer," *J. Adhes.*, 10, 237-253 (1979).
16. H. D. Conway and J. P. R. Thomson, "The determination of bond strength of polymeric films by indentation debonding," *J. Adhes. Sci. Technol.*, 2, 227-236 (1988).

17. U. Zorll, "Testing of the impact resistance of coatings under aspects of deformation energy," presented at the 15th FATIPEC Congress, Amsterdam, 1980. *Congress Book*, Vol. III, pp. 94-109.
18. R. D. Adams, P. Cawley, and C. C. H. Guyott, "Non–destructive inspection of adhesively–bonded joints," *J. Adhes.*, 21, 279-290 (1987).
19. P. A. Meyer and J. L. Rose, "Modeling concepts for studying ultrasonic wave interaction with adhesive bonds," *J. Adhes.*, 8, 107-120 (1976).
20. H. G. Moslé and B. Wellenkoetter, "Acoustic emission on painted steel sheets," presented at the European Federation of Corrosion Conference on Surface Protection by Organic Coatings, Budapest, 1979. *Proceedings*, 5, pp. 17-33.
21. W. Mielke, "Application of acoustic emission analysis for evaluation of the adhesion of coatings," presented at the 19th FATIPEC Congress, Aachen, 1988. *Congress Book*, pp. 471-486.
22. B. E. Dom, H. E. Evens, and D. M. Torres, "Thermographic detection of polymer/metal adhesion failures," in *Adhesion Aspects of Polymeric Coatings*, K. L. Mittal, Ed. New York: Plenum Press, 1983, pp. 597-621.

7

The Use of X-ray Fluorescence for Coat Weight Determinations

Wayne E. Mozer

Oxford Analytical, Inc., Andover, Massachusetts

1.0 INTRODUCTION

The technique of elemental analysis by x-ray fluorescence (XRF) has been applied to the quality control of coating weights at the plant level. Measurements by non laboratory personnel provide precise and rapid analytical data on the amount and uniformity of the applied coating. XRF has proved to be an effective means of determining silicone coating weights on paper and film, titanium dioxide loading in paper, and silver on film.

2.0 Technique

X-ray fluorescence is a rapid, nondestructive, and comparative technique for the quantitative determination of elements in a variety of matrices. XRF units come in a variety of packages; however, the type of unit most prevalent in the coating industry is described in this chapter.

The XRF benchtop analyzer makes use of a low level radioisotope placed in close proximity to the sample. The primary x-rays emitted from the excitation source strike the sample, and fluorescence of secondary x-rays occur. These secondary x-rays have specific energies, which are characteristic of the elements in the sample and are independent of chemical or physical state. These x-rays are detected in a gas–filled counter that outputs a series of pulses, the amplitudes of which are proportional to the energy of the incident radiation. The number of pulses from silicon x-rays, for example is proportional to the silicone coat weight of the sample. Since the technique is non destructive, the sample is reusable for further analysis at any time.

To ensure optimum excitation, alternate radioisotopes may be necessary for different applications. For silicone coatings and titanium dioxide in paper, an iron-55 (Fe-55) source is used. Fe–55 x-rays are soft (low energy) and do not penetrate far into a sample. For silver on film, a more energetic americanum–241 source has been used.

Placing samples just a few millimeters from the excitation source enables high sensitivities to be obtained. Irradiation of the sample is more efficient, the closer the sample is brought to the source. In the case of low energy x-rays, such as silicon, which are easily

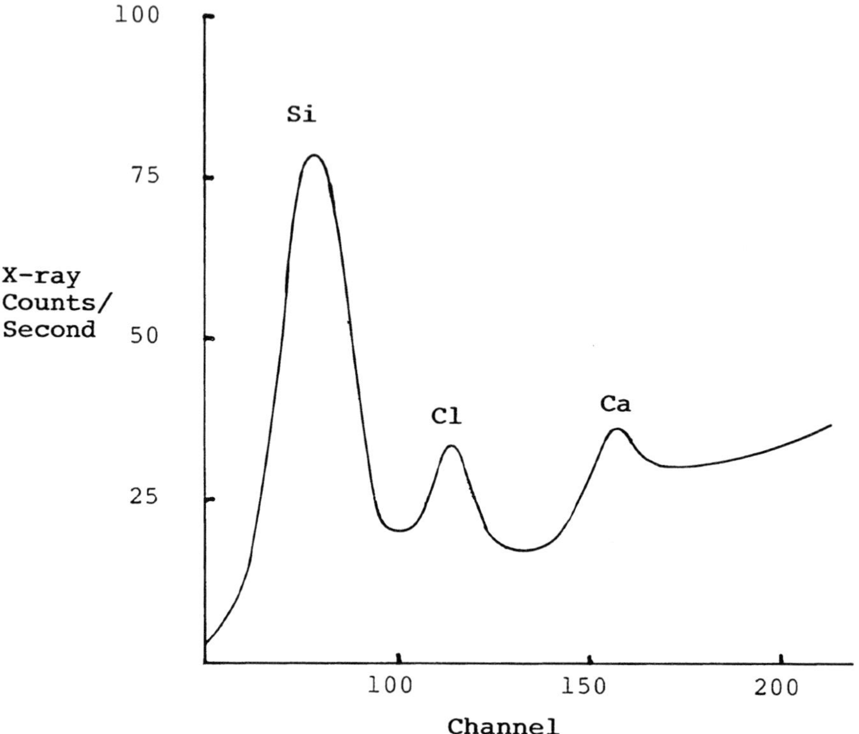

Figure 1   Spectrum scan of silicone coatings on super calendar craft paper.

absorbed by the atmosphere, a helium purge should be used. With optimum sample irradiation and helium purging when necessary, measurements within 2 minutes of counting time are typical for most samples.

Calibration curves for different elements and materials can be stored directly in the instrument and are available for recall. Curves are established by measuring a set of known samples or standards of the same material. Since XRF is a comparative technique, subsequent analyses are only as good as the quality of the calibration standards supplied.

The total accumulated intensity is actually a combination of signals from the analyte element and from the matrix in the form of background. Background intensities may vary depending on the thickness or the basis weight of the material. These differences may be determined by measuring uncoated or blank samples, which are automatically incorporated into the calibration.

3.0 METHOD

Three procedural steps must precede analysis. First, spectrum scan helps to set up various calibration parameters. Next a calibration is established. Almost all the time the instrument is in use, it is dedicated to analysis of production samples. Typically, spectrum scans are performed on representative standards to identify the occurrence of elemental peaks of interest in the spectrum, as illustrated in Figure 1. The scan is a qualitative tool and will show the presence of any potential interfacing element. For example, an Fe-55 source will excite

Table 1   Typical Results for a Silicone-on-paper Calibration[a]

Sample	Given concentration (g/m^2)	X-ray concentration (g/m^2)	Error
1	0.33	0.34	0.01
2	0.41	0.41	0.00
3	0.54	0.52	−0.02
4	0.62	0.62	0.00
5	0.70	0.72	0.02
6	1.50	1.50	0.00

[a]Standard error = 0.013 g/m^2.

the following range of elements in the periodic scale: aluminum to vanadium and zirconium to cerium. With the proper combination of sources, it is theoretically possible to measure from aluminum to uranium. Once any interference has been recognized by a spectrum scan, a calibration may be established. A sample is prepared by cutting a disk and placing it onto a nickel-coated paper holder. The holder ensures that the sample will be held as flat as possible, since an uneven surface will introduce errors in reproducibility, caused by scattering and vacancies at the incident surface. Table 1 gives a typical calibration for silicon coat weights.

Once a calibration has been established, two samples are selected which bracket the operational range of the samples measured. These samples will then be used to correct for changes in instrumental performance with time. The measurement of these two standards and of an uncoated sample for background correction purposes is a push-button procedure. Termed restandardization, is performed once per shift. Analyses may then be performed: an operator simple cuts a sample to be analyzed, inserts it into the instrument, and initiates the measurement. Within 23 minutes the instrument will determine the concentration of the element of interest, both printing and displaying the results.

4.0 ACCURACY

To ensure optimum accuracy, it is important to be aware of three sources of errors: suppliers of coating material, background changes, and interferences. It has been recognized, although not explained, that silicones of different suppliers give different sensitivities. When different silicones are analyzed on the same basis weight paper, lines with different slopes are obtained (Fig. 2). Thus, to avoid this source of error, separate calibrations are established for each supplier of silicone. Second, correction for changes in background will help reduce error.

Figure 3 illustrates two sets of standards of the same silicone on supercalendar kraft prepared on different dates. Quite evidently, the samples give the same slopes, they are offset. The analyzer will correct for this background change by having the operator insert an uncoated sample from the lot of paper to be coated. Thus, all subsequent analyses will be corrected for the apparent background change.

The final source of error is the presence of another material in high concentrations. Examples include silicone coatings on polyvinyl chloride (PVC) or titanium dioxide filled films. As the levels in these films change, the effect on the background for the silicon region

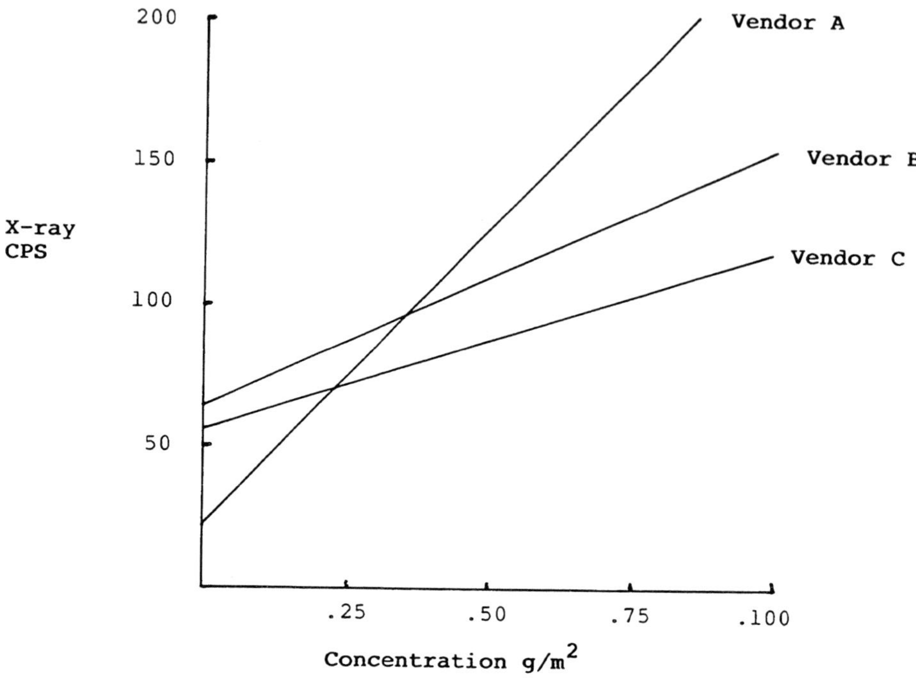

Figure 2 Differences in sensitivity in products of different suppliers of silicone.

of the spectrum will change. Thus, there will be raising or lowering of the apparent silicone coat weight for that film sample. This type of interference is easily corrected for by software that can be built into the analyzer. The interferences are recognized by using the spectrum scan, and automatically compensated for during calibration.

5.0   REPEATABILITY AND REPRODUCIBILITY

Repeatability in terms of precision of measurement is a statistical function determining the variability of repeat measurements on the accumulated intensity and x-ray counts. Since an iron-55 source excites titanium x-rays, more efficiently than silicon x-rays, it follows that sensitivities for titania coatings are higher, and therefore, measurements more precise. The precision may always be improved by increasing the coating time, but this does not always result in substantial improvements in accuracy. Typically, for silicone coatings, reported standard deviations correspond to ±0.01 g per square meter of silicone.

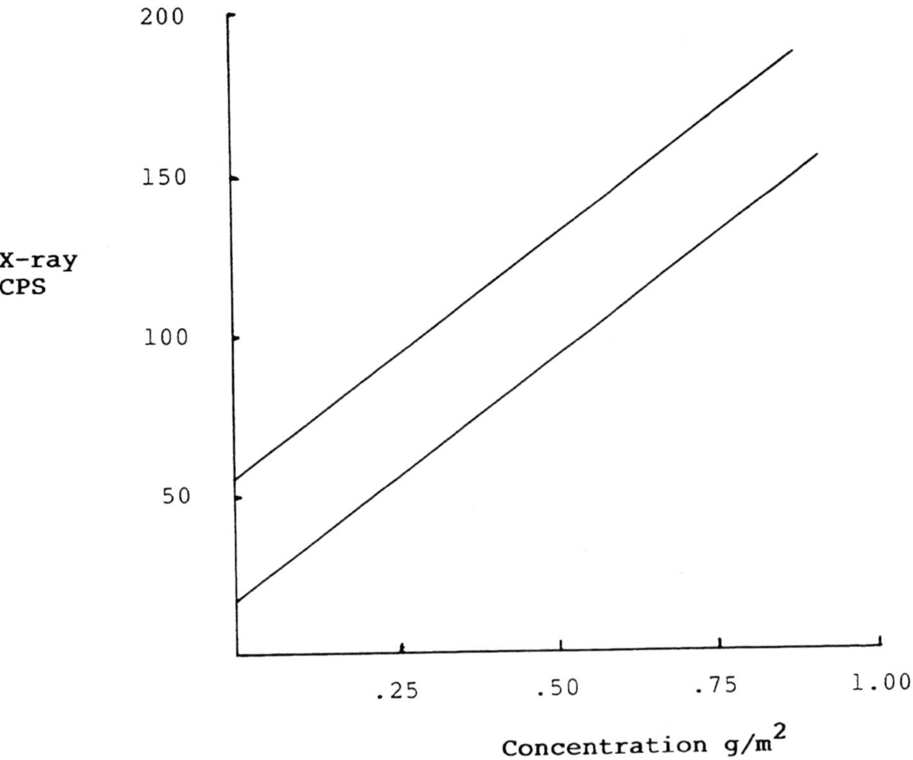

Figure 3 Differences in paper backings.

6.0 CONCLUSION

On-site XRF determinations are rapid enough and precise enough for effective quality control programs for elemental determinations on a variety of substrates. The use of XRF is not limited to measuring coatings; it is flexible enough to measure related products, such as tin, platinum, and rhodium catalysts, and other solutions.

8

Sunlight, Ultraviolet, and Accelerated Weathering

Patrick Brennan and Carol Fedor

The Q–Panel Company, Cleveland, Ohio

1.0 INTRODUCTION

Sunlight is an important cause of damage to coatings. Short wavelength ultraviolet light has long been recognized ad being responsible for most of this damage.[1]

Accelerated weathering testers use a wide variety of light sources to simulate sunlight and the damage that it causes. Comparative spectroradiometric measurements of sunlight and laboratory testers of various types show a wide variety of UV spectra. These measurements highlight the advantages and disadvantages of the commonly used accelerated light sources: enclosed carbon arc, sunshine carbon arc, xenon arc, and fluorescent UV. The measurements suggest recommendations for the use of different light sources for different applications.

2.0 SUNLIGHT

The electromagnetic energy from sunlight is normally divided into ultraviolet light, visible light, and infrared energy. Figure 1 shows the spectral energy distribution (SED) of noon midsummer sunlight. Infrared energy (not shown) consists of wavelengths longer than the visible red wavelengths and starts above 760 nanometers (nm). Visible light is defined as radiation between 400 and 760 nm. Ultraviolet light consists of radiation below 400 nm. The international Commission on Illumination (CIE) further subdivides the UV portion of the spectrum[2] into UV-A, UV-B, and UV-C as shown in Figure 1. The effects of the various UV wavelength regions are summarized in Table 1.

2.1 Variability of Sunlight

Because UV is easily filtered by air masses, cloud cover, pollution, and other factors, the amount and spectrum of natural UV exposure is extremely variable. Figure 2 compares the UV regions of sunlight, measured at Cleveland, Ohio, at noon on:

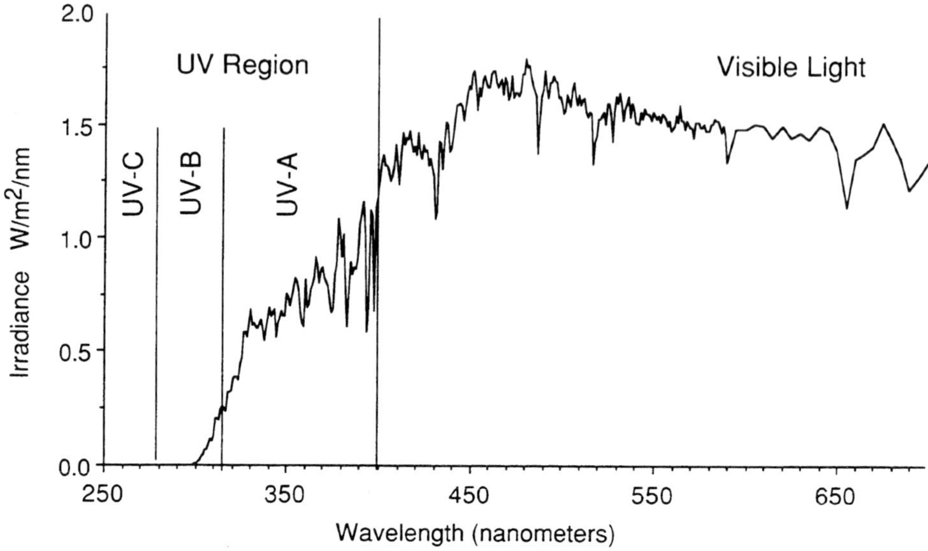

Figure 1   The sunlight spectrum.

- The summer solstice
- The winter solstice
- The spring equinox

These measurements are in essential agreement with data reported by other investigators.[4]

Because the sun is lower in the sky during the winter months, it is filtered through a greater air mass. This creates two important differences between summer and winter sunlight: changes in the intensity of the light and in the spectrum. Most important, much of the damaging, short wavelength UV light is filtered out during winter. For example, the intensity of UV at 320 nm changes about 8 to 1 from summer to winter. In addition, the short wavelength solar cutoff shifts from about 295 nm in summer to about 310 nm in winter. Consequently, materials sensitive to UV below 320 nm would degrade only slightly, if at all, during the winter months.

Table 1   Wavelength Regions of the UV

Region	Wavelength (nm)	Characteristics
UV-A	400–315	Causes polymer damage
UV-B	315–280	Includes the shortest wavelengths found at the earth's surface; responsible for severe polymer damage; absorbed by window glass
UV-C	280–100	Found only in outer space; filtered out by earth's atmosphere; germicidal

Source: Ref. 3.

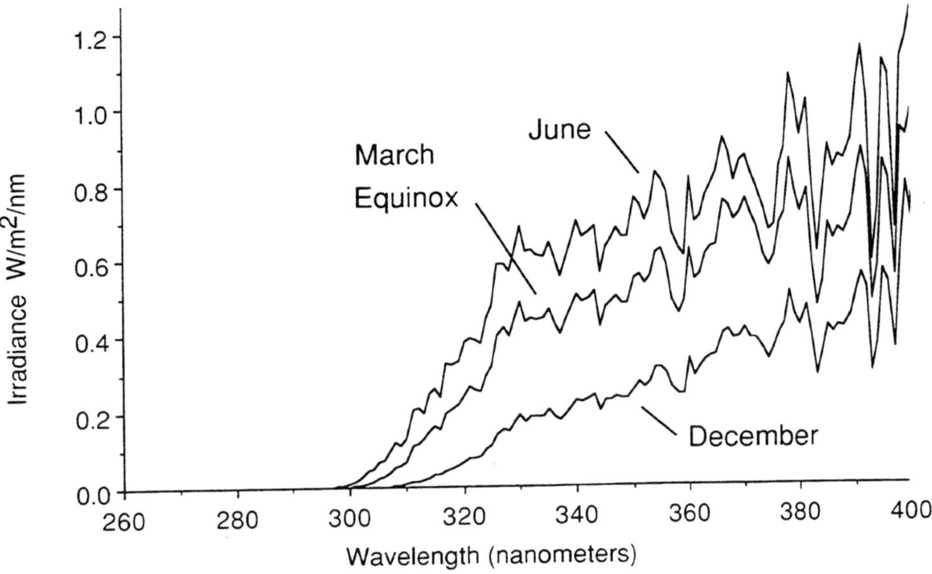

Figure 2   Seasonal variation of sunlight UV.

3.0   ACCELERATED LIGHT SOURCES COMPARED TO SUNLIGHT

The following discussion of accelerated weathering light sources confines itself to the question of UV spectrum. It does not address problems of light stability, the effects of moisture and humidity, the effects of cycles, or the reproducibility of results. For simulations of direct sunlight, artificial light sources should be compared to what we call the Solar Maximum condition: global, noon sunlight, on the summer solstice, at normal incidence. The Solar Maximum is the most severe condition met in outdoor service, and as such it controls which materials will fail. It is misleading to compare light sources against "average optimum sunlight," which is simply an average of the much less damaging March 21 and September 21 equinox readings. In this chapter, graphs labeled "sunlight" refer to the Solar Maximum: noon, global, midsummer sunlight. Despite the inherent variability of solar UV, our measurements show surprisingly little variation in the Solar Maximum at different locations. Figure 3 shows measurements of the Solar Maximum at three widely varied locations.

3.1   The Importance of Short Wavelength Cutoff

Photochemical degradation is caused by photons of light breaking chemical bonds. For each type of chemical bond, there is a critical threshold wavelength of light with enough energy to cause a reaction. Light of any wavelength shorter than the threshold can break the bond, but longer wavelengths of light cannot break it   regardless of their intensity (brightness). Therefore, the short wavelength cutoff of a light source is of critical importance. For example, if a particular polymer is sensitive only to UV light below 295 nm (the solar cutoff point), it will never experience photochemical deterioration outdoors. If the same polymer is exposed to a laboratory light source that has a special cutoff of 280 nm, it *will* deteriorate. Although light sources whose spectra include the shorter wavelengths produce faster tests,

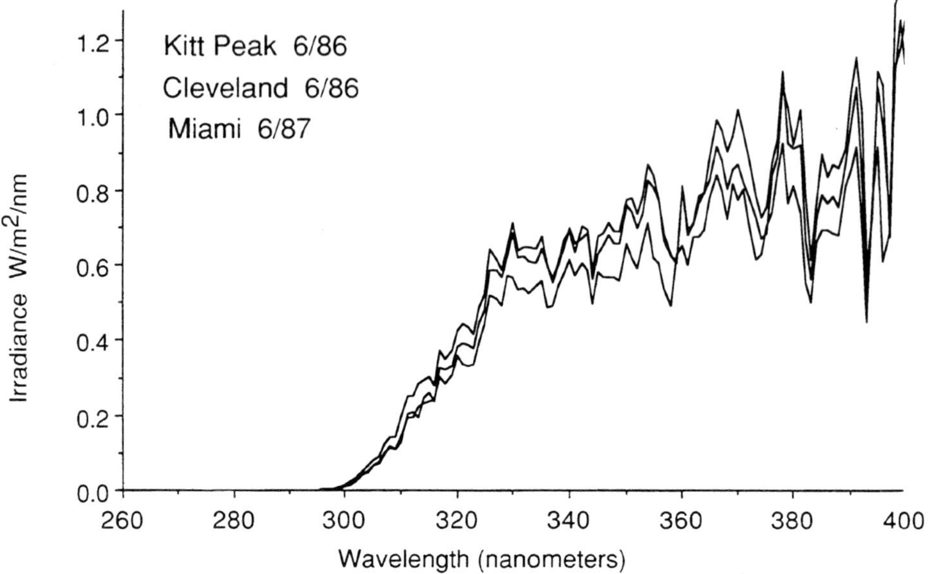

Figure 3 Solar maximum at three locations.

there's a possibility of anomalous results if a tester has a wavelength cutoff too far below that of the material's end–use environment.

4.0 ARC–TYPE LIGHT SOURCES

4.1 Enclosed Carbon Arc (ASTM G–23)

The enclosed carbon arc has been used as a solar simulator in accelerated weathering and lightfastness testers since 1918. Many test methods still specify its use. When the light output of this apparatus is compared to sunlight, some deficiencies become obvious. Figure 4 compares the UV spectral energy distribution of summer sunlight (Solar Maximum) to that of the enclosed carbon arc. The UV output of the enclosed carbon arc primarily consists of two very large spikes of energy, with very little output below 350 nm. Since the shortest UV wavelengths are the most damaging, the enclosed carbon arc gives very slow tests on most materials and poor correlation on materials sensitive to short wavelength UV.

4.2 Sunshine Carbon Arc (Open Flame Carbon Arc: ASTM G–23)

The introduction of the sunshine carbon arc in 1933 was an advantage over the enclosed carbon arc. Figure 5 plots the UV SED of summer sunlight against the SED of a sunshine carbon arc (with Corex D filters). While the match with sunlight is superior to the enclosed carbon arc, there is still a very large spike of energy, much greater than sunlight, at about 390 nm.

A more serious problem with the spectrum of the sunshine carbon arc is found in the short wavelengths. To illustrate this, a charge of scale is necessary to expand the low end of the graph. Figure 6 shows Solar Maximum versus sunshine carbon arc between 260 and 320 nm. The carbon arc emits a great deal of energy in the UV–C portion of the spectrum, well

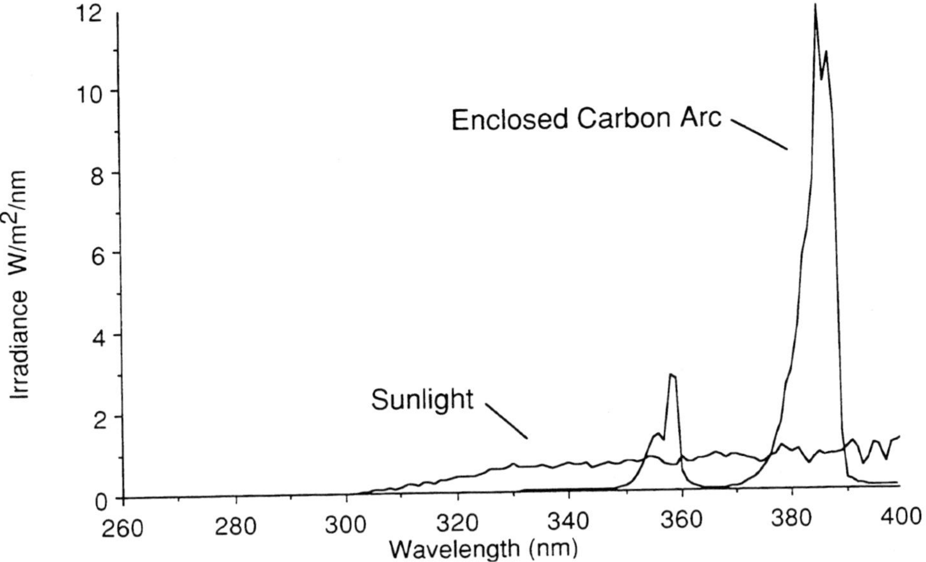

Figure 4 Enclosed carbon arc and sunlight.

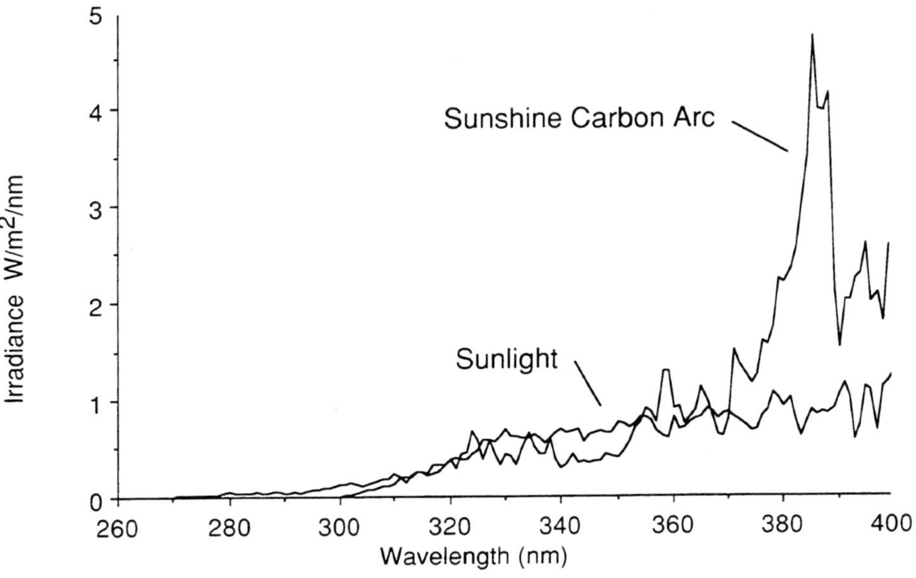

Figure 5 Sunshine carbon arc and sunlight: 260–400 nm.

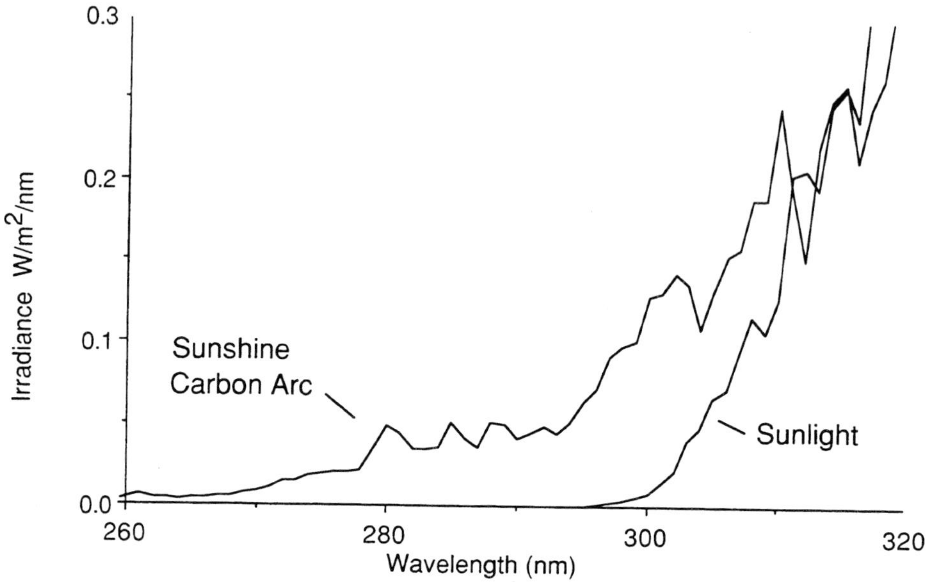

Figure 6   Sunshine carbon arc and sunlight: 260–320 nm.

below the normal solar cutoff point of 295 nm. Radiation of this type is realistic for outer space, but is never found at the earth's surface. These short wavelengths can cause unrealistic degradation when compared to natural exposures.

4.3 Xenon Arc (ASTM G–26)

The xenon arc was adapted for accelerated weathering in Germany in 1954. Our understanding of the xenon arc spectra is complicated by two variables: the effect of filters and the effect of irradiance settings.

Xenon arcs require a combination of filters to reduce unwanted radiation. Automotive test methods that call for xenon arc exposure usually specify quartz inner and borosilicate outer filters. This filter combination allows severe, unrealistically short wavelength UV (as low as 270 nm).

The most widely used filter combination foe the xenon arc is borosilicate inner and outer filters (boro/boro). This is a better simulation of sunlight than the quartz/boro filters, because the cutoff wavelength of the boro/boro is approximately 280 nm, somewhat closer to the sunlight cutoff of 295 nm. Figure 7 compares the xenon arc with the two filter combinations to the Solar Maximum.

Recent xenon arc models have a monitoring system to compensate for the inevitable light output decay due to lamp aging. The most common irradiance settings are 0.35 or 0.55 W/m² at 340 nm. Figure 8 shows how these two settings (with boro/boro) filters compare with the Solar Maximum. While 0.55 compares better with summer sunlight, 0.35 is more like winter sunlight. However, for practical reasons, 0.35 is the setting most commonly used.

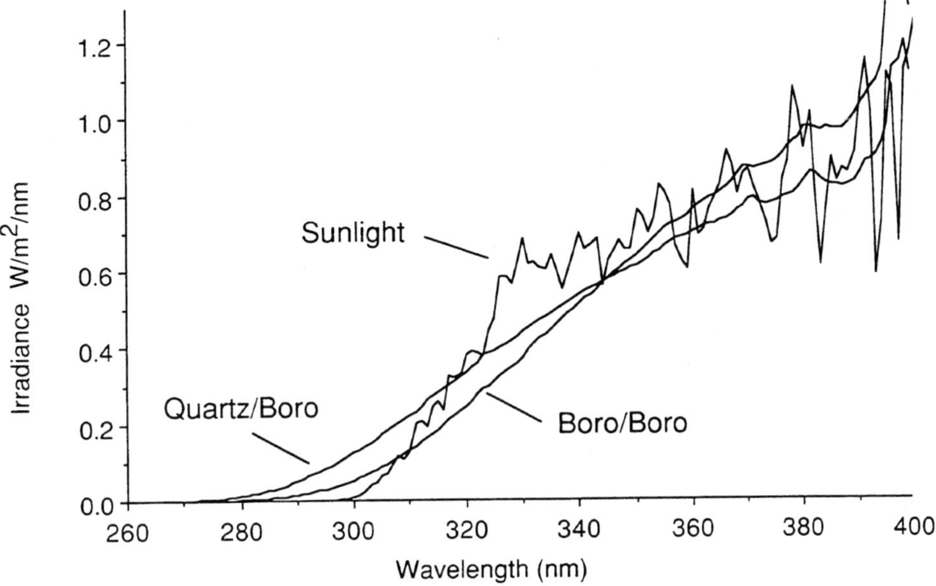

Figure 7 Xenon-effect of filters.

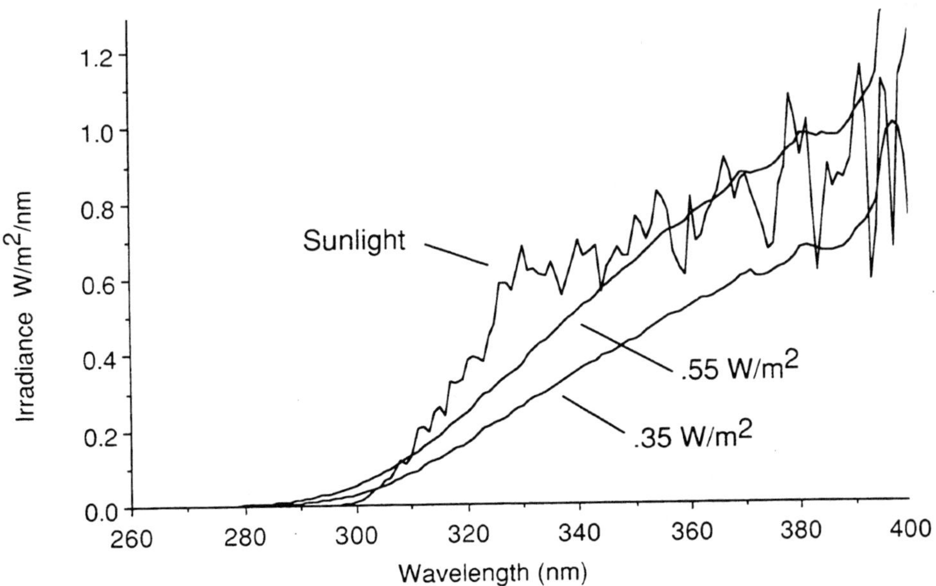

Figure 8 Xenon-effect of irradiance setting.

5.0 FLUORESCENT UV LAMPS

Since the 1970s, fluorescent UV and condensation testers have come into wide use. There are different lamps, with different spectra, for different exposure applications. The fluorescent UV testers and the arc testers use different approaches. The former do not attempt to reproduce sunlight itself, just the damaging effects of sunlight. This approach is effective because short wavelength UV causes almost all the damage to durable materials exposed outdoors. Consequently, fluorescent UV testers confine their primary light emission to the UV portion of the spectrum.

5.1 FS–40 Lamp (F40–UVB) (ASTM G–53)

In the early 1970s, the FS–40 became the first fluorescent UV lamp to achieve wide use. This lamp has been specified in some automotive test methods, particularly for coatings. Most of the FS–40's output is in the UV–B portion of the UV spectrum, along with some UV–A. This lamp has demonstrated good correlation to outdoor exposures for the gloss retention on coatings[5] and for the material integrity of plastics. However, the short wavelength output below the solar cutoff can occasionally cause anomalous results, especially for color retention of plastics and textile materials.[6]

5.2 UVB–313 Lamp (ASTM G–53)

Introduced in 1984, the UVB–313 is essentially a second–generation FS–40. It has the same SED as the FS–40, but its output is higher and more stable. Figure 9 plots the Solar Maximum against the UVB–313 and the FS_40. Because of its higher output the UVB–313 gives significantly greater acceleration over the FS–40 for most materials. With the exception of the automotive industry, the UVB–313 is the most widely used light source for the ASTM G–53 devices.

5.3 UVA–340 Lamp (ASTM G–53)

The UVA–340 was introduced in 1987 to enhance correlation in the G–53 devices. Figure 10 shows the UVA–340 compared to the Solar Maximum. This lamp is an excellent simulation of sunlight in the critical short wavelength UV region, from about 365 nm, down to the solar cutoff of 295 nm. Because the UVA–340 eliminates the short wavelength output (i.e., 5 output lower than sunlight), which can cause unnatural test results, it allows more realistic testing than many of the other commonly used light sources. The UVA–340 has been tested on both plastics and coatings and greatly improves the correlation possible with the fluorescent UV and condensation devices.[7]

6.0 CONCLUSIONS

The correlation between laboratory and natural exposures probably will always be controversial. As Fischer has indicated,[7] test speed and test accuracy tend toward opposition. Accelerated light sources with short wavelength UV give fast tests but may not always be accurate. Where they are wrong, however, they usually err on the safe side   they are too severe. Light sources that eliminate wavelengths below the solar cutoff of 295 nm will give better, more accurate results, but the price for increased correlation is reduced acceleration. Users must educate themselves to make this choice.

In addition we should point out that despite many chemists' fascination with light energy, the spectrum of a test device is only one part of the picture. With any accelerated test-

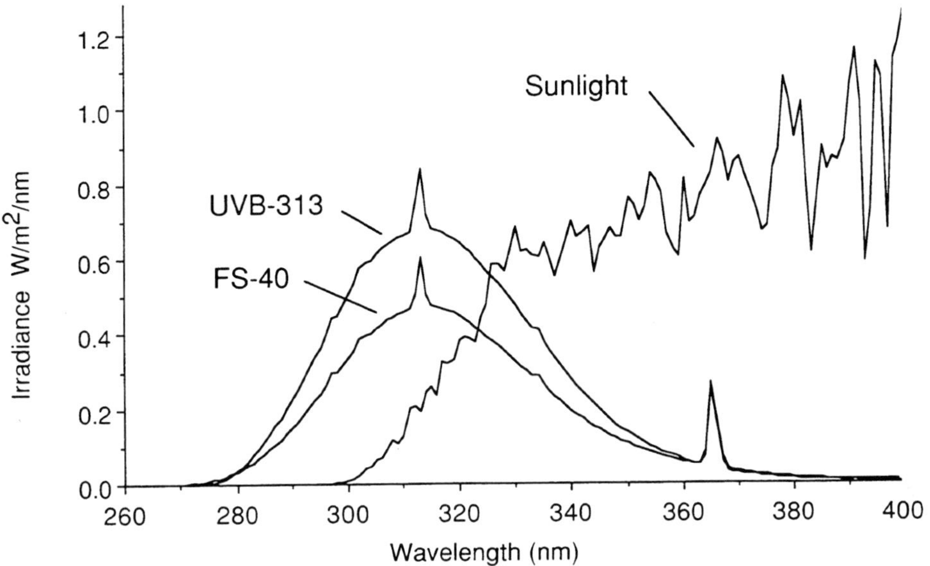

Figure 9 UVB-313 and FS-40.

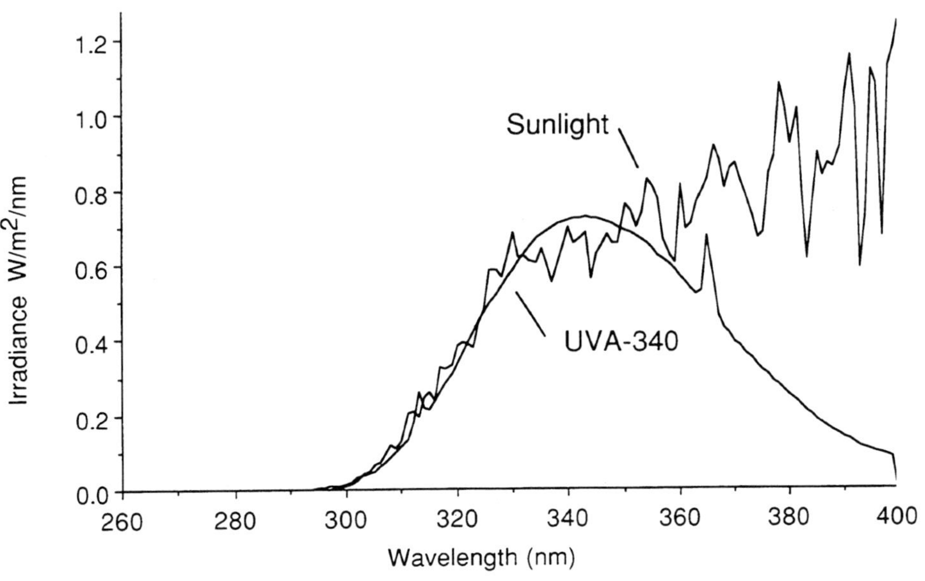

Figure 10 UVA-340 and sunlight.

er, there are a number of parameters that must be programmed: UV spectrum, moisture, humidity, temperature, and test cycle. Furthermore, the parameters that one chooses are, to a certain extent, arbitrary. No single test cycle or device can reproduce all the variables found outdoors in different climates, altitudes, and latitudes. Consequently, even the most elaborate tester is really just a screening device. Accelerated weathering data are comparative data. The real usefulness of accelerated testers is that they can give reliable, relative indications of which material performs best under a specific set of conditions.

ACKNOWLEDGMENTS

Most of the data in this paper were originally presented at the Society of Plastics Engineers Automotive RETEC, November 1987. The authors are grateful for the cooperation of Kitt Peak National Observatory, Kitt Peak, Arizona, and Ohio Spectrographic Service, Perma Ohio.

REFERENCES

1. N. Searle, and R. Hirt, "UV SED of sunlight," *J. Opt. Soc. Am.*, 55 (11) (1965).
2. CIE Standard No. 20, 19 .
3. D. Grossman, "Know your enemy: The weather," *J. Vinyl Technol.*, 3 (1) (1981).
4. G. Zerlaut, "Accelerated weathering & UV measurements," ASTM STP 781. Philadelphia: American Society for Testing and Materials, 1982.
5. G. Grossman, "Correlation of weathering," *J. Coatings Technol.*, 49 (633) (1977).
6. J. Dick, et al., "Weathering of pigmented plastics," SAE Technical Paper No. 850350, 1985.
7. R. Fischer, "Accelerated test with fluorescent UV-condensation," SAE Technical Paper No. 841022, 1984.

9

Cure Monitoring: Microdielectric Techniques

David R. Day

Micromet Instruments, Inc., Cambridge, Massachusetts

Developments in the area of microelectronics now enable the fabrication of microdielectric sensors that can analyze drying, curing, and diffusion phenomena in coatings.[1] Several types of microdielectric sensor have evolved in the past few years, the most sensitive being based on an interdigitated electrodes and field effect transistors fabricated on a 3 x 5 mm silicon chip.[2] The chip sensor is housed in a polyamide package and configured for ease of placement in various processing environments (Fig. 1).

1.0   THE DIELECTRIC RESPONSE

The dielectric response arises from mobile dipoles and ions within the material under test. As a coating cures, the mobilities of dipoles and ions are drastically reduced, sometimes by as much as 7 orders of magnitude. Microdielectric sensors are sensitive enough to follow those changes and are therefore useful for cure monitoring, cure analysis, and process control.[3]

The dielectric response is typically expressed by the quantities of permittivity or dielectric constant (E') and loss factor (E''):

$$E' = E_u + \frac{E_r - E_u}{1 + \omega\tau^2} \tag{1}$$

$$E'' = \frac{s}{e_0\omega} + \frac{E_r - E_u}{1 + \omega\tau^2}\omega\tau \tag{2}$$

where $(E_4-E_u)/(1+wt^2)$ is the dipole term, $se_0\omega$ is the conductivity term, and

E' = dielectric constant
E'' = loss factors
s = bulk ionic conductivity
e_0 = permittivity of free space (a constant)

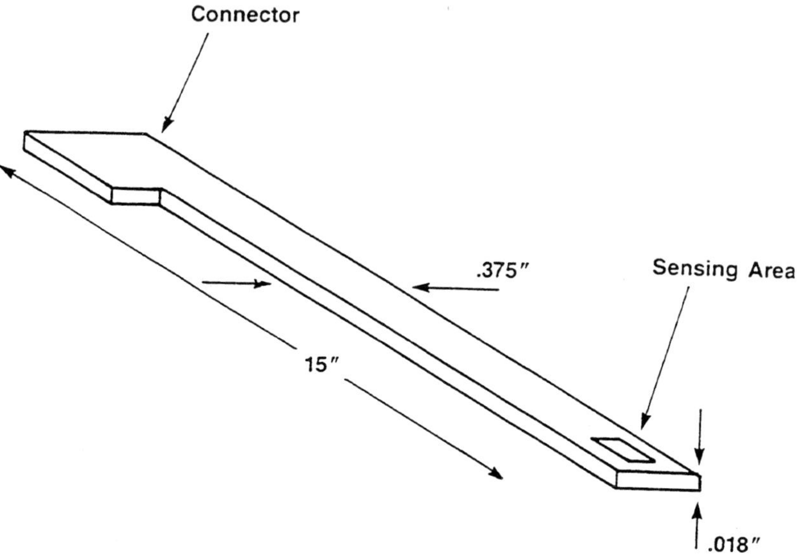

Figure 1   Schematic diagram of microdielectric sensor.

ω = frequency $\times 2\pi$

τ = dipole relaxation time

E_r = relaxed permittivity (low frequency E')

E_u = unrelaxed permittivity (high frequency E')

In addition to ions and dipoles, other factors, such as electrode polarization[4] or inhomogeneities, may influence the dielectric response; however, these generally play a minor role and are usually ignored.

The dielectric loss factor is the most useful quantity for monitoring cure reactions. Either dipole relaxation times or ionic conductivity levels may be monitored during cure. However, dipole relaxation times are usually difficult to determine because the conductivity response often dominates the dipole term in Equation 2. On the other hand conductivity usually can be determined throughout the entire cure process, especially if low frequencies (< 10 Hz) are monitored during the end of cure. The inverse of the measured conductivity, resistivity, is often proportional to viscosity of the material under test before gelation and related to rigidity after gelation.[5] Figure 2 shows loss factor data during an isothermal cure of an epoxy resin. Figure 3 compares the log (resistivity) calculated from the data in Figure 2 to degree of conversion as determined by the fractional generated heat method using differential scanning calorimetry (DSC). Figure 3 shows that the dielectric response can monitor the entire cure and is far more sensitive to last few percent of cure than DSC.

2.O   CHANGES IN RESISTIVITY DURING CURE

Figure 4 plots changes in resistivity during isothermal cure versus changes in the glass transition temperature. These data demonstrate the utility of microdielectric sensors for monitoring and process control in coatings.

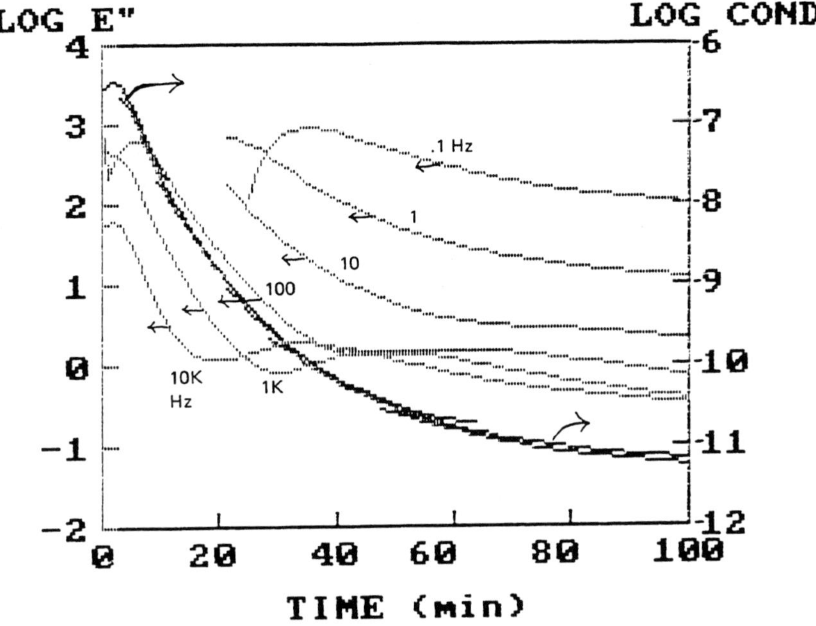

Figure 2 Dielectric loss factor data of isothermal (392°C) epoxy–amine cure. Frequencies range from 10^{-1} to 10^4 Hz.

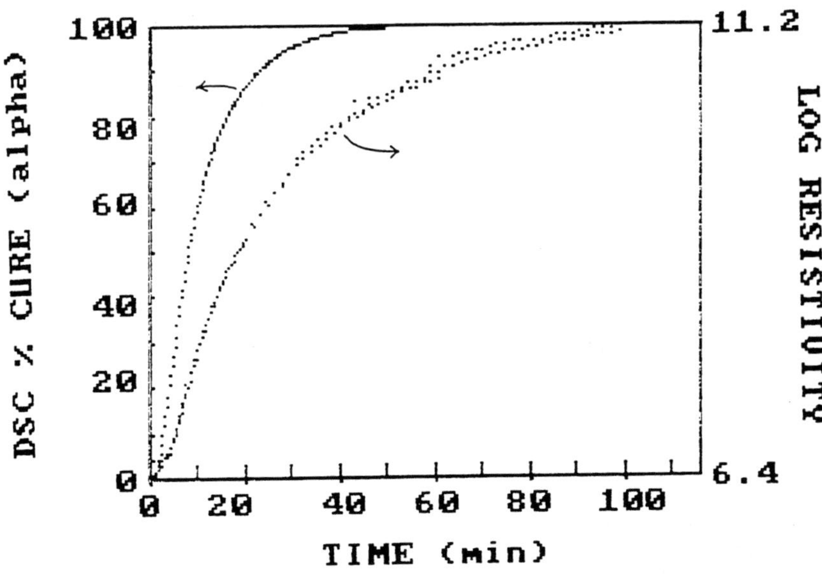

Figure 3 Ionic resistivity data from Figure 1 and degree of conversion as determined by DSC versus time.

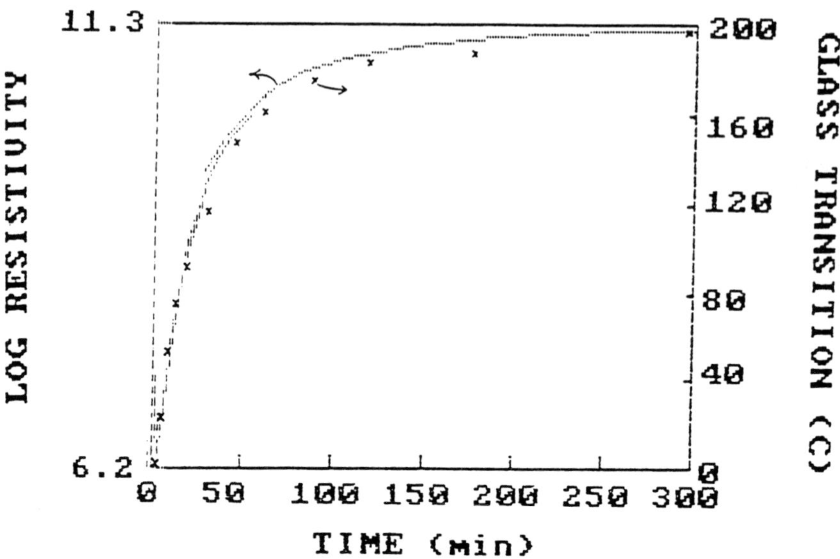

Figure 4   Ionic resistivity data and T_g during isothermal epoxy–amine cure.

Using the previously described techniques, ionic conductivity (or its reciprocal, resistivity) can be obtained in real time by continuously monitoring a range of frequencies. It can them be used to control the reaction through variation in either temperature or pressure. There exist several ways in which dielectric feedback may be used. Some of these are as follows.

Temperature may be held constant or controlled until a desired viscosity (measured dielectrically) is attained.

Viscosity may be held constant or varied at will through controlled variation in temperature.

Pressure, vacuum, or mold opening may be activated upon attainment of a critical viscosity of dielectric reaction rate.

Reaction may be terminated when the dielectric reaction rate decreases below some critical value.

2.1 Process Control Through Dielectric Feedback

Figure 5 is an example of a process-controlled cure of a graphite epoxy using microdielectric feedback. Control was achieved using an IBM PC with modified Micromet Instruments dielectric software and hardware. The cure was carried out in a hot press with temperature controlled from the computer. The process control software sequence was as follows.

1.   Heat and hold at 250°F until a log resistivity of 7.0 is reached (allows for degassing while preventing premature cure).

2.   Hold log resistivity (viscosity) at 7.0 until 350°F is reached (allows for controlled curing and prevents second viscosity minimum).

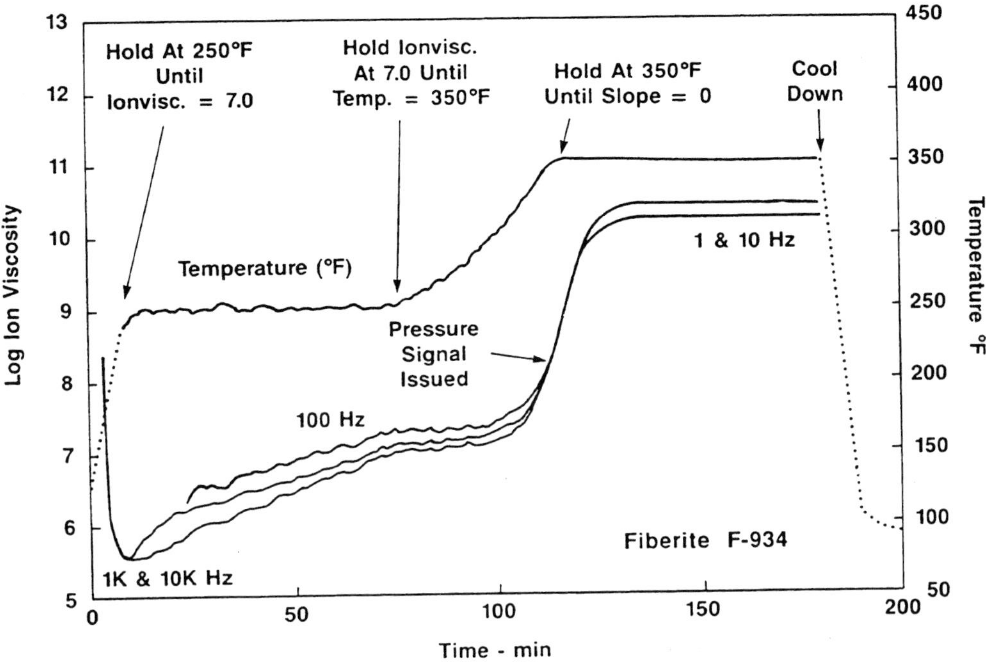

Figure 5 Process control of epoxy grafite cure utilizing microdielectric feedback.

3. Hold at 350°F until dielectric reaction rate is near zero (allows reaction to go to completion).
4. Cool down and notify operator that cycle has been completed.

2.2 Process Control Through Dielectric–Thermal Feedback

The cure in Figure 5 was completely controlled through both dielectric and thermal feedback from the microdielectric sensor in contact with the graphite epoxy material. Note that the second viscosity (resistivity) minimum typical of these materials was completely eliminated through the use of viscosity control. This technique is useful for limiting excessive bleed in composites that are prone to such problems. Finally the end point was detected and the reaction stopped, eliminating unnecessary overcuring time.

3.O SUMMARY

Microdielectric sensors are useful for monitoring cures of coatings under actual processing conditions. The ionic resistivity portions of the loss factor data correlate to viscosity during the early stages of reaction (before gelation or solidification). As the reaction progresses, the slope of the ionic resistivity can be used to monitor reaction rate and to detect when cure is complete. The microdielectric sensors provide a unique capability to correlate measurements made in the laboratory to those in the factory and to provide the necessary feedback information for adaptive process control.

REFERENCES

1. D. R. Day and D. D. Shepard, J. Coatings Technol., *60* (760);57 (1988).
2. (a) S. D. Senturia, N. S. Sheppard, Jr., H. L. Lee, and D. R. Day, J. Adhes., *15*, 69(1982).
   (b) N. F. Sheppard, Jr., D. R. Day, H. L. Lee, and S. D. Senturia, Sensors Actuators, *2*, 263(1982).
3. (a) W. E. Baumgartner and R. Ricker, SAMPE J., *19*(4), 6(1983).
   (b) D. R. Day, *Proceedings* of the SAMPE Symposium, Las Vegas, 1986, p.1095.
   (c) D. E. Kranbuehl, *Proceedings* of the SAMPE Symposium, Anaheim, CA, 1988.
   (d) D. R. *Proceedings* of the SAMPE Symposium, Anaheim, CA, 1988, p. 594.
4. D. R. Day, T. J. Lewis, H. L. Lee, and S. D. Senturia, J. Adhes. *18*, 73 (1985).
5. J. Gotro and M. Yandrasits, *Proceedings of the 45th* SPE ANTEC, Anaheim, CA, 1987, p. 1039.

Part 2
Coating and Processing Techniques

10
Wire-Wound Rod Coating

Donald M. MacLeod

Industry Tech
Oldsmar, Florida

1.0 INTRODUCTION

Wire-wound metering rods have been used for more than 75 years to apply liquids evenly to flexible materials. They were the first tools used to control coating thickness across the full width of a moving web. The 1980s saw a new popularity in rod use because of improved quality and the industry trend toward shorter converting runs. Wire-wound rods are used in a wide range of applications but find their greatest appeal in the manufacture of tapes, labels, office products, and flexible packaging. The first rods were made of ordinary carbon steel, wrapped with music wire. Today's metering rods use precision-ground core rods made of stainless steel, tightly wound with polished stainless steel wire at high speeds, on custom-designed winding machines. The resulting product is a laboratory-quality precision tool that can control coating thicknesses accurately within 0.0001 in. (0.1 mil). A typical wire-wound rod station is shown in Figure 1.

Also called applicator rods, Mayer bars, equalizer bars, coating rods, and doctor rods, this equipment has found uses in a wide variety of production applications, from the manufacture of optical films to wallboard panels. Wire-wound metering rods look deceptively simple. A stainless steel rod is wound with a tight spiral of wire, also made of stainless steel. The wire can be so small that it is almost invisible to the naked eye, or so large that the windings look like the coils of a hefty spring. Today, the industry has standardized on stainless steel rods because they can be used with almost every coating liquid. Earlier problems with rust and corrosion have been virtually eliminated. Where abrasive wear is a problem, some converters use chrome plating to prolong the life of the rod because of the hard surface presented by chromium. Chrome has its drawbacks however: it builds up unevenly at the extreme tops of the wires, changing the shape of the wires and the resulting coating thickness. Also, if not applied properly, chrome can acquire pitting marks or can flake off, contaminating the coating bath or causing uneven coating. Several new products introduced since 1985 have further expanded the market for rods. Where streaking or rod cleaning is a problem, rods with a Teflon surface are available. Particles that might wedge between stain-

103

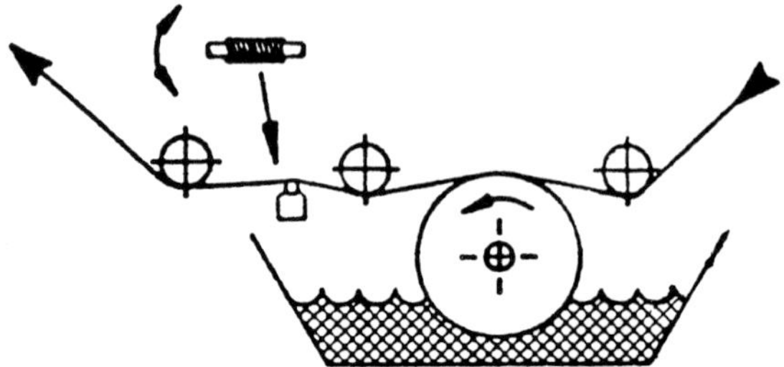

Figure 1   Typical coating station.

less steel wires tend to slide through, preventing buildups and subsequent streaks in the coating. Other users find the Teflon surface easier to clean, and a few claim increased rod life up to eight times that of stainless steel rods. Super-Coat * rods are conventional rods with a second, smaller wire wound around the primary wire of a normal rod. The shape of the grooves between the wires is altered drastically, which allows certain coaters to produce controlled coatings more than twice as thick as with conventional rods (see Fig. 2). Others users find that for any given coating thickness, the deeper grooves between the wires are spaced closer together and provide a smoother finish on the product.

Today's coating machines have wider webs and run at higher speeds, outproducing earlier equipment many times over. Meanwhile, the market for coated material requires more and more specialized products, hence more frequent changeovers and setups. Compared with other coating methods, changing coating rods requires little downtime for changeover and makes precise control of coating weight without changing formulations a practical consideration for many coaters.

2.0 HISTORY

In 1905 Charles W. Mayer founded the Mayer Coating Machines Company in Rochester, New York. The firm made equipment for manufacturers of carbon and wax papers, two new

---

*Product of Consler Scientific Design, Inc.

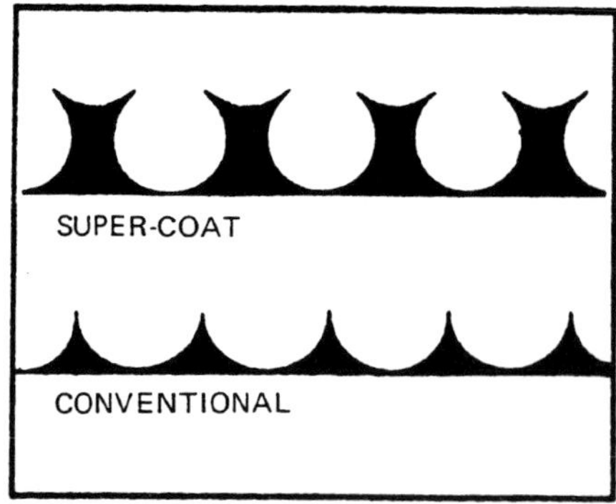

FIGURE 2   Super-Coat flow pattern.

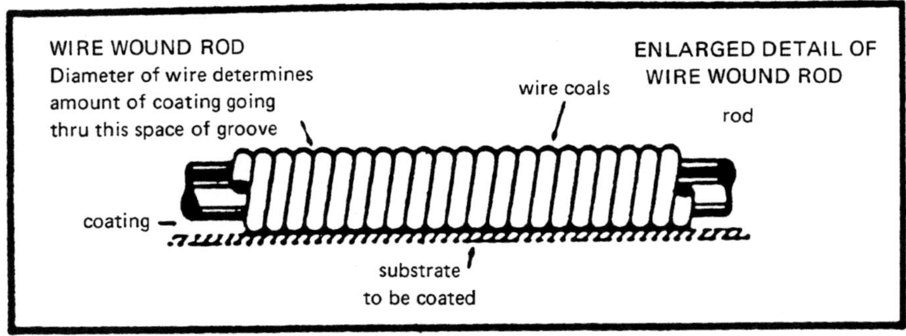

FIGURE 3   Rod design.

and growing industries. These machines used "equalizer bars" or "doctor rods," the forerunners of today's precision metering rods. The rods were made of carbon steel wound with different sizes of music wire. Mayer was issued a series of patents on his coating machines, including one in 1912 that covered his equalizer bars. Coaters of carbon and wax papers found that they could easily change the thickness of their coating by switching rods, so they began ordering rods with different wire sizes. Also, the early rods tended to rust and wear out, so Mayer found a ready and growing market for replacement rods, which became an important part of his business as more machines were shipped. Mayer's success in the coating machinery business became the target of federal antimonopoly laws, and he was forced to release his designs and patents to others in the 1930s. New companies sprang up, and their machines used doctor blades and roll coating methods, in addition to wire-wound rods. Under the federal guidelines, Mayer was not required to expose his proprietary techniques for making equalizer bars. The Mayer factory continued to produce these rods on special winding machines, which were developed for this purpose. As his machinery business declined, rod manufacturing continued to expand.

During this period, Mayer hired Ralph J. Consler on a contract basis, as a design engineer to develop new equipment for the still-growing carbon paper industry and the new "multiform" carbon coating market. Consler and another designer, Bruce DeNeve, formed DeNeve & Consler Company, a partnership in Rochester, in 1940.

3.0   THEORY AND PRINCIPLE

Coating thickness is governed by the cross-sectional area of the grooves between the wire coils of the rod, as shown in Figure 3. The geometry of this system creates a wet film thickness directly proportional to the diameter of the wire used. For example, you can double the coating thickness by doubling the wire size.

The groove between the wires determines the precise amount of coating material that will pass through. The initial shape of the coating is a series of stripes, spaced apart according to the spacing of the wire windings. Almost immediately, normal surface tension pulls these stripes together, forming a flat surface, ready for drying in air or under heat. Wet film coating thicknesses, accurate to 0.1 mil, can easily be achieved using metering rods. Mathematically, the average thickness of the area between the wires, and therefore the coating thickness, is 0.173 times the wire diameter. From a practical standpoint, most users use a 1:10 ratio to select the best theoretical rod size for a given wet coating thickness. A rod wound with an 0.015 in. diameter wire, for example, will ideally produce a wet film that is 0.0015 in. (1.5 mils) thick. In production coating, other physical factors can influence the actual coating thickness. The most important of these is the phenomenon of shearing action of the liquid. In effect, not all the coating material passes through the gaps; some liquid adheres to the surface of the wire. The impact of this is usually small, but it can be significant when small (0.025–.005 in.) wires are used, or when viscosities are high. In high volume coating operations, the speed of the web, web tension, penetration into the base material, and other factors will also affect the coating thickness. The sum of all these variations seldom exceed a 20% difference from the theoretical coating weight; however, the differences can be large enough to suggest a trial-and-error procedure for selecting the ideal wire size for each coating application. Because of the relatively low cost of rods, and the ease of changing from one wire size to another, the selection process is both economical and practical. Most coatings laboratories have a set of "lab rods," which are shorter versions of production coating rods in a variety of wire sizes. To test a coating for coverage, adhesion, opacity, or other characteristics, drawdown samples are made an examined. A piece of the substrate material is attached to the flat surface, and some coating liquid is puddled near the top of the substrate sample. A lab rod with a preselected wire size is chosen and pulled manually through the liquid, meeting a precise thickness of coating on the substrate.

The quality of test samples made by hand is largely dependent on the skill of the technician. Variations in rod pressure, angle of stroke, and speed can cause differences in the thickness and consistency of drawdown samples.

4.0 FILM THICKNESS

Wire-wound rod applies a finite thickness of liquid to a web of base material. In production coating operations, it is difficult to measure the actual thickness of a wet coating before it starts to dry through evaporation, although there are gages made for this purpose. In addition, many substrate materials are absorbent, so that there may be little or no actual increase in thickness after coating. If the coating is applied to a hard, nonabsorbent material and is allowed to dry, its dry thickness can be measured accurately with micrometers or other dry thickness gages. Table 1 compares the calculated wet film thickness to the wire diameter of metering rods. The 10:1 ratio between wire diameter and wet film thickness allows an easy selection of the required rod.

5.0 THE ROD COATING STATION

In production coating, the web must pass through a wetting station and then to the metering rod, where a measured thickness of coating is allowed to pass between the wires, and the excess liquid is returned to the reservoir. The coating liquid may be applied at the wetting station by several different methods. The web can be immersed directly into a tank; or an applicator can be rotated (Fig. 4) in the reservoir to transfer the liquid to the web at the top of its rotation (Fig. 5). It is important to apply an excess of coating liquid at this station, to let the metering rod to do its job.

(When an applicator roller is used in a rod coating system, the speed of the applicator is not a critical factor.) In addition, the machine operator can adjust the applicator roller speed within a side range, even while the machine is running. The web passes over the metering rod, which may be stationary or may be rotated slowly. The rotation may be either in the same direction as the web, or in the opposite direction. The choice of stationary rod, move-

TABLE 1 Wire Size Selection Chart

Wire Size	Wire Diameter (inches)	Wet film thickness (mils)	Wet film thickness (microns)	Wire Size	Wire Diameter (inches)	Wet film thickness (mils)	Wet film thickness (microns)
#0	None	0.00	0.0	#38	0.038	3.8	96.5
#2 1/2	0.0025	0.25	6.4	#40	0.040	4.0	101.6
#3	0.003	0.3	7.6	#42	0.042	4.2	106.7
#4	0.004	0.4	10.2	#44	0.044	4.4	111.8
#5	0.005	0.5	12.7	#46	0.046	4.6	116.8
#6	0.006	0.6	15.2	#48	0.048	4.8	121.9
#7	0.007	0.7	17.8	#50	0.050	5.0	127.0
#8	0.008	0.8	20.3	#52	0.052	5.2	132.1
#9	0.009	0.9	22.9	#54	0.054	5.4	137.2
#10	0.010	1.0	25.4	#56	0.058	5.6	142.2
#11	0.011	1.1	27.9	#58	0.058	5.8	147.3
#12	0.012	1.2	30.5	#60	0.060	6.0	152.4
#13	0.013	1.3	33.0	#62	0.062	6.2	157.5
#14	0.014	1.4	35.6	#64	0.064	6.4	162.6
#15	0.015	1.5	38.1	#66	0.066	6.6	167.6
#16	0.016	1.6	40.6	#68	0.068	6.8	172.7
#17	0.017	1.7	43.2	#70	0.070	7.0	177.8
#18	0.018	1.8	45.7	#72	0.072	7.2	182.9
#19	0.019	1.9	48.3	#74	0.074	7.4	188.0
#20	0.020	2.0	50.8	#76	0.076	7.6	193.0
#22	0.022	2.2	55.9	#78	0.078	7.8	198.1
#24	0.024	2.4	61.0	#80	0.080	8.0	203.2
#26	0.026	2.6	66.0	#82	0.082	8.2	206.3
#28	0.028	2.8	71.1	#84	0.084	8.4	213.4
#30	0.030	3.0	76.2	#86	0.086	8.6	218.4
#32	0.032	3.2	81.3	#88	0.088	8.8	223.5
#34	0.034	3.4	86.4	#90	0.090	9.0	228.6
#36	0.036	3.6	91.4	[a]			

[a]For coating up to 0.190 in. thick, use Super-Coat rods.

ment with different coaters and with different products. Establishing the ideal speed of rotation will also be different from job to job, and converters experiment to find the best procedure for each run. The most common procedure however, is to rotate the rod slowly in the opposite direction to the movement of the web. The rotation flushes the coating material between the wires, keeping the wire surfaces wet, and preventing setting up and hardening of some liquids. The rotation also distributes any abrasive wear evenly on the wires, and prevents flat spots from forming. The purpose of the metering rod is to remove excess coating liquid, allowing a measured amount to pass between the wire windings. The web should pass above the rod, to allow the excess liquid to fall back into the tank. The web, however, need not be perfectly horizontal, as long as the surplus coating can return to the tank through gravity. Metering rods for production coating can be made in a wide variety of sizes. The most common core rod diameters are quite small (3/16 and 1/4 in.), although sizes up to 1 in.

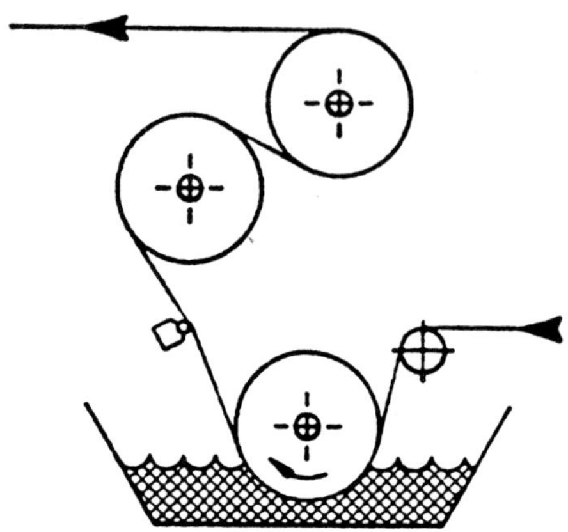

FIGURE 4 Web immersion method.

diameter are also used. The main advantages of small-diameter rods is their low cost and ease of storing and handling.

These thin rods must be supported in the coating machine, because they are not rigid and will deflect with pressure from the web. There are several types of rod holder in common use; the simplest is a square of rectangular steel bar, with a "V" groove machined into it. This rod holder is mounted between the side frames of the coating machine, and the metering rods are placed in the grooves. The "V" groove should be ground and polished to minimize wear on the rod and should be mounted accurately, at a right angle to the direction of web travel and parallel to the idler rollers of the machine (Fig. 6).

The design of a rod coating station should ensure that the web makes intimate contact with the wires of the metering rod. The wrap angle, the angle between the web direction as it approaches the rod and its direction as it leaves the rod, should be 15° for a heavy web tension, or up to 25° for a light web tension (Fig. 7).

Web tension is a critical factor in the design of a rod coating station. With a wrap angle of 15–25°, the web must be tight enough to ensure intimate contact with the metering rod, yet not so tight that the web is deformed by the wires. Adhesives and some other liquids can solidify between the wire windings of the rod whenever the coater is stopped. Many coating machines have a "throw-off" feature, a mechanical means of separating the web from the

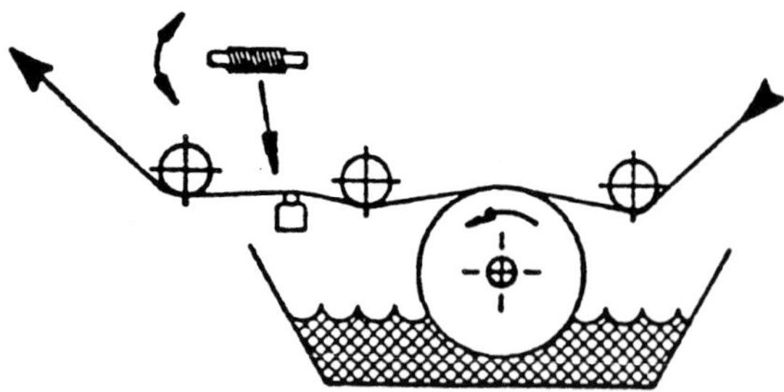

FIGURE 5 Applicator roller method.

FIGURE 6 Rod holder.

rod automatically, whenever the machine is turned off. This allows for quick removal of the rod for flushing and cleaning before the coating material has a chance to congeal between the wires (Fig. 8).

This automatic releasing feature also simplifies rod changing between production runs. One method used by coating machine manufacturers is a rocker arm throw-off system. A series of idler rollers presses the web against the metering rod while the coating machine is running. Whenever the coater is stopped, the idlers automatically rise, lifting the web up, away from the metering rod. At the same time, a water flushing system can be triggered to remove coating material from the rods before it can set up. Other techniques involve lowering the metering rod and its holder when the coater is turned off.

5.1 Rod Station Variations

Every coating application is different. Converters have created many unique systems for rod coating which they developed through experimenting in the lab or in actual production. A smoothing rod is a finely polished metering rod without wire, which is used several different ways. It can be used in the web path after printing with a gravure cylinder, to eliminate the etched pattern and enhance the coated surface. It also can be used after a wire-wound metering rod, when the viscosity of the coating liquid prevents it from leveling through surface tension effect (Fig. 9).

When a large amount of coating liquid must be metered off the web, two wire-wound rods are used in tandem, spaced 1 or 2 in. apart. The first rod has a larger wire size for doctoring off must of the excess liquid, and the second, smaller wire rod controls the finished coating thickness.

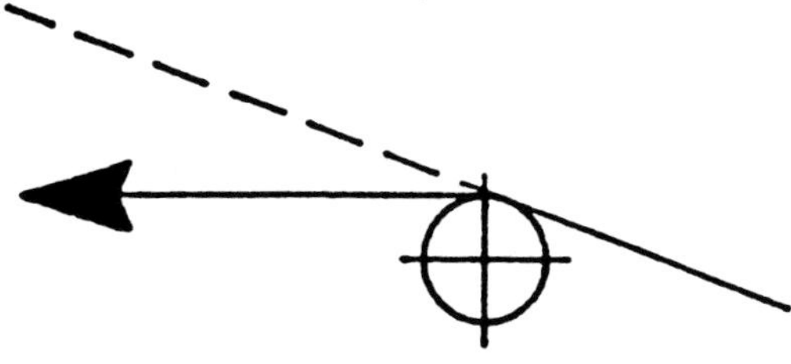

FIGURE 7 Wrap angle.

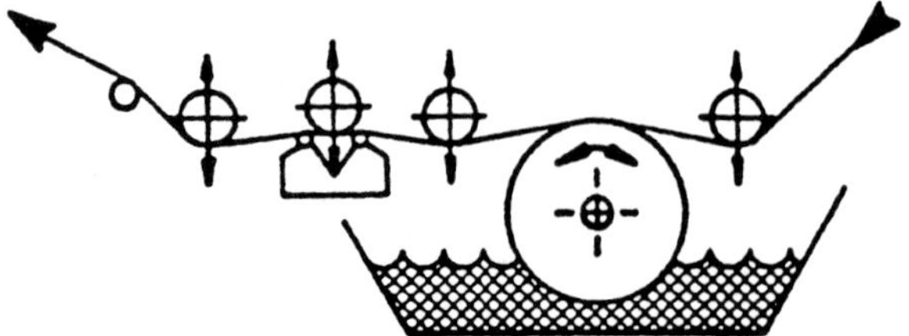

FIGURE 8 Automatic throw-off.

A typical coating station may use one or two metering rods and also have a position for a smoothing rod, which can be engaged for certain applications. The coating roll applies liquid to the moving web which is metered off by one or two rods; then it can be flattened by the smoothing bar. The complete coating station design must also consider draining, filtering, and pumping the coating liquid, controlling the wet edge of the web and disengaging the metering rods when the machine is stopped.

6.0 ADVANTAGES AND DISADVANTAGES

Metering rod coating is the third most popular method in use today, behind gravure and reverse roll coating.

6.1 Low Cost

Rod coating stations are relatively inexpensive and easy to add to existing coaters. Replacing worn or damaged rods, and changing from one coating weight to another, are both inexpensive and fast. The machine downtime for changing rods or cleaning them can be measured in minutes instead of hours, using much less labor than is required by other coating systems.

6.2 Precise Coat Weights

Metering rods can be selected to control the wet coating thickness in 0.1–0.2 mil increments. Without changing the coating formulation at all.

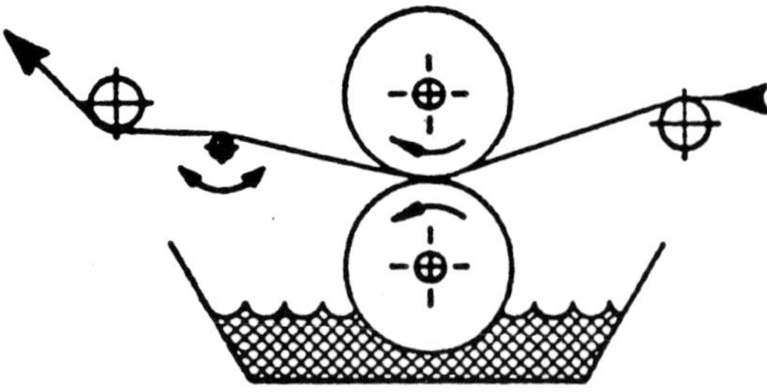

FIGURE 9 Smoothing rod.

6.3 Lower Setup Cost

Another important factor in the new popularity of rod coating is the worldwide trend toward shorter and shorter production runs. The faster setups inherent to rod coating allows the converter more productive running time, in addition to lower labor costs for changeovers.

6.4 Less Edge Wear

Rod coating offers the converter other advantages because of the method used to control wet edges. In a rod coating system, the dry edge is controlled by wipers or deckle straps on each edge of the applicator roll. Because the wipers are constantly wet with coating liquid, the tendency to scratch the roller is reduced. Even when scoring eventually occurs, it is on the applicator roll, not on the metering rod (which controls the final coating thickness). Also, because the wipers are easily moved, their position can be adjusted while the coater is operating, with no downtime at all.

6.5 Limitations

Standard metering rods work best with low viscosity liquids, which will flow easily between the wire windings. Two-wire Super-Coat rods can be used where viscosities are higher, however. Rod coating also can be used for a flat web, without tight or baggy edges. In production rod coating, the actual thickness of the coating can be affected by web speed, viscosity, and other factors. Depending on the application, rod coating speeds are usually limited to 1000 ft/min., although some coaters claim web speeds up to 2000 ft/min. The critical factor in the web speed of a rod coating system is the time required for the striations formed by the rod to level. The rod meters liquid by allowing a measured amount to flow through the spaces between the wires. Normal surface tension forces the raised portions to flow out and form a flat, even coating, but there is a time element involved, which is different for each coating material. The web speed must be controlled to allow time for leveling before the web is dried.

11

Slot Die Coating for Low Viscosity Fluids

Harry G. Lippert

Extrusion Dies, Inc., Chippewa Falls, Wisconsin

1.0 INTRODUCTION

Slot head coating has spawned a wide range of designs, some quite radical in their concept. This chapter discusses conservative manufacturing experience along with the experience of a wide variety of processors currently utilizing the proximity or wipe-on method.

2.0 MANIFOLD THEORY AND DESIGN

The primary purpose of a die is to define a width and provide an even coating in terms of cross-sectional thickness and smoothing. The manifold and coathanger section of the die is the main component in accomplishing uniform distribution. Smoothing will be addressed in a later section.

There are two basic styles of manifold design in use today: coathanger shaped, with a volumetrically reducing cross section (Fig. 1), and T-shaped, with a constant cross section (Fig. 2).

In either style, flow through the manifold is analogous to flow through a pipe in that there is an increasing resistance to material flow as the length increases. The wider the die (the longer the pipe), the greater the resistance to flow. It follows then that the primary criterion in a good die design is to ensure adequate flow to the ends of the die as the width requirements increase.

The coathanger-style die utilizes a slot section (preland) with a varying length downstream of the manifold to compensate for this pressure increase (see area *B*, Fig. 1). The pressure drop in the preland section must decrease at the same rate it increases in the manifold section. If the sum of these two components is equal at any point in the overall flow stream, the result is an even flow.

'Adapted from 1987 Polymers, Laminations and Coating Conference Book 1, December 1987. Copyright, TAPPI, Technology Park/Atlanta, Atlanta, GA 30348, 1987.

113

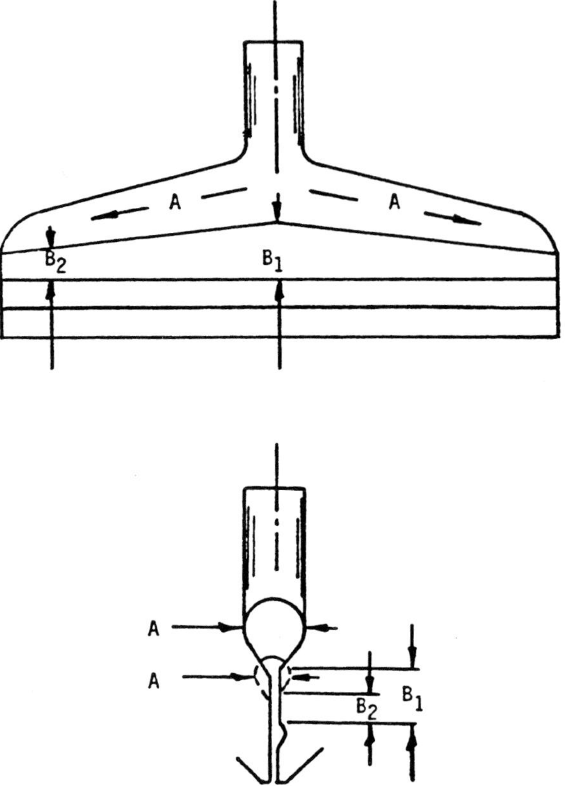

Figure 1 Coathanger volumetrically diminishing manifold.

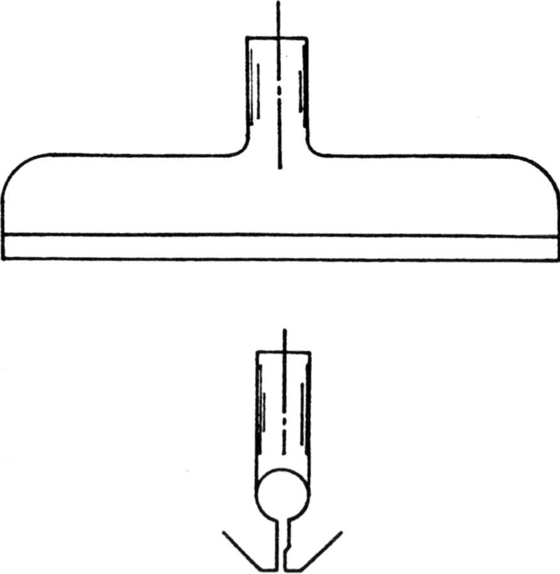

Figure 2 Constant cross section manifold.

It can be seen that while the generic coathanger style design is fixed, the overall dimensions may vary greatly, depending on a given die width, flow rate, or general coating material requirements. Generally as the die gets wider, the length of the preland section (B_1) must get longer and the manifold larger; as the flow rate increases, the manifold must get larger as well as the height of the gap at B_1. The compensating preland section downstream of the manifold allows the die design to be varied greatly to suit a given application.

These large internal designs are used for applications characterized by coating materials that vary greatly in viscosity levels or call for an extreme range of flow rates. Larger flow channels are less sensitive to rate and viscosity changes than small channels. Small internal designs are used for materials that require a low residence time in the die because of thermal degradation, or high shear rates to prevent gelation (thixotropic materials).

In analyzing the coathanger manifold, it must be emphasized that the manifold decreases in cross section as it approaches the ends of the die (dimensions A in Fig. 1); this rate of reduction may also be changed to suit a specific application. Because material is flowing out of the front of the die along its entire width, less material is presented to the manifold as it approaches the ends of the die. The reduction in manifold volume is an attempt to keep the velocity of the material at the ends of the die to a maximum, to compensate for the lower flow rate, and to prevent carbonization of the resin or changes in viscosity in a thixotropic or dilatant adhesive.

In summary, it can be seen that the coathanger manifold design can be modified to suit an application and still accomplish the primary criterion of even flow distribution.
To adequately design a coathanger die, the following information is required

1. Rheology curve (see Fig. 3). A rheology curve, a fingerprint of a particular resin, predicts its viscosity level at a given shear rate. This is required for all non-Newtonian or shear thinning fluids.
2. Flow rate or range of rates.
3. Material density at processing temperatures.
4. General material characteristics, such as heat degradability or thixotropicity.

The T-shaped manifold in the constant cross section style (see Fig. 2) has no compensating preland section; this is because of its inherent design. Rather, this style of die design relies on a larger manifold section to reduce the resistance to flow to the ends of the die; the larger the manifold, the less the resistance and the better the flow distribution. In theory, there can never be an even distribution, because no matter how large the manifold is, there will always be some pressure drop across it, and therefore less flow to the ends of the die when compared to the center.

The larger manifold has some drawbacks in that the residence time is greatly increased and flow at the ends of the die is nearly stagnant. The overall internal flow channel design cannot be increased or decreased to suit a given application, as it is restricted by the need for manifold size to achieve some flow distribution.

If the flow presented to the die is not constant over time, if the fluid is not homogeneous in terms of temperature and mix, and because there are inherent errors in viscosity measurement and theoretical flow calculations, thickness variations will sometimes occur. To adequately adjust these flow variations, a flexible lip is required as a fine-tune adjustment.

Having multiple entrances or a pump within a die simply represents attempts at producing uniform distribution, minimizing the effect of transverse pressure drop, and simplifying the job of manifold design.

SOLVENT A
Viscosity: 1100 + 250cps
(Ambient)
Solid content: 31.o+1.5%
MELTING VISCOSITY
CPS
800
600
400
200
SHEAR RATE
0
5,000
10,000
sec-1

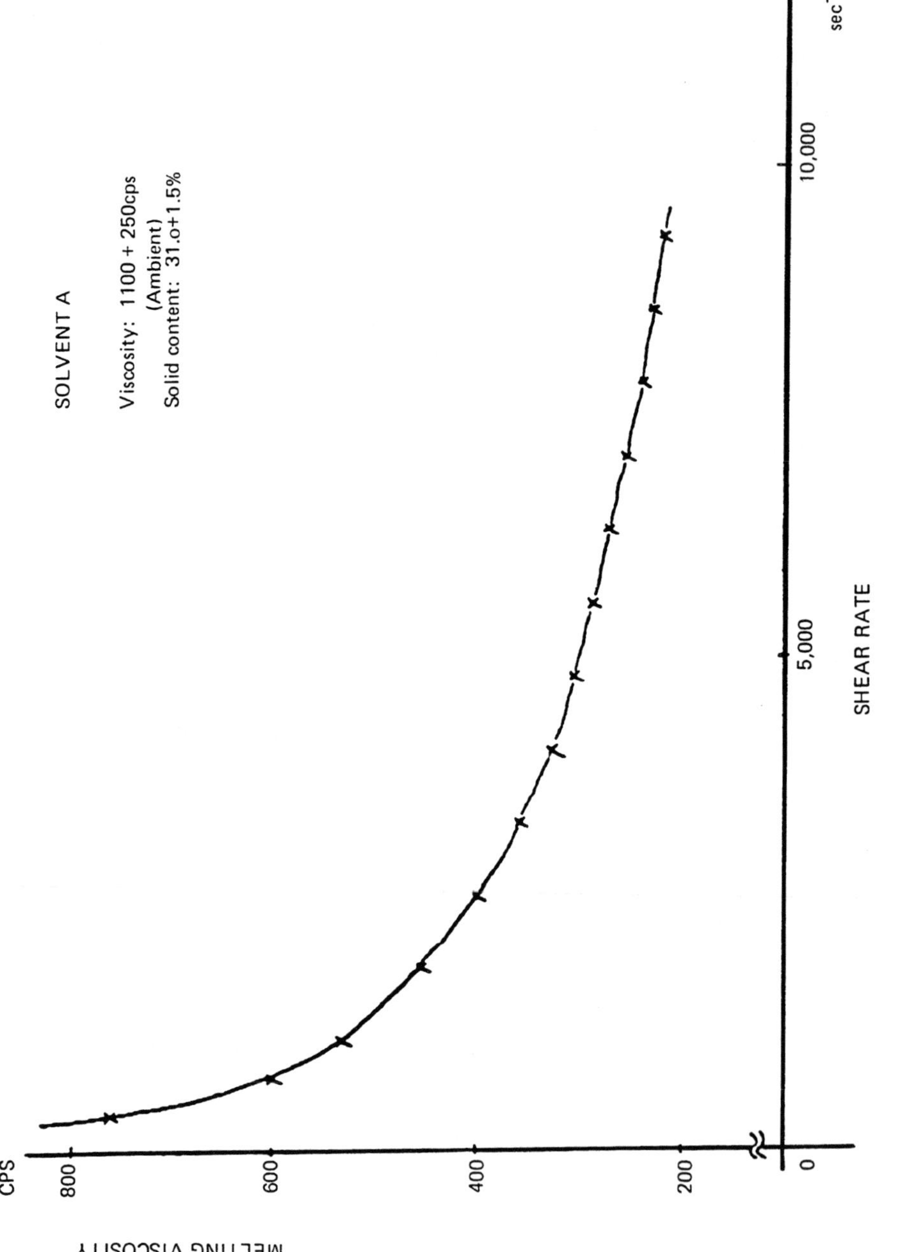

Figure 3

While a multiple-entrance design attempts to minimize the problem of flow distribution, it actually complicates it by introducing two pressure drops for every entrance installed. This can cause steps in the coating thickness, and with high viscosity fluids, parting lines, or "zippers," in the coating.

Pumping devices built into the die can improve flow distribution on simple T-slot manifold dies; however, they have some inherent disadvantages. Because of their complex design, streamlined flow and self-purging are compromised. With gear pumps built into a die, high shear causes a breakdown of some adhesives. The real question is, Why make tooling more complicated than necessary?

3.0 AIR ENTRAPMENT

Precautions to avoid air entrapment in the entire system must be considered carefully. Complicated piping runs or die manifolds will entrap air, resulting in excessive start-up times before complete purging is accomplished. It is also important to eliminate any point at which air can be induced into the system. We point out that a pan that is open to atmospheric pressure 5 placed just before the die manifold, such as on a gear pump die, can create air bubbles in the coating.

4.0 LIP DESIGN

4.1 Lip Adjustment Design

The two methods for adjusting lip opening are flexible lip (see Fig. 4), and sliding lip (see Fig. 5). Both designs have proved to be effective in wipe-on coating. Flexible lip die design is based on the hinge effect (see Fig. 4), allowing smooth, precise adjustment on relatively narrow centerlines. This single-piece concept also eliminates parting lines and flow-disrupting steps in the internal surface.

Sliding lip dies (see Fig. 5) have the lip bolted to the die body and use push–pull adjusting bolts to adjust lip position. This system distinct disadvantages. The clamping force must be great enough to seal between body and lip. During production, however, this force must be overcome when moving lip position, and the movement between lip and body will result in a

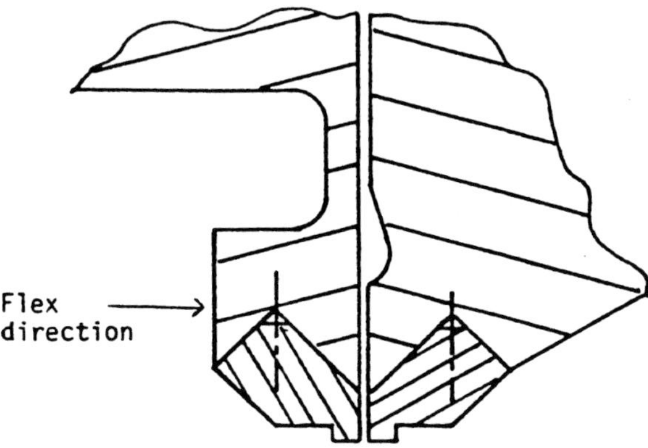

Figure 4 Flexible lip adjustment.

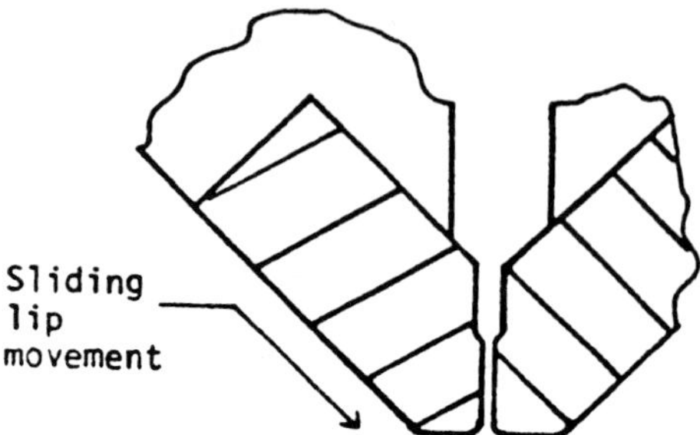

Figure 5　Sliding lip design.

series of unpredictable jumps. Furthermore, the cross-sectional size of the lip and the physical spacing of the web. The third disadvantage of the sliding lip is that lip face parallelism is distorted after adjusting.

4.2　Lip Wiping Face Design

Most low viscosity coatings will require some wiping shear to smooth the fluid onto the web. This wiping action is a balancing act of creating enough shear to force the coating to lay down without the lip face acting like a dam. This proximity coating.

Different fluids being coated on varying substrates at different coat weights and speeds require changes in wiping face design. These designs have been a takeoff from the technology used prior to slot die coating (roll smoothing bar and knife).

In general, four basic lip styles are being used today:

High shear flat wiping lip:

Medium shear flat wiping lip:

Medium shear rod wiping lip:

Sharp knife blade wiping lip:

5.0　DIE ADJUSTMENT AS IT RELATES TO MANIFOLD DESIGN

Precise coating weight control depend on the stability of the total system and on operator experience. Die performance will be at its best when a balance of pressure and flow is reached at the lip area. This balance is hard to define; changes in materials and rate will affect it. We can best describe this through the voice of an operator when he talks about the die being "jumpy or nervous" in one case and "lazy or unresponsive" in the other. In the case of the "jumpy" adjusting die, the pressure balance is too high; with the "lazy" one, pressure is too low. We will refer to this performance parameter as the operating window. When a die must be profiled excessively to improve end flow, the operating window will vary across the die; therefore, the die adjusting characteristics will differ (e.g., jumpy center, lazy ends). This effect is magnified in automatic control. When a die is set up at 0.010 in. opening in the center and 0.020 in. at ends to achieve balanced flow, 3% change in opening at the ends is 0.0006 in., and in the center it is a 0.0003 in. change. As we develop more dynamic computer programs to respond faster or to predetermine a target point, uniform die response becomes very important.

6.0   COAT WEIGHT ADJUSTMENT

Base coat weight is controlled by pump and line speed. Transverse area coat weight is a function of the lip gap adjustment. It is important to remember that the pump will always deliver a given amount of fluid to the web. When the operator adjusts the lip, material is not taken away, it is only moved from one place to another. This difference between roll or knife coating and slot die coating is often overlooked by the operator.

7.0   ADHESIVE SELECTION

Adhesive formulated for roll coating, in some cases, are not compatible with slot die heads. This is because roll coaters are high shear devices, whereas slot dies develop low shear. Any fluid that requires very high shear to create smoothing will not perform well on a low shear coating head. The shearing device or lip face can be modified to change shear level in two ways: changing the face-to-web gap (see Fig. 6 A) and changing the face length (Fig. 7). Provision should be made for ease of changing the shear level, as frequent changes of adhesive and formulation are common.

8.0   DIE STEEL AND PIPING SELECTION

Recent developments and reformulations of adhesives have led to highly corrosive or aggressive fluids. The die steel or plating must be carefully selected to ensure chemical compatibility, machinability, and reasonable cost. We find the majority of problems centering around highly acidic adhesives. A pH level of 4 will severely attack a chrome-plated die and render it unusable in a matter of 2–3 months. Several stainless steels offer reasonable prices and a highly corrosion-resistant makeup.

9.0   PROXIMITY VERSUS CONTACT COATING

Traditional die designs require being pushed into the backup roll to produce even coating distribution (contact coating). Utilizing modern manifold technology, we can reduce the roll contact pressure significantly, and therefore reduce roll damage and lip wear. Most important, when this application technique is chosen, the flex lip concept can be used.

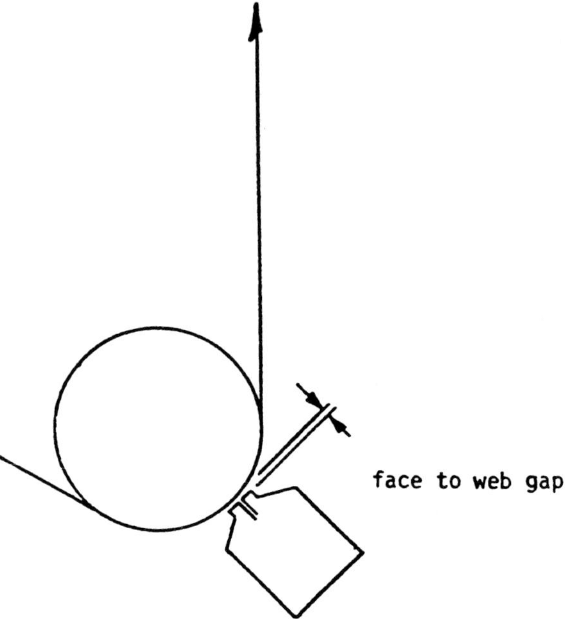

Figure 6   Changing gap between lip face and web.

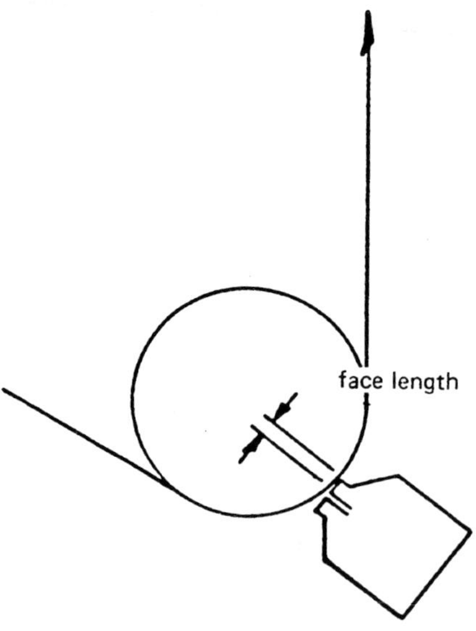

Figure 7   Changing lip face length.

In any system is of utmost importance to have inserts that are quickly replaceable, a lip step differential that can be easily changed, and the capability of increasing the shear without having the infeed lip rotate into and make contact with the web and roll (Fig. 8).

10.0 DIE POSITIONING

10.1 Slot Head-to-Roll Position and Angle of Contact

Several trends are evident within the industry.

1. In the past, a 12 o'clock (Fig. 9, B) or 6 o'clock contact position was most generally used. We find a 4 o'clock or 5 o'clock contact point to be in favor in most cases today (see Fig. 9, A). This is due to improved operator accessibility and the elimination of drool after pump shutdown.
2. The angle of contact depends on lip profile, die-to-roll gap, roll hardness, roll diameter, lip step difference, lip profile, and material spreading tendency. It is impossible to predict this angle without all these factors and in-depth experience.
3. Reduced contact pressure is necessary to reduce roll and die lip damage.
4. Since roll deflection is very hard and expensive to eliminate, a die that can be deflected or bent to conform to roll variance is required if roll deflection is a problem. It is common to use a steel roll backing up the elastomer roll to lessen the deflection problem, and with heated adhesives to help cool the elastomer roll.
5. Angle of attack of the die to roll must be pivoted around the pivot point indicated in Figure 8, for the best and most reproducible coating surface.

10.2 Lip Profiling

Lip shape and the relative position of the lead and trailing wipers to each other is of utmost importance in today's coating technology.

In some cases, using today's high technology coatings, a uniform and proper level of wiping action is required to produce satisfactory coating. As a result of the shear thinning characteristics in today's more difficult adhesives, proper profiling and angle of attack will produce a smooth and even coating, Any variance in lip profile will create differential wiping, causing an uneven appearance. Since, with modern coating heads, uniform distribution is not a function of the lip face, we can confine the lip face to a single function—namely, creating a proper environment at lay-down.

No hard data are available on lip face design as the interrelation between roll diameter, roll hardness, line speed, substrate, lip design, and adhesive viscosity characteristics come into play. There seem to be two technical camps. One group adheres to the flat, fixed wiping lip, with a differential step between the infeed lip and the wiping, or outfeed, lip. The other group tends to favor a rotating or fixed rod in the wiping area. This rod style design is as old as die coating, several patents have been issued.

10.3 Die Support Design and Operation

The interrelationship of die and mounting require that the two units act as one, both being equally important. Absolutely necessary to the successful operation of a slot coating head are the items in the series of design specifications indicated in Figure 10 and explained in Section 9.4.

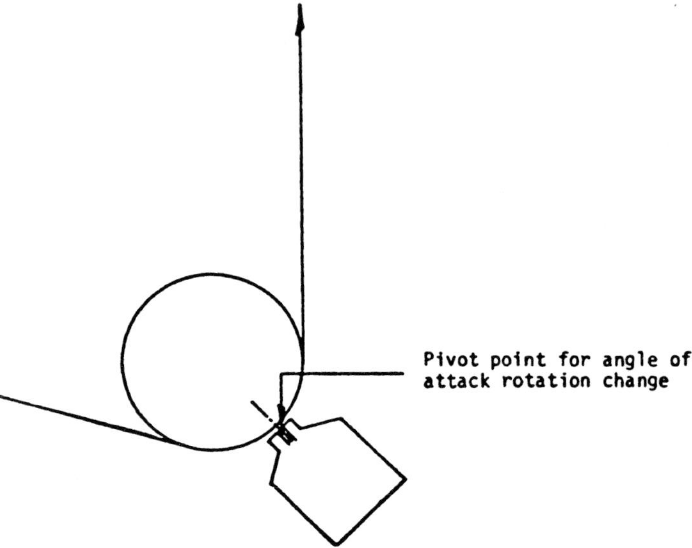

Figure 8 Proximity coating setup.

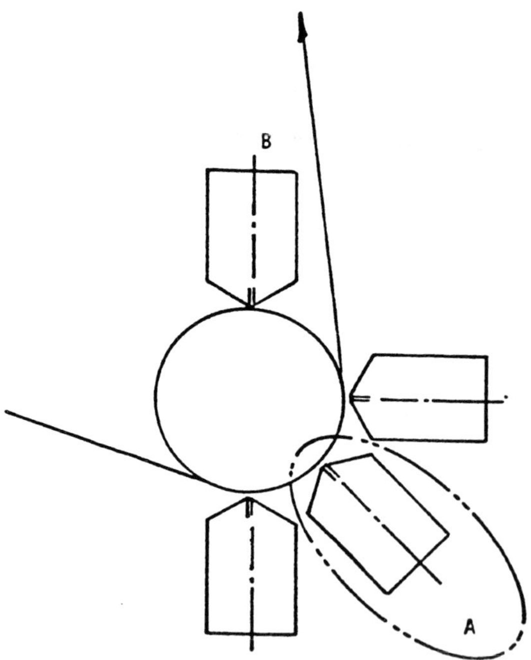

Figure 9 Die positions.

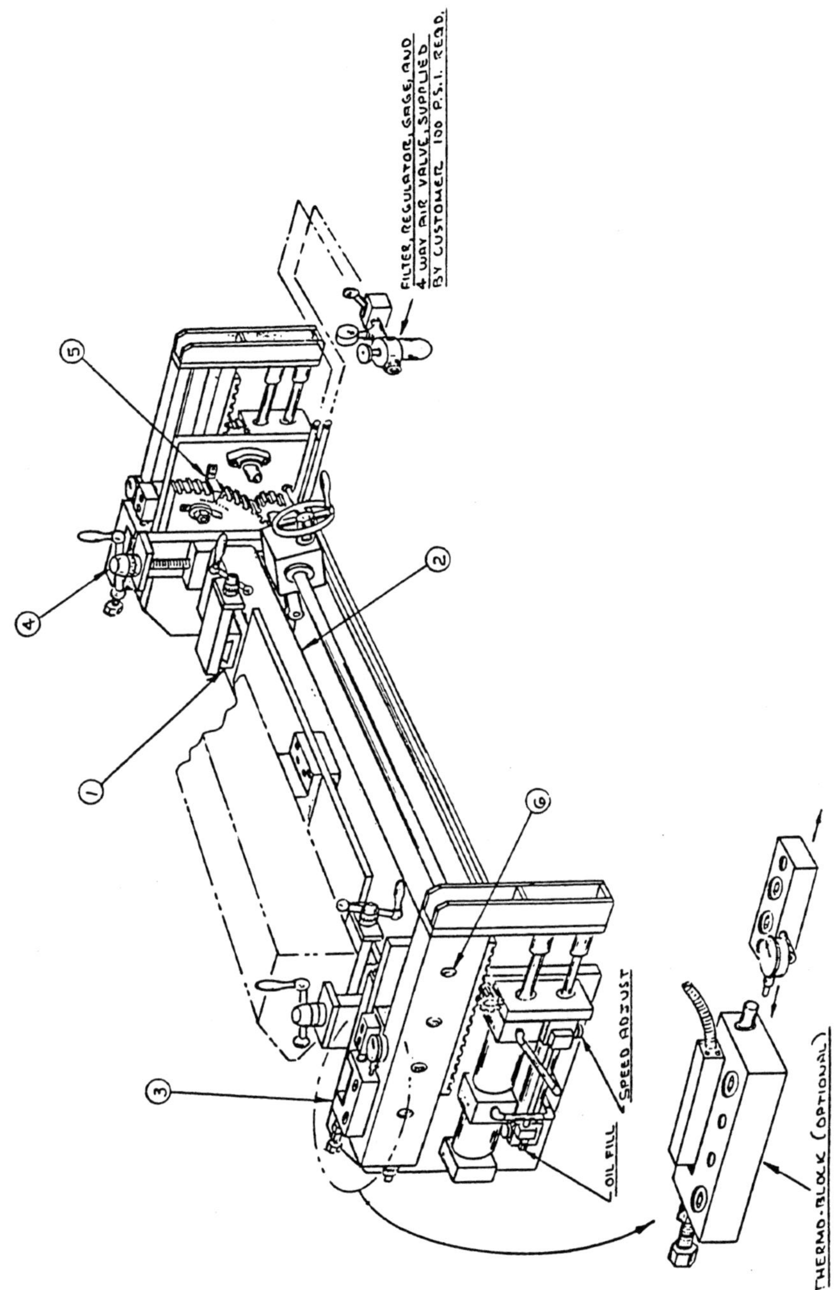

Figure 10 Coating die support; for explanation of numbered items, see Section 9.4.

Application and utilization of existing equipment designs will determine the best die-to-roll position. Operator convenience and the ability to view the point of laydown are very important. It is our opinion that mounting position *A* in Figure 9 is the most advantageous, and position *B* the least.

The support bed must be completely rigid and vibration-free. Any sag or bow in the die will create problems in the die-to-roll alignment. For this reason we do not recommend that the die be supported from its ends. On heated dies the designer must also allow for thermal growth while maintaining die straightness.

10.4 Support and Adjustment System Design Specifications

Item 1 in Figure 10: The die should be supported on precision-ground pads, with die straightness (contouring) adjustments for roll warp correction. On heated dies, insulation should be provided between die frame, and allowance made for die growth.

Item 2 in Figure 10: A heavy-duty rectangular tubing cross frame for maintaining straightness during operation and adjustment.

Item 3 in Figure 10: "X" in/out adjustment utilizing machine tool slideways with fast air actuation and hydraulic soft stop cushions. A micro stop correction adjustment with dial indicator or LVDT should be provided for each end.

Item 4 in Figure 10: "Y" adjustment provided for roll axis position alignment.

Item 5 in Figure 10: Angle of attack positioning adjusted through a rack-and-pinion gear arrangement, with indicator markings in degrees. Note that the angle of attack movement is centered around die lip (coating contact) point.

10.5 Die-to-Roll Positioning

Flexibility and repeatability are primary requirements. Difficult adhesives, speeds, and substrates will require different setup positions, and the ability to vary the die-to-web position easily with exact repeatability is of prime importance. The support frame must allow in/out movement and angle of attack adjustment (see Fig. 10)

In/out adjustment will have two functions: (a). Fast movement with 5–8 in. (130–200 mm) of travel, allowing lip cleaning, and (b). micro in/out to fine-tune roll-to-slot gap distance. This adjustment should allow differential end-to-end gapping.

The in/out adjustment must be on a straight line, always moving directly at the roll centerline. (Pivot-style mountings are not recommended.)

10.6 Angle of Attack Position Adjustment

To create different shear levels for proper film forming at varying speeds, the angle of attack must be adjustable. This movement should be accurate and smooth, and movement should cover about ±5°.

It is important that when one positioning point is adjusted, the other roll-to-die relationships remain unaltered. Angle of attack adjustment must pivot about the point indicated in Figure 8.

10.7 Lip Opening Setup

Depending on materials and laydown, a .008 to 0.012 in. even lip opening gap (Fig. 11) should be set before start-up, and adjusted for proper flow and laydown after start-up.

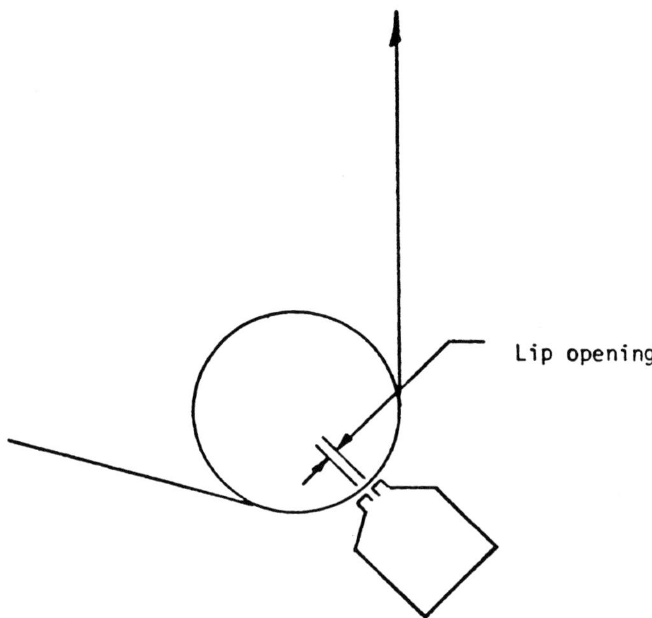

Figure 11 Lip opening setup.

The lip opening setting adjusts coat weight thickness, not die-to-roll gap. Roll coaters will have difficulty with this, as they have traditionally used roll-to-roll gap as the coat weight adjustment.

10.8 Die-to-Roll Gap Setup

The distance from die to roll or substrate is, in general, determined by web thickness and the viscosity of the fluid to be applied. The more clearance that can be maintained, the less damage there may be due to positioning or start-up mistakes.

We suggest, as a rule of thumb, that a clearance equal to the substrate thickness be set between substrate and infeed lip face. This distance may be less for materials of very low viscosity or for hard to smooth materials. Make sure the gap is from web to lip face, as shown earlier (Fig. 6), not from roll to lip face.

11.0 BACKUP ROLL DESIGN

Processors and equipment manufacturers alike would like to utilize steel backup rolls to improve runout (T.I.R) and to mitigate heat transfer problems. In some cases on lab or narrow production systems, steel rolls have been successful. In most cases, however, steel rolls are not as forgiving as an elastomer roll and therefore have not yet been accepted for production. We expect this to change as the technology matures.

Elastomer rolls have improved over the past several years to allow the precise roll-to-lip conformation absolutely required for proximity coating. When specifying an elastomer roll, the following items must be carefully considered:

runout tolerance
release characteristics
heat transfer
hardness
hardness uniformity
roll deformation (bow)
resistance to attack by the coating or cleaning agent

We see the use of rolls covered with urethane or Viton being most evident with diameters of around 300 mm and hardness of up to 90 durometer, Shore A. Runout tolerances of 0.0005 in. and better are being achieved.

The elastomer roll will be deformed by a given width web, and rolls must be provided to match web width changes. Spare rolls are also necessary, as damage often occurs.

12.0 AUTOMATIC CONTROL

12.1 Die Control

All commercially acceptable automatic die adjustment systems available today use flexible lip and heated bolt arrangements (see Fig. 12). Minor disagreements exist on details, but total overall performance does not seem to be greatly affected. The greatest confusion surrounds the effect of pure mechanical response time and how it relates to process analysis and its relation to response. All systems to date simply read the variation from target and

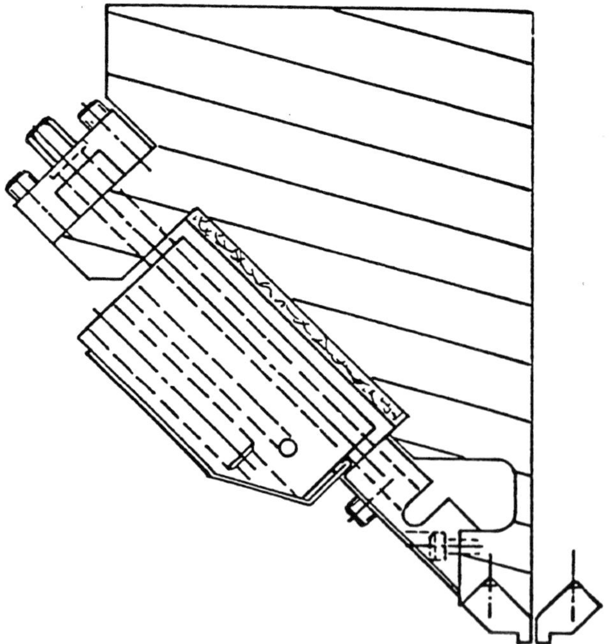

Figure 12 Automatic transverse coat weight control.

make a corrective response. Polymer lot-to-lot differences, temperature changes, in-plant drafts, and many other factors that affect gage have made it impossible to anticipate flow changes. Anticipation-based programs can be used in product changes, however, if known effects will happen over a relatively fixed period or on start-up.

During a production run, if a variance is seen, make sure that it is not a short-term effect, gone in the time it would take to make a change. We also may want to determine the variance trend. Only after careful analysis of the problem can we make a die adjustment. The time from discovery until an adjustment takes effect varies from line to line, however start-up to ±5% control, assuming the total system has reached some stability, will be 10–20 minutes, and ±3% control in 15–30 minutes.

Total and substrate weight separation, the first steps in the process control cycle, are achieved by a layout similar to that shown in Figure 13. Existing technology is available and has been proved in production for nearly a decade.

The Autoflex die features thermal expansion and contraction of the lip adjusting bolts to make finer adjustments in lip opening than are possible with strictly mechanical means.

Each lip adjusting bolt is fitted with a block containing a heater and air cooling path. Approximate gage uniformity is established in the conventional manner before the Autoflex system is engaged. Thickness variations are converted to lip opening correction by increasing or decreasing power to the individual lip bolt control blocks to trim variations to a minimum.

The key to the transverse thickness control is a microprocessor-based controller that is tied into the conventional computer control system.

On ambient operation dies, care must be taken to isolate the heat from the Autoflex bolts from the die body.

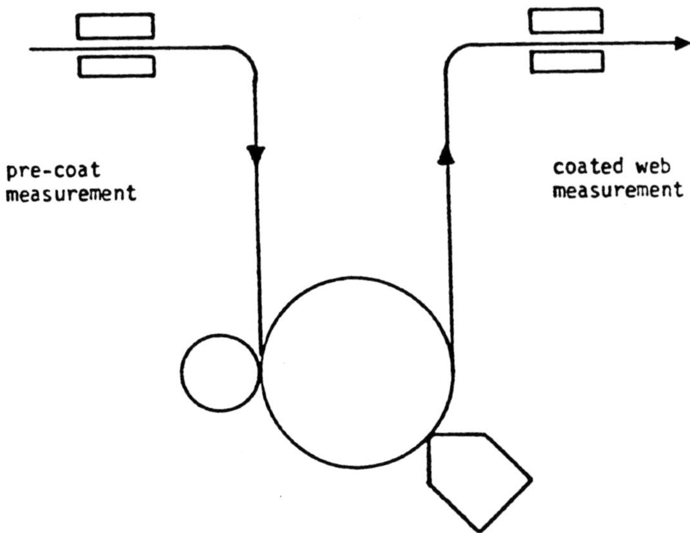

Figure 13 Schematic drawing of total and substrate weight separation in production control cycle.

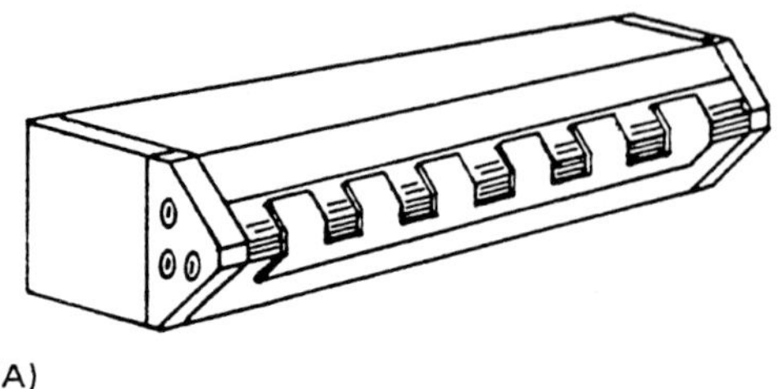

(A)

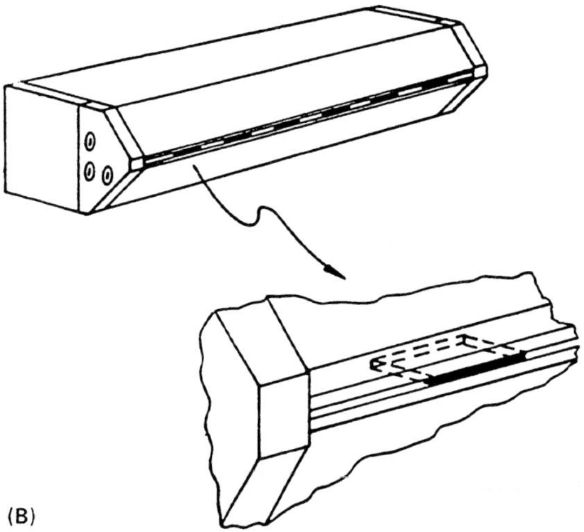

(B)

Figure 14 Deckling: (A) rake style and (B) shim style.

12.2 Die-To-Roll Position Adjustment System

The ability to repeat the original roll-to-die setup position is critical during start-ups and normal web splice coating interruption.

During operation, minor changes in die position may have to be made to accommodate roll expansion, changes in adhesive viscosity and smoothability, and substrate thickness variance. An automatic positioning device is available that will allow continuous adjustment, if necessary, of the web to the lip face. This is accomplished through a device similar to the Autoflex die bolt adjustment system. A heating and cooling device is installed in the manual adjustment system (U.S. Patent 3,940,221) for die-to-roll gap setup. Heating or cooling of this device will expand or contract the steel block and increase or decrease the die-to-roll gap. A usable method for monitoring smoothing must be employed and interfaced with the Autoflex computer.

13.0 DECKLING

Deckling may be necessary to reduce die width or to make stripe coatings. Be careful when attempting this, to make sure that shim materials are soft. We recommend Teflon/filled, Teflon/aluminum, foil/aluminum shim stock, or soft brass. For room temperature applications, an adhesive/foam/adhesive mounting tape works very well.

In many applications, a rake-type device may be used for stripe coating (Fig. 14A). This device does not clamp into lip and can be changed very quickly.

If excessive force is applied when the deckle is clamped into the lip, the lip will distort, causing lip wear or uneven coating at this point (Fig. 14B).

Deckling or stripe coating cannot be used in a system featuring the rotating rod lip design.

13.1 Air Entrapment Behind Deckling

When the die ends are deckled in, it is common for air to become trapped in this area. This air is slowly released, causing voids in the coating at the edges. A purge port with shutoff valve should be installed at the die ends to eliminate this problem.

14.0 DIE CLEANUP

In most cases, it is important to change coatings or coating formulation frequently. Therefore it is necessary to be able either to purge out the system or to clean the die completely. This holds true for the complete system, including pump filters and piping.

Purging is the easiest and most common method; but extreme care is necessary to streamline all flow areas to eliminate dead areas.

The opening and cleaning of a die can be an easy 30 minute experience or a 3–4 day nightmare. Slot die designs differ greatly; some are simple two-piece designs, while others have complex assemblies.

It is common to have a dual pumping and piping system, allowing quick changeover from one coating to the next, cleanup then taking place after the line is up and running.

12
Extrusion Coating with Acid Copolymers and Ionomers

Donald L. Brebner

E. I. du Pont de Nemours & Company, Wilmington, Delaware

1.0 PRODUCT CONSIDERATIONS

Acid copolymers and ionomers are high performance resins that offer adhesion, heat seal, and barrier properties markedly superior to those of conventional polyethylenes. Figure 1 shows the structure of acrylic and methacrylic acid copolymers, the two types of acid copolymer commercially available in the United States at this time. The presence of the methyl side group in the methacrylic acid copolymers results in some subtle differences between the two resin types, but they can be regarded as equivalent after allowing for the difference in molecular weight of the comonomers. For example, a 10 wt. % acrylic acid copolymer is equivalent to a 12 wt. % methacrylic acid copolymer in carboxyl group content.

Ionomers (Fig. 2)are derived from acid copolymers by partial neutralization of the carboxyl group with either sodium or zinc ions. Since the neutralization reaction results in a substantial increase in melt viscosity, it is commonly referred to as ionic cross-linking.

In both acid copolymers and ionomers, the melt and solid state properties ares strongly influenced by intramolecular hydrogen bonding, as illustrated in Figure 3. The forces involved in hydrogen bonding are almost 10 times stronger than the intramolecular forces in the nonpolar polyethylenes.

The ionomers are distinguished by the combination of hydrogen bonding and interchain ionic forces perhaps an order of magnitude stronger than the hydrogen bonds. As a consequence, the ionomers have a wide spectrum of melt and solid state properties, including better hot tack and grease resistance than acid copolymers of equivalent acid content.

With the acid copolymers, melt index and acid content are the major variables available to the resin producer. Since melt index is a measure of melt viscosity, it is related primarily to the processing characteristics of the resin. The resins now used for extrusion coating applications fall in the 5–15 melt index range to accommodate a broad field of processing needs.

Resins with acid contents of 3–15 % are currently on the market. The effects of increasing acid content are :

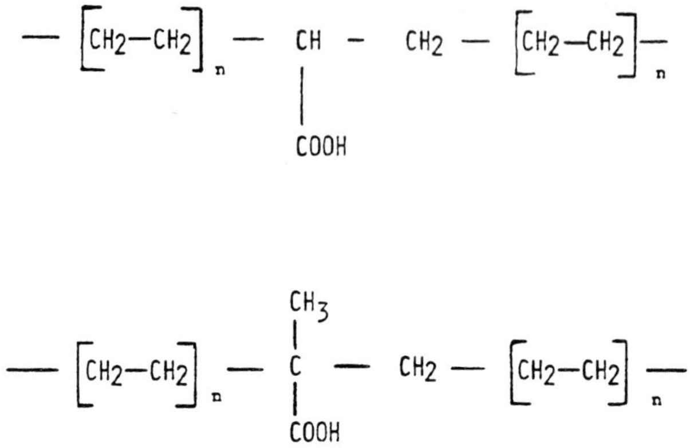

$$- \left[CH_2-CH_2 \right]_n - \; CH \; - \; CH_2 \; - \left[CH_2-CH_2 \right]_n -$$
$$\hspace{5.5cm} | \hspace{5cm}$$
$$\hspace{5.5cm} COOH$$

$$\hspace{5.5cm} CH_3$$
$$\hspace{5.5cm} |$$
$$- \left[CH_2-CH_2 \right]_n - \; C \; - \; CH_2 \; - \left[CH_2-CH_2 \right]_n -$$
$$\hspace{5.5cm} |$$
$$\hspace{5.5cm} COOH$$

Figure 1   Structure of ethylene acrylic (top) and methacrylic acid (bottom) copolymers.

better foil adhesion
better hot tack
lower seal initiation temperature
better oil and grease resistance
better abrasion resistance
higher gas and moisture permeability

The most important effect is the improved adhesion to unprimed aluminum foil, since most commercial applications of acid copolymers involved foil adhesion.

   The ionomers are characterized by additional variables, which can be manipulated to obtain a broader spectrum of properties in the final product. These variables include the following:

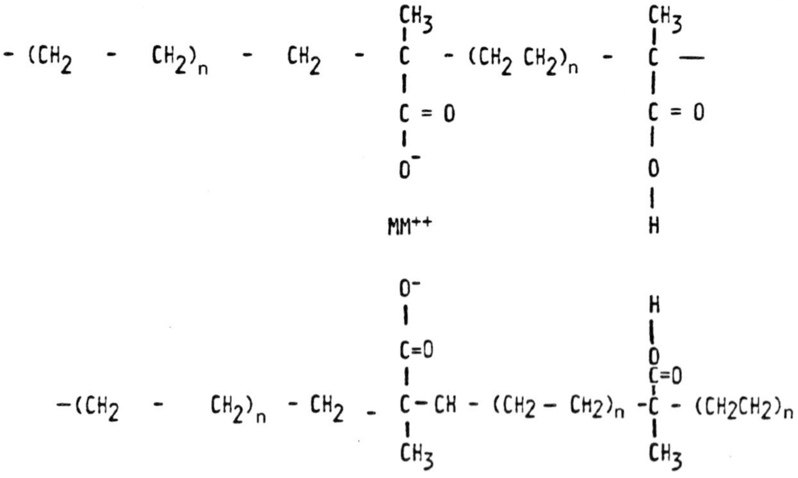

Figure 2   Ionomer structure.

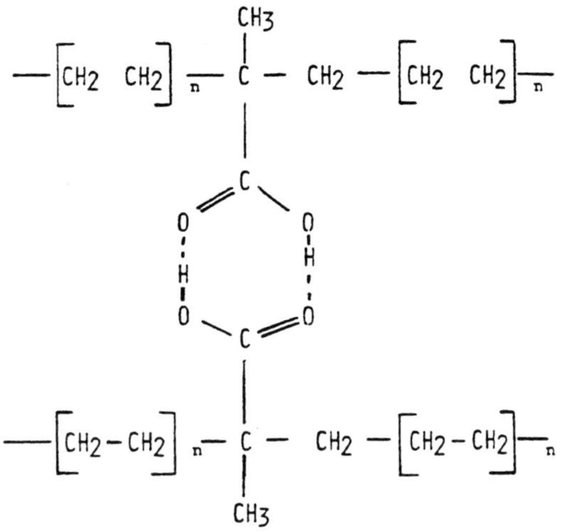

$$-\left[CH_2 \;\; CH_2\right]_n - \overset{\overset{\textstyle CH_3}{|}}{\underset{|}{C}} - CH_2 - \left[CH_2 \;\; CH_2\right]_n -$$

(chemical structure showing intramolecular hydrogen bonding between two carboxyl groups)

$$-\left[CH_2-CH_2\right]_n - \underset{\underset{\textstyle CH_3}{|}}{\overset{|}{C}} - CH_2 - \left[CH_2-CH_2\right]_n -$$

Figure 3 Intramolecular hydrogen bonding.

melt index (molecular weight of base resin
percentage of acid in base resin
degree of ionic cross-linking (percent acid neutralized)
type of ion used for neutralizatoion (Na or Zn)

In particular, the use of the sodium and zinc ions produces families of resins with distinctly different properties. The zinc ionomers are used for the majority of extrusion coating applications because of their excellent adhesion to unprimed foil.

The effects of composition on ionomer properties are shown in Table 1. In general, the ionomers based on high acid copolymers and with a high degree of neutralization have the best heat seal and product resistance characteristics. However, these advantages are balanced by a need for relatively high free carboxyl content for optimum foil adhesion. In practice, most extrusion coating resins used for coextrusion with nylon have a relatively high degree of neutralization. There are limited uses of sodium ionomers in extrusion coating applications that call for maximum grease resistance, but these applications are on either paper or primed foil.

2.0 END-USE CONSIDERATIONS

2.1 Foil Adhesion

One of the primary reasons for the selection of an acid copolymer or ionomer for use in extrusion coating is adhesion to unprimed aluminum foil. Extensive studies in Du Pont laboratories have found that acrylic and methacrylic acid copolymers of equivalent acid content have equal adhesion to foil. At a particular acid content, a zinc ionomer will have 15–20% lower foil adhesion. Resins with high acid content (10–13%) have 20–25% better adhesion than resins with low acid content (3–5%).

BREBNER

Table 1 Ionomer properties versus composition.

Higher Percentage of Acid	Higher Ion Linking	Higher free carboxyl: greater adhesion to polar substrates (foil, nylon)	
		Na	Zn
Greater stiffness	Greater stiffness	Better oil resistance	Less water sensitive
Better oil resistance	Better oil resistance	Higher gloss	Better adhesion (foil, nylon)
Higher gloss	Higher gloss	Lower haze	Less "neck-in"
Higher tensile and impact strength	Higher tensile and impact strength	Higher draw	
Lower haze	Higher toughness	Higher toughness	
Lower melt and softening points	Higher hot tack		

The acid copolymers and ionomers have additional utility because of the enduring nature of their adhesion to foil. High levels of adhesion of polyethylene to foil can be achieved with primers or high melt temperatures in extrusion, but the bonding is often deteriorated by extended exposure to aggressive environments. With acid copolymers and ionomers, most hard-to-hold products can be packaged without sacrificing long-term durability of the coating adhesion. Acid copolymers and ionomers of equal acid content are essentially equivalent in product resistance.

2.2 Heat Seal Characteristics

The heat seal properties of acid copolymers and ionomers constitute the second major area of superiority over low density polyethylene (LDPE). Seal initiation temperatures are lower and strengths higher than low density polyethylene. Figure 4 is a generalized plot of seal strength versus interface temperature for acid copolymers, ionomers, and LDPE. The acid copolymers and ionomers have similar seal initiation characteristics but the acid copolymers have a higher ultimate seal strength. Both have much higher seal strengths than LDPE.

Hot tack strength is the ability of a heat seal to remain together when a force is applied while it is still in the molten state. This is a critical property in vertical form-fill-seal applications, in which the product is loaded immediately after the seal is made. It is also critical in any high speed packaging operation in which the package is exposed to some form of abuse before the seal has cooled.

One method of measuring hot tack strength is the Du Pont spring test. A series of springs of different thicknesses and widths at the narrowest point provides varying levels of spring

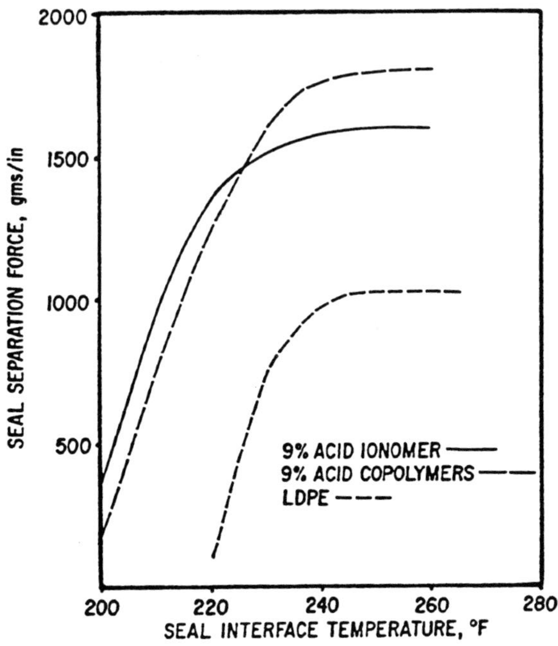

Figure 4 Heat seal strength versus bar temperature (1 mil coatings on 30 lb. kraft paper.

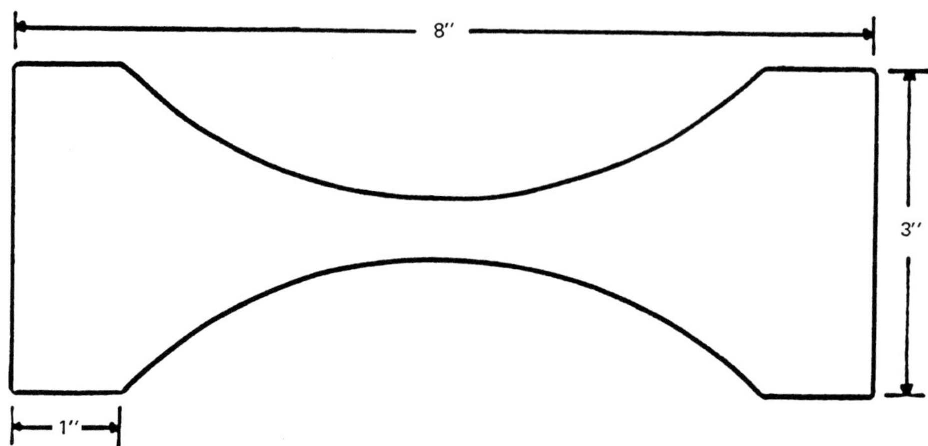

Figure 5   Spring for Du Pont hot tack tester.

tension (Fig. 5). Figure 6 shows how the spring is located inside the sample. The sample is heat sealed for 3 sec. at a pressure of 40 psi. The three-second dwell time allows an accurate measurement of the interfacial seal temperature by using a very fine thermocouple hooked up to a rapid response recorder.

When the heat seal bars are released, the springs apply an instantaneous force to the seal. The separation of the 1 in. wide seal measured to the nearest tenth of an inch. The results are plotted as the force required to obtain 20% seal separation as a function of temperature.

Figure 7 shows the hot tack characteristics of an ionomer and acid copolymers of equivalent acid content in comparison with low density polyethylene. LDPE, a completely nonpolar material, has very low hot tack strength. The acid functionality of the copolymers results in much higher hot tack strength because of interchain hydrogen bonding. The combination of hydrogen bonding and ionic bonding in the ionomers produces an even greater

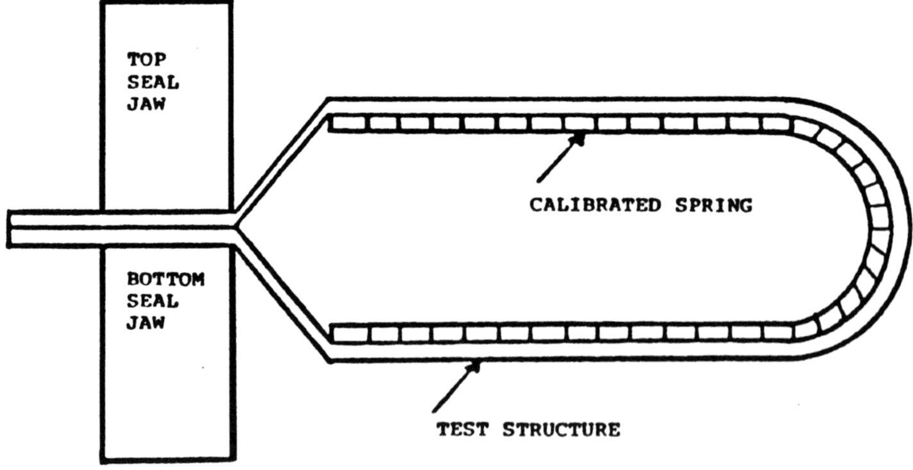

Figure 6   Hot tack spring test.

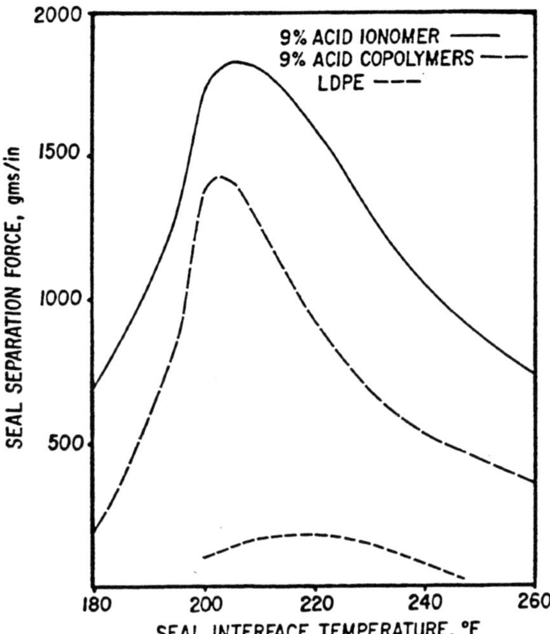

Figure 7   Hot strength versus interface temperature (1 mil coatings on 30 lb. kraft paper).

maximum hot tack strength and a broader hot tack range. In practice, this means that the ionomers will tolerate the broadest range of sealing conditions and offer the highest speeds on packaging lines.

3.0   PROCESSING CONDITIONS

3.1   Melt Temperatures

Low density polyethylene is typically processed at melt temperatures in the range of 600–630 °F. These high temperatures are necessary to promote surface oxidation of the resin and to ensure adequate adhesion to the substrate. However, it is not necessary to oxidize the acid copolymers or ionomers, since the acid functionality is responsible for adhesion to either foil or paper. These resins should be processed at the lowest melt temperature consistent with adequate product performance. One obvious reason to prefer a lower melt temperature is to reduce the chance of odor or taste problems in packaging sensitive products. Another reason is shown in Figure 8. Both the acid copolymers and ionomers can undergo a process called anhydride cross-linking, in which carboxyl groups in adjoining molecules react to form an anhydride cross-link and produce water as by-product. This reaction occurs at a significant rate in the temperature range at which the copolymers and ionomers are normally processed. The cross-linking reaction causes gel formation, while the water causes die buildup problems. As approximate guidelines, the melt temperatures shown in Table 2 are recommended for acid copolymers and ionomers, assuming that desired product properties can be achieved under these conditions.

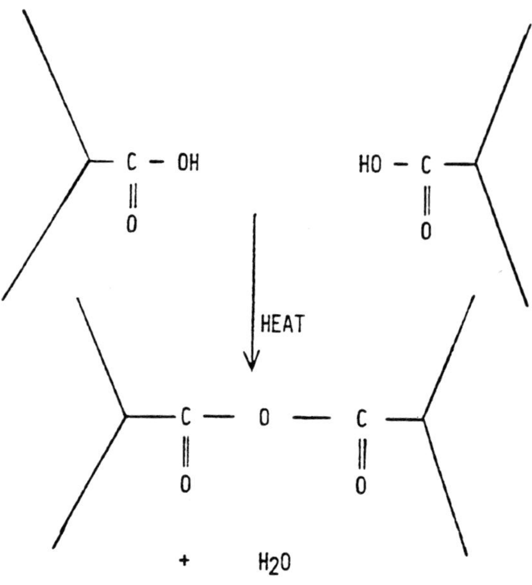

Figure 8   Anhydride cross-linking.

Table 2

Product	Melt temperature (F°)
High acid, high melt index	450–500
Intermediate acid and melt index	500–550
Low acid and melt index	550–590

3.2   Other Considerations

Both the acid copolymers and ionomers are slightly corrosive, so that metal parts exposed to these products at high temperature should be plated with nickel or chrome. Another factor to be considered is the moisture sensitivity of the ionomers. The ionomers are hygroscopic and as a consequence are packaged in foil-lined bags and boxes. When a box is opened, only a corner of the liner should be cut off to permit insertion of the transfer pipe. If this is done properly, it is easy to reseal the liner of a partially used box by folding it over and taping securely. With normal precautions, the moisture sensitivity of ionomers does not present an operational problem.

13

Porous Roll Coater

Frederic S. McIntyre

Acumeter Laboratories, Inc., Marlborough, Massachusetts

1.0 INTRODUCTION

Recent progress had been made in silicone-coated products that are curable by electron beam (EB) and ultra violet (UV). Advantages of UV and EB converting processes are shown in Figure 1. Of these new products, in which the release levels range between 25 and 50 g per 25 mm wide strip, typical viscosities of the materials tested are in the range of 500–1000 cp at room temperature. These silicone products are manufactured by Th. Goldschmidt, in West Germany, and Lord Corporation, in Pennsylvania, and other companies. The UV products are either one-part premixed/ready-to-use materials, or two-component products that require nitrogen inerting to overcome surface smear and to achieve a complete cure at web speeds up to 30 m/min per each UV lamp. A schematic diagram of nitrogen inerting process is shown in Figure 2. The UV lamps used for curing are rated at 300 W/in. (120 W/cm per lamp). The EB products are also premixed/ready-to-use, and as with the UV chemistries also require nitrogen inerting to obtain a full cure at typical web speeds of 200 m/min. The energy dosage is approximately 2 megarads (Mrad). A schematic diagram of an EB processor is shown in Figure 3.

2.0 EXTRUSION POROUS ROLL SYSTEM

2.1 Development

One recently developed extrusion roll coating system incorporates a slot nozzle coating head, located adjacent to a slow speed "nozzle roll" (see Fig. 4). The fluid is coated onto the "nozzle roll" and transferred to an adjacent "applicating roll," which in turn contacts the coating web for fluid transfer. The relative speed ratio between the "nozzle roll" and the "applicating roll" is approximately 1:30.

Reports on trials have noted that the curable coatings possess poor shear properties and that the speed ratio between the rolls is somewhat dependent on this limitation. The shear properties can be improved by heating the rolls, but this method is not always productive,

- ● NONPOLLUTING

- ● MORE EFFICIENT (LESS ENERGY REQUIRED)

- ● CAN BE USED ON BOTH FILMS AND PAPERS

- ● NON THERMAL CURE, ALLOWS MACHINE STOP & START
  WITH MINIMAL PRODUCT LOSS

- ● MULTIPLE PROCESS PERMITS SILICONE IN-LINE WITH
  HOT MELT PRODUCTS

- ● LOW HEAT ENERGY ALLOWS COATING OF THERMAL
  SENSITIVE WEB MATERIALS

- ● FULL CURE PROCESS MINIMIZES SILICONE MIGRATION
  AND CONTAMINATION

Figure 1 Advantages of radiation curing.

with the result that the coated web has the appearance of small blotches, 1–2 mm or larger, or lateral bands (chatter) of coating, rather than a smooth, uniform coating. Likewise, conventional roll coaters, which contain multiple rolls, experience a similar effect (see Fig. 5). This phenomenon is associated with fluids that have poor flow properties and dilatant characteristics. Viscosity of dilatant fluids increases with increasing shear rate as shown in Figure 6. The actual coat weight applied in either case is influenced by the web substrate material, surface conditions, silicone product properties, and the method of applications.

Most coating processes are able to apply these coatings between $1.3 - 1.6$ g/m^2 within industry standards; however, the type of web material greatly influences the final coating weight. Our research to reexamine the current methods for applying curable silicones enabled us to develop an alternative method for overcoming the shear problem. This design became Acumeter's Extrusion Porous Roll Applicating System. The system consists of a porous, stainless steel metal roll, an adjacent applicating rubber-coated roll, and a laminating roll. A schematic diagram of such a coater is shown in Figure 7. The porous roll receives a fluid supply from a positive displacement metering system, which is synchronous yet proportional to machine speed. A detailed diagram of a porous roll is shown in Figure 8.

2.2 Details and Disadvantages

The relative differential of the porous roll to the applicating roll can be either equal or slightly different. The applicating roll can operate at the same surface speed as the web, or slightly less, to create an additional smear of the coating on the web. The unique feature of the porous roll system is that shear is eliminated throughout the fluid transfer process. This means that the fluid material can be reasonably shear-sensitive and dilatant. The porous roll coating system simulates a printing process, similar to rotary screen ink printing (see Fig. 9). The major difference is that the porous roll system receives a fluid supply synchronous to machine speed, whereas the rotary screen printing process utilizes a doctor blade mechanism, located on the inside of the screen cylinder. The rotary screen printing process speed is influenced by the viscosity of the ink. This means that inks of high viscosity will limit machine speed.

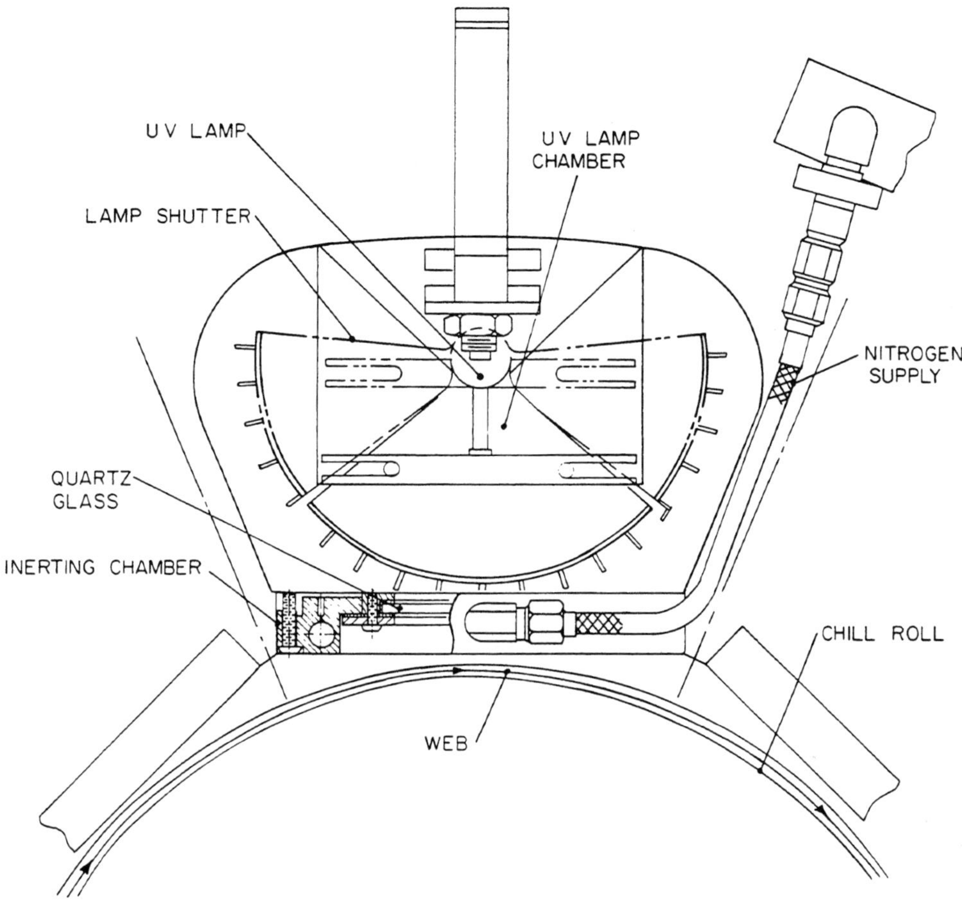

Figure 2 UV irradiation with nitrogen inerting.

In contrast, the porous roll system with synchronous positive displacement fluid supply is insensitive to viscosity change. The important element is synchronous fluid metering to obtain consistent and controlled coating wights applied at wide ranges of machine speeds.

Manufacturers of curable silicones have indicated that higher viscosity materials are desirable for obtaining special release levels. As mentioned earlier, conventional multiple-roll coating systems perform best when handling fluids less than 1000 cp. The porous roll system has been tested with coatings having viscosities to 10,000 cp.

Changing the micrometer openings in the porous roll applicator will permit handling of higher viscosity materials (see Fig. 10). For example, coatings of polyvinyl resin emulsion adhesives, having a viscosity of 5000 cp at a coating thickness of approximately 30 μm have been successfully applied without difficulty. Silicone products or other coating materials that have higher viscosities such as 10,000 cp may require a porous roll having larger openings, depending on flow properties. The larger porous openings will minimize the fluid back pressure and allow for consistent coat weights at various machine speeds.

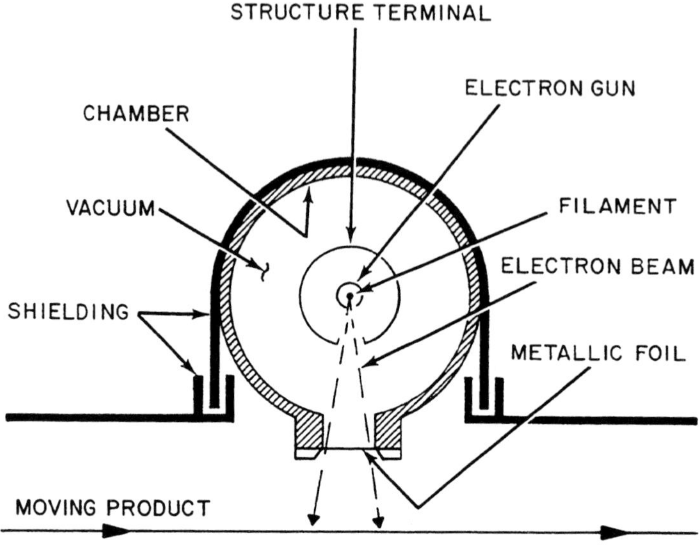

Figure 3 EB process.

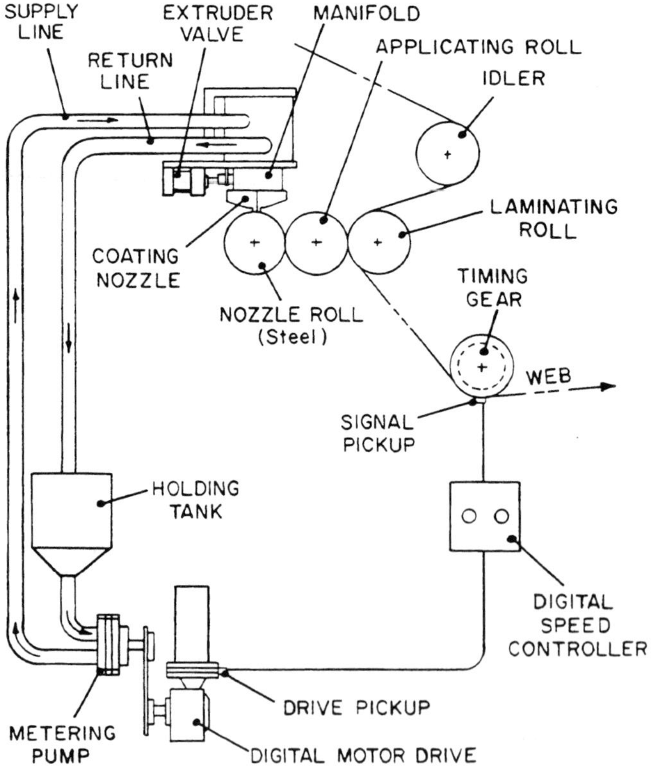

Figure 4 Roll coater with extrusion feed.

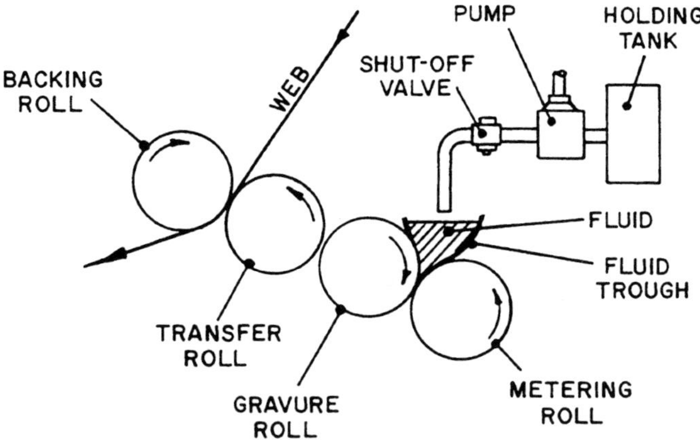

Figure 5 Transfer roll coater.

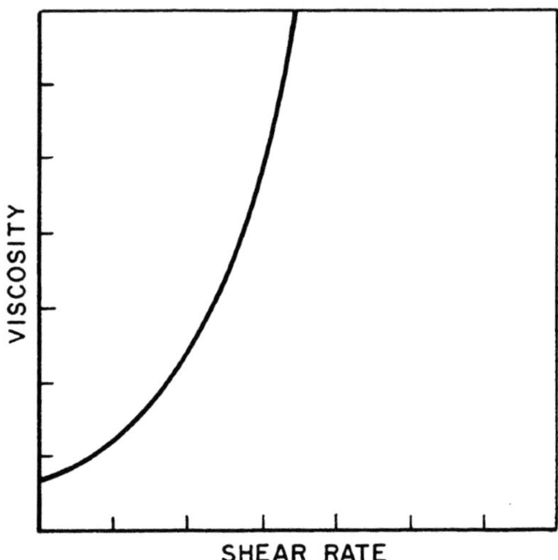

Figure 6 Viscosity dependence on shear rate of a dilatant fluid.

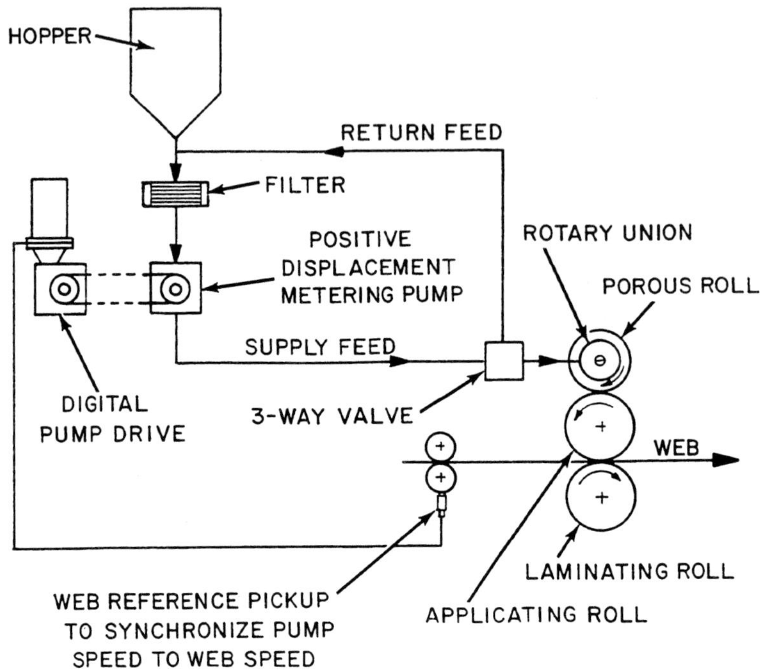

Figure 7   Porous roll coater.

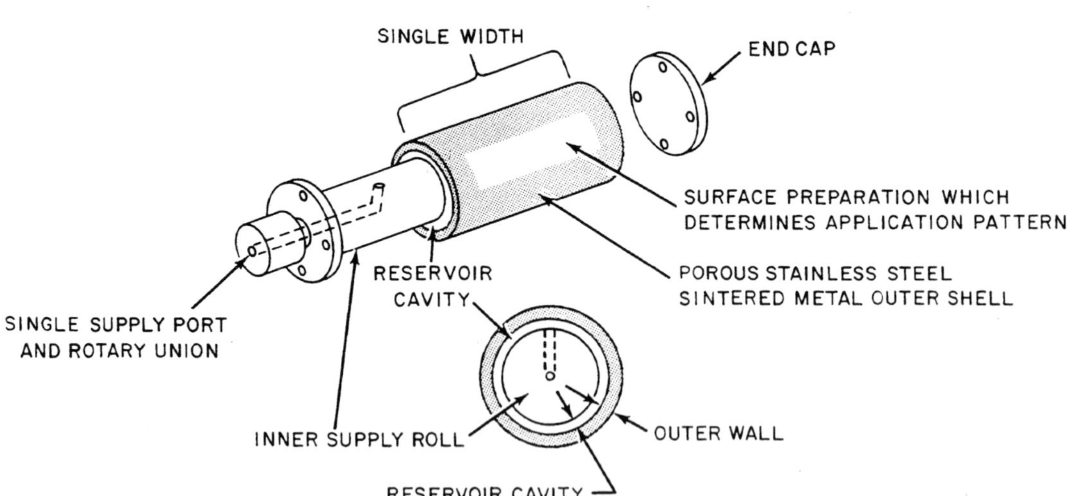

Figure 8   Diagram of a porous roll.

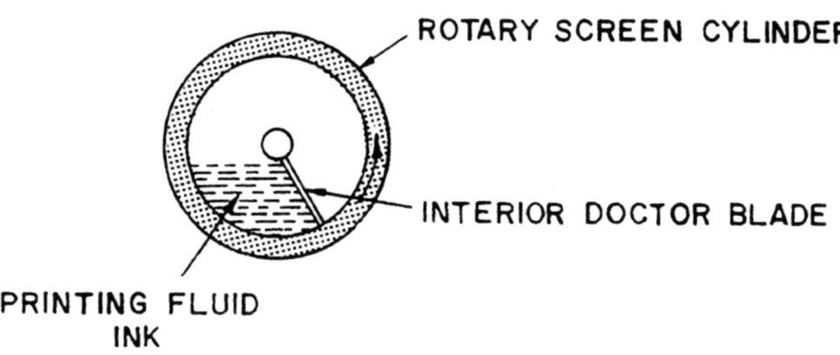

Figure 9 Rotary screen printer.

Figure 10 Porous roll applicator porosity.

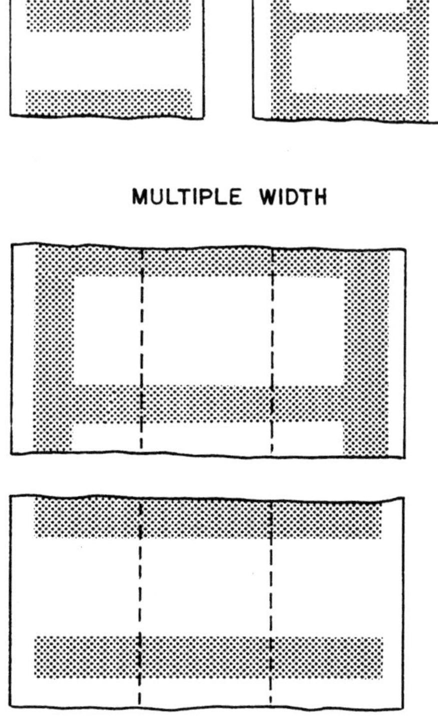

Figure 11 Pattern printing with porous roll.

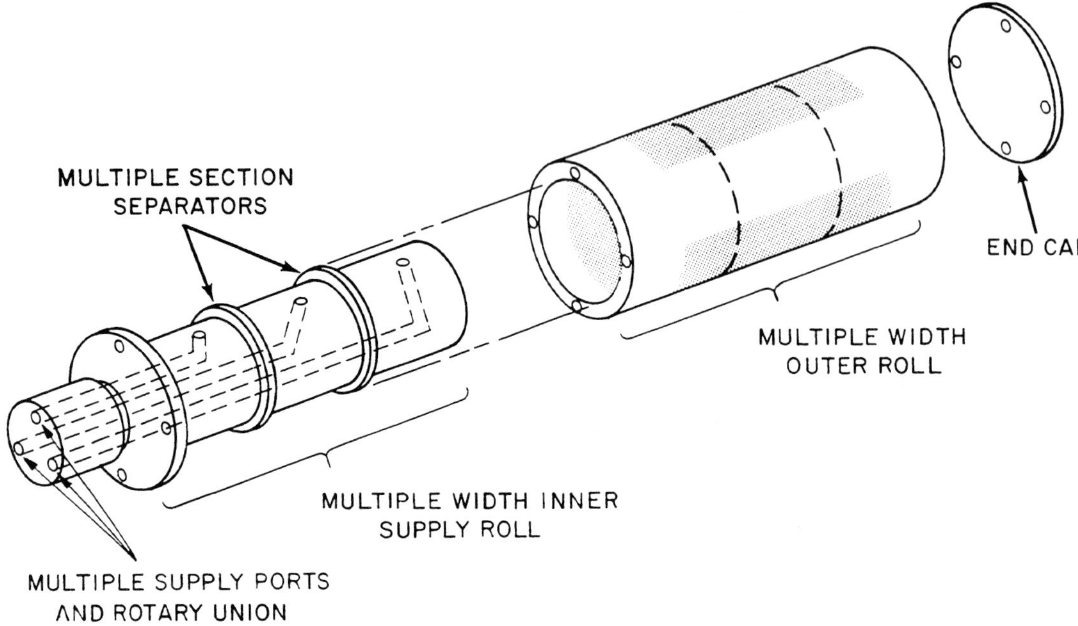

Figure 12 Cross section of a porous roll applicator.

The porous roll surface can be sealed for printing of fluids in patterns, as shown in Figure 11. This means that special coating patterns required in flexible packaging products, business forms, envelopes, tapes, and labels can utilize curable silicone coatings in the final converting process, rather than having pre-silicone-printed web materials.

The porous roll system also can be configured to handle multiple coating materials within the same applicator. A cross section of the applicator is illustrated (Fig. 12) to show the different chambers within which the multiple fluids are supplied. This feature permits the converter to coat different materials simultaneously, yet from the same applicator. Special tape products that require different release levels on the same web material can utilize this concept. For example, a pattern of 25 g/25 mm release can be applied on the left-hand side of the sheet, whereas a 100 g release coating can be applied in the center or adjacent location on the same sheet. Products such as double-sided, release-coated webs used in transfer tapes can be coated easily by utilizing two separate porous roll coating systems, located on either side of the web, as shown in Figure 13. UV or EB curing equipment is incorporated for cross-linking the coatings as dictated for completing the final products.

The porous roll system also allows for obtaining high release levels, such as 100, 200, and 400 g/25 mm, by using a two-component silicone system. This concept utilizes two porous rolls that are in direct contact with a single rubber coated applicating roll (Fig. 14). Component A is supplied through porous roll B. Component A represents the base silicone coating material, whereas component B is the diluent. The objective is to intermix the diluent into the base silicone. The proportional ratio of A to B determines the fluid release level. This concept is experimental at this stage, but tests have indicated that the two-component system is an alternative process for obtaining future "dial-in" release levels. A schematic diagram of a laboratory coater is shown in Figure 15, and diagrams of production-size units are shown in Figures 16–18.

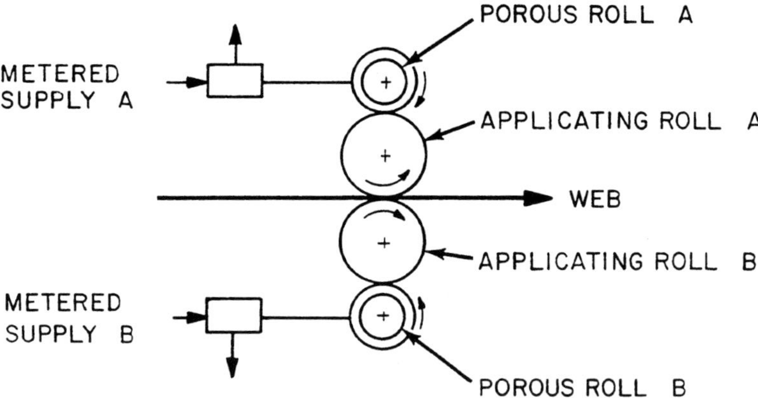

Figure 13 Two side coating with porous rolls.

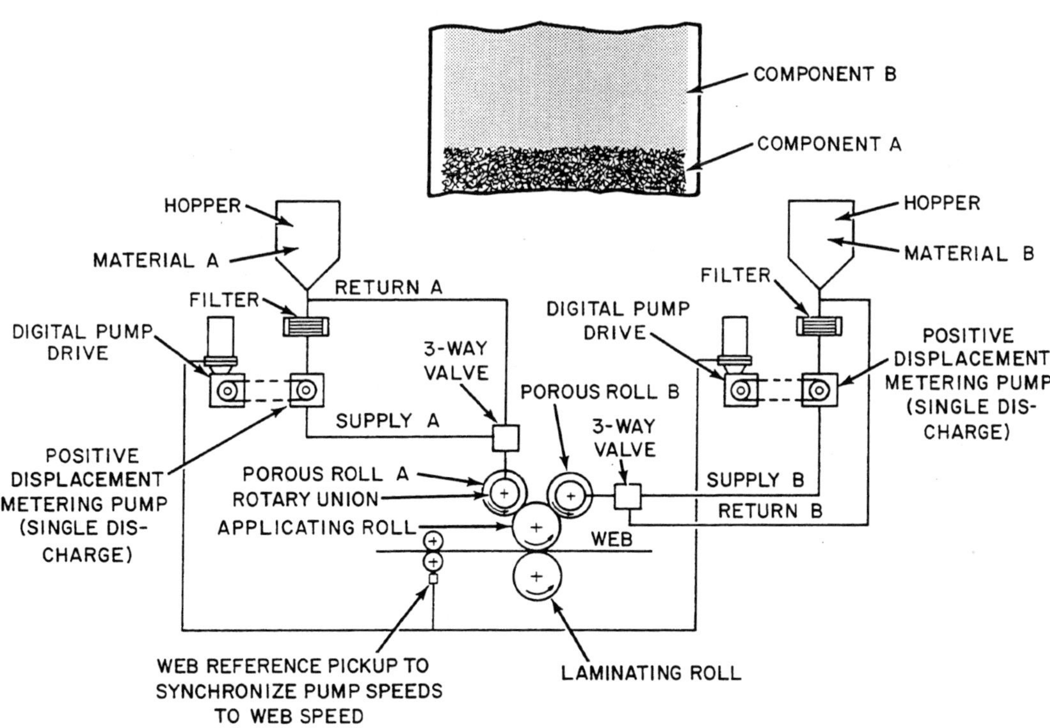

Figure 14 Application of two component silicone system.

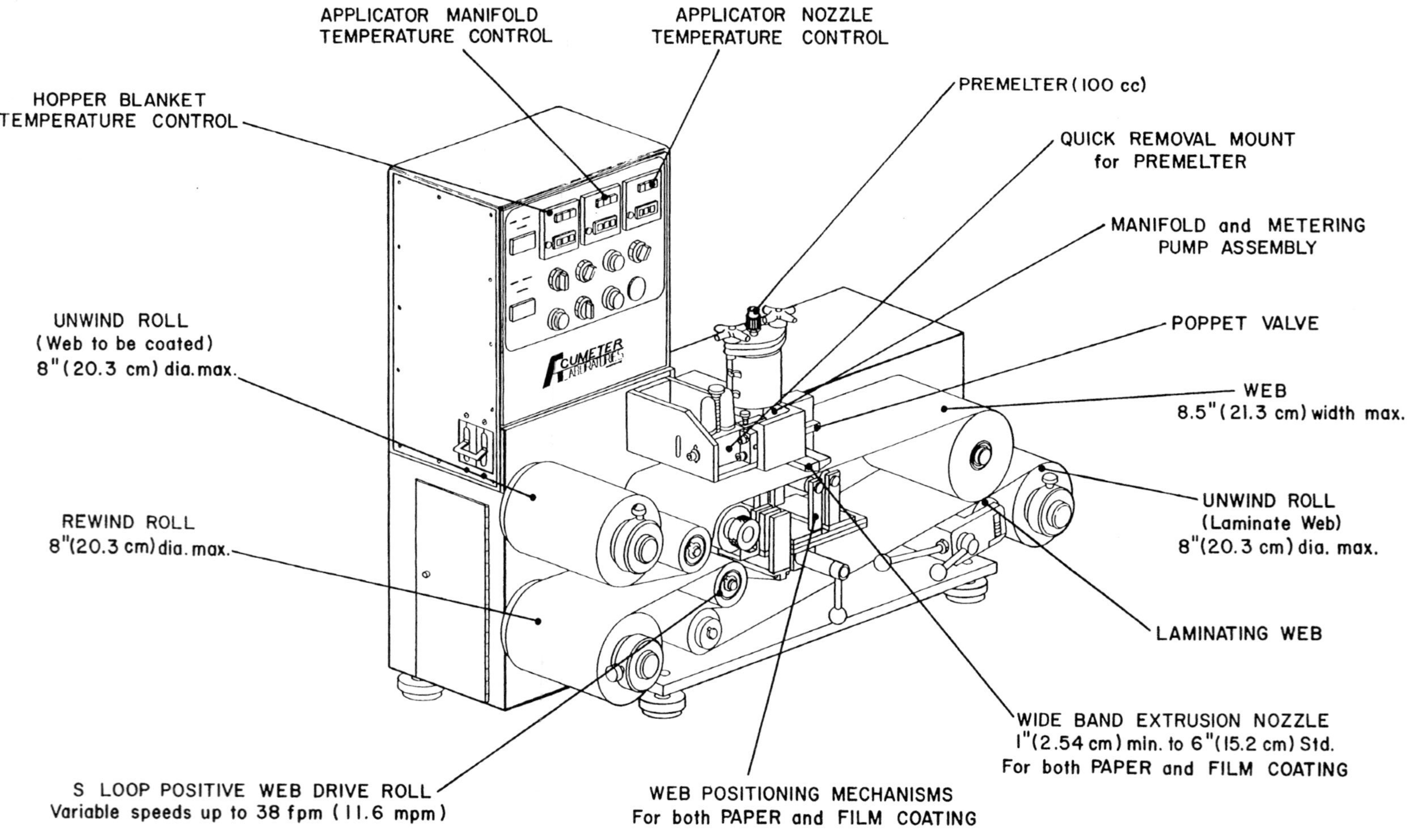

Figure 15 Schematic diagram of a porous roll laboratory coater.

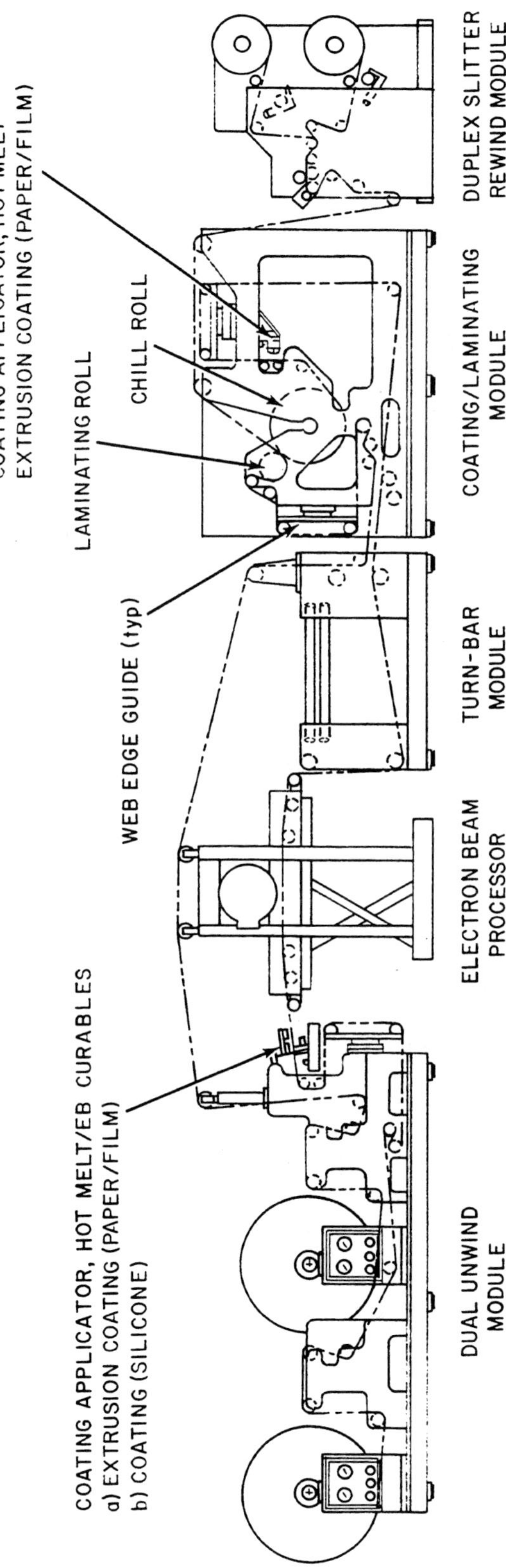

Figure 16 A coating line with a porous roll coater.

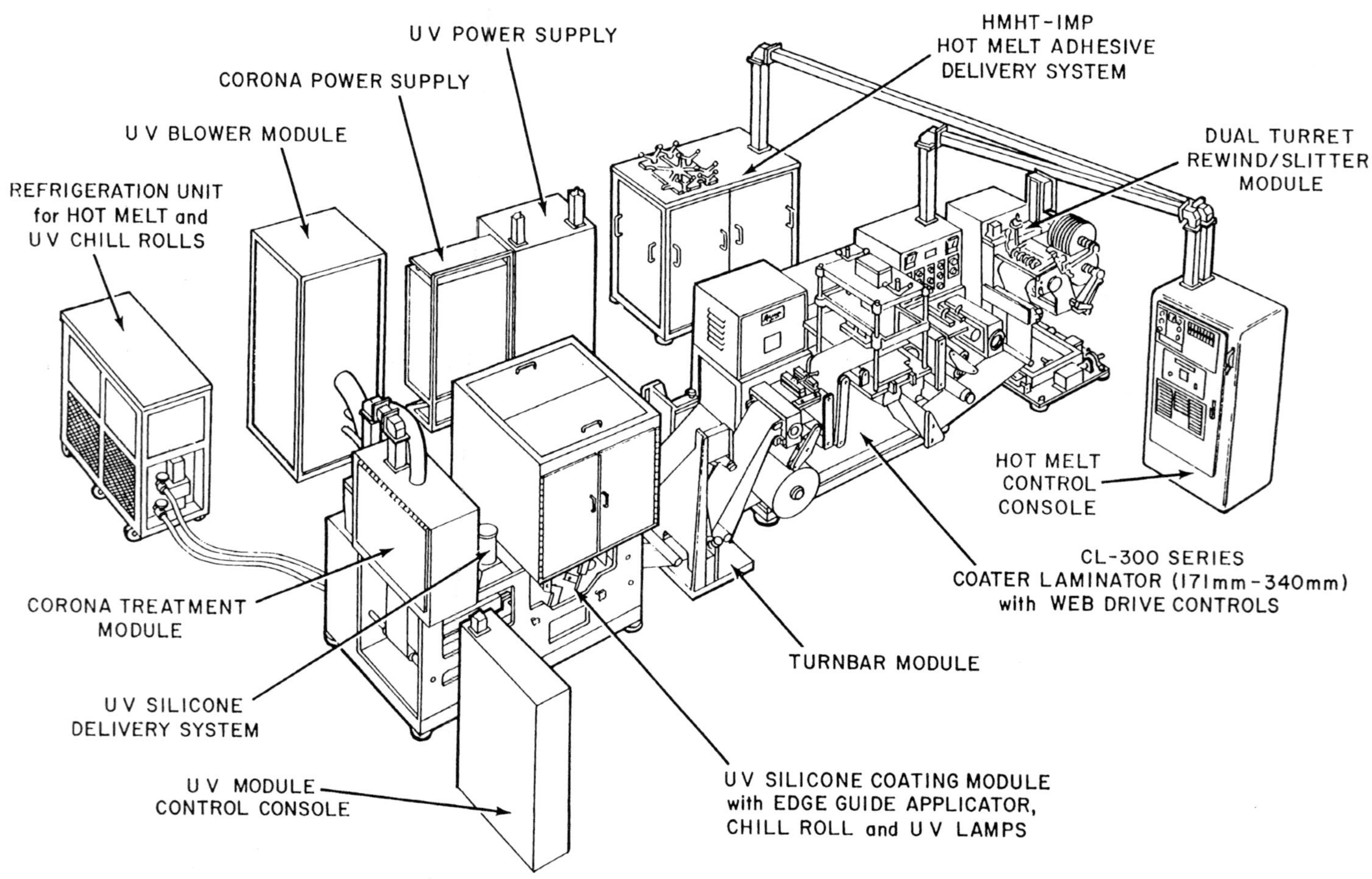

Figure 17 Label stock manufacturing line.

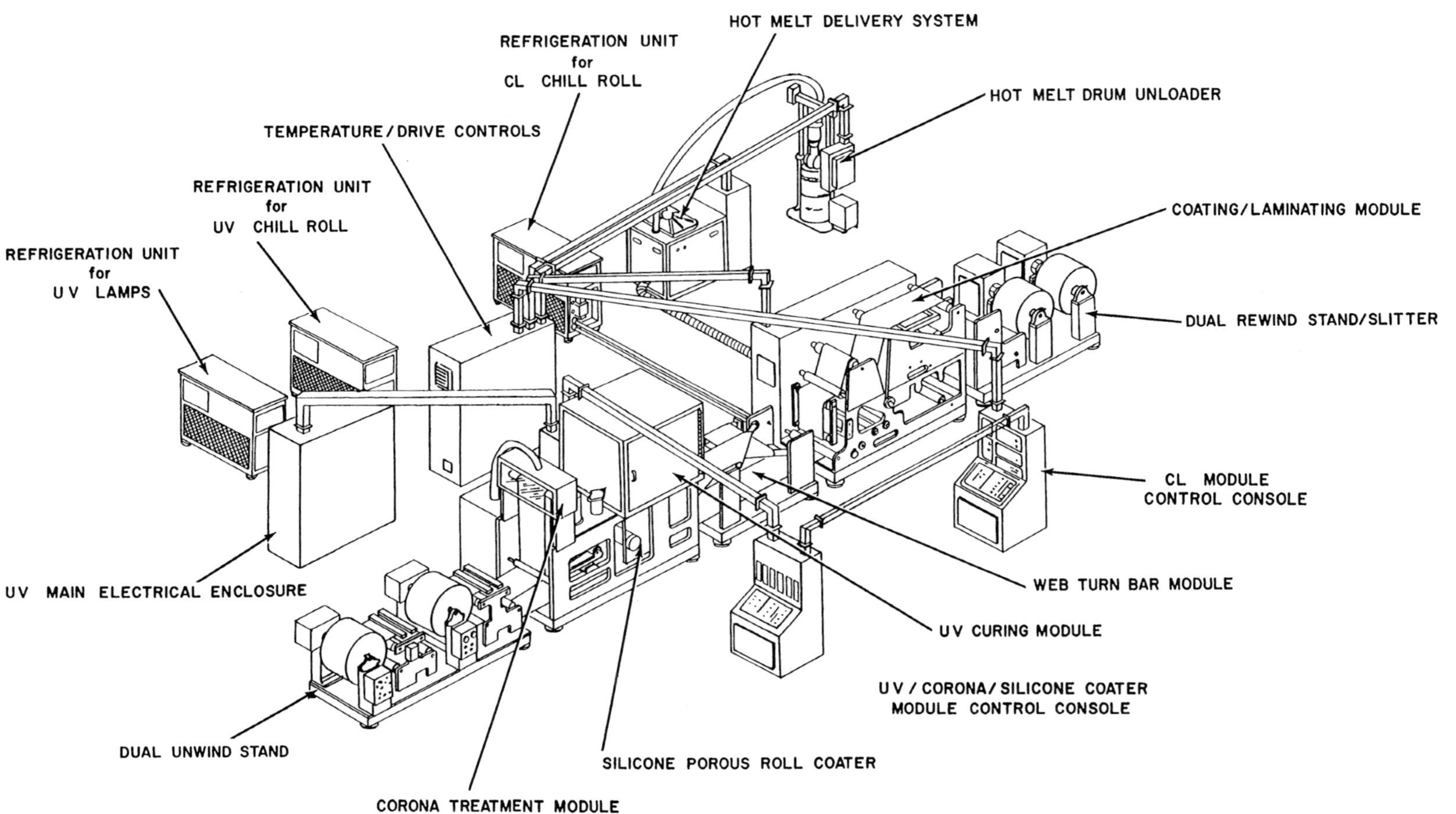

Figure 18 Coating line with a silicone porous roll coater.

14
Rotary Screen Coating

F. A. Goossens

Stork Brabant, Boxmeer, The Netherlands

1.0 INTRODUCTION

Not long after the introduction of multicolor printing machines, the one-color printing unit was also introduced, which through the years has found its way to a wide range of application areas such as wall cover printing, production of hard floor covering, technical coatings, artificial leather production, and last but not least as a one-color printing machine.

2.0 EQUIPMENT

Screen coating equipment (Fig. 1) consists of the following.

The screen, which is a seamless, perforated, nickel sleeve. The degree of perforation is expressed in the so-called mesh number, indicating the number of holes per linear inch.
The squeegee, which is mounted in the screen and serves as supply and distribution pipe of the paste. The squeegee blade, which is mounted to this pipe, pushes the paste out through the wall of the screen.
The whisper blade, which smooths the applied coating layer.

The amount of coating be applied is determined by four factors:

The choice of mesh number.
The squeegee pressure: that is, the angle formed between squeegee blade and screen. The smaller this angle, the higher the add-on.
The viscosity of the paste.
The squeegee setting with regard to the counter pressure roller.

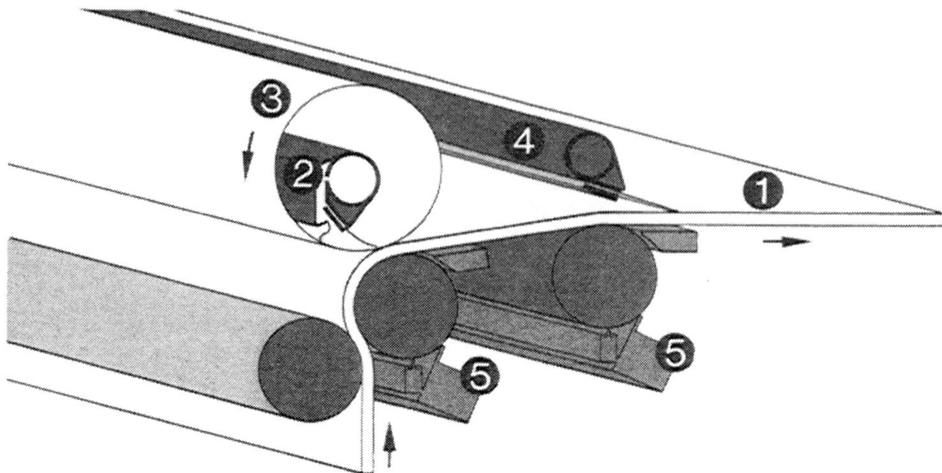

Figure 1   Screen coater. 1: web; 2: squeegee; 3: screen; 4: whisper blade; 5: doctor blade.

3.0 PRODUCTS

The coatings that can be applied by rotary screen are discussed in Sections 3.1–3.4.

3.1 Pattern-Type Coatings

In view of the quantities required for pattern-type coatings and flock glue application, frequent use is made of 125 and 40 mesh screens with a repeat size of 25 in. (64 cm).

End uses include women's fashions, wall-covering curtains, and partial chintz rainwear.

3.2 Dot Coating

Two types of screen are available to accommodate dot coatings: for regular dot coating, the dots are arranged in diagonal rows; for computer dot coating (CP), the dots are arranged at random.

For regular dot coating, the 17, 25, and 30 mesh screens find the widest application and the CP 30, CP 40, and CP 46 (where 30, 40, and 46 designate the number of perforations per square centimeter), are the most frequently used CP screens.

Because of the need to apply an oscillating squeegee on the outside surface of the screen (only used for dot-coating applications), which has been designated to suit 64 cm repeat screens, dot coating can be effected with 64 cm repeat screens only.

End uses for dot coatings include fusible interlinings (woven and nonwoven) and printbonding of nonwoven materials.

Advantages of screen coating compared to traditional coating by means of an engraved roller are as follows.

Screen costs are considerably lower than the cost of an engraved roller.
The time for exchanging screens is extremely short (~ 10 min).
The application amount is easy to control.

Light-sensitive nonwovens (20 g/m²) or knitted fabrics can be processed without problems thanks to the frictionless application system.
There is no penetration, resulting in effective use of paste and a good controllable hand.
The constant paste supply ensures precise level control by means of a past processor.

A new development is the application of *foam-dot coating*, in which the foam is applied with the so-called closed system (see Fig. 2). The foam, which has been made in the foam processor, is applied directly on the substrate via a "closed" squeegee, as shown (Fig. 2). The foam processor will be adjusted for the application weight wet.

For example, suppose we have a polyamide paste with 33% solids. To apply a dot coating with 10 g/m² dry add-on, the foam processor is adjusted for 30 g/m² wet add-on.

3.3 Overall Paste Coatings

Fine or thin coatings, in a range of 5–25 g/m² of dry application weight. Sorters used are 80 mesh 12% open area, 100 mesh SP (special) 20% open area and 125 mesh SP 12% open area, 130 µm thickness (indication only).

End uses include anoraks, skiwear, sportswear, rainwear, umbrellas, pressure-sensitive coatings (medical), and shift-fast coatings.

Medium coatings, in a range of 15–50 g/m² of dry application weight. Screens used are 60 mesh LR (long run) 14% open area, 125 µm thick and 40 mesh LR 20% open area, 130 µm thickness.

End uses include roller blinds, aluminum-coated curtains, bag material, and ironing bar covers.

Heavy past coatings, having a dry application weight of 50–130 g/m² are frequently applied with the 40 mesh HX (hexagonal) screen with 30% open area and 300 µm thickness.

End uses include vertical blinds, flame-retardant coatings on upholstery and technical coatings, and table cloths.
Screens for paste coating always have a 64 cm (25 in.) repeat size.
The squeegee blade is 28 mm × 0.2 mm.

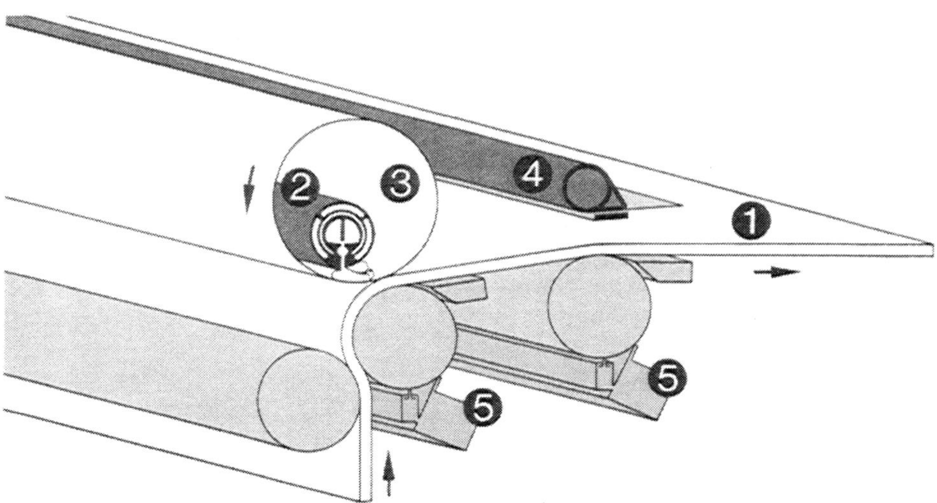

Figure 2   Foam coater. 1: web; 2: squeegee; 3: screen; 4: whisper blade; 5: doctor blade.

3.4 Foam Coatings

Only three type of screen are applicable when coatings are applied by means of thermally unstable foam (foam collapses in drier/stenter) or stable foam.

1. 40 mesh HX [50% open area, 300 μm thick, 91.4 cm (35.8 in.) repeat] for foam coatings having a dry weight of 20–60 g/m^2.
2. 14 mesh HX [40% open area, 450 μm thick, 91.4 cm (35.9 in.) repeat] for foam coatings having a dry weight of 50–100 g/m^2.
3. 25 mesh HX (30% open area, 350 μm thick, 91.4 cm (35.9 in.) repeat] for average foam coatings.

Stable foam is applied by means of a squeegee blade 40 mm wide and 0.2 mm thick.

End uses include black-out curtains or roller blinds, breathable coatings for anoraks and rainwear, and flock adhesive on curtains and shower curtains.

Thermally unstable foam coatings are applied directly by means of the "closed" system (see Fig. 2), via the foam processor and the closed squeegee through the screen on the fabric. The wet application weight can be adjusted on the foam processor.

For example, to apply 30 g/m^2 dry of an acrylic or polyurethane binder (50% solids) simply adjust the wet application weight on the foam processor at 60 g/m^2.

End uses include velour backcoatings, automotive backcoatings, mattress ticking, fixation of pile on pile fabrics, flame retardant and cigarette-proof coatings, imitation suede, and antislip coatings.

4.0 ADVANTAGES

1. Since substrate, screen, and counterpressure roller all have the same speed, coating is done without tension and friction. Thus, virtually all substrates can be processed on this system, including knitted fabrics, velours, nonwovens, and shift-sensitive materials such as skiwear, mattress ticking, and Lycra fabrics.
2. The user has penetration control: penetration into the substrate can be completely avoided or, if desired, controlled.
3. Thanks to the low system content, the coating method is clean, and fast changes are possible.
4. Coatings are exactly reproducible. Since parameters, squeegee pressure, squeegee setting, mesh number, and viscosity can be measured and read off, any given coating can easily be repeated.
5. Chemical savings (up to 20% of the coating weight) are realized in two ways: (a) through accurately controllable application and because the screen follows the web structure exactly (thus the textile character is maintained and an excess of paste, such as occurs with knife coating, is avoided) and (b) through great accuracy, in left/right and longitudinal directions, of the application amount.
6. Application is both tensionless and frictionless.
7. By means of the closed system, the user has total process control.
8. The knife coating option can be attached above the whisper blade roller, mentioned earlier.

This knife coating (see Fig. 3) can be used as a knife-on-air system for paste or unstable foam coatings and in the knife-over-roll coater made for foam applications. In both cases the apparatus is fitted with a paste or foam distribution system over the full working width. In this way it is possible to apply colored coatings with a totally even appearance.

The schematic diagram of a rotary screen coating line is shown in Figure 4.

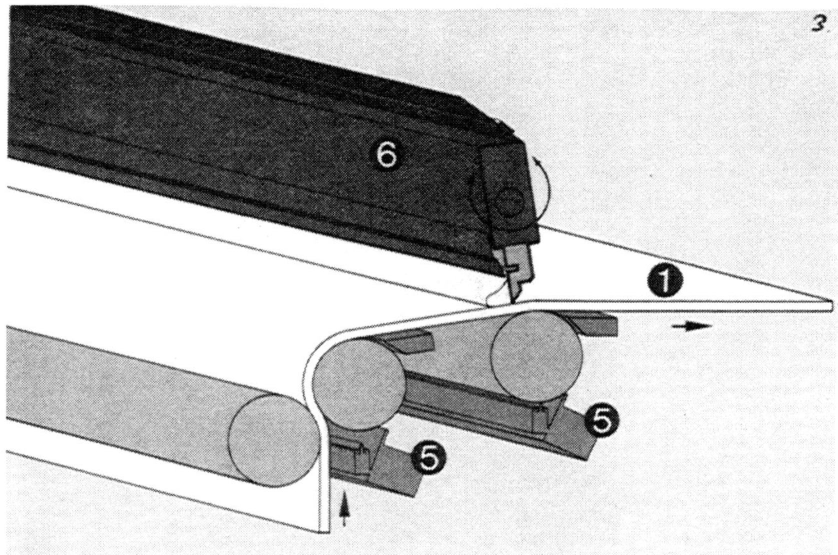

Figure 3   Adaptation of screen coater for knife coating.

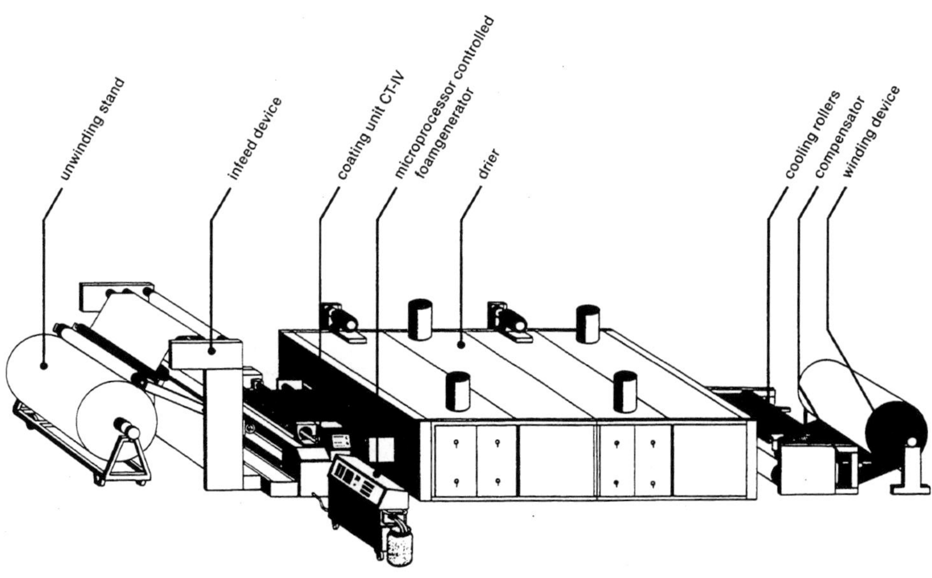

Figure 4   Screen coating line.

15
Screen Printing

Timothy B. McSweeney

Screen Printing Association International
Fairfax, Virginia

1.0 INTRODUCTION

The screen printing process is markedly different from most imaging processes generally associated with the graphic arts. First, the printing plate is actually porous, formed by a woven mesh of synthetic fabric threads or metal wire (or in at least one case, by a non-woven,, electroformed metal matrix), which is then combined with a selective masking material, commonly called a stencil. Because the coating material flows under pressure into an through this mesh or matrix before being deposited onto a substrate, the resulting coating has a thickness far greater than that of a material printed onto the substrate by offset lithography, gravure, flexography, xerography, or inkjet printing.

For this and other reasons, the screen printing process has many practical applications in industrial manufacturing areas in which other printing media have few or none.

The basic process steps are as follows (see also Fig. 1). The woven mesh (or matrix) is affixed to a rigid framework of aluminum or steel. In most applications, this framework forms a rectangular plane. However, variations are possible, including the cylindrical screen, which is affixed and sealed at both ends. In the case of mesh, whether of synthetic polyester monofilaments or stainless steel wire, tension is applied simultaneously in opposing directions to obtain a semirigid planar surface. This stretched printing screen then performs three distinct functions: (a) to meter the fluid coating (or ink) that flows through it under pressure (b) to provide a surface for shearing the viscous columns of coating material, which form during transfer to the substrate and (c) to provide support for the imaging elements (the stencil).

Ink or coating transfer is initiated by the imposition of pressure on the screen by means of a flexible plastic blade, the squeegee. Because of the flexibility of the blade material and its physical profile, a hydraulic action is caused by force exerted in two directions. The blade presses into the screen, and its inherent flexibility enables it to be put into direct contact with the substrate, thus effecting ink transfer. The blade also sweeps in a horizontal direction, thus applying the ink or coating as it moves, and causing the columns of material to shear as the printing screen rebounds after the squeegee has passed.

159

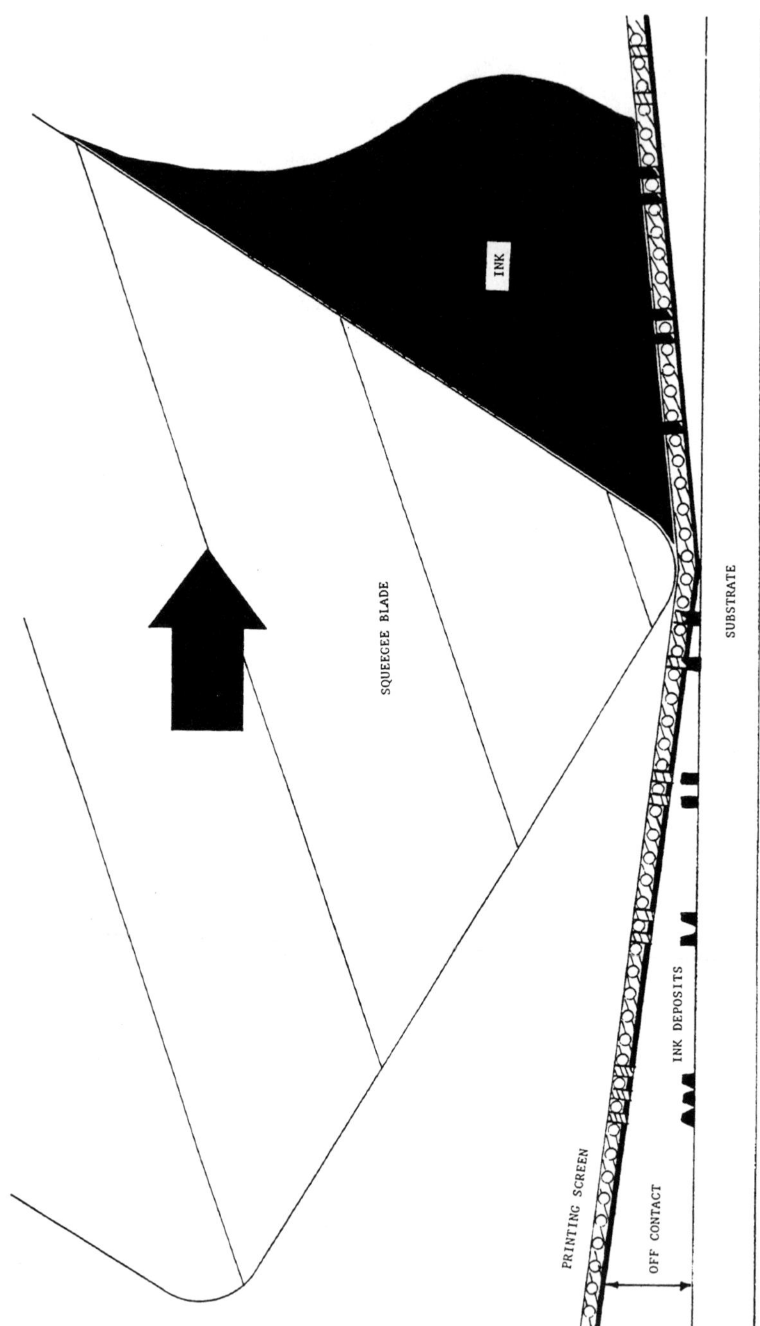

Figure 1 Ink transfer by screen printing.

It is the combination of the horizontal print stroke and the vertical (downward) pressure, effected by the squeegee glade, that forces the ink of coating into the ink "wells" created by the areas of intersection of woven mesh. (Similar wells, though shaped differently, are produced within the matrix of electroformed metal printing screens.) The elasticity of the printing screen then allows temporary contact with the substrate, and subsequent ink or coating transfer.

2.0 GEOMETRY OF THE PRINTING SCREEN

Due to a variety of materials and manufacturing methods, the geometry of the printing screen can vary considerably. Of most common use in industrial applications is the woven mesh, consisting of extruded polyester threads that have been woven into a precise and regular pattern known as "plain weave." For a given linear measurement, the number of thread crossings or intersections, called the mesh count, may be expressed, as produced, in either metric (European) or British (Japanese) increments. At 130 intersections (meshes) per centimeter and higher, it is common to weave the fabric in a twill weave, where one x-axis fiber crosses under and over every *other* y-axis fiber (instead of *every* y-axis fiber in the plain weave). Weaving produces thousands of roughly cubical apertures formed by fiber intersections at each four corners. It is the length, width, and depth of these apertures that form the ink wells and exert the primary influence on amounts of coating deposit.

Thread diameter though consistent within each woven fabric, is a variable that allows a wide choice of fabric thicknesses and thus substantial control over applied coating thicknesses. Threads available range from over 700 μm down to just over 30 μm (0.0275–0.0012 in.) in diameter. The fabric thickness, and therefore the ink well depth, is approximately 1.85 times thread diameter.

The woven mesh is tensioned to predetermined levels and affixed to a frame. For synthetic fabrics, the effects of tensioning can also reduce aperture depth slightly by reducing fabric thickness by up to 3%. For steel wire, however, there is no such reduction.

A stencil material, when present blocks the aperture either partially or totally, thus diminishing the amount of deposit, while adding depth to the walls of the aperture, thus increasing coating deposit.

The total volume of ink or coating deposited is thereby determined by the fabric thickness (the area of the screen/stencil that encompasses the image) minus the volume of the threads (or wires) within the image area.

The coating flow through the mesh aperture is also affected by its viscosity and the operation of the squeegee.

2.1 The Rotary Screen

In principle, the action of the cylindrical screen used in rotary screen printing is similar to that of flat screens, in that the coating material flows through open areas perforating a thin printing "plate." This differs, however, in the basic configuration, which is cylindrical rather than flat. The ink or coating is pumped into the interior of the cylinder, which is sealed at either end by printing heads. Also positioned within the cylinder is the squeegee.

The screen itself is composed (in one variation) of a seamless electroformed nickel mesh tube, which then receives an emulsion coating for producing the patterned stencil. Other methods for producing rotary printing screens achieve similar results.

Rotary screen printing has the advantage of continuous operation, without the intermittent sweep and return of the printing squeegee as in flat printing. Thus, the opportunity to increase speed of production is presented, particularly for substrates printed on a web.[1]

3.0 THE STENCIL

The masking of open mesh, known generically as the stencil, serves to control the pattern or shape of coating deposit (and to a lesser degree, the thickness of deposit). A variety of masking materials and applications are available, from CAD-generated cut films to photosensitive polymers and emulsions. Compatibility between the masking material and the ink or coating chemistry is necessary to preserve the stencil during printing.

Film stencils are adhered to the substrate side of the printing screen, while liquid emulsions are coated onto both sides. Positive-image photo film masters are then positioned on the screen prior to exposure of photosensitized masks. Exposed areas harden to form the mask, while unexposed portions are washed away in the development procedure.

Where the stencil blocks a mesh aperture, the ink or coating material is prevented from flow-through to the substrate, whereas partial blocking only restricts deposition. In this way, imaging is achieved.

4.0 DYNAMICS OF THE SQUEEGEE

The printing tool in screen printing is the squeegee, an instrument bearing some resemblance to the doctor blade used in gravure. However, the squeegee blade is much more flexible, consisting in most cases of a high density polyurethane elastomer, chosen for characteristic durability and solvent resistance. Other elastomers and rubber can also be used, depending on the application.

In flat screen printing, the squeegee makes intermittent and repetitive sweeps across the printing screen, thus forcing the fluid coating into the empty apertures of the mesh, and then into contact with the substrate (with the aid of the flexing of the screen). This action necessarily involves the application of force, applied in vertical (typically downward) and horizontal directions simultaneously; thus, the friction involved calls for wear resistance on the part of the blade.

Since the blade is made of a flexible elastomer, it also flexes, which produces an angle of attack with the mesh, typically about 80° from horizontal.

The properly tensioned screen itself has a force of several pounds per inch, more commonly expressed in newtons per centimeter, which resists the force of the squeegee. The ink or coating has a lesser force, dependent on its viscosity, which contributes to the hydraulic activity. Immediately following passage of the squeegee in the horizontal direction, the screen begins to separate from the substrate, and the coating shears away from the screen flowing into deposition on the substrate.

An additional tool, known as the floodbar, is a common component of automated flatbed screen printing presses. This thin, wide, and relatively flat metal bar passes in contact across the printing screen between squeegee print strokes, filling the open mesh with coating material. The flood stroke does not, however, exert sufficient pressure to cause the screen to flex into contact with the substrate. Therefore, no coating transfer out of the screen will take place. The flood stroke helps to ensure uniformity of the printing or coating application.

5.0 COATING TRANSFER

As a result of the application of the squeegee force, the mesh forms an area of contact with the substrate, and part of the ink or coating that fills the mesh aperture now adheres to the substrate. Cohesive forces within the viscous fluid (now in "columns") tend to hold the coating material together while it flows. As the squeegee passes, the flexing mesh separates from the substrate, and shearing forces cause the fluid to separate at the mesh. Even though the coating adheres to the aperture surfaces of the mesh, only relatively small layer of material actually remains within the mesh, as the coating separates from itself in obedience to shearing forces em thus the relatively thick deposit of coating material achieved by the screen printing process.

For proper ink or coating release from the screen, the rate of ink shearing must equal the speed of the printing squeegee and the rate of ink release from the mesh. Following transfer to the substrate, the ink or coating that has been released has a further tendency to flow, from a series of "column-shaped" deposits into a more uniform layer. This spreading phenomenon is largely dependent on the viscosity of the material that has been printed, the amounts of coating material deposited, and the time that elapses until the coating begins to dry.[2]

6.0 CONVERTING THE APPLIED COATING

Since the majority of inks and coatings applied by the screen printing process are liquid or at least in a semifluid state, complete conversion of the end product requires a drying or curing process. Among the methods used to achieve a solid film state are (from the slowest to the most rapid) evaporation, oxidation, catalytic curing, infrared radiation, ultraviolet radiation, and electron beam radiation. Each of these processes is most appropriate to specific coatings chemistries. In web systems and nonweb conveyorized systems, the speed of production depends not only on the ink or coating chemistry, but also on the stability of the substrate when under temperature (and other) conditions imposed by the conversion process, as well as the thickness of the applied coating film. Factors such as surface reflectivity, ambient humidity, and head absorption can also come into play.

7.0 CONCLUSION

The screen printing process is a unique method used for the application of inks or liquid coatings to a wide variety of substrate types, shapes, materials and surfaces. It is easily adapted to the problem at hand, capable of depositing relatively thick wet films in a short time in either a repetitive or continuous fashion. The process can be integrated with other production processes, whether specifically related to graphic arts or otherwise. Applications may be practical (e.g., conductive printed circuitry) or decorative, pigmented or transparent, finely detailed or broad in coverage. A comprehensive industry support system draws from the fields of photography, chemistry, industrial engineering, and manufacturing technology to provide a vast array of process variations and potential uses for this most flexible of printing techniques.

REFERENCES

1. M. Taylor, "Automatic printing of textile yard goods," *Technical Guidebook of the Screen Printing Industry* N4, Screen Printing Association International, Fairfax, Virginia, 1981.
2. E. Messerschmitt, *Screen Print.* 72(10) (1982).

16
Flexography

Richard Neumann

*Windmöller & Hölscher,
Lengerich/Westfalen,
Germany*

1.0 INTRODUCTION

Throughout the printing industry, flexography, or flexo, has established its reputation as a quality printing process bearing comparison with letterpress, gravure, and offset, which have been used industrially for many years. Today, the whole packaging sector and other areas of the printing industry would be unthinkable without this highly economical quality printing process. This is attributable primarily to the high flexibility flexo offers, its qualification for a wide range of materials, the large and variable range of print repeat lengths, the different press widths available, and the quite extraordinarily high production speeds. Other advantages include the highly diversified flexo press specifications and the possibility of using flexo in line with other printing techniques and processing operations.

Finally, the developments and improvements achieved in the field of press engineering, flexo printing plates, and flexo inks have recently contributed quite decisively to the position the process holds today.

1.1 Historical Development of the Aniline and or Flexographic Printing Process

Today's flexo process is far more than 100 years old. According to historians, extremely primitive aniline work was produced in the United States as far back as 1860.

The original name of this letterpress process, "aniline printing," is traceable to the aniline dyes used in the midnineteenth century, which were diluted in alcohol and had been used in printing for many years. This rubber printing process—until 1970, rubber printing plates were used exclusively—was initially employed for the printing of wrapping papers. The first aniline printing apparatuses are said to have been used in England and Germany beginning in 1890. From about 1910, some European machine manufacturers started to supply aniline printers in combination with paper bag machines to permit printed paper bags to be produced in a single pass. From the early 1920s till approximately 1940, aniline printing machines were slow speed units of simple design and light weight construction, and, for the most part, intended to operate in line with paper bag machines. However, the first four-color

165

roll-to-roll presses also were developed and introduced to the packaging industry at this time; in addition, printing inks were substantially improved, the first drying arrangements were installed on printing machines, and these were new packaging materials such as. cellophane and other nonabsorptive substrates, which had to be printed by this process.

In the course of the 1950s, this special letterpress technique gained in importance for a variety of packaging applications. The first pigment inks were developed, and machines providing improved stability, higher precision, and faster operation were designed, built, and even mechanized to some extent. It was also in the 1950s when—again in the United States—the name of "flexography" was created, which rather quickly became general usage in the trade worldwide. Flexography is defined as "a letterpress printing method utilizing rubber printing plates and liquid, quick-drying inks."

Over the past 25 years, this printing process has become a sidespread technology; it has been improved consistently, especially since 1972, when the photopolymer printing plate was developed and introduced. Over the past 5–8 years—and this is attributable first and foremost to the relatively low printing plate costs and the good print quality achieved, as well as to the versatile application possibilities—flexo has developed into an industrial printing process and is now a serious competitor of the established printing methods, offset and gravure.

1.2 The Flexo Process

Like the letterpress process, flexo utilizes raised printing plates of rubber or soft-elastic polymer, the raised areas of the plates transferring the ink onto the substrate. The low viscosity, quick-drying ink, which is diluted by means of solvents or water, is conveyed by a fountain roller or, more recently, by an ink chamber–doctor assembly and a transfer or anilox roller to the printing plates or stereos, as they are also called.

The most widespread mode of operation today is the roll-to-roll technology and printing in line with surface finishing machines. Whereas in former times only bags, sacks, and paper packaging materials were printed flexographically, the use of flexo today includes a great number of plastic packages in the food and hygiene sector, applications in the aluminum industry, and carrier bags, labels, wallpaper, corrugated board and cartons, newspapers, handkerchiefs, magazines, envelopes, stationery, and other articles requiring printing.

As a result of its high flexibility, flexo has succeeded in continually conquering new market areas and has penetrated into many a domain previously reserved for offset or gravure.

2.0 FLEXO PRESS SYSTEMS

The most usual flexo press layouts, which are known worldwide, can be subdivided into four main groups:

- end printers
- stack presses
- central impression machines
- in-line systems

Color decks and other machine components may often be identical for these different press versions. Nevertheless, to get the best possible final product, it is quite important and decisive to select the correct machine for a specific application.

The different machine versions are used for both printing and surface finishing of a wide range of products including packaging materials of paper, plastic films, and aluminum foil (i.e., from thinnest flexible polyethylene (PE) films to heavy-duty cartons). Number of colors, working width, print repeat length, and machine speed may vary quite substantially and may, accordingly, call for different machine layouts and conceptions.

2.1 End Printers

Today's flexo end printers have their origin in the aniline printing units developed around the turn of the century. Initially attached to paper bag machines, they served for one-or two-color "mechanized stamping" of paper bags, and later also for the printing of sacks and sheets. In the middle of our century, several thousand machine combinations were installed throughout the world, consisting of flexo end printers and paper converting machines for the printing and making of paper bags and sacks in a single operation. (Fig. 1). Following the development of polyethylene, this highly efficient printing technique was extended to include PE items as well.

These machine combinations dispense with a separate operation and reduce roll handling requirements; in addition, they save personnel, space, and capital expenditures, in addition to reducing waste. All these factors together generate improved economics. Today, end printers are available in working widths between 25 and 320 cm.

Depending on the capacity of the converting machines, speeds up to 400 m/min are achieved. Usually, three-or four-color stack printers with separate impression cylinders are used, but one-and six-color units are in use as well. Such end printers are employed exclusively for printing line drawings, solids, and type. They do not lend themselves to actual quality printing, which is reserved to roll-to-roll flexo presses featuring the appropriate optional equipment.

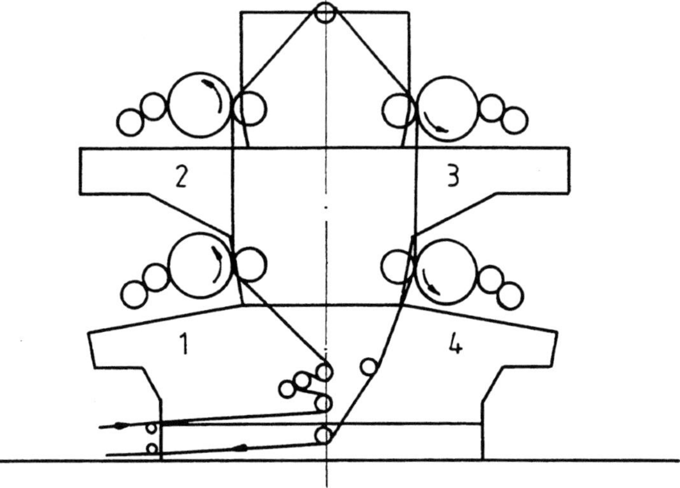

Figure 1 Flexo end printer with four color decks.

2.2 Stack Presses

The first flexo press working from roll to roll was a stack-type press developed on the basis of an end printer. Stack presses normally consist of four or six color decks with two or three color decks incorporated on top of each other on either side of the print unit frame. The configuration of such roll-to-roll presses is almost identical with all press manufacturers. Over recent years a high level of standardization of press layout has been reached (Fig. 2).

The web is fed from a simple or fully mechanized unwind unit for automatic reel splicing via a constant feed–draw assembly to the print unit frame, where it is printed in several colors, with a drier for drying the applied printing ink arranged after each deck. Leaving the last color deck, the web passes through a drying tunnel or other final drying arrangements for thorough drying of the ink and is then led via one or two chill rolls, combined with adjustable nip rolls, to the rewind unit or to a fully automatic, dc-motor-driven winding machine featuring roll splicing facility.

Stack presses are used for printing rigid materials. The between-color register accuracy in the machine direction is approximately ± 0.2 mm T.I.R. (total indicator runout).

Four color presses are used in the market, but the majority of central impression machines are six-color presses permitting four-or six-color halftone process printing using screen counts of the plate of 48, 60, or even 70 lines to the centimeter. This obviously calls for a perfect printing plate, appropriate printing inks, and a high precision machine. These requirements have been met, and as a result even flexible materials can today be printed flexographically with a quality approaching that achieved by the gravure and offset processes.

For the reasons outlined, the demand for central impression machines has drastically increased over recent years, this design being presently also used to print rigid materials such as papers and compounds. Apart from this, the application of these presses has been extended also to the wallpaper and corrugated board industry. The biggest machines have a 2–3 m diameter impression cylinder and cover a working width of approximately 2.5 m. They achieve printing speeds of approximately 300 m/min, and 400 m/min is not unusual with standard design presses.

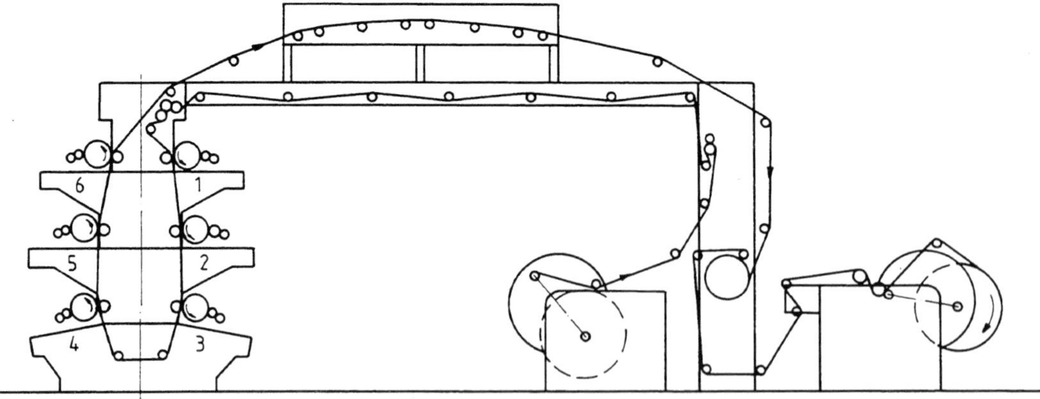

Figure 2 Stack-type flexo press with six color decks.

2.3 In-Line Systems

These machines are similar in configuration to gravure presses. For each color a separate frame is used, with the option of having any desired number of printing stations arranged one after the other, the drive connection between the individual frames being provided by universal joint shafts or the like. These machines call for a large floor space and involve substantial capital expenditure. This, in turn, means that they can be economically used only for long production runs at high speeds. Their main field of application is in the printing of aluminum foils, papers, and cartons (Fig. 3).

The essential advantage of this machine configuration lies in the long interunit drying paths, which permit wet-on-wet printing at high speeds. Modern high speed presses of this type achieve rates of up to 600 m/min (10 m/s) under production conditions. Stack presses allow the web to be printed with six colors on one side, with five colors on the face and one on the reverse or in configurations of 4:2 or 3:3. Stack presses normally achieve production speeds of 400 m/min on paper. They are built in working widths between 25 and 250 cm.

This press design, which was the best-selling machine version about 25 years ago, has been continuously superseded by central impression machines over the past decade, and today only 10–15% of all new web-fed presses are built on the stack principle.

2.4 Central Impression Machines

The first central impression flexo press (i.e., a press with the color decks arranged like satellites around a large-diameter common impression cylinder), was developed in 1953–1954 and introduced to the trade. The suggestion for the design of this press system came from the plastic film industry, where the need for a printing machine giving better web control had become obvious. And this need has been satisfied indeed: the web comes from the unwind, passing through a draw unit to the first color deck on the central impression cylinder, where it is pressed onto the cylinder surface and firmly held to it until it leaves the last printing deck. This working principle guarantees an immovably fixed position of the web so that no register variations whatsoever are liable to occur during printing. With the use of an appropriate drive system, maximum longitudinal register tolerances keep within 0.1 mm T.I.R. (Fig. 4). Hence it follows that this rotary press offers the best register accuracy without any mechanical or electronic aids and additional equipment.

The key component of the central impression machine is the common impression cylinder designed for maximum concentricity (5 ± T.I.R.) and perfectly constant temperature over the full width of the cylinder (± 0.5°C) during printing.

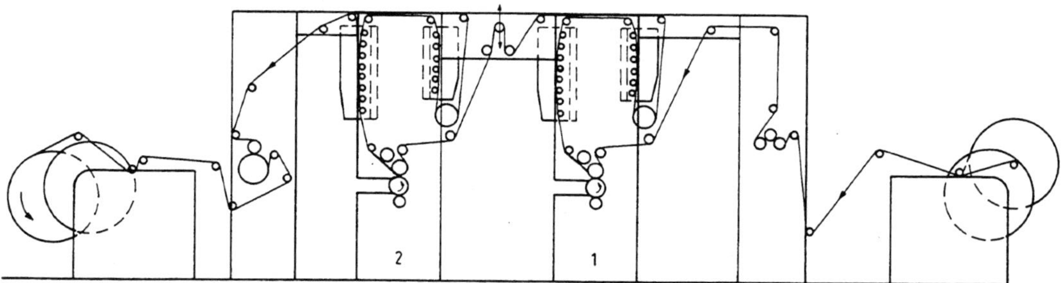

Figure 3 Flexo in-line machine with two printing stations.

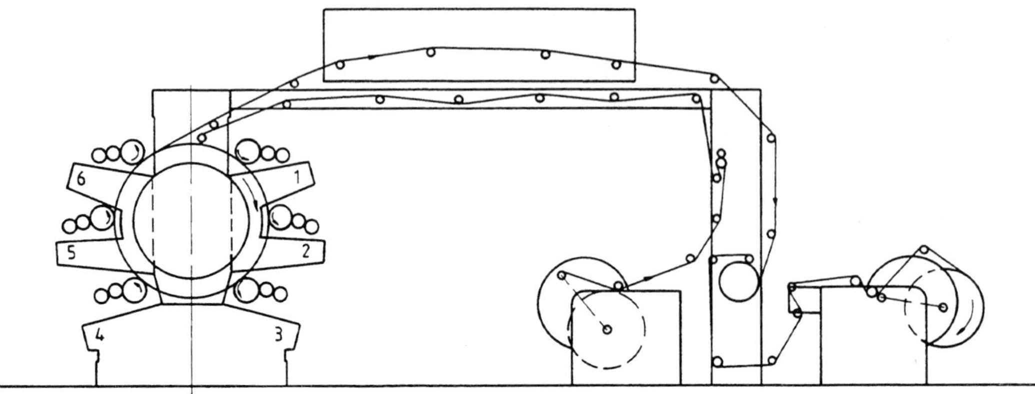

Figure 4 Central impression flexo press with six color decks.

Central impression machines are built in working widths of 60–160 cm. Apart from standard systems, there are also available today press versions utilizing so-called slide-in printing units or subunits, thereby reducing idle time on job changeover. Moreover, printing stations are being used which are designed for changing from flexo to gravure with or without the use of subunits. This concept adds to the application possibilities of the press and ensures optimal utilization.

3.0 THE MOST IMPORTANT FLEXO DECK SYSTEM

Having explained the four different flexo press versions today available in the marketplace, we turn to the most important and most interesting color deck system.

As a matter of fact, the color deck is the most important component of a flexo press. Today, only two printing deck systems are still of general and topic interest—the three roller system or fountain roller color deck, and the two-roller system or fountainless color deck. The latter has gained increasing importance over recent years and enjoys steadily growing popularity because it achieves higher print quality.

3.1 Three-Roller System (Fountain Roller Color Deck)

The three-roller printing deck was developed many decades ago and has been continually improved, it is today the most commonly used deck version found in flexo presses. It is relatively simple to operate and to set up, offers great flexibility and, until some years ago, was the appropriate design to meet all user requirements. The conventional three-roller printing deck is mounted on a pair of frames or consoles and normally consists of a pair of angular bearing blocks to take the plate cylinder and a pair of inking roller bearing blocks to accommodate the inking rollers (i.e., the anilox and the fountain roller). Also, an arrangement to receive the ink fountain would usually be incorporated. For changing the plate cylinder to print a different repeat length (or, for the purpose of impression adjustment), the plate cylinder, together with its angular bearing brackets, is horizontally loaded to or unloaded from the impression cylinder. This is done mechanically, hydraulically, or by motor. The same applies for the fountain and anilox rollers and for the ink duct (Fig. 5). This is accomplished through manual or motorized adjustment by threaded spindles or, more recently, by computerized systems providing different automation levels.

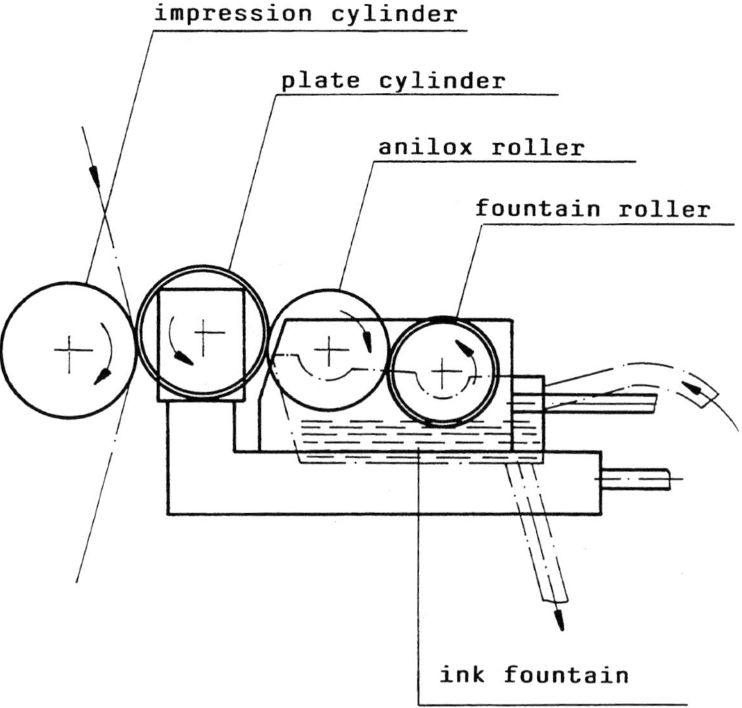

Figure 5 Three-roller system of fountain roller color deck.

At the moment the machine stops, (i.e., when printing is interrupted), the plate cylinder must be lifted clear of the impression cylinder and of the web, to prevent the plate and the substrate from sticking together. Thus the plate cylinder is slightly lifted or retracted in the horizontal and vertical directions. At the same time the plate cylinder must stop and come off the anilox roller so that no more ink is transferred to the plate. To prevent the ink from drying on the rollers while the press is standing still, the anilox and fountain rollers must continue rotating, powered by a separate motor. To prevent grinding and damage to the printing plates, the circumferential speeds of the anilox roller must always be equal to the surface speeds of the plate on the print cylinder.

With the three-roller printing deck, the amount of ink to be transferred to the substrate can be varied by altering the gap between fountain and anilox roller. This is an advantage with the three-roller system. The inherent disadvantage is, however, that changes in speed necessarily result in variations of the quality of ink transferred to the printing plates, which, in turn, results in color deviations in the printed image.

3.2 Two-Roller or Fountainless Printing Deck

The latest technology in flexo deck system design is relatively new and a great deal of engineering work and extensive experiments were necessary to accomplish an acceptable solution.

In the two-roller system, a plate cylinder and an anilox roller with doctor blade assembly has superseded the previously used fountain roller (Fig. 6). Almost all doctor blade units are

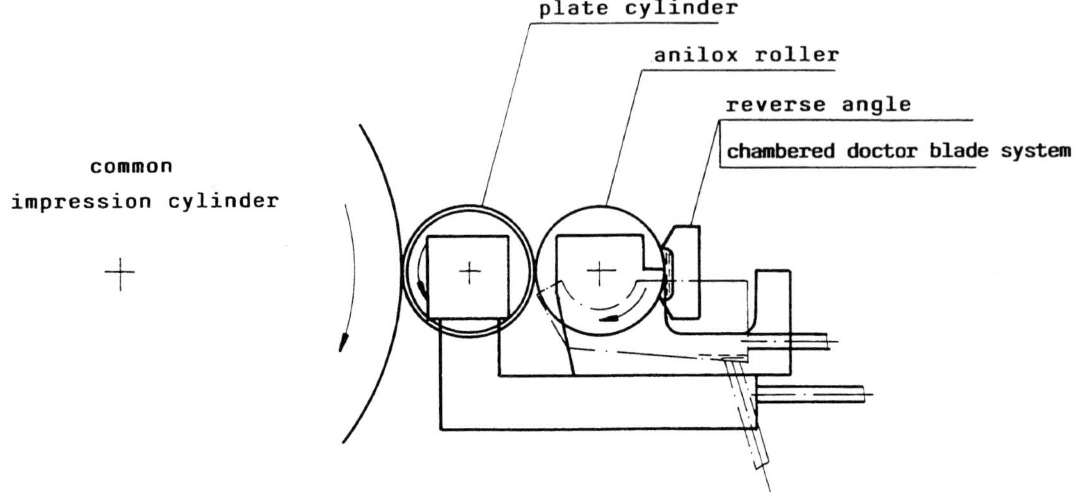

Figure 6 Two-roller or fountainless printing deck.

designed for laterally reciprocating doctor blade motion. Loading the doctor blade to the anilox roller is mostly done pneumatically, with fine adjustment of the contact pressure.

Initially, the anilox roller was being doctored only conventionally or positively, but over the past two years, negative or reverse-angle doctoring has been generally accepted and preferred. Press operators throughout the trade assert that especially when performing fine process work on high-speed wide-web presses, reverse-angle chambered doctor blades achieve more consistent and better results. This is attributable to the capability of such decks to provide even finer, more uniform, and exactly defined ink application over the full web width and throughout the machine's speed range. As an added benefit, anilox roller and doctor blade wear is minimized, provided the doctor blade is carefully set to the anilox roller with minimum contact pressure. Also, ink consumption is drastically reduced, no surplus ink being transferred from the anilox roller surface to the printing plates. This two-roller or fountainless printing deck with the reverse-angle chambered doctor blade system has been increasingly used throughout the industry and can be considered to be printing deck system of the future. Most of the attachments and options found with flexo presses are identical to those used with offset and gravure presses and do not feature any specific process-related characteristics.

4.0 PRINTING FORMS OR PLATES

Apart from the press design itself, printing plates are of paramount importance to quality flexo. To better understand flexo technology as a whole, some knowledge of printing plates and how they are made is indispensable.

Flexography owes its name to the exclusive use of flexible, elastic printing plates. These were made from natural and rubber compounds for decades, until the appearance on the market in 1972 of the first photopolymer plates for flexo, which have since succeeded in capturing quite a substantial market share. For many applications however, rubber plates are, still used today.

4.1 Rubber and Photopolymer Plate Making

As with any other printing process, good artwork is essential to making good rubber plates. For single-color printing one negative is needed, and for multicolor printing one negative for each color with continuous tones being separated into dots. From these negative films an original plate or a metal etching is prepared by photographic transmission using copper, magnesium, or zinc. On the basis of these metal blocks, so called matrices are molded from board soaked in phenolic resin; this is done in a special molding press. After cooling, the molded matrix is used for making the rubber plate in the same press, using natural rubber compounds in thicknesses from 2 to 10 mm (normally 2.76 mm) and of a Shore hardness, depending on intended application, between 40° and 70° Shore A. Minor corrections of plate thickness are achieved by grinding the back of the plate. For the printing of solids, even today, plates are cut from vulcanized rubber by hand.

It takes less time and work to make photopolymer plates than rubber plates, since the former are prepared directly from the negative, dispensing with metal etching and matrices.

There are two photopolymer systems, solid and liquid, the solid system being clearly preferred to the liquid technology. The solid system calls for a distinction between single-layer and multilayer plates. The latter utilize a soft base layer as a buffer layer, which serves to improve pressure compensation. On the back on in the middle, a layer of polyester film is provided to ensure dimensional stability. The light-sensitive polymer or relief layer is on top, covered with protective film, which is removed prior to exposure.

The plate-making process starts with back exposure to UV light for the purpose of sensitizing and polymerizing the plate. Next, UV main exposure takes place on the emulsion layer, through the negative, the latter being firmly held to the plate surface by vacuum. The decisive factors for proper polymerization of the printing area or image elements during exposure are the intensity of UV radiation and the time. Subsequent final exposure ensures perfect polymerization of the relief plate.

During the subsequent wash-out process using a special washout solution, the unexposed, nonpolymerized photopolymer is dissolved and removed from the plate. The plate is then dried by hot air and posttreated in an acid solution to get the final plate surface, final exposure of the plate increasing its resistance to solvents. Photopolymer printing plates are used in many different thicknesses and with variable relief depths to cover a wide range of applications. Hardness ranges from 45 to 60° Shore A.

Other plate-making methods have only minor importance and are not discussed here.

4.2 Printing Plate Mounting and Proofing

For production printing on the flexo press, the finished plates are pasted on the plate cylinder surface. This is mostly done using double-sided adhesive tape and whatever auxiliary equipment or machines will permit register-true plate mounting on the cylinders. The thickness of the adhesive film depends on the plate thickness; with a great number of flexo presses, the combined thickness is 3.0 mm.

Other mounting methods also are employed, but these are confined to special areas and therefore do not need to be discussed here.

Mounting the printing plates always takes place off the printing machine. With a view to avoiding unnecessary press downtime, the plate cylinders with the plates mounted on them are proofed before production begins. This is done using different methods and different apparatuses and machines. The proofing process is to demonstrate and ensure that the printing plates have been correctly and perfectly mounted, in accurate register and without blisters. Following this, the proofed plate cylinder can be installed in the flexo press for trouble-free production printing.

It is also possible to test the design of a new printing subject before plates are made. As far as rubber plates are concerned, a print proof can be produced from the metal block by letterpress printing, whereas with photopolymer plates a color proofing system can be used to prepare a high quality color proof from the negative.

If the printing plates are to be reused, they must be cleaned thoroughly after completion of the production run and carefully stripped off the print cylinder; they should be stored lying as flat as possible and, especially when photopolymer plates are involved, they must be wrapped in lightproof material. It is, of course, also possible to store the complete cylinders with the plates mounted.

5.0 PRINT SUBSTRATES AND PRINTING INKS

Because of the flexibility and versatility the flexo process offers, there is hardly any material that has not been printed flexographically, regardless of material thickness, elasticity or rigidity, surface properties, intended application, or many of numerous other material-related characteristics. Moreover, flexo permits printing of metal strips only a few centimeter wide or paper rolls as well as paper and film webs up to 3 m wide. Also, newspaper are printed by flexo today at production rates of approximately 700 m/min.

Obviously, then, there is a need for appropriate printing inks, bearing in mind that extremely varied demands are made on the ink to be used according to the specific material being printed, its proposed application, and the press speed. Also important in this respect are the very stringent statutory regulations that exist in the field of food packaging.

5.1 Print Substrates

The materials that can be flexo printed include paper, carton, cellophane, plastic films (both mono-and multilayer films), aluminum foils, and a wide range of special materials. Ranking first in the packaging industry, now as before, are wrapping papers, mostly consisting of blends of different pulps.

Sulfate kraft paper is a single-sided glazed pulp paper used primarily in the production of bags and sacks of all kinds. Sulfite and sulfate papers intended for similar applications are also flexographically printed. A drastic growth rate has recently been experienced in the field of kraft or test liner preprint for the corrugated board industry. Another area of flexo application includes tissue papers used as wrapping and toilet papers, hygienic papers for handkerchiefs and towels, as well as many other specialty papers. Also worth mentioning in this respect is the direct printing of corrugated board for use as outer package and the printing of cartons (e.g., liquid packs).

The first film to be printed flexographically, which raised quite substantial problems for all concerned, was cellulose film. Depending on its finish, cellulose film was mainly used in the packaging of foodstuffs and textiles. Since the arrival of plastic films, cellophane has lost a great deal of its former importance.

Polyethylene and polypropylene films possess many desirable properties and now rank first in the field of flexible packages. They are printed and processed in many different thicknesses ($\sim$ 10–250 μm), as either sheeting or tubing. Their main applications include carrier bags, food bags of all kinds, hygienic and sanitary articles and many other products, fertilizer and peat sacks, garbage sacks, cover films, shrink hoods, and the like. This wide selection of films is completed by polyvinyl chloride and polyvinyl dichloride films, and other special films used as mono films or in combination with other materials for packaging or to impart to the container specific properties desired for a certain product to be packaged. This is achieved by laminating different materials, by coating, and by coextrusion.

All these materials and material combinations are printed by the flexo process. This also applies for aluminum foil, which may be plain or laminated, lacquered or coated in thicknesses from 7 mm up. It is used primarily in the sweets and candy industry, as lid foil, and in the packaging of butter, soups, coffee, and bread, as well as for many other package applications.

5.2 Flexo Inks

The substrates named in Section 5.1 call for a great variety of flexo inks offering many different properties. Flexo inks are similar in composition to the inks used in gravure package printing; they always consist of colorants (dyes or pigments), binding agents (natural resins, artificial resins, or plastics), and a solvent or solvent blend. Whereas flexo inks used to be based on basic (soluble) dyes, pigment inks are used primarily today because of more exacting demands on the ink's fastness. To obtain the desired properties such as brilliance, adhesion, and qualification for laminating, the correct binding agents and additives must be selected.

Apart from the aforesaid properties, flexo inks are required to generate a quality end product. Fast and perfect drying of inks on the substrate during printing is another aspect of paramount importance, and in this respect the solvent of solvent mixture used is the decisive factor.

The drying system involves evaporation of solvents after the ink has been applied to the web. This drying process is substantially accelerated within the printing press using hot air, which is blown onto the web, and appropriate exhaust arrangements.

The most important solvents are hydrocarbons, alcohols, glycols, esters, and ketones. Recently, water-soluble pigment inks, once used exclusively in the printing of multiwall paper sacks, gift wrap, corrugated board, and wallpaper, have been playing an increasingly important role and also have been adopted in the fields of newspapers and plastic films. The obvious reason for the growing trend to use water-soluble inks in package printing and for plastic films, is the new set of laws calling for reduction of solvent emissions into the environment.

17
Ink Jet Printing

Naomi Luft Cameron

Datek Information Services, Newtonville, Massachusetts

1.0 INTRODUCTION

Ink jet printing refers to any system in which droplets of ink are ejected onto a printing surface to form characters, codes, or other graphic patterns. The ink jet concept dates from the 1860s, when Lord Kelvin developed the first practical jet for pattern generation. Early commercialization was in the oscillographic recorder area in the 1950s. Since the 1960s, ink jet developments have focused on computer output, with major contributions made by such scientists as Hellmuth Hertz in Europe (Lund Institute, Sweden) and Richard Sweet (Stanford University) and Steven Zoltan (Brush Instruments) in the United States.

Current commercial products range from printers for direct coding of packages, to high speed – low resolution direct mail printers (from Diconix), to graphic arts quality color plotters (from Iris Graphics), to graphic arts quality color plotters (from Hewlett Packard). To address this range of applications, several variants of the technology have been developed. Each approach involves tradeoffs among cost, speed, reliability, and print quality, determined by interactions between hardware and supplies.

Ink jet printing functions include:

creation of an ink stream or droplets under pressure
ejection of ink from a nozzle orifice
control of drop size and uniformity
control of which drops reach the paper
placement of drops on the recording surface

Control of these processes depends on several design variables, such as nozzle size, firing rates, drop deflection methods, and ink viscosity. Changing any variable typically requires adjustments to other system variables, making R&D advances slow and expensive.

Ink jet printers fall into two basic categories: continuous jet (synchronous) and impulse jet (drop-on-demand). Most early development took place in the continuous jet arena, but recent emphasis has shifted to the less complex (and therefore less costly) drop-on-demand approaches.

2.0 CONTINUOUS JET PRINTING

Continuous ink jet systems operate by forcing pressurized ink in a cylinder through nozzles in a continuous stream. Nozzle diameters range from 3 to 0.5 mil: the smallest nozzles can require up to 600 psi of pressure to eject the ink. The ink stream is unstable, breaking into individual droplets either naturally or through some applied stimulation such as ultrasonic vibration. Electrostatic deflection is used to control the droplets, which either reach the page in the desired pattern or are deflected into a "gutter" or "catcher" (see Fig. 1). Deflection can either be binary or variable.

Continuous jet printers either recycle or discard the unused ink, which can account for up to 98% of the generated droplets. Although preferable in terms of supply costs, recycling adds substantially to system complexity because of the need for pumps and ink purity safeguards (such as filters and solvent balance controls).

Continuous ink jet systems are capable of very high speed printing, with drop frequency rates sometimes in excess of 100 kHz. At these rates, it is difficult to control individual droplets: their tendency to combine into larger drops leads to print quality concerns. Ideally, however (i.e., when controlled), this tendency can yield enhanced image quality: because each printed pixel may be composed of multiple droplets, it is possible to control the pixel size, resulting in halftoning capabilities.High resolution halftoning cuts the speed of continuous ink jet systems considerably, however.

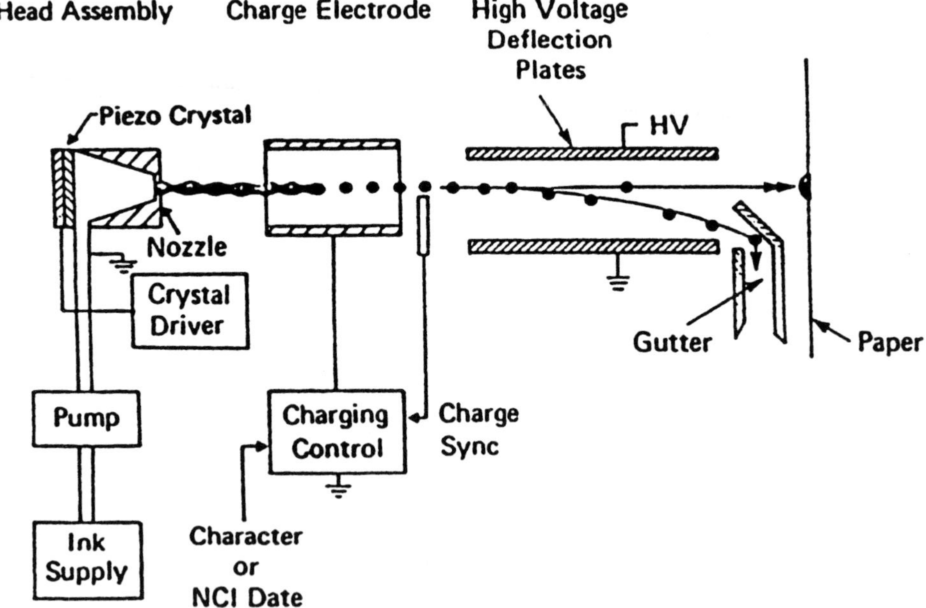

Figure 1   Continuous ink jet print system. (From: R. G. Sweet et al., U.S. Patent 3,373,437.)

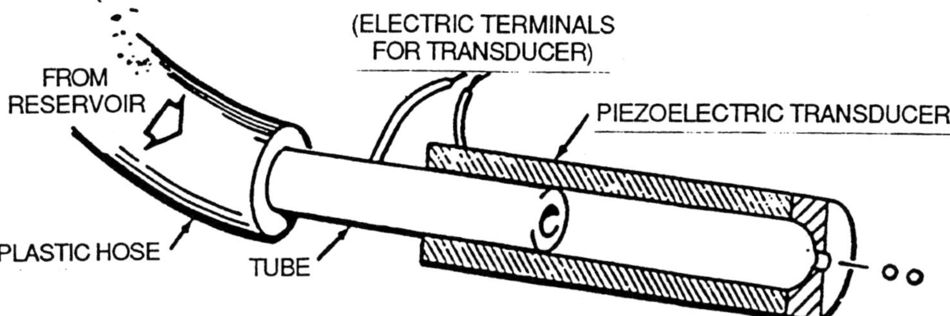

Figure 2 Piezoelectric impulse jet nozzle. (From: S. Zoltan, U.S. Patent 3,683,212.)

3.0 IMPULSE JET (DROP-ON-DEMAND) PRINTING

In contrast to continuous jet printers, drop-on-demand ink delivery systems create drops only as needed, thereby eliminating the need to control excess droplets. These systems are inherently binary: a drop either is ejected for placement on the receiver sheet or it's not.

Most of the system complexity of an impulse jet printer is in the printhead, since no recycling mechanism is required. Typically an impulse jet printhead has a pressurized reservoir of ink held directly behind a nozzle or orifice. When activated by electrical pulses, drops of ink are ejected and directed to page.

There are two basic methods of activating the ink droplets. The earliest impulse jet models used *piezoelectric* transducers, which squeeze the ink chamber or impulse the chamber at one end. Figure 2 shows a schematic of the piezoelectric approach. The alternative approach, *thermal activation* of ink drops, is gaining popularity as a result of developments by Hewlett Packard and Canon. A heater creates a bubble of ink vapor, which forces ink drops from the nozzle (Fig. 3).

Most newer ink jet printers use impulse jet technologies to address general-purpose printing applications. Advantages include mechanical simplicity, low hardware cost, and simplified logic. However, there are disadvantages as well: drop-on-demand printers are more sensitive to shock and vibration and have slower dot ejection rates (sometimes as low as 3 kHz). In addition, market acceptance has been slow, partly because of early reliability problems due to nozzle clogging from dried ink or paper dust.

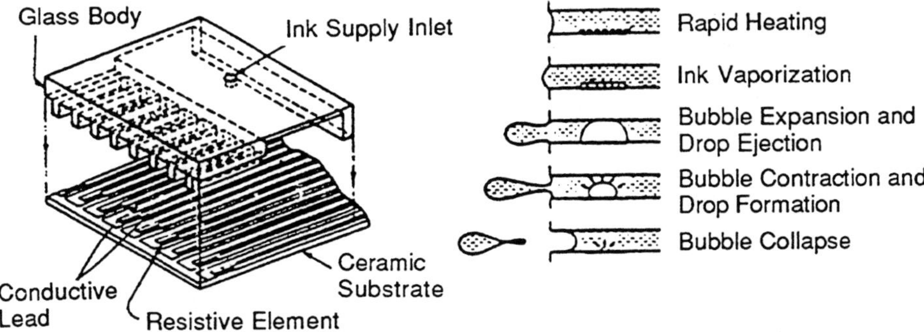

Figure 3 Thermally activated Canon impulse jet printhead. (From: A. B. Jaffe and R. N. Mills, from *"Color hardcopy for computer systems,"* SID Proceedings, 1983, Vol. 24/3, pp. 219–234.)

4.0 INK JET INKS

Ink jet printer design requires close matching of mechanical components, imaging inks, and, in many cases, the receiver materials. Ink chemistry is a critical link, determining diverse attributes such as viscosity, drop flight, corrosive properties, surface tension, drying time, dot shape, optical density and edge acuity, fade resistance, and compatibility with printing surfaces.

Traditionally inks have been liquid, with a water or solvent base. Continuous jet inks are usually based on one of a wide range of solvents, which permit fast drying on both porous and nonporous substrates. Impulse jet printers require high boiling point inks, usually water-based. Several recently commercialized or wax materials that are melted for ejection but solidify immediately on the receiver. Proponents claim that solid inks solve print quality problems from undesirable wicking of ink into paper fibers, although at the cost of considerable system complexity and embossed output, which is objectionable to some.

Ink formulation involves a series of tradeoffs, depending on type of printing system and the requirements of target applications. For instance, in impulse jet systems, where danger of nozzle clogging exists, inks must not dry within the nozzle; yet once ejected, they must dry fast enough on the paper to minimize feathering in the paper fibers. The difficulty of this undertaking is evident from the fact that few current ink jet printers offer true plain paper printing; most systems require special clay-coated papers to prevent wicking and this is a drawback in office environments.

Coloring agents in ink jet inks also present a challenge. Because pigment particles could make the ink too viscous and might cause clogging or undue wear, dyes are used instead. Resulting problems include lack of optical density (a grayish hue) and archival problems due to a tendency to fade.

The success of ink jet printing depends on developments that match systems' capabilities to printing requirements. Ink jet products have been highly successful for years in environments where speed and surface independence have been more important than image quality (e.g., for package coding and direct mail). Recent improvements in reliability, resolution, paper tolerance, and coloring agents could ultimately bring high quality office printing and color applications within reach.

Ink jet has been among the most challenging output technologies to perfect, with enormous R&D outlays and a long history of failed products. However, recent progress suggests that the allure of ink jet as a potentially elegant, low cost printing solution will finally be justified.

BIBLIOGRAPHY

"Advances in non-impact printing technologies for computer and office applications," in Proceedings from the First International Congress on Non-impact Printing, Venice, June 1981, Joseph Gaynor, Ed. New York: Van Nostrand Reinhold, 1982.

Ink Jet Printing—Its Role in Word Processing and Future Printer Markets. Newtonville, MA: Datek Information Services, August 1988.

Solid Ink Jet Color Imagining Technology. Newtonville, MA: Datek Information Services, August 1988.

18

Electrodeposition of Polymers

George E. F. Brewer

George E. F. Brewer Coating Consultants,
Birmingham, Michigan

1.0 INTRODUCTION

The electrodeposition of polymers is an extension of painting techniques into the field of plating and, like plating, is a dip coating process. The art of metal plating utilizes the fact that metal ions, usually Ni^{2+} or Cu^{2+}, can be discharged on the cathode to give well-adhering deposits of metallic nickel, copper, etc. The chemical process of deposition can be described as $1/2\ Me^{2+} + 1F$ (or 96,500 coulombs) of electrons gives $1/2\ Me^0$. In the case of electrodeposition of ionizable polymers, the deposition reaction is described as $R_3NH^+OH^- + 1F \rightarrow R_3N + H_2O$ or the conversion of water-dispersed, ammonium-type ions into ammonia-type, water-insoluble polymers known as cathodic deposition. Alternatively, a large number of installations utilize the anodic deposition process $RCOO^- + H^+$ less $1F \rightarrow RCOOH$. It should be mentioned that "R" symbolizes any of the widely used polymers (acrylics, epoxies, alkyds, etc.).

The electrodeposition process is defined as the utilization of "synthetic, water dispersed, electrodepositable macro-ions."

2.0 ADVANTAGES

Metal ions, typically $1/2\ Ni^{2+}$ show an electrical equivalent weight $1/2\ Ni^{2+}$ equal to approximately 29.5 g, while the polymeric ions typically used for electrodeposition exhibit a gram equivalent weight (GEW) of approximately 1600. Thus, $1F$ plates out of 30 g of nickel and deposits 1600 g of macroions. If we consider the thickness of the deposit, the advantage is even larger, since nickel has a specific gravity of 9, whereas pigmented organic coatings have a specific gravity of approximately 1.4.

Another advantage, and probably the reason for the rapid and still continuing industrial growth of the polymeric electrodeposition process, is the formation of uniformly thick coats on all surfaces of a formed workpiece, including such extreme recesses as the insides of car doors. This ability to extend coats into recesses is known as throwing power.

181

Still another advantage is the extremely small emission of vapors of volatile organic compounds (VOC), ranking electrodeposition with powder coating and radiation cure as the least polluting coating processes. Finally, the cost of operating an electrocoating tank is lower than that of any other painting method.

3.0 HISTORY

The earliest observation of migration in an electrical field was made in Moscow in 1808, by Reuss, who observed colloidal clay in water moving toward the anode. In 1861 Quinke in Berlin referred to migration observations made by Faraday and described the anodic migration of 30 substances in water and cathodic migration in turpentine. The first observation of an electrodeposition was made in London in 1905, by Picton and Linder, who found that ferric hydroxide in the presence of alcohol forms "hornlike" deposits on the cathode.

An electrophoretic separation of toxins and antitoxins was observed by Field and Teague, also in London, in 1907. The use of electrophoresis as an analytical tool culminated in 1948 when Tyselius received a Nobel prize.

The earliest industrial application for electrophoresis seems to have been made in 1919 when Davis, at General Electric, in Schenectady, New York, received a patent for the electrodeposition of bitumenous matter on wires. In 1923 Klein of London received a patent for the electrodeposition of rubber. A number of patents were granted after 1936 to Clayton, at Crosse & Blackwell, London for the deposition of waxes from emulsions. All of these processes utilized naturally occurring substances and none of them met with lasting success.

In 1965 Bogart, Burnside, and Brewer reported on the Ford electrocoating system which, by that time, had produced several million automotive wheels, hundreds of complete car bodies, and large numbers of samples of automotive, appliances, and general sample pieces.[1]

It is currently estimated that 1500 electrodeposition coating lines are in operation in the United States and that another 1500 lines operate overseas.

4.0 PROCESS

Spray painting and dip coating, use a carefully prepared and thinned batch of paint in a pot, agitated to ensure uniformity. Then the paint is transferred onto the workpiece until the pot is empty. Opposed to this, electrodeposition essentially transfers only the paint solids onto the workpiece. Thus, after the painting is done, the tank is still full of the aqueous portion of the paint, except for the paint bath increments that have been lifted out. These volumes of bath cannot be rinsed back into the tank by the use of water, because this would cause the tank to overflow. A fraction of the bath is, therefore, passed through an ultrafilter, which retains the paint solids, while the aqueous phase passes through filtration or permeation. The ultrafiltrate is then used to rinse the freshly coated workpieces and returns the lifted paint to the tank. For the same reason, paint of comparatively high concentration is used to replace the coated-out solids. The chemistry of the paint dispersion and deposition can be symbolized by the following equations:

$$RCOOH + BOH \xrightleftharpoons[\text{deposition}]{\text{solubilization}} RCOO^- + B^+ + H_2O \qquad (1)$$

anodic   solubilizer   dispersed   base
paint in-   external   anions
soluble

$$R_3N + HX \xrightleftharpoons[\text{deposition}]{\text{solubilization}} R_3NH^+ + X^- H_2O \qquad (2)$$

cathodic   typically   dispersed   acids
paint   lactic   cations
          acid

These two processes use bases or acids as external solubilizers, and the equations show that the bulk of the solubilizer stays in the bath. Thus, insufficiently solubilized paint has to be used as feed material, while the bath provides its excess solubilizer to balance the requirement.

A promising invention uses self-solubilizing cathodic materials called sulfonium bases.

$$\begin{array}{c} \circledR \\ R \\ R \end{array}\!\!\!\!> S^+ + OH^- \xrightarrow{\text{deposition}} \circledR H + \begin{array}{c} R \\ R \end{array}\!\!\!\!> SO$$

macro   paint   sulfoxide
cation

4.1 Throwing Power

Throwing power is measured as the depth of deposited paint in a standard cavity formed by test panels and narrow spacers. Indeed, most important feature of the electrocoating process is the ability to extend durable paint films into extreme recesses, for instance, automobile doors.

While throwing power is a property of the paint, it can be increased through higher applied voltage and higher bath conductivity (which helps transport electricity into recesses) lower coulomb per gram paint requirement (more efficient use of the electricity) larger openings for entrance into a cavity, and a larger perimeter of the opening. In other words, slits are more effective than round holes.

4.2 Maintaining a Steady State

Unlike other paint processes, the operation of an electrocoating tank, is similar to electroplating in that a balance of incoming and outgoing materials is necessary. Paint solids and evaporated water as well as other incidentally lost materials must be replaced. More precisely, the original tank fill is formulated to give one or more test pieces, meeting all the required solidities. For instance, a certain combination of a certain combination of resin, pigments, solvents, cross-linkers, and other materials gives the required deposited coating.

Yet the coating may contain these components in somewhat different proportions. Typically, a tank fill of 70 wt % resin plus 30 wt % pigment nay produce a deposited coating of 68 wt% resin plus 32 wt% pigment. If so, the paint replenishment (feed) must be richer in pigment than the original composition (fill). In general terms, the feed must replace all the materials leaving the tank. This is merely an analytical problem, but it can be cumbersome until solved.

4.3 Rupture Voltage

Of the factors that increase throwing power, increased voltage is the easiest to apply. There is, however, a limit called rupture voltage. It seems to be a dielectric breakthrough of the forming film and it depends on the surface of the metal being coated. At any rate, when voltages higher than the rupture voltage are applied, blemished (pockmarked) films and lower solidities (which reduce saltspray resistance, etc.) result.

5.0 EQUIPMENT

Figure 1 represents a typical electrocoating installation.

5.1 Conveyors

High volume production uses overhead conveyors entering the tank at a 30° angle. To save energy and space, free and power conveyors providing 90° vertical entry sometimes are combined with bulk–coating.

5.2 Metal Preparation

Most electrocoating installations use seven-to nine-stage zinc phosphate. Iron phosphate is also widely used. For nonferrous metals, special pretreatments are applied.

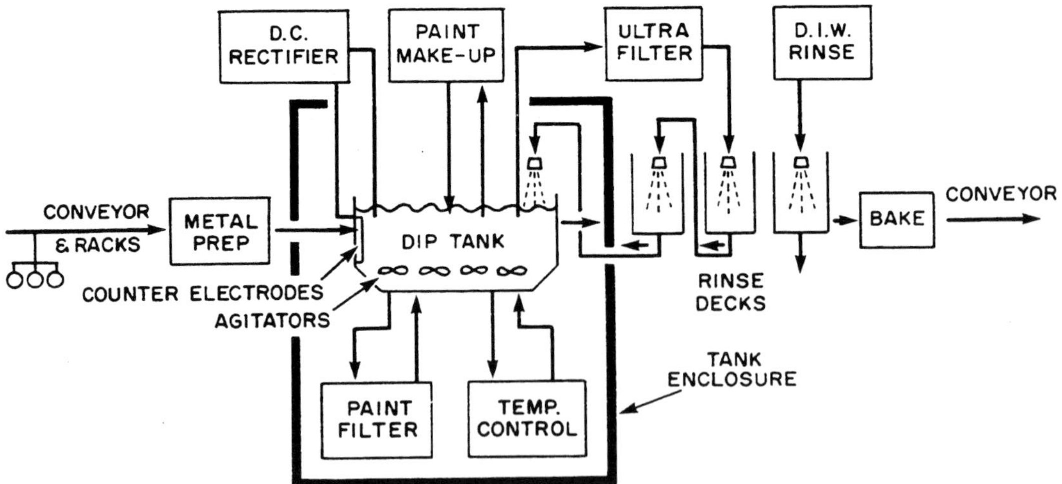

Figure 1   Schematic diagram of electrodeposition process; D.I.W. = deionized water.

5.3 Tank Enclosures

Safety is provided by a tank enclosure, which is interlocked with the power source and conveyor. Any entry de-energizes the system.

5.4 Dip Tanks

Tanks are operated on a continuous basis (conveyor or coil stock) or batch-type entry, in the latter case using "power and free" conveyors. The workplace is usually completely submerged 15 cm below the surface with 15 cm clearance on all sides. Tanks provide a workplace residence of approximately 3 minutes, though coating of cans and coils is done in seconds.

5.5 Rectifiers

Direct current power sources of less than 5% ripple factor over the full range are employed.

5.6 Counterelectrodes

Most electrocoating tanks are lined with an epoxy coat approximately 10 mils thick. In this case, the tank, the conveyor, the merchandise, and so on are all grounded: only the counterelectrodes, inserted below the surface of the paint, are of opposite (hot) polarity. The inserted electrodes are protected by plastic grates. If it is required to collect the counterions (acid or base groups), then the counterelectrodes are separated from the tank fill by membranes through which counterions pass by a process called electrodialysis. The tank itself can be used as counterelectrode unless separation of counterelectrode fluid is desired.

5.7 Agitation

Electrocoating baths of up to 500,000 liter volume contain from 10–20 wt % of solids and approximate the viscosity of water. Ejector nozzles and other agitating devices are used to move the entire bath volume in 6–30 minutes to prevent the solids from settling.

The dwell time in the bath is described as "turnover rate." For instance, the fill of an average tank of 100,000 liter volume contains about 10,000 kg of paint solids. One turnover is reached when 10,000 kg of solids (feed) has been added, which takes from 10 working days up to 10 months depending on the configuration of the merchandise and the production rate. At this time, by the law of probability, much of the original fill material still dwells in the tank. Considerable pumping stability is, therefore, required.

5.8 Temperature Control

Most tanks operate between 75° and 95°F. Practically all the required electrical and agitational energy is converted into heat and must be removed by use of chillers.

5.9 Ultrafilter

Paint savings and removal of dissolved impurities are accomplished by ultrafiltration, using membranes permeable to water, salts, and substances of less than 300 molecular weight. The ultrafiltrate (permeate) is used for usually three spray rinses and sometimes one dip rinse, followed by a deionized water rinse.

5.10 Paint Filters

Filter bags, wound filters, and indexing filters 5–50 μm pore size are used.

5.11 Paint Makeup

Symbolized in Equations 1 and 2, paint makeup is actually the last step in the manufacture of water-borne paints and is accomplished by mixing tank contents with all the needed paint components.

5.12 Deionized Water

Evaporated or otherwise lost water is replaced by the use of 75,000 Ω–cm of water, which is also used a final rinse.

5.13 Bake or Cure

Most electrocoats require a bake of 20 minutes at 350°F. However, lower baking materials are available; even ambient temperature curing materials are on the market.

6.0 LABORATORY

The tank control is carried out in about 3 hours per day by a technician, using the procedures found in the ASTM *Paint Testing Manual*.[2]

REFERENCES

1. *SAE J.*, pp. 81–83, August 1965.
2. G. E. F. Brewer and A. D. Hamilton, "Paint for electrocoating," ASTM Gardner-Sward *Paint Testing Manual*, 13th ed. Philadelphia: American Society for Testing and Materials; 1972, Section 8.10, pp. 486–489.

BIBLIOGRAPHY

1. Brewer, G. E. F., Ed. *Electrodeposition of Coatings*, ACS Advances in Chemistry Series, Vol. 119, Washington, DC: American Chemical Society, 1973.
2. Brewer, G. E. F., Chairman, "ACS Symposium on newer developments in electrocoating," *Org. Coatings Plast. Chem.*, 45 1–22, 92–113 (August 1981).
3. Chandler, R. H., "Advances in electrophoretic painting," *Bi– or Triannual Abstracts*. Braintree, Essex, England: R. H. Chandler, since 1966.
4. Duffy, J. I., Ed., *Electrodeposition Processes and Equipment*, Park Ridge, NJ: Noyes Data Corp., (1982).
5. Kardomenos, P. I., and J. D. Nordstrom, "Polymer compositions for catholic electrodeposition coatings," *J. Coating Technol.*, 54(686), 33–41 (March 1982).
6. Machu, W., *Handbook of Electropainting Technology*. (Electrochemical Publications, Ltd.,) Ayr, Scotland: 1978.
7. Raney, M. W., *Electrodeposition and Radiation Curing of Coatings*. Park Ridge, NJ: Noyes Data Corp., 1970.
8. Robinson, T., organizer of "Electrocoat" Conferences (even years, since 1982), Products Finishing, Cincinnati, OH.
9. Yeates, R. L., *Electropainting*. Teddington, England: Robert Draper Ltd., 1970.

19
Electroless Plating

A. Vaškelis

Lithuanian Academy of Sciences, Vilnius, Lithuania

1.0 INTRODUCTION

In electroless plating, metallic coatings are formed as a result of a chemical reaction between the reducing agent present in the solution and metal ions. The metallic phase that appears in such reactions may be obtained either in the bulk of the solution or as a precipitate in the form of a film on a solid surface. Localization of the chemical process on a particular surface requires that surface must serve as a catalyst. If the catalyst is a reduction product (metal) itself, autocatalysis is ensured, and in this case it is possible to deposit a coating, in principle, of unlimited thickness. Such autocatalytic reactions constitute the essence of practical processes of electroless plating. For this reason these plating processes are sometimes called autocatalytic.

Electroless plating may include metal plating techniques in which the metal is obtained as a result of the decomposition reaction of a particular compound; for example, aluminum coatings are deposited during decomposition of complex aluminum hydrides in organic solvents. However, such methods are rare and their practical significance is not great.

In a wider sense, electroless plating also includes other metal deposition processes from solutions in which an external electrical current is not used, such as immersion, and contact plating methods in which another more negative (active) metal is used as a reducing agent. However, such methods have a limited application; they are not suitable for metallization of dielectric materials, and the reactions taking place are not catalytic. Therefore, they usually are not classified as electroless plating.

Electroless plating now is widely used in modifying the surface of various materials, such as nonconductors, semiconductors, and metals. Among the methods of applying metallic coatings it is exceeded in volume only by electroplating techniques, and it is almost equal to vacuum metallization.

Electroless plating methods have some advantages over similar electrochemical methods. These are:

Table 1 Coatings obtained by electroless plating

Metal	$H_2PO_2^-$	N_2H_4	CH_2O	BH_4^-	RBH_3	Me ions	Others
Ni	Ni–P	Ni		Ni–B	Ni–B		
Co	Co–P	Co	Co	Co–B	Co–B		
Fe				Fe–B			
Cu	Cu	Cu	Cu	Cu	Cu	Cu	
Ag		Ag	Ag	Ag	Ag	Ag	Ag
Au		Au	Au	Au	Au		Au
Pd	Pd–P	Pd	Pd	Pd–B	Pd–B		
Rh		Rh					Rh
Ru				Ru			
Pt		Pt		Pt			Pt
Sn						Sn	
Pb			Pb				

1. Coatings may be deposited on electrically nonconductive materials (on almost any surfaces that are stable in electroless plating solutions).
2. Coatings have more uniform thickness irrespective of the shape of the product to be plated.
3. Deposition is simple—it is enough to immerse the (pretreated) product in the electroless plating solution.
4. It is possible to obtain coatings having unique mechanical, magnetic, and chemical properties.

Application of electroless plating, in comparison with electroplating techniques, is limited by the two factors: (a) it is more expensive, since the reducing agent costs more than an equivalent amount of electricity, and (b) it is less intensive, since the metal deposition rate is limited by metal ion reduction in the bulk of the solution.

2.0 PLATING SYSTEMS

To ensure chemical reduction of metal ions in a solution. The solution must contain a sufficiently strong and active reducing agent; that is it must have a sufficiently negative redox potential. The more easily the metal ions are reduced, the greater is the number of available reducing agents. Since only autocatalytic reduction reactions may be used successfully for deposition of coatings, the number of electroless plating Me–Red (metal–reducing agent) systems suitable for practice is not great (see Table 1).

Currently known electroless plating methods may be used to deposit 12 different metals; including metals belonging to the groups of iron, copper, and platinum (the well–known catalysts of various reactions) as well as tin and lead (only one solution has been published for deposition of the latter). Although deposition of chromium and cadmium coatings is described in the patent literature, autocatalytic reduction is not realized in these cases. Coatings are deposited on some metals by immersion plating only.

In some widely used processes, the deposition of metal is accompanied by precipitation of the reducing agent decomposition product—phosphorus and boron—and so the respective alloys are obtained. It is not difficult to deposit two or more metals at a time; electroless plating methods are known for deposition of more than 50 alloys of different qualitative composition, mostly based on nickel, cobalt, and copper.

The majority of reducing agents used in electroless plating are hydrogen compounds, in which H is linked to phosphorus, nitrogen, and carbon. It is in the reactions of these compounds that significant catalytic effects are possible, since in the absence of catalysts these reactions proceed slowly.

The most effective autocatalysis is obtained when the strongest reducer—hypophosphite—is used. In the absence of catalysts the reducer is inert and does not react even with the strong oxidants; only a few catalysts are suitable for it (e.g., Ni, Co, and Pd), but they provide for a catalytic process of the highest rate without reduction in the bulk of a solution. Other reducers are more versatile, for example, by using borohydride we may deposit coatings of almost all the metals mentioned. The reducing capacity of hydrogen compounds increases with an increase in pH of a solution. For this reason, the majority of electroless plating solutions are alkaline.

Such simple reducing agents as metal ions of variable valence (Fe^{2+}, Cr^{2+} and Ti^{3+}) usually are not suitable for deposition of coatings because noncatalytic reduction occurs rather easily. Recently, conditions have been established for autocatalytic deposition of tin and silver coatings using as reducing agents such metal complexes as $Sn(OH)_4^{2-}$ and $Co(NH_3)_6^{2-}$.

Depositions of some metals (ag, Au, Cu) by chemical reduction techniques was known as long ago as the nineteenth century, but it became popular after Brenner found (in 1945) a very efficient electroless nickel plating process using hypophosphite.[1] It was then that the term "electroless plating" was coined.

3.0 ELECTROLESS PLATING SOLUTIONS

The electroless plating solutions used in practice, in addition to the basic components (the salt of the metal to be deposited and a reducing agent). contain other substances as well. Usually, these are as follows.

1. Ligands, which form soluble complexes with metal ions, are necessary for alkaline solutions. Also, the use of stable complexes sometimes enhances the autocatalytic effect.
2. Substances controlling and maintaining a certain pH value of the solution: buffer additions are especially important, since in the course of metal reduction. Hydrogen ions are formed.
3. Stabilizers that decelerate reduction reaction in the bulk of a solution, hence enhance autocatalysis.

Sometimes, agents such as brighteners are also added to the solution.

The basic technological parameters of electroless plating solutions are discussed in Sections 3.1–3.4.

3.1 Deposition Rate

Deposition rate usually is expressed in micrometers per hour ($\mu m/h$; or mil/h, $\mu in./h$, mg/cm^2h). In the course of deposition, if the concentrations of reacting substances are not maintained at a constant level, this rate decreases. The values given in the literature are often averages, reflecting only the initial period. Such average rates depend on the ratio of the surface to be plated to the solution volume (dm²/liter).

The dependence of the deposition rate (v) on the concentration of reacting substances for a general case is rather complicated. It is often described by empirical equations, for example:

$$v = k \ [Me^{n+}]^a \ [Red]^b \ [H^+]^c \ [L]^d \tag{1}$$

where k is the rate constant (a constant value for a system of the given type) and $[L]$ is the concentration of a free ligand (not bound with metal ions in a complex). The exponents a and b are usually smaller than unity, while c is a negative value (in alkaline solutions OH⁻ ion concentration is used, and in such a case the exponent is often positive, $0 < c < 1$). Exponent d is usually close to zero; when the ligand is substituted, however, the deposition rate may change substantially. With constant concentrations of other solution components, the deposition rate decreases when the stability of a metal complex increases (when the concentration of free metal ions is lowered); however, this relationship for a general case is not rigorous.

The electroless deposition rate of most metals under suitable conditions is about 2–5 μm/h, and only electroless nickel plating rate may be as high as 20 μm/h (this corresponds to an electroplating process at current densities of 200 A/m²).

3.2 Solution Life

Solution life represents the maximum duration of solution usefulness. The beginning of metal ion reduction in the bulk of a solution may terminate its exploitation. In most modern electroless plating solutions, however, the reduction in the bulk usually does not occur under normal operating conditions, and the solution life is limited by the accumulation of reaction products or impurities. Thus, it is better to characterize the life of a solution not by time, which depends on the intensity of exploitation, but rather by the maximum amount of metal deposited from a volume unit of the solution (g/liter or μm/liter) or by turnover number showing how many times the initial amount of metal in the solution may be deposited in the form of a coating. This number may be as big as 10–20. After removal of undesirable substances accumulated in the solution, it may be used longer, just like electrolytes for electroplating.

After protracted exploitation of solutions, a certain amount of sediment may appear, since the bulk reaction may proceed on a limited scale even in fully stable solutions.

3.3 Reducing Agent Efficiency Factor

The amount of reducing agent (in moles or grams) that is consumed for deposition of a mole or gram of coating is indicated by the reducing agent efficiency factors. The required amount (according to the reduction reaction) of a reducing agent, which is equal, for example, to 2 moles for 1 mole of metal (nickel ion reduction by hyposphosphite or copper ion reduction by formaldehyde) is exceeded in real electroless plating processes as a result of the side reactions taking place.

3.4 Solution Sensitivity to Activation

The minimum amount of catalyst that must be present on the dielectric surface to initiate a reduction reaction is shown by the solution sensitivity to activation. This parameter is related to solution stability. The lower the stability of a solution, the easier is the initiation of a reaction, even on surfaces with low catalytic activity. A high sensitivity of a solution to activation is not always desirable, since metal from such solutions may be deposited even on surfaces that had not been activated; in such cases selective plating becomes impossible. When palladium compounds are used for activation, there should be no less than 0.01 and 0.03–0.05 µg of lead per square centimeter of a dielectric surface for nickel and copper plating, respectively. When silver is used as an activator, which is suitable only for some electroless copper plating solutions, it is necessary to have about 0.4 µg of silver per square centimeter.

4.0 PRACTICAL APPLICATIONS[2–4]

The applications of chemically deposited coatings may be divided into two groups. For decorative metallization of plastics, a thin (0.3–1.0 µm) layer of metal is chemically deposited on a dielectric surface, and its thickness is then increased by electroplating techniques. In this case, the properties of chemically deposited coatings and the nature of the metal are not of great significance; it is important only to ensure compactness and sufficient electrical conductivity of such coatings (a) for subsequent electroplating and (A) for providing the required adhesion of the metal layer. The metal for the chemically deposited underlayer is selected for process convenience and cost. For this purpose, nickel and copper coatings are used. Nickel is more convenient, since electroless nickel plating solutions are more stable and their compositions simpler than those of similar electroless copper plating solutions.

The adhesion of a coating to the nonconducting surface is essentially determined by the state of the surface, while the nature of the metal (at least for nickel and copper) usually has only a slight effect on adhesion. Copper coatings might be preferred because of their higher electrical conductivity. A copper underlayer is almost always used in the production of printed circuit boards.

Chemically deposited finished coatings, on the other hand, are thicker and their use is determined by their mechanical, electrical, and magnetic properties. The most popular are nickel (Ni–P and Ni–B) coatings deposited on metal products. Copper coatings 20–30 µm thick, deposited on plastics, exhibit good electrical conductivity and ductility and, therefore, are used in the production of printed circuit boards by additive processes. The entire circuit pattern is obtained by electroless techniques.

Coatings of cobalt and its alloys may be used to take advantage of their specific magnetic properties; silver and gold coatings are used because of their good electrical conductivity, optical properties, and inertness.

Electroless plating may be performed by using the plating solution once (until a greater part of any component in the solution is consumed and the reaction rate has sharply decreased) or by replenishing the substances that have been consumed in the course of plating. Long–term exploitation of solutions reduces the amount of plating wastes and ensures a higher labor productivity, but at the same time, it imposes more stringent requirements on plating solutions: they must be stable, and their parameters should not vary significantly with time. Besides, special equipment is required for monitoring and controlling the composition of such solutions. For this reason, long–term exploitation of solutions is applied only in large–scale production processes.

Single–use solutions are more versatile, but they are less economic and less efficient. A single–use method may be applied rather efficiently, however, when the solution has a simple composition and the basic components (first of all, metal ions) are fully consumed in the plating process, while the remaining components (such as ligands) are inexpensive and do not pollute the environment. In this case, single–use processes may be practically acceptable even in mass production.

An extreme case of single use of plating solutions is aerosol spray plating,[5] in which droplets of two solutions being sprayed by a special gun collide on, or close to, the surface being plated. One solution usually contains metal ions, while the other contains the reducing agent. Metal ion reduction in this case should be rapid enough to permit a greater part of the metal to precipitate on the surface before the solution film runs off it. This method is practical for deposition of such easily reducible metals as silver and gold, though such aerosol solutions are known for deposition of copper and nickel as well. The aerosol spray method is highly suitable for deposition of thin coatings on large, flat surfaces: this process is similar to spray painting.

Since the components of electroless plating solutions, first of all metal ions, may be toxic and pollute the environment, techniques have been developed for recovery of metals from spent plating solutions and rinse water. Other valuable solution components, such as ligands (EDTA, tartrate), may also be recovered.

Electroless plating usually does not require sophisticated equipment. The tank for keeping plating solutions must exhibit sufficient chemical inertness, and its lining should not catalyze deposition of metals. Such tanks are usually made of chemically stable plastics; metal tanks may be used as well—they can be made of stainless steel or titanium. To prevent possible deposition of metals on the walls, a sufficiently positive potential is applied to them using a special current source (anodic protection). Parts for plating may be mounted on racks; small parts may be placed in barrels immersed in the plating bath. Heating and filtration of solutions are carried out in the same way as in electroplating processes. Special automatic devices have been developed for monitoring and controlling the composition of plating solutions.

5.0 MECHANISMS OF AUTOCATALYTIC METAL ION REDUCTION

Autocatalytic metal ion reduction processes are highly complex: they contain many stages, and their mechanism is not understood in details. At present, it is possible to give an accurate description only of the basic stages of the catalytic process. Localization of the reduction reaction on the metal–catalyst surface (the cause of catalysis) is usually attributed to the requirement for a catalytic surface for one or more stages of the process to proceed. In accordance with one of the earlier explanations, only on a catalytic surface is an active intermediate product obtained, which then reduces metal ions. First, atomic hydrogen and, later, a negative hydrogen ionhydride—were considered to be such products. A reaction scheme with an intermediate hydride gives a good explanation of the relationships observed in nickel and cooper plating processes.[5] However, there is no direct proof that hydride ions are really formed during these processes. Moreover, the hydride theory explains only the reactions with strong hydrogen–containing reducers, which really may be H- donors.

A more versatile explanation of the causes of catalysis in these processes is based on electrochemical reactions. It is suggested that reducing agents are anodically oxidized on the catalyst surface and the electrons obtained are transferred to metal ions, which are cathodically reduced. The catalytic process comprises two simultaneous and mutually compensating electrochemical reactions. In this explanation of the catalytic process, electrons are the active intermediate product. However, electrons are fundamentally different from the conversational intermediate products of reactions. They may be easily transferred along the catalyst without transfer of the mass, and for this reason, the catalyst reaction, contrary to all other possible mechanisms (which are conventionally called "chemical mechanisms"), occurs not as a result of a direct contact between the reactants, or the reactants or the reactant and an intermediate substance, but because of the exchange of "anonymous" electrons via metal.

On the metal surface, when anodic oxidation of the reducer

$$\mathrm{Red} \rightarrow \mathrm{Ox} + n\mathrm{e} \tag{2}$$

and cathodic reduction of metal ions

$$\mathrm{Me}^{n+} + n\mathrm{e} \tag{3}$$

proceed simultaneously. A steady state in the catalytic system of electroless plating is obtained, in which the rates of both electrochemical reactions are equal, while;e the metal catalyst acquires a mixed potential E_m. The magnitude of this potential is between the equilibrium potentials E_c of the reducer and of the metal. The specific value E_m depends on the kinetic parameters of these two electrochemical reactions.

Electrochemical studies of catalytic metal deposition reactions have shown that the electrochemical mechanism is realized practically in all the systems of electroless plating.[4,6,7]

At the same time, it has become clear that the process is often not so simple. It appears that anodic and cathodic reactions occurring simultaneously often do not remain kinetically independent but affect each other. For example, copper ion reduction increases along with anodic oxidation of formaldehyde.[8] The cathodic reduction of nickel ions and the anodic oxidation of hypophosphite in electroless nickel plating solutions are faster than in the case in which these electrochemical reactions occur separately. This interaction of electrochemical reactions probably is related to the changes in the state of the metal–catalyst surface.

Electrochemical reactions may also hinder each other: for example, in reducing silver ions by hydrazine form cyanide solutions, their rate is lower than is separate Ag–Ag(1) and redox systems.

The electrochemical nature of most of the autocatalytic processes discussed enables us to apply electrochemical methods to their investigation. Only they must be applied to the entire system of electroless plating, without separating the anodic and cathodic processes in space. One suitable method is based on the measurement of polarization resistance. It can provide information on the mechanism of the process and may be used for measuring the metal deposition rate (both in laboratory and in industry).[9] The polarization resistance R_p is inversely proportional to the process rate i:

$$i = \frac{b_a \, b_c}{R_p(b_a + b_c)} \tag{4}$$

$$R_p = \left(\frac{dE}{di}\right)_{i=0} \tag{5}$$

where b_a and b_c are Tafel equation coefficients ($b = 1/\alpha nf$), α is the transfer coefficient, n is the number of electrons taking part in the reaction for one molecule of reactant, and $f = F/RT$ (F = Faraday number).

Autocatalytic metal reduction reactions also may not proceed in an electrochemical manner. Two courses of such reactions have been shown: (a) an intermediate metal hydride is formed, which decomposes to metal and hydrogen (reduction of copper ions by borohydride and, (b) the metal complex is hydrolyzed, resulting in precipitation of metal oxide on the surface, which then is reduced to metal by the reducer present in the solution (reduction of silver ions by tartrate).

6.0 STABILITY OF PLATING SOLUTIONS

Electroless plating solutions containing metal ions and reducing agents thermodynamically unstable systems. Metal ion reduction must proceed in the bulk of the solution.

The difference in the rate of metal ion reduction on the required surface (controlled catalytic reaction) and that of a reduction reaction in the bulk of a solution shows the effect of catalysis, and it determines, to a substantial degree, the practical usefulness of plating solutions. In an ideal case, the reaction in the bulk of a solution should not occur at all.

Formation of metal in the bulk of a solution is hindered by energy barriers: the activation barrier of homogenous reactions between metal ions and reducer and the barrier of the formation of a new phase (metal). The magnitude of the second barrier may be evaluated on the basis of thermodynamic principles.[10]

It was established empirically that the stability of plating solutions decreases with an increase in the concentration of reactants and temperature, with a decrease in the stability of metal ion complexes, and with the presence of solid foreign particles in the solution. Besides, it was found that stability decreases as the catalytic process rate and load increase. This may be attributed to the transfer of intermediate catalytic reaction products from the catalytic surface to the solution, where they may initiate a reduction reaction. To enhance the stability of solutions, it is recommended that lower concentration solutions and more stable metal complexes be used and that solid particles in the solution be removed by filtration. The most effective solution stabilization method is the introduction of special addition agents—that is, stabilizers.[4,11] Stabilizers, the number of which is very great, may be divided into two large groups: (a) catalytic poisons, such as S(II), Se(II) compounds, cyanides, heterocyclic compounds with nitrogen and sulfur, and some metal ions, and (b) oxidizers. It is assumed that stabilizers hinder the growth of fine metal particles, close to critical ones, by absorbing on them (catalytic poison) or passivating them (oxidizers).

Modern electroless plating solutions always contain stabilizers. Their concentration may be within the range of 1–100 mg/liter. Stabilizers, by hindering deposition of metal on fine particles, usually slow down the rate of the catalytic process on the surface being plated.

Table 2 Examples of Electroless Copper Plating Solutions

Components (g/liter) and parameters	Solutions		
	A	B	C
$CuSO_4 \cdot 5H_2O$	7	15	15
K–Na tartrate	25		
Na_2EDTA		30	45
NaOH	4.5	10	10
Formaldehyde (40%) ml/liter	25	20	10
Additives[a]	2	1–0.005	1–0.03
		2–0.03	2–0.05
pH	12.2–12.5	12.7	12.6
Temperature, °C	20	20	70
Deposition rate μm/h	0.4–0.5	2	3

[a]Solution A: $NiCl_2 \cdot 6H_2O$; solution B: sodium diethyldithiocarbamate, $K_4Fe(CN)_6$; solution C: 2.2′ – dipyridyl, polyethylene glycol (MW = 600).

This process may stop completely at a sufficiently high concentration of the stabilizers. In some cases, however, small amounts of stabilizers increase the deposition rate.

7.0 ELECTROLESS PLATING

7.1 Copper Deposition

Though copper coatings may be deposited using various reducers, only formaldehyde copper plating solutions are of practical importance. Autocatalytic reduction of copper ions by formaldehyde proceeds at room temperature in alkaline solutions (pH = 11–14); here copper ions must be bound into a complex. Suitable Cu^{2+} ligands for electroless copper plating solutions are polyhydroxy compounds (polyhydroxy alcohols, hydroxyacid anions) and compounds having a tertiary amine group and hydroxy groups (hydroxyamines, EDTA, and others). In practice, tartrate, EDTA, and tetraoxypropylethyl ethylenediamine (Quadrol) are used most often.

In the course of copper plating, along with the main reduction reaction

$$Cu^2 + 2CH_2O + 4OH^- \rightarrow Cu + 2HCOO^- + H_2 + 2H_2O \tag{6}$$

formaldehyde is consumed in the Cannizzaro reaction, and a total of 3–6 moles of CH_2O is consumed for the deposition of 1 mole of copper. During copper plating, much alkali is used including the Cannizzaro reaction. Consumption of OH– may be determined according to the equation (amounts of substances in moles):

$$\Delta OH^- = 3 \, \Delta \, Cu(II) + 1/2 \, \Delta \, CH_2O \tag{7}$$

Various formulations of copper plating solutions, which are totally stable and suitable for long exploitation (e.g., solution B in Table 2), have been developed. Three types of electroless copper plating solution have been distinguished in the literature: (a) low deposition rate solutions (0.5–1.0 μm/h), suitable for deposition of a copper underlayer; (b) solutions giving deposition rates of 4–5 μm/h (i.e., exhibiting a higher autocatalytic effect); and (c) solutions for deposition of highly ductile and strong copper coats (e.g., solution C in Table 2). All these solutions, essentially, have the same composition: they differ mostly by their additives. Besides, highly ductile coatings, which are used in the production of printed circuit boards by additives processes, are obtained at higher temperatures (> 40°C) and at a relatively low copper deposition rate.

7.2 Nickel Plating

Electroless nickel plating, in which hypophosphite is used as a reducer, is the most popular process.[12,13] Auto catalytic nickel ion reduction by hypophosphite occurs both in acid and in alkaline solutions. In a stable solution with a high coating quality, the deposition rate may be as high as 20–25 μm/h. This requires, however, a relatively high temperature, about 90°C. Since hydrogen ions are formed in the reduction reaction

$$Ni^{2+} + 2H_2PO_2^- + 2H_2O \rightarrow Ni + 2H_2PO_3^- + H_2 + 2H^+ \qquad (8)$$

a high buffering capacity of the solution is necessary to ensure a steady–state process. For this reason, acetate, citrate, propionate, glycolate, lactate, or aminoacetate is added to the solutions; these substances, along with buffering, may form complexes with nickel ions. Binding Ni^{2+} ions into a complex is required in alkaline solutions (here, besides citrate and aminoacetate, ammonia and pyrophosphate may be added); moreover, such binding is desirable in acid solutions because free nickel ions form a compound with the reaction product (i.e., phosphate), which precipitates and hinders further use of the solution.

Stabilizing additions for nickel plating solutions are less necessary than for copper solutions; nevertheless, they are added to ensure the stability of long–lived solutions.

Phosphorus is always present in the coatings when reduction is performed by hypophosphite. Its amount (in the range of 2–15 mass percent) depends on pH, buffering capacity, ligands, and other parameters of electroless solutions.

Borohydride and its derivatives may also be used as reducers for electroless nickel plating solutions. While temperatures of 60–90°C are required for the reduction of nickel ions by borohydride, dimethylaminoborane (DMAB) enables the deposition of Ni–B coatings with a small amount of boron (0.5–1.0 mass percent) at temperatures in the range of 30–40°C. Neutral and alkaline solutions may be used, and their composition is similar to that of hypophosphite solutions (Table 3).

7.3 Cobalt, Iron, and Tin Plating

Deposition of cobalt is very similar to that of nickel—the same reducers (hypophosphite, borohydride, and its derivatives) are used, and reduction relationships are similar.[14] Reduction of cobalt is more difficult, however, and cobalt deposition rates are lower than those of nickel; it should be noted that it is difficult to deposit cobalt from acid solutions. The Co–P and C0–B coatings obtained are of particular interest due to their magnetic properties.

Electroless iron plating is more difficult, and only one sufficiently effective iron plating solution is known, in which Fe ions form a complex with tartrate and $NaBH_4$ is used as a reducer. Fe–B coatings (about 6% B) are obtained in an alkaline solution (pH 12) at a temperature of 40°C and deposition rates of about 2 μm/h.

It is rather difficult to realize an autocatalytic tin deposition process. A sufficiency effective tin deposition method is based on the tin(II) disproportionation reaction in an alkaline medium.[15] In 1–5M NaOH solutions at 80–90 °C, it is possible to obtain a deposition rate of a few micrometers per hour.

7.4 Deposition of Precious Metals

Electroless silver plating is the oldest electroless metallization process; its present performance however, lags behind nickel or copper plating.[1] Unstable single–use ammonia silver plating solutions (with glucose, tartrate, formaldehyde, etc., as reducers) are usually employed. The thickness of coatings from such solutions is not great (< 1 μm). Such unstable solutions are more suitable for aerosol spray.

More effective electroless silver plating solutions have been developed using cyanide Ag(I) complex and aminoboranes or hydrazine as reducers: at temperatures of 40–50°C, the deposition rate is 3–4 μm/h, and in the presence of stabilizers these solutions are quite stable. Sufficiently stable electroless silver plating solutions may be obtained using metal ions such as Co(II) compounds as reducers.

Gold coatings may be deposited employing various reducers: however, the solutions are usually unstable. Solutions of sufficient stability have been developed with borohydride or DMAB as reducers using a stable gold cyanide complex.[16] At temperatures of 70–80 °C, the gold deposition rate reaches 5 Am/h and gold coatings of sufficient purity are obtained.

Thin gold coatings may be deposited on plastics by an aerosol spray method: gold complexes with amines are employed with hydrazine as a reducer, and a relatively thick coat (deposition rate as high as 0.4 μm/min) may be obtained.

Palladium coatings are easily deposited with hypophosphite as a reducer in alkaline solutions, in which Pd^{2+} ions are bound in a complex with ammonia, EDTA, or ethylenediamine. Palladium plating is performed at 40–50 °C, the deposition rate of the Pd–P (4–8 P) coat being in the range of 2–5 μm/h.

Coatings of platinum, ruthenium, and rhodium may be deposited using borohydride or hydrazine as a reducing agent. The process rate in a stable solution is low (0.5–2 μm/h).

Table 3 Examples of Electroless Nickel Plating Solutions

Components (g/liter) and parameters	Solutions		
	A	B	C
NiCl$_2$ · 6H$_2$O	30	30	25
NaH$_2$PO$_2$ · H$_2$O	10	20	30–40
Sodium acetate	8		
NH$_4$Cl			30
NH$_4$OH (25%), ml/liter			30–35
Glycine		20	
NaNO$_2$			0.02–0.1
pH	5	6	9
Temperature, °C	90	80–90	30
Nickel deposition rate μm/h	15	7–15	1.8

7.5 Deposition of Metal Alloys

About 60 coatings of a different qualitative composition containing two or more metals may be deposited. Such metals as copper, iron, zinc, tin, rhenium, tungsten, molybdenum, manganese, thallium, and platinum group metals may be introduced into nickel and cobalt coats, and nickel, cobalt, tin, zinc, cadmium, antimony, bismuth, lead, and gold into copper coats.

In the electroless deposition of metal alloys, the same thermodynamic relationships as those of alloy deposition by electroplating techniques are valid; it is clear that it is difficult to introduce into coatings metals that are difficult to reduce, such as chromium and manganese. Besides, in the case of chemical reduction, an additional factor—catalytic properties of metals—becomes apparent. Great amounts of additional metal may be introduced into a coat of nickel, copper, and so on, only when that metal is catalytic or, at least, inert with respect to oxidation of the reducer. The amount of metals–catalysts in the alloy may be as high as 100%, that of catalytically inert metals up to 50%, and the amount of metals–inhibitors may be only 10–20%. When a less catalytically active metal is introduced, the deposition rate decreases.

8.0 PROPERTIES OF CHEMICALLY DEPOSITED METAL COATINGS

Only in rare cases are chemically deposited metal coatings so pure, and have so regular a structure, that their properties are the same as those of the corresponding chemically pure substance. Very different properties may be exhibited by coatings containing a nonmetallic component—phosphorus or boron.

The density of coatings is a little lower than that of bulk metal. This is related to a rather irregular coatings structure: they contain more defects (pores and inclusions of foreign matter). For example, chemically deposited copper usually has a great number of microscopic voids 20–300 Å in diameter, formed by the hydrogen occluded in the coating. Ni–P and Ni–B coatings usually have a layered structure, which results from the nonuniform distribution of phosphorus and boron in the coatings.

Mechanical properties of the coatings may vary within a wide range depending on the electroless plating conditions, plating solutions composition, and deposition rate.

For chemically deposited finish copper coatings, such as those on printed circuit boards, sufficient resistance and ductility are of great importance. Coatings that have a tensile strength of about 40–50 kg/mm^2 may be obtained at a temperature of 50–70°C. Their ultimate elongation, which characterizes ductility, may be as high as 6–8%. Copper coatings obtained at room temperature are more brittle. Highly ductile coatings may be obtained only from solutions containing special additives. Ductility increases when deposited coatings are heated in an inert atmosphere at temperatures of 300–500°C.

Ni–P and Ni–B coatings are relatively hard; after deposition, their hardness, which depends on the amount of P and B, is 350–600 kg/mm^2 (3400–5900 MPa) of Ni–P coatings and 500–750 kg/mm^2 (4900–7400 MPa) for Ni–B coatings, while after heating at about 400°C, it is 800–1000 kg/mm^2 for Ni–P and 1000–12500 kg/mm^2 for Ni–B. Hence, such coatings have the same hardness as that of chromium coatings. The tensile strength of Ni–P coatings is in the range of 40–80 kg/mm^2.

The ductility of nickel coatings, if their hardness is taken into account, is rather high: their ultimate elongation is less than 2%. Such a combination of hardness, wear resistance, and ductility is unique.

The electrical conductivity of chemically deposited coatings is usually lower than of the respective pure metals. Resistivity of thin copper coatings (0.5–1.0 μm) deposited at room temperature is 3–4 × 1.0^{-8} Ω·m—twice as great as that of pure bulky copper. The surface resistance of such coatings is 0.03–0.07 Ω/□. However, ductile copper coatings that are obtained at temperatures of 50–70°C have a resistivity of 2 × 10^{-8} Ω·m, close to that of pure copper.

The resistivity of Ni–P and Ni–B coatings depends on the amount of nonmetallic component, and it is usually in the range 3–9 × 10^{-7} Ω·m; that is much higher than that of pure bulky nickel (0.69 × 10^{-7} Ω·m). Heating causes reduction in resistivity.

Magnetic properties of the coatings of such ferromagnetic materials as nickel and cobalt may vary within a very wide range. With an increase in the amount of phosphorus in nickel coatings, their ferromagnetism decreases, and coatings containing more than 8 mass percent of phosphorus or 6.5 mass percent of boron are nonmagnetic.

Coatings of Co–P, Co–B, and cobalt alloys with other metals have highly different magnetic properties. These depend on the composition of the coatings, their structure, and thickness, and they may be controlled by changing the composition, pH, and temperature of electroless plating solutions. Usually, cobalt coatings exhibit a high coercivity (15–80 kA/m); however, soft magnetic coatings (0.1–1.0 kA/m) may be deposited as well.

Optical properties of coatings are less varied and do not differ so much from those of pure metals. Chemically deposited coatings are usually dull; when special additives are introduced, bright coatings are obtained. Since they are not used as finish decorative coatings, properties of appearance and brightness usually are not essential.

Silver and gold coatings are often used as mirrors, but the light–reflecting surface is usually the inner surface, which is adjacent to the smooth glass surface. Chemically deposited thin gold films are employed as optical filters; they pass visible light but reflect infrared rays and radio waves.

Chemically deposited coatings are usually less porous than the respective electroplates; therefore, they provide better protection of the basis metal against corrosion. Corrosion resistance of the coatings themselves may be different depending on structure and composition. NI–P and Ni–B coatings are more resistant to corrosion than nickel electroplates; this may be due to their fine crystalline structure.

REFERENCES

1. A. Brenner, and G. Riddell, *J. Res. Natl. Bur. Stand.*, 37, 31 (1946).
2. W. Goldie, *Metallic Coating of Plastics* Hatch End, Middlesex, England: Electrochemical Publications Ltd., Vol. 1, 1968; Vol. 2, 1969.
3. F. Pearlstein, "Electroless plating," *Modern Electroplating*, 3rd ed., F. A. Lowenheim, Ed. New York; 1974, p. 710.
4. M. Šalkauskas, and A. Vaškelis, *Khimicheskaya Metallizatsiya Plastmass.* Leningrad: Khimiya, 1984.
5. R. M. Lukes, *Plating*, 51, 969, 1066 (1964).
6. M. Paunovic, *Plating*, 55, 1161 (1968).
7. F. M. Donahue, *Oberfläche–Surface*, 13, 301 (1972).
8. A. Vaškelis, and J. Jačiauskiene, *Elektrokhimiya*, 17, 816 (1981).
9. I. Ohno, and S. Haruyama, *Surface Technol.*, 13, 1 (1981).
10. A. Vaškelis, *Elektrokhimiya*, 14, 1970 (1978).
11. E. B. Saubestre, *Plating*, 59, 563 (1972).

12. K. M. Gorbunova, and A. A. Nikiforova, *Physicochemical Principles of Nickel Plating*, translated from Russian, 1963, TT 63– 11003.
13. G. Gawrilov, *Chemische/Stromlöse/Vernickelung*. Saulgau, Wurt.: Eugen Leuze Verlag, 1974.
14. K. M. Gorbunova et al., *Fiziko–Khimicheskiye Osnovy Processa Khimicheskogo Kobaltirovaniya*, Moscow: Nauka, 1974.
15. A. Molenaar, and J. J. C. Coumans, *Surface Technol.*, 16, 265 (1982).
16. Y. Okinaka, *Gold Plating Technology*, H. Reid and W. Goldie, Eds. Hatch End, Middlesex, England: Electrochemical Publications, Ltd., 1974, p.82.

20

The Electrolizing Thin, Dense, Chromium Process

Michael O'Mary

The Armoloy Corporation, DeKalb, Illinois

1.0 GENERAL DEFINITION

The electrolizing process uniformly deposits a dense, high chromium, nonmagnetic alloy on the surface of the basic metal being treated. The alloy used in Electrolizing provides an unusual combination of bearing properties: remarkable wear resistance, an extremely low coefficient of friction, smooth sliding properties, excellent antiseizure characteristics, and beneficial corrosion resistance. Electrolized parts perform better and last up to 10 times longer than untreated ones.

The solution and application processes are carefully monitored at all Electrolizing facilities. The result is a fine-grained chromium coating that is very hard, thin, and dense and has absolute adhesive qualities. The Electrolizing process deposits a 99% chromium coating on the basis metallic surfaces, whereas normal conventional chromium plating processes tend to deposit 82–88% chromium in most applications.

Electrolizing calls for the cleaning and removal of the matrix on the basis metal's surface by multicleaning process, using a modified electrocoating process that causes the chromium metallic elements of the solution to bond to the surface porosity of the basis metal. It is during this process that the absolute adhesive characteristics and qualities of Electrolizing are generated. The Electrolizing coating will not flake, chip, or peel off the basis metal substrate when conventional ASTM bend tests and impact tests are performed.

Three basic factors are always present after applying Electrolizing to metal surfaces:

- Increased wear (Rockwell surface hardness of 70–72 R_c)
- Added lubricity characteristics
- Excellent corrosion resistance

2.0 APPLICATIONS

2.1 General

Electrolizing can meet a variety of engineering needs: chrome plating according to specifications, flash chrome plating, repair or salvage work, and heavy chrome application before grind. Electrolizing increases wear resistance, reduces friction, prevents galling and seizing, minimizes fretting corrosion, offers resistance to erosion, and provides corrosion resistance. It can be applied to all commonly machined ferrous and nonferrous metals, including aluminum, titanium, stainless steels, coppers, brass, and bronze. The coating is not recommended for magnesium, beryllium, columbium, lead, and their respective alloys.

On aluminum, Electrolizing increases wear life and surface strength, reduces oxidation and corrosion, enhances appearance, and prevents galling, seizing, and erosion. Electrolizing is also conductive, whereas other treatments to aluminum are nonconductive. The result with Electrolizing is no static buildup.

Because it is a thin, dense coating, Electrolizing exhibits its best wear and lubricity properties on hardened surfaces. It is most effective when the basis metal is 40 Rc or harder. In severe wear applications, a basis metal should be hardened to the 50–62 Rc range before Electrolizing is performed. Electrolizing will improve performance on any basis metal, but it is not a substitute for heat treating.

Table 1 lists appropriate uses for Electrolizing at various basis metal Rockwell hardness ranges.

2.2 Specific

Electrolizing is used across a variety of industries for a multitude of purposes. With Electrolizing, dies and molds for both rubber and plastics have better release characteristics and reduced wear (especially when abrasive materials are involved); cutting tolls experience longer wear life; nuclear components exhibit better antigalling and corrosion resistance properties; dies used in stamping, drawing, forming, and blanking have sharper cuts and increases in work life; engine and transmission parts (such a valves, valve guides, pis-

Table 1 Uses for Electrolizing

Basis metal hardness range (R_c)	Application	Thickness of Electrolizing recommended (in.)
18–30	Electrolizing will handle low-loaded or stress conditions and provide basic corrosion resistance.	0.0001
30–40	Electrolizing increases wear resistance where the metal will basically support the wear factor. Corrosion resistance is good in this range.	0.0007–0.0009
40–50	Corrosion resistance increases as basis metal gets harder Wear resistance properties begin to improve greatly.	0.0003–0.0005
≥ 50	This is the most common and recommended range for an Electrolizing application. Corrosion resistance is superior, and the maximum wear resistance benefits exhibited.	0.00005–0.0003

tons, gears, and splines) are protected without detrimental tolerance changes or variations; standard pumps and meters handle corrosive liquids and materials much better; and bearing and bearing surfaces can run longer and cooler and are superior to a 400 series stainless steel for corrosion protection. In fact, Electrolizing makes it possible to use standard ferrous steels in place of stainless steel in many applications, including food processing, medical environments, and ball or roller bearing applications.

Other applications include drive transmissions, power transmissions, gears, molds, screws, sleeves, threaded parts, and valves. In the automotive industry, applications include not only engine parts, but also large metal form dies for auto body parts. Electrolizing has been utilized on applications in these and other industries:

aerospace	gauges and measuring equipment
aircraft and missiles	pharmaceutical
armament	photography (motion and still)
automotive	medical instruments
business machines	metalworking
cameras and projectors	molds (plastic and rubber)
computers	motor industry
cryogenics	nuclear energy
data processing	refrigeration
electronics	textile industry
food processing	transportation

Specifically, Electrolizing is approved and meets the following aerospace, nuclear, and commercial specifications:

Air Research Company, Garrett, CO
American Can Company
AMS 2406
AMS 2438
AVCO Lycoming—AMS 2406
Bell Helicopter
Bendix Company
   Utica, NY, division
   Teterboro, NJ, division
   Kansas City, MO, division
   South Bend, IN, division
Boeing
   BAC 5709 Class II, Class IV
   QQC 320
Cleveland Pneumatic Tool-CPC Specs (Chromium), QQC320
Colt Industries,
   Menasco, TX, division
DuPont
Fairchild Camera
Fairchild Republic
General Dynamics
General Electric
   Lynn, MA

Cincinnati, OH (aircraft)
Wilmington, MA
Wilmington, NC (nuclear)
Fitchburg, MA
Gillette Company, Boston
Grumman Aircraft
IBM, 40 to 45
Johnson & Johnson, New Jersey
Kaman Aircraft
    QQ–C–320
    AMS 2438
McDonnell/Douglas
    PS 13102
    QQC320
MIL–C–23422
Nabisco Company
Ozone Industries
Perkin Elmer Company, most divisions
Pratt & Whitney
    QQC320
    PWA 48
    AMS 2406
Procter & Gamble, Cincinnati, OH
QQ–C–320 B Class II
Raytheon
U.S. Navy
    Newport News, Electric Boat, Portsmouth Naval Shipyard
    Hamilton Standard
    HS 332, HS246
    QQ–C–320B
Western Gear Company
Western Electric Company
Westinghouse

3.0 SURFACE PREPARATION

Substrate surfaces must be free of oil, grease, oxides, and sulfides. Parts should be surface finished before shipment to an Electrolizing facility, where they will undergo further cleaning and surface preparation through a multicleaning process. Surfaces will no be changed significantly in configuration by the Electrolizing process.

A clean surface is very important. When conventional chrome is utilized, component failure is often erroneously attributed to "spalling." The effect is actually due to a residual contaminant, which was covered or surrounded by electrodeposited chrome and subsequently dislodged as a result of sliding or other mechanical action. This leaves a void in the chrome plating which, in turn, causes further deterioration of the surface.

The surface must be free from scale and soils to secure the best corrosion resistance. Any scale left on the surface will gradually acquire a rusty appearance and act as a nucleus for additional rust to form. On articles that are to have a "stainless" appearance, every trace of scale must be removed by grinding, pickling, or polishing.

The solution of Electrolizing, Inc., is to use a multicleaning procedure. Multicleaning is not a trade name, nor is it a novel, or supercleaning, technique. It is merely a designation for a carefully planned cleaning program prescribed for each individual part, utilizing every possible cleaning method available—such as vapor degreasing; solvent, detergent, diphase and alkaline cleaning; dry vapor honing; electrolytic and ultrasonic cleaning; vibratory cleaning; and hydroblasting. Maintained throughout are high level quality control standards, consisting of microscopic before-and-after inspection of each part, intermittent inspection between processing phases, and disciplined handling procedures.

The significant surfaces to be coated should be completely finished prior to processing. For best results, the surface finish should be 32 RMS or better. The finer the surface, the better it is to Electrolize. This is true to a finish of 2–4 RMS or better. Surface finishes that are received by Electrolizing with 4 RMS or less may show some slight roughing of the surface after application. In most cases these finishes can be lapped or polished back to their original condition. All surfaces should be free from plated coatings. Parts that have been nitrided should be machined or vapor-blasted before processing. The Electrolizing coating will not adhere to brazed or welded areas unless these are machined or vapor-blasted to remove all impurities before processing. Unless otherwise specified, the coating should be applied after all basis metal processing has been completed. This includes heat treatment, stress relieving (when required), machining, brazing, welding, and forming. The coating should be applied directly to the basis metal without any intermediate coating.

Electrolizing is a low temperature process. The temperature of parts during the cleaning and coating process does not exceed 180°F. There is no distortion or annealing of the basis metal.

Generally, parts should be less than 20 ft. long, 30 in. in diameter, and 4000 lb in weight. Parts exceeding these limits should be discussed with Electrolizing, Inc.

4.0 SOLUTION

Electrolizing is a blend of some of the best chrome salts and selected proprietary catalysts and additives that create its unique proprietary features. The solution is carefully monitored and its quality maintained through the use of exacting, sophisticated equipment and laboratory controls. The materials are distributed only of accredited licensees within the Electrolizing licensee program.

5.0 PROPERTIES

5.1 Thickness

Electrolizing is applied in a very thin, dense layer on the basis metal. The Electrolizing process does not create a buildup on corners or sharp edges; it conforms exactly to the surface. There is no change in either the conductivity or magnetic properties of the basis metal.

Electrolizing deposits range from 0.000010 to 0.003 in. per side. The practical coating range is from 0.000025 to 0.001 in. The average deposit is 0.0004–0.0008 in. per surface. Thickness tolerances of ±0.000010 to ±0.000050 in. can be maintained, depending on thickness specified and the quality level of the part. Coating thickness that may exceed 0.001 in. can be applied, and tolerances can be maintained to eliminate post-Electrolize grinding op-

Table 2 Thickness Ranges for Most Metals

Coating/side (in.)	Tolerance (in.)
0.000050″	± 0.000010
0.0001	± 0.000020
0.0002	± 0.000050
0.0003	± 0.000050
0.0004	± 0.000050
0.0005	± 0.000050
0.0006	± 0.000075
0.0007	± 0.0001
0.0008	± 0.0001
0.0009	± 0.0001
0.001	± 0.0002

erations. Tables 2 and 3 list thickness ranges and tolerances for most metals. It is recommended that the coating thickness and tolerance be established by consulting with Electrolizing Inc.

Electrolizing thickness is always in direct relation to the basis metal and will vary from one basis metal to another. However, the thickness established for each will remain constant and predictable. In all applications calling for a close inspection to monitor deposit thickness, Electrolizing, Inc., recommends the use of a exacting reverse etchant method of ascertaining deposit thickness. This method is nondestructive to the basis metal. If necessary, destructive microphotos will verify deposit thickness and consistency.

5.2 Adhesion

Electrolizing has outstanding adhesive characteristics. The adhesion of the coating is such that when examined at four diameters, it will not show separation from the basis metal on test specimens bent repeatedly through an angle of 180°, on a diameter equal to the thickness of the specimen, until fractured.

Table 3 Thickness Ranges for Aluminum

Coating/side (in.)	Tolerance (in.)
0.0001	± 0.000050
0.0002	± 0.000050
0.0003	± 0.0001
0.0004	± 0.0001
0.0005	± 0.0002
0.0006	± 0.0002
0.0007	± 0.0002
0.0008	± 0.0003
0.0009	± 0.0003
0.001	± 0.0003

The Electrolizing coating forms a lasting bond with the surface by permeating the surface porosity of the basis metal, but it can be removed by licensed Electrolizing plants without detrimental effects to the basis metals.

5.3 Corrosion

Electrolizing resists attack by most organic and inorganic compounds (except sulfuric and hydrochloric acids). The Electrolizing coating is typically more noble than the substrate; therefore, ti protects against corrosion by being free of pores, cracks, and discontinuities, and by providing a uniform structure and chemical composition. Porosity, hardness, and imperfect surface finishes of basis metals will affect the corrosion-resistant properties of Electrolizing; however, all basis metals that are Electrolize coated will have enhanced corrosion-resistant characteristics.

Samples can be subjected to standard ASTM B–117 and B–287 salt spray tests. The Electrolizing process also meets the following specifications: QQ–C–320, AMS–2406, Mil–C–23422, Mil–P–6871, ANP–39, and NASA ND–1002176.

5.4 Wear Resistance (Surface Hardness)

Electrolizing is one of the hardest chromium surfaces available, measuring 70–72 R_c, as applied. "As applied" refers to the measurable hardness of the Electrolizing coating when measured on the basis metal to which it is applied. The basis metal plays an important role in determining how wear resistant the Electrolizing surface will be. Generally speaking, Electrolizing increases measurable hardness 10–15 points, as shown in Table 4.

In all cases, Electrolizing is 70–72 R_c. However, the basis metal directly affects the measurable hardness that can be achieved. The harder the basis metal, the higher the Electrolizing measurable hardness will be. Test measurements of the surface hardness should be made using the Knoop or Vickers method with a 5–10 gram load on a diamond point.

High hardness values indicate good wear resistance, but there are other factors to consider, such as coating surface texture, coating density, substrate cleanliness before coating application, interfacial energy of adhesion between the coating and the substrate, energy of adhesion between coating surface and the opposing sliding material, type of lubricant used, and the combination of opposing materials.

Increased density improves wear resistance because it results in fewer cracks, inclusions, and voids, which in turn reduces the rate of corrosive attack and provides more resistance to fragmentation, spalling, and wear.

Surface fatigue wear—the category to which true spalling belongs—also affects wear resistance in conventional QQ–C–320 chrome plating. The stress concentrations here are

Table 4 Measurable Hardness (R_c)

Basis metal	Electrolizing coating
≤ 18	18–25
18–35	30–50
35–50	50–70
50+	70+

generated during the electrodeposition, an integral phenomenon when chromium crystals are formed. The Electrolizing coating—essentially a chromium alloy—is virtually devoid of these internal stress concentrations, and therefore has minimal tendency to spall. This is one of the reasons for the substantial superiority in wear resistance under mechanical impact conditions of Electrolizing to most conventional chrome platings.

The Electrolizing coating also has a lower kinetic friction coefficient, and in wear resistance it is superior to all other chrome platings (including electroless nickel and a tungsten-carbide coating), as measured on three different test machines: the Taber, the Falex, and the LFW–1.

Finally, the Electrolizing coating exhibits extremely low adhesive wear coefficients: 1.72×101^{-7} on steel against a copper alloy, and 1.19×10^{-7} on steel against steel with a petroleum oil (10^{-8} representing the ultimate or best wear coefficient ever measured).

5.5 Lubricity

An Electrolized coating has a coefficient of friction of 0.11. Electrolizing also produces kinetic friction coefficients as low as 0.045 under unidirectional sliding test conditions with a fluorosilicone oil on the LFW–1 test machine, and a 0.069 with an additive-free white mineral oil on the Falex lubricant tester. Electrolizing's low friction factor is invaluable when extreme temperatures are involved.

5.6 Conformity

Electrolizing works best when applied to a relatively smooth surface (12–32 RMS or finer). Below 4 RMS, the process may deter slightly from the fine finish of the part, requiring post-Electrolize operations.

Internal and external surfaces on nearly all shapes and configurations can be uniformly processed. Slots or grooves less than 0.187 in. wide, having a depth greater than width, and bores less than 0.187 in. diameter will require special engineering to assure a uniform coating. Electrolizing, Inc., recommends test runs for such parts and special discussions with Electrolizing engineers on dimensions less than 0.187 in.

5.7 Heat Resistance

The maximum operating temperature recommended for the Electrolize coating is approximately 1600°F (710°C). Time at temperature should be reviewed with Electrolizing, Inc., before testing or specifying Electrolizing. In general, oxidation occurs around 1100°F (430°C), progressing to 1650°F (740°C), and then to diffusion.

5.8 Brightness

The Electrolizing coating is smooth, continuous, fine-grained, adherent, uniform in thickness and appearance, and free from blisters, pits, nodules, porosity, and edge buildup. Electrolizing is used as the final coating on parts and equipment. The Electrolizing coating is shiny by application. However, satin (matte) finishes can be attained, if specified.

5.9 Hydrogen Embrittlement

During conventional chrome plating processes, a detrimental side effect occurs: hydrogen occlusion. Hydrogen penetrates into the substrate and causes embrittlement of the metal part with subsequent reduction of mechanical properties, particularly fatigue strength. Most conventional chrome plating control documents, therefore, specify a final 375°F bake to remove hydrogen gas.

The longer the plating cycle, the more likely it is that hydrogen embrittlement will occur. Embrittlement is also more likely to occur after an acid clean. Shot peening and/or liquid honing can be used to relieve embrittlement stress.

Hydrogen embrittlement is extremely unlikely with Electrolizing because the Electrolizing processing avoids most of the causes of a true hydrogen embrittlement. Electrolizing, Inc., does not include postprocess baking in its process technique. However, if postprocess baking is required by a customer, Electrolizing, Inc., is able to include it as a standard procedure.

21

The Armoloy Chromium Process

Michael O'Mary

The Armoloy Corporation, DeKalb, Illinois

1.0 GENERAL DEFINITION

The Armoloy process is a low temperature, multistate, chromium alloy process of electrocoating based on a modified chromium plating technology. However, instead of the customary chromium plating solutions, the Armoloy process uses a proprietary chemical solution. The solution and application process are carefully monitored at all Armoloy facilities. The result is a satin finish chromium coating that is very hard, thin, and dense and has absolute adhesive qualities. Armoloy deposits a 99% chromium coating on the basis metallic surfaces, whereas conventional chromium plating processes tend to deposit 82–88% chromium in most applications.

The Armoloy process involves cleaning and removing the matrix on the basis metal's surface by special proprietary means and using a modified electrocoating process that causes the chromium metallic elements of the solution to permeate the surface porosity of the basis metal. It is during this process that the absolute adhesive characteristics and qualities of Armoloy are generated. The Armoloy coating actually becomes part of the basis metal itself, and the result is a lasting bond and a continuous, smooth, hard surface. The surface will not chip, flake, crack, peel, or separate from the basis metal under conditions of extreme heat or cold, or when standard ASTM bend tests are involved.

Three basic factors are always present after applying Armoloy to metal surfaces:

increased wear (70–72 R_c surface hardness)
added lubricity characteristics (including the ability to utilize Armoloy against Armoloy)
excellent corrosion resistance

211

2.0 APPLICATIONS

2.1 General Applications

All ferrous and most nonferrous materials are suitable for Armoloy application. Service life of parts has been increased to 10 times normal life and even higher in certain applications. However, basis metals of aluminum, magnesium, and titanium are not good candidates for the Armoloy process.

Because it is a thin, dense coating, Armoloy exhibits its best wear and lubricity properties on hardened surfaces. It is most effective when the basis metal is 40 R_c or harder. In severe wear applications, a basis metal should be hardened to the 58–62 R_c range before Armoloy application. Armoloy will improve performance on any basis metal, but it is not a substitute for heat treating.

Table 1 lists appropriate uses for Armoloy at various hardness ranges.

2.2 Specific Applications

Armoloy is used across a variety of industries for a multitude of purposes. With Armoloy, dies and molds for both metal and plastics have better release characteristics and reduced wear (especially when abrasive materials are involved); cutting tools experience longer wear life; nuclear components exhibit better antigalling and corrosion resistance properties; dies used in stamping, drawing, forming, and blanking have sharper cuts and increases in work life; engine and transmission parts (such as valves, valve guides, pistons, gears, and splines) are protected without detrimental tolerance changes or variations; standard pumps and meters handle corrosive liquids and materials much better; and bearings and bearing surfaces can run longer and cooler and are superior to a 440C stainless steel for corrosion protection. In fact, Armoloy makes it possible to use standard ferrous steels in place of stainless steel in many applications, including food processing, medical environments, and ball or roller bearing applications.

Other applications include drive transmissions, power transmissions, gears, molds, screws, sleeves, threaded parts, and valves. In the automotive industry, applications not only include engine parts, but also large metal form dies for auto body parts.

Table 1 Use for Armoloy

Basis metal hardness range (R_c)	Application
18–30 R_c	Armoloy will handle low-loaded or stress conditions and provide basic corrosion resistance.
30–40	Armoloy increases wear resistance where the metal will basically support the wear factor. Corrosion resistance is good in this range.
40–50	Corrosion resistance increases as basis metal gets harder. Wear resistance properties begin to improve greatly.
>50	This is the most common and recommended range for an Armoloy application. Corrosion resistance is superior, and the maximum wear resistance benefits are exhibited.

3.0 SURFACE PREPARATION

Substrate surfaces must be free of oil, grease, oxides, and sulfides. Parts should be surface finished before shipment to Armoloy, where they will undergo further cleaning and surface preparation through proprietary means. Surfaces are not changed significantly in configuration by the Armoloy process. Armoloy does not fill scratches, pits, or dents, but rather conforms to such imperfections and may highlight them.

The preparation by Armoloy of the surface conditions tends to improve overall surface finish on most parts submitted for processing. The special mechanical methods utilized by Armoloy do not feature any acids, etching, or reversing methods used in conventional chromium plating; consequently, there is no detrimental effect to the basis metal. Surface finishes that are received by Armoloy with 8 RMS or less may show some slight roughing of the surface after application. In most cases these finishes can be lapped or polished back to their original condition. It is important be aware that Armoloy is a silver-satin finish and does not exhibit the maximum reflective properties of some conventional chromium processes.

4.0 PROPERTIES

4.1 Thickness

Armoloy is applied in a very thin, dense layer on the basis metal. The coating will range from 0.000040 to 0.0006 in. (1–15 µm) per side. Normal proven average deposits per side are in the 0.0001–0.0002 in. (2.5–5 µm) range. Tolerances can be held to ± 0.000050 in. (±1 µm) on deposits up to 0.0002 in. (5 µm) per side. If required, Armoloy can be controlled to ±0.000025 in. (±0.5 µm) per surface. There is no change in either the conductivity or magnetic properties of the basis metal.

Armoloy thickness is always in direct relation to the basis metal, and it will vary from one basis metal to another. However, the thickness established for each will remain constant and predictable. In all applications calling for a close inspection to monitor deposit thickness, the Armoloy Corporation recommends the use of a exacting reverse etchant method of ascertaining deposit thickness. This method is nondestructive to the basis metal.

4.2 Adhesion

Armoloy will not chop, flake, crack, peel, or separate from the basis material under standard ASTM bend tests or under conditions of extreme heat or cold (from − 440 to 1600°F). The Armoloy coating forms a lasting bond with the surface by permeating the surface porosity of the basis metal. Armoloy can be removed by licensed Armoloy plants without detrimental effects to the basis metals.

4.3 Corrosion

Armoloy resists attack by most organic and inorganic compounds (except sulfuric and hydrochloric acids). The Armoloy coating is typically more noble than the substrate; therefore, it protects against corrosion by being free of pores, cracks, and discontinuities, and by providing a uniform structure and chemical composition. Porosity, hardness, and imperfect surface finishes of basis metals will affect the corrosion-resistant properties of Armoloy; however, all basis metals that are Armoloy coated will have enhanced corrosion-resistant characteristics. Samples are subjected to standard ASTM B–287 salt spray tests. Armoloy also conforms to Mil QQC–320B, AMS 2406, and AMS 2438.

4.4 Wear Resistance

Armoloy is one of the hardest chromium surfaces available, measuring 70–72 R_c, as applied. "As applied" refers to the measurable hardness of the Armoloy coating when measured on the basis metal to which it is applied. The basis metal plays an important role in determining how wear resistant the Armoloy surface will be. Generally speaking, Armoloy increases measurable hardness 10–15 points, as shown in Table 2.

Armoloy itself is always 70–72 R_c. However, the basis metal directly affects the measurable hardness that can be achieved. The harder the basis metal, the higher the Armoloy measurable hardness will be. Test measurements of the surface hardness should be made using the Knoop or Vickers method with a 5–10 gram load on a diamond point.

High hardness values indicate good wear resistance, but there are other factors to consider. Corrosion and lubrication also affect wear. Armoloy improves wear resistance not only by being extremely hard, but also by resisting corrosion and by improving lubricity with its nodular microsurface.

4.5 Lubricity

Special techniques used in its application render Armoloy self-lubricating, creating a nodular surface. Armoloy's low friction factor is invaluable under condition of extreme temperatures. Table 3 gives coefficients of friction on various materials at 72°F with no lubricants.

An added feature of the Armoloy coating is its ability to be run against itself in friction-related applications. The special nodular (orange peel) surface that is created by the Armoloy process allows for the special "Armoloy-to-Armoloy" feature to reduce frictional characteristics. The nodular surface creates excellent retention of lubricants and a continuous dispersion of them, which help to provide other low friction features. Special attention must be paid in the engineering of tolerances when Armoloy will be operated against another Armoloy surface.

4.6 Conformity

Armoloy conforms exactly to the basis metal surface. All threads, flutes, and even scratches are reproduced in detail. Armoloy works best when applied to a relatively smooth surface (12–32 RMS). RMS finish will improve slightly, to a low of about 8 RMS. Below 4 RMS the process may deter slightly from the fine finish of the part.

Table 2 Measurable Hardness (R_c)

Basis metal	Armoloy coating
≤ 18 R_c	18–25
18–35	30–50
35–50	50–70
> 50	> 70

Table 3 Comparison of Coefficients of Friction

Materials			Coefficients	
1	versus	2	Static	Sliding
Steel		Steel	0.30	0.20
Steel		Babbitt metal	0.25	0.20
Steel		Armoloy	0.17	0.16
Babbitt metal		Armoloy	0.15	0.13
Armoloy		Armoloy	0.14	0.12

Internal and external surfaces on nearly all shapes and configurations can be uniformly processed. Slots or grooves less than 0.187 in. wide, having a depth greater than width, and bores less than 0.187 in. in diameter will require special engineering to assure a uniform coating.

4.7 Heat Resistance

Armoloy will withstand temperatures of −440 to 1800°F. At temperatures above 1600°F, Armoloy will react with carbon monoxide, sulfur vapor, and phosphorus and begin to soften. At bright red heat, oxidation occurs in steam or alkaline hydroxide atmospheres. Hardness, wear resistance, and corrosion properties will be reduced at temperatures above 1600°F. Conversely, the Armoloy coating will remain basically stable at temperatures below −200°F.

4.8 Brightness

Armoloy is used as the final coating on parts and equipment. It has a very attractive satiny-silver matte and "micro-orange peel" finish. If necessary, Armoloy can be polished after application to enhance its surface finish and reflectivity.

4.9 Hydrogen Embrittlement

In all electrochemical plating processes, free ions of hydrogen ar created and released. Frequently, these ions entrap themselves in the molecular structure of the basis material, resulting in hydrogen embrittlement. The following conditions make hydrogen embrittlement extremely unlikely with the Armoloy process:

No acids are used in the preparation process.
The vapor blast (liquid hone) or dry hone procedure aids in relieving residual surface stress.
No "reverse clean" or etchant is used before the part is Armoloy processed.
The plating cycle times are very short and the Armoloy chrome is deposited so rapidly that Armoloy seals the surface porosity of the basis metal before hydrogen ions can invade the surface of the basis metal. However, if required, Armoloy can be and will be postplate heat treated to specification.

22

Sputtered Thin Film Coatings

Brian E. Aufderheide

W. H. Brady Company, Milwaukee, Wisconsin

1.0 HISTORY

Sputtering was discovered in 1852 when Grove observed metal deposits at the cathodes of a cold cathode glow discharge. Until 1908 it was generally believed that the deposits resulted from evaporation at hot spots on the cathodes. However, between 1908 and 1960, experiments with obliquely incident ions and sputtering of single crystals by ion beams tended to support a momentum transfer mechanism rather than evaporation. Sputtering was used to coat mirrors as early as 1877, finding other applications such as coating fabrics and phonograph wax masters in the 1920s and 30s. The subsequent important process improvements of radio frequency (rf) sputtering, allowing the direct deposition of insulators, and magnetron sputtering, which enables much higher deposition rates with less substrate damage, have evolved more recently. These two developments have allowed sputtering to compete effectively with other physical vapor deposition processes such as electron beam and thermal evaporation for the deposition of high quality metal, alloy, and simple inorganic compound coatings, and to establish its position as one of the more important thin film deposition techniques.

2.0 GENERAL PRINCIPLES OF SPUTTERING

Sputtering is a momentum transfer process. When a particle strikes a surface, the processes that follows impact depend on the energy of the incident particle, the angle of incidence, the binding energy of surface atoms, and, the mass of the colliding particles (Fig. 1).

In sputtering, the incident particles are usually ions, because they can be accelerated by an applied electrical potential. If the kinetic energy with which they strike the surface is less than about 5 eV, they will likely be reflected or absorbed on the surface. When the kinetic energy exceeds the surface atom binding energy, surface damage will occur as atoms are forced into new lattice positions. At incident ion kinetic energies above a threshold, typically 10–30 eV, atoms may be dislodged or sputtered from the surface. At normal incidence,

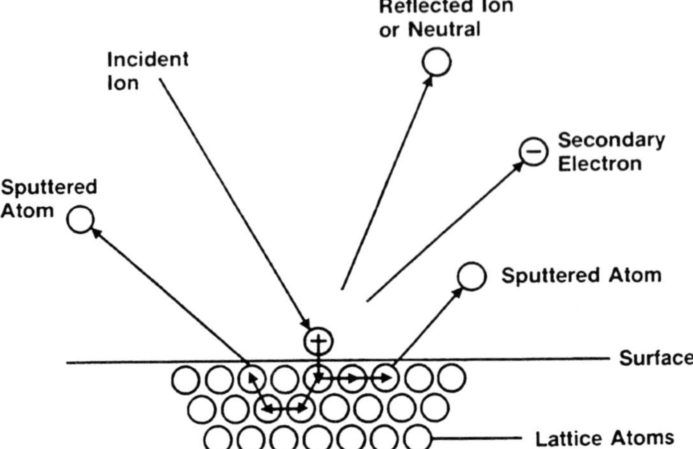

Figure 1 Schematic representation of some of the processes that follow ion impact during sputtering. (Courtesy of W. H. Brady Co.)

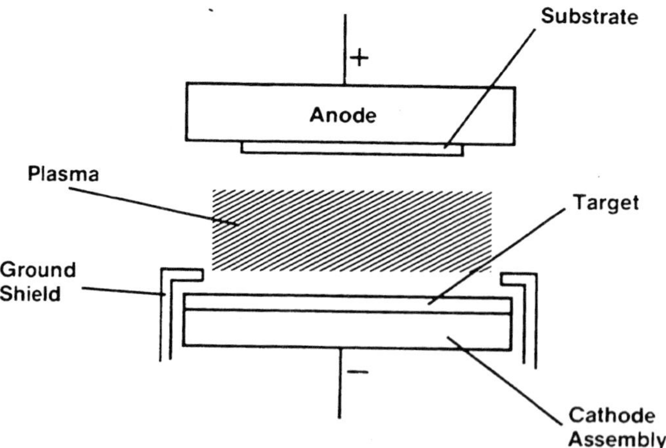

Figure 2 Schematic representation of a diode sputtering assembly. (Courtesy of W. H. Brady Co.)

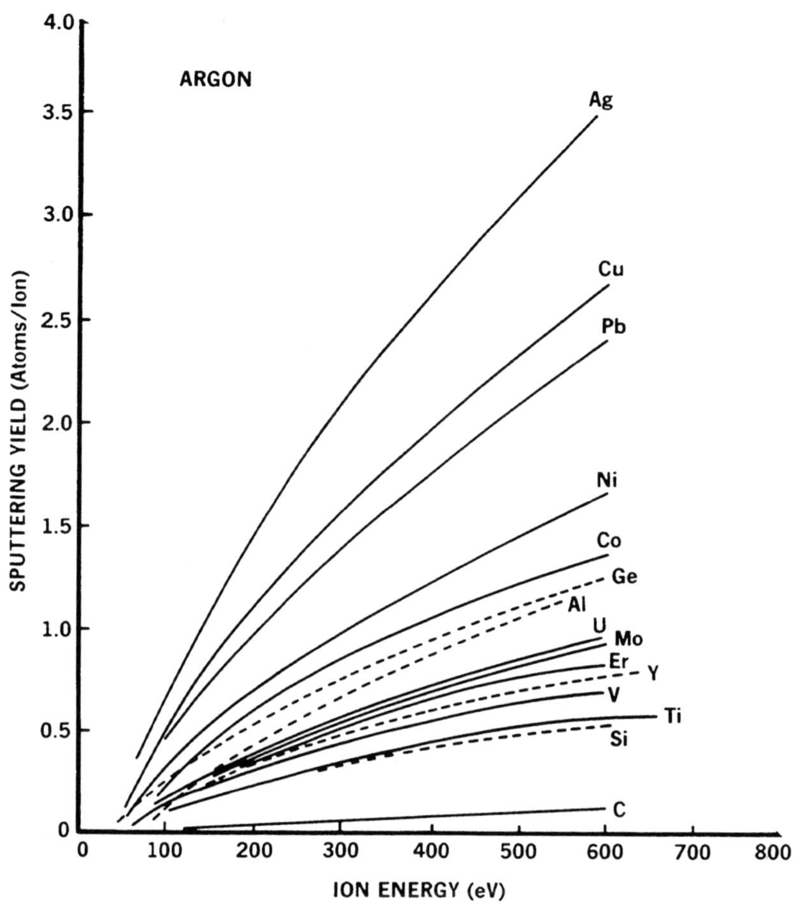

Figure 3 Variation of sputtering yield with ion energy at normal angle of incidence. (From J. A. Thornton, in *Deposition Technologies for Films and Coatings*, R. K. Bunshah, Ed. Park Ridge NJ: Noyes Publications, 1982, p. 179.)

multiple internal collisions are required, but at lower angles, sputtered atoms can be produced directly. These sputtered atoms and ions can be condensed on a substrate to form a thin film coating.

Energetic ion bombardment is usually achieved by a low pressure process of the glow discharge type. The basic process configuration, in this case diode, is shown in Figure 2. The vacuum chamber is equipped with a target (cathode), the source of coating material, and a substrate to be coated. To dissipate the considerable heat generated in the target during the sputtering process, it is usually bonded to a water–cooled metal (copper) backing plate with solder or conductive epoxy. The target also may be directly cooled by water for greater cooling capacity. The chamber is evacuated and then backfilled with an inert gas, usually argon, to a pressure of 10^{-3} to 10^{-1} torr.* An electrical potential is applied between the target (cathode) and substrate holder (anode). This produces a low pressure glow discharge or

---

* 1 pascal (Pa) = 1 N/m^2 = 0.075 torr.

plasma between the two electrodes. Grounded dark space shields are used to prevent a discharge from forming in undesirable areas. In a dc glow discharge of this type current is carried by electrons that are collected from the plasma by an anode and by positive ions leaving the plasma as they are accelerated toward the target. A continuous supply of additional ions and electrons must be available if the discharge is to be sustained. Some of the ions striking the target surface generate secondary electrons, which are accelerated by the cathode potential. These electrons, with energies approaching the applied potential, enter the plasma and ionize gas atoms, producing the necessary additional ions and electrons to sustain the discharge.

The relative rates of deposition for different materials depend largely on the sputter yield (Fig. 3), defined as the number of target atoms ejected per incident particle. Sputter yield depends on the target material, silver showing the highest yield, and generally increasing with incident ion energy and mass.

3.0 SPUTTER DEPOSITION SOURCES

3.1 Direct Current Diode Sputtering

The simplest and oldest sputter deposition source is dc diode. The two electrodes are usually parallel to each other, spaced 4–8 cm apart and the substrate is placed on the anode as in Figure 2. The applied potential is typically 1000–3000 V dc with argon pressures of about 0.075–0.12 torr. The dc diode configuration has important disadvantages, including low deposition rate (~ 400 ô/min for metals), high working gas pressure, targets limited to electrical conductors, and bombardment of the substrate by plasma electrons, resulting in substrate heating. The cathode systems discussed next can be used to improve on the performance of the dc diode.

3.2 Triode Sputtering

A heated filament (Fig. 4) is used as a secondary source of electrons for the discharge; an external magnet can also be used to confine electrons and increase isolation probability. Triodes can produce much higher deposition rates, up to several thousand angstroms per minute, at lower pressures, $(0.5–1 \times 10^{-3}$ torr) and voltages (50–100 V). The usefulness of triodes has been limited by difficulties in scaling up to large cathode sizes and corrosion of the emitter filament by chamber gases.

3.3 Radio Frequency Sputtering

Nonconducting materials cannot be directly sputtered with an applied dc voltage because of positive charge accumulation on the target surface. If an ac potential of sufficiently high frequency is applied, an effective negative bias voltage is produced such that the number of electrons that arrive at the target while it is positive equals the number of ions that arrive while it is negative. Because the mass of the electron is very small relative to ions present, the target is positive for only a very short time, and deposition rates for rf diode are almost equivalent to dc diode. This resulting negative bias allows sputtering of an insulating target. The frequency used in most practical applications is usually 13.56 MHz, a radio frequency band allocated for industrial purposes by the Federal Communications Commission. Rf sputtering allows insulators as well as conductors and semiconductors to be deposited with the same equipment and also permits sputtering at a lower pressure $(5–15 \times 10^{-3}$ torr). One major disadvantage of rf sputtering is the need for electromagnetic shielding to block the rf

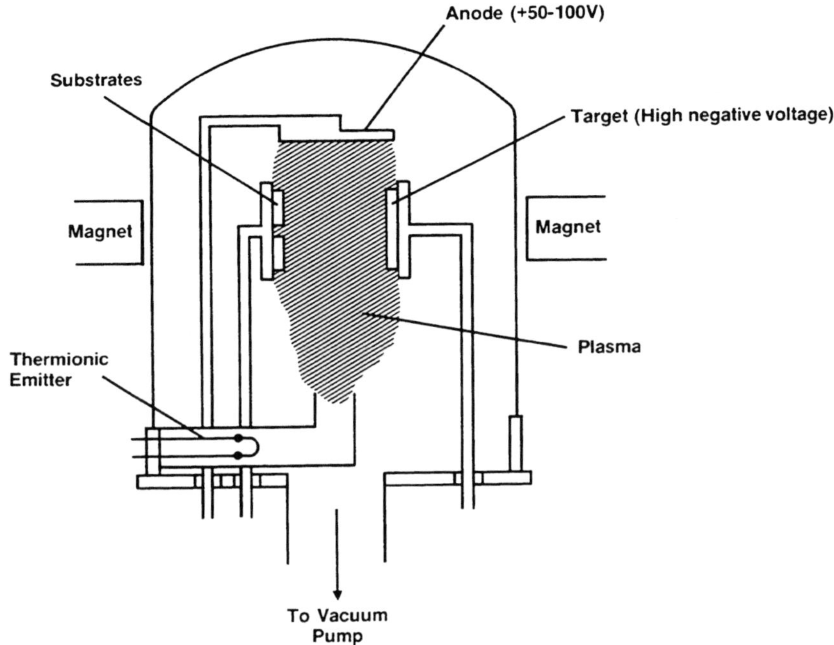

Figure 4 Schematic representation of a triode sputtering process. (Courtesy of W. H. Brady Co.)

radiation. Also, the power supplies, matching network, and other components necessary to achieve a resonant rf network are very complex.

3.4 Magnetron Sputtering

The magnetron cathode is essentially a magnetically enhanced diode. Magnetic fields are used to form an electron trap which, in conjunction with the cathode surface, confines the E X B electron drift currents to a closed–loop path on the surface of the target. This "race-track" effectively increases the number of ionizing collisions per electron in the plasma. The magnetic confinement near the target results in higher achievable current densities at lower pressures ($10^{-3} - 10^{-2}$ torr), nearly independent of voltage. This manner of cathode operation is described as the magnetron mode and is capable of providing much higher deposition rates (10 times dc diode) with less electron bombardment of the substrate and therefore less heating. Factors affecting deposition rate are power density on the target, erosion area, distance to the substrate, target material, sputter yield, and gas pressure. Dc is usually used for magnetron sputtering, but rf can be used for insulators or semiconductors. When magnetic materials are sputtered, a thinner target is often necessary to maintain sufficient magnetic field strength above the target surface. The three most common magnetron cathode designs, described below, are illustrated in Figure 5.

3.4.1 *Planar Magnetron*

An array of permanent magnets is placed behind a flat, circular,or rectangular target. The magnets are arranged such that areas in which the magnetic field lines are parallel to the target surface from a closed loop on the surface. Surrounding this loop, the magnetic field

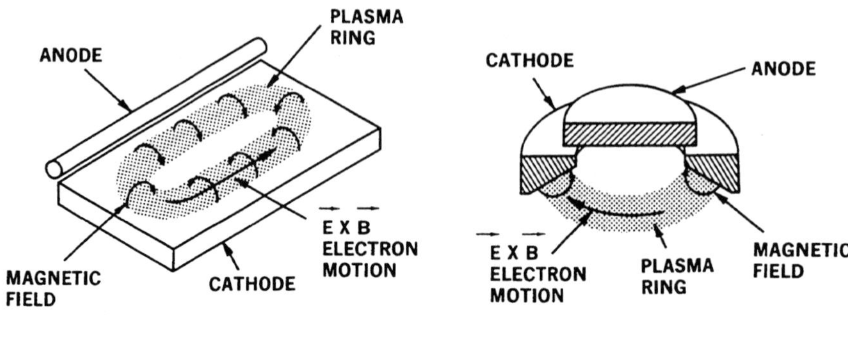

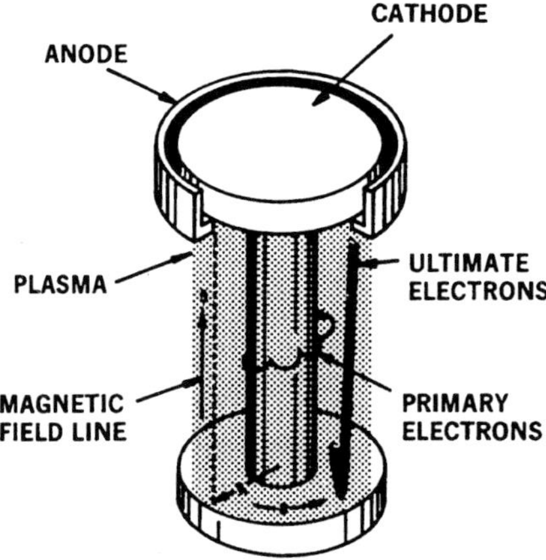

Figure 5 Clockwise from upper left: schematic representations of planar magnetron, gun–type magnetron, and cylindrical post magnetron sputtering sources. (Adapted from J. A. Thornton, in *Deposition Technologies for Films and Coatings*, R. F. Bunshah, Ed. Park Ridge NJ: Noyes Publications, 1982, pp. 194–195.)

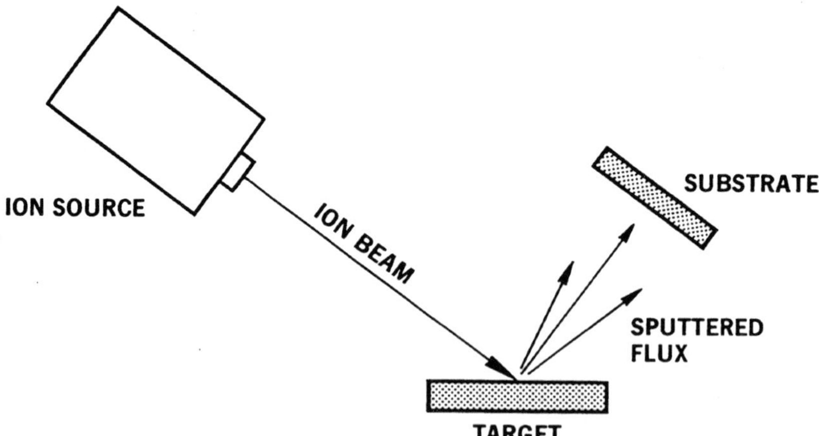

Figure 6 Schematic representation of ion beam sputtering source showing relative locations of target and substrate. (From J. A. Thornton, in *Deposition Technologies for Films and Coatings*, R. K. Bunshah, Ed. Park Ridge, NJ: 1982, p. 211.)

lines generally enter the target, perpendicular to its surface. This produces an elongated electron racetrack and erosion pattern on the target surface. Because of the nonuniformity in target erosion, utilization of target material is poor, typically 26–45%. This also results in nonuniform deposition on a stationary target. Uniformity is provided by substrate motion, usually linear or planetary, combined with uniformity aperture shielding. Planar magnetron cathodes are usually operated at 300–700 V providing a current density of 4–60 mA/cm^2 or a power density of 1–36 W/cm^2.

Deposition rates are generally proportional to the power delivered to the target. Much of this power is dissipated as target heating. The primary factor limiting magnetron deposition rates is the amount of power that can be applied to the target without causing it to melt, crack, or warp. This is controlled by the cathode water cooling design, and the thermal conductivity of the target, the backing plate, and the interface between them. The planar magnetron cathode has been scaled up in production applications to several meters in length and serves as an important industrial coating tool.

3.4.2 *Cylindrical Magnetron*

Two variations on a cylindrical cathode design can be used to coat large surface areas: the cylindrical post magnetron, which sputters outward from a central post target, and the cylindrical hollow or inverted magnetron, which has target erosion on the inner wall of a cylindrical target. Operating parameters are similar to the planar magnetron. The E X B (electric field strength X magnetic flux density) current closes on itself by going around the post or cylinder. Electrostatic or magnetic containment is often used to minimize end losses. Erosion is uniform along the post or inside of the cylinder. This enables fairly uniform coating without substrate movement. Hollow cathodes are especially effective at coating objects of complex shapes. Another cylindrical cathode, the rotatable magnetron, uses a magnet array similar to a planar magnetron and rotates the target or magnets to obtain uniform erosion.

3.4.3 *Ring or Gun Magnetron*

The ring or gun magnetron source includes a circular cathode and a concentric centrally located anode. As with other magnetrons, high deposition rates are possible with little substrate heating. Because of the circular design, planetary substrate motion is necessary for deposition uniformity. This design is extensively used for small–scale applications but has not been scaled up to larger dimensions. Arrays of these cathodes have been used to coat large areas.

3.5 Beam Sputtering

A separate ion beam source (Fig. 6), as opposed to a glow discharge, may be used to erode the surface of a target. The energy, direction, and current density of the ion beam may be controlled independently, and it is possible to work at background pressures lower than other sputter deposition methods used. Unique film properties can sometimes be obtained using ion beam deposition, bit it is generally limited to coverage of rather small areas and lower deposition rates.

3.6 Reactive Sputtering

Argon is usually employed as the working gas in sputter deposition processes. It is relatively inert, being incorporated in the growing film only when trapped or embedded in its surface. Other more reactive gases such as water vapor, oxygen, and nitrogen are normally present in the deposition chamber as low level contaminants, which have been outgassed from the substrate, target, and chamber walls. These gases may be incorporated into the growing film

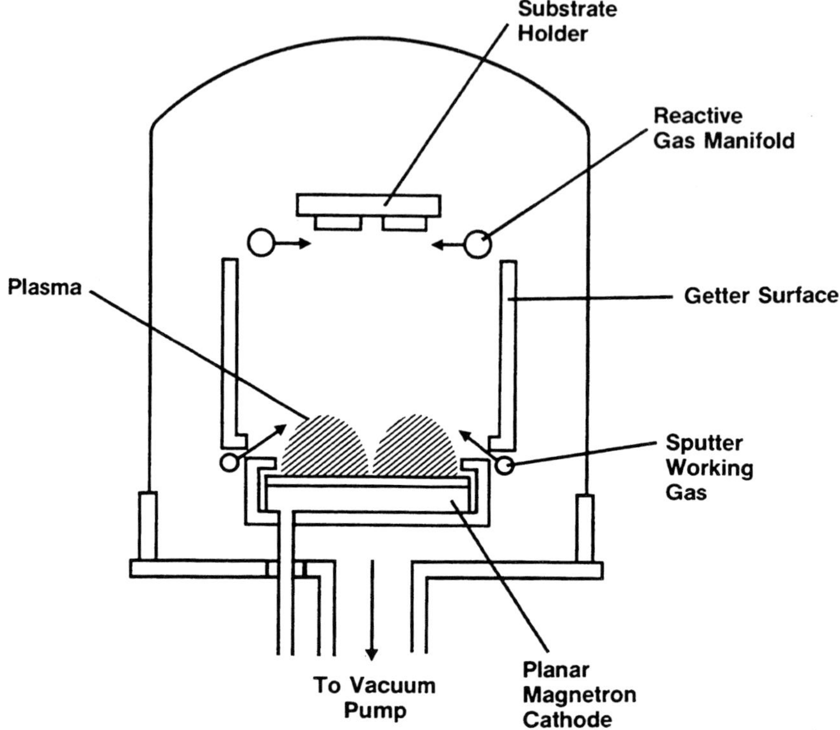

Figure 7 Schematic representation of a reactive sputtering apparatus. (Courtesy of W. H. Brady Co.)

by reacting with condensed atoms on the substrate surface, forming small amounts of oxides, nitrites, carbides, and other similar compounds of the sputtered material.

In reactive sputtering, gases are intentionally introduced into the deposition chamber to completely react with the forming thin film. A gas manifold system (Fig. 7), is often used to provide uniform distribution of the reactive gas at the substrate and minimize reactive gas at the target surface.

Reactive sputtering is a very nonlinear process. The growing film on the substrate acts like a getter pump for the reactive gas up to the pressure at which a stoichiometric compound is formed. At this point, the pumping rate of the substrate decreases substantially, and reactive gas pressure increases in the chamber. This gas may react with the surface of the target, resulting in decreased deposition rates due to lower sputter yields of compounds and other factors. Consequently, most reactive deposition processes attempt to work near the transition region of the target, where stoichiometric compounds are formed at the substrate and the cathode target is metallic. Reactive sputtering therefore permits the formation of compounds from simple metallic targets in the dc mode or rf mode. This process is widely used for deposition of oxides and nitrites such as silicon oxides, silicon nitrite, titanium nitride, and indium tin oxide.

4.0 OTHER PROCESS CONSIDERATIONS

Vacuum chambers used for sputtering have to be capable of producing a vacuum better than 5×10^{-4} torr to minimize contamination from background gases. A load lock is used on some devices when contamination of the sputter chamber by gas absorbed when the chamber is opened is unacceptable, or to decrease pump down time. A wide variety of pumping systems are used including diffusion, turbomolecular, and cryogenic pumps. Throttling of the pumps is necessary in many cases due to high working pressures.

Most target materials have considerable surface contamination accumulated during manufacture and storage. Targets are presputtered to a depth sufficient to remove these contaminants before deposition. A shutter is often used to protect the substrate from the contamination while the target is being cleaned.

The sputtering process can be very dependent on such equipment design factors as gas distribution systems, position of pumps, and distance from target to substrate. These factors and others must be carefully considered when transferring a process from one chamber to another or when scaling up to production levels.

A major factor controlling equipment configuration is the nature of the substrate. The substrate may be small rigid pieces such as microelectronic circuitry or much larger rigid items such as architectural glass. If the substrate is flexible, roll–to–roll processing is often used. Equipment is designed to handle the substrate in a manner that will promote the achievement of coating uniformity.

5.0 PROPERTIES OF SPUTTERED THIN FILM COATINGS

The electrical, optical, and other properties of thin films often vary from the properties measured in bulk materials. Conditions present at the substrate during deposition can have a significant influence on these properties. When a sputtered atom condenses on a substrate, it transforms its kinetic energy to the surface lattice. The resulting loosely bonded atom has mobility on the surface and will migrate over the surface, interacting with other absorbed atoms until it finds a permanent low energy site, or is desorbed. As the film thickness builds, atoms within the lattice move to more stable positions by bulk diffusion.

For many metals, the mobility of surface and bulk atoms is related to the ratio of substrate temperature (T) and the melting point of the metal (T_m). At low T/T_m values, surface and bulk diffusion play little role because atoms have insufficient energy to move from their initial position. In films deposited under these conditions (Fig 8), the internal crystal structure is poorly defined and has many defects; voids are also present due to shading effects. Conditions tending to produce lower T/T_m are higher melting temperature target materials (e.g. refractory metals), substrate cooling, higher working gas pressures (which "thermalize" or reduce sputtered atom energies), and reactive gases adsorbed on the surface. Higher T/T_m values result in more surface and bulk diffusion in the growing film, producing denser columnar grains with fewer defects and defined boundaries.

In addition to substrate heating, T can effectively be raised by bombarding the substrate with ions or electrons during deposition. In a diode system, the substrate is often in the plasma, resulting in film heating due to high energy electron impact. A bias voltage can also be applied to the substrate to accelerate incoming ions, increasing the kinetic energy transferred to the surface. When depositing alloys or compounds, higher T values may decrease the sticking coefficient of one component, changing the overall film composition. An important point to remember is that the kinetic energy of sputtered atoms typically is 10 times

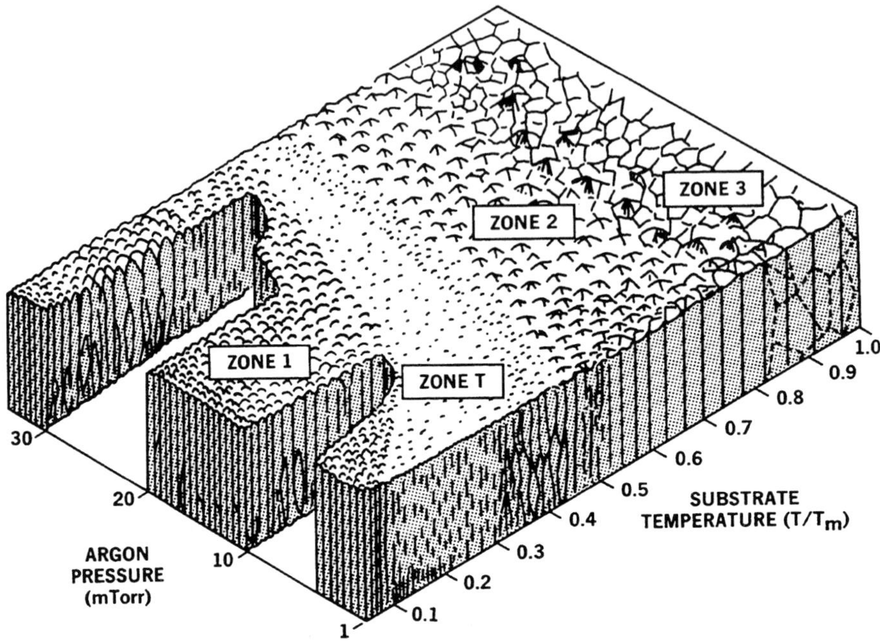

Figure 8 Schematic representation of the relationship between substrate temperature and argon pressure on the structure of metal coatings deposited by sputtering using a cylindrical magnetron source. T is the substrate temperatures and T_m is the melting point of the coating material in absolute degrees. (From J. A. Thornton, in *Deposition Technologies for Films and Coatings*, R. F. Bunshah, Ed. Park Ridge, NJ: 1982, p. 214.

that of evaporated species. This factor plays a significant role in determining the difference in properties between sputtered and evaporated films.

Most sputtered thin film are in a state of compressive or tensile stress. The stress is due to the mismatch in thermal coefficients of expansion between the substrate and the thin film and to internal stresses built up within the film as a result of imperfections in the crystal lattice. In higher T_m materials, deposited at low T/T_m, internal stresses usually dominate and may reach the yield strength of the film, resulting in fracture. Both magnitude and type of internal stress have been found to be influenced by working gas pressures. Thermal stress usually predominates in higher T/T_m coatings.

6.0 THIN FILM MATERIALS

Most metals have been deposited by sputtering; some of the more commonly used metals are aluminum, chromium, copper, gold, molybdenum, nickel, platinum, palladium, silver, tantalum, titanium, tungsten, vanadium, and zirconium. This group of metals represents an extremely broad range of electrical, magnetic, mechanical, optical, and other physical properties. If a single metal does not provide the necessary properties, alloys such as stainless steel, nichrome, and cobalt chrome may also be sputtered. The different elements comprising an alloy may have widely ranging sputter rates, but because of concentration at the target surface of the slower sputtering species, alloys can generally be sputter deposited with the same composition as the target. Semiconductors such as silicon ands carbon can also be sputtered, but if conductivity of the target is too low, rf power may be necessary. Oxides, nitrites, carbides, sulfides, and other compounds of the metals, alloys, and semiconductors listed above may be made via reactive sputtering with added oxygen, nitrogen or ammonia, methane or other gaseous hydrocarbons, and hydrogen sulfide, respectively.

7.0 APPLICATIONS FOR SPUTTERED THIN FILMS

Sputtered coatings are used for a wide variety of applications, some of which are listed below, according to the function performed by the thin film.

7.1 Electrical

Metals and alloys are sued as conductors, contacts, and resistors, and in other components such ass capacitors. Transparent conductors such as indium tin oxide and thin metals serve as electrodes for LCDs touch panels, other display devices, and solar cells. Thin films are also extensively used in microelectronic devices.

7.2 Magnetic

Some high performance magnetic data storage media are deposited via sputtering. Cobalt alloys such as cobalt–chromium and to a lesser extent, nickel, iron, and samarium alloys are typically used.

7.3 Optical

Thin metal and dielectric coatings are used to construct mirrors, antireflection coatings, light valves, laser optics, and lens coatings, and to provide architectural energy control and optical data storage.

7.4 Mechanical

Hard coatings such as titanium carbide, nitride, and carbon produce wear–resistant coatings for cutting tools. Molybdenum sulfide serves as a solid lubricant.

7.5 Chemical

Thin film coatings can be used to provide high temperature environmental corrosion resistance for aerospace and engine parts, catalyst surfaces, gas barrier layers, and lightweight battery components.

7.6 Decorative

Titanium nitride is deposited on watch bands and jewelry as a hard gold–colored coating. Metals are deposited for weight reduction in automotive and decorative graphics applications.

8.0 ADDITIONAL RESOURCES

The following professional societies include sections dealing with sputtered coatings: American Vacuum Society (offers short courses in sputtering and coatings), Society of Vacuum Coaters, Electrochemical Society, and Materials Research Society. Journals that cover developments in sputtered coatings include *Journal of Vacuum Science and Technology,*

Thin Solid Films, Journal of Applied Physics, Vacuum, Progress in Surface Science, and the *Journal of the Electrochemical Society*.

BIBLIOGRAPHY

1. Bunshah, R. F., et al., *Deposition Technologies for Films and Coatings*. Park Ridge, NJ: Noyes Publications, 1982.
2. Chapman, B., *Glow Discharge Processes, Sputtering and Plasma Etching*. New York: Wiley, 1980.
3. Coutts, T. J., *Active and Passive Thin Film Devices*. New York: Academic Press, 1978.
4. Maissel, L. I., and R. Glang, *Handbook of Thin Film Technology*. New York: McGraw–Hill, 1970.
5. Vossen, J. L., and W. Kern, *Thin Film Processes*. New York: Academic Press, 1978.

23

Cathodic Arc Plasma Deposition

H. Randhawa

Vac–Tec Systems, Inc., Boulder, Colorado

1.0 INTRODUCTION

The cathodic arc plasma deposition (CAPD) method[1,2] of thin film deposition belongs to a family of ion plating processes that includes evaporative ion plating[3,4] and sputter ion plating.[5,6] However, the cathodic arc plasma deposition process involves deposition species that are highly ionized and posses higher ion energies than other ion plating processes. All the ion plating processes have been developed to take advantage of the special process development features and to meet particular requirements for coatings, such as good adhesion, wear resistance, corrosion resistance, and decorative properties.

The cathodic arc technique, having proved to be extremely successful in cutting tool applications, is now finding much wider ranging applications in the deposition of erosion resistance, corrosion resistance, decorative coatings, and architectural and solar coatings.

2.0 CATHODIC ARC PLASMA DEPOSITION PROCESS

In the cathodic arc plasma deposition process, material is evaporated by the action of one or more vacuum arcs, the source material being the cathode in the arc circuit (Fig. 1). The basic coating system consists of a vacuum chamber, a cathode and an arc power supply, an arc ignitor, an anode, and substrate bias power supply. Arcs are sustained by voltages in the range of 15–50 V, depending on the source material; typical arc currents in the range of 30–400 A are employed. When high currents are used, an arc spot splits into multiple spots on the cathode surface, the number depending on the cathode material. This is illustrated in Figure 2 for a titanium source. In this case, an average arc current/arc spot is about 75 A. The arc spots move randomly on the surface of the cathode, typically at speeds of the order of tens of meters per second. The arc spot motion and speed can be further influenced by external means such as magnetic fields, gas pressures during coatings, and electrostatic fields.

Materials removal from the source occurs as a series of rapid flash evaporation events as the arc spot migrates over the cathode surface. Arc spots, which are sustained as a result of

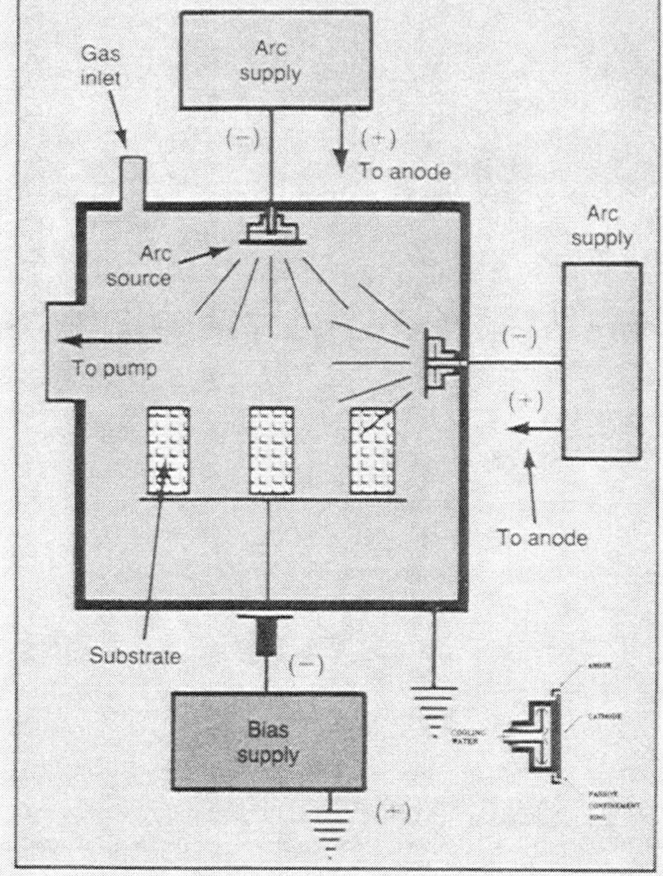

Figure 1   Schematic of a cathode arc deposition system.

the material plasma generated by the arc itself, can be controlled with appropriate boundary shields and/or magnetic fields.

CAPD is markedly different from physical vapor deposition process. Some of its characteristic features are as follows:

1. The material plasma is generated by one or more arc spots.
2. A high percentage (30–100%) of evaporation material is ionized.
3. The ions exist in multiple charge states (e.g., in case of Ti, Ti^+, Ti^{2+}, Ti^{3+}, etc.).
4. The ions possess very high kinetic energies (10–100 eV).

These characteristics of CAPD result in deposits that are of superior quality in comparison to other plasma processes. Some of these advantages are as follows:

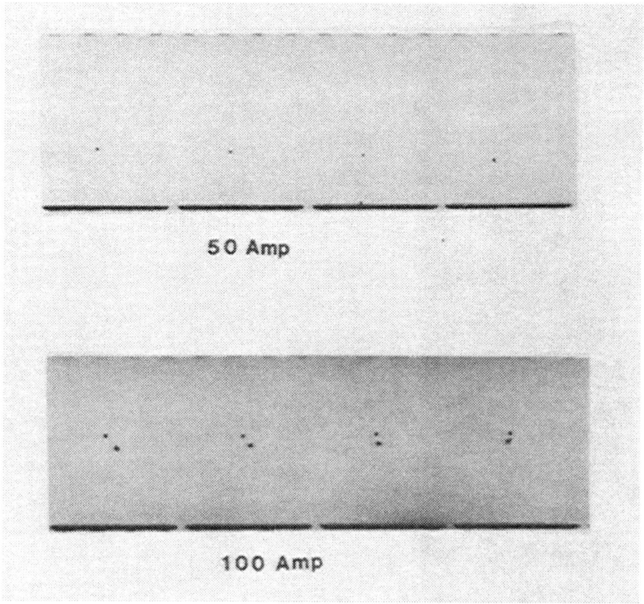

Figure 2   Number of arc spots on the titanium cathode arc source as a function of arc current.

1. Good quality films over a wide range of deposition conditions (e.g., stoichiometric reacted films with enhanced adhesion and film density can be obtained over a wide range
   of reactive gas pressures and evaporation rates).
2. High deposition rates for metals, alloys, and compounds with excellent coating uniformity.
3. Low substrate temperature.
4. Retention of alloy composition from source to deposit.
5. Ease of producing reacted compound films.

3.0 CATHODIC ARC SOURCES

A schematic cross section of a cathodic arc source is shown as an inset in Figure 1; ia a
photograph of a typical large area source. The arc source comprises a cathode (source material), an anode, an arc ignitor, and a means of arc confinement.

The method of arc confinement is a key factor in arc source design and configuration.
Cathodes using magnetic fields or boundary shields are limited to small sizes of the order of
a few inches in diameter. This limits the uniformity attained from such sources. It is generally necessary to use a multitude of such sources to obtain a reasonable coating quality. Arc
sources employing confinement passive boarders (Fig. 3) with predetermined electronic
characteristics may be built much larger and over a wide range of sizes. Such cathode
sources provide good uniformity over a wide range of substrate sizes in various industrial
applications. A typical thickness uniformity observed using an 8 in. × 24 in. titanium cathode at a source to substrate distance of 10 in. was approximately 10% over a flat area meas-

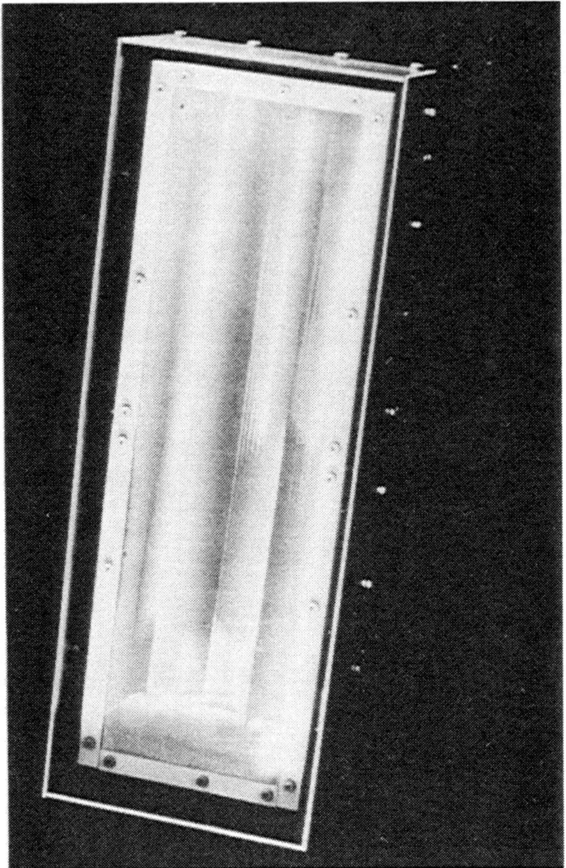

Figure 3   Typical large–area cathodic arc source.

uring 5 in. × 20 in. Furthermore, the target utilization of such arc sources exceeds 70%—much higher in comparison to the magnetron sputtering source (N 40%).

4.0   CATHODIC ARC EMISSION CHARACTERISTICS

The cathodic arc results in a plasma discharge within the material vapor released from the cathode surface. The arc spot is typically a few micrometers in size and carries current densities as high as $10A/\mu m^2$. This high current density causes flash evaporation of the source material and the resulting evaporant consists of electrons, ions, neutral vapor atoms, and microdroplets. Emissions from the cathode spots are illustrated in Figure 4. The electrons are accelerated toward the cloud of positive ions. The emissions from the cathode spots split into a number of spots. The average current carried per spot depends on the nature of the cathode material. The extreme physical conditions present within cathode spots are listed in Table 1.

It is likely that almost 100% of the material may be ionized within the cathode spot region. These ions are ejected in a direction almost perpendicular to the cathode surface. The

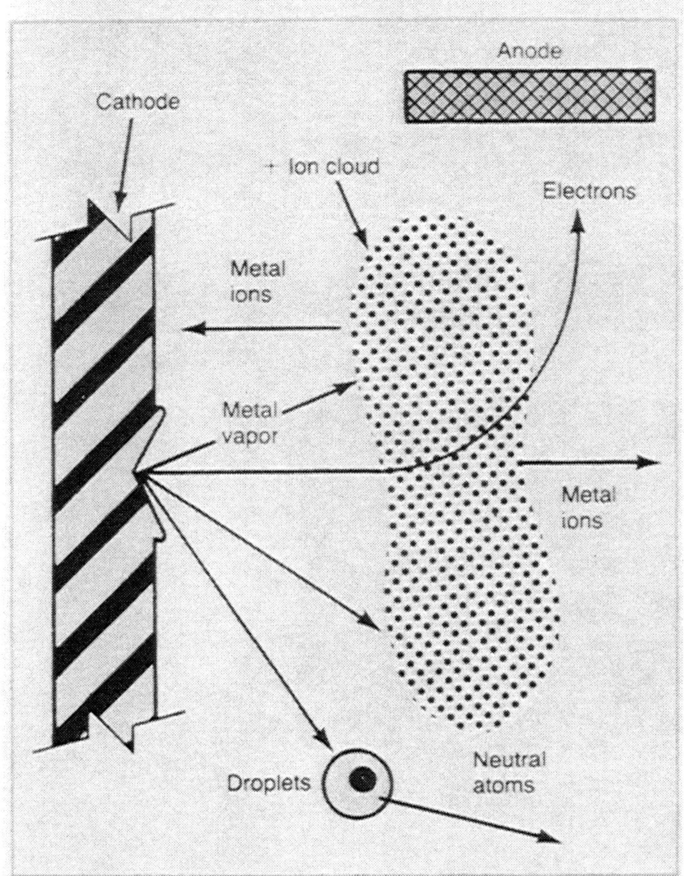

Figure 4   Emission characteristics of a cathodic arc source.

Table 1   Physical Conditions: Cathode Arc Spot

Temperature, K	$4 \times 10^3 - 10^4$
Pressure, Mpa	$0.1 - 10$
Power density, W/cm^{-2}	$10^7 - 10^9$
Electric field, V/cm^{-1}	$10^4 - 10^5$
Current density, A/cm^{-2}	$10^6 - 10^8$

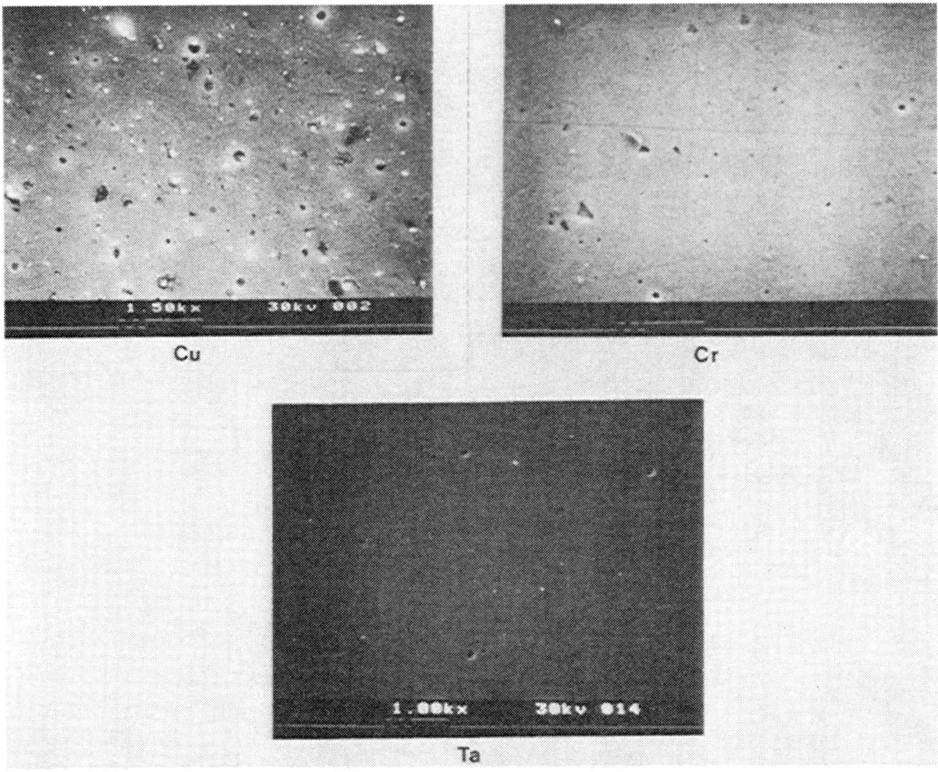

Figure 5   Microdroplet emission from metals having different melting points.

microdroplets, however, have been postulated to leave the cathode surface at angles up to about 30° above the cathode plane. The microdroplet emission is a result of extreme temperatures and forces that are present within emission craters. The microdroplet emission is greater for metals with low boiling points. Figure 5 shows such results for copper, chromium, and tantalum.

5.0 MICRODROPLETS

Microdroplets are emitted as one of the products of the flash evaporation events. In an uncontrolled situation, very high microdroplet densities may be produced and deposited onto the substrates. The microdroplets are found to be metal–rich in composition in the case of reacted compound films. Microdroplet size and density can be controlled in the arc deposition process. Parameters and source design are the key factors that influence the density and size of the microdroplets.

As previously reported, microdroplet density and size vary with the material. Zirconium nitride films, deposited under the same conditions as titanium nitride, exhibit a much lower density of much smaller microdroplets (N 0.1–0.2 μm). It is believed that the smaller microdroplets result from the higher melting point and low vapor pressure of zirconium coupled with the higher arc spot velocity observed on a zirconium cathode surface. The higher arc spot velocity results in a low mean residence time of the arc spot on a given localized area; thus, it minimizes localized overheating, hence the size and density of the microdroplets.

The arc motion of a conventional arc source was studied using a very high speed photographic technique. The arc speed was measured to be approximately 8 m/sec. The application of suitable external magnetic fields was found to enhance the arc speed to 17 m/sec. A corresponding reduction in macroparticles was observed. The source design as well as the operating gas pressure during deposition had an effect on the microdroplet emission. A new arc source using these modified microdroplets could be totally illuminated. This is illustrated for films of titanium and zirconium nitrides and titanium dioxide, as shown in Figure 6.

6.0 RECENT DEVELOPMENTS

A major interest in the cathodic arc process till recently has been in the deposition of hard coatings for tribology, wear, and decorative applications. Deposition, characterization, and performance evaluations of nitrites, carbides, and carbonitrites of several materials [Ti, Zr, Hf, (Ti–Al), (Ti–Zr)(Ti–6Al–4V) etc.] using cathodic arc deposition process have been investigated in detail and will not be discussed here. Some of the most recent developments

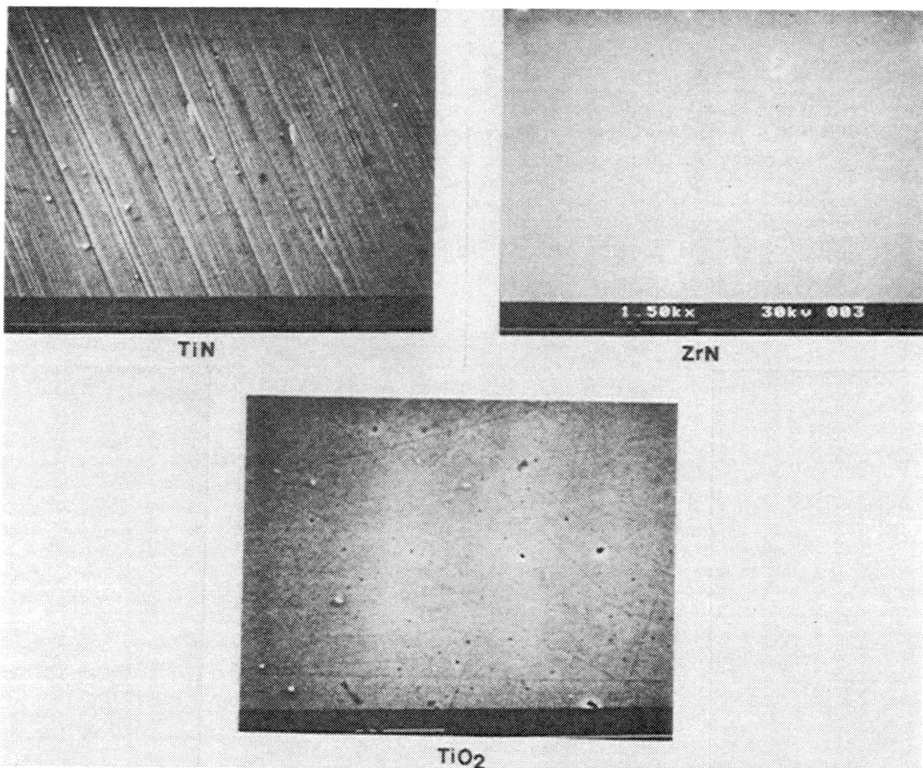

Figure 6 Scanning electron micrographs showing surface topography of various films using modified arc technology.

involve deposition of oxides and multicomponent materials for architectural glass, solar reduction applications, barrier films, and so on. Thin films of tin, zirconium nitride, titanium dioxide, zirconium dioxide, oxide of copper, and other metallic materials have been investigated for these applications. Films of TiO_2, and ZrO_2, were deposited in a reactive mode using an oxygen–gas mixture. ZrO_2 and TiO_2 films with very low absorption ($< 5\%$) in the visible light range, and excellent adhesion to plastics and glass substrates have been deposited. In fact TiO_2 films with a sharp cutoff at approximately 400 nm were found to be very suitable for UV filters.

TiN and ZrN films deposited by cathodic arc have also been investigated for architectural glass coatings. The deposition rates and stoichiometry control were found to be superior to magnetron sputtering. A deposition rate as high as 10 times that of magnetron sputtering for production scale was demonstrated.

Multicomponent films consisting of Inconel and Ni–Cr–Al–Y alloys have been also successfully deposited at rates as high as 1 mm/min. The film composition as analyzed by spectroscopic techniques (e.g., ESCA and AES) was found to be within 10–15% of the source material. This makes cathodic arc an excellent choice for multicomponent materials.

The cathodic arc deposition process has proved to be capable of fulfilling the most exacting demands in applications as diverse as tool coatings, decorative coatings, architectural glass coatings, and turbine engine coatings. Developments are continuing to broaden the range of various potential applications of the cathodic arc.

REFERENCES

1. H. Randhawa and P. C. Johnson, "A review," *Surface Coatings Technol.* 31 303 (1987).
2. J. L. Vossen and W. Kem, Eds., *Thin Film Processes*. New York: Academic Press, 1978.
3. J. A. Thornton, in *Deposition Technologies for Films and Coatings*, R. F. Bunshah, Ed. Park Ridge, NJ: Noyes Publications, 1982.
4. W. M. Mullarie, U.S. Patent No. 4,430,184 (1984).
5. H. Randhawa and p. C. Johnson, *Surface Coatings Technol.*, 33, 53 (1987).
6. H. Randhawa, presented at ASM International Strategic Machining and Materials Conference, Orlando, FL, 1987.

24

Industrial Diamond and Diamondlike Films

Arnold H. Deutchman and Robert J. Partyka

BeamAlloy Corporation, Dublin, Ohio

1.0 INTRODUCTION

The mechanical, electrical, thermal, and optical properties of diamond make it attractive for use in a variety of different applications ranging from wear-resistant coatings for tools and engineered components to advanced semiconductor structures for integrated circuit devices.[1] Until recently, diamond "coating" was done by bonding single-crystal diamond grits to the surfaces of the components to be coated. Applications for diamond coatings were limited therefore to tooling used for cutting and grinding operations.

Recent advancements in plasma-assisted chemical vapor deposition (PACVD) and ion beam enhanced deposition technologies make it possible to form continuous diamond and diamondlike carbon films on component surfaces. There new films have many of the mechanical, thermal, optical and electrical properties of single-crystal diamond, and they make possible the diamond facing of precision tools and wear parts, optical lenses and components, and computer disks, as well as the production of advanced semiconductor devices. The most flexibility, in terms of properties of the deposited diamond films and types of material coatable, is found when the films are formed using ion beam deposition techniques.

2.0 DIAMOND AND DIAMONDLIKE FILMS

The ability to diamond-coat tools and engineered components required that diamond precursor material be condensed from a vapor phase as a continuous film onto the surface of the component to be coated. Furthermore, the deposition must proceed so that the vapor-deposited material condenses with the structure and morphology of diamond. Diamond is a metastable form of carbon; as such, when condensed from a vapor or from a flux of energetic particles, it will tend to assume its most thermodynamically stable state or form-graphite. With advanced processes like chemical vapor deposition (CVD) and ion beam enhanced deposition, it is possible to influence, to a certain degree, the energy and charge states of the particles in the vapor phase, thus allowing some control over the energy state (stable or metastable) and crystallographic and stoichiometric form of the deposited films. Thus it is feasible to synthesize a variety of diamond and diamondlike films with a range of mechanical, chemical, optical, electronic, and thermal properties.[2–6] Practical applications for the various films on actual engineered components will be determined by the film properties desired (i.e., hardness, resistivity, optical transmission, etc.) and the nature of the deposition process used to produce the films.

3.0 FILM DEPOSITION TECHNIQUES

The development of techniques and technologies capable of the deposition of continuous thin diamond and diamondlike carbon films has been sparked by advances in the semiconductor and thin film deposition industry.[7–8] These new techniques enable diamond film deposition on the surfaces of both semiconductor and nonsemiconductor materials for potential use in a wide variety of applications, both electronic and nonelectronic. Two distinct coating methodologies have been developed, one relying on deposition from an excited plasma discharge sustained in a low pressure atmosphere of hydrocarbon gases (plasma-assisted CVDD), and one relying on the direct deposition of carbon films, either without or with simultaneous bombardment by an intense flux of high energy ions (ion beam enhanced deposition).

3.1 Plasma-Assisted Chemical Vapor Deposition (PACVD) Techniques

Deposition of diamondlike carbon films with the plasma-assisted CVD technique proceeds by exciting hydrogen–hydrocarbon–argon gas mixtures either in a glow discharge[9–15] or with microwave radiation[16–18] (Fig. 1). In both cases a plasma is produced, and free carbon atoms are generated by the thermal decomposition of the hydrocarbon gas component. The carbon atoms liberated in the plasma have enough energy to allow tetragonal carbon–carbon (diamond) bonding, making possible the condensation of diamond and diamondlike carbon films. The films produced are actually mixtures of trigonally bonded carbon (graphite), tetragonally-bonded carbon (diamond), and other allotropic crystalline forms.[7,8]

To dissociate the hydrocarbon starting gas and provide enough thermal energy to allow formation of trigonal and/or tetragonal carbon bonding, temperatures in the plasma discharge must exceed 2000°F. Deposition rates on the order of 1 μm/hr are achievable. The presence of free hydrogen in the processing gas helps to promote film growth with higher concentrations of tetragonally bonded diamond versus trigonally bonded graphite. This occurs because graphitic bonds are much more chemically reactive than are diamond bonds, resulting in selective etching of the graphite component of the films by free hydrogen gas in the plasma. However the presence of hydrogen trapped in the deposited diamond films can produce high levels of tensile stresses, leading to embrittlement and buckling.

Applications for diamond films formed by the PACVD techniques are limited to those in which the substrate can be raised to temperatures in excess of 2000°F. Also, the diamond films deposited by this technique grow epitaxially and therefore condense on and adhere best to crystalline substrates like silicon and germanium. Therefore diamond films deposited by PACVD techniques are best suited for applications in semiconductor structures and devices. Field effect transistors have been made by forming ohmic and Schottky contacts on semiconducting diamond substrates deposited by the CVD process. The films can also be useful as passivation layers on integrated circuits, since they can be made very hard, as well as resistant to attack by acid, alkali, and organic solvents. Formation of diamondlike carbon films on metallic, ceramic, and certainly plastic substrates calls for techniques that do not

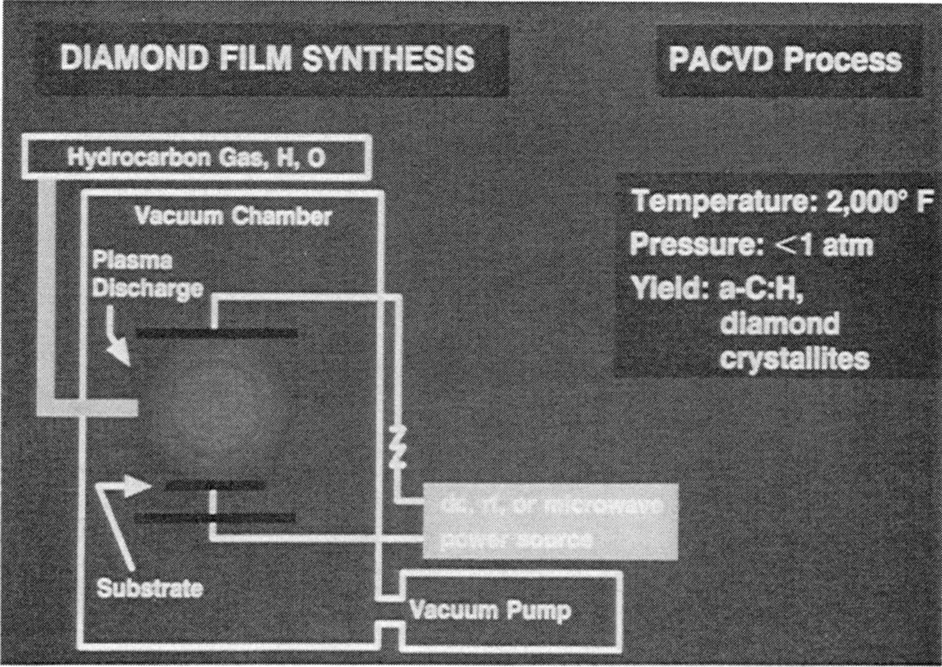

Figure 1 Diamond film synthesis with the plasma-assisted chemical vapor deposition process (PACVD). A range of hydrogenated diamond and diamondlike film structures can be produced at low pressure, and temperatures in the 2000°F range.

require high temperatures and are not as sensitive to the metallurgy of the substrate to be coated.

3.2 Ion Beam Enhanced Deposition (DIOND)

The high temperatures and pressures required for the formation of tetragonally bonded diamond structures can be provided on a microscopic level by the bombardment of films of carbon atoms with energetic ions. Bombarding ions with energies of only 100 eV produce picosecond temperature and pressure spikes at the target surface in excess of 7300°F, and 120,000 atm. The thermal agitation and shock wave accompanying 100 eV ion impingement is capable of forming single-crystal diamond nuclei approximately 1 nm in diameter. Thus under ion bombardment, carbon atoms deposited in continuous films on a substrate surface can combine at the surface to form all possible combinations of carbon–carbon bonding, including the uniform tetragonal bonding exhibited by natural diamond.

Ion beam based deposition processes can be implemented either by direct acceleration and implantation of carbon atoms[20–23] or by deposition of thin carbon films with simultaneous bombardment of the films by an additional energetic ion beam[24–27] termed ion beam enhanced deposition. The latter technique is more flexible in that a wider latitude of film morphologies and properties can be produced, higher deposition rates are achievable, and a wider variety of materials can be coated. During the ion beam enhanced deposition process, carbon films are deposited on the surface to be diamond coated while the surface is at the same time illuminated with a secondary energetic ion beam (Fig. 2). The energy of the car-

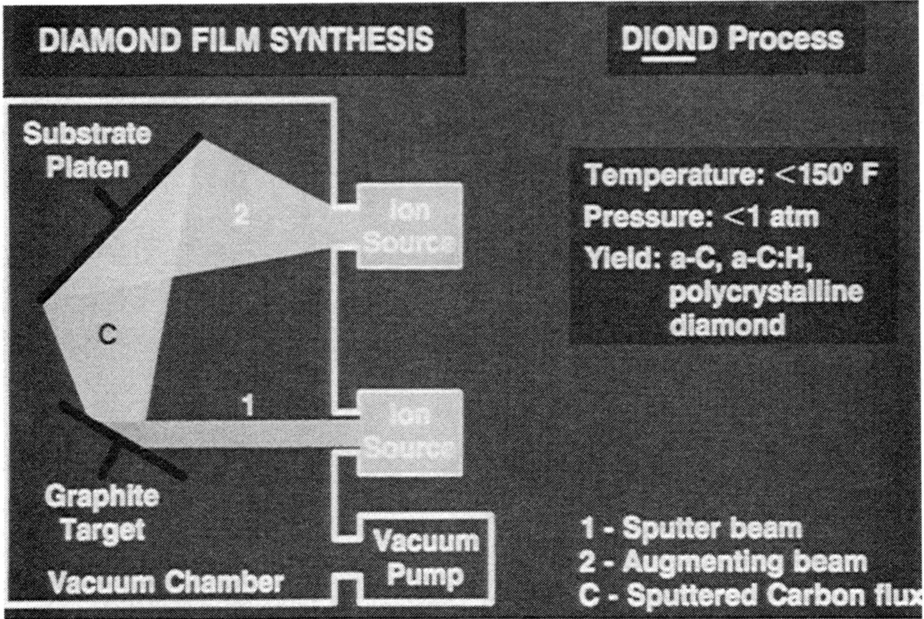

Figure 2   Diamond film synthesis with the ion beam enhanced deposition process (DIOND). A range of diamond and diamondlike film structures can be produced at low pressure, and temperatures that do not exceed 150°F .

bon atoms can be varied during the deposition, as can the energy of the secondary ion beam. No additional energy in the form of heat is required, so coating temperatures can be held below 150°F. In addition, the films are initially mixed into the surface being coated, thereby allowing them to be formed easily on virtually and substrate material, and optimizing film adhesion.

4.0   DIAMOND AND DIAMONDLIKE FILM PROPERTIES

The operating parameters of both the PACVD and the DIOND deposition processes can be adjusted to optimize the mechanical, electrical, optical, and thermal properties of the deposited films. For resistance to wear and erosion, film hardness can be optimized. Films with hardnesses in the 3000–4500 (DPHN) range are produced routinely, and even though they are thin (0.1–5.0 μm), they produce dramatic increases in wear resistance on wear components and tools.

The surfaces of natural diamond are lubricous, with extremely low coefficients of friction. The lubricity of diamondlike carbon films can be further enhanced by adjusting process parameters to increase concentrations of trigonally bonded graphitic structures. Coefficients of friction as low as 0.001 have been produced in diamondlike films rubbing against steel.

Optical properties can also be optimized by adjusting process parameters. Diamondlike films that are optically transparent in the visible and infrared regions of the electromagnetic spectrum, as well as films that are colored, can also be produced.

Films with an extremely wide range of electrical resistivities can be deposited by varying process parameters. By adjusting the relative concentration of trigonally and tetragonally bonded carbon atoms, the resistivities of the films can be varied from as low as $1\Omega\cdot$ cm to as high as 1×10^{12} cm.

5.0 POTENTIAL APPLICATIONS

The development of the PACVD and the DIOND ion beam enhanced deposition techniques enable the deposition of a range of diamond and diamondlike films with a variety of operating properties. The applications for each technique will be determined by the film properties required and the temperatures that components to be treated can sustain during coating (Fig. 3).

The high temperature PACVD process is finding primary applications in the semiconductor device area, since semiconducting and ceramic substrate materials can sustain high processing temperatures. Applications in the semiconductor electronics area include production of active semiconducting elements (amplifiers, oscillators, and electro-optic elements) that operate at higher switching speeds and higher temperatures (and thus power

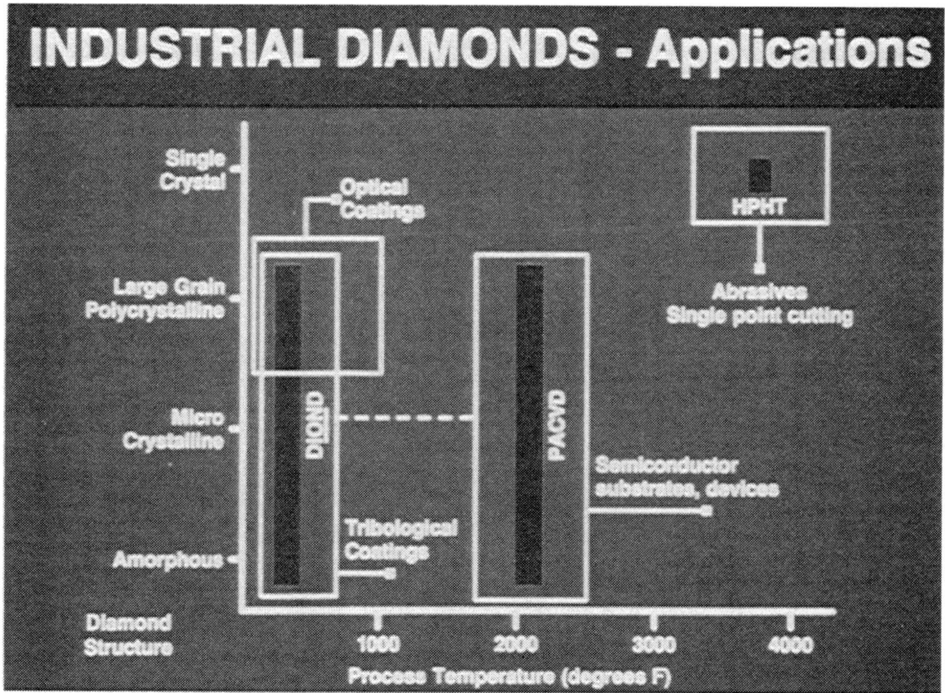

Figure 3 Potential application areas for diamond and diamondlike films shown on a continuum of film properties and processing temperatures. The PACVD process will find primary use in semiconductor electronic applications, in contrast to the ion beam enhanced deposition (DIOND) process, which will find broader use in the deposition of trbological coatings for engineered mechanical and optical components.

levels), high efficiency heat sink layers, and insulating layers for protection against oxidation and chemical contamination.

The availability of the DIOND ion beam enhanced deposition process, which is able to produce a range of diamond and diamondlike carbon films at low temperatures, and depositable on virtually any substrate material, makes a number of applications for diamond coatings feasible. Wear-resistant surfaces for a wide variety of precision tools and wear components (e.g., gears and bearings) are now being produced. The dielectric properties of the films allow them to be used as protective coatings for the magnetic recording media on fixed disks for computers, as well as the magnetic recording heads themselves. Since the films can be made optically transparent, optical components like lenses and mirrors can be surfaced for resistance to erosion and scratches. The films can also be used to reduce surface friction on components as well as to produce graphitic self-lubricating layers for applications in which conventional lubricants cannot be used.

The availability of processes capable of depositing diamond and diamondlike films on the surfaces of engineered components offers opportunities to improve the operating performance of many existing products and to develop a wide variety of new ones in the semiconductor, electronic, optical, computer, and mechanical components industries. Key to the successful use and application of these new industrial diamond and diamondlike films are a solid understanding of the range of film properties that can be produced and proper selection of the techniques used to deposit them.

REFERENCES

1. G. Graff, *High Technol.*, 7, 44 (1987).
2. J. C. Angus, et. al., *Thin Solid Films*, 118, 311 (1984).
3. C. Weissmantel, in *Thin Films from Free Atoms and Particles*, K. J. Klabunde, Ed. Orlando, FL: Academic Press, 1985, p. 153.
4. J. C. Angus, *Thin Solid Films*, 142, 145 (1986).
5. D. B. Kerwin et. al., *Thin Solid Films*, 148, 311 (1987).
6. R. Messier et. al., *Thin Solid Films*, 153, 1 (1987).
7. R. C. DeVries, *Annu. Rev. Mater. Sci.*, 17, 161 (1987).
8. J. C. Angus et. al., *Science*, 241, 913 (1988).
9. D. S. Whitmell et. al., *Thin Solid Films*, 35, 255 (1976).
10. B. V. Spitsyn et. al., *J. Cryst. Growth*, 52, 219 (1981).
11. K. Enke, *Thin Solid Films*, 80, 227 (1981).
12. S. Matsumoto et. al., *J. Appl. Phys.*, 21, 1183 (1982).
13. E. T. Prince, *J. Vac. Sci. Technol. A3*, 3, 694 (1985).
14. J. D. Warner et. al., *J. Vac. Sci. Technol. A3*, 3, 900 (1985).
15. A. R. Nyaiesh et. al., *J. Vac. Sci. Technol. A3*, 3, 610 (1985).
16. O. Matsumoto et. al., *Thin Solid Films*, 128, 341 (1985).
17. O. Matsumoto et. al., *Thin Solid Films*, 146, 283 (1986).
18. N. Fujimori et. al., *Vacuum*, 36, 99 (1986).
19. S. Aisenberg et. al., *J. Appl. Phys.*, 42, 2953 (1976).
20. E. G. Spenser et. al., *Appl. Phys. Lett.*, 29, 228 (1976).
21. J. H. Freeman et. al., *Nuclear Instrum. Methods*, 135, 1 (1976).
22. T. Miyazawa et. al., *J. Appl. Phys.*, 55, 188 (1984).
23. J. W. Rabalais et al., *Science*, 239, 623 (1988).
24. C. Weissmantel, *Thin Solid Films*, 92, 55 (1982).
25. M. J. Mirtich et. al., *Thin Solid Films*, 131, 245 (1985).
26. C. Weissmantel et. al., *J. Vac. Sci. Technol. A4*, 6, 2892 (1985).
27. A. H. Deutchman, et. al., *Ind. Heating, LV(7)*, 12 (1988).

25

Tribological Synergistic Coatings

Walter Alina

General Magnaplate Corporation
Linden, New Jersey

1.0 INTRODUCTION

The solution of wear and many related problems for any application is very much experience dependent. A scientific basis for resolving these problems unfortunately has not yet been found. Using experience and history, it is possible to recommend a number of potential solutions; however, the ultimate proof is in the actual trial of the application. This is because there are so many variables within each application that the slightest change could make a difference in the selection of the appropriate coating. Even though applications appear to be identical, there are always slight differences such that the same coating selection will not always perform in the same manner.

The production of synergistic coatings on steel (Nedox) or aluminum (Tufram) is based on the principle of infusion of a dry lubricant or polymer into the coatings. General Magnaplate has developed a family of such coatings (Nedox), each one representing specific properties, such as hardness, lubricity, corrosion protection, and dielectric strength. The standard hardfacing for steel is an electroless nickel coating. There are a number of electroless nickels that vary the phosphorus content and consequently have differences in hardness and corrosion resistance. Choice of such a coating varies and is based on the application requirements.

Synergistic coatings for aluminum (Tufram) have been used successfully for many years. The system can accommodate almost all aluminum alloys, provided a copper content of 5% and a silicon content of 7% are not exceeded. Higher percentages of these constituents (set up too great a change in substrate resistivity, hence) prevent the buildup of required film thickness.

The prime purpose of the Tufram system is to produce films having properties such as improved wear resistance, better surface release (lower coefficient of friction), good corrosion resistance, and high dielectric strength.

The principle of these coatings is based on a hardcoat after which a polymer or dry lubricant is infused into the coating substrate.

All coatings are used in a wide variety of industries. Some are in compliance with the regulations of the U.S. Food and Drug Administration and can be used in food and medical applications.

The improvement in wear resistance to aluminum ranges from 5 to 25 times that without the coating. It is difficult to put an exact number on the improvement, since it varies from one application to another.

2.0 WHAT ARE SYNERGISTIC COATINGS?

Synergistic coatings are not really "coatings" in the conventional sense of the word. They are created during multistep processes that combine the advantages of anodizing or hardcoat plating with the controlled infusion of low friction polymer and/or dry lubricants. These "coatings" become an integral part of the top layers of the base metal rather than merely a surface cover. Since the resulting surfaces are superior in performance both to the base metal and to the individual components of the coatings, the proprietary processes that produce them are identified as "synergistic." Why do they work?

The Tufram process, a family of coatings for aluminum alloys, converts the hydrated aluminum oxide $Al_2O_3 \cdot H_2O$ and replaces the H_2O of the newly formed ceramic surface with inert polymeric material that provides a self-lubricating surface. In the process, the aluminum crystals expand and form porous anchor crystals that remain hygroscopic for a short period of time.

The particles of the specific polymer selected are then introduced under controlled conditions of properly balanced solutions, time, and temperature, to permanently interlock with the newly formed crystals.

This results in a harder-than-steel continuous lubricating plastic–ceramic surface of which the polymeric particles become an integral part (Fig. 1).

Synergistic coatings are a great asset in solving many wear problems, but how were these problems resolved years ago? Their solution was expensive, time-consuming, and very frustrating.

Normally a surface was built up with 0.005–0.020 in. of hard chrome, then precision machined to a specification. After this surface had been polished to 4 RMS or better, it basically qualified as a wear surface because it was hard and the 4 RMS microfinish reduced the coefficient of friction. If the coefficient of friction of such a surface were compared to the coefficients for synergistic coatings (see Table 1), the latter would be seen to have much lower numbers, consequently lower frictional values.

It is important to have some familiarity with the coefficient of friction between materials. Table 1 gives both static and dynamic coefficients of friction of many basic coatings. Each fractional number implies the frictional forces applied against two surfaces.

The coefficients in Table 1 are to be considered only as a guide. Many factors can alter figures derived from these laboratory constants, including:

Applied loads
Point loading
Loading stresses
Substrate hardness
Temperature conditions
Ambient moisture present

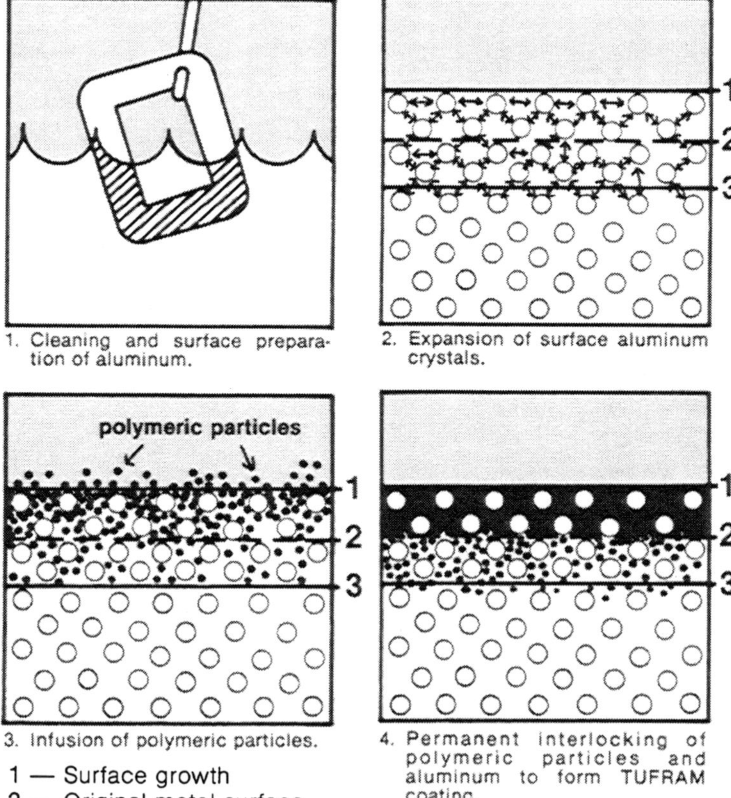

1. Cleaning and surface preparation of aluminum.

2. Expansion of surface aluminum crystals.

3. Infusion of polymeric particles.

4. Permanent interlocking of polymeric particles and aluminum to form TUFRAM coating.

1 — Surface growth
2 — Original metal surface
3 — Penetration into metal

Normally you can estimate ½ penetration and ½ growth for most alloys, i.e., thickness 0.002 means 0.001 into metal and 0.001 growth above the original surface.

Figure 1 The Tufram process sequence.

Environmental variations
Microfinish components

Because of the complexity of the wear and friction phenomena, all these conditions must be considered when designing any system.

All testing was standardized as follows:

1. Specimens were ground and polished to 4–6 RMS.
2. Laboratory temperature was maintained at 72°F, with a relative humidity of 54%; absolute humidity of Hg. 10.6.
3. An average of five readings per test run.
4. A T.M.I. Slip and Friction Tester (model 98.5) was used, with a constant load of 200 g.
5. The materials and coatings were listed in Table 2.
6. The friction tester used was the T.M.I. slip and friction tester Model No. 98–5, using a constant 200 gram load. It is shown in Figure 2.

TABLE 1 Friction Data by Materials

Upper plate[a]	Lower plate[a]	Coefficients of friction	
		Static	Kinetic
Ice	Ice	0.000	0.000
Hi-T-Lube	Hi-T-Lube	0.251	0.217
Hi-T-Lube	Steel	0.056	0.049
Hi-T-Lube [b]	Hi-T-Lube [b]	0.034	0.034
Lectrofluor 604	Glass	0.190	0.172
Lectrofluor 604P	Teflon	0.106	0.089
Magnagold	Magnagold	0.245	0.211
Magnagold	Magnagold + Ni	0.484	0.357
Magnagold	Teflon	0.150	0.123
Magnagold	Steel	0.300	0.246
Magnagold	Nickel	0.326	0.259
Magnagold	Glass	0.177	0.155
Magnagold	Aluminum	0.248	0.220
Magnagold	Chromium[a]	0.193	0.174
Magnagold	Steel	0.313	0.285
Magnagold	Nickel	0.367	0.329
Magnagold	Titanium P	0.559	0.494
Magnagold	Copier paper	0.518	0.497
Magnagold	MOS/2	0.304	0.270
Magnagold	Hi-T-Lube	0.264	0.244
Magnagold	Graphite over paper	0.260	0.234
Magnagold TFE	Magnagold + TFE	0.225	0.174
Magnaplate HCR	Magnaplate HCR	0.198	0.174
Magnaplate HCR	Glass	0.138	0.125
Magnaplate HCR	Aluminum	0.346	0.289
Magnaplate HCR	Teflon	0.142	0.120
Magnaplate HMF	Hi-T-Lube [b]	0.032	0.031
Magnaplate HMF	Magnaplate HMF	0.160	0.147
Magnaplate HMF	Teflon	0.059	0.053
Magnaplate HMF	Glass	0.251	0.192
Magnaplate HMF	Steel	0.212	0.181
Nedox S/F 2	Steel	0.301	0.260
Nedox S/F 2	Teflon	0.103	0.090
Nedox S/F 2	Nedox S/F 2	0.179	0.123
Nedox S/F 2	Glass	0.137	0.130
Tufram 604	Aluminum	0.429	0.372
Tufram H–2	Tufram H–2	0.171	0.139
Tufram H–2	Glass	0.203	0.169
Tufram H–2	Aluminum	0.377	0.264
Tufram H–2	Teflon	0.134	0.120
Tufram H–0	Tufram H–0	0.249	0.222
Tufram H–0	Glass	0.180	0.150
Tufram H–0	Aluminum	0.251	0.219

TABLE 1 Continued

		Coefficients of friction	
Upper plate[a]	Lower plate[a]	Static	Kinetic
Tufram H–0	Teflon	0.121	0.103
Tufram L–4	Tufram L–4	0.184	0.173
Tufram L–4	Aluminum	0.353	0.294
Tufram L–4	Glass	0.256	0.189
Tufram L–4	Teflon	0.142	0.130
Tufram R66	Glass	0.162	0.149
Tufram R66	Tufram R66	0.148	0.115
Tufram R66	Aluminum	0.329	0.272
Tufram R66	Teflon	0.133	0.100
Aluminum	Titanium A	0.413	0.376
Aluminum	Titanium P	0.614	0.531
Aluminum	Teflon	0.237	0.186
Aluminum	Glass	0.175	0.137
Aluminum	Aluminum	0.646	0.563
Aluminum	Magnagold	0.304	0.263
Aluminum	Chromium	0.199	0.185
Aluminum	Nickel	0.258	0.233
Aluminum	Steel	0.466	0.375
Chromium	Chromium A	0.176	0.159
Chromium	Aluminum	0.266	0.216
Chromium	Magnagold	0.176	0.149
Chromium	Nickel	0.405	0.356
Chromium	Steel	0.254	0.210
Copier paper	Copier paper	0.275	0.259
Graphite over paper	Graphite over paper	0.322	0.302
Hard chromium	Titanium P	0.344	0.304
Hard chromium	Teflon	0.095	0.078
Hardcoated aluminum	Glass	0.151	0.127
Hardcoated aluminum	Teflon	0.178	0.157
Hardcoated aluminum	Hardcoated aluminum	0.264	0.220
MOS/2	MOS/2	0.433	0.418
Nickel	Teflon	0.148	0.120
Nickel	Chromium	0.192	0.174
Nickel	Aluminum	0.330	0.253
Nickel	Magnagold	0.308	0.267
Nickel	Nickel	0.317	0.279
Steel	Titanium P	0.493	0.410
Steel	Hi-T-Lube	0.254	0.218
Steel	Graphite over paper	0.245	0.225
Steel	Aluminum	0.349	0.247
Steel	Magnagold	0.377	0.308
Steel	Magnagold +Ni	0.675	0.607
Steel	Teflon	0.269	0.269
Steel	Nickel	0.723	0.553
Steel	Glass	0.127	0.116

TABLE 1 Continued

Upper plate[a]	Lower plate[a]	Coefficients of friction	
		Static	Kinetic
Steel	Chromium	0.202	0.174
Steel	Nickel	0.431	0.333
Steel	Magnagold	0.218	0.194
Steel	Nickel	0.353	0.315
Steel	Steel	0.423	0.351
Teflon	Titanium A	0.232	0.209
Teflon	Titanium P	0.291	0.240
Teflon	Hard chronium	0.210	0.191
Teflon	Magnagold + ni	0.209	0.160
Teflon	Magnagold	0.161	0.114
Teflon	Magnaplate HCR	0.178	0.167
Teflon	Magnaplate HMF	0.172	0.154
Teflon	Nedox SF2	0.149	0.120
Teflon	Tufram H–2	0.137	0.127
Teflon	Tufram H0	0.167	0.138
Teflon	Tufram L4	0.149	0.131
Teflon	Tufram R66	0.180	0.149
Teflon	Teflon	0.083	0.070
Teflon	Steel	0.184	0.157
Teflon	Nickel	0.223	0.190
Teflon	Hardcoated aluminum	0.207	0.183
Teflon	Glass	0.097	0.097
Teflon	Aluminum	0.194	0.177
Titanium A	Steel	0.358	0.317
Titanium A	Magnagold	0.264	0.226
Titanium A	Hard chromium	0.375	0.332
Titanium A	Aluminum	0.345	0.288
Titanium P	Steel	0.369	0.283
Titanium P	Magnagold	0.290	0.258
Titanium P	Hard chromium	0.328	0.288
Titanium P	Aluminum	0.430	0.321
Titanium A	Titanium A	0.359	0.303
Titanium A	Teflon	0.174	0.142
Titanium P	Titanium A	0.415	0.370
Titanium P	Teflon	0.223	0.193

[a]A superscript "a" indicates additional polish after coating; a "b" indicates postburnishing—comparable to breaking in the surface.

Table 2   Key to Materials and Coatings Listed in Table 1

Designation in Table 1	Description
Aluminum	6061 T6 grade (0.250) thickness
Steel	1032 grade H32 (0.250) thickness
Titanium A	6Al/4V (0.250) thickness
Titanium P	Vacuum deposited at 10–5 torr, 2 μm thickness, purity 99.99%
Glass	Tempered (0.250) thickness
Teflon	White, virgin grade (0.250) thickness
Nickel	Autocatalytic 6/8% phosphorus (0.001)
Hard chromium	Industrial grade (0.0003)
Hard anodize	6061 T6 (0.002)
Tufram	Proprietary aluminum coating
Nedox	Proprietary treatment for steel and stainless steels and nonferrous metals
Hi-T-Lube	Proprietary solid film metal alloy lubricant
Magnagold	Proprietary method for vacuum coating of titanium nitride
Magnaplate HMF	Proprietary ultrahard, high microfinish for most base metals
Magnaplate HCR	Proprietary ultrahard and exceptionally corrosion-resistant coating for aluminum

3.0   WEAR TESTING

Wear tests are difficult for most people to comprehend. However, a system that is considered a reasonably good standard is the Taber Abraser (Fig. 3). This instrument requires a panel 4 in. square, coated with the material that is being tested. It can vary loads (grams), the number of cycles (generally 10^3–10^4) and the type of abrasive wheel (either CS–10 or CS–17) being used.

After a set of conditions has been established, it is easy to use them as a standard for new coatings, since there now is a basis for comparison.

The principle of this wear method is a weight loss of coating based on the degree of wear.

Note the conditions and weight loss in Figures 4 and 5 for aluminum and steel, respectively.

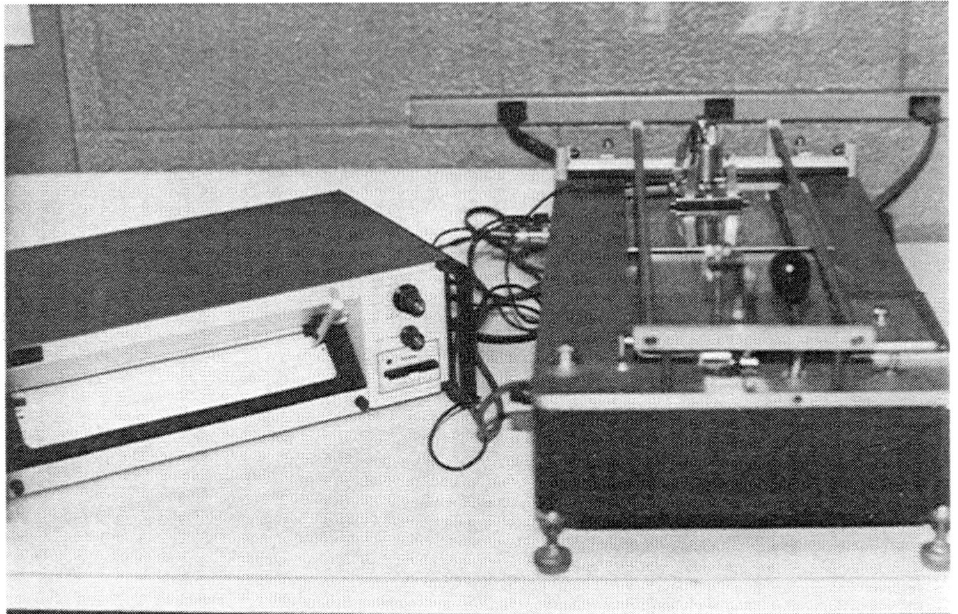

Figure 2   T.M.I. Slip and friction tester.

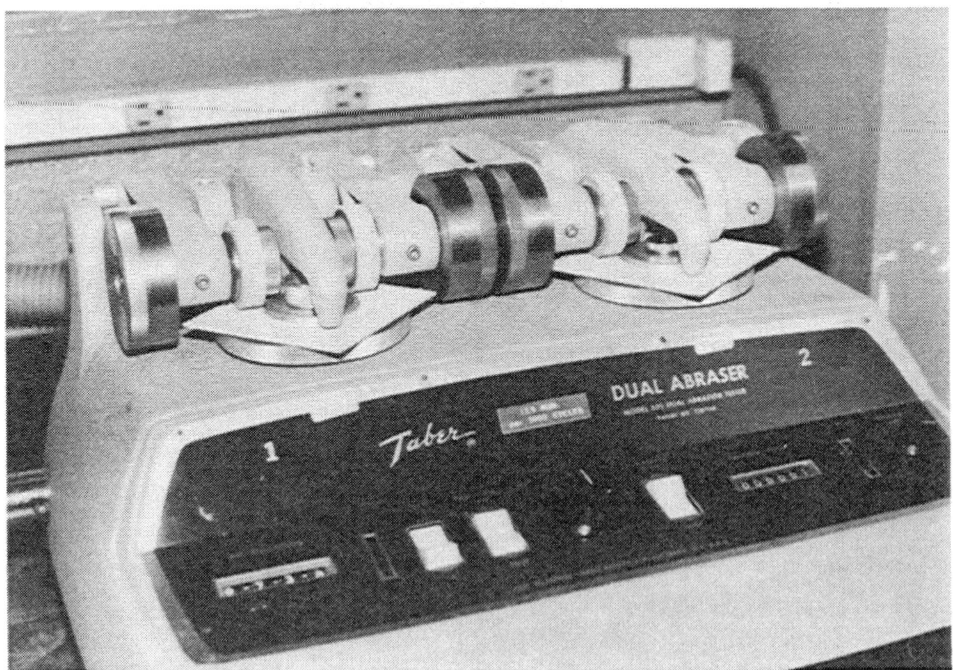

Figure 3   The Taber Abraser.

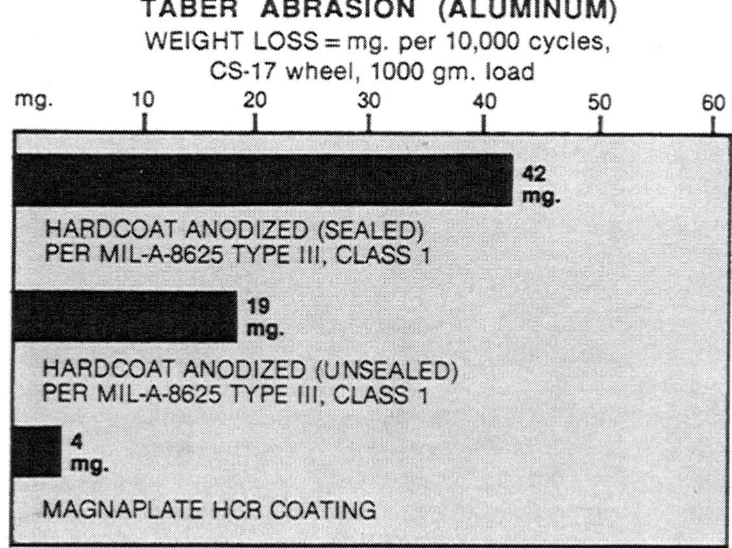

Figure 4   Weight loss following Taber abrasion for aluminum samples with various coatings.

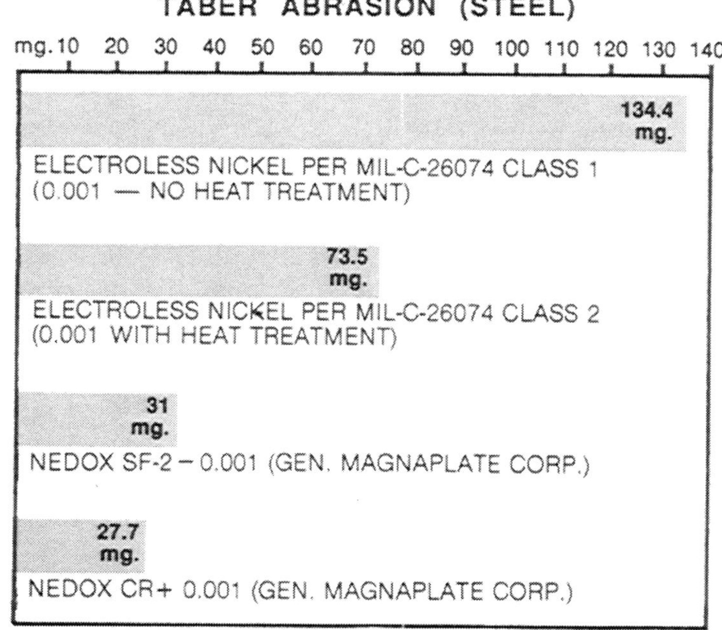

Figure 5   Weight loss following Taber abrasion for steel samples with various coatings.

4.0 COATING FAMILIES

4.1 Polymer Coatings (Lectrofluor)

The properties that can be easily achieved using polymer or organic coatings are dielectric, chemical corrosion, and radiation resistance; in addition, special organic coatings can be used with food and pharmaceutical applications in compliance with regulations of the U.S. Food and Drug Administration.

All coatings are proprietary, since they are new developments, and specific nomenclature (601,604,615,611, etc.) is used for identification.

Coatings can be applied by spraying, by dipping, or by an electrostatic powder process. In most cases, a curing temperature will be in the range of 300–750°F will achieve maximum surface hardness and minimize porosity. Polymer coatings normally range from 0.001 to 0.015 in. in thickness.

Salt spray resistance of these coatings is excellent: approximately 2000 hours (6 years in the atmosphere). The chemical resistance is the main property of the 601 coating, which can withstand strong acids and alkalies at temperatures up to 2000°F. The 604 material has similar resistance but can be utilized in food applications. The 615 is a tough, durable coating (D-80 durometer) that can withstand temperatures from –400 to 500°F, and can also be used in food applications. The 611 coating, which has good release characteristics, also is used in applications in the baking industry and comes in a variety of colors.

4.2 Magnesium (Magnadize) and Titanium (Canadize)

Since magnesium and titanium are widely used materials, especially in the field of aerospace and computers, their users often are confronted with very severe problems.

A system for hardfacing each material has been developed which is basically an electrochemical process. The synergistic coating system is still far the best for a wear application, however, and special fluoropolymers or dry lubricants are infused into the hardfacing. For magnesium the synergistic coating is called Magnadize and for titanium it is called Canadize.

For magnesium the thickness can vary from a minimum of 0.0002 in. to a maximum exceeding 0.0015 in. Normal application thickness is approximately 0.0005–0.001 in.

It is more difficult to build up thickness for titanium, so the normal application thickness is between 0.0002 and 0.0005 in.

A typical application for Magnadize would be a magnesium engine mount for aircraft. The entire mount would be hardcoated, and the gear spline would receive dry lubricant to improve the efficiency of the part.

A classic application for Canadizing is titanium hardware for aircraft. Such components are anodized with an infusion to a thickness of 0.0002–0.0004 in. to prevent the titanium from seizing.

4.3 Titanium Nitride (Magnagold)

Since the introduction of titanium nitride to the industrial sector, coating companies have been offering this service without sufficient knowledge of its engineering properties. The coating was known to be extremely hard, and it was being applied to cutting tool exclusively. A study of this coating material, however, revealed some interesting properties, noted in Tables 3–5.

One of the unique features of the titanium nitride process is the uniformity of the coating. This is critical for many applications involving missile, computer, and semiconductor appli-

Table 3   Some Physical Properties of Magnagold Coatings

Hardness	R_c80–85
Chemical resistance to 30% concentrations of nitric and sulfuric acids on copper and steel substrates at ambient temperature	Virtually no attack
Alkali resistance	Virtually no attack
Taber abrasion test, CS 10 wheel, 1000 g load 10,000 cycles	Average weight loss > 0.5 mg
Coating thickness	1–3 μm
Uniformity of thickness	15×10^{-5} 0.000015 in. (max)
Crystal lattice	Body-centered cubic, a = 4.249 ô
Density	5.44 g/cm^4
Thermal conductivity, cal/cm/sec/°C	~ 0.162 (at 1500°C) ~ 0.167 (at 1600°C) ~ 0.165 (at 1700°C) ~ 0.136 (at 2300°C)
Coefficient of thermal expansion	$\times 10^{-6}$ cm/°C 9.35± 0.04 (at 25 ~1100°C)
Electrical resistivity	40 μΩ(at 27°C)
Thermonic emission work function	3.75 eV
Microhardness (Hu)	2050 kg/mm^2 (load 100 g)

cations. Other excellent applications are found in the plastic extrusion industry (which utilizes the superior release properties in molds and dies); the medical industry (for coating delicate surgical instruments, and preventing solder adherence when special heat treatments are employed in the complex multistep process).

In this Magnagold process, a uniform coating, held to within a few millionths of an inch thickness, can be applied safely to even the most critical, closest tolerance parts via special cleaning fixturing mechanisms and techniques that enable the part to be rotated 360° while traveling through the vapor stream. The process sequence is outlined in Figure 6; an production setup appears in Figure 7.

Because the parts to be coated are rotated both radially and axially within the unique ion bombardment chamber, there are no restrictions on shapes. In addition, all surfaces, except clamp/and fixture areas, are uniform in coating thickness. Normally, the thickness ranges from 0.00003 and 0.0002 in., with final determination being edge sharpness and/or wear. (Load life is the ultimate criterion for the final decision.)

Since the condition of the substrate surface is a key element in the effectiveness of any physical vapor deposition (PVD) process, General Magnaplate has engineered an exclusive, proprietary predeposition surface preparation. A combination of specialty designed equipment and chemical cleaning techniques prepares the component surface to assure permanently interlocked anchoring of the "coating." Conventional vapor deposition applica-

Table 4 Wettability of Various Metal Surfaces

Wettability Metal	Contact	(Temperature/Condition)
Cu	180°	(1100°C vac)
	126°	(1180°C vac)
	148°	(1560°C NH$_3$)
	136°	(1130°C Ar)
Al	147°	(850–1000°C vac)
	135°	(900°C Ar)
Cd	139°	(450°C vac)
Pb	102°	(450°C vac)
Sn	140°	(350°C vac)
Bi	147°	(370°C Ar)
Fe	100°	(1500°C vac)
	132°	(1550°C Ar)
Co	104°	(1550°C vac)
Ni	~70°	(1550°C vac)
	110°	(1450–1500°C N$_2$)

tors are not equipped with the extensive facilities that permit the meticulous care and attention required in the precleaning phase of the process. The parts are mounted on a specially designed cylindrical fixture, and then the entire work cylinder enters the vacuum chamber. A vacuum (1 x 10^{-6} torr) is achieved, after which the system is purged with argon gas as an additional cleaning step. Titanium metal (99.9%) is then vaporized by a plasma energy source. This is followed by the precise introduction of nitrogen, the reactive gas, into the chamber. The parts to be coated are cathodically charged by high voltage (dc), thereby attracting accelerated ions of titanium. Simultaneously, they combine with nitrogen to produce the tightly adhering, highly wear-resistant titanium nitride PVD coating.

Note: Some alloys are sensitive to temperatures up to 900°F and can be reduced in hardness if (The substrate material selected is not heat-compatible with this process.) It is possible to lower processing temperature to prevent certain steels from annealing; however, there may be a slight reduction in the hardness of the titanium nitride coating.

Table 5 Static (S) and Kinetic (K) Coefficients of friction for Variously Coated Components of a Panel Assembly

Lower panel	Upper panel	
	Steel (4–8 μm in. RMS)	Teflon
Steel (4–8 RMS)	S: 0.534 ± 0.079	0.184 ± 0.029
	K: 0.400 ± 0.093	0.157 ± 0.029
Magnagold (4–8 RMS)	S: 0.218 ± 0.028	0.161 ± 0.014
	K: 0.194 ± 0.030	0.114 ± 0.017

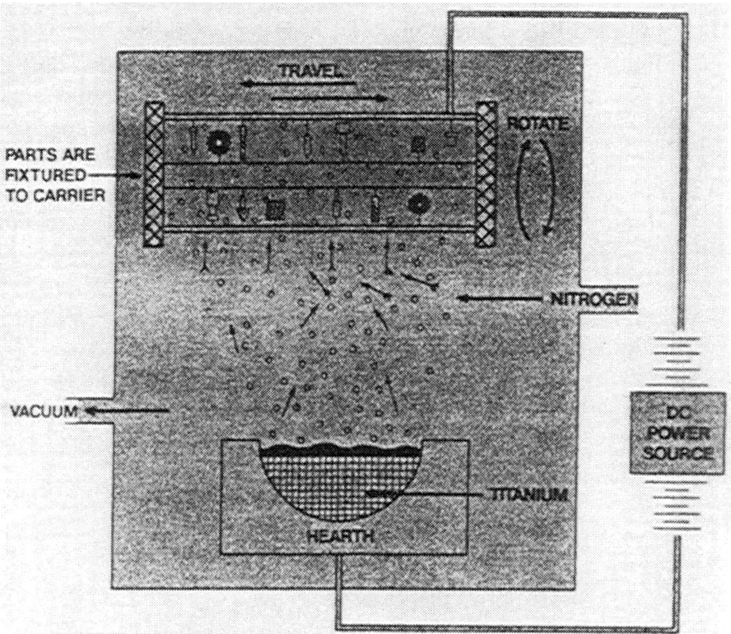

Figure 6 The Magnagold process sequence.

Figure 7 The Magnagold production setup.

sive, proprietary predeposition surface preparation. A combination of specialty designed equipment and chemical cleaning techniques prepares the component surface to assure permanently interlocked anchoring of the "coating." Conventional vapor deposition applicators are not equipped with the extensive facilities that permit the meticulous care and attention required in the precleaning phase of the process. The parts are mounted on a specially designed cylindrical fixture, and then the entire work cylinder enters the vacuum chamber. A vacuum (1×10^{-6} torr) is achieved, after which the system is purged with argon gas as an additional cleaning step. Titanium metal (99.9%) is then vaporized by a plasma energy source. This is followed by the precise introduction of nitrogen, the reactive gas, into the chamber. The parts to be coated are cathodically charged by high voltage (dc), thereby attracting accelerated ions of titanium. Simultaneously, they combine with nitrogen to produce the tightly adhering, highly wear-resistant titanium nitride PVD coating.

Note: Some alloys are sensitive to temperatures up to 900°F and can be reduced in hardness if (The substrate material selected is not heat-compatible with this process.) It is possible to lower processing temperature to prevent certain steels from annealing; however, there may be a slight reduction in the hardness of the titanium nitride coating.

26

Chemical Vapor Deposition

Deepak G. Bhat

GTE Valenite Corporation, Troy, Michigan

1.0 INTRODUCTION

Chemical vapor deposition (CVD) is a technique of modifying properties of surfaces of engineering components by depositing a layer or layers of another metal or compound through chemical reactions in a gaseous medium surrounding the component at an elevated temperature. In formal terms, CVD may be defined as a technique in which a mixture of gases interacts with the surface of a substrate at a relatively high temperature, resulting in the decomposition of some of the constituents of the gas mixture and the formation of a solid film of coating of a metal or a compound on the substrate.

2.0 PROCESS

A modern CVD system includes a system of metering a mixture of reactive and carrier gases, a heated reaction chamber, and a system for the treatment and disposal of exhaust gases. Figure 1 shows the basic arrangement of various components of industrial CVD systems.

The gas mixture (which typically consists of hydrogen, nitrogen or argon, and reactive gases such as metal halides and hydrocarbons) is carried into a reaction chamber that is heated to the desired temperature by suitable means. The various techniques include resistance heating with Kanthal, Globar (SiC) or graphite heating elements, or induction. In some cases, the substrate is heated directly by passing an electric current through it.

Typical operating parameters for a conventional CVD process are shown in Table 1. Different variations of the conventional method have been developed over the last few decades. These include moderate-temperature CVD (MTCVD), plasma-assisted CVD (PACVD) and laser CVD (LCVD). In the MTCVD process, the reaction temperature is reduced to below about 850°C by the use of metalorganic compounds as precursors. Therefore, this technique is also referred to as metalorganic CVD (MOCVD). In the microelectronics field

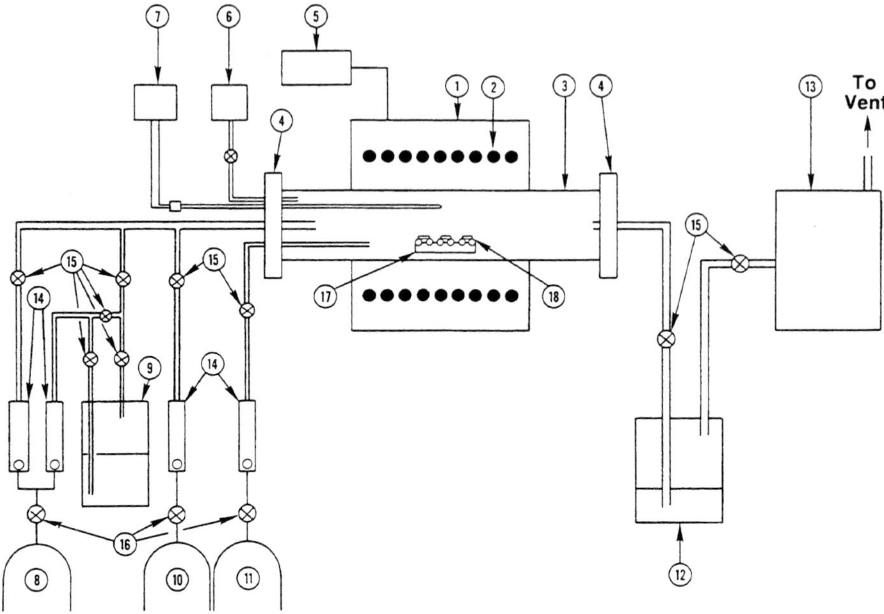

Figure 1 Schematic diagram showing various components of a typical chemical vapor deposition system operating at one atmosphere pressure: 1: Reactor, 2: Heating elements, 3: Reaction chamber, 4: Water-cooled end flanges, 5: Power controller, 6: Pressure gauge, 7: Temperature sensor and controller, 8, 10 and 11: Precursor gas sources, 9: Metal halide (liquid) vaporizer, 12: Particulate trap, 13: Gas scrubber, 14: Flow meters, 15: Flow control valves, 16: Gas tank regulators, 17: Substrate support, 18: Substrate. For a system operating at low pressures, the vacuum pumping system is connected between 12 and 13.

Table 1 Typical Parameters in Conventional Chemical Vapor Deposition

Temperature	Pressure	Precursors
$\geq 800°C$; typically up to $2000°C$	$\leq 10^{-6}$ torr to 1	Reactive gases: metal halides, carbonyls Reducing gases: H_2 Inert gases: Ar, N_2 Other gases: CH_4, CO_2, NH_3, other hydrocarbons

Table 2 Chemical Reactions in CVD

Reaction	Equation
Thermal decomposition, or pyrolysis	$CH_3SiCl_3 \rightarrow SiC + 3HCl$
Reduction	$WF_6 + 3H^2 \rightarrow W + 6HF$
Oxidation	$SiH_4 + O_2 \rightarrow SiO_2 + 2H_2$
Hydrolysis	$2AlCl_3 + 3H_2O \rightarrow Al_2O_3 + 6HCl$
Coreduction	$TiCl_4 + 2BCl_3 + 5H_2 \rightarrow TiB_2 + 10HCl$

where this technique of widely used, it is also commonly referred to as organometallic vapor phase epitaxy (OMVPE). In the PACVD technique, the heating of the gas mixture is accomplished by creating a high energy plasma that activates the chemical reactions at considerably reduced temperatures as compared to the conventional CVD. In the case of LCVD techniques, the same effect is achieved by using a laser beam to heat the gas volume or the substrate.

All CVD systems require a mechanism by which the products of the chemical reaction are treated. These products contain various reactive and potentially hazardous constituents, as well as particulate matter, which must be trapped and neutralized before the gases are exhausted to the atmosphere. In addition, since most CVD processes are carried out at subatmospheric pressures, the pumping equipment must be protected from the relatively hot, corrosive gases. This is usually done by using nonreactive materials for the pump components.

The deposition of coatings by CVD can be achieved in a number of ways. The chemical reactions utilized in CVD are shown in Table 2. These reactions between various constituents occur in the vapor phase over the heated substrate, and a solid film is deposited on the surface. The coatings are, therefore, termed "overlay" coatings. On the other hand, a surface film can also be deposited by causing a reaction between the substrate surface and one or more of the constituents of the vapor phase. One example of this technique is the formation of a nickel aluminide film on the surface of nickel by a reaction of aluminum trichloride and hydrogen from the vapor with the surface of nickel at a high temperature. Such a coating may, therefore, be called a conversion coating.

3.0 APPLICATIONS

The CVD technique is applicable for the deposition of a wide variety of materials, such as metals, compounds, ceramics, powders, and whiskers. Typical materials deposited by CVD, and applications of the CVD technique are summarized in Tables 3 and 4. One of the earliest applications of CVD was in the manufacture of pigments. Powders of TiO_2, SiO_2, carbon black, and other materials such as Al_2O_3, Si_3N_4, and BN, have been routinely made by CVD. In a variation of the conventional CVD technique, powders of nuclear fuel materials from the fuel rods used in nuclear reactors have been coated in a fluidized bed with coatings of SiC, graphite, and ZrC for containment of fission products.

Because of the nature of the process, CVD is used to deposit high-purity metals from their halide or carbonyl precursors, and the technique has been especially useful for synthesizing refractory metals. The most commonly used precursors for CVD are metal halides. For a successful application of CVD, it is necessary to be able to decompose the halides at

Table 3 Typical Materials Deposited by CVD

Material	Example
Metals	Al, As, Be, Bi, Co, Cr, Cu, Fe, Ge, Hf, Ir, Mo, Nb, Ni, Os, Pb, Pt, Re, Rh, Ru, Sb, Si, Sn, Ta, Th, Ti, U, V, W, Zr; also carbon and boron
Compounds	II–VI and III–V compounds, borides, carbides, nitrides, and silicides of transition metals, as well as sulfides, phosphides, aluminides, etc.
Ceramics	Al_2O_3, AlN, B_2O_3, BN, SiC, Si_3N_4, UO_2, Y_2O_3, ZrO_2, etc.

relatively moderate temperatures (e.g., $\leq 1000°C$). Thus, many metals whose halides are stable in this temperature range are difficult to deposit successfully. In many of these cases, organometallic compounds have been used successfully. Typical examples of metallic components deposited by CVD are shown in Figure 2.

In recent years, another interesting application of the CVD technology has achieved prominence. Deposition of whiskers of metals and refractory compounds is of significant technological importance because of the potential for development of composite materials. Composites have become a very important new class of engineering materials for aerospace structural applications.

Whiskers are needle-shaped single crystals of materials, typically 1 μm or less in diameter and several micrometers long. It has been demonstrated that the addition of whiskers to ceramics, which are inherently brittle, significantly improves their fracture toughness. Various refractory compounds have been deposited in the form of whiskers by CVD. These include Al_2O_3, Cr_3C_2, SiC, Si_3N_4, TiB_2, TiC, TiN, ZrC, ZrN and ZrO_2. Figure 3 shows an example of TiC whiskers deposited by CVD. It is to be expected that with the increasing prominence of the composite materials in the advance engineering components, many more materials will be synthesized in the whisker and fiber forms for these applications.

As stated earlier, the conventional CVD calls for relatively high temperatures. This requirement imposes certain limitations on the type of substrate that can be successfully used for deposition. Typically, most ceramic materials, graphite, and refractory metals such as tungsten and molybdenum are found to be quite suitable because of their high thermal and chemical stability in typical CVD process environments. Steels have also been used suc-

Table 4 Applications of the CVD Technique

Tribological coatings	Decorative films
Wear-resistant coatings	Superconducting films
High-temperature coatings for oxidation resistance	Emissive coatings
	Coatings for fiber composites
Dielectric insulating films	Free-standing structural shapes
Optical/reflective films	Powders and whiskers
Photovoltaic films	

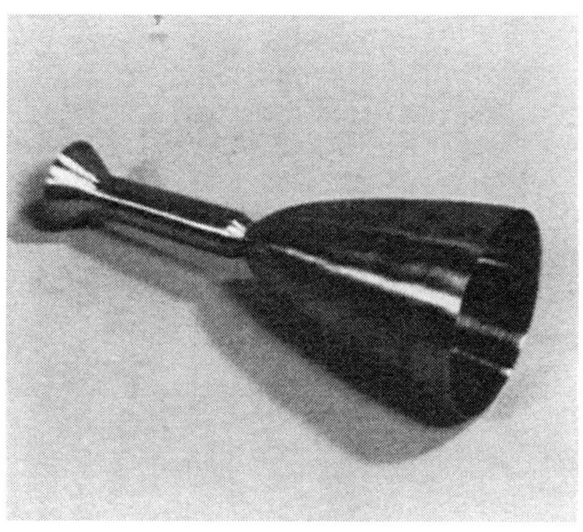

A B

C

Figure 2 Metallic components deposited by CVD. (A) Iridium-coated rhenium thrust chamber for liquid rockets, 75 mm major diameter × 175 mm length × 0.75 mm wall thickness; (B) Tungsten crucible, 325 mm diameter × 575 mm height × 1.5 mm wall thickness; (C) Tungsten manifold, about 175 mm long. (Photographs courtesy of Ultramet Corporation, Pacoima, California; reprinted with permission.) Figure 2 of "A Review of Chemical Vapor Deposition Techniques, Materials & Applications," by D. G. Bhat, Surface Modification Technologies, pp. 1–21, The Metallurgical Society, 420 Commonwealth Drive, Warrendale, PA 15086.

Figure 3   Scanning electron micrograph showing whiskers of TiC deposited by CVD from a gas mixture containing TiCl$_4$, H$_2$, and CH$_4$ at about 1100°C. The formation of whiskers is enhanced by nickel, which acts as a catalyst during the growth of whiskers by the vapor-liquid-solid (VLS) mechanism.

cessfully, but certain precautions must be taken for best results. For example, most steels other than austenitic or ferritic steels undergo solid state phase transformation in the 700–800°C temperature range. This transformation is accompanied by changes in microstructure, physical properties, and dimensions that could be detrimental for the coating or the component in the intended application. In addition, the chemical stability of steel may be compromised in some CVD coating operations, as in the case of tungsten deposition as a result of the reaction of steel with the fluoride gases. Methods to avoid these possibilities exist; for example, one can deposit a film of nickel by electrolytic or electroless means to protect the substrate. Figure 4 shows an example of a stainless steel compressor blade coated with a CVD tungsten carbide coating by a moderate-temperature CVD process. The component was electroless nickel plated before CVD.

One of the most widely known and practiced application of CVD is in the manufacture of coated cemented carbide cutting tools. These tools are made of tungsten carbide-cobalt alloys on which a wear-resistant coating of a refractory compound is applied by CVD. The commonly used coatings include TiC, TiN, and Al$_2$O$_3$, and their combinations. Other coat-

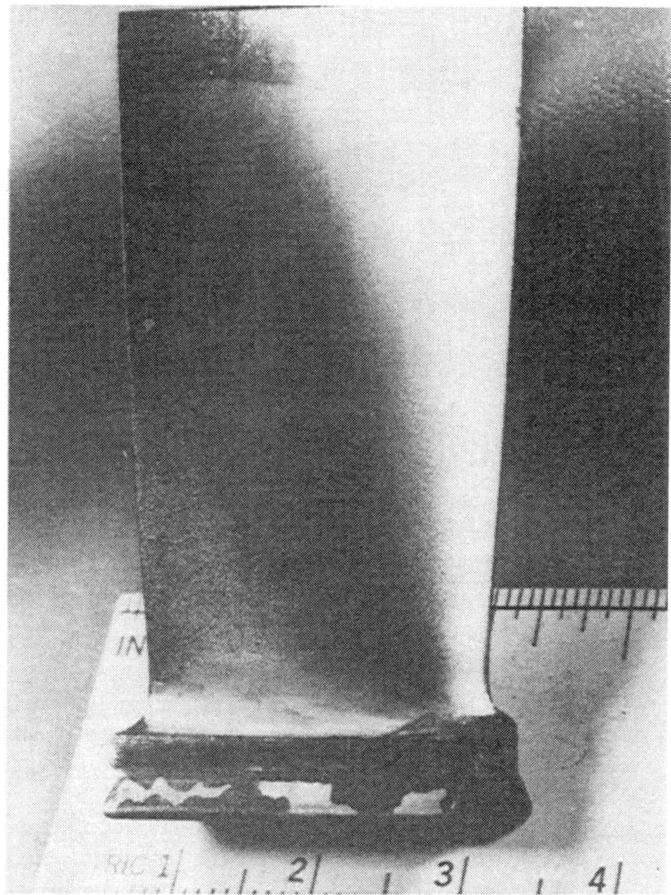

Figure 4   Photograph showing a 17–4 PH stainless steel compressor blade coated with a tungsten carbide coating in a MTCVD process. The blade if first coated with an interlayer of nickel by electrolytic or electroless plating techniques to protect it from the corrosive action of hydrofluoric acid gas generated during the deposition reaction.

ings include Ti(C,N), TaC, HfN, and ZrN. Figure 5 shows several cemented carbide cutting tool inserts coated with TiC and TiN.

Another application of wear-resistant CVD coatings is in areas involving erosion and abrasion, such as sand blast nozzles, slurry transport and handling equipment, coal gasification equipment, and mining equipment. In these applications, air-or waterborne particles of sand, fly ash, or other particulate matter traveling at speed and under pressure can cause abrasion and erosion of surfaces they contact. Conventionally, many of these surfaces are protected by applying wear-resistant coatings by various means, such as electroplating, flame or plasma spraying, laser cladding, and weld overlay coating techniques. Chemical vapor deposition has also been used successfully in many of the applications. The graph

Figure 5 Photograph showing cemented tungsten carbide cutting tool inserts coated with TiC and TiN coating in a conventional CVD process. These coatings impart improved wear resistance to the carbide tools, allowing them to run at higher speeds and chip loads in the machining of various materials.

(Fig. 6) shows the relative wear rates of various coatings, other hardfacing materials and ceramics against sand, indicating that CVD coatings can be successfully used in these applications.

Tribological coatings present another use for CVD coatings: thus improve the coefficient of friction between sliding or rolling surfaces in contact, thereby reducing wear due to adhesion, abrasion, or other causes. Typical coatings used in these applications include refractory compounds such as carbides, nitrides, and borides of transition metals. The important properties of coatings in these applications include hardness, elastic modulus, fracture toughness, adhesion, grain size, and to a certain extent, chemical stability depending on the service environment.

One of the elegant applications of CVD tribological coatings is for ball bearings. As shown in Figure 7, the wear resistance and service life of a steel ball bearing improved dramatically when either the balls or the races were coated with CVD TiC. Other applications of tribological coatings include various steel components such as coatings on dies used in molding, extrusion, and similar metalworking operations.

Coatings used for high-temperature applications require high thermal stability. Refractory compounds having low vapor pressure and high decomposition temperature are generally suitable in these cases, depending on service environment. Other properties, such as abrasion resistance, oxidation resistance, thermal shock resistance, and compatible thermal expansion characteristics are also important. Thus, typical coatings used in these applications include certain refractory metals, Al_2O_3, B_4C, SiC, Si_3N_4, SiO_2, and ZrO_2, and refractory metal silicides. Composite coatings such as $Al_2O_3+ZrO_2$ and $Al_2O_3+Y_2O_3$ have also been studied. Most of these coatings can be deposited by CVD. Typical applications for these coatings include rocket nozzles, reentry cones, ceramic heat exchanger components, afterburner parts in rocket engines, and gas turbine and automotive engine components. An-

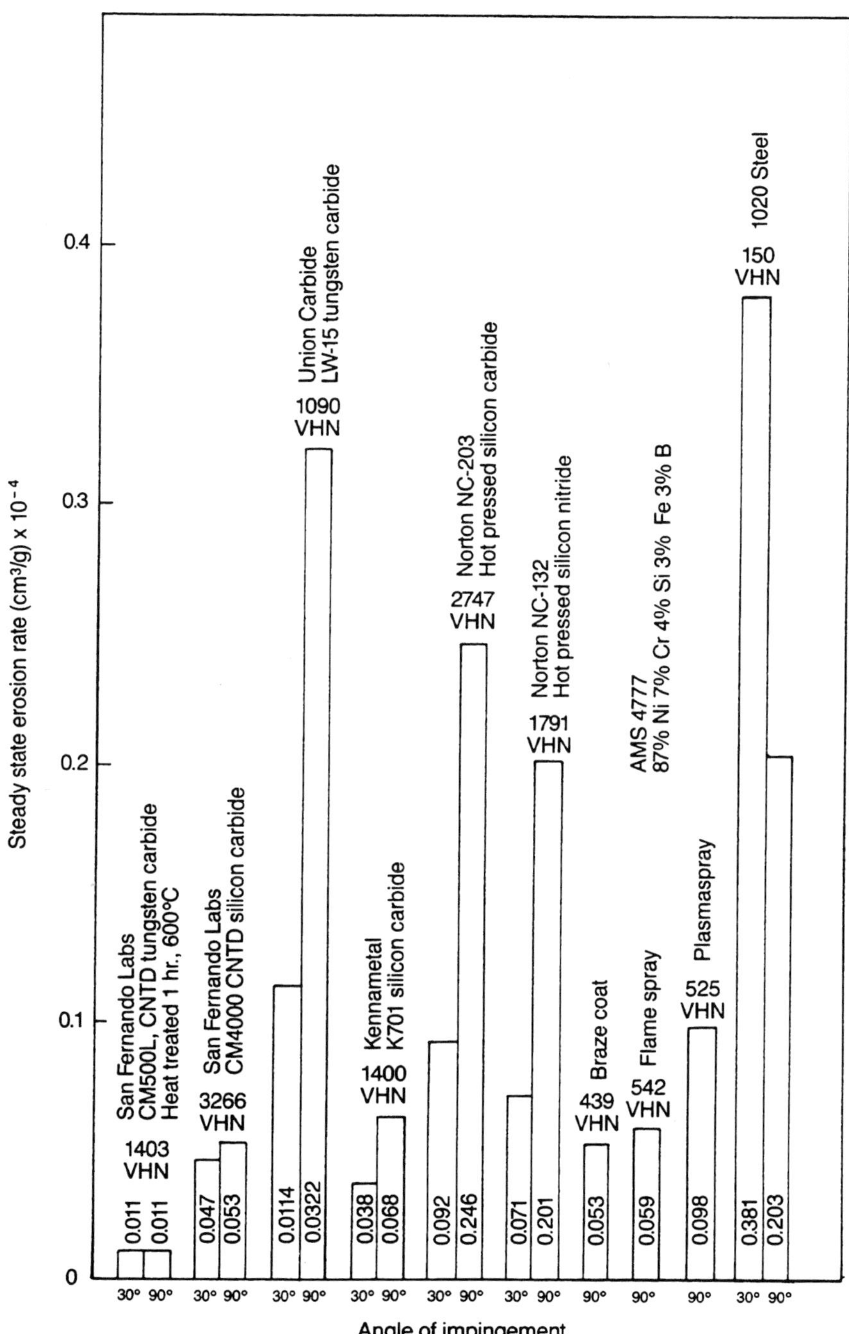

Figure 6 Steady state erosive wear rate of ultrafine-grained CVD tungsten-carbon (CM 500L) and SiC (CM 4000) coatings and other hardfacing materials, coatings, and ceramics. The eroding medium is 200 micron SiC particles impinging at a velocity of 30 ms⁻¹ at room temperature. (Data from Hickey, et al., *Thin Solid Films*, vol. 118, p. 321 (1984). Reprinted with permission from Elsevier Sequoia, S.A., Switzerland.)

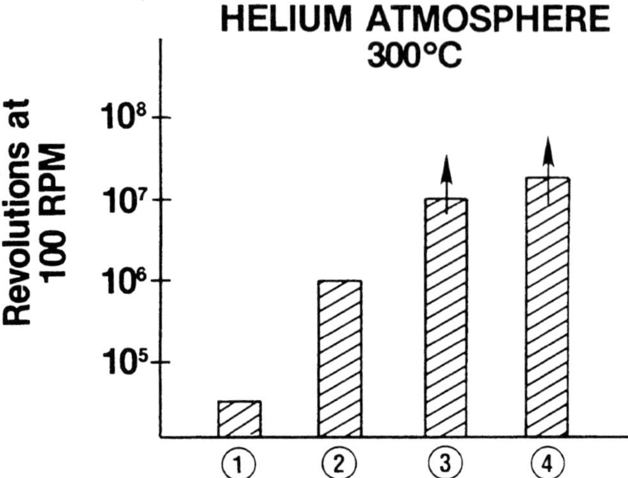

(1) **Standard steel ball bearing –
complete seizure.**

(2) **Standard steel ball bearing with MoS₂
spray – complete seizure.**

(3) **Ball bearing with TiC-coated races – wear
of balls but no seizure.**

(4) **Ball bearing with TiC-coated balls – wear
of races but no seizure.**

Figure 7   Effect of CVD TiC coating on the frictional and wear behavior of steel ball bearing components. (Data from Hintermann, H.E., *Thin Solid Films*, vol. 84, p. 215 (1981). Reprinted with permission from Elsevier Sequoia, S.A., Switzerland.)

other well-known example of a protective refractory coating is the SiC-coated hardware used in the microelectronics field for manufacturing coated silicon wafers. Figure 8 shows typical examples of graphite susceptor components coated with SiC. An iridium-coated rhenium thrust chamber for spacecraft was shown in Figure 2.

In recent years, advances in the technology of carbon-carbon composites have led to the fabrication of components out of these materials, which are then coated by CVD or the new technology of chemical vapor infiltration (CVI) with various refractory compound coatings, most notably SiC. Other ceramic fiber composites based on alumina and silica have also been coated in a similar manner for high temperature service. Figure 9 illustrates one of the techniques used for coating of porous fiber preforms by CVI.

The more exotic CVD techniques that were mentioned earlier, such as PACVD and LCVD, have found important applications for the deposition of new types of coatings. One of the most interesting applications is the deposition of diamond films by PACVD. The diamond films have unique properties and application potential ranging from wear-resistant coatings for cutting tools to coatings for laser mirrors, fiber-optics, dielectric films and heat

Figure 8 Photograph of graphite susceptor components coated with CVD SiC. These components are used in the microelectronic industry as supports for wafers during deposition of various thin films.

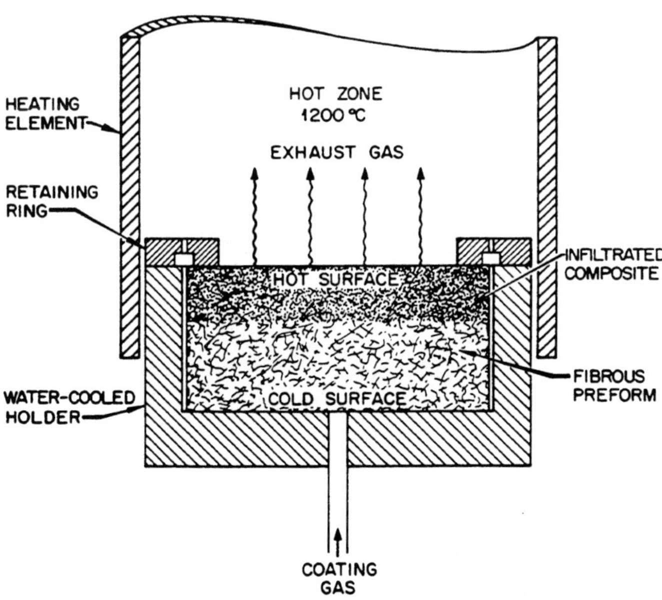

Figure 9 Schematic diagram showing a technique of chemical vapor infiltration of porous fiber preforms, in which a coating of a protective material such as SiC is deposited. In this method, a thermal gradient across the preform allows diffusion of the reactive gas mixture progressively from the hot surface to the cold surface, uniformly coating the preform. (Data from Stinton, et al., *Ceramic Bulletin*, vol. 65, p. 347 (1986). Reprinted with permission from the American Ceramic Society, Westerville, OH.) American Ceramic Society.

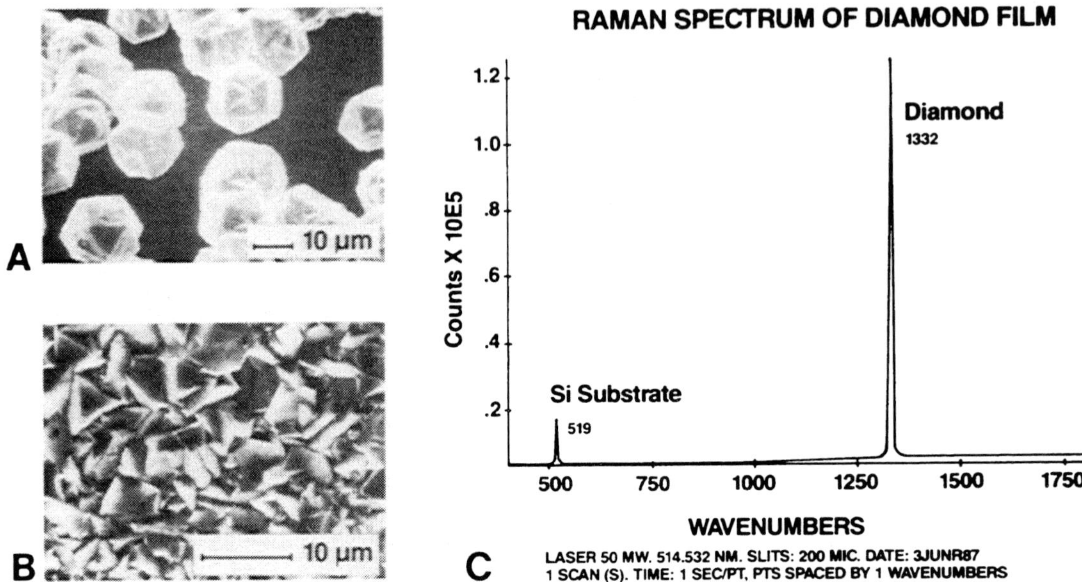

Figure 10   Photographs showing diamond crystals (A) and polycrystalline diamond film (B) deposited on a silicon wafer substrate by microwave plasma-enhanced CVD, from a gas mixture containing hydrogen and methane. The Raman spectrum (C) clearly shows the characteristic Raman shift for diamond at 1332 cm^{-1}. (Data courtesy of Drs. R. Messier, Materials Research Laboratory, The Pennsylvania State University, and P. K. Bachmann, Philips Research Laboratories, Eindhoven, The Netherlands.)

sinks in microelectronic circuits. Figure 10 shows an example of a diamond film deposited on silicon, with the characteristic Raman peak at 1332 cm^{-1}. Coatings deposited by the LCVD technique find applications in laser photolithography, repair of VLSIC masks, laser metallization, and laser evaporation-deposition.

4.0 SUMMARY

The chief characteristics of CVD may be summarized as follows:

1.   The solid is deposited by means of a vapor phase chemical reaction between precursor compounds in gaseous form at moderate to high temperatures.
2.   The process can be carried out at atmospheric pressure as well as at low pressures.
3.   Use of plasma and laser activation allows significant energization of chemical reactions, permitting deposition at very low temperatures.
4.   Chemical composition of the coating can be varied to obtain graded deposits or mixtures of coatings.
5.   Controlled variations in density and purity of the coating can be achieved.
6.   Coatings on substrates of complex shapes and on particulate materials can be deposited in a fluidized-bed system.
7.   Gas flow conditions are usually laminar, resulting in thick boundary layers at the substrate surface.

8. The deposits usually have a columnar grain structure, which is weak in flexure. Fine-grained, equiaxed deposits can be obtained by gas phase perturbation of chemical reactions by various techniques.
9. Control of vapor phase reactions is critical for achieving desirable properties in the deposit.
10. A wide variety of metals, alloys, ceramics, and compounds can be manufactured as coatings or as free-standing components.

It is clear that chemical vapor deposition is a versatile technique for the deposition of coatings of a wide variety of materials, as well as for the fabrication of free-standing structural components.

BIBLIOGRAPHY

T. M. Besmann, D. P. Stinton and R. A. Lowden, "Chemical Vapor Deposition Techniques," *MRS Bulletin*, November 1988.

D. G. Bhat, "Chemical Vapor Depositor," Chapter 2 in *Surface Modification Technologies—An Engineer's Guide* (T. S. Sudarshan, Ed.), Marcel Dekker, Inc., New York, NY (1989), p. 141.

J. M. Blocher, Jr., "Chemical Vapor Deposition," Chapter 8 in *Deposition Technologies for Films and Coatings-Developments and Applications* (R. F. Bunshah, et al., eds.), Noyes Publications, Park Ridge, NJ (1982), p. 335.

W. A. Bryant, "Chemical Vapor Deposition," Chapter 6 in *Surface Modification Engineering, Vol. I: Fundamental Aspects* (R. Kossowsky, Ed.), CRC Press, Boca Raton, FL (1989), p. 189.

R. A. Holzl, "Chemical Vapor Deposition Techniques," *Techniques of Materials Preparation and Handling—Part 3* (Techniques of Metals Research Series, vol. 2) (R. F. Bunshah, ed.), Interscience Publishers, New York, p. 1377 (1968, p. 1377.

H. O. Pierson (ed.), *Chemically Vapor Deposited Coatings*, American Ceramic Society, Columbus, OH, (1980).

C. F. Powell, J. H. Oxley, and J. M. Blocher, Jr. (Eds.), *Vapor Deposition*, John Wiley and Sons, Inc., New York (1966).

K. K. Yee, *International Metals Reviews*, Review No. 226 (1978).

27

Solvent Vapor Emission Control

Richard Rathmell

Londonderry, New Hampshire

For business operations that include the wet coating of a surface, followed by drying, the amount of volatile organic compound (VOC) released to the atmosphere is important. Increased awareness of ambient air quality, and various regulations affecting solvent vapor emissions, do not change the need to make a business economically profitable.

1.0 REGULATORY BACKGROUND

For a perspective on the VOC regulations, the government now monitors ambient air quality to measure several contaminants; particulates (dust), sulfur dioxide (SO_2), ozone, and others. The amount of ozone is associated with "smog" and volatile organics in the air; it is most noticeable on hot summer days and in metropolitan areas. Industrial coating operations are important point sources that may emit tons of VOC. Automotive traffic and refueling release much more VOC, but the thousands of smaller sources are not as easy to control.

The federal Clean Air Act of 1961 promulgated an important set of regulations that establish limits and also require the states to act to meet ambient air quality standards. State regulations may be more stringent than federal regulations, but not less. Also, local regulations, such as county, municipal, ore regional authority, may be more stringent. In some areas, the state or local authorities are judges by some to be too lenient toward emissions and by others to be antibusiness in enforcement of regulations. In many areas, the industrial emissions have been reasonable well controlled, but the ambient ozone standard of 0.12 ppm ozone has not been attained. (This is unrelated to the "ozone depletion" problem at high altitudes.)

The federal government now discriminates between "attainment areas" and "nonattainment areas." Regulations also discriminate between New Sources and Existing Sources. New source performance standards may be based on a cost–benefit analysis, but in some nonattainment areas a more stringent LAER (lowest achievable emission rate) may be required, to be negotiated on a case-by-case basis. Existing sources, and some new sources may be subject to RACT (reasonable available control technology).

Typically a state agency will control the permit applications and approval for an industrial operation. However, it must be determined locally what regulations and limits are applicable, and who will control the permitting process.

2.0 ALTERNATIVE CONTROL PROCESSES FOR VOLATILE ORGANIC COMPOUNDS

The business and process decisions facing those who apply and dry a wet surface coating to a product require consideration of several factors including:

1. Product quality needs
2. Safety
3. Regulatory and air quality restrictions
4. Costs
5. Ease and reliability of operation

The process alternatives include the following.

1. Reformulation
   A. Conversion from solvent-based to a water-based wet formulation.
   B. Conversion to a high solids formulation with little or not volatile component
2. Vapor destruction
   A. Thermal oxidation of the organic compound to produce carbon dioxide and water vapor
   B. Catalytic oxidation
3. Vapor recovery
   A. Absorption–desorption using activated carbon
   B. Direct vapor condensation

Reformulation has been the subject of considerable research, which is beyond the scope of this chapter. Much success has been achieved, but in many cases the use of volatile organic compounds gives a superior end product quality or advantage. Safety in handling solvents and costs require examination.

Vapor destruction, or oxidation, typically requires less capital expense than vapor recovery, but the extra cost for recovery facilities may be cost justified if an appreciable amount of solvent can be reused, reducing the required annual purchase of new solvent. If oxidized, some heat energy may be recovered from the solvent, but in all cases additional energy must be purchased to operate the vapor oxidizer.

Where carbon adsorption–desorption facilities are used to recover solvents, it sometimes costs more to recover the solvent than to purchase new solvent; such extra costs are justified by the need to avoid air pollution.

There is no control process that is preferred for all situations. It is necessary to consider specific cases to determine what will be preferred. Typically, however, thermal oxidation will be preferred for smaller rates of vapor emissions or where the solvent cannot be reused, and catalytic oxidation will be considered only where the airflow required for fume capture leads to relatively dilute vapor concentrations. Solvent recovery by direct vapor condensation will be preferred where larger amounts of vapor can be reused and new dryers are planned.

2.1 Safety

Process safety while handling volatile organic compounds requires sure methods of avoiding fire and explosion damage.

The conventional procedure, in which solvent-based coatings are applied and dried, is to sweep the organic vapors away with an excess airflow, such that the vapor concentration is too low to sustain combustion, should an ignition temperature develop.

Typically at lease 300–400 volumes of air are used to dilute one volume of vapor that it cannot be ignited, and it is not uncommon to use several thousand volumes of air per volume of vapor.

It should be understood, however, that fire and explosive hazards also can be controlled by maintaining and oxygen concentration too low to sustain combustion. The common solvents will not ignite or sustain combustion in an atmosphere containing less that 10–13% oxygen (versus air with 21% O_2). This control procedure is a key to an economical condensation solvent recovery process.

2.2 Operating Costs

The cost for controlling vapor emissions is very largely dependent on the amount of air that is admixed with the vapor. Excess air can provide the advantage of more assured safety and/or greater percentage of vapor capture in a fume hood, but the excess air greatly adds to the cost of removing the vapor from the air.

In industrial operations that incorporate a vapor incinerator or vapor recovery system, it is important to minimize the flow of dilution air to a degree that does not sacrifice safety. For example, toluene vapor in concentrations below 1.2% vapor in air are too lean to sustain combustion if ignited, and toluene vapor concentrations above 7.1% are too rich to sustain combustion if ignited. There limits are called LEL (lower explosion limit) or LFL (lower flammable limit), or UEL or UFL for the upper limits. Other solvents have slightly different limits, which are applicable under "normal" dryer conditions up to 200°F.

It is common practice to ensure that a dryer exhaust flow is not more than 25% LEL (a fourfold safety factor) to minimize the risk of combustible mixtures within the dryer. The National Fire Protection Association standards will permit vapor concentrations "not to exceed 50% of the LEL," provided a continuous indicator and alarm set up is arranged to shut the system down at a somewhat lower wet point. However, when the average exhaust concentration is in the range of 40–50% LEL, the absence of higher concentrations within the dryer cannot be guaranteed. Relatively few dryers are normally operated with 40% or higher LEL vapor concentrations.

On the other hand, where printing is applied to a small percentage of a surface, such as in a multicolor rotogravure line or flexographic line, it is typical for substantially more air to be used per unit of solvent applied, and for the total exhaust mixture to be relatively low in LEL concentration.

Effective vapor capture sometimes requires more airflow than the LEL safety consideration.

3.0 VAPOR OXIDATION

Most of the common solvent vapors may be oxidized and converted into clean and harmless carbon dioxide and water vapor. However, chlorinated solvents would produce objectionable hydrochloric acid vapors if oxidized.

A thermal oxidizer typically is operated in the 1200–1500°F range, with a hot gas retention of 0.3–0.6 second to achieve substantially complete oxidation of the organic materials.

Catalytic oxidizers typically are operated several hundred degrees cooler than thermal oxidizers, with temperatures depending on the specific catalyst used and the concentration of vapors oxidized.

The more expensive noble metal catalysts, such as platinum, will tolerate temporarily higher temperatures than cheaper catalysts, which are susceptible to thermal deactivation. Some impurities in the air may poison any catalyst.

The heat energy released by the vapor oxidation may, in some cases, be useful for heating the process dryers or ovens. Usually, the high temperature gases from the oxidizer are used to preheat the cooler vapor-laden air, and residual heat is still sufficient to process needs.

The cost of oxidizing a given amount of vapor depends on how much dilution air is present or, for a given amount of air, how much vapor is present. More air requires a larger incinerator to retain the hot gases for the minimum time required to complete the oxidation reaction and more energy is required to bring the air to the combustion (oxidation) temperature.

However, if the vapor concentration is maintained close to 40% LEL or above, the solvent vapor can supply substantially all the energy required. At lower concentrations it becomes increasingly necessary to supply auxiliary fuel *or* to provide more air–air heat transfer to preheat the vapor laden air.

For example, one cubic foot of toluene vapor diluted with more or less air in the exhaust flow to be incinerated will be as shown in Table 1.

From Table 1 it can be appreciated that a reduction in airflow (for a given flow of solvent vapor) will proportionately reduce the size of the vapor incinerator, but the size of heat exchanger or the amount of added fuel required is affected to a much greater degree.

It is theoretically possible to provide enough heat exchanger capacity to obviate the need for additional fuel for normal operation. In practice, an auxiliary fuel burner is needed for start-up, and it must be kept ignited and ready to heat the air when the vapor concentration decreases.

Heat exchangers for vapor thermal oxidizers usually are either the shell-and-tube type, using stainless steel tubes, or ceramic beds. Some metal plate–plate exchangers also are used, but in every case it is important to prevent leakage or short-circuiting of vapor-laden air to the exhaust gases, or bypassing the combustion zone. Such leakage or bypassing can generate objectionable odors from partially oxidized organics.

The ceramic bed heat exchangers operate by periodically reversing the flow direction through at least two or more beds, which are alternately heated and cooled. Outgoing hot combustion gases flow through a bed until the ceramic pieces reach a set temperature, then the flow is reversed and vapor-laden gases are heated as they flow through the hot bed into the combustion zone. There is no problem if the vapor-laden gases ignite in the bed prior to the combustion space, but before flows are switched back it is desirable to first purge vapor-laden gases from the cooling bed into the combustion zone. Non oxidized vapors should not be pushed out with exhaust flow. With relatively large beds it is practical (but not inexpensive) to provide the high heat transfer area needed to accommodate relatively dilute vapor flows. The bed size required can be minimized by a high frequency of flow switching; the airtight dampers may be switched every few minutes. The ceramic pieces must be selected to tolerate frequent temperature changes and to accommodate the thermal expansion–contraction cycle that occurs. If dust is released by thermal movements or abrasion, it may prevent direct usage of the residual hot gases in the dryers and ovens.

Metal surface heat exchangers, with hot combustion gases in one side and the cooler vapor-laden gases on the other side, operate continuously, without flow reversal or switching dampers. Thermal expansion–contraction can be a problem, leading to torn welds or fractures and to leakage of the higher pressure vapor-laden air into the lower pressure oxidized discharge flow. Such leakage can generate objectionable odors by the scorching of the vapors.

In heat exchangers of the shell-and-tube type, longer tubes with baffled counterflow over the tubes are more efficient than short tubes with cross-flow.

A recent development, patented by the Wolverine Corporation (Merrimac, MA), overcomes the expansion problem with long-tube heat exchangers; the tube is free to expand and contract at one end within a slip tube that acts as an air aspirator. In this arrangement, a small amount of lower pressure oxidized vapor is allowed to leak back into the oxidation zone; leakage in this direction is acceptable.

Table 1 Operating Variables[a] for Thermal and Catalytic Incinerators

Variable	Type of Incinerator				
	Thermal			Catalytic	
LEL in exhaust, %	40	25	10	10	5
Volume of exhaust, ft^3/min	208	333	833	833	1667
Assumed exhaust temperature	200	200	200	100	100
Assumed combustion temperature	1400	1400	1400	900	900
Temperature rise required	1200	1200	1200	800	800
Temperature rise resulting from vapor oxidation	1160	725	290	290	145
Temperature rise required from preheater or auxiliary fuel	40	745	910	510	655
Requirement from preheater or other fuel, Btu $\times 10^3$/h	9	170	820	460	1180
Available temperature differential across heat exchanger (with no other fuel used)	1160	725	290	290	145
Ratio of heat exchanger area[b] required (to avoid auxiliary fuel comsumption)		1	12	6.75	35

[a]All temperatures in degrees Fahrenheit.
[b]Assuming equal coefficient; $A = Q/\Delta T$, where A = the heat transfer area, Q = heat flow, and ΔT = temperature differential.

4.0 SOLVENT RECOVERY

Solvent recovery may be preferred to vapor oxidation if the solvent can be reused, allowing a company to save substantially on the purchase of new solvent. Although vapor oxidation can return some energy value from the solvent, usually the "chemical" value is appreciably more than the energy value.

There are two important approaches to solvent recovery: activated carbon and direct condensation. The carbon approach is substantially more costly. The condensation approach is not always applicable, but it can be very much less expensive where the dryers for a coated web and condensers are designed as a system.

4.1 Carbon Adsorption

When a carbon bed is installed to capture solvent vapors exhausted with air from a dryer, it is usual to provide two or more carbon beds and to periodically desorb from one while adsorbing in the other(s). Desorption is accomplished by heating the bed and sweeping the vapors out, typically with steam.

The steam and vapors are then condensed, and the steam condensate (water) must be separated from solvent. Typically about 3–4 or up to 10 lb of steam is used per pound of solvent recovered. If the recovered solvent is soluble in water, (alcohols, ketones, etc.), the need to separate the water presents additional expenses. Some vapors, such as methyl ethyl ketone, absorb onto carbon with so much heat release that precautions are needed to prevent spontaneous ignition from burning up the air-swept carbon bed.

There are many carbon adsorption systems, in which the value of solvent recovered exceeds the cost of operation. Often, however, it costs more to recover the solvent than to buy new solvent. In such cases, the higher recovery cost is charged to the need to avoid air pollution.

4.2 Direct Vapor Condensation

Direct vapor condensation can be more economical than using carbon beds for solvent recovery. The energy consumption is usually less than 10% of the energy to operate a conventional air-swept dryer and carbon bed, and there is no water added to the solvent by the process. Also, there is no exhaust flow to the atmosphere in the preferred designs. The vapor condensers are much smaller than required for carbon bed recovery system, and there is another advantage in that when two or more dryers are operated with different solvents or blends of solvents, separate vapor condensers avoid the mixing of solvents and the costs of separating them. Usually a water chiller type of refrigeration system is required, but one system can serve all the condensers.

However, there are limitations on the applicability of this process: the process dryers must be substantially "airtight" to permit the contained atmosphere be recirculated countless times with a minimum of interchange with outside atmosphere. Indirect heaters, such as steam coils, are necessary. Also, the oxygen level in the contained atmosphere should be kept below the limit (10–13% O_2) that would sustain combustion; otherwise it is necessary to use very cold condensers and relatively high recirculation rates to keep the vapor levels safely below the vapor LEL limits. There is a need for a relatively small flow of inert (low oxygen content) gas to offset the tendency for the wet web to drag air (21% O_2) into the dryer, and it is necessary to maintain a ready reservoir of pressurized inert gas (a) to facilitate a rapid air purge and dryer start-up after the dryer has been opened for any reason, and (b) to provide a safety cushion for a fail-safe shutdown during an electrical power outage.

The sources of inert (low oxygen) gas required include the flue gas of a gas-fired steam boiler and purchased liquid nitrogen or carbon dioxide. Where flue gases are used, the gas burner must be of the type that can maintain a low ratio of excess air to fuel for various fuel firing rates. A compressor and pressurized storage tank can provide the ready reservoir for fast start-up and fair-safe shutdowns, or a tank of liquid nitrogen with vaporization facilities can be used.

In some important respects, the operation of an inerted airtight dryer is inherently safer than a conventional air-swept dryer. In an air-swept dryer there is a transition zone between a flammable wet interface and a nonflammable exhaust, and there is the potential for a temporary excess solvent loading into the dryer to produce a large volume of combustible mixture. In an inerted dryer, there is no flammable interface, and any temporary excess solvent loading will not make a combustible mixture. When the coating process and wet web is stopped for any reason, there is no tendency for outside air to exchange with the atmosphere contained in the dryer, except as air may be drawn in to replace the volume of vapor condensed, or to make up for gas volume contraction as the contained gas cools down. In the Wolverine systems, the normal operating vapor concentration in the dryer is designed to prevent the condensation due to vapor volume of any unsafe air inhalation into the dryer. Normally, the vapor condenser temperature is selected to draw the vapor concentration below the organic LEL level when the coating process is stopped. This then provides a double safety factor with both a safe O_2 LEL level and a safe organic-in-air LEL level.

When a dryer is shut down overnight or for a weekend, it is not necessary or desirable to purge the contained atmosphere to the outside atmosphere.

28

Surface Treatment of Plastics

William F. Harrington

Uniroyal Adhesives and Sealants Company, Inc.
Mishawaka, Indiana

1.0 INTRODUCTION

No single step in the coating process has more impact on film adhesion than surface preparation. Film adhesion to a plastic is primarily a surface phenomenon and requires intimate contact between the substrate surface and the coating. However, intimate contact of that plastic surface is not possible without appropriate conditioning and cleansing.

Plastic surfaces present a number of unique problems for the coater. Many plastics, such as polyethylene or the fluorinated polymers, have a low surface energy. Low surface energy often means that few materials will readily adhere to the surface. Plastic materials often are blends of one or more polymer types or have various quantities of inorganic fillers added to achieve specific properties. The coefficient of thermal expansion is usually quite high for plastic compounds, but it can vary widely depending on polymer blend, filler content, and filler type. Finally, the flexibility of plastic materials puts more stress on the coating, and significant problems can develop if film adhesion is low due to poor surface preparation.

2.0 FUNCTIONS OF SURFACE PREPARATION

Treatment of the plastic surface performs a great many functions depending on the individual polymer type involved.

2.1 Removal of Contamination

Any substance that interferes with the contact between the coating film and the plastic must be removed. Process oils, dirt and grime, waxes, mold release agents, and poorly retained plasticizers must all be removed.

2.2 Control of Surface Roughness

No surface is perfectly smooth, and various techniques enable the coater to match the finish on the plastic component to the coating viscosity for improved film adhesion.

279

2.3 Matching of Surface to Adhesive

Conversion of the outside surface of the plastic can provide an interface that is more like, or more compatible with, the chemical structure of the coating.

2.4 Providing a Boundary Layer

Plating or coating the surface with a primer often provides a more suitable surface for coating adhesion.

2.5 Control of Oxide Formation

Polymers generally exhibit good resistance to oxidation, but since an oxidized surface has a higher level of surface activity, several procedures promote oxide formation.

2.6 Control of Absorbed Water

Many plastics readily absorb moisture from the surrounding atmosphere. This moisture can interfere with film adhesion, especially for coatings that require heat cure cycles.

3.0 FACTORS IMPACTING PREPARATION INTENSITY

Often the goal of surface preparation is to accomplish several of the function listed in Sections 2.1 to 2.6. A sequential series of cleaning techniques is appropriate where a single step or process might prove inadequate. In fact, the intensity of surface preparation procedures used depends on a variety of factors, all of which must be considered in the product design process.

3.1 Type of Plastic

Higher surface activity reduces the need for surface preparation. Some molded thermoplastic parts that require a high melt temperature and a fast cool-down cycle are more easily coated. Many thermoset plastics, such as phenolic, are readily coated.

3.2 Surface Contamination

The type and level of contamination will dictate the degree of cleaning required. Silicone release agents on molded parts require very intense cleaning procedures.

3.3 Initial and Ultimate Strength Requirements

Film adhesion develops faster on prepared and clean surfaces. In addition, higher adhesion values are achievable with higher levels of surface cleaning.

3.4 Service Environment

Better surface adhesion generally indicates a more permanent coating, one more adaptable to end-use conditions.

3.5 Time

Delays between surface preparation and coating application often require repetition of preparation steps to reclean, or higher levels of cleaning to start with, to overcome possible recontamination.

3.6 Component Size

Typically, large parts are more difficult to process than small parts, and certain processes are not suitable or available for large parts.

3.7 Cost

Within the manufacturing/product cost cycle, certain limitations have an impact on the degree of preparation utilized. Limitations on surface preparation procedures can affect returned goods and warranty claims, cost often overlooked in the design phase.

4.0 SURFACE PREPARATION TECHNIQUES

As may be expected, there are nearly as many procedures for cleaning and preparing plastic surfaces as there are polymer types. Some procedures are inexpensive and easy to accomplish; others are not. Most plastic types have a few highly recommended procedures for achieving the best finish for coating, but often the user is faced with many equally valid choices or, worse, conflicting advice on the level of intensity.

Plastics can be obtained in variety of forms: molded, sheet, shaped part, foam, and film. Many polymers can be readily blended together to achieve specific properties. Fillers and plasticizers are also included in the resin matrix to yield certain characteristics. Obviously, the same surface preparation procedure may not be appropriate in each of these cases, even though the major ingredient remains the same.

The manufacturers of both the plastic and the coating should be prepared to give advice on appropriate surface preparation techniques to use with the product. However, since user needs differ so widely, even the manufacturer may not have a definitive answer for an individual circumstance. Considerable experimentation may be required to identify the appropriate technique for a given operation.

The procedures outlined here in the remainder of this chapter concentrate on the technique, rather than the plastic. Where appropriate, however, certain techniques have a listing of the polymer types for which they are most suitable.

Finally, many of the processes utilize hazardous, corrosive, toxic, flammable, or poisonous chemicals. Safe handling methods, worker training, and appropriate control procedures are essential to minimize risk in the work environment.

4.1 Solvent Cleaning

Solvent action removes surface contamination by dissolving the unwanted substance. This is the easiest and most commonly used procedure an often serves as a first step in more complex procedures.

Organic solvents and water are the solvents recommended. The organic solvents can be flammable or nonflammable. Most commonly used for plastics are acetone, methyl ethyl ketone, isopropyl alcohol, methyl alcohol, toluol, 1,1,1-trichloroethane, naphtha, and sometimes Freon—either by itself or blended with another solvent. Water is cheap and readily available but often has trace levels of impurities, which can also contaminate surfaces. Most often recommended is distilled or deionized water. Water is frequently used as a rinse for other surface preparation procedures.

These solvents can be used in processes as simple as the wiping of a dampened cloth across the plastic. Immersion is a technique usually accomplished by submerging the part in a swirling bath with heat applied to speed the solvent action. Spray cleaning has the advantage of washing off the contamination with the force of the spray. Vapor degreasing requires that the plastic part be suspended over a boiling tank of solvent. As the vapors condense on the part, the constant flow over the surface washes it clean. Ultrasonic degreasing utilizes high frequency vibration from sonic waves to dislodge the contaminants in a solvent bath.

In all cases, frequent changing or filtering of the solvents is recommended to prevent residue buildup and recontamination. Compatibility of the solvent with the plastic should be thoroughly checked. Since some plastics absorb water, care should be taken. Heat is often used to dry the parts after washing, but heat can readily distort thermoplastics.

4.2 Detergent Cleaning

Soaps and detergents do for plastic components what they do in the dishwasher and washing machine at home.

Emulsification of oils, greases, and some mold releases is easily achieved in either hot or cold water solutions. Detergent cleansing is often used as a preliminary step to mechanical treatments.

The names of the cleaners are familiar, since they are frequently items found in the home. Ivory soap, Ajax, Borax, and trisodium phosphate are all recommended in various cleaning operations and do quite well.

An immersion wash is quite effective, unless the plastic is water sensitive. Scouring with a medium-to-stiff bristle brush is excellent for dislodging many contaminants.

Solution cleanliness must be monitored. A clean water rinse is essential, since soaps can act as contaminants if not washed off. Thorough drying at elevated temperatures is recommended.

4.3 Mechanical Treatment

Physical scrubbing of plastic surfaces removes oxides and contaminated layers. A solvent or detergent cleaning must precede the mechanical treatment to avoid scrubbing surface contaminants into the roughened surface. This process often comes before chemical treatments.

Sanding, either wet or dry, is a commonly used procedure, using a grit size of 40 to 400, depending on the amount of surface to be removed and the surface finish desired. Softer plastics are obviously more susceptible to damage. For parts that have a complex configuration, grit blasting (wet or dry) or wire brushing is more appropriate. Grit size and type can be varied to obtain the proper finish. Grit size and type can be varied to obtain the proper finish. Machining can be used to expose fresh layers of plastic to coat.

In all cases, the roughened surface should be vacuumed or air blasted to remove residual dust or grit. A solvent wipe or water rinse followed by elevated temperature drying is also recommended.

4.4 Chemical Treatment

Generally, the most effective surface preparation is a chemical etch of the plastic to be coated. Both physical and chemical characteristics can be altered to improve wet-out and film adhesion. Nearly always, chemical treatment is preceded by one or more cleaning operations to remove surface contamination. There prior operations reduce solution contamination and ensure optimum interaction between the solution and substrate.

Most often, the surface is washed or immersed in a bath containing an acid, base, oxidizing agent, chlorinating agent, or other highly active chemical. Each procedure requires control of the parts by weight of active ingredients, the temperature of the solution, and the elapsed time of immersion. Some procedures have a wide range of ingredient ratios, while others are quite specific. Temperature of the solution is inversely related to time of the immersion; that is, the higher the temperature, the shorter the exposure.

Virtually all chemical etch procedures require water rinsing (once or twice), and an elevated temperature drying is recommended. With active ingredient treatments, it is imperative that solution strength be monitored and renewed at appropriate intervals.

4.4.1 Sulfuric Acid–Dichromate Etch

By far the most commonly recommended chemical treatment for plastic parts, the sulfuric acid–dichromate etch is used on acrylonitrile–butadieme–styrene (ABS), acetal, melamine or urea, polyolefins, polyphenylene oxide, polystyrene, polysulfone, and styrene–acrylonitrile (SAN). For each plastic, a different ingredient ratio and immersion temperature and time may be recommended.

The following list is offered as a guide to a possible range of parameters:

Ingredient	Parts by weight	Range
Potassium or sodium dichromate	5	0.5–10.0
Concentrated sulfuric acid	85	65.0–96.5
Water	10	0–27.5
Time:	10 seconds to 90 minutes	
Temperature:	Room temperature to 160°F	

While the ranges are extremely wide, experimental trials coupled with test results will allow the use to identify the most appropriate values for a given plastic.

4.4.2 Sodium Etch

For truly difficult surfaces to coat, such as the various fluoroplastics and some thermoplastic polyesters, highly reactive materials must be used. Metallic sodium (2–4 parts) is dispersed in a mixture of naphthalene (10–12 parts) and tetrahydrofuran (85–87 parts).

Immersion time is approximately 15 minutes at ambient temperatures, followed by thorough rinsing with solvent (ketone) before water rinsing.

4.4.3 Sodium Hydroxide

A mixture of 20 parts by weight of sodium hydroxide and 80 parts of water is an effective treatment of thermoplastic polyesters, polyamide, and polysulfone. Heating the solution to 175–200°F and immersing for 2–10 minutes is appropriate.

4.4.4 Satinizing

Satinizing is a process developed by Du Pont for their homopolymer grade of acetal (U.S. Patent 3,235,426). Parts are dipped in a heated solution of dioxane, paratoluene sulfonic acid, perchloroethylene, and a thickening agent. After the dip cycle, parts are heat treated, rinsed, and dried according to a prescribed procedure.

4.4.5 Phenol

Nylon is often etched with a 80% solution of phenol in water. Generally, the treatment is conducted at room temperature by brushing onto the surface and drying for about 20 minutes at approximately 150°F.

4.4.6 Sodium Hypochlorite

A number of plastics, particularly the thermoplastic types and the newer thermoplastic rubbers, can be chlorinated on the surface by applying a solution of the following ingredients (parts by weight).

Water	95–97
Sodium hypochlorite, 15%	2–3
Concentrated hydrochloric acid	1–2

Parts can be immersed for 5–10 minutes at room temperature, or the solution can be brushed onto the surface for the same period.

4.5 Other Treatments

A variety of other cleaning and preparation techniques are available. Many of these procedures are unique to plastic processing and were developed to overcome the low surface activity many plastics exhibit. For best results in most cases, these treatments require the prior removal of surface contamination by solvent or detergent cleaning.

4.5.1 Primers

Primers are an adhesion-promoting coating used to develop better adhesion of the final coating to the plastic substrate. The primer can be any number of chemical types, including silanes, urethane polymers, isocyanates, nitrile phenolic, or vinyl. To protect the substrate from recontamination, a primer is usually coated onto the surface soon after other surface preparation procedures have been completed.

4.5.2 Flame Treatment

The impingement of a flame on many plastics, such as the polyolefins, acetal, fluoropolymers, and polycarbonate, oxidizes the surface to provide a higher level of surface energy and better film adhesion. This process can be especially effective on complex shapes and molded parts. Superheated air (1000°F) performs much the same function as a flame.

4.5.3 Exposure to Ultraviolet Radiation

Highly intense radiation from an ultraviolet source provides an ionized or highly polar surface.

4.5.4 Drying

Since many plastics readily absorb atmospheric moisture, simple oven drying can be effective.

4.5.5 Plasma Treatment

Most plastics benefit from a plasma treatment. Parts are exposed to gases, which are ionized by radio frequency or microwave discharge. Effective gases include neon, helium, oxygen, and moisture vapor. Although very effective in improving surface conditions that promote better film adhesion, this process usually is limited by equipment restrictions to smaller components and parts.

4.5.6 Corona Discharge
For film and other thin gage plastics, surface tension can be improved by passing between two electrodes. This treatment is suitable for high speed operations.

5.0 EVALUATION OF SURFACE PREPARATION

In the early stages of developing an appropriate surface preparation technique, it is necessary to have an adequate screening test to determine the effectiveness of the process. There are several methods and tests available.

Water break test.   Water poured on a clean surface will be smooth and will sheet across the face. On oily or poorly treated surfaces, however, water will bead. This test has limited effectiveness on low polarity surfaces such as polyolefins and fluoroplastics, and it is not recommended for surfaces that absorb moisture.

Contact angle test.   A drop of a standardized reference fluid will have a defined contact angle at the edge of the drop. Well-prepared surfaces are easier to wet and will exhibit a lower contact angle than an unprepared surface.

Tape test.   A flexible tape can be applied to the surface and a peel test conducted. The tape should be applied under controlled conditions, with standardized pressure and a defined dwell time. For consistent results, the type, width, and brand of tape should remain constant.

Quick strip test.   After a part has been coated and properly conditioned, a grid pattern of cuts through the coating is made. A standard tape is applied with constant pressure to the surface, then quickly stripped. Counting the number of squares of the grid that were removed gives an indication of film adhesion.

Environmental testing.   Once parts have been treated, coated, and conditioned, conducting a series of heat cycling, weatherometer, stress tests, and various exposure tests will determine the effectiveness of the coated part to meet end-use conditions.

BIBLIOGRAPHY

Cagle, Charles V., *Adhesive Bonding*, New York: McGraw-Hill, 1968.
Military Standardization Handbook: Adhesive Bonding, MIL-HDBK– 691B, U.S.Department of Defense, 1987.
Snodgren, Richard C., *Handbook of Surface Preparation*. New York: Parmerton Publishing, 1974.

29

Flame Surface Treatment

H. Thomas Lindland

Flynn Burner Corporation
New Rochelle, New York

1.0 INTRODUCTION

The need for surface treatment was recognized shortly after the development of polyolefin materials, resulting in the evolution of various treatment methods. W. H. Kreidl pioneered the process of using an oxidizing flame on polyolefins to produce a surface receptive to printing and coating. At the same time, Kreidl's assistant, Kritchever, was developing the concept of using an electrical corona to produce the same result. The two men went their separate ways, and both methods have been widely used for the past 30 years for various surface treating applications.

2.0 SURFACE TREATMENT

The exact mechanism of surface treatment is still unknown. In spite of repeated efforts, using sophisticated instrumentation and complex laboratory methods, we still do not have an accurate understanding of the process. Fortunately, we do not need this information to use the process.

Solid surface have a surface energy specific for various materials. For a liquid drop to spread on a given surface, the liquid surface tension must be lower than the critical surface tension of the solid. Metal and glass exhibit a high surface energy, whereas plastics have a low surface energy. Pretreatment increases the surface energy and therefore its wettability. It may also eliminate a weak boundary layer, thus improving adhesion.

In flame treating, the high temperature of the combustion gases causes oxygen molecules to become disassociated, forming free, highly chemically active oxygen atoms. In addition, because of the energy in the high temperature combustion process, oxygen atoms may also lose electrons to become positively charged oxygen ions. Such an electrically neutral gas made of equal amounts of positively and negatively charged particles is known as a plasma. Plasma may be hot or cold.

In flame treating, these high speed, energetic, very reactive oxygen ions or free oxygen atoms bombard the plastic surface and react with the molecules. This process oxidizes the surface and requires an oxidizing flame, which is a flame with an excess of oxygen.

In corona treating, high voltage fields cause the oxygen molecules to break into free atoms, which can react with the plastic. Those that do not react with the surface recombine into molecules of normal diatomic oxygen or into unstable molecules of ozone—triatomic oxygen. In addition, surface treatment involves many other complex reactions.

In many cases, the objective is to treat the surface to a predetermined critical surface tension expressed in dynes per centimeter. ASTM specification D-278-84 describe a method of evaluating the level of surface treatment. An increase in surface energy is usually related to an improved adhesion. However, sometimes a substrate may be wettable and still not provide the desired adhesion level. Flame treatment to a consistent level is required. This calls for maintaining the operating parameters constant and having a uniform substrate surface.

3.0 BURNERS

Several burner designs are available.

3.1 Atmospheric Burners

In an atmospheric gas burner, part of the air required for combustion is premixed with the gas is prior to the point of ignition. The air premixed with gas is referred to as primary air. The remaining air, called secondary air, is drawn from the surrounding atmosphere at atmospheric pressure; hence the name atmospheric burner. The forerunner of the atmospheric burner system is the well-known Bunsen burner.

To deliver the air–gas mixture to the burner head, the principles developed by Venturi and Bunsen were utilized (Fig. 1). Delivering gas through an orifice (1) at a pressure ranging from 3 to 11 inches of water column would produce about 40–60% of the primary air (2) required. As long as sufficient secondary air (3) is present, proper combustion can take

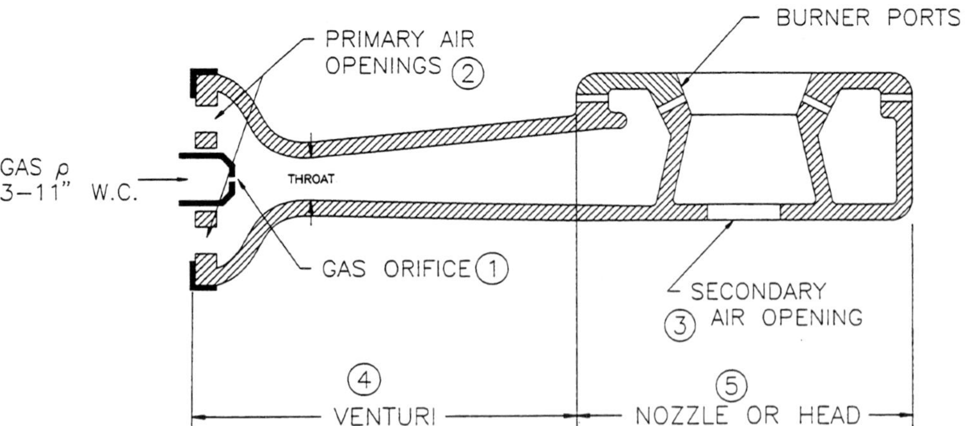

Figure 1 Atmospheric burner.

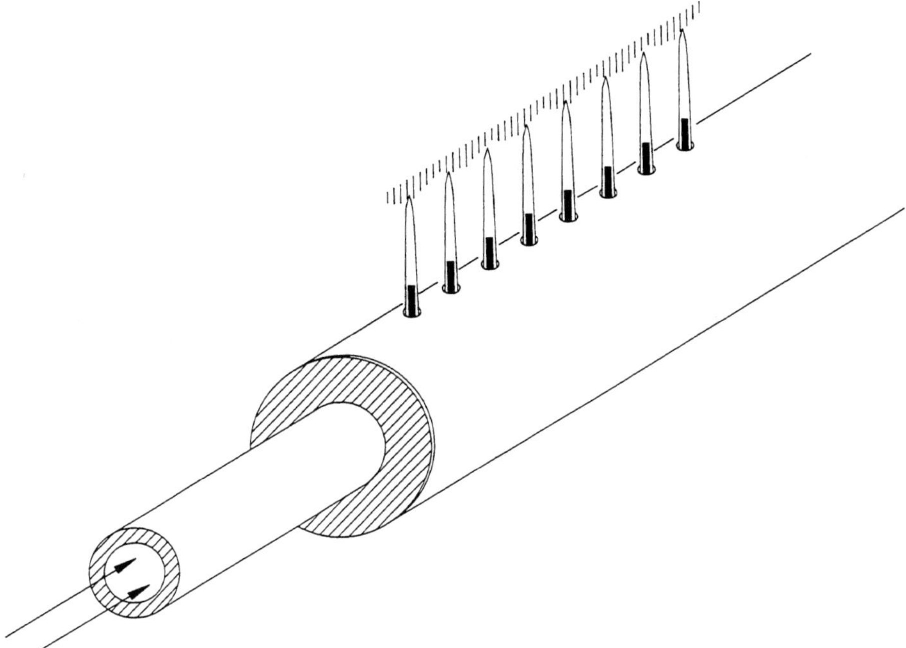

Figure 2 Pipe burner with drilled holes.

place. The venturi mixer (4) coupled with a burner nozzle (5) or head became the mainstay of early gas applications. Many varieties of such burners exist today.

The need to distribute the flame in a wider pattern was resolved by providing a pipe with drill holes (Fig. 2). Such burners work well but are limited to low energy output.

3.2 Power Burners

Once the flame has been established, the flame front burns back toward the source of the air–gas mixture. The front is stable only if the air–gas velocity matches the rate at which the mixture burns. If the velocity is too high, the flame will lift off the ports. If the velocity is too low, the flame will retract toward the source of the air–gas mixture and will cause a flash-back. This tendency can be minimized by keeping the port areas small. The ignition temperature is also important. It takes approximately 630°C to ignite natural gas, but only 480°C to ignite propane. The cool metal surrounding a port will quench the flame, provided there is sufficient mass to absorb the heat. This explains the need for a certain port depth.

The key in resolving these problems is to lower the velocity of a portion of the air–gas mixture or to limit the capacity of the burners. Gun type nozzles were developed, as illustrated in Figure 3. A portion of the air–gas mixture is diverted into the small protected area (1). The velocity of this portion of the mixture is reduced until the piloting will provide continuous ignition to the main air–gas stream emitting from the large center port (2).

This allows the rate of mixture velocity out of the nozzle to be increased, increasing the heat output. This feature, known as flame retention, is inherent in all power burners. Unlike atmospheric burners, power burners utilize a power source of combustion air.

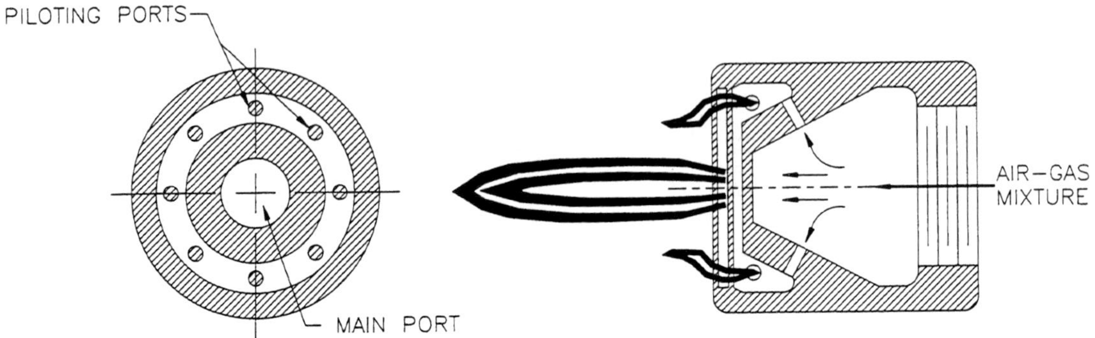

Figure 3 Gun-type nozzles.

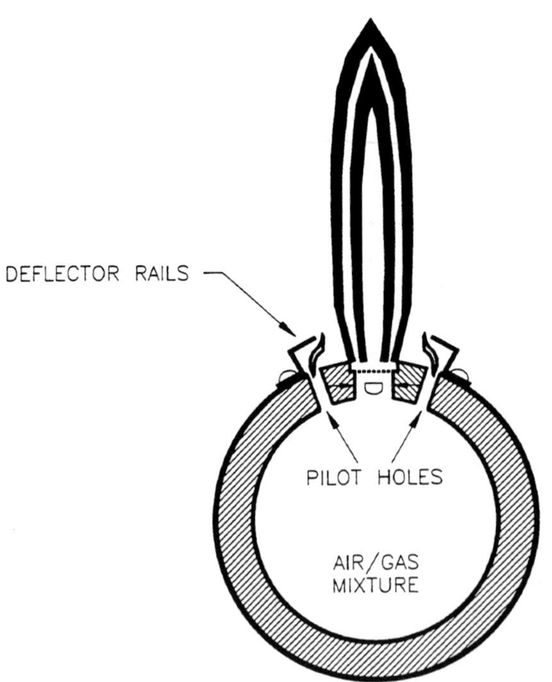

Figure 4 Line burner with deflector rails.

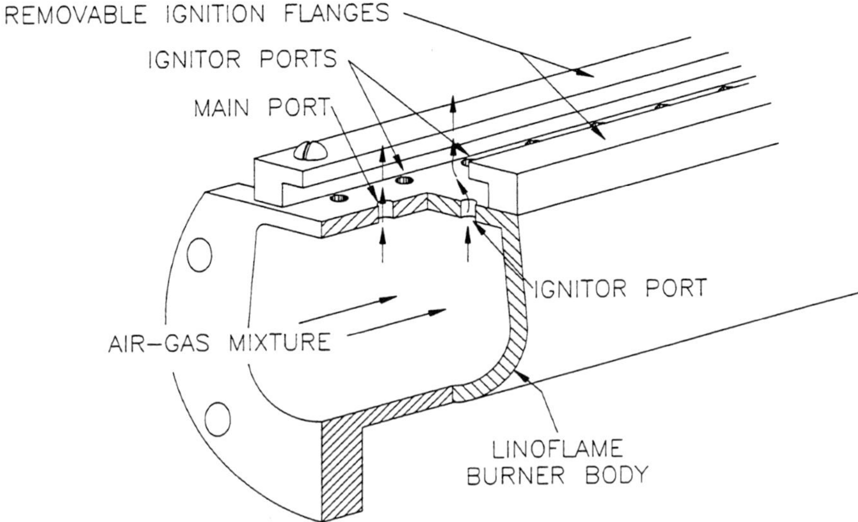

Figure 5 Drilled port line burner.

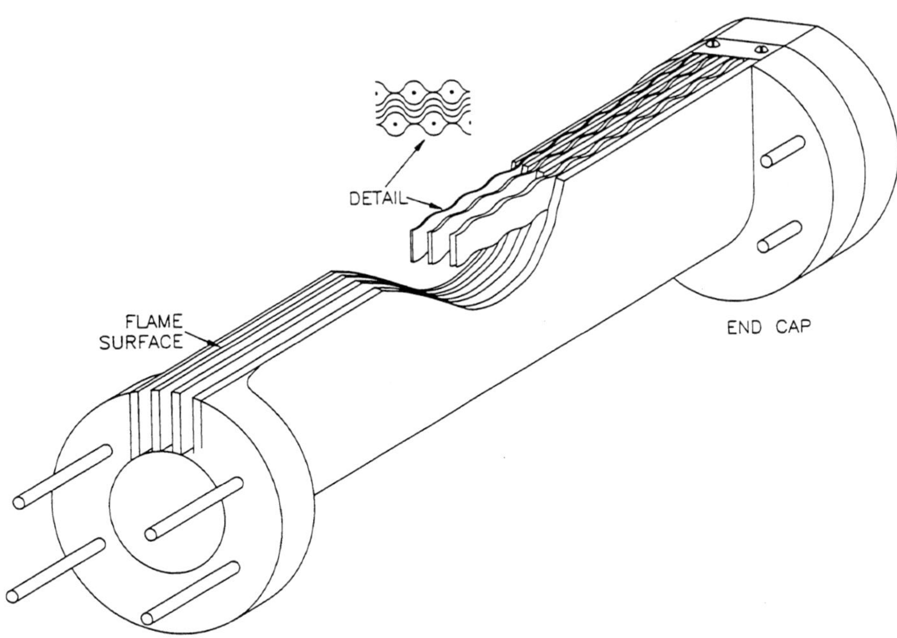

Figure 6 Ribbon burner.

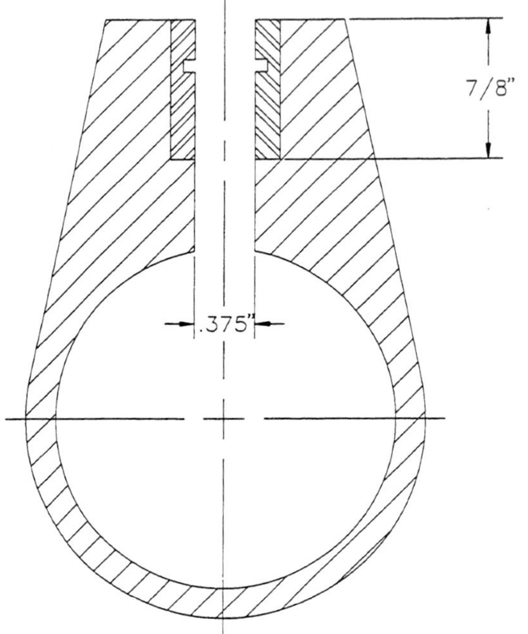

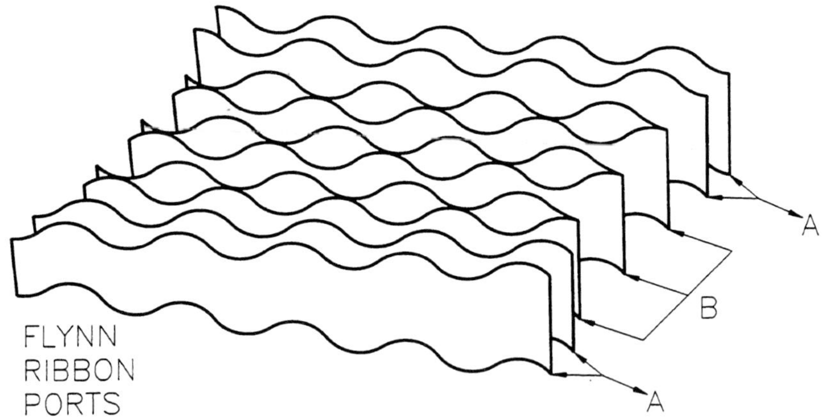

Figure 7   Ribbon burner details.

Patented under the following: United States Pat. No. 3,047,056 3,053,316 2,647,569 Canadian Pat. No. 665,440 662,919 503,393 Canadian Pat. No. 932,688 934,339 Other Foreign Patents

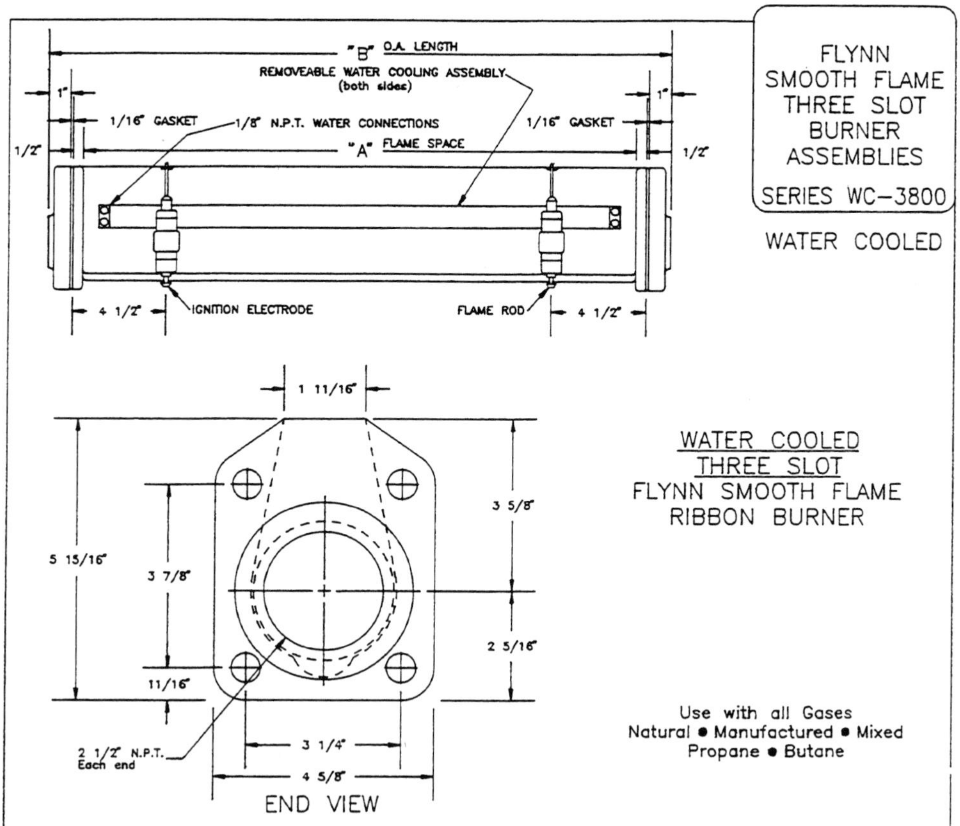

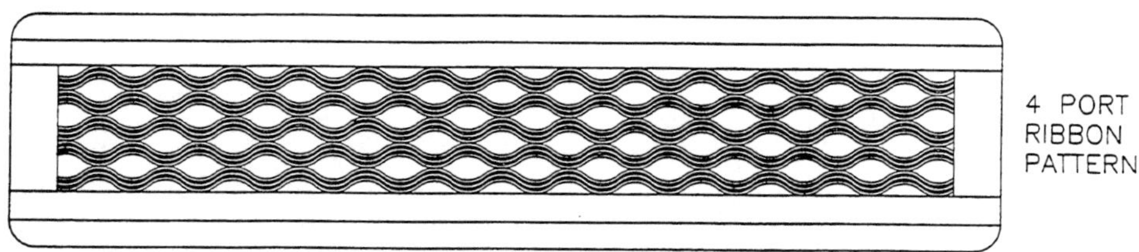

CATALOG No.	"A" FLAME SPACE	"B" O.A. LENGTH	BTU / HR	SHIPPING WEIGHT
WC-3800-60	59 3/4"	62 7/8"	900,000	132 lbs
WC-3801-72	71 3/4"	74 7/8"	1,080,000	155 lbs
WC-3802-80	80 1/4"	83 3/8"	1,200,000	176 lbs
WC-3803-84	84 3/8"	87 1/2"	1,260,000	203 lbs
WC-3804-48	47 5/8"	50 3/4"	720,000	105 lbs

Figure 8 Type 3800 water-cooled smooth flame burner.

Figure 9 Details of four-port ribbon burner.

The flame retention feature was introduced to line burners, such as drilled pipe burners. A single row of ports was drilled down the center and, following the pattern of the flame retention nozzle, rows of small holes were drilled on each side. Deflectors or ignition rails were placed over the two rows of piloting holes (Fig. 4). Figure 5 illustrates a cross section of a typical drilled port line burner that is widely used in the industry today.

Ribbon burners (Fig. 6)were developed by Harold Flynn. A suitable slot was milled out on a casting and the ribbon stack inserted. The velocity of a portion of the air–gas mixture was reduced, establishing piloting along each side of the rows of main ports, which were produced by the ribbon configuration (Fig. 7). This eliminated any external deflectors or rails. Further refinements of ribbon line burners have made it possible to manufacture various slot widths and ribbon configurations to produce a variety of flame patterns.

4.0 FILM TREATMENT

One of the problems associated with the attempt to utilize line burners for the treatment of polyolefin films was the requirement of increasing the flame velocity enough to penetrate the boundary air layer on the web surface. To treat the substrate with direct flame, sufficient thermal energy had to be developed to penetrate this layer. Further more, the exit velocity of the burner flame had to be adjusted to varying web speeds.

The principle of a burner designer to meet these requirements is shown in Figure 8. The ribbon channels of piloting or ignition flames firing at reduced exit velocities provide a constant supply of ignition to the center ribbon channel, enabling it to fire at exit velocities far in excess of the normal speed of flame propagation, greatly increasing the energy output of the burner.

Figure 9 shows a burner designed for film posttreatment application. It has a single slot due to the lower energy output requirement. The water cooling is built into the face. The side bars assure a smooth flame, which is required for the posttreatment process. The capacity, flame velocity, and flame retention capability are all related to the ribbon configuration. By varying the number of ports and the width of the ribbon stack, it is possible to produce a customized flame pattern for each application.

30

Plasma Surface Treatment

Stephen L. Kaplan and Peter W. Rose

Plasma Science, Inc.
Belmont, California

1.0 INTRODUCTION

Gas plasmas make up 99% of our universe, existing mainly as stars. Although rare on earth, natural plasmas include lightning, the aurora borealis, and St. Elmo's fire. Table 1 lists certain plasmas and characterizes them by particle density and temperature.

Plasmas can be produced and controlled by ionizing a gas with an electromagnetic field of sufficient power. One useful form of gas plasma is made by introducing gas into a reaction chamber, maintaining pressure between 0.1 and 10 torr, and then applying radio frequency (rf) energy. Once ionized, excited gas species react with surface of materials placed in the glow discharge.

The physical and chemical properties of plasmas depend on many variables: chemistry, flow rate, distribution, temperature, and pressure of the gases. Additionally, rf excitation frequency, power level, reactor geometry, and electrode design are equally important. Dissociated gas molecules quickly recombine to their natural state when the plasma's power source is shut off.

2.0 TYPES OF PLASMA

Plasmas occur over a wide range of temperatures and pressures, however all plasmas have approximately equal concentrations of positive and negative charge carriers, so that their net space charge approaches zero.

In general, all plasmas fall into one of three classifications. Elements of high-pressure plasmas, also called *hot plasmas*, are in thermal equilibrium (often at energies > 10,000°C). Examples[1] illustrated in Table 1 include stellar interiors and thermonuclear plasmas. *Mixed plasmas* have high temperature electrons in midtemperature gas (~ 100–1000°C) and are formed at atmospheric pressures. Arc welders and corona surface treatment systems use mixed plasmas. *Cold plasmas*, the focus of this chapter, are not in thermal equilibrium. While the bulk gas is at room temperature, the temperature (kinetic energy) of the free elec-

295

Table 1   Ranges of Particle Densities and Temperatures for Plasmas of Various Types

Plasma type	Particle density (no./cm^3)	Temperature (K)
Natural plasmas		
Stellar interiors	10^{22}–10^{25}	~10^8
Stellar atmospheres	10^{10}–10^{16}	10^4–10^6
Earth's ionosphere	10^{10}–10^{12}	10^2–10^3
Manmade plasmas		
Thermonuclear plasma	10^{12}–14^{14}	10^8–10^9
Constructed arc		
Plasma jets	10^{16}–10^{18}	1–5×10^4
Free-burning		
Electric arcs	10^{16}–10^{17}	7–10×10^3
Combustion flames	10^{16}–10^{15}	3–5×10^3
Low pressure arcs	10^{10}–10^{12}	1–3×10^3
Cold gas plasma	10^{10}–10^{12}	300–600

trons in the ionized gas can be 10–100 times higher (as hot as 10,000°C),[2] thus producing an unusual, and extremely chemically reactive environment at ambient temperatures.

There are two types of cold plasma, as determined by electrode configuration. *Primary plasmas* are generated directly by rf energy between the electrodes of a reaction chamber. *Secondary plasmas* exist downstream of the energy field, carried by gas flow and diffusion. Secondary plasmas are less desirable for surface modification because the farther downstream from the rf field the parts to be treated are, the less reactive the plasma becomes. One part may shield another, creating nonuniformity, and less surface area can be treated before all active species are locally depleted, reducing effectiveness with larger loads.

3.0   A TYPICAL PLASMA PROCESS CYCLE

Modern plasma processing equipment is entirely automated. Typical continuous treatment equipment is pictured in Figure 1.

To prepare for treatment, the process engineer enters set points into a microprocessor controller. The conditions of the process have been determined previously to be effective for the specific application. A key switch locks out the set points, so that the operator cannot inadvertently modify the process. A process may consist of single or multiple process steps, each comprising a complete plasma cycle.

Loading the chamber, the machine operator places parts on the electrode/shelves. After closing the door, the operator starts the process by pressing the start button. The controller monitors and controls the process until it is complete.

The first step the controller takes is evacuating the reaction chamber by opening the isolation valve to the vacuum pump. At a predetermined base pressure, process gases enter the chamber. A delay lets the reaction chamber's pressure stabilize, and then the power is switched on, creating a plasma. The controller begins the process timer. An impedance-matching network continuously and automatically minimizes power mismatch between generator and chamber as long as the rf power is on.

Figure 1 Typical continuous treatment equipment.

The step ends after the process time has expired, or after an optional temperature set point has been achieved, at which time the rf power and the process gases are shut off. The vacuum pump evacuates the process gas and by-products from the chamber, and the system repeats the entire cycle for the next step. If the last step has been completed, the chamber vents to atmosphere and the controller alerts the operator that the process is complete.

4.0 PLASMA CHEMISTRY

Three properties of the cold gas plasma—chemical dissociation, kinetic energy from ionic acceleration, and photochemistry,—make this unique environment effective for surface treatment.

Exposing gases to sufficient electromagnetic power dissociates them, creating a chemically reactive gas that quickly modifies exposed surfaces. At the atomic level, plasma contains ions, electrons, and various neutral species at many different energy levels. One of the excited species formed is the free radical, which can directly react with the surface of organic materials, leading to dramatic modifications to their chemical structure and properties. Modification sites also occur when ions and electrons bombarding the surface have gained enough kinetic energy from the altering electromagnetic field to knock atoms or groups of atoms from surfaces. Furthermore, gas phase collisions transfer energy—forming more free radicals, atoms, and ions.

Combining of dissociated species gives off photons as they are returning to their ground state. The spectrum of this glow discharge includes high energy UV photons, which will be absorbed on the top surface layers of the substrate, thus creating even more active sites. The color of the glow discharge depends on the plasma chemistry, and its intensity depends on the processing variables.

The plasma process modifies only several molecular layers, thus appearance and bulk properties are usually unaffected. In addition, plasma changes the molecular weight of the surface layer by scissioning (reduction in molecular length), branching, and cross-linking organic materials. The chemistry of the plasma determines its effects on a polymer.

5.0 SURFACE TREATMENT APPROACHES

Like many polymer processes, plasma is a chemical process. Three types of cold plasma treatments are used in processing polymers.

1. *Activating plasmas* use a gas or gases that react with the product to modify its chemistry. Such plasmas use oxygen, ammonia, air, halogens, and other gases for cleaning surface contaminants, microablating the surface, and substituting various chemical groups onto the polymer chain. Activating plasmas are discussed below.
2. *Grafting plasma* treatment first activates the surface by exposure to a chemically inert plasma, then bathes the surface in a vapor of an unsaturated monomer (without plasma generation). The free radicals previously formed on the polymer surface initiate grafting reactions with the reactive monomers.
3. *Plasma polymerization* utilizes plasma energy to initiate gas phase polymerization reactions causing the deposition of organic onto surfaces within the plasma chamber.[2]

6.0 PLASMA ACTIVATION OF PLASTICS

Activating plasmas have three competing molecular reactions that alter the plastic simultaneously. The extent of each depends on the chemistry and the process variables. They are as follows.

1. *Ablation* (microetching), or removal by evaporating surface material either for cleaning or for creating surface topography.
2. *Crosslinking*, or creating covalent bonds or links between parallel long molecular chains.
3. *Substitution*, the act of replacing atoms in the molecule with atoms from the plasma.

Ablation is an evaporation reaction in which the plasma breaks the carbon-to-carbon bonds of the hydrocarbon polymer. As long molecules become shorter, their volatile monomers or oligomers boil off (ablate) and they are swept away with the exhaust. Ablation is important for surface cleaning, and where desired, for surface etching. Cleaning removes from polymer surfaces such external organic contaminants as hydraulic oils and mold releases. Equally important is the removal of internal contaminants such as processing aids and internal lubricants that have bloomed to the surface. Often an oxygen-containing plasma is selected to facilitate rapid breakdown of the suspected contaminant into a volatile by-product. Cleaning by plasma is more effective than cleaning by vapor degreasing or by other methods. Plasma produces a "superclean" surface; but if gross contamination exists, parts may be precleaned by ultrasonic cleaning, or solvent–vapor degreasing so that the plasma process time is kept to a minimum and thus remains cost effective.

Once cleaned, the plasma begins ablating the top molecular layers of the polymer. Amorphous, filled, and crystalline portions will be removed at different rates, giving a technique effective for increasing surface topography with a view to increasing mechanical adhesion or for removing weak boundary layers formed during molding.

Cross-linking, on the other hand, is done with an oxygen-free noble gas(argon or helium). After the plasma has generated surface free radicals, these react with radicals on adjoining molecules or molecular fragments to form cross-links. This process increases the strength, the temperature resistance, and the solvent resistance of the surface.

Unlike ablation or cross-linking, substitution replaces one atom or group from the surface with active species from the plasma. In this case, free radical sites on the surface are free to react with species in the plasma, including but not exclusively free radicals, thus altering surface chemistries by the addition of covalently bonded functional groups. The selection of the process gas determines which groups will be formed on the modified polymer. Gases or mixtures of gases used for plasma treatment of polymers include nitrogen, argon, oxygen, nitrous oxide, helium, tetrafluoromethane, water, and ammonia. Each gas produces a unique plasma chemistry. Surface energy can be quickly increased by plasma-induced oxidation, nitration, hydrolyzation, or amination.

Very aggressive plasmas can be created from relatively benign gases. For example, an oxygen and tetrafluoromethane (Freon 14) plasma contains free radicals of fluorine. Oxidation by fluorine free radicals is known to be as effective as oxidation by the strongest mineral acid etchant solutions, with one important difference: hazardous and corrosive materials are not used. As soon as the plasma is shut off, the excited species recombine to their original stable and nonreactive form. In most cases treatment of the exhaust effluent is not required.

Gases that contain oxygen are generally more effective at increasing the surface energy. For example, plasma oxidation of polypropylene increases the initial surface energy of 29 dynes/cm to well over 73 dynes/cm in just a few seconds. At 73 dynes/cm, the polypropylene surface is completely water wettable. Increased surface energy results in a plasma that yields polar groups such as carboxyl, hydroxyl, hydroperoxyl, and amino. A higher energy (hydrophilic) surface translates to better wetting and greater chemical reactivity of the modified surface to adhesives, paints inks, and deposited metallic films, providing for improved adhesion and permanency.

The enhanced surface reactivity is characterized in the laboratory by studying water wettability. Wettability describes the ability to spread over and penetrate a surface; it is measured by the contact angle between the liquid and the surface. The relationship between contact angle and surface energy is inverse—the contact angle decreases with increasing surface energy. Wettability can easily be induced on normally nonwettable materials such as polyolefins, engineering thermoplastics, fluoropolymers, thermosets, rubbers, and fluoroelastomers.

Noble gases(argon, helium, etc.) generate surface free radicals that react either with other radicals on the surface, yielding molecular weight changes, or with the air, when the part is removed from the chamber, thus increasing the surface energy.

Process gases such as fluorocarbons will generally provide a lower energy, or hydrophobic, surface by substitution of abstracted hydrogen with either fluorine or trifluoromethyl radicals to form a fluorocarbon surface. Fluorination is favored in some medical applications, where it undesirable to have catheters be wetted by blood. The nonwettable barrier layer also inhibits chemical penetration, a consideration that is important for specialty packaging.

Table 2 Typical Bond Improvement[a] After Surface Treatment: Solvent-Washed Plasma

Material		Shear strength (psi)	Failure mode[b]
Valox	(polyester)	522	Adh
		1644	Coh
Noryl	(polyphenylene oxide)	617	Adh
		1799	Coh
Durel	(polyarylate)	250	Adh
		2161	Coh
Vectra	(LCP)	939	Adh
		1598	Coh
Lexan	(polycarbonate)	1705	Adh
		2242	Coh
Delrin	(polyacetal)	165	Adh
		857	Adh

[a]Lap shear strength 3M Weldbond #2256.
[b]Adh = adhesive failure, Coh = cohesive failure of adherent.

Table 3 Paint Adhesion on Polyethylene: Urethane Heat-Cured Automotive Paint

Test condition	Failure	Liftoff(%)
Control (no plasma)	OB[a]	100
Plasma treated	5B	0

[a]ASTM Test Method D3359–83, "Measuring Adhesion by tape test."

7.0 ADHESION

Bonding in manufacturing processes is a specialized field, but generally, cleanliness and wettability are necessary for good adhesion.

High surface energy alone does not guarantee better adhesion; however, the versatility of the process enables tailoring of the surface chemistry for optimal adhesion or superior product performance. It is not uncommon to move the focus of failure from the bond line into the adherent or into the adhesive with a many-fold increase in the adhesion. Examples of typical plasma improvement on a range of materials for epoxy bonding[3,4] and for coating[5] are shown in Tables 2 and 3.

Extremely mobile polymers, like elastomers, have shelf lives measured in minutes or hours. The flexible molecular chains rotates the high energy functional groups into the bulk. Once an active surface has been treated and properly coated with adhesive, the modification is permanently tied to the surface. Thus priming of a treated elastomer fixes the surface chemistry.

8.0 SUMMARY

Plasma surface treatment is an effective, efficient method of modifying a wide variety of polymeric and elastomeric surfaces. Adhesion strength of treated materials often exceeds that of the adherent. The plasma process is not operator-sensitive; its other significant characteristics include reproducibility, cleanliness, and the ability to more consistently provide high reliability bonds.

REFERENCES

1. B. Chapman, *Glow Discharge Processes*, New York: Wiley, 1980.
2. J. R. Hollahan and A. T. Bell, *Techniques and Applications of Plasma Chemistry*. New York: Wiley, 1974.
3. S. L. Kaplan and P. W. Rose, "Plasma surface treatment of plastics," presented at the SPE 46th Annual Technical Conference, Atlanta, April 1988.
4. S. L. Kaplan, and P. W. Rose, in *Plastics Finishing & Decorating*, D. Satas, Ed., New York: Van Nostrand Reinhold, 1986.
5. S. L. Kaplan and S. Rines, "Plasma pretreatment for plastics," *Prod. Finish.*, January 1988.

31
Surface Pretreatment of Polymer Webs by Fluorine

R. Milker

Lohmann GmbH, Neuwied, Germany

Artur Koch

Ahlbrandt System GmbH, Lauterbach, Germany

1.0 INTRODUCTION

Bonding, coating laminating, painting, and printing require good substrate adherence. This requires above all, surface polarity, which permits mechanical and, in particular, chemical bonds. For this reason, polymeric materials are treated by means of oxidation processes entailing three main groups: corona discharge, flame treatment, and chemical.

All processes are more or less disadvantageous. The chemical methods have been proved only in narrow, limited fields of application (e.g., as liquid pickling agents), or they require high supervision and a substantial maintenance effort (e.g., ozone treatment). State-of-the-art fluorination, is troublesome because it is a discontinuous process and not feasible in many cases, especially for web-shaped substances. Corona pretreatment requires high investment and is strongly liable to interference. In the area of the dielectric material, fires occur frequently, causing short circuits in the pretreatment station. Additionally, only one side of the web material can be activated by the corona discharge.

This chapter describes an attractive new pretreatment method implemented by Lohmann GmbH, featuring the continuous fluorination of web materials.[1]

2.0 THE FLUORINATION PROCESS

The surface of web-shaped, polymeric substrates is subjected continuously, for a short time, in a suitable reaction chamber, to elemental fluorine—attenuated with an inert gas. Thus, the surface energy of the polymeric material is increased to such an extent that excellent adherence to other polymers (e.g., lacquers and adhesive agents) is attained.[2]

The fact that this fluorination technique is feasible at all, on an industrial scale, is due to the astonishing chemical–technological developments realized during the past 20 years, which having made fluorine not only an important initial substance but even a component in large-scale technical operations.

2.1 Fluorine

Fluorine, an almost colorless gas, is one of the strongest oxidizing agents; it is surpassed only by a few other oxidants (e.g., chlorine fluoride, chlorine trifluoride). It reacts with almost all organic and inorganic substances; the few exceptions include, first of all, nitrogen and the inert gases helium, neon, and argon, plus some metal fluorides in the highest valence state, as well as other fully fluorinated compounds (e.g., CF_4 or SF_6). Fluorine's great reactivity can be explained by the interaction of the low dissociation energy of the molecule itself and the very strong bond fluorine forms with other atoms. Moreover, since the fluorine atom is rather small, the spatial relations in fluorine compounds admit high coordination numbers of the relevant central atom.

For a long time, the extreme aggressiveness of fluorine has limited its use in industrial applications.

Indeed, as late as 1936 a technical encyclopedia stated: "Due to the difficult manufacture and storage, fluorine has no practical importance for industry." It was only a little more than two decades ago that all difficulties impairing the manufacture of fluorine on a large scale could be considered to have been overcome.

Nowadays, elementary fluorine in the liquefied state is transported even in fuel trucks. It is mainly used for the manufacture of the highly volatile uranium hexafluoride, which is known to serve for the separation of the uranium isotopes U^{235} and U^{238}. Thus fluorine has become a key product for the nuclear industry. In Germany, Kali Chemie AG, which has had a leading part in fluorochemistry in general, is the only manufacturer of this extremely reactive substance.[3]

2.2 Plant for Continuous Surface Fluorination

Figure 1 diagrams the construction of a plant for the pretreatment of web-shaped materials with fluorine.[4] Figure 2 illustrates an actual set up. Fluorine supplied in steel pressure bottles is attenuated to a certain concentration with inert gas (nitrogen, rare gases, and compressed air) in a corresponding dose discharge devise. This fluorine–inert gas compound is available for the system via the line 9 in Figure 1. The attenuation step can be dispensed with by connecting instead to a steel pressure bottle filled with a fluorine–nitrogen compound.

The fluorine delivery amounts for a 10 vol% standard compound—due to the higher filling pressure—to 50% of that obtained with the pressure bottle containing elementary fluorine. After the reaction chamber 2: Fig. 1 has been filled with attenuated fluorine, the fluorine–inert gas compound is circulated through reaction chamber 2 by the pump (5) via pipe (6) and the two stop valves (7 and 8). The web-shaped material to be pretreated is fed into chamber 2 via the intake airlock (3) and leaves it via the outlet airlock (4).

The parameters time of contact and concentration of fluorine influence the surface energy of the polymer. The surface effect attainable depends additionally on the chemical quality of the polymer. Because of the extreme reactivity of the elemental fluorine, all polymeric materials disposing of substitutable hydrogens can, in principle, be activated. Airlocks 3 and 4 are supplied with inert gas via pipe 11. The airlocks, as well as the series-connected flow resistors (12) prevent, to a great extent, the escape of fluorine and hydrogen fluoride (HF) into the ambient air. Such a system of airlocks is described in details in Ref. 5. From pipe 9 and via dosing valve 10, the escaped portion of fluorine is replaced, but a small quantity of it is lost as HF by virtue of the chemical conversion with the reactant. Since a partial fluorination causes surface activation, which is sufficient for many purposes, the quantity of fluorine actually consumed or the resulting quantity of HF is rather small.

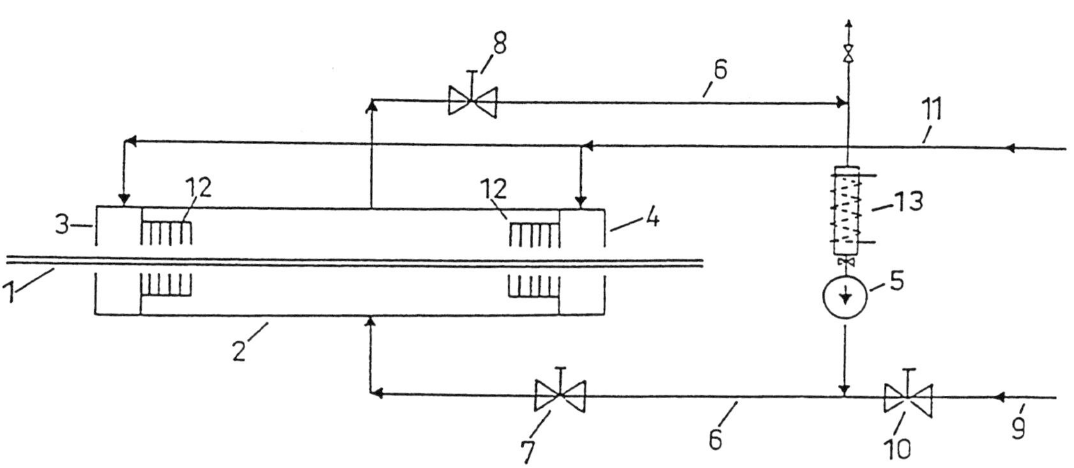

Figure 1 Schematic diagram of a continuous surface fluorination pretreatment plant. Numbered components are identified in the text.

Figure 2 The surface fluorination pilot plant of Lohmann GmbH, Neuweid, Germany.

The low portions of HF in the system are removed by means of a hydrogen fluoride absorber (13 in Fig 1). Such a cleaning is also necessary if the fluorine is to be dosed by a glass rotameter. The absorber comprises a pipe made of Monel, nickel, or steel (diameter, 50 mm; length, 450 mm), and filled with granular, porous sodium fluoride. Welded to both sides of the pipe are caps that carry supply and drain pipes. The absorber can be heated up to 300°C by means of an electric tube furnace. The HF absorption ensues at room temperature. For regeneration, the absorber is heated in the nitrogen flow to +300°C. A suitable porous sodium fluoride can be produced by heating grained or pelleted sodium bifluoride to 250–300°C in a nitrogen flow.

If required, the entire system can be rinsed via a gas washer: the F_2 and HF portions in the rinsing gas are absorbed in counterflowing diluted potash lye and are made innocuous. The exhaust air is completely free from pollutants.

2.3 Safety Precautions

The hazard associated with the gases fluorine and hydrogen fluoride being delivered through the airlock to the open air is comparable to the one caused by ozone released, which occurs inevitably as a result of the corona pretreatment. The safety measures known and approved for ozone apply also to fluorine and hydrogen fluoride. For these gases, a threshold limit value of 0.1 ppm in air was fixed, as for ozone. Biological assays have shown, that the toxicity of fluorine and hydrogen fluoride is many times less than that of ozone.

Since fluorine is perceptible by its smell rather like heavily chlorinated water, even in rather low concentrations, cases of poisoning are extremely rare.

On-the-job safety is safeguarded by installing a chemical detector giving an acoustic alarm and interrupting the fluorine supply if the threshold contents of fluorine admitted to the air is exceeded.

3.0 PRETREATMENT WITH FLUORINE: APPLICATION EXAMPLES

Extensive serial tests on a pilot plant as described in Section 2.2 illustrate that:

A concentration of 5–10 vol % fluorine in the plenum chamber is fully sufficient to raise the surface energy of the polymer above 50 mN/m.

Average fluorine consumption is only 2.5 kg for the pretreatment of approximately 100,000 m^2 of polymer surface. (The fluorine costs are thus practically negligible.)

All tests described below were performed with a working width of 1 m; the reaction chamber to which the fluorine–nitrogen was admitted had a length of 1 m and a height of 30 cm, for a volume of 300 liters.

3.1 Polyethylene–Vinyl Acetate Copolymer Foam

A closed-cell foam, 1 mm thick, is to be pretreated on both sides at a web speed of 5 m/min.

After having passed the fluorine atmosphere, the surface tension of the foam material increases from 30 mN/m to 54 mN/m. If the foam material is used as for example, supporting material for double-sided, pressure-sensitive adhesive tapes, the adhesive bond is as good as with the corona-pretreated material. This is shown in Table 1.

Electron spectroscopy for chemical analysis (ESCA) research [8] indicates that activation of the foam selected in this example is predominantly based on the partial fluorination of the terminal methyl group of the comonomeric vinyl acetate. As with the corona pretreatment,

TABLE 1 Adhesive Bonding Versus Pretreatment

Foamed plastic	Surface tension (Mn/m)	Pebra test[6]	Force test[7] (N/625 mm^2)
Not pretreated	< 30	3 sec	99
Corona pretreated	50	> 2 min	129–149
Fluorine pretreated	54	> 2 min	133

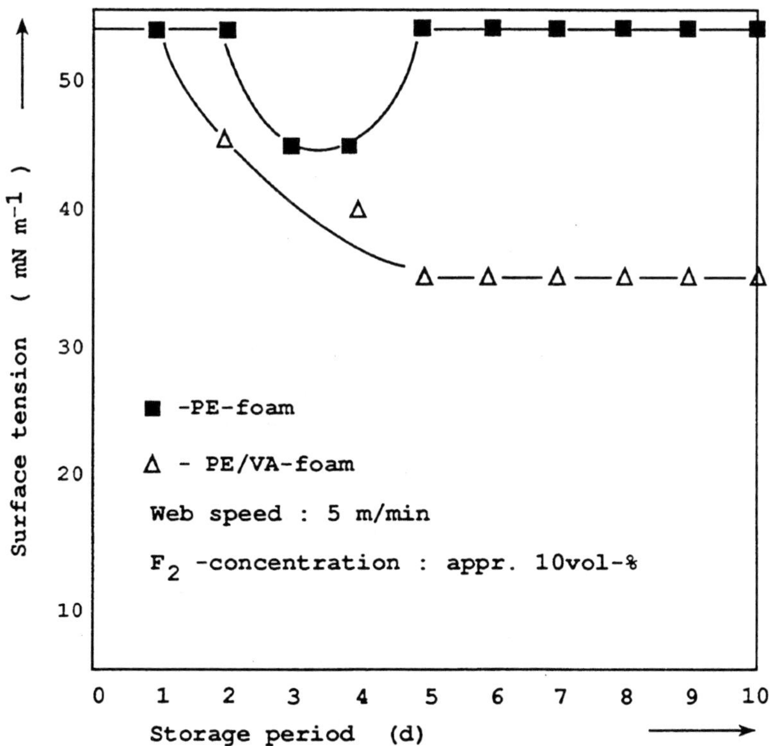

Figure 3 Storage stability of the surface energy of polythylene (PE) and PE–vinyl acetate (VA) copolymer foams subjected to fluorine pretreatment.

TABLE 2   Surface Tension Values Obtained with Corona and Fluorine Pretreatments

Type of sheeting	Surface tension values (mN/m)		
	Initial	Corona	F_2
Low density polyethylene	30	35	> 54
High density polyethylene	32	< 32	> 54
Biaxially oriented polypropylene	32	35	> 54
Polyethylene sulfide	35	> 54	> 54
Polyvinyl chloride	40	> 54	> 54
Polyurethane resin	32	< 32	> 45
Polyethylether ketone	35	> 54	> 54

the surface tension degrades gradually, but it stabilizes on a higher level, as indicated in Figure 3. Pure polyethylene foam, however, maintains for weeks
the conferred surface tension of more than 50 mN/m.

3.2 Plastic Sheeting

With the maximum web speed of 24 m/min (F_2 concentration 5 vol %) attained in the pilot plant, common film or sheeting qualities of differing widths were successfully pretreated on one or two sides, as shown in Table 2. The pretreatment effect remained constantly high throughout an observation period of 6 weeks.

3.3 Air Cushion Sheeting

For certain applications an adhesive agent coating that is applied on the burl side has to be firmly bonded to the surface of the sheeting or film, even in the interstices.. Here, the fluorination technique is in fact predestined to be a solution for problems. As expected, such a polyethylene air cushion sheeting or film can be pretreated as well as other polyethylene sheetings.

3.4 Terpolymer Rubber

Ethylene propylene rubber monomer (EPDM) belongs to the materials that will be used to an ever-increasing extent in the future—in particular in the automobile industry.

The negative side of this versatile, almost indestructible polymer body is its obstinate pretreatment behavior. So, after the conveyance of EPDM sectional cuts on an endless belt through the fluorination chamber, only 45 mN/m is measured. This surface energy, however guarantees satisfactory substrate adherence.

4.0 ADVANTAGES OF SURFACE PRETREATMENT WITH FLUORINE

Compared with other pretreatment processes, the fluorination method not only features a wide spectrum of applications but unlike corona pretreatment and ozonization, it does not require electrical apparatus, which is susceptible to interference and requires intensive maintenance. Additionally, the basic investment for fluorine fittings (discharge station) and for a reaction chamber is lower than the prime cost of a corona pretreatment station.

A particular advantage of continuous fluorination can be seen in the application to web-shaped bodies, which—when wound in a compact roll—are not accessible for fluorination by present state-of-the-art means. Such materials can be continuously surface activated now from roll to roll, independent of roll length.

Another very important argument for the fluorination technique is the fact that the surface treatment effect, unlike corona pretreatment, is long-lasting—at best, irreversible—on both sides. This is of enormous importance for practical applications in industry, since subsequent surface refining processes need not follow immediately the surface activation.

Furthermore, the fluorination process provides effective pretreatment of the following materials:

Burled, embossed, or otherwise irregularly formed surfaces of sheetings or films.
Narrow fabrics.
Foams wider than 2 mm.
Biaxially stretched polypropylene sheeting of any width.

5.0 SUMMARY

Now that fluorination technique has become controllable, it offers promising and beneficial prospects for the pretreatment of polymeric, web-shaped materials. It cannot and will not displace corona pretreatment, which has proved its usefulness for decades, but is surely an attractive solution for a number of special applications.

REFERENCES

1. R. Milker, *Coating*, in 11, 294–298. (1985.)
2. EP 0 214 635
3. KC-Fluor, Kali-Chemie AG
4. G 87 03 823.4
5. DOS 30–38–741
6. Test instruction AW 14, Pebra
7. Test instruction 0400, Lohmann GmbH, Neuwied, West Germany.
8. Dr. Ruppert, Chemical Institute of the University of Bonn, personal communication.

32

Calendering of Magnetic Media

John A. McClenathan

IMD Corporation
Birmingham, Alabama

1.0 INTRODUCTION

Calendering is a continuous web process: two or more cylindrical rolls are nipped together under mechanical load and driven via a system motor. This chapter discusses the calendering of coatings in the manufacture of magnetic media products.

2.O CALENDERING MAGNETIC MEDIA PRODUCTS

In the magnetic tape industry, the calendering process, relates to improving performance of the manufactured product. Improvements to the surface of the applied coatings on the substrate, usually polyester film, can be varied depending on the end use of the magnetic media product; they include:

Improved surface finish.
Densified coating.
Improved aesthetics.
Reduced product thickness.
Improved product electrical output.
Reduced recorder head wear.
Promotion of coating durability.

Some of the product improvements listed are interrelated and can be optimized. As an example, improvement of the surface finish will be achieved by the calendering process and the coating will be densified. Iron oxide in the coating binder system must be maintained at the coating surface and not caused to migrate toward the base film. The surface can become very smooth if a percentage of binder system is caused to migrate to the coating surface, displacing the iron oxide particles away from the eventual recording head. The condition of "too smooth" can also be achieved, and the lack of surface boundary lubrication air will allow contact between the tape and recording head, resulting in excessive drag on the tape.

The surface conditions required on the magnetic media products are critical. The hydrodynamic lift clearances are 5–10 microinches (μin.)(0.12–2.5 μm), as compared to a coated surface roughness (after calendering) of 0.6–1 μin. (0.014–0.025 μm) for the magnetic media products. To achieve this smooth finish, the coating to be calendered is placed in contact with the polished surface of a steel hot roll. The basic nip area, as shown in Figure 1, is the physical contact area between two rolls. One roll is the compliant roll and has a lower modulus of elasticity than the second roll, which is usually steel. Calendering results from the process, which consists of a controlled nip pressure and controlled heat applied to the product. Optimization of these two basic parameters will, in most cases, cause product quality improvements.

The most noted factors to be controlled during calendering are as follows.

1. *Variable and controlled nip pressure.* The calenders are usually hydraulically loaded to achieve loadings from 500 to 4000 PLI, where PLI is pounds per inch of roll face, (89–714 kg/ cm).
2. *Steel (or hot) roll surface finish.* Roll surfaces are hardened and chrome plated to provide optically perfect surfaces of better than 0.5 μin. (0.012 μm).
3. *Temperature of the product.* The temperature of the hot steel roll and the compliant roll surfaces must be controlled. The steel roll is heated internally by water, oil, or electricity and controlled to ±1°C.
4. *Cleanliness.* Compliant rolls must not be a source of particle contaminants. Any foreign particles or dirt will mark the product when passed through the high pressure nip area.
5. *Rolls must be of good physical design.* Deflection must be minimum. Roll runout of 0.001 in. (0.025 mm) TIR or better must be maintained. The face surfaces on all rolls must be straight. Tolerances depend on the compliant roll material's ability to conform to the steel roll surface.
6. *Compliant roll.* A variety of compliant roll materials with different surface finishes and hardnesses are available and include the following: paper filled—cottons, polyester, nylon, polyimide, rubber, steel, and epoxy–glass. The hardness range of most of the compliant rolls (excluding steel) that are used is 80–90 Duro Durometer "D."
7. *Temperature.* Control of the web temperature during passage through the calender must be held at an acceptable level to provide plasticity to the coating binder system without distorting the base film. Roll diameters, number of nips, web speed, web thickness, and the web wrap angles on the rolls provide controllable variables for calendering success.

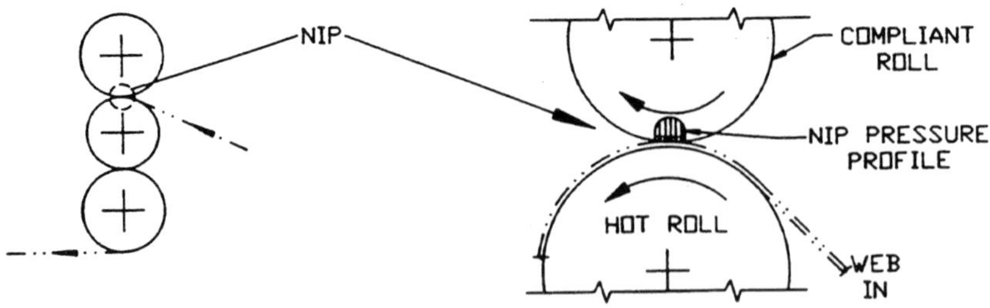

Figure 1 Nip definition.

3.0 CALENDER DESIGN

Calender configurations vary from a simple two-roll configuration to a multiroll stack of eleven rolls. Unsupported compliant rolls require larger diameters to provide high nip loading capability. This results in the most common arrangement, featuring the two external rolls made of steel (not compliant) to provide the required nip stiffness. This concept creates stacks of three, five, seven, nine, or eleven rolls. The arrangement can be horizontal or vertical (Fig. 2). The loading methods can vary but most often are from the bottom on a vertical stack to allow easy opening for emergencies.

Intermediate and end steel rolls can be cooled as well as heated, if required, to control the process web temperature. The sizes of the intermediate rolls also are often reduced in a multiroll stack machine to reduce the required overhead installation space and reduce the nip area, thereby providing a high unit area loading.

The loading system is usually hydraulic but can also be pneumatic or mechanical. Direct loading requires larger loading cylinders than a moment arm mechanical advantage system.

The calender stack power requirements vary with the speed, web width, pressure, and type of compliant rolls used. The usual power source consists of a motor and reducer driving through a coupling directly to the bottom steel roll in a vertical stack, or the fixed—end steel roll in a horizontal stack. All rolls can be driven with separate drive motors if space allows. If only one prime mover is provided on the stack, a means of rotating all rolls is required to allow transporting the product with the nips open and also allow closing of the nips without product damage.

Calenders are provided with emergency stop cables, nip guards, and drive stop interlocks to open all nips.

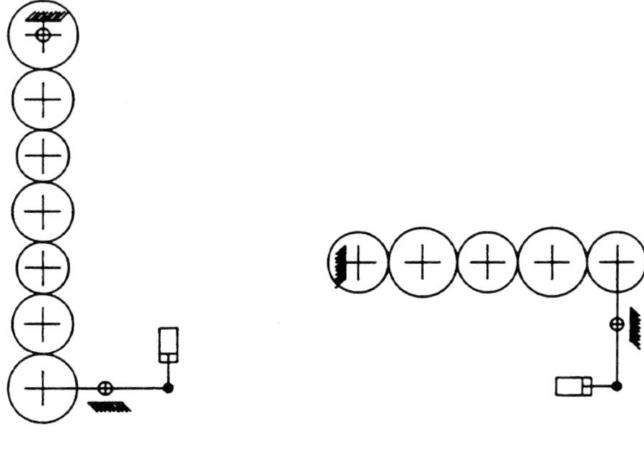

Figure 2 Typical calender arrangements.

REFERENCES

1. J. J. Brondijk, P. E. Lvierenga, E. E. Feekes, and W. J. J. M. Sprangers, "Roughness and deformation aspects in calendering of particulate magnetic tape," *IEEE Trans. Magn. MAG–23*(1), (January 1987).
2. Alex N. Tabenkin, "The growing importance of surface finish specs," *Machine Design*, Sept. 20, 1984.
3. J. A. McClenathan, *The Design of a Magnetic Tape Calender*. Birmingham, AL: IMD, Feb. 17, 1970.
4. Harold Bredin, "Tribology and the design of magnetic storage system." *Mech. Eng.* March 1985.
5. A. J. Holloway, "Polyester film surface definition and control," presented at the Symposium of Magnetic Media Manufacturing Methods, Hawaii, May 1983.

33
Embossing

John A. Pasquale III

Liberty Machine Company, Paterson, New Jersey

1.0 GENERAL

Embossing is a method by which a web is textured by the use of a pattern roll pressing against a backup roll under controlled conditions. One can emboss both thermoplastic and nonthermoplastic webs by choosing the proper roll arrangement to deform the web.

To emboss thermoplastic materials, the web is deformed by preheating and pressing it with a cooled embossing (pattern) roll to set the pattern and cool the web to retain that pattern.

The degree of preheating to soften the web must be carefully controlled so that no melting or degrading of the web will take place. To make the heat removal process as efficient as possible, no more heat should be applied than is needed to satisfactorily emboss the product.

To emboss a nonthermoplastic web such as paper, textile, or foil, one must apply pressure that exceeds the elastic limit of the substrate and imparts the pattern. This type of embossing involves the use of male and female rolls, either two rolls with matched patterns made of steel or other metal, or a steel pattern roll that comes in contact with a filled backup roll, which takes a permanent deformation for a given pattern by running the steel embossing roll in contact with the backup roll during a "running in" period. In some cases, special rubber-covered rolls can be used; their behavior eliminates the need for "running in."

2.0 THERMOPLASTIC WEBS

Embossing thermoplastic webs is achieved by using an engraved metal roll pressing against a rubber-covered backup roll. The metal roll is cooled with a refrigerated solution to remove heat from the product and to set in the embossed pattern. The backup roll is internally cooled, mainly to increase the life of the rubber covering. The roll may also be cooled externally by a water bath and squeeze roll system, especially if the web is an unsupported thermoplastic film having a tendency to adhere to a hot rubber surface.

Embossing of a web depends on many variables such as:

315

Degree of preheating and rheological properties of the product.
Sheet thickness.
Hardness of the rubber backup roll.
Embossing roll pattern and its cooling capacity.
Postemboss cooling.

A fine balance exits between the preheating and the removal of heat to set the pattern. Applying the appropriate amount of preheat but insufficient cooling results in the inability to deform the web, which has not been softened enough. The best embossing system is one that optimizes heat input and removal for a given thickness and speed.

The hardness of the backup roll plays a role in the finished product's texture. If a roll is too hard, a good definition of fine surface patterns might be achieved, however, displacement of material within the product for deep embossings may not be possible. If the roll is too soft, it will allow deep embossings to show through the back of the product, an effect that is objectionable in some applications.

The need to cool the product after it has left the embossing roll is another important factor. Appropriate postcooling facilities, usually cooling rolls, are used to bring the sheet temperature as close to ambient conditions as possible before the product is rewound into a roll. When cooling a thin sheet, the problem of retained heat is minor because a thin sheet releases heat easily. In heavy sheet embossing, although the surface of the web might feel cool, the heat is retained in the body of the sheet. This heat, if it remains, will cause a loss of embossed grain when the product is later rewound for further processing.

Embossing units can be placed in various geometric positions. They are usually either vertical, where the web path enters in a horizontal manner, as shown in Figure 1, or they can be placed in a horizontal fashion, as shown in Figure 2. Under special conditions and for certain applications, it might also be advantageous to find them disposed at a particular angle, as shown in Figure 3.

The preheated web should enter the embossing nip perpendicular to the line of action of the embossing and backup rolls, to ensure that the web is not prematurely cooled by striking either the embossing roll or the backup roll first. It is acceptable and sometimes desirable, however, that the backup roll be contacted first. Many webs are unstable in their preheated condition and will be easily creased if they enter the nip unsupported. A short arc of contact before the embossing action of the nip allows the web to be flattened. It is also preferable to

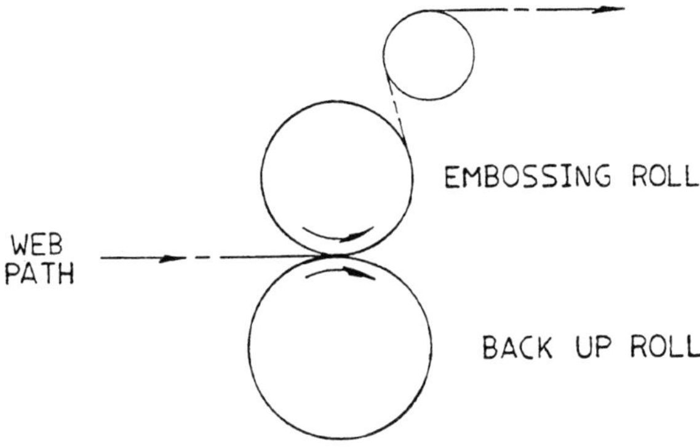

Figure 1 Vertical embossing unit.

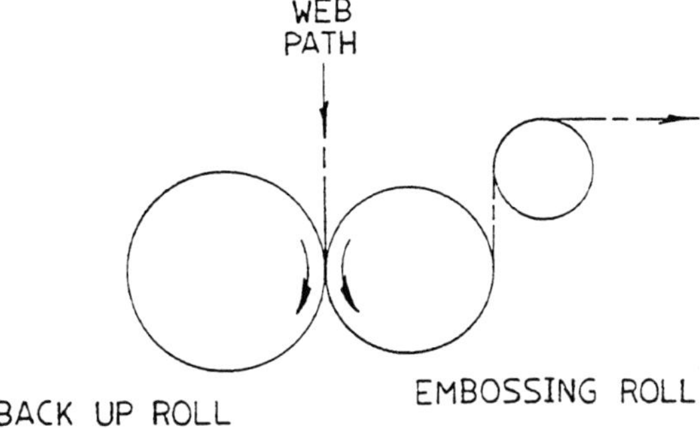

Figure 2 Horizontal embossing unit.

contact the backup roll first because the surface temperature of the backup roll is higher than that of the cool embossing roll; thus the amount of heat removed from the preheated web is minimized. Furthermore, the heat is removed from the back side rather than the face that is to be embossed.

2.1 Embossing Machines for Thermoplastic Webs

Figure 4 shows an embossing machine that incorporates a preheating drum with the addition of a surface radiant heater, an embossing section consisting of a metal embossing roll and a rubber-covered backup roll, and appropriate cooling rolls.

Polyvinyl chloride (PVC) film having a moderate thickness of 0.2–0.3 mm (0.008–0.012 in.) is unwound from a tension-controlled unwind to assure that it is not stretched during the process. It enters the preheating area by passing over a spreader roll and is applied to the heat drum by use of a lay-on roll. The lay-on roll is a rubber-covered, pneumatically operated roll, whose function is to lay the web onto the heated drum surface, allowing for intimate contact and preheating.

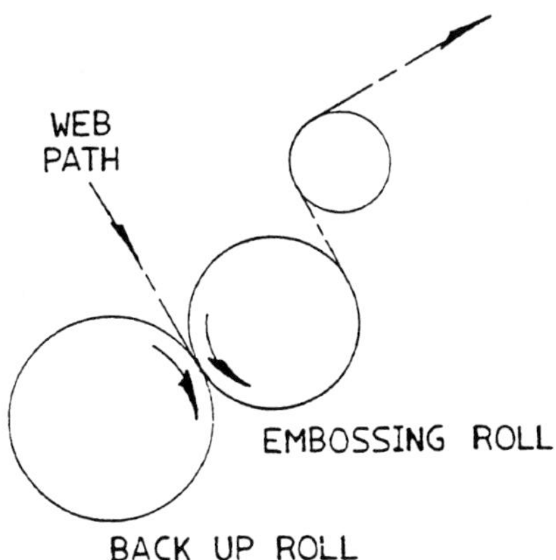

Figure 3 Embossing unit at an angle.

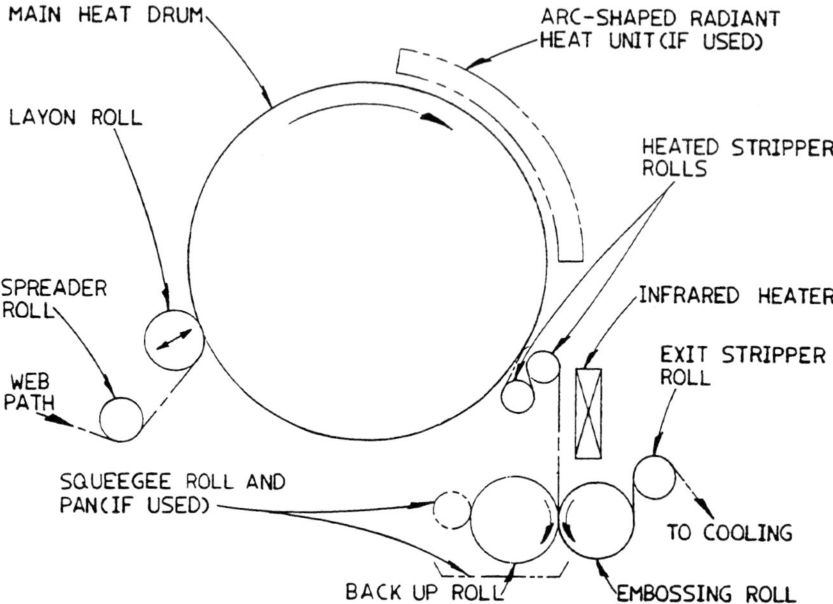

Figure 4 Embossing machine.

The drum is heated by steam or hot oil and is usually 1– 1.5 m (36–60 in.) in diameter. Steam heat is preferred, since it responds to temperature changes more readily than does hot oil. For this thickness of web, the arc-shaped radiant heat unit around the periphery of the drum is not required.

The steam-heated drum has a double-shell construction as shown in Figure 5. Saturated steam enters from one end and passes through the center shaft of the drum; then it is fed through passages to the annular section formed by the inner and outer shells of the drum. As the steam fills the annular chamber and performs its work, condensation takes place and is removed through similar passages at the opposite end of the drum by a syphon tube and exits through the shaft opposite the steam inlet. Both the steam and the condensate enter and leave the drum through rotary joints furnished with bronze hoses to withstand the temperature of the steam. The rotary joints are pipe connections, which allow the drum to rotate freely while they remain stationary to provide a solid connection to the steam pipe and condensate return system.

The embossing section consists of a 250–300 m (10–12 in.) diameter, double-shelled embossing roll, which is designed to allow a cooling solution to pass through it in an efficient manner to remove heat as rapidly as possible, thereby setting in the embossing.

Embossing rolls typically have a double-shell design with a spiral wrap for the most efficient passage of the cooling medium.

For drums heated by hot oil, the construction is similar, but spiral windings forming passageways or channels are provided in the annular space between the inner and outer shells to allow the oil to flow in a prescribed path and in the most efficient manner, to promote good heat transfer. The hot oil is pumped through the center shaft of the drum and enters through conduits similar to the steam-heated drum design; after it has done its work, it leaves by similar passages at the opposite end of the drum. It is appropriate to pump the hot oil at rates

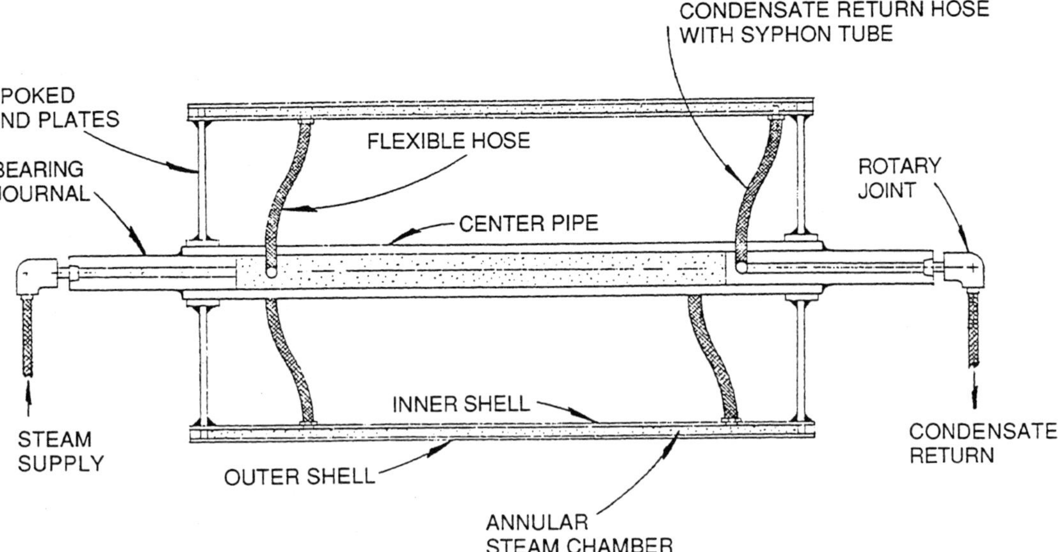

Figure 5 Steam-heated drum.

that will cause it to flow in a turbulent manner through the drum annulus passageways, to provide the best heat transfer coefficient for optimal results.

The preheated web is stripped away from the main heating drum by use of the heated stripper rolls. These rolls help to maintain the temperature of the web and allow it to be passed into the embossing section. The stripper rolls are of single-shell construction and vary in size from 100 to 150 mm (4–6 in.) for steam-heated systems.

The infrared heater shown on the drop into the embossing section is used to maintain the ambient temperature around the web and to assure that the surface to be embossed will be entering the nip at approximately 160°C (320°F).

The embossing roll design is also important, since pressure is exerted between the embossing roll and the rubber-covered backup roll. The double-shell construction, along with the spiral winding, produces a roll that is structurally strong while still having excellent heat transfer capability.

The cooling solution is supplied to the roll through the use of rotary joints with rubber hoses; the solution goes in one end, flows through he spiraled annular construction of the roll, and is discharged at the other end.

Roll design becomes very important from the standpoint of heat transfer efficiency, and care must be taken to properly size the entrance and exit ports (to permit the required flow rate) and the annulus (to assure that the cooling solution is passing through it in the turbulent region of flow), thereby promoting the highest heat transfer coefficient and maximizing the cooling effect.

The backup roll is rubber covered and is cored for internal cooling, but primarily to keep the interface of the steel and rubber covering cool to avoid premature bond failure. The covering used on the roll should be abrasion resistant and able to withstand heat.

The two rolls are brought together hydraulically or pneumatically, with the embossing roll usually being the nondriven roll, while the backup roll is motor driven. To prevent stretching the product as it passes from one section to the other, it is important that synchronization between the main heating drum and the embossing section be accurately maintained.

It is important that the embossing roll be wrapped by the thermoplastic web. This enables cooling to take place while the web is still in intimate contact with the textured embossing roll.

Postcooling is more important on heavier gage materials, where residual heat in the web will later result in temperature rise in the rewound roll, causing a loss of embossing. In a sheet of this thickness, most of the cooling is achieved by the embossing roll and exit cooling roll. For the heavier sheets, it is recommended that dual-shell cooling rolls averaging 460–610 mm (18–24 in.) diameter and cooled through rotary joints be used to complete the cooling system.

3.0 NONTHERMOPLASTIC EMBOSSING

Nonthermoplastic materials are embossed in a manner similar to the thermoplastic materials except that the force between the embossing roll and the backup roll is greater, so that the elastic limit of the material being deformed can be exceeded.

The predominant factor in embossing nonthermoplastic materials is the use of male and female rolls. This roll combination includes either metal male and female rolls, a metal embossing roll operating against a filled roll, or a metal embossing roll acting against a special compound, rubber-covered roll.

The use of metal male and female embossing rolls gives the most concise and definitive texture possible. The engraved patterns on both rolls are carefully matched to allow for insertion of the respective male parts into their counterparts.

The most commonly used combination of rolls is the metal embossing roll and the filled backup roll. This backup roll has either a saturated paper or a textile fabric composition, both of which are densely packed to form a very solid homogeneous structure. The roll is then finished to a specific diameter and when placed in the embossing machine is usually "run in" by bringing the metal embossing roll into contact with the filled roll under load. By continually running the two rolls in contact with each other, the embossing roll pattern is imparted to the surface of the backup roll, giving a male–female characteristic without the use of two mating metal rolls. This combination is excellent for thin as well as heavier materials, since embossing roll–backup roll contact is inconsequential with one metal roll and a composition backup roll.

The use of a rubber-covered backup roll for nonthermoplastic materials depends strictly on the embossing load. There are certain rubber compositions, such as cast urethanes, which will act similar to a composition roll. They do not require "running in" but will deform under the load of the embossing roll.

The backup rolls for nonthermoplastic materials have a much larger diameter than those used for the thermoplastic webs. This is due to embossing loads and wear. The larger surface gives less revolution of the roll, hence less wear of the surface as compared to a similar roll. The larger diameter gives a better structural member to resist deflection.

34

In-Mold Finishing

Robert W. Carpenter

Windsor Plastics, Inc.
Evansville, Indiana

1.0 INTRODUCTION

In the past few years a lot of attention has been focused on the utilization of in-mold finishing of thermoplastic parts. In-mold foiling, available since the early 1970s, has recently developed into an important market because of refinements in tool design and improvements in foils, and through the expansion of in-mold capabilities. Insert molding has also been available for a number of years and has continued to develop as manufacturers' search for technologies that will provide them with superior part quality.

In-mold foiling offers the customer a product that benefits from the economics of having the part finished simultaneously with the injection molding process. The part is also more durable than a component finished with one of the conventional techniques because of foil quality and the inherent benefits of transferring the finish at the melt temperature of the thermoplastic. In-mold foiling also provides the designer with new opportunities, since this application of finishes is not restricted to flat surfaces.

Developing technologies include the application of specialized films such as wood, vinyls, and leathers, and the expansion of design capabilities through interfacing in-mold components with other techniques like electroplating.

With the opportunity to look at designs in their earliest stages, instead of trying to fit the process to an awkward design, the processors began to fully utilize the benefits afforded by in-mold applications. The obvious benefits lie in the reductions in labor and burden because the part is partially or completed finished in the molding cycle, and in many cases without a drastic change in the cycle time required. An other positive aspect of the in-mold processes is the ability to conform to geometries that would be impractical or impossible to finish by more conventional techniques. A third benefit is the variety of materials that may be utilized in the process, which allows for efficiencies of scale when a single mold is utilized to produce a variety of versions with different colors or patterns; moreover, the materials selected can be customized without regard to physical or chemical properties.

321

In-mold decoration has a long history. Inserts, preformed and trimmed, have been utilized for decorative components with deep draws for at least two decades, and commercial applications of in-mold foiling began in the early 1970s.

2.0 PROCESS

2.1 Laminates

Insert molding consists of using pressure or heat to form a laminate with matched metal dies and inserting the formed laminate in the mold before each shot. Tooling to remove excess laminate from the edges is normally required and, where openings are required to match the molded part, cutting dies must be capable of locating these openings in the laminate within 0.002 in. Through the development of melt flow and laminate construction, extremely deep draws and complicated geometries have been accomplished with this technique. The laminate is a very important consideration in this technique, since the physical properties of the finished part are totally reliant on its construction. Utilizing heavy clear films as top coats, these laminates have developed systems capable of high abrasion resistance. In some of the earliest laminate films, tradeoffs existed between abrasion and weathering. In these cases the laminate film would pass 1000 cycles on a Taber Abrader with a CS–17 wheel and a 1000 g load, but would be subject to delamination when subjected to severe weathering conditions involving heat, humidity, and UV exposure. Through years of development, present systems have been improved, and now specialized laminate formulations utilizing fluorocarbon-based systems are being designed for use on exterior automotive applications.

2.2 Foiling

In-mold foiling has been gaining more acceptance in the past 5 years, although as mentioned, the technology existed in the early 1970s. In-mold foiling differs from insert molding in that the foil is transferred from a carrier that is continuously fed through the molding machine; moreover, the carrier is not preformed, but is formed in the cavity during the injection cycle. To accomplish this a foil must be selected with a carrier substantially different from that conventionally used for hot stamping. This is necessary because of the potential for stretching and wrinkling as the foil is moved through the tool and as it is formed during injection. In addition, the release must not allow premature flaking when the melt contacts the foil surface, and the bonding coat must be compatible with the temperatures and pressures associated with injection molding. Foil development for in-mold applications has resulted in dramatic improvements for this process throughout the 1980s. Not only have many of the concerns over splitting, stretching, and wrinkling been resolved, but more advanced top coat systems have improved resistance to abrasion exposure.

The in-mold foiling process produces an excellent bond between the substrate and the foil, because the transfer occurs at the melt temperature of the substrate material. Because the foil is formed in the tool, there is no requirement for preforming, trimming, or cutting, as in insert applications. The foil can be fed on a continuous carrier without interrupting the molding cycle, and the relatively thin film will conform to surface patterns utilized to enhance the visual characteristics of the part, such as ticking for woodgrain applications. In-mold foiling can be successfully incorporated where windows and slots are required, since the film is relatively thin (0.0003 in.), and after the transfer any excess covering the opening can be removed with a simple blowoff operation at the molding machine.

To further enhance the capabilities of in-mold techniques, processors have begun to look into ways to accurately register foils to allow for the application of patterns and graphics in specific areas. Such registration can be accomplished through a guidance mechanism based on holes punched in the side of the foil, or by utilizing an electronic scanning device that locates voids along the edge of the foil and registers the foil accordingly. Since this system requires locating a specific pattern at a point on the molded surface, accuracy in both lateral and transverse alignments is more critical, and the use of statistical process controls for gaging and appearance values is highly recommended. Generally, the development work in this area is performed with the application of logos on flat surfaces.

Successful experiments have been run in which the logo is transferred to a nonflat surface, such as radiused sidewalls or recessed graphics, however. In conjunction with these attempts, work has been done to accomplish the simultaneous transfer of registered foils on both the first and second surfaces of a clear methacrylate part. By starting with a preprinted foil and utilizing this latter technique, some sample parts have been produced by in-mold means that would have required as many as 10 additional secondary operations to achieve the required finish. Still in its early stages of development, this process holds promise for providing the domestic supplier with a competitive technique for decorative acrylic logos.

2.3 New Materials

In addition to the processing innovations for in-mold applications, the use of unconventional materials offers opportunities for the market expansion of in-mold products. Although some restrictions do apply, vinyls, leathers, and even wood materials have been successfully applied to plastic substrates. These materials not only offer the designer a much greater variety of colors and textures, but also provide the opportunity to add a tactile element to the design through the feel of the materials themselves and through the use of techniques such as backing the selected material with a compatible carrier. Testing to determine the effect of environmental exposures is under way at this time to analyze the properties of these new materials and thereby their application potentials. Work is also being done at present to apply conductive films in-mold to effect shielding from electromagnetic interference, and basic development work is under way to electroplate a film onto a mold surface and then injection mold the part structure behind the plate surface to produce an electroplated part in the mold.

One important note to both the designer and the processor: the in-mold processes must be viewed as integral parts of system designs. It is difficult to retrofit an in-mold process for an existing component, and the conversion is less cost effective if the job is at a mature market stage. Practice early involvement, determine whether an in-mold application will be cost effective, design the part to accommodate the technique, and interface the in-molded part with other decorative components to maximize the appearance and economic values. A case in point is a door trim panel.

The interior trim panel assembly consists of the chrome-plated back bezel, a custom color-injection-molded support bezel, and an overlay that has been produced with an in-mold foil process. With respect to its impact on the electroplating-on-plastics industry, this application broadens the scope for decorative interiors because of its unique design and cost effectiveness.

Before this three-component system was developed, such parts were produced by an insert molding process followed by masking an vacuum metallization. The major benefit was that the component part was of a one-piece construction. It was felt that this was the most cost-effective way of producing interior trim panels. As a result, less chrome plating was required for interiors.

In this new application, chrome plating the backup bezel adds rigidity to the overall construction, and through this rigidity, large areas of plastic may be removed without affecting the integrity of the overall assembly. Once assembled, the chrome-plated backup bezel also serves to form the bright bead to frame the color and foil applications utilized in the assembly.

These three components are essentially assembled by a series of snap fits. This assembly arrangement produces a tight, secure assembly and since each decorative element is a separate component, the lines of separation between the color, foil, and the brightwork are very crisp and distinct.

A benefit of multiple-component assembly over one-piece assembly is that a variety of color combinations can be achieved with the basic set of tools by changing the custom color or by changing the foil. A large number of color and design variations can be achieved. This enables the manufacturer to utilize his tooling over a number of years. An additional advantage is the natural appearance and inherent integrity of the chrome-plated bead, which are superior to these properties in products made by conventional metallization or foiling. Because no top coating is required and no individual masking, secondary tooling costs can also be reduced. This means that the assembly is cost competitive as well. This product is an example of what early involvement and preengineering can accomplish.

3.0 CONCLUSION

One of the most formidable arguments for the use of in-mold products is the potential requirement in the future for the reduction of volatile organic compounds (VOCs) in the finishing of plastics. The printing of the carriers produces far less VOC emissions than does the painting of the part, and since the in-mold process does not degrade the transferred film, release of volatiles at the molding machine is minimal. In the final analysis, in-mold finishing techniques offer great promise because they simultaneously address the areas of quality improvement, environmental control, and cost effectiveness, and these are precisely the factors that have plagued the domestic processor for the past decade.

Part 3
Materials

35

Acrylic Polymers

Ronald A. Lombardi and James D. Gasper

ICI Resins US, Wilmington, Massachusetts

1.0 INTRODUCTION

Since their introduction decades ago, acrylic polymers have gained a strong foothold in the coatings and allied industries as a result of their improved flexibility and adhesion compared to polyvinyl acetate emulsions, phenolics, and styrene-butadiene latex combined with their moderate cost. In addition, their significantly improved outdoor durability, including resistance to UV degradation, has mandated their use in several applications. In many respects the name "acrylic" has become synonymous with a high performance level in a polymer system.

Presently, acrylics are available in three physical forms: solid beads, solution polymers, and emulsions. The emulsion form is by far the dominant form in use today. This is due generally to the ease of tailoring properties, and the lower hazards and manufacturing costs compared to the solid and solution polymers.

2.0 CHEMISTRY AND MANUFACTURE

2.1 Monomers

Acrylic monomers are esters of acrylic and methacrylic acid. Some common esters are methyl, ethyl, isobutyl, *n*-butyl, 2-ethylhexyl, octyl, lauryl, and stearyl. The esters can contain functional groups such as hydroxyl groups (e.g., hydroxyethyl methacrylate), amino groups (e.g., dimethylaminoethyl methacrylate), amide groups (acrylamide), and so on, in addition to the carboxylic acid functionality of the unesterified monomer. Acrylic monomers can be multifunctional (e.g., trimethylolpropane triacrylate, or butylene glycol diacrylate, to mention two). The polymer chemist has a wide range of monomers to select from when designing a specialty polymer system.

Typically mixtures of comonomers are chosen for the properties they impart to the polymer. Adhesive strength, for example, is increased by using monomers with low glass transition temperatures such as butyl acrylate or 2-ethyl hexylacrylate. Members of the carbox-

327

ylic acid group of acrylic and methacrylic acids also tend to increase the adhesive properties of polymers. Cohesive strength is usually imparted by the harder acrylic monomers such as methyl methacrylate and methyl acrylate. Molecular weight is also a significant contributing factor, and these two parameters must be carefully balanced by the polymer chemist.

When functional groups are needed for postreaction, monomers such as hydroxyethyl methacrylate or N-methylolacrylamide are incorporated. Hydroxyl groups can be used in combination with melamine and epoxy curing agents to achieve cross-linking. Similarly, other functional groups (acid groups, amines, amides, etc.) can be incorporated.

2.2 POLYMERIZATION METHODS

2.2.1 Bulk Polymerization

As the name implies, bulk polymerization is accomplished by initiating acrylic monomers in the absence of solvents other than the monomers themselves. Typically peroxide or azo initiators are used. The major problem of course is the viscosity increase realized after approximately 30% conversion. Generally, heavy-duty mixers and temperatures exceeding 150°C are used to control viscosity. The well-known Tromsdorf gel effect is often observed at high conversion, leading to rapid exotherms, high molecular weight fractions, and increased polydispersity. Processes for the bulk polymerization of acrylic copolymers are described in the patent literature [1–3].

2.2.2 Solution Polymerization

Adding a solvent to a bulk polymerization recipe allows much easier control of viscosity at high conversions. Again peroxide (e.g., benzoyl peroxide, lauryl peroxide) or azo (azobisisobutyronitrile) initiators are used. Mercaptans and halogenated hydrocarbons are used to regulate molecular weight. Most solvents also act as chain transfer agents to some extent. Generally acrylic polymers prepared by solution polymerization are less than 100,000 molecular weight units.

2.2.3 Suspension Polymerization

The process of suspension polymerization, as the name implies, involves the use of a dispersing agent to stabilize monomer droplets in a continuous phase, usually water. With continued agitation, the suspended droplets are polymerized using oil-soluble initiators, again peroxides and azo compounds. In fact, the polymerizing droplets exhibit bulk polymerization kinetics, as each droplet is actually a small bulk reactor with the continuous aqueous phase acting as a heat sink. Typical suspending agents include polyvinyl alcohol, polyacrylic acid, and hydroxyethylcellulose. As might be expected, suspension polymerization is usually limited to monomer compositions whose glass transition temperature Tg is near to or greater than ambient temperature; otherwise, gelation or blocking easily results in the dry bead. Following polymerization, the suspension polymer particles, or beads, are dewatered, washed to remove impurities such as suspending agents and electrolytes, and dried.

2.2.4 Emulsion Polymerization

Perhaps the most complex of the polymerization processes, emulsion polymerization has been well studied and described. As the name implies, the acrylic monomers are emulsified using a surfactant and suspended in a continuous phase, water. Water-soluble initiators are used to initiate polymerization. Some common examples are ammonium or potassium persulfate, hydrogen peroxide, and redox pairs such as t-butyl hydroperoxide–sodium formaldehyde sulfoxylate.

Polymerization is generally thought to initiate in the water phase. A growing radical precipitates to a micelle containing monomers and continues to polymerize. The growing micelle is resupplied with fresh monomer by diffusion from monomer droplets. For a thorough discussion of the emulsion polymerization process, see reference 4.

3.0 VERSATILITY OF ACRYLICS

3.1 Glass Transition Temperature

By selection of the proper monomers, the glass transition temperature of the polymer and therefore the likely application area can be varied. The glass transition temperature of a polymer is the simple average value in degrees Celsius representing a range of temperatures through which the polymer changes from a hard and often brittle material into one with soft, rubberlike properties. Although these average Tg values sometimes vary with the test method used, they are reproducible within certain limits and represent specific polymer characteristics. The glass transition temperature is useful as a guideline for softness of hand, low temperature flexibility, and room temperature hardness and softening point. The glass transition temperature should be used to compare hardness and softness of latex only within a simple polymer group.[5]

Table 1 illustrates the wide range in Tg resulting from different monomer compositions.

3.2 Emulsion Acrylics

Emulsions have become the dominant technology in acrylic polymers, and we therefore focus much of our discussion on this category. Table 2 correlates Tg ranges of emulsion acrylics with specific application areas. There is obviously overlap to be expected among the ranges. By varying the monomer compositions, we have a clear indication of the versatility of acrylics.

The wide range for adhesives encompasses pressure-sensitive polymers and heat activating polymers, which dry to a tack-free state.

3.2.1 Physical Properties

On a physical basis, emulsions can be characterized in the following fundamental areas: solids content, viscosity, pH, particle size, minimum film forming temperature (MFT), and particle charge.

TABLE 1 Glass Transition Temperature

Homopolymer	Tg (°C)
Acrylic acid	112
Methyl methacrylate	106
Methyl acrylate	8
Isopropyl acrylate	− 8
Ethyl acrylate	− 24
N–Butyl acrylate	− 56
2–Ethyl hexylacrylate	− 65

Source: Ref. 6.

TABLE 2 Glass Transition Temperature Versus Application Area

Tg (°C)	Suggested application area
80–100	High heat resistant coatings
50–65	Floor care coatings
35–50	General industrial coatings
10–40	Decorative paints
25–35	Binders for inks
– 60–25	Adhesives

Solids content is determined by drying latex to constant weight, viscosity is determined by use of the Brookfield Viscometer, and pH is determined by pH meter. Particle size usually ranges from about 0.05 to 0.5 μm, depending on the type and amount of surfactant employed; measurement is usually made by laser light scattering technique, electron microscopy, or ultracentrifuge.

Particle charge is discussed in the paragraphs that follow.

As mentioned previously, surfactants stabilize liquid monomer droplets formed by agitation. Polymerization takes place in monomer miscelles to form the solid dispersed particles of polymer in water. Surfactants are classified into three types: anionic, nonionic, and cationic.[7]

The most common type, anionics, ionize in water to leave a negative charge on the latex particle. Nonionics do not ionize but stabilize by a combination of hydrophobic and hydrophilic regions on the molecule. Cationics, which are not commonly used, ionize to give the particle a positive charge.[8]

Surfactants help to improve mechanical and chemical stability of the emulsions but also tend to increase the water sensitivity of the dried film.

3.2.2 Minimum Film Forming Temperature

Unlike solution polymers, which have one homogeneous phase, emulsions are polymer particles dispersed in water, the continuous phase. Upon drying, the particles must combine or coalesce to form a continuous film. If the process is carried out at room temperature (25°C) with a polymer whose Tg is above 25°C, film formation will not occur. In this case the polymer must be heated to above its Tg or a coalescing agent must be added to the latex to soften or plasticize the particles so that they will combine to form a continuous film. Coalescents often used are high boiling liquids that have a solvating effect on the polymer but later evaporate so that the full physical properties of the polymer alone are achieved. Generically many of these coalescents are classified as glycol ethers. Their relative toxicology has recently been studied.[9]

4.0 APPLICATION AREAS

4.1 Coatings

The use of acrylic polymers for both emulsion and solution coatings is very diverse. Before discussing the areas in which acrylics might be used in coatings, let us first identify the fundamental reason for considering the use of a coating, namely, to protect the substrate being coated from the environment. There are additional reasons, depending on the end-use area.

Once the need for a coating has been decided, a further requirement is apparent: namely, the coating must adhere to the surface that has been coated, since to decorate or protect any surface, the coating must remain in position. This phenomenon is known as adhesion. The reader is directed to an excellent review of the methods to determine satisfactory adhesion of a coating.[10] An additional point to consider is the air dry nature of properly formulated waterborne acrylics. Their full properties are developed not through extensive heat cure cycles but by room temperature drying through the use of coalescing agents, as discussed earlier. Where heat is available, increased solvent and heat resistance can be attained by the addition of cross-linking agents.

Each end-use product area requires certain specific property characteristics to achieve the feature needed for the application; the type of substrate to be coated as well as the manner in which the part is used determine the property requirements. It is clear that plastics do not require a corrosion-resistant coating; therefore a coating designed for plastics has property requirements different from a coating designed for metals. At the same time, adhesion of the coating, which is important for any substrate, becomes a key characteristic.

However, achieving adhesion to any specific substrate may require a totally different product requirement. Materials that adhere to metal may fail to bond to plastic. Therefore coatings for metals may have to differ significantly from coatings for plastics or wood to achieve the desired performance. Each substrate area, therefore, requires the design of a specific coating to meet its individual end-use requirement for the intended application.

Returning to application of acrylics in coatings, we will center our discussion on five major areas:

Automotive.
General metal.
Maintenance.
Wood.
Business machines.

4.1.1 Automotive
With the increased use of plastics in automotive applications has come a surge in the use of coatings. Plastics are coated to improve resistance to chemicals, solvents, ultraviolet light, and abrasion, as well as exterior durability.[11] Other applications for coatings include the engine enamels area, underbodies, and auto refinishing work. Performance properties required include grease resistance, durability, and adhesion to oily metals.

4.1.2 Coatings for Metal
Coatings are needed for metal casings and transformers. The requirements for such coatings include chemical resistance, hardness, and corrosion and humidity resistance. Since steel is still commonly used, the ability to protect this metal from corrosion remains an important requirement.[12]

4.1.3   Maintenance Coatings

Coatings are required for bridges and storage tanks, where again, properties of corrosion and humidity resistance are required.

4.1.4   Wood Coatings

Coatings for boards destined to be used in furniture and kitchen cabinets require blocking and detergent resistance, sandability, and resistance to grain raising.

4.1.5   Business Machines

Coatings for calculators, typewriters, copy machines, and analytical instrumentation are common examples in the business machine category. Performance required includes chemical and solvent resistance and adhesion to plastics [e.g., polycarbonate, polyphenylene oxide, acrylonitrile–butadiene–styrene (ABS)].

4.2   Adhesives

Both solvent and emulsion acrylic adhesives are extensively used in the industry, but before discussing adhesives, we need to address the fundamental difference between a coating and an adhesive. A coating must adhere to only one substrate; and adhesive must adhere to one substrate, then to a second substrate. A coating, once applied, is exposed to the elements and must withstand abrasion, marring, solvents, water, and heat. It may require high gloss, and other special properties, as well.

An adhesive is protected to a certain degree by being sandwiched between two substrates. It, therefore, does not have to have some of the performance properties that must be built into a coating. It must ideally have a bond strength high enough to fracture or tear at least one of the substrates. In many cases, the bond strength should not be materially affected by heat, solvents, or water. Therefore, an adhesive must not only have good anchorage to both substrates (adhesive strength), it must also have high enough cohesive strength to fracture or tear one of the substrates upon delamination. Thus an adhesive must balance adhesive strength with cohesive strength.

Another basic difference between emulsions (coatings and adhesives) is in their film formation properties. To have hard, tack-free, and heat-resistant coatings, the glass transition temperature of the polymer is intentionally designed to be higher than room temperature. The coating then requires a coalescing agent to form a clear continuous film. Adhesives form films at room temperature without the need for coalescing aids. A soft flexible polymer film is desired for an adhesive and this film should be thermoplastic (i.e., able to soften and flow repeatedly upon the application of heat). The film can subsequently be cross-linked through functional groups if heat and solvent resistance is desired.

Acrylic-based adhesives are normally employed where improved specific adhesion and/or resistance to yellowing from exposure to ultraviolet rays is required. Acrylics are used in three main areas: heat-sealable adhesives, laminating adhesives, and pressure-sensitive adhesives. These are discussed separately.

4.2.1   Heat-Sealable Adhesives

Heat sealing is used for bonding two substrates where one or both are impervious to water. Typically, the more heat-and solvent-resistant substrate is coated first and the solvent or water is driven off in an oven. At this point, the web can be wound up on itself (if the dried adhesive is tack free or nonblocking) and heat sealed to the secondary substrate at a later date. Alternatively, the adhesive-coated substrate can be laminated simultaneously to the second substrate to form the finished product. The latter case is the laminating adhesive technique, discussed later. When heat is applied to activate or soften the adhesive, two conditions must be met for adequate bonding:

1. The adhesive must have sufficient flow at the activation temperature to properly wet out the secondary substrate.
2. The adhesive must have a chemical affinity or specific adhesion to the particular substrate. This comprises the secondary chemical bonding forces, which give rise to what we call adhesive bonding.

Many applications involving food packaging fall into this category. Examples including lidding-type adhesives for coffee creamers and jams and the blister packaging of pharmaceuticals.

4.2.2 Laminating Adhesives

Laminating adhesives function much the same as heat-seal adhesives except that the temperature necessary to activate the adhesive is much lower. Where heat-seal adhesives may require an activation temperature of 120°C (250°F), laminating adhesives can be designed to function anywhere between room temperature and 90°C (200°F). Recently, new low temperature curing types of adhesive have, to some degree, replaced two-component solvent–polyurethane adhesives in the flexible packaging area.[13] Designing a room temperature curing mechanism into an acrylate system further enhances opportunities to replace solvent systems.[14] Considerable progress has also been made using low temperature curing acrylics in the industrial area.[15]

Typical applications include vinyl-to-wood laminating and bonding vinyl to ABS for automotive interiors.

4.2.3 Pressure-Sensitive Adhesives

The pressure-sensitive method of bonding is often called the "one-way" bonding method. The adhesive is coated onto the substrate either directly or by transfer coating. The adhesive is protected by a release liner until it is ready to be used. When application is desired, the liner is removed and the adhesive-coated substrate is bonded to the other substrate using pressure alone. Pressure actually activates the adhesive, hence the name of this method. Upon firm pressure, the tacky adhesive mass actually flows and bonds itself both mechanically and chemically to the other surface.

However, to be functional, a pressure-sensitive adhesive must be more than just very tacky. Fly paper has a very high degree of tack but lacks internal strength or cohesiveness. A functional pressure-sensitive material will have high tack or adhesive strength combined with high cohesive strength, the basic requirement for any adhesive.

Pressure-sensitive adhesives are used in many tape areas such as packaging, masking, electrical mending, medical, and mounting. The graphic arts area includes vinyl decals and special decorative films such as clear and metallized polyester films. These areas generally require an outstanding balance of properties that must be retained under severe outdoor exposure and temperature extremes. Such properties as outstanding resistance to ultraviolet light degradation and plasticizer migration are required. Adhesives for this area must produce clear and colorless films, dictating the use of acrylics exclusively.[16]

Solvent-based polystyrene acrylics (PSAs) have been traditionally used in these areas. More recently they have been replaced in varying degrees with their emulsion counterparts.[17] However, there still exist technical areas in which the solvent systems prevail.[18] These include applications requiring high levels of heat, water, and solvent resistance.

4.3 Inks

The use of acrylic polymers to formulate waterborne inks has grown steadily in the past 10 years. The major cause for this growth has been government regulations imposing limits on solvent emissions by printers. The same reasons, of course, sponsored the growth of waterborne coatings and adhesives. The choices left to the printers are:

Install solvent recovery–incineration systems.
Develop compliant high solids solvent inks.
Develop adequately performing waterborne inks.

Since many printing companies have selected the third option, our discussion will focus on it.

In addition to the pigment, waterborne acrylic inks are usually composed of two types of acrylic polymer, an emulsion-polymerized acrylic and an alkali-soluble solid acrylic type of resin. The emulsion acrylic contributes to finished ink properties such as film formation, adhesion to plastic films, and scuff, heat, and alkali resistance. However, emulsion polymers are very poor pigment wetters. In addition, they have poor flow characteristics compared to their solvent-borne counterparts. The alkali-soluble resins are good pigment wetters, display good rheology, and prevent drying of the ink in the cylinders when the line is stopped during printing runs.[19] Many different types of alkali-soluble resin can be used in combination with the emulsion acrylics.

Table 3 compares the properties of acrylic polymers to four other generic classes: maleic rosinates, shellac, protein, and styrenated maleic copolymers.

By combining the best properties of emulsion-polymerized and alkali-soluble acrylics, reasonably performing waterborne inks are gradually replacing the traditional solvent types. These waterborne inks are being used in such diverse applications as stamps, cigarette cartons, beer cans, milk cartons, food packaging, wallpaper, grocery bags, newspapers, and corregated boxes.[19]

Despite these successes, there still exist drawbacks and technical problems with the waterborne inks.[20] Among these are the following:

Drying problems. To maintain solvent line speeds, more efficient drying ovens are required.
Foaming. Use of suitable antifoam agents has considerably minimized this problem.
Freeze–thaw stability. Depending on formulation, the printer must also consider storage conditions during winter months.
Resolubility. Sometimes referred to as drying out on the press, this requires a careful balance between the alkali-soluble resin and the emulsion polymer.

With increased research and development efforts, the outlook for improved performance and growth of waterborne acrylic-based inks is quite optimistic.

5.0 COATING TECHNIQUES

A variety of coating techniques are used for applying adhesives, coatings, and inks. The specific method chosen depends on certain criteria, including:

Viscosity/rheology.
Film thickness desired.

TABLE 3 Property Comparison of Alkali-Soluble Acidic Resins

Property	Resins				
	Acrylic polymers	Maleic rosinates	Shellac	Protein	Styrenated maleic copolymers
Dispersibility	Fair–good	Fair	Fair	Fair	Good
Wet rub	Fair–good	Poor	Fair	Poor	Poor
Dry rub	Good	Fair	Good	Good	Fair
Heat resistance	Fair–good	Poor	Poor	Good	Good
Foam	Fair–good	Poor	Fair	Fair	Poor
Print quality	Fair–good	Fair	Fair	Poor	Good
Gloss	Fair–good	Poor	Good	Poor	Good
Drying rate	Fair	Poor	Poor	Poor	Good
Shelf life	Good	Fair	Fair	Fair	Fair
Resolubility	Fair	Fair	Good	Good	Good

Source: Ref. 19.

Shape of substrate.
Cost.
Versatility of use.
Thickness tolerances.

Table 4 illustrates the coating techniques that have been used successfully in three applications areas: adhesives, coatings, and inks. (The list in not exhaustive.)

The following methods are widely used and will be discussed in greater detail:

Gravure.
Flexographic.
Wire-wound rod.
Knife over roll.
Reverse roll.

5.1 Gravure Coaters

Gravure coaters are used typically when coating weights under 0.2 mil are desired. The gravure coater is used in the ink industry when a precise, reproducible amount of ink has to be applied to a substrate. Gravure cylinders are engraved with various fine patterns called cells. The patterns are designated by the number of cells per inch, followed by the cell name, which represents the shape of the cell (e.g., quadrangular, helical, pyramidal). After application, the applied ink pattern will flow together on the substrate to form a continuous film.

The major advantage of gravure coaters is that coating weight is independent of web tension and line speed. However, once running, the coating weight can be varied only by

TABLE 4 Commercial Coating Techniques

Technique	Usefulness with		
	Adhesives	Coatings	Inks
Gravure	+	+	+
Flexographic	−	−	+
Letterpress	−	−	+
Lithographic offset	−	−	+
Screen	+	−	+
Wire-wound rod	+	+	−
Knife over roll	+	+	−
Reverse roll	+	+	−
Direct roll	+	+	−
Floating knife	+	+	−
Curtain coating	+	+	−
Air spray	+	+	−
Airless spray	+	+	−
Electrostatic spray	+	+	−
Brush	+	+	−
Dip coating	−	+	−

changing the percent solids of the ink (or adhesive). This method, therefore, is costly because of the need to maintain many cylinders to accommodate varying coating weights.

Laminating adhesives for the flexible packaging industry use gravure coaters. Typical coatings weights for these applications are about 0.2 mil. These are applied primarily using a quadrangular cell pattern.

5.2 Flexographic Coaters

In flexographic printing, a cylinder similar to a gravure cylinder picks up ink from the fountain and, through a series of rolls, deposits a film of wet ink on a raised printing surface. This raised printing surface then applied ink to the substrate. The process is ideal for dimensionally unstable substrates such as thin polyolefin films, because the printing plate will deform slightly when it comes in contact with the film and not cause deformation of the film. Application viscosity of flexographic inks is between 35 and 200 cp. Typical dry film thickness applied is 0.1–0.3 mil.[20]

5.3 Wire-Wound Rod Coaters

Wire-wound rods are used when relatively low coating weights (0.2–0.8 mil) are required. Adhesive or coating viscosities of 200– 1000 cp would be typical. Simplicity of operation, low maintenance, and inherent low shear characteristics make this technique a popular choice. Rod coaters apply an excess of coating to the substrate and then utilize a wire-wound rod for metering to the desired coating weight. The final amount of coating remaining on the substrate is a function of the area of the cavity between the tightly wrapped wires on the rod. The number given to the rod indicates the thickness of the wire wound on it. For example, a #16 rod would have 0.016 in. diameter wire wound on it. As the number increases, so does the thickness of the wire and volume of the cavity between adjacent wires; thus higher numbers deposit greater amounts of adhesive. Additional factors that affect coating weight deposition for a given rod are web tension and uniformity, adhesive solids, viscosity, and rheology. When applying low coating weights, wire-wound rods are second only to gravure coaters in depositing accurate and uniform coating thicknesses. Ease of changeover and low cost give wire-wound rods an advantage over gravure coaters.

5.4 Knife Over Roll Coaters

Knife over roll coaters have long been used to apply adhesives for high viscosity, high coating weight, pressure-sensitive products such as masking tapes. A knife over roll coater consists of a rubber or steel backing roll, a knife, and a coating hopper. Since the substrate passes between knife and backing roll, any variation in backing thickness will cause a variation in adhesive applied. This is the major drawback to the knife over roll method. If backing thickness increases, the amount of adhesive deposited in that area decreases. Therefore, the uniformity of the adhesive laydown is only as good as the profile or caliper of the backing material being coated.

Typically viscosities of 10,000–100,000 cp can be applied by this method, which does give it some attractiveness. Depending on adhesive solids and viscosities, coating weights of 0.5–2.0 dry mils can be applied.

5.5 Reverse Roll Coaters

Reverse roll coaters have been the primary method for the production of high quality, pressure-sensitive decals for the graphic arts industry. The reverse roll coater consists of a rotating steel applicator roll, a stationary steel metering roll, and a rotating rubber backup roll. The applicator roll moves in the opposite direction of the web, which indicate the derivation of the name. Coating weight is controlled by the speed of the applicator roller relative to the speed of the backup roller. Speeding up the applicator roll increases coating weight, whereas slowing of the applicator roll decreases coating weight. The primary reason for using a reverse roll coater is that a very precise, uniform coating weight across the web can be attained. Dried film tolerances of ~0.0001 in. are possible. The other desirable feature is that coating weight thickness is independent of the backing weight variation. This is exactly opposite to the relationship on a knife over roll coater. Coating thickness in a reverse roll coater is generally a function of:

Gap between applicator and metering roll.
Applicator roll speed.
Adhesive solids, viscosity, and rheology.

Two primary coating supply systems are used when applying adhesives or coatings by reverse roll. For higher viscosities (10,000–20,000 cp), a nip-fed arrangement is utilized. For lower viscosities (3000–10,000 cp), a pan-fed system is preferred.

The principal drawback to the reverse roll coater is the expense, since high precision rolls and bearings are required. However, the initial investment is usually outweighed by the improvement in quality of the finished product.

REFERENCES

1. J. A. Brand and L. W. Morgan, U. S. Patent 4,546,160; S. C. Johnson & Son, Inc.
2. K. W. Free, U. S. Patent 3,821,330; E. I. Dupont de Nemours and Co.
3. K. Shimada et al., U. S. Patent 3,968,059; Mitsubishi Raon Co. Ltd.
4. *Encyclopedia of Polymer Science and Technology*, Vol. 5, pp. 801–859.
5. B. F. Goodrich Chemical Company, Latex Product Date, Bulletin L–12, p. 14.
6. Rohm and Haas Company, Bulletin SP197, Special Products Department.
7. B. F. Goodrich Chemical Company, Latex Bulletin L–10, p. 4.
8. B. F. Goodrich Chemical Company, Latex Bulletin L–19, p. 12.
9. R. A. Heckman. *J. Prot. Coatings Linings*, *3*:10 (1986).
10. J. E. Fitzwater, Jr. *J. Water-Borne Coatings*, *8*:3 (1985).
11. J. E. Fitzwater, Jr. *J. Water-Borne Coatings*, *7*:3 (1984).
12. G. Pollano and A. Lurier, *J. Water-Borne Coatings*, *9*:1 (1986).
13. P. Foreman, *Paper Film and Foil Converter*, November 1982.
14. R. A. Lombardi, *Adhes. Age*, February 1987.
15. A. H. Bealieu and F. P. Hoenisch, *J. Water-Borne Coatings*, *10*:1 (1987).
16. D. Satas, *Handbook of Pressure Sensitive Adhesive Technology*, 2nd ed. New York, Van Nostrand Reinhold, 1989.
17. R. A. Lombardi, *Higher Performance Water-Borne Pressure Sensitive Adhesives*. Itasca, Il. Pressure Sensitive Tape Council Seminar, 1987.
18. F. T. Koehler and J. A. Fries, *Acrylic Emulsion PSAs: Performance vs. Acrylic Solutions*. Spring Seminar, Adhesive and Sealant Council, Atlanta, 1985.
19. G. Sen, *Am. Ink Maker*, *65*:12 (1987).
20. M. T. Nowak, *Am. Ink Maker*, *62*:10 (1984).

36

Vinyl Ether Polymers

Helmut W. J. Müller

BASF AG, Ludwigshafen/Rhein, Germany

1.0 GENERAL

The general formula for vinyl ether polymers used in the production of adhesives and coatings is as follows:

$$-[CH_2-CH]_n-$$
$$|$$
$$OR$$

where R is —CH_3 (methyl)

or —CH_2—CH_3 (ethyl)

or —CH_2—CH—CH_3 (isobutyl)
$$|$$
$$CH_3$$

The consistency of these polymers depends on their molar mass and ranges from viscous oils to rubbery solids.

    Vinyl ether was first converted into a resinous polymer more than a century ago. Between 1920 and 1930, it became readily accessible by the techniques of Reppe chemistry and thus attracted industrial interest. Means were then investigated for polymerizing it. In 1938 the large-scale production of vinyl ether polymers commenced in the Ludwigshafen works of the former IG-Farbenproduktion (now the main production site of BASF AG). GAF Corporation started production after 1940, followed by Union Carbide Corporation, which relinquished the field in 1976.

2.0 MONOMERS

The Reppe reaction between acetylene and an alcohol gives rise to vinyl ether:

$$HC \equiv CH + H{-}OR \rightarrow CH_2 = CH{-}OR$$

It is still the only method of producing vinyl ethers that has acquired industrial significance. The properties of the constituent monomers are listed in Table 1.

3.0 PRODUCTION OF POLYMERS

Vinyl ethers can be easily polymerized in bulk or in solution, by batch or continuous techniques. Owing to the considerable heat of reaction, careful control and elaborate equipment are essential. The monomers and the initiator are metered continuously into the reactor, and the polymerization reaction sets in within a few minutes. After the reactor has been completely charged, it is closed, and polymerization proceeds further under pressure. The boiling point of the monomer or the solvent governs the rate at which the temperature rises during the reaction. The heat of reaction is removed by means of a reflux condenser and/or the reactor cooling system.

4.0 PRODUCTS ON THE MARKET

Table 2 lists vinyl ether polymers on the market. As suggested in the table, it is common practice to describe the degree of polymerization by the K value, which is a measure for the average molar mass.

The form in which the polymers are offered depends on the production method and the demands imposed in application. Thus they can be supplied in the solvent-free form or, to facilitate handling, as solutions in common solvents. Lutonal I 60 D and I 65 D are highly viscous 55% secondary aqueous dispersions and are produced from an 80% solution of Lutonal I 60 in mineral spirit (b p 60–140°C) with different protective colloids.

5.0 PROPERTIES

The degree of polymerization governs the physical form of polyvinyl ethers (i.e., whether they are viscous oils, tacky plasticizing resins, or rubbery substances). The products on the market are normally colorless. They may occasionally display a slight yellow to brown discoloration, owing to side reactions of the initiator system, but this seldom affects their performance in any way. The polymers offer good resistance to hydrolysis. Products with a low to medium molecular mass (i.e., K 65) may have a slight odor that emanates from residual monomers or oligomers.

The solubility of vinyl ether polymers is shown in Table 3. It depends on the alkyl group. The fact that polyvinyl methyl ether is listed as a water-soluble polymer can be explained by the hydrogen bond between the liquid medium, water, and the oxygen atom in the ether group. Thus the polymer, which is itself insoluble, is solvated by water to form a soluble associate. On the application of energy, the hydrogen bonds are again loosened—that is, the polyvinyl methyl ether loses its hydrate envelope and is thus precipitated. This occurs, for

Table 1 Monomers Used for Producing Vinyl Ether Polymers

Product	Structure	Molecular weight	Density at 20°C (g/cm^3)	Flash point (°C)	Melt point (°C)	Boiling point (°C)
Vinyl methyl ether	CH_2=CH—O—CH_3	58.1	0.747	− 60	− 122	∼ 6
Vinyl ethyl ether	CH_2=CH—O—CH_2—CH_3	72.1	0.754	− 45	− 115	∼ 36
Vinyl isobutyl ether	CH_2=CH—O—CH_2—CH—CH_3 CH_3	100.2	0.769	− 15	− 112	∼ 83

Table 2 Coventional vinyl ether polymers

Polymer based on	Commercial name	Manufacturer	K value	Nature of polymer	Glass transition temperature
Vinyl methyl ether	Lutonal M 40	BASF	∼ 50	Soft resin	∼ − 25
	Gantrez M	GAF	∼ 50	Soft resin	∼ − 25
Vinyl ethyl ether	Lutonal A 25	BASF	∼ 12	Viscous oil	∼ − 45
	Lutonal A 50	BASF	∼ 60	Soft resin	∼ − 30
	Lutonal A 100	BASF	∼ 105	Tacky rubber	∼ − 25
Vinyl isobutyl ether	Lutonal I 30	BASF	∼ 25	Viscous oil	∼ − 25
	Lutonal I 60	BASF	∼ 60	Soft resin	∼ − 25
	Lutonal I 60 D	BASF	∼ 60	Soft resin	∼ − 20
	Lutonal I 65 D	BASF	∼ 60	Soft resin	∼ − 15

Table 3 Solubility of vinyl ether polymers

Polymer	Solvents[a]								
	Al H	Ar H	Chl H	LA	HA	E	Lk	Hk	Water
Polyvinyl methyl ether	−	+	+	+	+	+	+	+	+
Polyvinyl ethyl ether	+	+	+	+	+	+	+	+	−
Polyvinyl isobutyl ether	+	+	+	−	+	+	−	+	−

[a]Al H = aliphatic hydrocarbons, Ar H = aromatic hydrocarbons, Chl H = chlorinated hyrocarbons, LA = lower alcohols, HA = higher alcohols, E = esters, LK = lower ketones, HK = higher ketones.

instance, in Lutonal M 40 solutions at about 28°C. The precipitation point is influenced by the concentration of the aqueous solution and the presence of any solvents.

6.0 APPLICATION

Vinyl ether polymers constitute a classical group of starting materials for the production of adhesives and coatings. They are primarily used in combination with other raw materials or—in one case in the pressure-sensitive adhesive sector—with polymers of the same type but different molecular mass.

Blending with vinyl ether polymers improves a number of the properties of the unblended products, including anchorage, adhesion or tack on difficult substrates, resistance to plasticizers, and resistance to aging.

The main applications in the adhesives sector are blends of polyvinyl methyl ether with starch or dextrin for label adhesives or blends of medium molecular weight polyvinyl ethyl ether with resin solutions for bonding carpets. Secondary dispersions of medium molecular weight polyvinyl isobutyl ether lag somewhat behind in terms of volume but not in importance. They are blended with acrylic dispersions in the production of pressure-sensitive adhesives. Blends of vinyl ethyl ether polymers of various molecular weights are used as pressure-sensitive adhesives on the medicinal sector.

In the field of surface coatings, polyvinyl methyl ether and medium molecular weight polyvinyl ethyl ether are formulated together with cellulose nitrate, chlorinated binders, and styrene copolymers for coating metal foil, plastics, film, paper, and other flexible substrates, and for antifouling paints.

BIBLIOGRAPHY

Müller, H. W. J., Vinyl ether polymers, in *Handbook of Pressure Sensitive Adhesive Technology*, 2nd ed., D. Satas. Ed. New York: Van Nostrand Reinhold, 1989.

37

Poly(styrene–butadiene)

Randall W. Zempel

Dow Chemical Company
Midland, Michigan

1.0 INTRODUCTION

Styrene–butadiene (SB) polymers are used in coatings formulations primarily to improve coating strength, printability, gloss, pigment and fiber binding, and substrate bond strength. The styrene–butadiene polymers are produced primarily by emulsion polymerization to form a latex. A latex is a dispersion of finely divided spherical particles of polymer in water. The monomer ratios of styrene and butadiene can be adjusted to give the desired amount of flexibility or stiffness. Polystyrene latex contains hard polymer particles that will not form a continuous film upon drying at room temperature. Polybutadiene latex contain very soft, gummy polymer particles that will produce a weak, sticky film when dried. By copolymerization of styrene and butadiene monomers and adjustment of the monomer ratios, it is possible to obtain a copolymer with a wide range of intermediate properties. Thus, a copolymer latex with a high styrene content will produce a very tough, durable film but will not have the flexibility and adhesive performance of a latex with a high butadiene content.

The temperature at which the copolymer changes from a brittle to a rubbery state is known as the glass transition temperature (T_g). Figure 1 illustrates the effect of the concentration of styrene on the T_g of SB copolymers. As the concentration of styrene in the copolymer increases, the T_g also increases. This property is important in coatings for determining durability of the dried film and optimizing curing conditions.

A typical latex will contain 45–55% polymer, with the balance being water.

Table 1 illustrates the typical properties of a styrene–butadiene latex. The particle size, composition, cross-linking, and molecular weight can all be controlled, and this is accomplished through free radical emulsion polymerization.

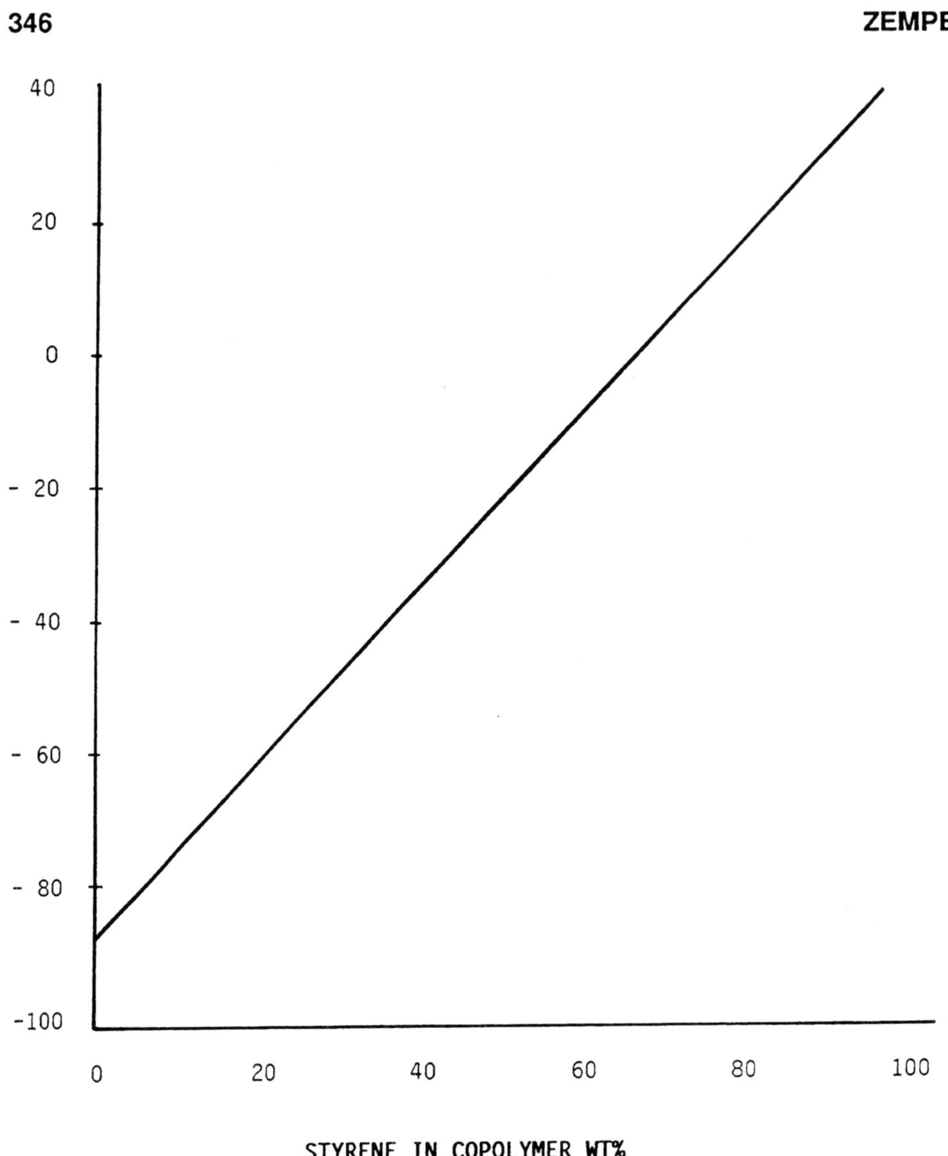

Figure 1 The effect of copolymer styrene content on glass transition temperature.

Table 1 Typical Properties of an SB Latex

Property	Value
Color	Milky white
Solids	50% wt.
pH	7–9
Brookfield viscosity	< 500 cp at 25°C
Average particle diameter	2000ô
Surface tension	45 dynes/cm
Specific gravity (at 25°C)	1.01
Styrene/butadiene ratio	50:50
Film properties	
tensile strength (at break)	550 psi
elongation	520%

2.0 EMULSION POLYMERIZATION

Emulsion polymerizations require an emulsifier or surfactant composed of a hydrophobic and a hydrophillic portion to give the latex colloidal stability. As low levels of surfactant are added into the aqueous solution in a reactor vessel, the surfactant will saturate the water phase and as the concentration is increased, the surfactant molecules will aggregate to form micelles. This concentration is known as the critical micelle concentration. As the styrene and butadiene monomers are added, they will diffuse through the water phase and into the micelles until an equilibrium is obtained. The majority of polymerization occurs within these monomer-swollen micelles. Figure 2 is a representation of the process.

Polymerization begins when water-soluble free radical initiators are added. Initiators, such as persulfate salts, are used for reactions above 50°C and redox systems are used for reactions at lower temperatures. The initial reactions occur in the water phase to form a free radical. The free radical will react first with monomer double bonds in the water phase and chain growth will begin. As the molecular weight increases, the chain becomes more hydrophobic and migrates to the swollen micelles, where the majority of polymerization occurs. The polymerization occurs in random fashion, and residual double bonds have the capability of further chain growth. This gives the SB latex the ability to form a three dimensional polymer matrix. This cross-linking or branching can have a strong influence on the polymer's mechanical properties.

Chain transfer agents are added to control the molecular weight. They will lower the molecular weight of the polymer through chain termination and will result in improved wetting and binding characteristics. The surface of the latex particle can also be modified by functional modifiers such as amides, amines, and carboxylic acids, which provide improved colloidal stability and increased adhesion properties.

3.0 CHARACTERISTICS OF STYRENE–BUTADIENE LATEX

In coatings applications, several physical properties are key to determining performance of the latex polymer. These include glass transition temperature, particle size, minimum film formation temperature, rheology, and surface energy.

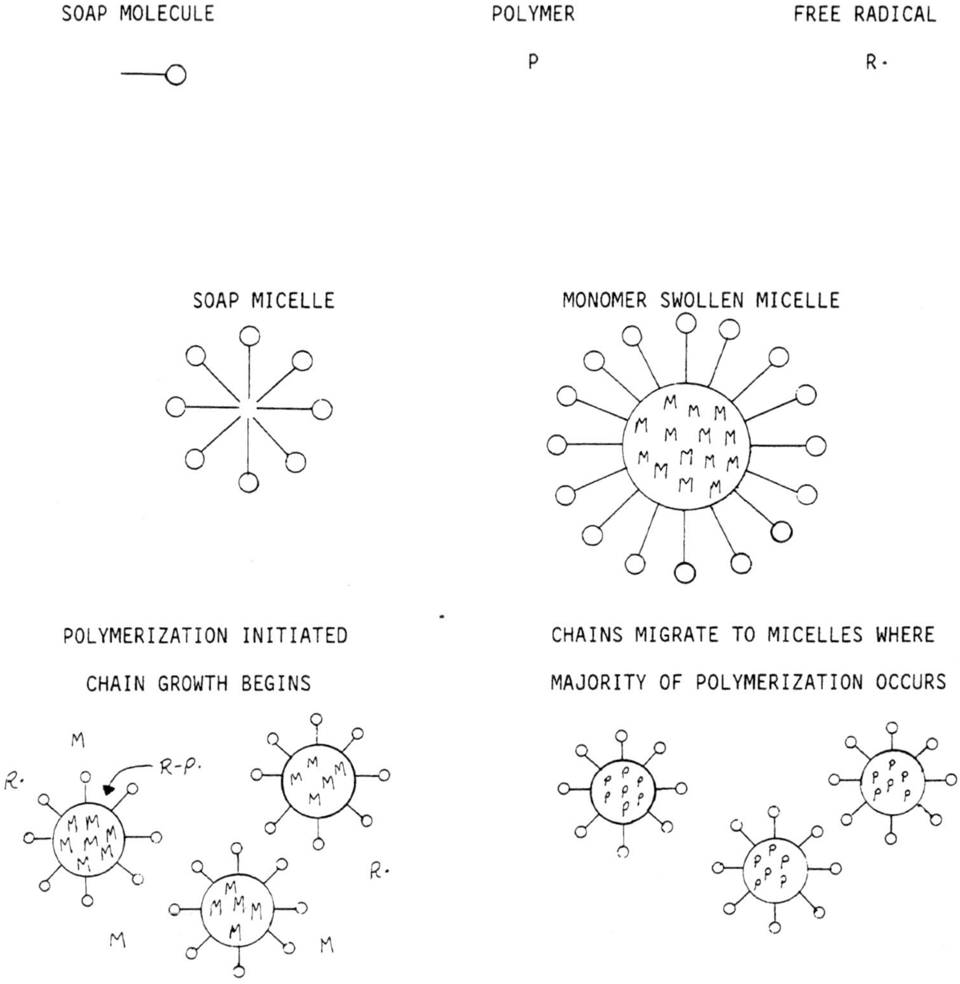

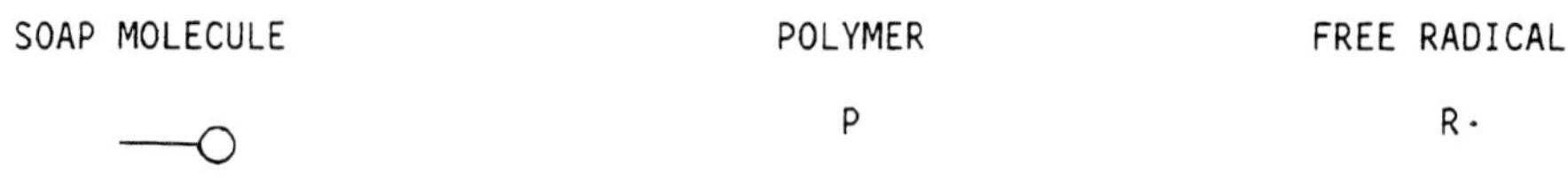

Figure 2 Free radical emulsion polymerization.

The S/B ratio and glass transition temperatures are useful as a basis for comparing performance characteristics of different SB latexes. End-use properties can be plotted against the S/B ratio and T_g to help determine the range of applicable polymers. Latex particle size and distribution affect high and low shear viscosity, gloss, and pick strength. Minimum film formation temperature (i.e., the minimum temperature at which a continuous film will form) is essential for determining process drying conditions. The minimum film formation temperature of a typical SB latex is slightly higher than its T_g. The reactions of the latex to increased shear stress and shear rate are important rheological properties because they will affect machine runnability. As the shear stress is increased, shear thinning, shear thickening, or Newtonian flow can occur. The surface tension or surface energy of a latex will affect wettability and adhesion of the coating to substrates.

4.0 USES

Styrene–butadiene latex is used in paper and paperboard coatings, textile coatings, as binders and coatings for flooring felts, as carpet backing, and in many other coatings application. The material can be produced with a wide variety of properties and can be tailored to specific end uses. Advantages of SB latex in paper coatings are excellent mechanical stability, pigment hiding power, improved gloss and ink holdout, and better printability. In textiles and carpets, SB latex provides excellent fiber adhesion, high filler loading, and a degree of dimensional stability and flexibility. Many other coatings applications use these basic features of SB latexes to improve end-use performance.

BIBLIOGRAPHY

Blackley, P.C., *Emulsion Polymerization, Theory and Practice*. London: Applied Science Publishers, 1975.

Klun, R. T., "Fundamentals of latex technology", presented at 1986 TAPPI Coatings Conference, Washington, DC.

Miller, F. A., Chapter 1, in *Treatise on Coatings, Film Forming Compositions*, R. R. Meyers and J. S. Long Eds. New York: Dekker, 1968, pp. 1–54.

38
Liquid Polymers for Coatings

Robert D. Athey, Jr.

Athey Technologies
El Cerrito, California

1.0 INTRODUCTION

High solids coatings are an alternative to solvent-borne coatings because of the coating industry's concerns for energy costs and the environmental impact associated with solvents. The use of 100% solids coatings, wherein a polymer binder precursor has functional groups, was accepted by the industry.

Synthesis techniques and monomers used control the binder properties through the influence on type and concentration of functional groups, chain configuration, and synthesis by-products. The impact on functionality is emphasized in several reviews.[1-3] The effect on fluid properties is also discussed. One paper on current commercial polymers includes recommended cure and uses.[2]

2.0 FLUID PROPERTIES

Liquid polymers, to be useful, must be capable of blending with additives, such as pigments, curatives, and stabilizers, Berry and Morrell discuss some of these variables in connection with liquid rubber.[4] The factors that control this intermixing capability contribute to the viscosity of the coating system. These factors include:

Interchain entanglement—a function of molecular weight.
The glass transition temperature of the polymer.
Inter-or intrachain interactions (e.g., hydrogen bonding).

Hydrogen bonding is important in many aspects of the flow and mixing of telechelic (end group functional) polymers and has been discussed by many authors.[5-11]

3.0 COMMERCIAL LIQUID POLYMERS

The commercial liquid addition polymers fall into several categories. Some of these products are listed in Table 1. Among the materials shown are some oligomers for incorporation into ultraviolet or electron beam (EB) coating systems.

3.1 Polymers with Random Functionality

Polymers having random functionality are available, but have limited usage. Cured poly(1,2-butadiene) was used for an experimental can coating (Firestone Tire & Rubber Co.). A solvent-based solution composed primarily of a polymer of trans-1,4-butadiene, which is thermally cured, is a commercial can coating (E. I. Du Pont de Nemours & Co.). Thermal cure depends on oxidative cross-linking in air, although additives, such as phenolic

Table 1 Commercial Liquid Polymers

Type of material	Manufacturer brand name
Polyether toluene diisocyanate (TDI) capped polymer	Upjohn Polymer Chemicals /Castethane Du Pont / Adiprene series Spencer Kellog / Many
Polyester TDI capped	Thiokol / Solithane Spencer Kellog / Many
Alkyd–urethane	Spencer Kellog / Spenkel series
Aliphatic oil–urethane	Spencer Kellog / Spenlite series
dimercapto liquid polymer	Phillips Chemical Co. / PM Polymer
dimercapto liquid polysulfides	Thiokol / LP–2, LP–3
telechelic polybutadiene	Arco Chemical Co. / Poly BD, also have BD/St and BD/AN
Saturated polymer	Witco Chemical /Formerz series
Diols	Inolex Chemical / Lukorez series Hooker Chemical / Rucoflex
Bishydroxyethyl dimerate	Emery Industries / Emery 9360–A
Hydroxyl telechelic block poly(ethylene oxide–copropylene oxide)	BASF Wyandotte / Pluronic series (have triols and quadrols
UV-cured epoxy resins	Union Carbide / Cyacure series
Acrylic end-capped urethanes	Freeman Chemical / Chempol Morton Thiokol, Inc. / Uvithane Polymer Systems Corp. / Purelast
Acrylic end-capped polybylene glycol	Alcolac Chemical / DV–1402
Acrylic end-capped epoxy resins	Interaz / Celrad series and others
Epoxy end-capped polyisobutylene	Amoco / Actipol series
Polyisoprene, natural and synthetic	Hardman Inc. / Kalene and Isolene series

resins, are occasionally used. Since curing consumes considerable energy and produces gaseous by-products, alternative methods are desirable. Synthetic nonfunctional butadiene–acrylonitrile copolymer (Hycar 1312) is used with sulfur cure as a castable elastomer and as a plasticizer for polyvinyl chloride (PVC) or nitrile rubber compounding.

Nonfunctional liquid polymers based on natural products have been used commercially as plasticizers and castable elastomers for many years. They are based on natural rubber that has been depolymerized by oxidative degradation.[12]

Liquid polymers without functional groups are also used as tackifiers in adhesive applications. These include polymers mentioned in the preceding paragraphs and also liquid polybutene (Chevron Chemical Co.). This product has also been used as a stir-in additive for roof coating latexes.

A liquid polymer having random carboxyl functionality is recommend as an epoxy resin modifier to toughen cast plastic moldings. Its structure suggests that it is an emulsion copolymer of butadiene–acrylonitrile–acrylic acid in ratios 83:13:4, 78:17:5, and 62:32:6. Its application is similar to that of the liquid randomly epoxidized polybutadiene polymers (FMC Corp.).

3.2 Telechelic Polymers

Polymers having end group functionality are termed telechelic, and a recent review has been published.[76] The commercial telechelic polymers have a variety of available functionalities.

Amine-terminated butadiene–acrylonitrile is promoted as a flexibilizing agent for epoxy resins (B. F. Goodrich Chemical Co.). Liquid resins with hydroxyl end groups in free radical polymerized polybutadiene, styrene, or acrylonitrile are available (Arco Chemical Co.). Vinyl-terminated butadiene homo-copolymers with acrylonitrile are used as UV-or EB-cured coating modifier (Goodrich). A polymer with mercaptan functional terminal groups is used in sealant applications (Phillips Chemical Co.). A bromine-terminated liquid polymer is also being developed (Polysar Ltd.).

Carboxyl-terminated butadienes have been available from several sources (Goodrich, Phillips, Thiokol Chemical Corp.). These products have been recommended as epoxy resin modifiers and rocket propellant binders. Carboxyl-terminated poly(butadiene–acrylonitrile) copolymers with 10, 18, and 28% acrylonitrile are available (Goodrich), as well as developmental hydroxyl-terminated copolymers, claimed to be diols (Scientific Polymer Products).

3.3 Curing

Curing of liquid rubbers for such applications as caulks, sealants, and O-rings, has been practiced for many years. These cures depend on the reaction of the rubber double bond. Yanko[13] showed that a standard rubber vulcanization recipe does not cause curing of a liquid poly(styrene–butadiene). It has been shown that the formulation of hard rubber required 10 to 40 times the normal amount of sulfur to cure the liquid polyacrylobutadiene. Hardman and Lang[14] developed a lead peroxide quinone dioxime cure system for liquid depolymerized natural rubber. The American Synthetic Rubber Corporation[15] claims a sulfur for their liquid poly(styrene–butadiene) rubber in road tar and bitumen modification applications. Argabright and Kuntz[16] used polythiols with UV light or peroxide to cure a liquid poly(styrene–butadiene). Friedmann and Brossas[17] used bis-silyl coupling agents to react with the double bonds of a liquid polybutadiene. With the exception of hard rubber, these systems, are normally low in tensile strength and elongation, as the cure sites are located at random along the polymer chain—the cross-linking density is low and the resultant gel is somewhat cheesy.

The main problem with the random cure on the liquid polymer chain is the distance between cross-links. The development of strength requires such a high degree of cross-linking that little capability for molecular motion is retained and a hard plastic product results. However, most applications require flexibility of elastomers. The solution is to use the liquid polymer's telechelic functions in a combined chain extension and cross-linking, retaining the segmental mobility, while still attaining the desired strength.

Examples of the types of cure for telechelic polymers of varying functionality are shown in Table 2. Just about any chemical reaction may be used for curing or chain extension.

Table 2 Telechelic Polymer Cure Systems

Curative	Telechelic Species	Ref.
Peroxide	Polybutadiene diol–urethane end-capped reacted with unsaturated glycidyl ester	55
Polymerizable oligomer	Carboxylated telechelic rubber	56
Polyisocyanates	Polyisoprene, polybutadiene, polystyrene diols	57
Tetramethyldiamine	Halogenated telechelic polybutadiene	58
Diisocyanates	Hydroxylated telechelic polybutadiene	59, 60
Hexachloro *p*–xylylene	Carboxylated telechelic polybutadiene	61
Epoxy + *tert*–amine	Carboxylated telechelic polybutadiene of poly(acrylonitrile–co-butadiene	62
Rubber	Peroxyl-terminal liquid rubbers	63
4, 4′–Diamino-diphenyylmethane	Isocyanate-terminated liquid polymer	22
Epoxy + *tert*-amine	Carboxylated telechelic polyisobutylne	64
Diamines	Bromominated telechelic polybutadienes	65
Rubber + diisocyanates	Hydroxylated telechelic polybutadienes or poly(propylene oxide)	66
Chlorinated elastomer	Amineated telechelic polystyrene	67
tert-Polyamines	Halogenated telechelic polybutadiene	68
Iron	Carboxylated telechelic polybutadiene	69
Urethane	Isocyanate end-capped diols	70
tert-Amines	Mercapto ester capped liquid polymer	71
Mercaptan-terminated polymer	Liquid polymer with terminal unsaturation	72
Metal oxides	Carboxylated telechelic polychloroprene	73
Amine + iron	Mercaptan-terminated butadiene copolymer	74
Organotin and organotitanate	Hydroxy-terminated polysiloxane	75

French[18] has shown how to assess cross-link density in telechelic polymers by evaluation of gel content. Determinations of thermooxidative effects[19] and variation in swelling of a variety of poly(butadiene urethanes)[20] have been used to evaluate the cross-link density.

Britain reported observations on catalyst effects on the urethane end-capped diols, which sometimes gel upon storage depending on the catalyst used. Reaction conditions for telechelic urethane polymer cures were modified by van Glulick and coworkers by including salts that coordinated to the amine curative.[22] Comparison of random cure versus telechelic cure in carboxylated polybutadienes is discussed by Mastrolia and Klager.%[23] Telechelically cured polymers showed improved mechanical behavior and storability. Molecular weight relation to tensile strength, tear strength, and hardness were discussed by Ono and coworkers.[24] A thermal cure based on styryl end groups requiring no added chemicals was demonstrated by Garcia.[25]

4.0 APPLICATIONS

The utility of liquid, especially telechelic, polymers was widely investigated.

The perfluoroalkyl iodide modified liquid polybutadiene may be used as a furniture polish or as an oil and water repellent textile coating.[26,27] MgO-carboxylated telechelic polybutadiene combinations are useful when a thixotropic plastic coating is called for.[28]

Not all the liquid polymer coatings are based on telechelic functionality. Gray[29] described a randomly functional hydroxyacrylic oligomer synthesis for use in melamine resin cured coatings. Hamann and coworkers[30] devised a coating composed from a phenolic adduct to a liquid polybutadiene. The liquid poly(1,2-butadienes) used in metal coating have already been noted. However, these liquid polymers are not capable of chain extension.

Vaughn[31] devised metal or textile coating polymers having terminal carboxyl groups. There are thermosetting acrylic powder coatings containing telechelic functionality.[32] The Ford Motor Company[33] devised a powder coating based on carboxylated telechelic polymer and an epoxy curative.

The acrylic–methacrylic end-capped oligomers are viscous liquids used in the radiation-cured higher solids formulations. The lower portion of Table 1 shows some commercial varieties of such products. Morris[34] compared several acrylated oligomers in wood finishing applications, testing for adhesion and abrasion resistance to choose the optimum oligomers. Christmas[35] compared monomer-diluted, radiation-curable oligomers on a variety of metal substrates to obtain formulations optimized for impact resistance, adhesion, and flexibility in a "T" bend test.

Siebert et al.[36] used carboxylated telechelic polybutadienes to toughen epoxy coatings. Jones and coworkers[37] devised an epoxy coating cured by a mercaptan-terminated polymer. McPherson and Gillham[38] characterized epoxy resin coatings cured by carboxylated telechelic polybutadiene by torsional pendulum analysis, noting effects of molecular weight and compatibility of the components. Drake et al.[39] reviewed elastomer-modifier, epoxy-based coatings.

A steel coating was devised from a hydroxy telechelic poly(styrene–butadiene and polyisocyanate curative.[40] A solventless coating capable of air curing was composed of an acryloyl-terminal polymer, a product of an acryloyl chloride esterification of a hydroxylated telechelic polymer.[41]

Acetylene-terminated polyphenylenes were reported as coatings by French.[42] McDonald[43] used a methacrylate-terminated epoxy resin in a radiation-cured coating. Pechiney St. Gobain[44] devised a coating based on styrene (or other monomers) and a maleic-terminated unsaturated polyester. Vranken and Dufour[45] increased the number of electron beam cross-linking sites in low molecular weight polymers by converting mercaptoethanol chain transfer termini to acryloyl esters on acrylic polymers.

Ko et al. used oligomers glycols converted to the acryloyl or methacryloy esters in a catalyzed peroxide cure steel coating.[46] Moisson-Frankhauser et al.[47] compared the use of maleinized polybutadiene to a carboxylated telechelic polybutadiene in alkyd coatings. A peroxide-cured unsaturated terminal polybutadiene gave better impact strength than poly(1,2-butadiene).[48]

An aqueous nail varnish consisting of a polythiol, an olefin-terminal liquid polymer and hardening agent, was claimed.[49] A concrete coating was made from a mercaptan telechelic polychloroprene with ZnO, MgO, and an inorganic peroxide and an amine.

Moisson-Frankhausen et al.[51] used carboxylated telechelic polybutadiene to produce aqueous electrophoretic coatings. Maleic or phthalic half-esters of hydroxylated telechelic polybutadienes were also used for electrophoretic coatings.[52] Anderson and Dowbenko[53] used a mercaptan-terminal polymer as a chain transfer agent in the synthesis of a block copolymer intended for coatings. A non-solvent coating was made from a block copolymer having an end group functional polymer as a precursor.[54]

5.0 CONCLUSIONS

The variety of coating compositions described shows the interest of the coating industry in liquid and especially in telechelic polymers. Problems related to rheology, cure rates, and curing chemistry remain. The factors governing rheology, compatibility with pigments and additives, cure type and extent, and the final coating properties lie in the composition and the polymerization conditions of telechelic polymer synthesis.

REFERENCES

1. D. M. French, *Rubber Chem. Technol.*, 42, 71 (1969).
2. R. D. Athey, *Prog. Org. Coatings*, 7(3), 289 (1979).
3. R. D. Athey, *J. Coatings Technol.*, 54(690), 47 (1982).
4. J. P. Berry, and S. H. Morrell, *Polymer*, 15, 521 (1971).
5. H. Onuma, et al., *Bull. Inst. Chem. Res. Kyoto Univ.*, 44, 123 (1966).
6. O. A. Olkhonikov, et al., *Vysokomol. Soed. A*, 15(11), 2512 (1973).
7. L. P. Semenova, et al., *Vysokomol. Soed. A*, 17(2), 318 (1975).
8. N. Okui, et al., *Kobunski Robunshu, Engl. Ed.*, 3(8), 1671 (1974).
9. A. A. Berlin, et al., *Vysokomol. Soed. B*, 17(7), S43 (1975).
10. A. N. Dunlop, and H. L. Williams, *J. Appl. Polym. Sci.*, 17, 2945 (1973).
11. *Specialty Polymers*. A multiclient study. Skeist Laboratories, Inc., Livingston, NJ, 1975.
12. P. Thiron, *Ploym. Age*, 264, October 1974.
13. J. A. Yanko, *J. Polym. Sci.*, 3, 576 (1984).
14. K. V. Hardman, and A. J. Lang, *Rubber Age*, 39(12), 431 (1961).
15. American Synthetic Rubber Corp., Flosbrene 25 (technical brochure).
16. P. A. Argabright, and I. Kuntz, U.S. Patent 3,030,344.
17. G. Friedmann, and J. Brossas, *ACS Polym. Div. Prepr.* 26(1), 268 (Apr. 1985).
18. D. M. French, *J. Macromol. Sci., Chem.*, A11(3), 643 (1972).
19. I. M. Erlikh, and E. L. Rubinshtein, *Kauchuk Rezina*, 1976 (4), 33.
20. V. B. Zabrodin, et al. *Vysokomol. Soed. B*, 17(8), 590 (1975).
21. J. W. Britain, *Acs Org. Coatings Plast. Chem. Div. Prepr.*, 22(1), 290 (1962).
22. N. M. Van Glulick, et al., *Rubber Chem. Technol.*, 49(2), 320 (1976).

23. E. J. Mastrolia, and K. Klager, Propellants Manufacture, Hazards and Treating. *Advances in Chemistry Series* No. 88, 122. Washington, DC. American Chemical Society, 1969.
24. K. Ono, et al. *J. Appl. Polym. Sci.*, 21, 3223 (1977).
25. D. Garcia, *ACS Polym. Div. Prepr.*, 26(1), 268 (1985).
26. Thiokol Chemical Corp., British Patent 1,403,649.
27. Thiokol Chemical Corp., British Patent 2,325,561.
28. C. H. Sheppard, and R. J. Jones, U.S. Patent 3,931,354.
29. R. A. Gray, *J. Coatings Technol.*, 57(728), 83 (Sept. 1985).
30. K. Hamann, et al., U.S. Patent 3,770,688.
31. W. L. Vaughn, U.S. Patent 3,803,087.
32. Union Chemique Belgique, British Patent 1,441,814.
33. Ford Motor Co., British Patent 1,424,968.
34. W. J. Morris, *J. Coatings Technol.*, 56, (715), 49 (Aug. 1984).
35. B. K. Christmas, *Mod. Paint Coatings*, 152, October 1984.
36. A. R. Siebert, et al. *ACS Org. Coatings Plast. Chem. Div. Prepr.* 34(91), 759 (1974).
37. F. B. Jones, et al., U.S. Patent 3,803,089.
38. C. A. McPherson, and J. K. Gillham, *ACS Org. Coatings Plast. Chem. Div. Prepr.*, 38(1), 229 (1978).
39. R. S. Drake, et al., *Epoxy Resin Chemistry II.* ACS Symposium Series No. 221, 1. Washington, DC: American Chemical Society, 1983.
40. DeSoto Inc., U.S. Patent 3,933,760.
41. Mitsubishi Gas Chem. Ind., German Patent 2,533,846.
42. J. E. French, *ACS Org. Coatings Plast. Chem Div. Prepr.*, 35(2), 72 (1975).
43. W. H. McDonald, U.S. Patent 4,091,050.
44. Pechiney St. Gobain. German Patent 2,052,961.
45. A. Vranken, and P. Dufour, German Patent 2,450,621.
46. K. Ko, et al., German Patent 2,533,846.
47. J. M. Moisson-Frankhauser, et al., British Patent 1,361,547.
48. Nippon Zeon Co. Ltd, Japanese Patent 49,052,851.
49. Bristol-Myers Co., Belgian Patent 49,052,851,
50. Electrochem. Ind. Co. Ltd, Japanese Patent 74,037,410.
51. J. M. Moisson-Frankhauser, et al. German Patent 2,305,912.
52. Atlantic Richfield Co., British Patent 1,358,234.
53. C. C. Anderson, and R. Dowbenko, U.S. Patent 4,016,332.
54. Union Chemique Belgique, German Patent 2,450,621.
55. Matsushita Electrical Works, U.S. Patent 3,957,903.
56. E. V. Khabarova, et al., *Kauchuk Rezina*, 1976(3), 45.
57. C. Pinazzi, et al., *Makromol. Chem.*, 1Z5, 705 (1974).
58. D. C. Edwards, *ACS Polym. Div. Prepr.*, 15(2), 56 (1974).
59. G. L. Staton, et al., *ACS Polym. Div. Prepr.*, 15(2), 52 (1974).
60. L. M. Sergeyev, et al., *Polym. Sci. USSR*, 12, 2339 (1970).
61. Z. N. Nudelman, et al., *Int. Polym. Sci. Technol.*, 1(4), T/27 (1974).
62. A. E. Kalaus, et al. *Int. Polym. Sci. Technol.*, 2(2), T/60 (1975).
63. N. M. Kuzmina, and L. S. Chuko, *Int. Polym. Sci. Technol.*, 1(10), A/12 (1974).
64. V. F. Antipova, et al., *Kauchuk Rezina*, 1975(1), 20.
65. Z. P. Chiruikova, et al., *Kauchuk Rezina*, 1975(12), 5.
66. I. M. Tunkel, et al., *Kauchuk Rezina*, 1974(3), 8.
67. H. Okamoto, et al., *Nippon Gomu Kyokaishi*, 49(6), 520 (1976).
68. B. N. Pronin, et al., *Vysokomol. Soed. A*, 19(3)5, 455 (1977).
69. M. Roux-Michollet, et al., *ACS Polym. Div. Prepr.*, 19(2), 369 (1978).
70. A. H. Frazer, and E. J. Goldberg, U.S. Patent 2,888,440.
71. Phillips Petroleum Co., U.S. Patent 3,860,566.
72. Product Research Chemical Corp., Belgian Patent 821,959.
73. K. Marabushi, U.S. Patent 4,205,150.

74. R. W. Ireland, and D. E. Skillicorn, U.S. Patent 3,607,845.
75. P. A. E. Guinet, and R. P. Puthet, U.S. Patent 3,409,573.
76. E. J. Goethals, Ed., *Telechelic Polymers: Synthesis and Applications*. Boca Raton, FL: CRC Press, 1988.

39

Polyesters

H. F. Huber and D. Stoye

Hüls Troisdorf AG,
Troisdorf/Marl, Germany

1.0 INTRODUCTION AND SCOPE

Polyesters are polymers containing the recurring ester unit ~ COO ~. They are formed by a condensation reaction between carboxyl-and hydroxyl-containing compounds. Historically, the first polyesters were condensation products of glycerol and phthalic anhydride. Those resins were used in the paint industry as early as 1910, but they were of low molecular and brittle. Later on, that is by 1925, they were modified by the inclusion of monofunctional fatty acids, which compensated the third functional group of the glycerol and elastified the resins. If unsaturated fatty acids were used, they rendered the resulting resins cross-linked by air oxidation. All these resins are now generally referred to as "alkyd resins"; they are treated in Chapter 40.

During the second half of this century a large number of di-and trifunctional acids and glycols became commercially available, thus creating new possibilities for they synthesis of polyesters. It was the commercialization of terephthalic acid—or rather the dimethyl ester—that spurred the development of polyethylene terephthalate, which today plays an important role for the manufacture of fibers, and films, and as an engineering plastic. In contrast to this polymer, however, which is made up of one pair of reactants only and is, therefore frequently referred to as homopolyester, it is the combination of more than two reactants that yields copolyesters. For the manufacture of copolyesters, the chemist can today draw from a large selection of monomers. This makes copolyesters a class of polymers, which can be varied over a very wide range:

Low or high molecular weight.
Saturated or unsaturated.
Linear or branched.
Liquid or solid.
Amorphous or crystalline.

359

Soft/elastic or hard.
Tough or brittle.

The glass transition temperature of copolyesters can be varied over a broad span. They can be modified in the main chain by other polymeric blocks (e.g., polyester units), and they can be prepared with varying terminal groups: hydroxyl, carboxyl, epoxy, acrylic functions, and others. Consequently, they can undergo further chemical reactions, especially cross-linking, to render a coating duroplastic.

The versatility of this class of polymers, combined with good stability toward light, heat, oxygen, and many chemicals, has contributed to an impressive and continuing gain in importance over the past 30 years.

Unsaturated polyesters also should be mentioned. They contain maleic, fumaric, or other unsaturated acids and, thus, have double bonds in the main chain. They are used as coatings—generally dissolved in monomeric styrene—in the furniture industry. Without styrene, they may be used in conjunction with other base polymers to enhance and broaden adhesion properties. This chapter deals with saturated copolyesters only.

2.0 CLASSIFICATION OF SATURATED POLYESTERS

Saturated polyesters for use as binders in paints and coatings may be classified according to their molecular weight and their functionality. High molecular weight polyesters are predominantly linear, thermoplastic polymers with molecular weights from 10,000 to 30,000. Generally, they are copolyesters containing terephthalic and/or isophthalic acid and aliphatic diacids and a blend of diols. In contrast to terephthalic acid homopolyesters, they exhibit better solubility in solvents.[1] In coatings they impart a high degree of flexibility paired with excellent surface hardness and stability. High molecular weight linear polyesters may be used as physically drying binder components in paints, although the majority of uses are in baking enamels for highly flexible coatings, such as coil and can coatings in combination with amino resins or other suitable hydroxyl-reactive cross-linkers.[2] Certain special grades of high molecular weight polyesters are ground and used as thermoplastic powder coatings.

Low molecular weight polyesters range from 500 to 7000 g/mol and are, in general, not suitable as physically drying binders.[3] Because of their low degree of polymerization, they carry a great many functional terminal groups.[4] Low molecular weight polyesters may be linear or branched; by variation of the manufacturing process, it is possible to incorporate mostly either hydroxyl or carboxyl end groups, or both kinds.

By themselves, low molecular weight polyesters are not satisfactory film formers. They require a reaction partner, which is capable of reacting with the end groups of the polyester and causes the formation of a cross-linked, duroplastic film.[5] Amino resins[6] and polyisocyanates[7] are suitable as such cross-linking agents for hydroxyl polyesters, whereas epoxy resins and polyoxazolins may be used for carboxyl polyesters. By proper selection of the reactants, the formulator can design products ranging from two-component or one-pack solvent-borne baking enamels with amino resins or blocked polyisocyanates to powder coatings or, via the salt formation of carboxyl polyesters, water-soluble stoving paints.

By reacting the native terminal groups of polyesters (i.e., hydroxyls or carboxyls) with at least bifunctional monomers or oligomers, saturated polyesters may be further modified in many ways. Aminoplasts may be employed to prepare thermosetting precondensates with hydroxyl polyesters. Partially blocked polyisocyanates, when blended and reacted with excess amounts of hydroxyl polyesters, will yield thermosetting binders. Similarly, one can obtain silicone-, epoxy-, or acrylic-modified polyesters.[8–11] For the purpose of further modification, polyesters with molecular weights between 1000 and 5000 are best suited, given their higher content of functional hydroxyl or carboxyl groups and their better reactivity as compared to higher molecular weight counterparts. In that respect it is frequently desirable not to convert all end groups, in order to leave residual cross-linkable functions for a coating or merely to improve the adhesion of the coating to certain substrates.

Table 1 lists the various ways to cross-link high or low molecular weight or of various modifications. Still the most important groups of cross-linking agents are amino resins, especially melamine–formaldehyde condensation resins.[12–15] Other cross-linkers are gaining in importance, however. With the rapid growth of the market for powder coatings, polyester–epoxy hybrids and blocked isocyanate curing agents are becoming increasingly popular.[16] Polyester–isocyanate based binder systems are increasingly used in solvent-borne paints, such as coil and can coatings, where a high degree of elasticity and resistance to weather or other attacks is required. Silicone-modified polyesters are known for excellent weather resistance and—with sufficiently high silicone content—good heat and chemical stability also.[17] Epoxy-modified polyesters are suitable for cross-linking with acidic resins or acid anhydrides and offer, thus, a formaldehyde-free and less toxic way to cross-link polyesters. Finally, acrylated polyesters serve as binders for radiation-cured coatings for varnishes, painting inks, and adhesives.

Table 1   Types of Polyester, Cross-linkers and Applications

Type of polyester	Crosslinker	Field of application
Linear high molecular weight polyesters	Melamine resins, benzo-guanamine	Coil and can- coating paints, primers, coatings for collapsible tubes
Linear low molecular weight hydroxyl polyesters	Melamine resins, benzo-guanamine resin, polyiso-cyanates	Industrial stoving paints, two-component paints
Branched low molecular weight polyesters	Melamine resins, benzoguan-amine resins, urea resins, polyisocyanates	Industrial stoving paints, two-component powder coatings
Carboxylated polyesters	Melamine resins, epoxy resins, triglycidylisocyanurate, polyoxazolin cross–linker	Waterborne coatings, powder coatings
Silicone-modified polyesters	Siloxane melamine resins	Coil coating paints, heat-resistant paints
Isocyanate-modified polyesters	Isocyanates, blocked polyiso-cyanates	Coil coating paints, industrial paints
Epoxy-modified polyesters	Polyanhydrides, carboxylated polyesters	Primers, industrial paints
Acrylated polyesters	Melamine resins, radiation cure	Industrial paints, adhesives printing inks

3.0 MANUFACTURING PROCESSES

3.1 Reaction Components

Aromatic and aliphatic polycarboxylic acids and various polyols may be used for the penetration of linear or branched polyesters of varying molecular weight. The nature of the monomers and their relative amounts determine the properties of the resulting copolyester. The most important raw materials for the preparation of copolyesters are listed in Table 2 (18–20). Whereas for the manufacture of linear polyesters strictly bifunctional acids and alcohols must be used,for branched polyesters a certain amount of tri-or higher functional monomers is required. To warrant manufacture in commercial quantities on a consistently high quality level, a high degree of purity is required for the raw materials and, where applicable, a constant ratio of isomers. The polycondensation reaction is very sensitive to impurities, and even small amounts of foreign matter will effect discoloration, a change in molecular weight distribution, or even gelling.

3.2 Technical Manufacturing Processes

Saturated polyesters are made by esterification and transesterification. Both are reversible equilibrium reactions, which yield the desired product by continuing removal of condensation water. Under the conditions of the esterification reaction, transesterification occurs at the same time. Consequently, when more than two diols or decarboxylic acids are being used, a statistical distribution of the monomers must be expected. The rate of reaching equilibrium is catalytically accelerated by acids, bases, or transition metal compounds. Commonly, small amounts of titanium, tin, or antimony compounds are used as catalysts.

Low molecular weight linear or branched polyesters are generally prepared by a one-step process. This involves blending all monomers and reacting them at temperatures from 125 to 240°C. The condensation products are removed either by the application of vacuum (vacuum-melt process) or by passing through a stream of inert gas such as nitrogen or carbon

Table 2 Monomers for Producing Saturated Polyesters

o-Phthalic anhydride	Ethylene glycol
Isophthalic acid	Propanediol 1–, 2
Terephthalic acid	Diethylene glycol
Adipic acid	Butanediol 1–, 3
Sebacic acid	Butanediol 1–, 4
Trimellitic anhydride	Hexanediol 1–, 6
Pyromellitic anhydride	Neopentyl gylcol (NPG)
Trimesic acid	Trimethyl pentanediol (TMPD)
Hexahydrophthalic anhydride	Cyclohexane dimethanol (CHDM)
Hexahydrophthalic acid	Tricyclodecane dimethanol
5–*t*–Butyl isophthalic acid	Hydropivalic acid–NPG–ester
Dodecanoic acid	Bisphenol A bishydroxyethyl ether
Dimerized fatty acids	Trimethylol propane (TMP)
	Pentaerythritol

dioxide (driving gas en melt process) or by the addition and subsequent azeotropic distillation of a solvent (azeotrope process).

Higher molecular weight copolyesters are made in two steps. The first step involves the preparation of a polyester precondensate either by transesterification or by esterification of the acids with excess diols. The second step, referred to as the polycondensation consists of further removal of diol from the molten precondensate in vacuum at temperatures above 250°C.

Other manufacturing processes for polyesters are known, but these are not normally used for commercial purposes.[1]

4.0 PROPERTIES OF POLYESTERS

4.1 Morphology

Depending on the choice of components used to build up the polyester, the morphology of the polymer may be varied continuously over a wide range from completely amorphous to highly crystalline. Monomers imparting regularity—for stance, by a sequence of methylene groups or by symmetrical substituents on a ring—will lead to crystallinity in the polymer. Accordingly, polybutylene terephthalate or a copolyester consisting of terephthalic and adipic acids and butanediol will exhibit a high degree of crystallinity. Less regular molecules, such as isophthalic or orthophthalic acid or neopentyl glycol will yield only weakly crystalline or, more likely, completely amorphous polymers.

As a rule, for use in coatings, amorphous polyesters will prevail; however, there may well be instances in which crystallinity can be used to advantage—for instance, in thermolastic powder coatings,[21] where crystallinity helps to reduce the permeability of the coating.[22]

4.2 Solubility

Naturally, chemical composition will be the dominant factor in determining the solubility of a polyester; however, this dominance may be exhibited either directly or indirectly (by affecting the morphology). The solubility of polyesters decreases rapidly with increasing crystallinity.[23] Highly crystalline copolyesters are insoluble in any common solvents. They can be dissolved only in blends of phenol and *o*-dichlorobenzene at higher temperatures. Polyesters of medium crystallinity are soluble in methylene chloride; weakly crystalline polyesters will also dissolve in aromatic hydrocarbons, such as toluene. Amorphous polyesters will dissolve in a variety of polar solvents, such as esters, etheresters, ketones, and aromatic and chlorinated hydrocarbons.

Within the amorphous polyesters, chemical composition will have a direct effect on solubility. Thus shifting the glycol ratio in a polyester from ethylene glycol toward neopentyl glycol will enhance the solubility of the resulting polyester. Detailed information on the solubility of a variety of polyesters may be found in vendors' literature.[24,25]

Polyesters are essentially insoluble in water, unless special measures are taken to impart hydrophilic sites. This may be achieved in one or more of the following ways: Incorporation of polyethylene oxide segments or anionic groups, using either carboxyls (e.g., by means of trimellitic anhydride[26]) or salts of sulfoisophthalic acid.[27] Detailed treatments on waterborne polyesters may be found elsewhere.[28,29] True solubility of the polymers is attained only rarely. Rather, a gradual increase in water dispersibility is observed until, by way of forming polyelectrolyte salts, a colloidal solution is obtained. Frequently, auxiliary solvents such as butyl glycol ethers are also employed.

4.3 Molecular Weight Distribution

Polyesters are prepared by polycondensation, a fully reversible reaction. The desired degree of polymerization is obtained by shifting the reaction equilibrium in that excess glycol is removed. A concise treatment concerning low molecular weight linear polyesters is available in German.[4] Essentially, assuming comparable reactivity of end groups, the molecular weight distribution is a purely statistical one. For higher molecular weight linear polyesters, it tends to drop off fairly rapidly toward higher molecular weights and to tail off toward lower ones. For a typical polyester the following data were found:

$$M_w = 22,200 \qquad M_n = 8300 \qquad \text{dispersity} = 2.7$$

By gel permeation chromatographic analysis, a content of 3% (by area) of low molecular weight (> 1000) species was measured. One complication is the possible formation of cyclic oligomers—for instance, the one consisting of 2 moles of terephthalic acid and 2 moles of ethylene glycol or several others. Most of these oligomers are crystalline and tend to cause turbidity in resin solutions and, occasionally, in paint films. Much work has been devoted to the elimination of these oligomers, resulting in ways and means to avoid this interference factor.[30]

Much more complicated is the situation when trifunctional monomers are used and branched polymers are obtained.[28] As Carothers[31] and Flory[32] have shown, the molecular weight distribution broadens rapidly with increasing degree of polymerization and soon a critical conversion point is reached, beyond which gelation of the polymerization batch occurs. This can be avoided by a two-step process, which introduces additional functional groups in a controlled manner. A typical example for this technology is the preparation of a branched hydroxyl polyester in a first step, followed by reaction with trimellitic anhydride as a second step.[26,28,30]

4.4 Functionality and Reactivity

In contrast to other polymerization processes (e.g., radical polymerization), polycondensation does not tend to cause imperfections due to branching or chain termination. Therefore a polyester consisting of bifunctional components only will be just about perfectly linear. Only the incorporation of trifunctional monomers such as trimethylol propane or trimellitic acid causes a controlled increase in functionally above 2. As the polycondesation is generally carried out with an excess of glycol, the acid number declines more rapidly with increasing degree of polymerization than the hydroxyl number; thus, except for low molecular weight polyesters, the reactivity of the polyester is primarily linked to its terminal hydroxyl groups. Since secondary hydroxyls are less reactive than primary ones, a terminal primary hydroxyl group generally remains. This, in turn, reduces the reactivity of a polyester primarily to a function of its molecular weight, in accordance with the decreasing concentration of end groups with increasing size of the polymer molecule.

4.5 Transition Temperatures

These transition temperatures are of interest in conjunction with polyesters. In order of importance for coatings, there are:

Glass transition temperature, T_g.
Softening point, T_f.
Crystalline melting point, T_m.

The theory of the relationship between thermodynamic equilibrium melting points and copolymer compositions was developed by P. J. Flory.[33] On the basis of the free volume theory, Flory and Fox[34] proposed a general theory on glass transition theory, which led to the formula suggested by Fox,[35] which permits the calculation of the glass transition temperature from the contributions of the individual monomers. Polyesters obey the formula quite well. Thus, from knowledge of the glass transition temperatures of the homopolyesters, copolyesters may be calculated with good accuracy. For the purpose of this chapter, however, a few basic guidelines may be helpful.

Purely aliphatic polyesters will—with increasing ratio of methylene to ester groups—asymptotically approach the glass transition temperature of polyethylene (N 160 K). Accordingly, a polyester of azelaic acid and hexamethylene diol has a glass transition temperature of about 190 K, whereas one made of terephthalic acid and bisphenol A has a glass transition temperature 480 K. More detailed data on glass transition temperatures of homopolyesters may be found in the literature.[36,37]

For amorphous polyesters, the softening point—usually determined by the ring and ball method—is generally approximately 50–90°C above the glass transition temperature. As one would expect, this difference increases with increasing molecular weight.

Crystalline melting points depend largely on the chemical nature of the crystalline species. Thus, a highly aliphatic, crystalline polyester will melt at approximately 340 K, whereas more aromatic crystalline species will melt at considerably higher temperatures (e.g., polyethylene terephthalate, 540 K). More information on melting points of polyesters is available.[38]

4.6 Compatibility of Polyesters

The compatibility of polyesters with other resins and polymers depends on many parameters, which also are partly interrelated, including molecular weight, glass transition temperature, morphology, and chemical composition of the polyester. Among the most prominent examples for the use of polyesters in combination with other polymers are blends of low molecular weight polyesters with cellulose acetobutyrate in base coats of two-coat metallic automotive finishes. High molecular weight polyesters are used as coresins for nitrocellulose lacquers and for vinyl copolymers.[24,25]

4.7 Chemical Properties

Generally speaking, polyesters have excellent stability toward light, oxygen, water, and many chemicals. The weakest spot in the polymer chain is the ester group, with its potential sensitivity to hydrolysis. Accordingly, degradation may occur, provided there is an environment that combines moisture with acids or alkali or other catalytically active materials, plus the possibility for the moisture to permeate into the polyester. The latter is largely suppressed in the case of crystalline polyesters or for cross-linked polyester paint films. Degradation of polyethylene terephthalate was investigated by Buxbaum.[39] Because this author was using low molecular weight model compounds, this study may, to some extent be applicable to other polyesters as well.

5.0 ANALYTICAL PROCEDURES

In the characterization and quality control of saturated polyesters, the end groups are analyzed according to procedures of the International Standards Organization (ISO) and the American Society for Testing and Materials (ASTM): for hydroxyl number, ISO 4629; for acid number, ISO 3682 or ASTM D 2455. For each type of polyester there will be narrow limits in those two parameters. Further test methods for polyester solutions are the non-volatile matter content (ISO 3251, ASTM D1259), color number (Gardener Scale, ISO 4630, ASTM D 1544), flow time (Ford cup, ISO 2431, ASTM D 1200), viscosity (ISO 3219, ASTM D 1725) and flash point (ISO 1523, ASTM D 1310). For solid polyesters the ring and ball method is used to determine the softening point.

Molecular weight is frequently determined indirectly as inherent or reduced viscosity. The respective formulas are:

$$\eta_{inh} = \frac{t/t_o}{c} \qquad \eta_{red} = \frac{t/t_o - 1}{c}$$

Where
t = elution time of solution
t_o = elution time of solvent
c = concentration

Frequently, the control of the foregoing analytical data is not sufficient to ensure consistent quality for the intended use. In such cases, application technical tests must be carried out to learn more about polyester reactivity, the mechanical properties of the coating, gloss, compatibility, and adhesion, and resistance to water, chemicals, weather, etc.

6.0 PREPARATION OF POLYESTER COATINGS[20]

For the preparation of paints based on saturated hydroxyl polyesters, amino resins are the most commonly chosen cross-linkers. Melamine–formaldehyde condensates are preferred over urea–formaldehyde resins in view of the superior properties of the former in weathering stability as well as the balance of surface hardness and elasticity. For further improvement of adhesion, gloss, and water resistance, benzoguanamine resins are used. Selection among the many commercially available melamine resins is primarily determined by the compatibly of the melamine resin with the type of polyester it is to be combined with. Monomolecular, etherified condensation products such as hexamethoxymethyl melamine or only partially etherified products exhibit a very broad range of compatibility. Higher molecular weight resins containing several melamine units in the molecule are less compatible and can be used with selected polyesters only.

To meet the commonly established stoving conditions of 120–270°C with acceptably brief reactions times, the cross-linking reaction between hydroxyl polyesters and amino acid resins—especially the etherified variety—requires the use of acidic catalysts. A common choice is p-toluolsulfonic acid or its salts with volatile amines, such as morpholine, or nonionically blocked acidic compounds. Hetero-or homopolar-blocked acids are latent catalysts, which become effective only during the baking cycle.[40,41] Thus, a premature reaction between polyester and melamine resin at room temperature in the liquid paint is effectively suppressed and the shelf life of the paint appreciably extended.

The properties of polyester coatings are essentially influenced by the molecular weight and the composition of the polyester as well as by the nature and the amount of the melamine resin. With increasing amounts of amino resin, baked coatings will become harder and less elastic, whereas resistance to solvents and chemicals increases. In general, good results are obtained with 10– 35 wt% of total binder content. In coil and highly elastic can coating paints, the amino resin content may be reduced to as little as 5%.

Of less commercial importance than melamine resin cross-linked polyesters are those with other cross-linking agents such as isocyanate or epoxy cross-linked or siloxane-or radiation-cured systems. Naturally, to avoid side reactions and insufficient shelf life when designing a paint formulation, solvents, catalysts, pigments, additives, and fillers must be chosen with due regard for the chemistry involved.

6.1 Solvent-Borne Polyester Coating[20]

For the preparation of solvent-borne polyester paints, all the binder components are first dissolved; then the pigments, fillers, and additives are added, thoroughly dispersed, and milled. For reasons of economy and to alleviate the milling procedure, it is advantageous to conduct the milling step in a fraction of the total binder. To avoid difficulties in pigment wetting when using binders of different polarity, the fraction used for milling should contain all the binder components in the proper proportion. Upon completion of the milling step, the retained binder portion is added together with leveling or flow agents or other additives and the required amounts of solvents, Paints based on saturated copolymers can be applied with all common techniques. Those most widely practiced are roller coating for can and coil coating, followed by various spray applications (air assisted, airless, or electrostatic spray) with, for instance, rotating discs or belts.

6.2 High Solids Paints

High solids paints formulated with saturated copolyesters have total solid contents between 65 and 80 wt%. They require low viscosity—that is, low molecular weight polyesters and monomolecular melamine—formaldehyde resins.[42–49] Due to the low molecular weight of the polyesters, the elasticity of paint films thus obtained is inferior to that of conventional solvent-borne polyester paints. Therefore, high solids polyesters paints are primarily used as spray-applied or dip paints. When formulating such paints, one should take into account the high polarity of the binder and use wetting agents for better pigment wetting, as well as effective flow agents and preferably nonionically blocked catalysts for improved electrostatic spray applicability.

Polyester–urethane two-component paints allow the manufacture of products with higher solid content and lower emission as compared to polyester–melamine combinations.

6.3 Waterborne Paints

Waterborne polyester–melamine paints are made from saturated polyesters containing an increased mount of carboxyl groups,[5,28,50–52] having acid numbers between 45 and 55 mg KOH/g and molecular weights of approximately 2000 g/mol. The polyesters are usually combined with water-soluble melamine–formaldehyde resins such as hexamethoxymethyl melamine in a ratio of 70:30 to 85:15. The use of a catalyst is not necessary, as the carboxyl groups accelerate the reaction to a sufficient degree. Before blending these polyesters with the melamine resin, the carboxyls of the polyester must be neutralized with amine, usually dimethylethanolamine. Auxiliary solvents such as butyl glycol ether may be used to reduce viscosity, improve pigment wetting, enhance shelf life, and improve the thinning characteristics of the paint. Depending on the nature of the binder, the application viscosity and the stoving cycle, 5–15 wt % solvent based on total paint formula is used.

Lately, it has been possible to develop polyesters with acid numbers of only about 20 mg KOH/g and molecular weights of 4000 g/mol, which upon neutralization are thinnable with water. The higher molecular weight allows a further reduction in emissions when curing with melamine resins. Paints thus formulated may be applied by roller; they display a high degree of flexibility and impact resistance in the cured film.

6.4 Solvent-Free Coatings

Saturated polyesters play an important role for the formulation of thermosetting powder coatings. They can be combined as carboxyl polyesters with epoxy resins or triglycidyl isocyanurate (TGIC),[7,16] or as hydroxyl polyesters with blocked solid isocyanates, predominately isophorone diisocyanate (IPDI). Thermoplastic powders are used mainly to cover welding seams in cans. The preparation of powder coatings involves blending of the binder resins with pigments, additives, and catalysts, homogenizing in a kneader–extruder, then grinding and screening. Electrostatic spray is the prevailing application method.

For adhesive coatings, special, low viscosity polyesters were developed.[53] The low viscosity in spite of comparatively high molecular weight is attributed to the special structure of these polyesters, consisting of an essentially linear main chain with a large number of alkyl side chains attached to it. By reacting the terminal groups further, the polyesters may be functionalized to carry terminal acrylic bonds, which render them reactive to high energy radiation such as ultraviolet light or accelerated electrons. Products thus formulated are solvent-free. They must be applied warm by roller or slot die coating and radiation cured.

7.0 PROPERTIES AND APPLICATIONS OF POLYESTER COATINGS

Polyester coatings, especially cross-linked ones, exhibit excellent flexibility or even elasticity. they show very good impact, scratch, and stain resistance. Their good adhesion properties, especially to metals, combined with good corrosion protection and weather resistance, have made them indispensable in a number of fields.

7.1 Sheet and Coil Coatings

Precoated sheet and coil have enjoyed remarkable growth over the past two decades, and polyesters have played an ever-increasing role in this industry. Polyesters are being used for primers and top coats with equally good success.

Polyester-based coil coating primers have been discussed concisely by Schmitthenner[54] and by Robertson.[55] The tough requirements that coated coil stock is expected to meet—for instance, for use as façade sheeting,—necessitate the use of primers, which have three main functions:

Promotion of adhesion to pretreated metal.
Corrosion protection.
Elastification of the two-coat system consisting of primer and top coat.

The two-coat system permits the optimization of adhesion of the primer to the pretreated metal and, at the same time, intercoat adhesion. High molecular weight polyesters with relatively high glass transition temperatures and in films approximately 5 μm thick have yielded excellent results on all counts. Corrosion-protective action is enhanced by the use of anticorrosive pigments. For the formulation of top coats, the end use will influence the choice of polymers. Paints with outstanding weather resistance can be formulated from medium molecular weight polyesters.[56] The UV absorption of polyesters used for this purpose ends at wavelengths below those at which the UV irradiation of sunlight begins. Schmitthenner has given another example of high molecular weight, highly elastified polyesters, that can be used to formulate correspondingly elastic enamels for end uses such as home appliances.

7.2 Can Coating

The term "can coating" embraces paints for use on a vast variety of decorated metals for packaging purposes. This includes cans of all kinds, including food cans and aerosol cans, collapsible tubes, and caps and closures of many kinds. Generally, the paint is applied to flat sheet stock, followed by printing, stacking up for storage, then stamping and forming. Especially difficult are the requirements for drawing and redrawing cans and closures, where the degree of forming is too high to be achieved in a single draw. The most common substrates are tin-plated steel, aluminum, and to a smaller extent, directly chromated steel (tin-free steel,TFS).

Accordingly, the requirements for can coatings are extreme in many ways. They include printability and block and scratch resistance, yet sufficient elasticity to permit forming and drawing without damage to the paint. On the finished article, the paint must be stable with respect to its contents, it must be stain resistant, sometimes heat sealable, and for food preservation, sterilization able to withstand, nontoxic, and neutral in taste and odor. The composition of can coatings for use in conjunction with foodstuffs is regulated in many countries individually, however, the U.S. Food and Drug Administration (FDA) is recognized internationally. The FDA provides a list of components, [57] that may be used to prepare polyesters intended for use as coatings in direct contact with food.

In conclusion, polyesters offer an excellent combination of physical properties especially concerning the balance of elasticity and surface hardness, paired with excellent adhesion, resistance to yellowing, and stability against the majority of materials now packaged in cans and tubes.

7.3 Automotive Paints

In the area of automotive paints, polyesters compete with a variety of other polymers. Acrylics, epoxy esters, and even alkyds have their share of the market. As automotive paints are spray applied, low molecular weight, and frequently branched polyesters tend to be used predominantly. two areas have developed as special grounds for polyesters. These are base coats for two-layered metallics and chip-resistant fillers.

For the former, low molecular weight, branched polyesters are used in conjunction with cellulose acetobutyrate and polymeric melamine resins. This combination permits optimum alignment of the metallic flake pigments immediately after application. The use of polyesters for chip-resistant fillers makes use of the excellent elasticity. Special polyester fillers made of upgraded polyurethane resin are used on areas of the car body where stone chipping occurs most frequently (e.g., rocker panels).

7.4 Industrial Paints

Polyester paints are gaining importance for the formulation of industrial baking enamels as a result of their weathering resistance, good abrasion and chemical stability, well-balanced elasticity, and surface hardness. This applies especially to high solids products and for waterborne sprayable and dip paints. Texturized paints are used for metal covers and housings for appliances, machinery, and data processing equipment primarily because of their good abrasion resistance and favorable hardness–elasticity properties. Automotive uses include metal parts in the engine compartment and fixtures for rearview mirrors and windshield wipers. Polyester paints are attractive here because of their corrosion protection and resistance to oil, fuel, and cleaning agents. Office furniture, such as steel desks and filing cabinets require high impact resistance, which is well met by polyester paints. Similarly, household appliances—if not made from precoated sheet—are also frequently coated with polyester-based paints.

7.5 Two-Component Paints

Saturated polyesters play an important role as hydroxyl-functional binder components in the formulation of two-component polyurethane paints. Compared to other hydroxyl-functional materials, polyesters permit complete adjustment of the elasticity of the coating of substrate and coating is required, to avoid crack formation upon temperature fluctuations. In addition, polyesters are used in isocyanate-cured, two-component automotive repair paints because of their weathering and chip resistance.

7.6 Powder Coating

Polyesters are finding use in thermosetting and thermoplastic powders. The latter are a specialty application only. The overwhelming majority are thermosetting powders. They may be either hydroxyl polyesters cured with blocked isocyanates or carboxyl polyesters cured with di-or triepoxides.

The most important advantage of *thermoplastic powders* is the speed of film formation, which is accomplished merely by melting and does not require reaction time for a chemical cure. For this purpose, high molecular weight, partially crystalline polyesters are best suited.[21] The use is predominantly for coating the welding seams of three-piece cans.[22]

Polyesters for use in *thermosetting powders* are generally amorphous, with molecular weights between 2000 and 6000, glass transition temperatures above 55°C, and softening points between 100 and 120°C. They must have good pigment compatibility. In general, they have a functionality higher than 2, which means branching in either the main chain or terminal components of higher functionality, as achieved by capping with trimellitic anhydride.[58,59]

Hydroxyl polyesters are generally cured with blocked isocyanates, such as isophorone diisocyanate blocked with ε-caprolactam[7,60] or uretdiones.[61]

Carboxyl polyesters may either be cured with epoxy resins (polyethylene sulfide–epoxy hybrides) or with TGIC. Polyester–epoxy hybrides enjoy the largest volume. They are very versatile and are, thus, finding use in many and diverse applications. Their advantage over hydroxyl polyesters–blocked isocyanate materials is the absence of blocking agents, which have to be driven off during the curing cycle. In turn, they require catalysts, such as quaternary ammonium compounds,[62] to achieve satisfactorily short curing times at acceptable cure temperatures. Owing to the relatively high content of epoxy resins (60–100 parts per 100 parts polyester) they have a tendency to chalking, hence are limited with respect to weather resistance. The latter can be essentially improved by using TGIC at a level of approximately 10 parts per 100 parts polyester.[63] Weather resistance of TGIC-cured carboxyl polyesters is excellent; however, powder coatings formulated on this basis have a tendency to display orange peel upon longer storage of the powder.[64]

7.7 Radiation-Curable Coatings

Saturated polyesters, of course, are not reactive to UV light or accelerated electrons such as used generally for radiation cure. There are basically two ways to render them reactive:

Incorporation into the polymer backbone of unsaturated acids, such as maleic or fumaric acid.
Capping with reactive, preferably acrylic or methacrylic vinyl double bonds.

The former method includes the unsaturated polyesters, such as those used as fiberglass-reinforced polymers in vehicle construction with or without the use of monomeric styrene as a reactive diluent. In the field of coatings, this material combination has found only very limited acceptance for a variety of reasons. The only type of use worth mentioning is for furniture finishes.

Vinyl-terminated polyesters, or more specifically acrylated polyesters, may be prepared by direct esterification of hydroxyl polyesters with acrylic acid or by reaction of carboxyl polyesters with glycidyl methacrylate.[65]

The most widely practiced method of attaching acrylic double bonds to polyesters is via the reaction of hydroxyl polyesters, diisocyanates, and hydroxyalkyl acrylates leading to acrylated polyester urethanes. Such materials are finding growing acceptance either by themselves or in combination with reactive diluents for a variety of uses. The reactive diluents are predominantly di-and triacrylates. The uses are diversified; overprint varnishes in the graphics and packaging sectors are presumably the largest group.

7.8 Adhesives

Adhesives should be mentioned for the sake of completeness, especially the kinds that are predominantly used in the form of continuous coatings. These include:

Heat-seal lacquers and hot melt coatings.
Laminating adhesives.
Pressure-sensitive adhesives.

For heat-seal lacquers and hot melt coatings, saturated, high molecular weight polyesters can be very useful. They offer good blocking resistance and swift bonding combined with attractive physiological properties (e.g., FDA conformity).[66]

For laminating adhesives, high molecular weight polyesters may be used in conjunction with isocyanates to effect high bond strength, good temperature stability, and even sterilization resistance.[67]

Pressure-sensitive adhesives, a totally new field for polyesters, may well gain considerable importance in the near future.[68,69]

REFERENCES

1. J. Rüter, K. König, and K. H. Seemann, *Ullmanns Encyclopädie*, Technische Chemie Vol. 19. Weinheim: Verlag Chemie, 1980, pp. 61–88.
2. D. Stoye, *Materialprüfung*, 15, 410 (1973).
3. E.-C. Schütze, *Oberfläche-Surface*, 11, 203 (1970).
4. J. Dörffel, J. Rüter, W. Holtrup, and R. Feinauer, *Farbe und Lack*, 82, 796 (1976).
5. J. Dörffel and U. Biethan, *Farbe und Lack*, 82, 1017 (1976).
6. W. L. Hensley, *Paint Varn. Prod.*, p. 68 (1966).
7. R. Gras and H. Riemer, U.S. Patent 4,528,355 (1985); Hüls.
8. R. Capanni et al., *Farbe Lack*, 78, 831 (1972).
9. V. Mirgel and K. Nachtkamp, *Farbe Lack*, 89, 928 (1983).
10. K. Schmitt, J. Disteldorf, and F. Schmitt, U.S. Patent 4,151,152 (1975); Hüls.
11. D. Berger, U.S. Patents 4,097,465 4,097,466 (1975); Hüls.
12. K. H. Hornung and U. Biethan, *Farbe Lack*, 76, 461 (1970).
13. U. Biethan, K. H. Hornung, and G. Peitscher, *Chem. Ztg.*, 96, 208 (1972).
14. U. Biethan, and K. H. Hornung, in *Tenth FATIPEC Congress Book*, 1970, p. 277.
15. U. Biethan, J. Dörffel, and D. Stoye, *Farbe Lack*, 77, 988 (1971).
16. A. Schott, R. Gras, and E. Wolf, U.S. Patent 4,258,186 (1977); Hüls.
17. F. O. Stark, in *Organic Coatings, Science and Technology*, Vol. 6. New York: Dekker, 1984, pp. 87–100.
18. D. L. Edwards et al., *Paint Varn. Prod.*, p. 44 (1966).
19. M. F. Koistra, *J. Oil Colour. Chem. Assoc.*, 62, 432 (1979).
20. U. Biethan et al., *Ullmanns Encyclopädie Technische Chemie.*, Vol. 15. Weinheim: Verlag Chemie, 1978, pp. 625–628.
21. K. Brüning and K. G. Sturm, U.S. Patent 4,054,681 (1977); Dynamit Nobel.
22. H. J. Hölscher, *Neue Verpack.*, 1, 46 (1984).
23. *Dynapol Copolyesters for Coatings and Adhesives* (product information leaflet), Hüls AG, Marl, West Germany.
24. *Dynapol Coatings* (product information ringbook), Hüls AG, Marl, West Germany.
25. *Vesturit—Saturated Polyester Resins for Stoving Enamels* (brochure), Hüls AG, Marl, West Germany.
26. Technical Service Report GTSR 22 and GTSR 36, Amoco Chemicals Co., Naperville, Il.
27. G. Schade, U.S. Patent 4,104,262 (1978); Dynamit Nobel.
28. J. Dörffel, *Farbe Lack*, 81, 10 (1975).
29. J. Dörffel and W. Auf der Heide, *Farbe Lack*, 86, 109 (1980).
30. D. W. Sartorelli and R. R. Smith, EP Patent 0105016 (1983); Goodyear.
31. W. H. Carothers, *Trans. Faraday Soc.*, 32, 39 (1936).
32. P. J. Flory, *J. Am. Chem. Soc.*, 63, 3083 (1941).
33. P. J. Flory, *Principles of Polymer Chemistry*. Ithaca, NY: Cornell University Press, 1953, pp. 568–576.
34. T. G. Fox and P. J. Flory, *J. Appl. Phys.*, 21, 581 (1950).
35. T. G. Fox, *Bull. Am. Phys. Soc.*, 1, 123 (1956).
36. D. W. van Krevelen, *Properties of Polymers*. New York: Elsevier, 1972.
37. W. A. Lee and R. A. Rutherford, *Polymer Handbook*, Vol. III. New York: Wiley, (1975), p. 139.
38. R. Hill, *Fibres from Synthetic Polymers*. London: Elsevier, 1953.
39. L. Buxbaum, *Angew. Chem.*, 80, 225 (1968).
40. W. Andrejewski and D. Stoye, British Patent. 1 481 182 (1973); Hüls.
41. W. J. Mijs, W. J. Muizebelt, and J. B. Reesink, *J. Coatings Technol.*, 55, 45 (1983).
42. D. Stoye, W. Andrejewski, and J. Dörffel, *Thirteenth FATIPEC Congressbook*, 1976, p. 605.
43. D. Stoye and J. Dörffel, *Adv. Org. Coatings Sci. Technol.*, 2, 183 (1980).
44. D. Stoye and J. Dörffel, *Pigment Resin Technol.*, 9(7), 4; 9(8), 8 (1980).
45. L. Gott, *J. Coatings Technol.*, 48, 52 (1976).

46. R. Buter, *Farbe Lack*, 86, 307 (1980).
47. D. Stoye and J. Dörffel, in *Organic Coatings, Science and Technology*, Vol. 6. New York: Dekker, 1984, pp. 257–275.
48. L. W. Hill and K. Kozlowski, *J. Coatings Technol.*, 59, 63 (1987).
49. F. N. Jones and D. D.-L. Lu, *J. Coatings Technol.*, 59, 73 (1987).
50. K. H. Albers, A. W. McCollum, and A. E. Blood, *J. Paint Technol.*, 47, 71 (1975).
51. M. R. Olson, J. M. Larson, and F. N. Jones, *J. Coatings Technol.*, 55, 45 (1983).
52. K. L. Payne, F. N. Jones, and L. W. Brandenburger, *J. Coatings Technol.* 57, 35 (1985).
53. H. Müller et al., U.S. Patent 4,668,763 (1987); Dynamit Nobel.
54. M. Schmitthenner, paper presented at the ECCA Annual Congress, Brussels, 1983.
55. R. R. Robertson, paper presented at the 1982 Fall Technical Meeting, NCCA.
56. M. Schmitthenner, paper presented at the Fourteenth International Conference on Organic Coatings Science and Technology, Athens, 1988.
57. Federal Register, Vol. 21, Section 175.300 [Food and Drug Administration].
58. S. Harris, *Polym. Paint Color J.*, 174 (4114), 162–164 (1984).
59. E. Bodnar and P. Taylor, *Pigment Resin Technol.*, 15(2), 10– 15 (1986).
60. R. Gras, F. Schmitt, and W. Wolf, U.S. Patent 4,246,380 (1981); Hüls.
61. R. Gras et al., U.S. Patent 4,413,079 and 4,483,798 (1984); Hüls.
62. P. L. Heater Jr., EP Patent 0 056 356 (1985); Goodyear.
63. Ciba-Geigy Bulletin, Araldit PT 810. Ciba-Geigy, Basel, Switzerland.
64. W. Marquardt and H. Germeller, *Farbe Lack*, 86, 696–698 (1980).
65. J. W. Saracsan, EP Patent 0,089,913 (1983); Goodyear.
66. *Dynapol Coatings Bulletin L-650* (product Information), Hüls AG, Marl, West Germany.
67. *Polyester Resins, Adhesives and Coatings*, (brochure), Dayton Chemical Division of Whittaker Corp., West Alexandria, OH.
68. H. F. Huber and H. Müller, *Conf. Proc. Radcure*, 86, 12–23 (1986).
69. H. F. Huber and H. Müller, *Conf. Proc. Radcure*, 87, 8–35 (1987).

40
Alkyd Resins

Krister Holmberg

Berol Kemi AB
Stenungsund, Sweden

Alkyd resins represent a class of polymers that are used in surface coating formulations because of their low cost and versatility. The term "alkyd" was coined by Kienle and Ferguson and is derived from "al" of alcohol and "cid" of acid; "cid" was later changed to "kyd."

Alkyd resins in a broad sense refer to polymers. By convention, however, polyesters with unsaturation in the backbone are not referred to as alkyds but are termed unsaturated polyesters.

The specific definition of alkyds that has gained wide acceptance is that alkyds are polyesters modified with fatty acids. The nonmodified resins are then called saturated polyesters. Terms like "oil–free alkyd" and "oil–modified polyester" can also be found in the literature

1.0 CLASSIFICATION

Alkyds are synthesized from three basic components, polybasic acids, polyols, and (except for oil–free alkyds) fatty acids. The nature and proportions of these components control the properties of the resin. The amount of combinations is enormous, and specification of an alkyd resin must involve several parameters. The most important ways of classification are given below.

1.1 Oil Length and Type of Oil

Depending on the weight percentage of fatty acid in the resin, alkyds are referred to as short oil (< 45%), medium oil (> 55%). However, some confusion exists regarding the terminology. Sometimes oil length refers to percentage of triglyceride, in which case fatty acid content has to be recalculated into triglyceride. The second approach can be converted into the first by division by 1.045.

375

The type of fatty acid used also governs the properties of the alkyds. The resins are classified as drying, semidrying, and nondrying, depending on the degree of unsaturation in the fatty acid residues (iodine number of > 140, 125–140, and < 125, respectively).

Oxidative drying of alkyds, which involves air oxidation of polyene structures in fatty acid residues, is at maximum around 50% oil length. After drying, film hardness is inversely proportional to the degree of fatty acid modification.

Short oil alkyds generally give films of high quality with regard to color and gloss retention but low flexibility and with poor adhesion. Long oil alkyds are usually superior in terms of pigment dispersion, rheological properties, and storage stability.

Examples of properties of alkyd resins related to oil length and type of oil are shown in Table 1.[2]

1.2 Percentage of Phthalic Anhydride

Phthalic anhydride is the most commonly used raw material in alkyd compositions, and the weight percentages usually stated. There is an inverse relationship between percentage of phthalic anhydride and degree of fatty acid modification, short oil alkyds having above 35%, medium oil alkyds between 20 and 35%, and long oil alkyds below 20% phthalic anhydride.

1.3 Acid Value and Hydroxyl Number

The acid value is defined milligrams of potassium hydroxide required to neutralize 1 g of resin. For alkyd resins, 0.1 M KOH in ethanol is normally employed.

The hydroxyl number (sometimes called hydroxyl value) is the milligram of potassium hydroxide equivalent to the amount of acyl groups reacted in the acylation of 1 g of resin. A known amount of acylating reagent (often acetic anhydride or phthalic anhydride in pyridine) is added to the resin sample, and the hydroxyl number is obtained by back–titration with alkali.

Usually the hydroxyl number is considerable higher than the acid value. Baking alkyds require a certain concentration of hydroxyl groups to react with the amino acids, and for air drying alkyds the concentration of hydroxyl groups determines pigment wetting properties. The concentration of carboxyl groups is of particular interest in alkyds for water borne coatings. To achieve water solubility without excessive use of cosolvents, these resins are processed to a high acid value and the carboxyl groups are subsequently neutralized with ammonia or an amine (see Section 6.2).

2.0 PRINCIPLE OF ALKYD SYNTHESIS

Polyesters may be prepared either by direct condensation of at least difunctional acids and at least difunctional alcohols, or by employing reactive derivatives of the acid or the alcohol component. The first type of reaction is reversible, and to shift the equilibrium toward the product side, the water formed must be removed from the reaction zone. In practice, such removal may be carried out in various ways, such as azeotropic distillation using an organic solvent, sweeping the vapor away by means of a stream of inert gas, or applying a vacuum.

Reactions with reactive derivatives may be regarded as irreversible. Anhydrides and epoxides are the most common type of reactive derivatives of acids and alcohols, respectively.

Table 1 Effect of Oil Length and Type of Oil on the Properties and Uses of Alkyds

Oil type	Oil length (%)	Typical oil	Properties
Oxidizing	≥ 60	Linseed, safflower, soybean, tall oil fatty acids; wood oil in blends with other oils; dehydrated castor oil	Soluble in aliphatic solvents. Compatible with oils and medium oil length alkyds; good drying characteristics. Films are flexible, with reasonable gloss and durability
Oxidizing	45–55	Linseed, safflower, soybean, tall oil fatty acids; wood oil in blends with other oils	Soluble in aliphatic or aliphatic-aromatic solvent mixtures. Good drying characteristics, durability and gloss
Oxidizing	≤ 45	Linseed, safflower, soybean, tall fatty acids; wood oil in blends with other oils; dehydrated castor oil	Soluble in aromatic hydrocarbons. Low tolerance for aliphatic solvents. Usually cured at elevated temperatures either by heating with manganese driers or with urea or melamine formaldehyde resins
Nonoxidizing	40–60	Coconut oil, castor oil, hydrogenated castor oil	Soluble in aliphatic–aromatic solvent blends. Usually used as a plasticizer for thermoplastic polymers such as nitrocellulose
Nonoxidizing	≤ 40	Coconut oil, castor oil, hydrogenated castor oil	Soluble in aromatic solvents. Used as a reactive plasticizer that chemically combines with other resin entities (e.g. melamine–formaldehyde resin)

Source: Ref. 2.

Polyesterification is one of the prime examples of step–growth polymerization. This is a type of reaction in which each polymer chain grown at a relatively slow rate over a much longer period of time than in an addition polymerization reaction, and in which the initiation, propagation, and termination reactions are approximately identical in both mechanism and rate. [3]

The equilibrium constants of polyesterification is normally equal to that of the analogous model reaction between monofunctional compounds. This has been explained by the proposition that the reactivity of the functional end groups in the growing polyester chain is independent of the degree of polymerization. In other words, at all stages of polymerization the reactivity of every functional group is the same. This principle of equal reactivity of functional groups, first demonstrated by Flory [4, 5] is important for alkyd synthesis since it permits the application of statistical considerations to the problem of distribution of the bonds formed during polymerization.

Polyesterification carried out in the absence of an added catalyst has been found to follow third–order kinetics. [5, 6] The carboxyl groups act as catalyst and the mechanism involved is the following:

$$2\ RCOOH \rightleftharpoons RCOO^- + RCOOH_2^+$$

$$RCOOH_2^+ + R^1OH \rightleftharpoons RCOOHR^{1+} + H_2O$$

$$RCOOHR^{1+} + RCOO^- \rightleftharpoons RCOOR^1 + RCOOH$$

The second step is believed to be rate determining. [7] In reaction media of low dielectric constant, such as esters and polyesters, the ions are probably associated as ion pairs. The decrease in concentration of carboxyl groups can be expressed:

$$-\frac{d[COOH]}{dt} = k[COOH]^2 \cdot [OH]$$

If the polyesterification is performed in the presence of an acid catalyst, the reaction becomes second order. [6, 7] At high degrees of conversion, however, the reactions become sluggish. This has been ascribed to depletion of the catalyst; at low concentration of remaining carboxyl groups, a catalyst, such as p–toluenesulfonic acid, may compete favorably in reacting with hydroxyl groups, thus acting as a chain–terminating additive. [8, 9]

3.0   FUNCTIONALITY AND PREDICTION OF GEL POINT

In the condensation of bifunctional reactants, the functionality of the reaction product is always 2, regardless of the extent of reaction. In the reaction between a triol and a dibasic acid, using equivalent amounts of hydroxyl and carboxyl groups, on the other hand, the functionality of the molecules formed increases as the reaction proceeds. This is illustrated below for the reaction between glycerol and adipic acid; the tetraester produced has a functionality of 4.

$$2CH_2\!-\!CH\!-\!CH_2 + 3HOOC\!-\!(CH_2)\!-\!COOH \longrightarrow$$
$$\quad|\quad\ |\quad\ |$$
$$\ OH\ \ \ OH\ \ \ OH$$

$$CH_2\!-\!CH\!-\!CH_2\!-\!OCO\!-\!(CH_2)_4\!-\!COO$$
$$\ |\quad\ |\qquad\qquad\qquad\qquad\ \ |$$
$$OH\ \ \ OH\qquad\qquad\qquad\qquad CH_2$$
$$\qquad\qquad\qquad\qquad\qquad\qquad\ |$$
$$\qquad HOOC\!-\!(CH_2)_4\!-\!COO\!-\!CH$$
$$\qquad\qquad\qquad\qquad\qquad\qquad\ |$$
$$\qquad\qquad\qquad\qquad CH_2\!-\!OCO\!-\!(CH_2)_4\!-\!COOH$$

3.1 Actual Functionality

In the example above, the actual functionality equals the maximum functionality of both the triol and the dibasic acid; a high cross–linked structure will form and gelation will eventually occur.

If the glycerol and the adipic acid are, instead, used in equimolecular amounts, a linear polymer will be obtained. The actual functionality of the triol is now 2, instead of 3, since on the average only two hydroxyl groups from each glycerol molecule will react.

$$CH_2\!-\!CH\!-\!CH_2\!-\!O\!-\![-CO\!-\!(CH_2)_4\!-\!COO\!-\!CH_2CH\!-\!CH_2\!-\!O\text{-}]_n\!-\!CO\!-\!(CH_2)_4\!-\!COOH$$
$$\ |\quad\ |\qquad\qquad\qquad\qquad\qquad\qquad\quad\ |$$
$$OH\ \ \ OH\qquad\qquad\qquad\qquad\qquad\qquad OH$$

In the alkyd syntheses an excess of hydroxyl over carboxyl groups is used. The actual functionality of the acid component is then equal to its maximum functionality. The actual functionality of the polyol, on the other hand, is lower than the maximum functionality and can be calculated[10] from the following formula:

$$F_{actual} = \frac{F_{maximum}}{1+n}$$

where n is the fraction of the hydroxyl groups present in excess of carboxyl groups.

For example, in the case above where glycerol and adipic acid are used in equimolecular amounts, hydroxyl groups are present in 50% excess. Hence, the actual functionality will be:

$$F_{glycerol} = \frac{3}{1+0.5} = 2$$

The overall maximum functionality of an alkyd composition, including both acid and alcohol components, is expressed as

$$F_{\text{overall, max}} = \frac{\text{Total equivalents}}{\text{Total moles}}$$

A more useful expression is obtained if the excess of one component (usually excess polyol) is disregarded:

$$F_{\text{overall, max}} = \frac{\text{Total equivalents} - \text{excess equivalents}}{\text{Total moles}}$$

The following example illustrates the procedure. An alkyd as prepared according to the formulation of Table 2.

Total equivalent polyols: $1.0 + 4.5 = 5.5$
Total equivalent acids: $0.8 + 3.6 = 4.4$
Excess equivalent polyol: $5.5 - 4.4 = 1.1$

$$F_{\text{overall,act}} = \frac{(5.5 + 4.4) - 1.1}{4.6} = 1.91$$

3.2 Gel Point

The overall actual functionality can be used as a measure of whether an alkyd composition will gel. In doing this, Carothers's definition of gel point (i.e., that gelation corresponds to an infinite number average molecular weight) is normally used.[11–13]

It can be shown that the extent of reaction at the gel point P_{gel} can be written:

$$P_{\text{gel}} = \frac{2}{1.91} = 1.05$$

Now, if P_{gel} is less than 1, the composition will gel before reaction has gone to completion. A value of P_{gel} exceeding 1 indicates that the reaction will not gel (assuming ideal conditions).

Using the value $F_{\text{overall, act}}$ obtained from the values of Table 2 gives

Table 2 Alkyd Formation

Raw materials	Moles	Maximum functionality	Equivalents
Diethylene glycol	0.5	2	1.0
Glycerol	1.5	3	4.5
Fatty acid	0.8	1	0.8
Adipic acid	1.8	2	3.6

$$P_{\text{gel}} = \frac{2}{1.91} = 1.05$$

This indicated that 105% conversion is needed to give gelation. Consequently, the composition will not gel.

3.3 Alkyd Calculations

The forgoing theory has been the basis for computerized calculations on alkyd formulations. It is well known that alkyds should be processed close to the gel point; gelation during cooking is a disaster, however, a proposed method of saving such runs notwithstanding. [14] In the calculation, the relative amounts of the ingredients can be varied to arrive at a gel point at a certain acid value. Furthermore, the formulation can be modified by incorporating other acids and alcohols into the recipe, as well as by excluding original ones. The only demand made on a new raw material is that its molecular weight and maximum functionality be known.

The calculations give not only the relative amounts of the starting materials to arrive at gel point at a certain acid value, but also the molecular weight of the alkyd at a given value, higher than the acid value corresponding to gelation.[15]

There is usually a discrepancy between theoretical and experimental P_{gel}. The reason for this is that the polyesterification is not an ideal process, but is subject to a number of side reactions. Furthermore, the true functionality of the starting materials may not always correspond to the theoretical values.

The most important correction factors to be taken into account in alkyd synthesis calculations are[16]:

1. Intramolecular reactions, particularly esterifications, are known to take place in most alkyd syntheses. The true functionality will be reduced.
2. The reactivity of one of the functional groups of a multifunctional reactant is reduced because of steric and/or electric effects. One example of this is glycerol, the secondary hydroxyl group of which reacts slower than the two primary groups.
3. The true functionality of a component is higher than the theoretical value. This applies to polyols in general, since etherification invariably occurs parallel to the esterification reactions. Etherification of a polyol having n hydroxyl groups leads to a new polyol with a functionality of 2 n minus 2. Unsaturated fatty acids also have a true functionality higher than the theoretical value, since cross–linking between the chains often occurs during alkyd processing.[17]

4.0 RAW MATERIALS

4.1 Polybasic Acids

Due to ease of handling, good balance of properties, and economy, orthophthalic acid is the most important polybasic acid for alkyds. It is almost exclusively used in its anhydride form. Isophthalic acid is used as a replacement for phthalic anhydride when a tougher, faster drying, and more chemical– and heat–resistant coating is required. The meta position of its carboxyl groups makes the formation of intramolecular cyclic esters more unlikely with isophthalic acid, leading to higher molecular weights and high viscosities of the alkyds.

Maleic anhydride is sometimes used in limited amounts in alkyd formulations. The olefinic bond of this compound enables it to form Diels–Alder and ene adducts with unsaturated acids in drying oils.[18, 19] As a consequence, the total functionality of the system will increase, leading to high viscosity and risk of gelation. The incorporation of maleic anhydride into alkyd formulations generally improve color and water resistance.

Longer aliphatic dibasic acids, in particular adipic and azelaic acid, may be used as minor ingredients to impart flexibility in the alkyd structure. Tri– and tetrafunctional acids or anhydrides, such as trimellitic and pyromellitic anhydride, are incorporated to produce alkyds of high acid value. Chlorinated and brominated compounds (e.g., tetrachloro– and tetrabromophthalic anhydride) are used to impart fire–retardant properties to the resin.

4.2 Polyols

Usually a mixture of polyols having a functionality of 2 to 4 is used in an alkyd formulation. Ethylene glycol, diethylene glycol, propylene glycol, and neopentyl glycol are the most important diols; glycerol and trimethylol propane are commonly used triols, and pentaerythritol is the tetraol of choice. The choice of polyol components is mainly responsible for the degree of branching of the alkyd. The flexibility of the resin is also influenced by the distance between the hydroxyl groups—diethylene glycol, for instance, gives a more flexible product than ethylene glycol. Neopentyl glycol, because of its branched structure, gives heat– and hydrolysis–resistant films.

4.3 Fatty Acids and Oils

The vast majority of the nonbasic acids used in alkyd compositions are derived from vegetable oils. Fish oil also is used to a small degree. Rosin, which is a mixture of resinous monobasic acids having a molecular weight of about 330, is sometimes incorporated to improve hardness, initial drying, and water resistance. Synthetic aromatic acids, such as benzoic acid, can improve greater hardness and improved gloss retention.

The degree of unsaturation in the fatty acid governs the drying properties of alkyds. In general, the higher the iodine number of the oil or the fatty acid, the faster the drying (see Section 1.1). The position of the double bonds is also of importance, conjugated bonds being much more reactive in autoxidation than nonconjugated.[20]

Soybean oil is the workhorse among the vegetable oils. Linseed oil is used for fast drying alkyds. Tall oil, sunflower oil, and safflower oil are also common as raw materials for drying alkyds. Coconut oil and castor oil are extensively used in nondrying alkyds. The fatty acid composition of the most important vegetable oil is given in Table 3.

5.0 MANUFACTURING PROCESS

Alkyds are prepared by polycondensation of the acid and alcohol components until a predetermined acid value–viscosity relationship has been achieved. The reaction is normally under inert gas or solvent vapor to minimize oxidation of unsaturated components. In the initial phases of the reaction, the drop in acid value is rapid and the increase in viscosity is slow. Toward the latter stages of the reaction, the drop in acid value is true. Preparation may be performed by a solvent–free process, by the fusion method, or by using a small amount of solvent that forms an azeotrope with water (i.e., the solvent method). Furthermore, in the preparation of fatty acid modified polyesters, either the triglyceride or the fatty acid derived

Table 3 Typical Fatty Acid Composition (%) of Vegetable Oils

Fatty acid		Castor	Coconut	Linseed	Olive	Palm	Palm kernel	Peanut	Safflower	Soybean	Sunflower	Tall	Tung
						Vegetable oils							
C_8	Caprylic		6				3						
C_{10}	Capric		6				4						
C_{12}	Lauric		44				51						
C_{14}	Myristic		18			1	17						
C_{16}	Palmitic	2	11	6	16	48	8	6	8	11	11	5	4
C_{18}	Stearic	1	6	4	2	4	2	5	3	4	6	3	1
C_{18}	Oleic	7	7	22	64	38	13	61	13	25	29	46	8
C_{18}	Linoleic	87	2	16	16	9	2	22	75	51	52	41	4
C_{18}	Linolenic	3		52	2				1	9	2	3	4
C_{18}	Eleostearic												3
C_{18}	Ricinoleic												80
C_{20}	Arachidic							2				2	

Source: Ref. 25.

from it may be employed as starting material. These two procedures are referred to as the monoglyceride and the fatty acid process, respectively.

5.1 Fusion Methods Versus Solvent Method

The *fusion method* is the older process but still widely used, especially for alkyds of an oil length of 60% or more. The reaction is carried out at a temperature of 220–250 °C , and the inert gas sparge, which is used for dewatering, also causes some loss of volatile polyols and of phthalic anhydride.

In the *solvent method* the esterification is performed in the presence of a small quantity of water–immiscible solvent, usually xylene. The process is carried out under continuous azeotropic distillation of the solvent. The xylene–water vapor mixture is condensed, the water is separated, and the organic distillate is returned to the reactor. The reaction temperature is governed by the refluxing temperature which, in turn, depends on the amount of xylene used, 5% being a normal value. In general, the solvent method offers better control of the resin composition, as there is virtually no loss of raw materials by volatilization or sublimation.[21]

5.2 Monoglycerides Versus Fatty Acid Process

When a triglyceride oil is heated together with polyols and polybasic acids, the polyols react preferentially with the acids and a heterogeneous mixture of triglyceride and unmodified polyester is obtained. The way to overcome this problem is to perform a controlled transesterification of the fatty acids prior to the condensation step. This is usually done by reacting 1 mole of triglyceride with 2 moles of glycerol (or another polyol) at a temperature of 220–250 °C until *monoglyceride stage* is obtained. The preferred catalysts are Pbo, $Ca(OH)_2$, and Ca- soaps.[22]

$$
\begin{array}{ccc}
CH_2OCOR & CH_2OH & CH_2OCOR \\
| & | & | \\
CHOCOR \;+\; CHOH & \xrightarrow[\text{catalyst}]{\Delta\;\;\text{alkaline}} & 3CHOH \\
| & | & | \\
CH_2OCOR & CH_2OH & CH_2OH
\end{array}
$$

After formation of monoglycerides, the polybasic acids and the rest of the polyols are added and the condensation is carried out until the desired viscosity–acid value relationship is reached.

The *fatty acid process* uses not the triglyceride but the fatty acid itself. Apart from giving a greater freedom in the choice of polyol components, this process is more reproducible and gives better control over molecular weight and molecular weight distribution of the resin, 12 In addition, this process enables the condensation to be carried on to lower acid values, which is advantageous for the drying properties.

The choice of manufacturing process is not only a matter of finding the most suitable synthesis procedure. There exists distinct differences in film properties between alkyds of the same composition but prepared by different processes.[23, 24]

6.0 ALKYDS FOR REDUCED SOLVENT EMISSION

During the past decades a number of new technologies, all derived from the demand to reduce or eliminate the organic solvent from the coating system, have been developed.

6.1 High Solids Coatings

By synthesizing alkyds of narrow molecular weight distribution and by reducing the mean molecular weight of the resin, coatings with a considerably higher solids content can be attained.[25] Strictly speaking, the term "high solids" refers to coatings with more than 80% nonvolatiles by volume. In practice, paints with 70% and even 60% are often included in the high solids category.

6.2 Water Borne Coatings

Alkyds with high acid values can be formulated to give isotropic solutions by neutralization with ammonia or an amine and by using an alcohol or a glycol ether cosolvent.[26] Such formulations may be regarded as a type of microemulsion, with the alkyd resin constituting both the oil component and the anionic surfactant.

Alkyd emulsions are prepared through post emulsification techniques. The surfactants used for these emulsions are usually non ionics having hydrophile–lipophile balance (HLB) values of 12–14.[27] The technique has mainly been applied to long oil alkyds.

6.3 Powder Coatings

Powder coatings may be regarded as the ultimate solution to the problem of solvent emission. Oil–free alkyds (saturated polyesters) are being used in thermosetting powder coatings in combination with epoxy or isocyanate cross–linkers.[28]

7.0 MODIFIED ALKYDS

7.1 Polyamide Modification

Thixotropic alkyd resins are made by combining alkyds with polyamide resins and heating the mixture so that amide and ester interchange reactions occur. The thixotropic behavior of these resins is attributed to hydrogen bonding between ester carbonyl and amide groups distributed along the polymer chain.[29]

7.2 Vinyl Modification

A vinyl monomer, usually styrene or vinyl toluene, can be polymerized onto an alkyd containing unsaturated groups. The products obtained are mixtures of graft copolymers and polymer blends.[30] Vinyl–modified alkyds are characterized by very fast drying and by hard, water–resistant films.

7.3 Other Modifications

Alkyds may be modified with *silicones* to give copolymers with excellent durability and heat resistance. *Alkyl methacrylate* modified alkyds are fast curing and have good gloss retention and durability. Modification with *epoxy resins* yields products that have excellent adhesion properties and improved water and chemical resistance.

8.0 USES

Alkyd resins as a group are characterized by good adhesion and drying properties. The films produced have good flexibility and durability. By various modifications, specific properties can be improved. A weak point of alkyds is the susceptibility to alkaline hydrolysis.

Long oil alkyds are soluble in aliphatic solvents. Normally applied by brush, they are used in exterior trim paints and wall paints, as well as in marine and metal maintenance paints. They are also widely used in clear lacquers.

Medium oil alkyds are soluble in aliphatic–aromatic solvent blends. The air drying type is used as the standard vehicle for industrial applications, such as primers and undercoatings, maintenance paints, and metal finishes. The nonoxidizing type is often used as external plasticizer in nitrocellulose lacquers.

Short oil alkyds are soluble in aromatic but not aliphatic solvents. The air drying type is used in baking primers and enamels, either as the sole binder or together with other resins, such as urea or melanine resins. The nondrying type is mainly used as plasticizing resin in nitrocellulose lacquers and in combination with urea or melanine resins in stoving and acid curing finishes.

Oil–free alkyds are widely used together with melanine resins in high performance stoving coatings.

REFERENCES

1. R. H. Kienle and C. S. Ferguson, *Ind. Eng. Chem.*, *21*, 349 (1929).
2. D. H. Solomon, *The Chemistry of Organic Film Formers*. New York: Krieger, 1977, p. 91.
3. R. W. Lenz, *Organic Chemistry of Synthetic High Polymers*. New York: Wiley Interscience, 1967, p. 6.
4. P. J. Flory, *J. Am. Chem. Soc.*, 64, 2205 (1942).
5. P. J. Flory, *Principles of Polymer Chemistry*. Ithica, NY: Cornell University Press 1953.
6. S. D. Hamann, D. H. Solomon, and J. D. Swift, *J. Macromol. Sci. Chem.*, A2, 153 (1968).
7. J. H. Saunders and F. Dobinson, *Comprehensive Chemical Kinetics*, Vol. *15*: in the series *Non–Radical Polymerization* C. H. Bamford and C. F. H. Tipper, Eds, Amsterdam: Elesier, 1976, pp. 505, 507.
8. P. J. Flory, *Chem. Rev.*, 39, 137 (1946).
9. C. E. H. Bawn and M. B. Huglin, *Polymer*, 3, 257 (1962).
10. W. C. Spitzer, *Off. Dig.*, 36, 16 (1964).
11. W. H. Carothers, *Trans. Faraday Soc.*, 32, 39 (1936).
12. D. H. Solomon, *The Chemistry of Organic Film Formers*. New York: Kreiger, 1977, pp. 65, 82, 88, 108.
13. C. W. Johnson, *Off Dig.*, 32, 1327 (1960).
14. A. Gal and A. Hardof, *Org. Coatings Sci. Technol.*, 8, 55 (1986).
15. L. A. Tysall, *Calculation Techniques in the Formulation of Alkyds and Related Resins*. Teddington, United Kingdom: Oaint Research Association, 1986.
16. D. H. Solomon, *The Chemistry of Organic Film Formers*. New York: Kreiger, 1977, pp. 84–86.
17. D. Firestone, *J. Am. Oil Chem. Soc.*, 40, 247 (1963).
18. A. E. Rheineck and T. H. Khoe, *Fette Seifen Anstrichm.*, 71, 644 (1969).
19. K. Holmberg and J.–A. Johansson, *Acta Chem. Scand.*, B36, 481 (1982).
20. D. H. Solomon, *The Chemistry of Organic Film Formers*. New York: Kreiger, 1977, pp. 38–74.
21. E. G. Bobalek, E. R. Moore, S. S. Levy, and C. C. Lee, *J. Appl. Polym. Sco.*, 8, 625 (1964).
22. R. A. Brett, *J. Oil Colour Chem. Assoc.*, 41, 428 (1958).

23. J. R. Fletcher, L. Polgar, and D. H. Soloon, *J. Appl. Polym. Sci.*, 8, 659 (1964).
24. D. H. Solomon and J. J. Hopwood, *J. Appl. Polym. Sci.*, 10, 993 (1966).
25. K. Holmberg, *High Solids Alkyd Resins*. New York: Dekker, 1987.
26. S. Paul, *Surface Coatings*. New York: Wiley, 1985, pp. 561–568.
27. B. G. Bufkin and J. R. Grawe, *J. Coatings Technol.*, 50 (647), 65 (1978).
28. S. Paul, *Surface Coatings*. New York: Wiley, 1985, pp. 680–682.
29. A. G. North, *J. Oil Colour Chem. Assoc.*, 39, 695 (1956).
30. E. F. Redknap, *J. Oil Colour Chem. Assoc.*, 43, 260 (1960).

41

Phenolic Resins

Kenneth Bourlier

Union Carbide Corporation
Bound Brook, New Jersey

PHENOLIC RESINS

Synthetic phenolic resins were developed and commercialized in the early 1900s by Leo Baekland.[1] The reaction of phenol and formaldehyde produces a product that forms a highly cross–linked three–dimensional polymer when cured. The resins have found use in various applications in the coatings industry because of their excellent heat resistance, chemical resistance, and electrical properties. They also offer good adhesion to many substrates and have good compatibility with other polymers.

Phenolic resins have two basic classifications: resoles and novolaks. Resoles, or heat–reactive resins, are made using an excess of formaldehyde and a base catalyst. The polymer that is produced has reactive methylol groups that form a thermoset structure when heat is applied.

Novolaks are made using an excess of phenol and an acid catalyst. Reaction occurs by the protonation of the formaldehyde,[2] and the intermediate is characterized by methylene linkages rather than methylol groups. These products are not heat reactive, and they require additional cross linking agents such as hexamethylenetetramine to become thermosetting. These reactions can be thought of as nucleophillic attack of phenoxide ion on the formaldehyde, or electrophillic substitution by protonated formaldehyde on the aromatic ring.[3]

Both polymerization reactions evolve water during cure. This condensation reaction serves to limit film thickness to approximately 3 mills because the volatiles will cause blistering while curing takes place. Baking temperatures are generally 300–400° F with a bake time of 10–30 minutes.

Phenolic resin polymers have been used in the coatings industry for many years because of their excellent performance properties and their relatively low cost. Some of the major applications are in rigid packaging, maintenance primers, printing inks, and epoxy hardeners. A brief discussion of these applications follows.

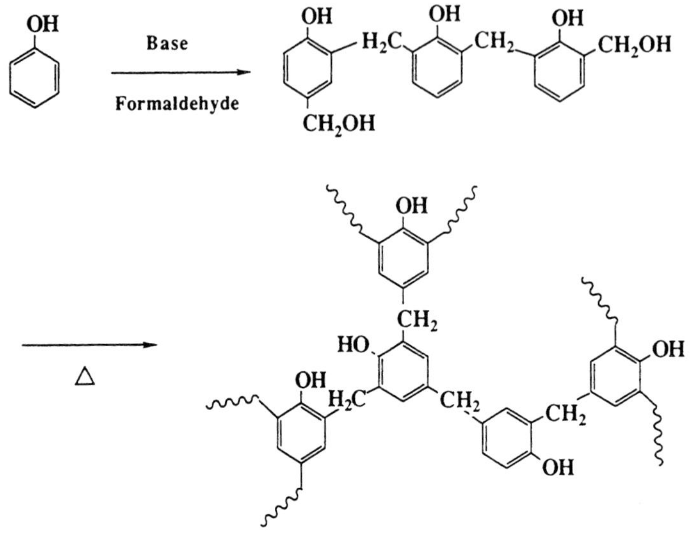

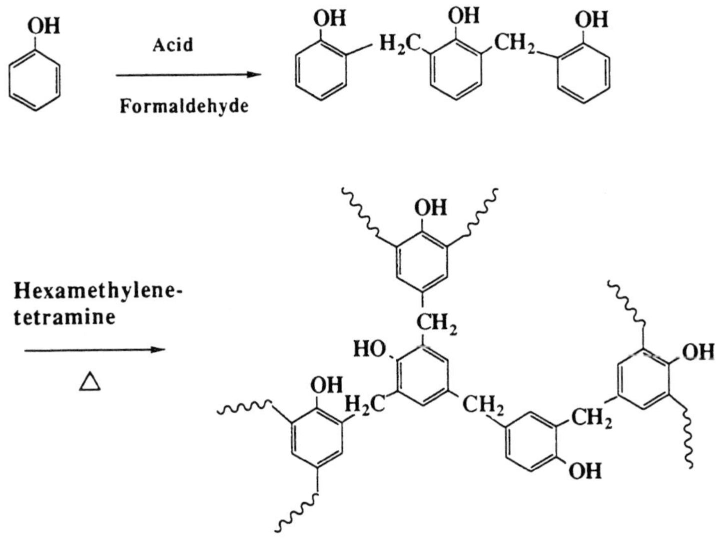

1.0 RIGID PACKAGING

Phenolic resins have found use in metal containers because they offer outstanding corrosion protection and excellent adhesion, along with chemical and solvent resistance. Typical end-use applications are food container coatings, drum and pail lining, pipe coatings, tank linings, and coil coating primers, or certain polyester resins.

Unmodified phenolics are often used in drum, pail, and pipe coatings. Since the substrate is usually rigid, the brittleness associated with these phenolic formulations does not present a problem. The excellent solvent resistance of phenolic resins make them ideal for this application. A film thickness of 0.2–0.3 mills often provides sufficient protection. Thicker

TABLE 1 Drum and Pail Lining

Ingredients	Parts by Weight
Phenolic resin[a]	73.30
Acetone[b]	7.00
Methyl ethyl ketone	9.00
Ethanol	55.00
Propyl Propasol solvent	14.00
Cyclohexanone	15.00
Leveling agent	0.05

[a]BKS-2600, Union Carbide Corp.
[b]Silwet L-7602, Union Carbide Corp.
Source: Ref. 4.

films can also be used if coatings are applied in several layers to prevent blistering during cure.[3] Bake schedules generally range from 300–400 °F with times varying from 10 minutes to several hours. The coatings may be clear or pigmented with iron oxide. A typical formulation is listed in Table 1. Formulations may be applied by spray, brush, or roller coating.

The requirements of food can linings are quite different from those of drum coatings. Food can coatings must have sufficient flexibility to survive any impact that could fracture the coating and thus expose the metal surface to the contents of the can.[5] The phenolic resins are based on blends of phenol and cresols that offer varying degrees of flexibility and chemical resistance. Resins based on bisphenol A are used in critical taste applications. High molecular weight epoxies are used as modifiers to further enhance the flexibility of the cured films.

Another consideration for food can formulations is sulfur staining. Foods high in protein such as meats and fish contain sulfur, which may permeate to the substrate and cause discoloration.[5] Phenolics, because of their high degree of cross–linking, offer outstanding resistance to sulfur containing foods. A food container formulation is listed in Table 2.[6]

TABLE 2 Food Container Formulation

Ingredients	Parts by Weight
Phenolic resin[a]	22.90
Epoxy resin[b]	34.00
n-Butanol	11.30
Methyl Propasol acetate	11.30
85% Phosphoric acid	0.25
Leveling agent[c]	1.00

[a]Ucar BKS-7570, Union Carbide Corp.
[b]Epon 1007, Shell Chemical Co.
[c]Silicon resin SR 882M, General Electric.
Source: Ref. 6.

2.0 MAINTENANCE PRIMERS

Non–heat–reactive phenolics are used in combination with drying oils or alkyds to produce air drying maintenance coatings. The phenolic resins supply solvent and moisture resistance while the oils or alkyds impart flexibility and good filming characteristics.[7] In the past, parasubstituted phenolic novolak resins were actually cooked with unsaturated oils such as linseed or castor oil.[8] The mixtures were boiled at 400–500 °F until a thick resinous solution was produced. The reaction mechanism is not completely understood, but it is believed that quinone methylide structures in the phenolic resin react with the double bond in the oil through a Diels–Adler type of reaction.[9] The resin solution is then let down in mineral spirits or aromatic solvents along with metal driers such as cobalt or manganese. Usually a combination of driers gives the best overall performance.[10] The result is an air dry finish that has excellent weathering resistance. Today, it is increasingly popular to simply cold–blend phenolic resins[11] based on *p–ter–* butyl phenol along with solvents, driers, and highly reactive oils such as tung oil. Coating properties are excellent, and the procedure takes less energy and time.

3.0 PRINTING INKS

Phenolic resins are used in numerous printing in formulations. Lithographic, letterpress, and gravure inks are oleoresinous and have phenolic resins added to the ink systems to improve film hardness and gloss.[12] Parasubstituted phenolic novolak resins that have been modified with rosin make excellent ink additives. Rosin modification of the phenolic resin raises the melting point and gives the phenolic excellent oil solubility. The formulations consist of a phenolic resin, oils, pigments or dyes, driers, and lubricants or plasticizers.[13] The proper balance of ingredients gives the desired combination of hardness, viscosity, penetration, and drying rate.

4.0 EPOXY HARDENERS

Phenolic resins may be combined with epoxy resins for use in protective coatings. Phenolic–epoxy products are also used in laminates, prepreg manufacturing, molding materials, and electrical insulation coatings. The phenolic resin is used as a coreactant to produce thermoset systems with improved heat and chemical resistance. Non–heat–reactive (novolak) resins are used to cross–link the epoxies. The epoxy resins are typically epoxy–phenolics or bisphenol A–based resins,[14] The reaction mechanism is different from the resole–epoxy reactions. The coating is heat–activated and used a base catalyst such as an amine, dicyandiamide, or an imidazole.[15] The phenolic hydroxyl group reacts with the epoxy group to form a polyether structure,which has an advantage because no volatiles are released during cure. This allows for thick films to be produced with low shrinkage and no voids from volatile emission. The phenolic–epoxy reaction using a base catalyst can be demonstrated as follows:

For critical electrical applications, "high purity" resins are used. These products are made according to stringent specifications limiting the amount of water, ions, and free monomers present in the resin.

$$\left(\begin{array}{c} \text{OH} \quad \text{OH} \\ \text{CH}_2 \quad \text{CH}_2 \end{array} \right)_n + \begin{array}{c} \text{O} \qquad\qquad \text{O} \\ \text{H}_2\text{C}-\text{CH}\cdot\text{CH}_2\text{-O-R}-\text{OCH}_2\cdot\text{CH}-\text{CH}_2 \end{array}$$

$$\xrightarrow[\triangle]{\text{Base}}$$

$$\left[\begin{array}{l} \text{—OCH}_2\,\text{CH(OH)CH}_2\text{-O-R}-\text{OCH}_2\cdot\text{CH(OH)CH}_2\text{O—} \\ \text{CH}_2 \\ \\ \text{—OCH}_2\text{CH(OH)CH}_2\text{-O-R}-\text{OCH}_2\cdot\text{CH(OH)CH}_2\text{O—} \\ \text{CH}_2 \end{array} \right]_n$$

5.0 SUMMARY

Phenolic resins were one of the first synthetic polymers to have widespread commercial importance. Their outstanding performance properties have given them a permanent role in the coatings industry. They are used in applications ranging from railroad tank cars to carbonless copy paper. Phenolic resins are also used on a wide variety of substrates including metal, wood, paper, and ceramics. There are so many types of phenolic resin available to the market–place that a particular resin can be selected for virtually any application.

Phenolic resins should not be overlooked when choosing a high performance polymer. Phenolics currently do not receive the same attention as some of the more recently developed polymers, but for many applications there is simply no substitute for phenolic resins. Phenolics will continue to have an important role in the coatings industry because of their versatility, coatings properties, and reasonable price.

REFERENCES

1. J. S. Fry, C. N. Merriam, and W. H. Boyd, *Chemistry and Technology of Phenolic Resins and Coatings*, in ACS Symposium Series. *Applied Polymer Science*. Washington, DC: American Chemical Society, 1985, p. 1147.
2. R. T. Morrison, and R. N. Boyd, *Organic Chemistry*. 3rd ed. Boston: Allyn and Bacon, 1973, p. 1147.
3. J. S. Fry, C. S. Merriam, and W. H. Boyd, *Chemistry and Technology of Phenolic Resins and Coatings*, ACS Symposium Series, *Applied Polymer Science*. Washington, DC: American Chemical Society, 1985, p. 1149.
4. Union Carbide Corp., *Formulation Suggestions—Durable Phenolic Baking Coatings for Rigid Metal Substrates*. (F–60675), 1988.
5. A Knop, and L. Pilato, *Phenolic Resins*. Berlin: Springer–Verlag, 1985, p. 247.

6.   Union Carbide Corp., *Ucar Phenolic Resin BKS–7570*. (F–60689), 1988.
7.   R. W. Martin, *The Chemistry of Phenolic Resins*. New York: Wiley, 1956, p. 203.
8.   R. R. Myers, and J. S. Long, *Film Forming Compositions*. New York: Dekker, 1972, p. 155.
9.   R. W. Martin, *The Chemistry of Phenolic Resins*. New York: Wiley, 1956, p. 205.
10.   R. G. Middlemiss, *J. Water Borne Coatings*, November 1985.
11.   S. H. Richardson, *Paint Varnish Prod.*, August 1955.
12.   A. Knop, and W. Scheib, *Chemistry and Application of Phenolic Resins*. Berlin: SPringer–Verlag, 1979, p. 192.
13.   National Association of Printing Ink Manufacturers, *Pattern Printing Ink Formulae*. NAPIM, 1974.
14.   J. S. Fry, *Ucar Phenolic Resins for Epoxy Hardeners*. Union Carbide Corp. (P–)–3)57).
15.   U. S. Patent 3,493,630, Union Carbide Corp.

42

Olefinic Thermoplastic Elastomers

Jesse Edenbaum

Consultant
Cranston, Rhode Island

1.0 INTRODUCTION

Olefinic elastomers based upon ethylene–propylene and/or ethylene–propylene–diene monomer rubbers (EPR or EPDM) are thermoplastic by virtue of their alloying with isostatic crystalline polypropylene and/or high density polyethylene (HDPE). These products are produced by high intensity mixing in Banburies continuous mixers, and extruders.

The olefin plastics are either pellets or reactor beads, while the rubber can be in bale form for mixing in a Banbury. For mixing in an extruder or continuous mixer, the rubber must be converted to a pellet or granular particle. The high intensity mixing results in a simultaneous comminution of the polymers, with the olefin as the continuous phase and the rubber the dispersed phase. Thus the blend is thermoplastic: the blend viscosity is largely controlled by the choice of polyolefin, and the elasticity by the rubber segment of the blend.

Thermoplastic olefinic elastomers (TPOEs) can be manufactured by blending alone, which limits the elevated temperature properties of the mix, or by blending and cross–linking in situ during the compounding operation. When the compounds are cross–linked, the elevated temperature properties are enhanced. The processing "nerve" of the blend is reduced, and thinner, more complex extruded products are possible.

The polymer compound is made broadly versatile by the inclusion of a great variety of additives. In addition to the initial choice of polymers, the ratio of plastic to rubber (hard to soft segment) controls the hardness of the compound to some degree. The use of high permamence petroleum oils that function as permanent plasticizers assists in the control of hardness. Flexural modulus or toughness is more readily controlled by the rubber polymer. The combined use of these ingredients results in a wide variety of physical properties.

Fillers, such as fine particle calcium carbonates, clays, talc, and silicas are all usable. TPOE compounds cannot be made to be clear; but very pale, pastel colors are possible. Translucent colors are possible in thin sections. For outdoor use, protection against UV radiation is needed. The general–purpose compounds are not flame retardant inherently and require a package of halogen donor additives to pass any necessary specifications.

395

2.0 PROPERTIES

TPOE compounds are inherently of high elongation. They retain their elongation over a wide range of hardness, while the tensile properties vary with hardness; That is, the higher the hardness, the higher the tensile strength. Low temperature properties are excellent over a wide range of hardness. Brittle point is well under –60 °C. Permanent set tends to be fair, that is, it varies with temperature and increases at elevated temperature, which limits its usefulness under those environments. The cross–linked compounds, which have superior permanent set, form an exception. These compounds generally have excellent ozone resistance, are superb in electrical properties, and have no processibility problems. TPOE compounds, like their derivative compounds, are inherently heat stable and for general–purpose use require a small dose of antioxidant to protect the polymers during processing. When complex end products are specified, the protective package is custom designed into the finished compound.

2.1 Limitations of TPOE Compounds

Chemical and oil resistance of TPOE compounds is poor. The blended versions are low in compression set at elevated temperatures. Highly cross–linked versions are hygroscopic and must be dried before processing. Because of their high olefinic contents, these polymer blends are not readily decorated. Finishes of any kind require special surface preparation.

3.0 USAGE

TPOE compounds are widely used for the injection molding of automotive underchassis parts. Air dams, stone deflectors, and sight shields are some of the parts that are used throughout the world. These parts are colored by pigmentation. The inertness of the compounds allows the use of low cost pigments. Surface coating on an exposed product like sight shields requires priming or the use of a conductive compound before electrostatic paint can be applied.

Wire and cable compounds are typical of a broad use segment of extruded products. The low specific gravity, high dielectric properties, ease of processing, excellent low temperature properties, and good high temperature ratings (90 °C UL) allows use for battery booster cables, appliance wire, mining cable, and other primary wire and cable jacketing applications.

High moisture resistance and generally good outdoor use properties and temperature range recommend the use of these compounds for sporting goods such as swim fins, goggles, snorkeling tubes, and bike and other sports grips, as well as various industrial and automotive flaps, knobs, and handles, and many miscellaneous molded product applications.

Other parts that can be extruded include weatherstrip, flashing, tubing, and a variety of edge finishings. The inertness of these polymers allows contact with many other materials without bleed or transfer, but evaluation before use must always be performed.

3.1 Coating Applications

TPOEs do not lend themselves to coating use. These polymer compounds are, like most polyolefins, insoluble in most solvents, and therefore cannot be solution coated onto substrates. They can be coextruded, hardcore backbone with a flexible edge type of application. They can be hot melt applied as a coating where the properties described are desired. They will stick to most if not all other polyolefins. Thin films can be extruded and then laminated to other olefin plastics or onto paper, board, or cloth.

3.2 Primer Systems

For adhesion to TPOE products, flash primer systems such as Seibert Oxidermo or du Pont are recommended. Priming systems also are available from PPG and from many of the industrial paint producers. TPOE products can also be primed by activation of the surface using a benzophenone treatment followed by exposure to ultraviolet light. It is always necessary to evaluate each compound individually. Compounds vary widely, and caution must be used before a finish is accepted.

43

Ethylene Vinyl Alcohol Copolymer (EVOH) Resins

R. H. Foster

Eval Company of America
Lisle, Illinois

1.0 POLYMER

Ethylene vinyl alcohol (EVOH) resins are hydrolyzed copolymers of vinyl acetate and ethylene. The vinyl alcohol base has exceptionally high gas barrier properties, but it is water soluble and difficult to process. By copolymerizing ethylene with vinyl alcohol, the high gas barrier properties are retained and significant improvements are achieved in moisture resistance and processibility.

A typical reaction for producing EVOH resins is shown on page 398.
Under proper conditions this reaction yields a copolymer that is more than 99% hydrolyzed.

Table 1 lists the range of EVOH resins presently available.

2.0 BARRIER PROPERTIES

EVOH copolymers are highly crystalline, and their properties are highly dependent on the relative concentration of the comonomers. Generally speaking, as the ethylene content increases, the gas barrier properties decreases, the moisture barrier properties improve, and the resins are processed more easily (see Table 2).

The presence of a hydroxyl group in the molecular chain, renders the gas barrier properties of the EVOH resins sensitive to moisture. As the moisture content of the resin increases, the gas barrier properties decrease. However, by proper use of companion materials in a multilayer structure, this effect can be minimized.

Table 3 shows how the combination of different materials can be used to minimize the moisture content of the EVOH and maintain superior gas barrier properties. In this article, any aqueous–based food is considered to be 100% relative humidity; the storage environment is shown at relative humidities of 65 and 85%.

In addition to outstanding gas barrier properties, EVOH resins offer excellent barriers to a variety of flavors, aromas, and solvents (see Tables 4 and 5).

$$x\ \mathrm{CH_2=CH_2} \ +\ y\ \mathrm{CH_2=CH\!-\!O\!-\!CO\!-\!CH_3} \longrightarrow -(CH_2\!-\!CH_2)_x\!-\!(CH_2\!-\!CH)_y-$$

ethylene monomer   +   vinyl acetate monomer   ethylene vinyl acetate copolymer

$$(CH_2\!-\!CH_2)_x\!-\!(CH_2\!-\!CH)_y \ \xrightarrow[\text{catalyst}]{\text{heat}}\ (CH_2\!-\!CH_2)_x\!-\!(CH_2\!-\!CH)_Y$$

ethylene vinyl acetate copolymer   ethylene vinyl alcohol copolymer
                                                                   (EVOH)

Table 1 Properties of Ethylene Vinyl Alcohol Resins

Property	Range
Melt index	0.7–15.0 g/10 min
Density	1.12–1.21 g/cm^3
Melting point	156–191 °C
Crystalline temperature	134–167 °C
Ethylene content	27–48 mol %

Table 2 Barrier Properties

EVOH	Gases[a]				Moisture[b]
(Mol % ethylene)	O_2	CO_2	N_2	He	
27	0.01	0.04		7.7	6.8
32	0.02	0.05	0.002	15.5	3.8
38	0.03	0.10	0.003	25.5	2.1
44	0.07	0.20	0.005	40.0	1.4
48	0.11	.032	0.007	52.0	1.4

[a]In cubic centimeter-mils per 100 in.^{2} 24 hours per atm at 20°C and 65% relative humidity.
[b]In gram-mils per 100 in.^{2} 24 hours per atm 40°C and 90% relative humidities.

It is these exceptional barrier properties that permit the use of extremely thin layers of EVOH resins in coextrusion–coated structures for gable top drink cartons, single–serve drink cartons, dry mixes, etc.

In addition to excellent barrier properties, EVOH resins have very good mechanical and thermal properties (see Table 6).

3.0 REGULATORY APPROVAL

EVOH resins meet US Food and Drug Administration requirements for direct food contact as specified in The Code of Federal Regulations (21 C F R, Section 177.1360), issued September 21, 1982, for all temperatures through 100 °C. The use of EVOH resins as an indirect food contact layer also meets the FDA requirement for high temperature retort applications.

Table 7 lists packaging structures produced using various fabrication methods.

When using EVOH resins in multilayer structures, it is necessary to user an adhesive to gain adequate bonding strength to the other polymers or substrates. Commercially available adhesive resins such as Plexar, Admer, Modic, or Bynel are suitable for use with EVOH resins. EVOH resins are produced domestically by EVAL Company of America (EVAL resins) and E. I. DuPont (Selar OH resins).

4.0 Fabrication Methods

There are three basic methods of using EVOH to provide the barrier layer in multilayered structures. These are:

Coextruded structures in which EVOH resins are combined with polyolefins or polyamides to form the structure.
The use of EVOH films that are laminated to other substrates or coated with other materials.
The coating of various substrates or monolayer containers with EVOH resins.

Generally speaking, coextrusion or lamination is used for structures in which the EVOH layer must be protected from the effects of moisture. When packaging aqueous–based products, (tomato ketchup, barbeque sauce, etc.), various combinations of less costly polyolefins

Table 3 Oxygen Transmission (O_2TR) Rate Versus Barrier Layer Relative Humidity

Film structure[a]			Temperature (°C)	Inside package			
				Outside (65%RH)		Outside (85%RH)	
Outside: 0.8 mil	Barrier: 0.6 mil	Inside: 2 mils		RH of barrier (%)	O_2 TR^b (barrier %)	RH of barrier (%)	O_2TR^b (barrier %)
PP	27 mol % EVOH	LLDPE	5		0.03		0.05
				79		88	
			20		0.04		0.10
PP	27 mol % EVOH	PP	5		0.02		0.04
				75		86	
			20		0.03		0.07
PET	27 mol % EVOH	PP	5		0.02		0.04
				72		84	
			20		0.03		0.07
Nylon	32 mol % EVOH	LDPE	5		0.01		0.02
				67		81	
			20		0.03		0.05
PP	32 mol % EVOH	LDPE	5		0.05		0.05
				79		88	
			20		0.04		0.12
PS	44 mol % EVOH	LDPE	5		0.05		0.05
				70		83	
			20		0.17		0.27

[a] PP, polypropylene; PET, polyethylene terephthalate; PS, polystyrene.
[b] In cm^3 per 100 in^2 per 24 hours per atm.

are used to provide structural strength and to prevent excess moisture from reaching the EVOH barrier layer and a lowering of gas barrier properties.

Coating techniques can also be used to produce multilayered structures by the use of either multiple coatings or coextrusion coating. The resulting structure will be very similar to a coextruded structure.

Spray, dip, and/or roller coating of EVOH resins are used to produce containers for carbonated beverages or to establish a barrier to solvents, aromas, or odor. Manufacturers of such products include Kuraray Company Ltd. (Eval resins) and Nippon Gohsei (Soarnol resins) in Japan and Solvay (Clarene resins) in Europe.

Table 4 Flavor Barrier Data (Citrus and Tropical Flavors)[a]

Film[b]	Thickness (μm)	Duration					
		1Hr	2Hr	15Hrs	1Day	4Days	35 Days
EVOH–32	15	A	A	A	A	A	A
EVOH–44	15	A	A	A	A	A	A
BO EVOH	15	A	A	A	A	A	A
BO Nylon	15	A	A	B	B	B	B
OPP	20	B	B	C	C	C	C
PVDC	25	A	A	B	B	B	B
LDPE	50	D	D	D	D	D	D

[a]Key A, no detection; B, faint flavor; C, partial flavor; D, flavor clearly distinguished.
[b]BO EVOH, biaxially oriented EVOH; BO Nylon, biaxially oriented nylon; OPP, oriented polypropylene; PVDC, polyvinylidene chloride; LDPE, low density polyethylene.

Table 5 Flavor, Aroma, and Solvent Barrier Properties[a]

Resin[b]	Allyl sulfide: Garlic-food type (Croutons snacks, salad dressing)	Acetic acid Vinegar-food type (Cheddar cheese, snacks, condiments)	Ethylene Acetate: Laminating adhesive solvent residual	Toluene: Printing ink solvent residual	Methyl ethyl ketone Printing ink solvent residual
HDPE-Nylon-EVA	0.00008	0.92	0.03	0.02	0.005
HDPE-EVOH-EVA	0.00075	0.035	0.0043	0.007	0.035
PVDC-PP-PVDC	0.0068	1.98	0.34	0.22	3.09
OPP-HDPE-EVOH-EVA	0.00076	1.40	0.15	0.00003	0.09
EVA-Glassine-PVDC	0.50	4.18	6.47	3.15	15.1

[a]g/24 hrs, m2, 100 ppm at 70°F.
[b]HDPE, high density polyethylene; EVA, ethylene vinyl acetate; PVDC, polyvinylidene chloride; OPP, oriented polypropylene.

Table 6 Properties of Three Commercial Resins

	Eval EVOH Resin Grade		
Property	F101	H101	E105
Maximum service temperature °C	240	240	250
Melting point °C	181	175	164
Glass transition temperature °C	69	62	55
Tensile strength, psi	11,220	9,400	8,520
Tensile yield strength, psi	10,365	8,300	7,385
Elongation at break, %	230	280	280
Young's modulus, psi	385,000	345,000	298,000
Ethylene content, mol %	32	38	44

Table 7 Barrier Applications

Fabrication process	Application	Structure[a]
Lamination or extrusion coating	Food kits, pouches, dry mixes snacks	BON/ barrier/ LLDPE Nylon/ PET/ barrier/ LDPE OPP/ barrier/ LDPE Paper/ HDPE/ barrier/ EVA
Coextrusion coating	Aseptic packaging	LDPE/ Paper/ barrier/ PP/ Surlyn
	Juices and other drink products	LDPE/ Paper/ LDPE/ barrier LDPE/ Paper/ LDPE/ barrier LDPE/ Barrier

[a]BON, biaxially oriented nylon; LLDPE, linear low density polyethylene; PET, polyethylene terephthalate; LDPE, low density polyethylene; OPP, oriented polyethylene; HDPE, high density polypropylene; EVA, ethylene vinyl acetate; PP, polypropylene.

44
Elastomeric Alloy Thermoplastic Elastomers

Charles P. Rader

Monsanto Chemical Company

Akron, Ohio

A thermoplastic elastomer (TPE) is a material with the functional performance of a thermoset rubber, but the processibility of a conventional thermoplastic. Elastomeric alloys (EAs) are a generic class of TPEs composed of two or more polymer systems between which a synergistic interaction has arisen, giving rise to properties significantly better than those expected from a single blend of these polymer systems. EAs are medium performance, moderate cost TPEs.

Figure 1 compares the generic classes of TPEs by performance and by cost. Thus, EAs have higher performance and cost more than either the styrenic or olefinic blend TPEs whereas thermoplastic polyurethanes, copolyesters, and polyamides cost more and give higher performance than EAs.

1.0 PROPERTIES

Elastomeric alloys may have either two phases or a single phase. A two–phase EA is an alloy of vulcanized rubber with a thermoplastic polyolefin. The alloying of these two polymeric phases gives rise to higher ultimate tensile strength, improved retention of physical properties at elevated temperature, improved resistance to hydrocarbon fluids, lower compression set, and lower tension set. These properties thus qualify EAs for applications for which a simple rubber–polyolefin blend would be inadequate.

The need for a compatibilizer to stabilize intermingling of the rubber and polyolefin phases of a two–phase EA will be determined by their relative solubility parameters. Having essentially equal solubility parameters for the two phases eliminates the need for a compatibilizer whereas a significant difference between the solubility parameters requires a compatibilizer.

Single–phase EAs and are said to consist of an ethylene vinyl acetate–acrylate ester–polyolefin blend with significant plasticizer content. They may or may not contain carbon black.

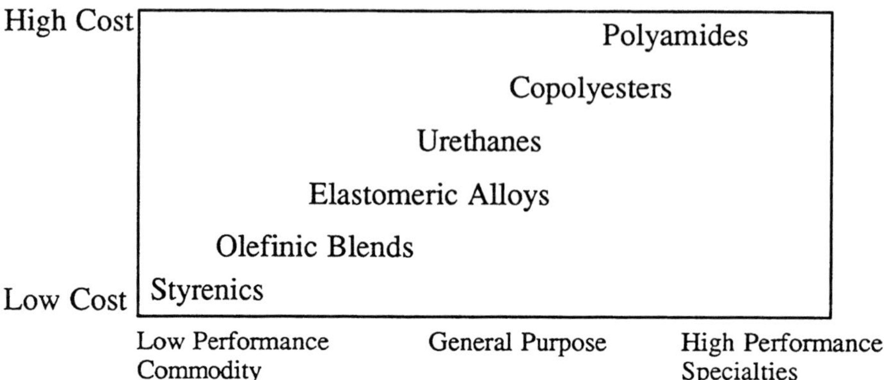

Figure 1   Comparison of performance and cost of the generic classes of thermoplastic elastomers.

Two–phase EAs range in hardness from 55 Shore A to 50 Shore D. EAs derived from ethylene propylene diene monomer (EPDM) rubber and polypropylene have a service temperature range from –60°C to 135°C in air. Two–phase EAs from nitrile rubber and polypropylene have a service temperature ranging from 40°C to 125°C in air. The specific gravity of single–phase EAs ranges from 1.2 to 1.3, and that of two–phase EAs from 0.9 to 1.0. Although the properties of an EA are quite competitive with those of a thermoset rubber, the ultimate tensile strength is generally significantly lower relative to a thermoset rubber of the same hardness. This difference is generally of little consequence in actual service, since very few rubber articles are used at an elongation anywhere close to the ultimate limit.

The fatigue resistance of two–phase EAs is truly outstanding, and the compression and tension set are competitive with those of thermoset rubber stocks specially compounded for these properties. This is an unusual combination of properties for a specific rubber composition.

Fluid resistance of EPDM rubber based two–phase EAs is quite broad. These materials are extremely resistant to water, aqueous solutions (including acids and bases), and a variety of polar organic fluids. In nonpolar, hydrocarbon fluids, these materials retain their physical properties quite well; however, they do experience significant swell. If low swell is a criterion for adequate service with an EA, the use of one based on nitrile rubber and polyolefin is recommended. These alloys give much lower swell in hydrocarbon fluids than those based on EPDM rubber.

2.0 PROCESSING

EAs capitalize on the numerous processing advantages of a TPE. For example, in addition to being fully compounded, EAs are simpler to process, have shorter processing times, permit recycling of scrap, have lower energy consumption, allow tighter, more economical quality control, and make possible the use of processing methods not available for thermoset rubber fabrication.

EAs require drying before processing, in contrast to thermoset rubbers. The equipment and methods used for fabricating thermoplastic are commonly used with EAs whereas thermoset rubber fabricating equipment is generally unsuitable. These materials are highly non–Newtonian in flow properties, being extremely sensitive to shear. Two–phase EAs are normally processed with a melt temperature in the 180–220°C range, and MPRs in the 160–190°C range.

Extrusion of an EA requires a thermoplastics extruder (L/D > 20 = 1) with the capability of achieving melt temperatures up to 220°C. These materials can readily be injection molded with the same equipment used for polyethylene and polypropylene, with cycle times in seconds, rather than the minutes required for a rubber vulcanization.

Blow molding, a method commonly used for molding hollow articles from thermoplastics, is unsuitable for molding thermoset rubbers. It is highly useful for molding a variety of hollow articles (e.g., convoluted bellows) from EAs. Thermoforming and heat welding are other fabrication methods, unsuitable for use with thermoset rubber, which have been found to be extremely useful for fabricating articles from EAs.

3.0 USES OF ELASTOMERIC ALLOYS

Introduced commercially in 1981, EAs have experienced spectacular growth during the 1980s. In 1987, their use exceeded 10,000 metric tons worldwide. By the mid 1990s, it is projected that their applications will reach 50,000 metric tons per year.

These materials have been especially well received by the automotive industry, with typical uses being hoses, jacketing, grommets, seals, convoluted boots, and weatherstripping. They are prime candidates for under–the–hood uses, where the temperatures are progressively increasing.

EAs function nicely as a material for window glazing and expansion joints in architectural applications. More than 50 major North American buildings have utilized the unique properties of EAs. Mechanical rubber goods include those featuring a rubber article as a component part of a useful assembly. Typical commercial uses of EAs in this area include household appliances, office equipment, toys and other articles requiring the use of seals, boots, tubing, and bushings, and other rubber articles produced by extrusion, injection molding, or blow molding.

EAs from EPDM rubber and polypropylene have excellent electrical properties—dielectric strength, resistivity, power factor, dielectric constant—which render them highly suitable for use as a primary electrical insulator, as well as a jacketing material. Electrically conducting wire can be coated by crosshead extrusion of an EA and used in automotive, appliance, construction, and many other applications.

The low toxicity of two–phase EAs recommends them highly for applications involving direct contact with foods and potable water. These materials also offer much promise for medical applications embracing direct contact with pharmaceutical preparations to be taken orally or injected into the bloodstream.

45

Polyvinyl Chloride and Its Copolymers in Plastisol Coatings

Jesse Edenbaum

Consultant
Cranston, Rhode Island

1.0 INTRODUCTION

PVC plastisols are liquid dispersion systems of polyvinyl chloride and/or PVC copolymer resins in compatible plasticizers. The liquids vary in viscosity from thin milklike fluids to heavy pastes have the consistency of molasses.

The lowest viscosity products are generally used for spray coatings and some paper and fabric coatings, while the higher viscosity products are often utilized in dipping, slush molding, rotocasting, and other more specialized procedures.

2.0 FORMULATION

Viscosity of plastisol is controlled by formulation techniques, and it is often kept low in by the addition of inactive diluents such as odorless mineral spirits. If more than minor amounts of diluents are used, the product is often referred to as an organosol.

These products share a common compounding technology. The primary components are the dispersion–grade resin, plasticizers, PVC stabilizers (which are common to all PVC), and assorted fillers, pigments, and a wide variety of additives to control properties of the product in storage, during processing, and in the finished state. A typical plastisol formulation is shown in Table 1.

2.1 Resins

PVC dispersion resins are very fine particle size products made by emulsion polymerization and finished by the spray drying technique. These paste resins are characterized by their molecular weight, particle size, and shape. They are predominantly homopolymers, but there are also a wide variety of copolymers made with polyvinyl acetate as the comonomer. The comonomer content will normally vary from 3 to 10%. Other comonomers are sometimes used.

Table 1　Formulation for a Typical Plastisol

Component	Phr level range[a]
PVC primary dispersion resin	60–100
PVC blending resin	0–40
Plasticizers	
Primary, secondary, stabilizing	30–120
Fillers	0–100
Pigments (color)	0.1–5
Stabilizers	0.5–5
Others	As needed

[a]Phr (Parts per hundred resin) is the normal manner of expressing PVC formulations.

High molecular weight resins give products with high physical properties and good viscosity stability, depending on the plasticizer used. The particle size of the resins is generally between 0.5 and 2.0 μm. Particle size and shape control the surface area, and a highly irregular surface area allows for faster absorption and more rapid viscosity increase. Copolymer resins speed fusion and assist adhesion. The molecular weight of the resins relates directly to the speed of fusion and physical property development. The higher the molecular weight, the slower the fusion and the longer the time and/or the higher the temperature needed for maximum physical properties.

Blending resins are made by the emulsion process, by modified emulsion, and by standard suspension procedures, some are made by the mass polymerization method. Blending resins are characterized by larger particle size and very low plasticizer absorption. They are used to reduce cost, decrease plastisol viscosity, and aid in the release of air. They help control the finish and are used where a mat surface is desirable. They are usually lower in molecular weight than the primary resins, and one disadvantage is their tendency to settle out if the plastisol viscosity is too low.

2.2　Plasticizers

Plasticizers are liquids that provide mobility to the plastisol system. They are of primary significance, and they are selected first when formulating. Plasticizers differ in the permanence characteristics they impart to the finished product. The blend of plasticizers also will assist in the control of viscosity and its stability, and in the fusion characteristics of the finished plastisol. The plasticizers are the same as used in dry (pellet or dry blend) compounding. The finished compounds are the same with regard to permanence, weathering, and chemical and electrical properties.

Plasticizers are high boiling, chemically stable, organic esters. They are derivatives of phthalic anhydride and adipic, sebacic, trimellitic, or phosphoric acids. They are esters prepared in condensation reactions with alcohols of chain lengths from hexyl through tridecyl and mixtures thereof.

Initially, the viscosity of the plasticizer strongly correlates with the viscosity of the plastisol. Upon aging, however, the viscosity increases depending on the resins chosen and the activity or solvency of the plasticizer. The higher the plasticizer/resin ratio, the lower is the rate of viscosity rise.

These ester plasticizers can be rated according to their activity in the following manner. Phthalate esters are the fastest fusing and solvating. The lower the alcohol chain, the more rapid is the fusion and the poorer the viscosity stability. Straight chain or Alfol alcohol esters have greater viscosity stability at the same carbon chain length. Thus, normal octyl phthalate will provide greater viscosity stability than dioctyl phthalate (DOP), the di–2–ethylhexyl phthalate. Fusion time and temperature comparisons are much closer and virtually indistinguishable.

Adipic acid esters follow the same alcohol chain length rules as do all other plasticizers. Aromaticity plays a significant role in fusion and viscosity stability. Comparing butylbenzyl phthalate (BBP) to DOP and diisodioctyl phthalate (DIDP) results in the following order: BBP fuses faster than DOP, which is faster than DIDP. The higher the aromaticity, the poorer is the viscosity stability.

Polymeric plasticizers based on linear dibasic acids such as adipic, sebacic or azelaic acids and esterified with dihydric glycols, such as propylene and butylene, of varying molecular weight, serve to supply high performance and oil and solvent resistance to plastisols. Care must be taken in formulation, since these plasticizers are normally of medium to very high viscosity.

The epoxy types of plasticizer such as epoxidized soybean oil (EPO), and butyl epoxy tallate are widely used for stabilizer boosters. The tallate provides better viscosity control than the EPO. The EPO, however, is nontoxic.

Secondary plasticizers of petroleum oil derivation and other origins are utilized to control viscosity and viscosity stability, and to provide special effects such as harder exposed surfaces by allowing for fugitive plasticizers in top coats. Such volatility can also be provided by solvents. These pastes are those referred to as organisols.

2.3 Solvents

Solvents and diluents can be the aliphatic mineral spirits previously mentioned. They lower viscosity and extend shelf life by retarding the viscosity increase due to the solvation of the resin, but sometimes these agents may reduce viscosity to a point at which settling can occur. Blends of diluents and secondary plasticizers are advantageous for control of those conditions.

2.4 Other Additives

Stabilizers must always be liquids or pastes when used in plastisols. They are barium, cadmium, or zinc containing products. For nontoxic products, calcium zinc soap pastes are used. Leads are rarely used, and octyl tins are used for some special applications.

Fillers are non–oil–absorbing inert minerals such as calcium carbonate, talc, and some clays. Fillers are used to lower cost by extension of the total formulation. They are usually very fine particle products. They also can add dullness to the surface finish, import a dry hand, reduce tackiness, and be used to control the viscosity. Fillers must be capable of being readily deagglomerated by the shear exerted in the plastisol preparation process.

Pigments are used to color plastisols and organosols, and any pigment suitable for general PVC use is usually satisfactory for paste technology. However, pigments must also be added as dispersions. They must be dispersed in plasticizer carriers on three–roll mills, sand mills, or at least, using a high speed disperser, such as a Cowles dissolver, where the high shear at the blade tip will deagglomerate and disperse the pigment.

Other compounding ingredients used in plastisols are of lower volume usage but serve to expand the versatility of the products. Surfactants are often added to control the viscosity and to help the release of air bubbles during coating procedures. They also assist in fabric penetration, where this is required. They are mostly nontoxic surfactants derived from polyethylene glycol.

Thixotropic agents control strike–through and fabric penetration, as well as viscosity in various processes. They alter the flow for spray applications and dipping by increasing low shear viscosity and allowing high flow under high shear. Lubricants are added to function as release agents during casting on paper and slush molding. They also control surface tack of cast films.

Blowing agents are added (as pastes) in upholstery and flooring applications. The blowing agents are catalyzed by the use of some of the stabilizers that contain zinc. The most widely used blowing agent is azodicarbonamide. This is produced in various particle sizes, which help control the blowing temperature. Density of the finished product is controlled by using combinations of blowing agent level, varying the particle size of the blowing agent and then blowing agent catalyst, and compound gelation control by careful selection of the plasticizers and resins. This procedure results in closed–cell foams. Open–cell foams are produced when the blowing occurs at the gelation temperature.

In addition, UV and weathering protection agents, flame retardants, and fungicides are all useful where required.

3.0 PLASTISOL MANUFACTURING PROCEDURES

The basic mixing procedure for plastisols can be performed in various forms of equipment. The procedures are generally alike regardless of the size or type of equipment used. In the typical process procedure that follows, all nonliquid ingredients must be added as pastes dispersions except for the bulk fillers and resins. Care must be exercised to rinse containers with some of the weighed, withheld plasticizer.

1. Add three–quarters of the total liquids to the mixer.
2. Start the agitator.
3. Add the solids in the following order:
    Fillers
    Blending resins
    Dispersion resins and other dry ingredients
    The dry powders should be added slowly onto the surface, allowing them to be pulled into the vortex.
4. The temperature of the mixture should not be allowed to rise above 90°F.
5. The mixture must go through a high shear stage to deagglomerate all clumps. When a smooth creamy state has been reached, the balance of the paste materials is added and, lastly, the balance of the liquids. Some additional mixing maybe necessary.
6. Deaeration is usually required, except for some foam applications. Total time consumed depends on formula, volume, and equipment.

3.1 Equipment

Simple mixers, from a 55 gallon drum, high speed mixer with an intensive mixing blade, to 1000 gallon tank mixing equipment with multiple variable speed mixers with built–in deaeration systems are used for plastisol preparation

3.2 Quality Control

Quality control tests have to be performed on the liquid systems as well as the fused products. Tests on the paste are unique to liquid systems but controls of the fluxed PVC product are the same as those performed on other solid polymers. Table 2 lists the tests typically performed on liquid plastisols and organosols.

Tests are performed as necessary to control the manufacturing processes as well as the product. Finished product specifications are agreed upon between the producer and the customer. Physical properties of the finished films or coatings can be tensile strength, elongation, hardness, low temperature flexibility, high temperature performance, coating and/or processing speed, yield, color, finish, and many more. These liquid systems are complex due to their conversion from a liquid state to a solid during the manufacturing process.

3.3 Coating Application

Plastisols and organosols are widely used as coatings. They can be applied by many procedures. Heavy–walled products such as dip–coated tool handles, dishwasher racks, plating racks, and decorative or protective coatings for work and garden gloves can be produced by hot and cold dipping. In the hot dip process, a permanent coating is made by priming the metal to be coated and then preheating the product before dipping it into the plastisol. Viscosity control and pseudoplasticity of the dipping plastisol allow the application of a single heavy coat, which is then oven fluxed, resulting in a permanent coated product. Dishwasher racks and plating racks are made this way.

Gloves are cold dipped and the weight of the coating and penetration are controlled by formulation and viscosity. Thixotropic plastisols control the weight of the coatings. Gelation and fluxing times and temperatures are controlled by the resins and plasticizers selected.

Specialty products such as certain catherers, and odd–shaped tubular products can be hot dipped into plastisol by preheating a dipping form, dipping repeatedly as needed, and then fluxing, cooling in water, stripping the product from the dipping form, and repeating the process, reusing the form.

In slush molding, conveyorized molds are heated and filled with plastisol. After a controlled time span, the plastisol is poured from the molds into a recirculating supply tank. The molds have a gelled coating on the inside walls. They go through an oven to complete the fluxing or fusion and are water cooled; then the product (a rain boot typically) is stripped from the mold, trimmed, inspected, and packed. The mold is cleaned and the cycle repeated.

Encapsulation coating is done on small electrical parts, metal products, and other applicable items. Thin to heavy coatings are available. Products can be primed if good adhesion is required. Products are cold dipped, and the fluxing is performed according to the coating thickness.

Spray coating is used for products of irregular shape. Automobile kick panels are plastisol spray coated. Again, priming is performed when needed. The product can be preheated or not. Spray plastisols are often diluted with solvent. Viscosity controls sag during spraying and gelation. Other applications are metal outdoor furniture, appliances, and many protective coating applications.

3.4 Continuous Thin Film Applications

The largest use of plastisols and organosols in coatings is for fabric and other porous substrate coatings and impregnation. There are myriad applications for clothing, such as raincoats, pocketbooks, shoe fabrics, and gloves. Coated fabrics serve as leather replacements for luggage and upholstery for automobiles and homes, and for wall covering with and without textile backing. Industrial protective coverings such as awnings are made from PVC paste coated fabrics in many different weights.

There are many applicable procedures fro coating these materials onto flexible substrate. In knife coating, for example, a liquid is dispensed onto a moving material web, and the clearance between the knife edge and the web controls the application thickness. Knife coating equipment has many variations that contribute to the versatility of the process.

Direct roll coating, three–roll, nip–fed reverse roll coaters, and variations on these are also used. The web can be embossed and also laminated in line. Plastisols perform as lamination adhesives using these procedures. Plastisols can also be used to totally impregnate heavy webs by the dip saturation procedure.

Almost all the products described can be in–line embossed, or further decorated by printing, flocking, and laminating. Most of the products described are fused or fluxed to their final physical properties in line, except those that are to be laminated or assembled into a multilayered product.

There are many film products made by spread coating of these pastes. Adhesive–coated films for decorative uses, decals, wall covering, mat casting, and floor coverings can be manufactured this way. Again, these products can be printed, embossed, decorated further, or laminated.

In every instance the properties of the finished product are controlled by the plasticizer system used. The application technique is further controlled by the viscosity characteristics, which are, in turn, controlled by the resins selected, and the balance of the formulation enhances the manufacturing process efficiency.

46

Polyvinyl Acetal Resins

Thomas P. Blomstrom

Monsanto Chemical Company
Springfield, Massachusetts

1.0   CHEMISTRY AND MANUFACTURE

The polyvinyl acetals are a family of high molecular weight polymers prepared by the mineral acid catalyzed acetalization of polyvinyl alcohol. The chemistry of preparation is outlined in Equation 1, where R can be H or any of a wide variety of organic radicals.

$$\sim\!\!\!CH_2\text{---}CH\text{---}CH_2\text{---}CH\!\!\!\sim + RCHO \xrightarrow{H^+} \sim\!\!\!CH_2\text{---}CH\text{---}CH_2\text{---}CH\!\!\!\sim + H_2O$$

$$\underset{OH}{|} \qquad \underset{OH}{|} \qquad\qquad\qquad \underset{O}{|} \qquad \underset{O}{|}$$

$$CH$$

$$|$$

$$R$$

$$(1)$$

While many polyvinyl acetals have been prepared,[1,2] the only commercially important polymers are the formal (R = H) and the butyral (R = n-C^3 H^7), derived from the acetalization of polyvinyl alcohols with formaldehyde and n-butyraldehyde, respectively. The generalized structure of the polyvinyl acetal molecule is shown in Figure 1. Molecular weights, as weight average molecular weight (Mw), range from about 25,000– 150,000 for the formal and from 40,000 to 250,000 K for the butyral, depending on grade. The three chemical moieties are randomly distributed along the polymer chain. While simple in concept, the actual commercial preparation of these polymers is quite complex, requiring many processing steps. For the formal and the majority of the butyral resins currently of importance in the coating industry, these steps include extensive solvent and by-product purification and recovery operations. A schematic process flow diagram for the manufacture of a typical polyvinyl acetal, polyvinyl butyral (PVB), is shown in Figure 2.

417

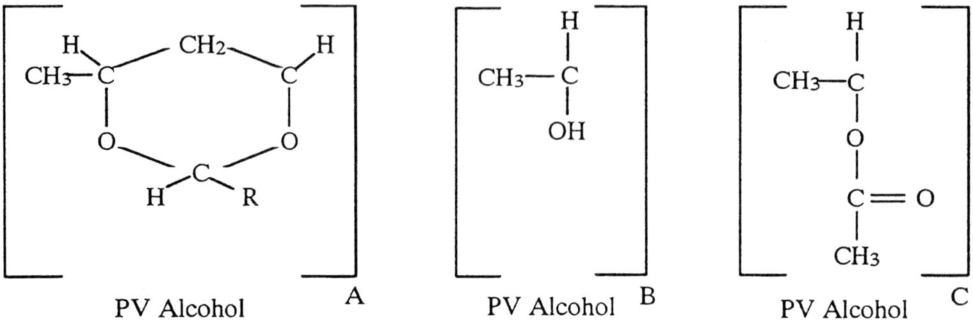

Figure 1   Generalized structure of the polyvinyl acetyl molecule.

2.0   AVAILABILITY, ECONOMICS

Monsanto is the only U.S. producer of polyvinyl formal, under the brand name of Formvar. The resins are available in the United States as white to pale buff-colored, free-flowing powders, in fiber drums at about 150 lb net, costing \$5.60–\$6.80 per kilogram (10 April 88) depending on grade and quantity. Outside the United States polyvinyl formal is supplied by Shin Nippon Chisso in Japan as Vinilex F, by Wacker Chemie in Germany as Pioloform F, and Siva in Italy.

Monsanto, which produces a full line of PVB's has production facilities both in the United States and in Europe. In addition to the grade used exclusively for Monsanto's safety glass interlayer sheet, Saflex, a wide range of specialty grades known as Butvars are available for a variety of coatings applications. Also in the United States, Union Carbide supplies a limited range of coatings grade polyvinyl butyral known as Vinyl Butyral Resins XYHL and XYSG. Du Pont produces the resin for captive use in its safety glass interlayer sheet, Butacite. In Europe, PVB is available from Monsanto, Hoechst, Wacker Chemie, Dynamit Nobel, and Rhône Poulenc. In Japan, a range of grades is produced by Sekisui Chemical Company, while Chang Chun produces the resin in Taiwan. In the United States, PVB resins are available as white, free-flowing powders at 135–155 lb net in fiber drums at about \$6.60–\$9.00 per kilogram, depending on grade and quantity.

Monsanto also produces an aqueous dispersion of a plasticized high molecular weight polyvinyl butyral known as Butvar Dispersion BR. This material air dries to a tough, transparent film exhibiting low to moderate adhesion to most surfaces; thus it is useful for, among other applications, removable protective coatings and textile sizing. This material is available in steel drums at about \$4.45–4.70 per kilogram (wet), 50–52% solids basis.

Worldwide production of polyvinyl formal resin is estimated at 3500–4500 metric tons annually. Approximately, 45,000–50,000 metric tons of polyvinyl butyral resin is produced in the free world each year, the majority going into the production of safety glass interlayer. Production figures for polyvinyl butyral dispersion are not available.

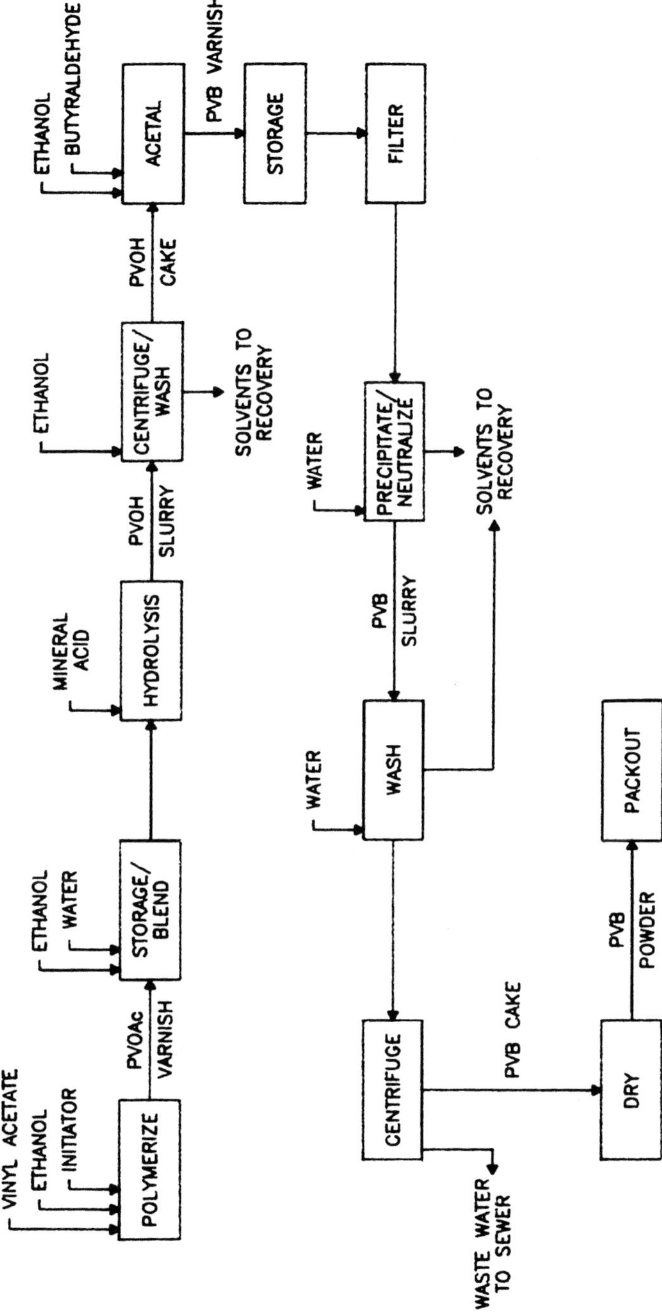

FIGURE 2 Process flow diagram of a typical polyvinyl acetyl, polyvinyl butyral molecule.

3.0 PROPERTIES

3.1 Reactivity and Compatibility

In general, the polyvinyl acetals are characterized by toughness and excellent adhesion to a wide variety of surfaces. They are resistant to most nonpolar solvents and to attach by both acids and bases, although they are slowly attacked by strong aqueous acids. Their resistance to hydrocarbons and mineral oils is outstanding. Because of the residual hydroxyl groups in the polymers (Fig. 1), they are readily cross-linked with a variety of widely available cross-linking reagents that react with hydroxyl. Examples of the more common cross-linking reagents are dialdehydes and phenolic, amino, isocyanate, and epoxy functional resins. The butyral shows limited compatibility with other resins (e.g., alkyd, polyvinyl chloride, silicones, urea– and melamine–formaldehyde, cellulose acetate butyrate, and ethyl cellulose) and excellent compatibility with nitrocellulose, epoxy, phenol–formaldehyde, isocyanate, and some rosin ester derivatives. The formal, being more polar than the butyral, shows more limited compatibility with other resins. It is fully compatible with most common isocyanate and epoxy resins, and shows limited compatibility with alkyd, phenolic, melamine– and urea–formaldehyde resins, and silicones.

3.2 Physical and Chemical Properties

The physical and chemical properties of Butvar resins are shown in Table 1, and the mechanical, thermal, and electrical properties are given in Table 2. Table 3 gives solubility data for the Butvar resins in a variety of solvents. The physical and chemical properties of the Formvar resins are shown in Table 4; the mechanical, thermal, and electrical properties are listed in Table 5. Solubility data for the Formvar resins are given in Table 6 [3]. Properties of Butvar Dispersion BR are shown in Table 7 [4]. The data given in Tables 1–7 are typical properties, not the manufacturer's specifications. For actual product specifications, the manufacturer should be consulted. The polyvinyl formal grades are all chemically identical; they differ only in molecular weight. The polyvinyl butyrals fall into two chemical groups, high (17.5–21 wt %) residual polyvinyl alcohol, and low (10.5–13 wt %) polyvinyl alcohol. Within each of these groups, molecular weight is also varied. In general, as the percentage of polyvinyl alcohol increases, toughness, modulus, and adhesion to polar surfaces will increase, while the range of useful solvents decreases (conversely, solvent resistance increases). As the percentage of polyvinyl alcohol decreases, the resins become more broadly soluble, softer, and lower in glass transition temperature.

Within a given chemical group, increasing molecular weight results in a modest increase in tensile strength and modulus, increased solution viscosity, and a modest increase in solvent and chemical resistance. In general, even the lowest molecular weight grades of these resins are high enough in molecular weight to be well within the chain entanglement region of the E' spectrum such that physical properties do not change drastically as molecular weight increases. Molecular weight variations can be used effectively to control the rheology/viscosity of coatings formulations more than the properties of the final coating.

3.3 Solution Viscosity

Solution viscosity is a function of chemical composition, molecular weight, and solvent composition. Within a given chemical class, solution viscosity in a given solvent will increase with molecular weight. The effect of solvent composition on solution viscosity is more complex and not within the scope of this discussion. Solvent systems are often chosen for a variety of reasons other than viscosity: drying rate, toxicity and other environmental concerns, utility with other resins needed in a formulation, etc. Usually by judicious experimentation, a compromise solvent system can be found, that will give a reasonable mix of the required properties.

3.4 Plasticizers

Plasticizers are often used to soften the films prepared from the polyvinyl acetals. Among the suggested plasticizers for polyvinyl butyral are butyl benzyl phthalate, 2-ethylhexyl diphenyl phosphate, dihexyl adipate, and a variety of other phosphate, phthalate, adipate, sebacate, and ricinoleate esters. Rosin-based polyester, and blown linseed oil plasticizers are also used.

Diethyldiphenyl and dicyclohexyl phthalates, butyl benzyl phthalate and phosphate esters including 2-ethylhexyl diphenyl phosphate polyester, chlorinated naphthalenic, and adipate diesters are useful for polyvinyl formal. By proper choice of plasticizer type and level, the physical–mechanical, chemical, and adhesion properties of these resins can be tailored for a wide variety of applications. Fitzhugh and Crozier [5] have studied the effect of a number of plasticizers on the mechanical properties of polyvinyl acetal resins.

3.5 Toxicology

Butvar resins are regulated by the U.S. Food and Drug Administration under the Code of Federal Regulations, as indirect food additives. Butvar resins have also been subjected to acute toxicity studies on laboratory animals. Subject to the appropriate regulations, they are useful in a number of packaging applications for both fatty and aqueous foods.

Both Butvar and Formvar resins have flash points in excess of 370°C. The lower explosive limit (LEL) for Butvar dust in air is 20 g/m^3. Although these materials are considered to be nontoxic in ordinary everyday handling, good industrial hygienic practices should be observed when using them.

4.0 SURFACE COATING APPLICATIONS

4.1 Polyvinyl Butyral [2, 8]

The inherent properties of adhesion to a wide variety of surfaces, film toughness and chemical/solvent resistance, and film clarity of the polyvinyl butyral resins makes them the vehicle of choice in a wide variety of specialty coating applications. They adhere tenaciously to most polar surfaces: wood, glass, metals, ceramics, pigments, etc. Their high binding efficiency allows them to be used at very high pigment loadings. Ceramic films are typically coast in thicknesses from fractions of a mil to several millimeters, using 4–5% Butvar B–76 as the sold binder [6]; most permanent coatings applications will require considerably higher binder levels than this for adequate film strength. Polyvinyl butyral resins have been used as the binder of choice in thermophotographic and photographic coatings, [7] in photoconductive coatings, in electrophotographic coatings, and as a coating on optical recording disks, all of which depend on film toughness, binder efficiency, and optical clarity. Their toughness, chemical and heat resistance, and binding efficiency render them useful in powder coatings and photothermographic coating applications.

Polyvinyl butyrals are used extensively as wood coatings, where their resistance to natural wood oils makes them a primary choice for sealers and wash coats. An example of an application on the Western Pine Association Knot Sealer number WP578 is given in Table 8. A well-known application of polyvinyl butyrals is in the manufacture of wash primers for

TABLE 1 Typical Physical and Chemical Properties of Butvar Polyvinyl Butyral Resin

Properties	ATSM method[a]	Butvar resins						
		B–72	B–74	B–73	B–76	B–79	B–90	B–98
Physical Form				White, free-flowing powder				
Volatiles, maximum as packed,%		3.0	3.0	3.0	5.0	5.0	5.0	5.0
Molecular weight (weight average x 10^{-3})	(1)	170–250	120–150	90–120	90–120	50–80	70–100	40–70
Solution viscosity 15% by wt, mPa·s	(2)	7000–14000	3000–7000	1000–4000	500–1000	100–400	600–1200	200–400
10% by wt, mPa·s	(2)	1600–2500	800–1300	400–1000	200–450	75–200	200–400	75–200
Specific gravity 23°/23° (± 0.002)	D792–50	1.100	1.100	1.100	1.083	1.083	1.100	1.100
Burning rate, cm/min	D635–56T	2.5	2.5	2.5	2.5	2.3	0.9	2.0
Refractive index (± 0.0005)	D542–50	1.490	1.490	1.490	1.485	1.485	1.490	1.490
Water absorption (24 h), %	D570–59aT	0.5	0.5	0.5	0.3	0.3	0.5	0.5
Hydroxyl content, as % polyvinyl alcohol	D1396–58	17.5–20.0	17.5–20.0	17.5–20.0	11.0–13.0	10.5–13.0	18.0–20.0	18.0–20.0

Acetate content, as % polyvinyl acetate	D1396–58	0–2.5	0–2.5	0–2.5	0–1.5	0–1.5	0–1.5
	0–2.5						
Butyral content, as % polyvinyl butyral, approx	80	80	80	80	80	80	80
Chemical[b]							
Resistance to							
weak acids	D543–56T	E	E	E	E	E	E
	E						
strong acids	D543–56T	E	E	E	E	E	E
	E						
weak bases	D543–56T	E	E	E	E	E	E
	E						
strong bases	D543–56T	E	E	E	E	E	E
	E						
Organic solvents							
alcohols	D543–56T	P	P	P	P	P	P
	P						
chlorinated	D543–56T	G	G	G	F	F	G
	G						
aliphatic	D543–56T	E	E	E	F	F	G
	G						
aromatic	D543–56T	F	F	F	P	P	F
	F						
esters	D543–56T	F	F	F	P	P	F
	F						
ketones	D543–56T	F	F	F	P	P	F
	F						

[a]All properties were determined by ASTM methods except:

(1) Molecular weight was determined via SEC/LALLS in tetrahydrofuran.

(2) Solution viscosity was determined in 15% solutions in 60:40 toluene/ethanol at 25°C, using a Brookfield viscometer. Also in 10% solution in 95% ethanol at 25°C. using an Ostwald viscometer.

[b]Key: E, excellent; G, good; F, fair; P, poor

Source: Ref. 3.

TABLE 2 Typical Mechanical, Thermal, and Electrical Properties of Butvar Polyvinyl Butyral Resin

Properties	ASTM method	Butvar resins						
		B–72	B–74	B–73	B–76	B–79	B–90	B–98
Mechanical[a]								
Tensile strength, MPa								
yield	D638–58T	47–54	47–54	45–52	40–47	40–47	43–50	43–50
break	D638–58T	48–55	48–55	41–48	32–39	32–39	49–46	39–46
Elongation, %								
yield	D638–58T	8	8	8	8	8	8	8
break	D638–58T	70	75	80	110	110	100	110
Modulus of elasticity								
(apparent), GPa	D638–58T	2.3–2.4	2.3–2.4	2.2–2.3	1.9–2.0	1.9–2.0	2.1–2.2	2.1–2.2
Flexural strength, yield, MPa	D790–59T	83–90	83–90	79–86	72–79	72–79	76–83	76–83
Hardness, Rockwell								
M	D785–51	115	115	115	100	100	115	110
E	D785–51	20	20	20	5	5	20	20
Notched Izod Impact strength								
(1.25 cm x 1.25 cm), J/m	D256–56	59	59	55	43	43	48	37
Thermal (°C)								
Flow temperature,	D569–59	145–155	135–145	125–130	110–115	110–115	125–130	105–110

Glass temperature[a]		72–78	72–78	72–78	62–72	62–72	72–78	72–78
Heat distortion temperature	D648–56	56–60	56–60	55–59	50–54	50–54	52–56	45–55
Heat-sealing temperature[b]		105	105	99	93	93	205	200
Electrical								
Dielectric constant								
50 Hz	D150–59T	3.2	3.2	3.0	2.7	2.7	3.2	3.3
1 kHz	D150–59T	3.0	3.0	2.7	2.6	2.6	3.0	3.0
1 MHz	D150–59T	2.8	2.8	2.6	2.6	2.6	2.8	2.8
10 MHz	D150–59T	2.7	2.7	2.5	2.5	2.5	2.7	2.8
Dissipation factor × 10^3								
50 Hz	D150–59T	6.4	6.4	5.8	5.0	5.0	6.6	6.4
1 kHz	D150–59T	6.2	6.2	5.5	3.9	3.9	5.9	6.1
1 MHz	D150–59T	27	27	22	13	13	22	23
10 MHz	D150–59T	31	31	22	15	15	23	24
Dielectric strength, V/m (3.2 mm thickness)								
short time	D149–59	17	17	19	19	19	18	16
step by step	D149–59	16	16	16	15	15	15	15

[a]Glass temperature (T_g) was determined by differential scanning calorimeter.

[b]Heat-sealing temperature was determined on a 25 m dried film on paper, cast from a 10% solution in 60:40 toluene/ethanol. A dwell time of 1.5 seconds at 414 kPa (60psi) line pressure was used on the heat sealer.

Source: Ref. 3.

TABLE 3 Solubility of Butvar Resins[a]

Solvent	Butvar solutions agitated at room temperature for 24 h		
	5% Solid Solution	10% Solids Solution	
	B–72, B–73, B–74	B–76, B–79,	B–90, B–98
Acetic acid (glacial)	S	S	S
Acetone	I	S	SW
Butyl acetate	I	S	PS
n–Butyl alcohol	S	S	S
Butyl Cellusolve	S	S	S
Cyclohexanone	S	S	S
Diacetone alcohol	PS	S	S
Diisobutyl ketone	I	SW	I
N, N–Dimethylacetamide	S	S	S
N, N–Dimethylformamide	S	S	S
Dimethyl sulfoxide	S	S	S
Ethyl acetate, 99%	I	S	PS
Ethyl acetate, 85%	S	S	S
Ethyl alcohol, 95%, or anhydrous	S	S	S
Ethylene dichloride	SW	S	SW
Ethylene glycol	I	I	I
Isophorone	PS	S	S
Isopropyl alcohol, 95%, or anhydrous	S	S	S
Isopropyl acetate	I	S	I
Methyl acetate	I	S	PS
Methyl alcohol	S	SW	S
Methyl ethyl ketone	SW	S	PS
Methylene chloride	PS	S	S
Methyl isobutyl ketone	I	S	I
Naphtha (light solvent)	I	SW	I
N–Methyl–2–pyrrolidone	S	S	S
Propylene dichloride	SW	S	SW
Tetrachlorethylene	SW	SW	SW
Tetrahydrofuran	S	S	S
Toluene	I	PS	SW
Toluene/ethyl alcohol (95%) (60:40 by weight)	S	S	S
1, 1, 1–Trichloroethane	SW	S	SW
Xylene	I	PS	SW

[a]Key: S Completely soluble; PS partially soluble; I insoluble, SW, swells (hazy, turbid).
Source: Ref. 3.

the priming of metal surfaces to be used in hostile environments (e.g., the hulls of naval vessels). There are a number of formulations available, both single-and two-package systems. Another example of a metal coating based on polyvinyl butyral is Metal Coating 2009, described in Table 9.

Other coating applications of the polyvinyl butyrals include solder masks for printed circuit board manufacture, heat-fusible wire coatings, and zinc oxide-based photosensitive paper coatings; they also serve as the binder/vehicle for iron oxide in the production of magnetic recording tapes. They are widely used as toughness/flexibility/adhesion promoters in the production of inks for letterpress, flexographic, and gravure printing, and as a component in toners for reprography.

4.2 Polyvinyl Butyral Dispersions

The dispersion of plasticized polyvinyl butyral in water, marketed by Monsanto as Butvar Dispersion BR,[4] is widely used as a permanent surface size in critical textile applications e.g., seat belt and parachute webbing), where the toughness of the butyral film lends outstanding abrasion resistance to the fabric. The relatively high surfactant levels used in the manufacture of this dispersion reduce the inherent adhesion of the resin to most highly polar surfaces such as metals and glass. This property is used to advantage in the preparation of removable coating for temporary protection of sensitive surfaces. The dispersion is typically formulated with other dispersions/latex as extenders to reduce cost. Pigments and plasticizers can be incorporated, as can small amounts of solvents to enhance adhesion properties.

4.3 Polyvinyl Formal [2, 8]

Polyvinyl formal, available in several viscosity grades from Monsanto as Formvar, is widely used as an oil-resistant insulating coating for magnetic wire. For this application it is normally cross-linked by formulating with phenolic, epoxy, or urethane resins to enhance properties. These coatings are tough, strongly adhering, abrasion resistant, and totally impervious to hydrocarbon oils and lubricants. Polyvinyl formal serves as the base vehicle for coatings for printed circuit boards, both permanent and fugitive (e.g., solder masking), for the surface treatment of glass fibers and boron filaments (abrasion resistance), and as an antifog coating on glass. It finds application in photothermographic and electrographic coatings. It is used as a binder in magnetic tape coatings, particularly for the high performance chromium oxide coated tapes. It finds use as an ink binder and as a lacquer to protect the surface of lithographic plates.

In general, polyvinyl formal is higher in modulus and less susceptible to solvent attack than is polyvinyl butyral. Conversely, solvent choices are more limited for the formal. Films are somewhat yellow, reducing their utility in applications calling for a colorless film.

TABLE 4 Typical Physical and Chemical Properties of Formvar Polyvinyl Formal Resin

Properties	ASTM Method[a]	Formvar resins			
		5/95E	6/95E	7/95E	15/95E
Physical Form		White, free-flowing powder			
Volatiles, maximum, %		2.2	2.2	2.2	2.2
Molecular weight $\times 10^{-3}$ (weight average)	(1)	25–35	35–45	40–60	70–150
Solution viscosity 15% by wt, MPa·s	(2)	140–280	250–500	325–675	1800–3500
Resin viscosity	(2)	8–12	12–15	15–20	37–53
Specific gravity, 23°/23° (± 0.002)	D792–50	1.227	1.227	1.227	1.227
Burning rate, cm/min	D635–56T	2.0	2.3	2.3	2.5
Refractive index (± 0.0005)	D542–50	1.502	1.502	1.502	1.502
Water absorption (24hs), %	D570–59aT	1.2	1.2	1.2	1.2
Hydroxyl content, as % polyvinyl alcohol	D1396–58	5.0–6.5	5.0–6.5	5.0–6.5	5.0–6.5
Acetate content, as % polyvinyl acetate	D1396–58	9.5–13.0	9.5–13.0	9.5–13.0	9.5–13.0

Formal content, as % polyvinyl formal, approx		82	82	82	82
Chemical[b]					
Resistance to					
weak acids	D543–56T	E	E	E	E
strong acids	D543–56T	E	E	E	E
weak bases	D543–56T	E	E	E	E
strong bases	D543–56T	E	E	E	E
Organic solents					
alcohols	D543–56T	G	G	G	G
chlorinated	D543–56T	P	P	P	P
aliphatic	D543–56T	E	E	E	E
aromatic	D543–56T	G	G	G	G
esters	D543–56T	G	G	G	G
ketones	D543–56T	G	G	G	G

[a]The ASTM method noted for hydroxyl content and acetate content rafers specifically to polyvinyl butyral resins. However, the same method is applicable to the polyvinyl formal resins. All other properties were determined by ASTM methods except:

(1) Molecular weight was determined by SEC/LALLS in hexafluoroisopropanol.

(2) Solution viscosity was determined in 15% by weight solutions in 60:40 toluene/ethanol At 25°C, using a Brookfield viscometer. Resin viscosity—5 g resin made to 100 ml with ethylene dichloride—measured at 20°C using an Ostwald viscometer.

[b]Key: E, excellent; G, good; F, fair; P, poor.

Source: Ref. 3.

TABLE 5 Typical Mechanical, Thermal, and Electrical Properties of Formvar Polyvinyl Formal Resin

Properties[a]	ASTM Method	Formvar resins			
		5/95E	6/95E	7/95E	15/95E
Mechanical					
Tensile strength, MPa					
yield	D638–58T	59–66	59–66	59–66	59–66
break	D638–58T	52–59	52–59	52–59	52–59
Elongation, %					
yield	D638–58T	7	7	7	7
break	D638–58T	50	50	50	50
Modulus of elasticity					
(apparent), GPa	D638–58T	2.7–3.1	2.7–3.1	2.7–3.1	2.7–3.1
Flexural strength, yield, MPa	D790–59T	117–124	117–124	117–124	117–124
Hardness, Rockwell					
M	D785–51	150	150	150	150
E	D785–51	65	65	65	65
Notched Izod impact strength					
(1.25 cm × 1.25 cm), J/m	D256–56	70	70	70	70
Thermal (°C)					
Flow temperature, 6.9 MPa	D569–59	140–150	140–150	140–150	160–170

		103–113	103–113	103–113	103–113
Glass temperature	[b]	103–113	103–113	103–113	103–113
Heat distortion temperature	D648–56	83–87	85–90	85–90	87–93
Heat-sealing temperature	[c]	96	96	99	107
Electrical					
Dielectric constant					
50 Hz	D150–59T	3.2	3.2	3.2	3.4
1 kHz	D150–59T	3.3	3.3	3.3	3.0
1 MHz	D150–59T	3.1	3.1	3.1	2.8
10 MHz	D150–59T	3.0	3.0	3.0	2.8
Dissipation factor $\times 10^3$					
50 Hz	D150–59T	8.1	8.1	8.1	8.7
1 kHz	D150–59T	10	10	10	10
1 MHz	D150–59T	21	21	21	21
10 MHz	D150–59T	19	19	19	18
Dielectric strength					
(3.2 mm thickness), V/m					
short time	D149–59	24	13	13	12
step by step	D149–59	12	12	12	13

[a]Conversion factors: MPa × 145 = psi; GPa × 145 × 10^3 = psi; J/m × 53.38 lb–ft/in.

[b]Glass temperature (T_g) was determined by differential scanning calorimeter.

[c]Heat-sealing temperature was determined on a 25 m dried film on paper, cast from a 10% solution in 60:40 toluene/ethanol. A dwell time of 1.5 seconds at 414 kPa (60 psi) line pressure was used on the heat sealer.

Source: Ref. 3.

TABLE 6 Solubility of Formvar Resins[a]

Solvent	Formvar resins 15/95E, 7/95E, 6/95E, 5/95E
Acetic acid (glacial)	S
Acetone	I
Aniline	S
Benzene	I
Butyl alcohols	I
Butyl acetate	I
Carbon disulfide	I
Cresylic acid	S
Cyclohexanone	I
Diacetone alcohol	I
Diisobutyl ketone	I
Dimethyl sulfoxide	S
N, N–Dimethylacetamide	S
N, N–Dimethylformamide	S
Ethyl acetate, 99%	I
Ethyl acetate, 85%	I
Ethyl alcohol, 95%, or anhydrous	I
Furfural	S
Hexane	I
Isopropyl alcohol, 95%, or anhydrous	I
Methyl acetate	I
Methyl alcohol	I
Methyl benzoate	S
Methyl butynol	S
Methyl Cellosolve acetate	I
Methyl ethyl ketone	I
Methyl isobutyl ketone	I
Methyl pentynol	S
N–Methyl–2–pyrrolidone	S
Nitropropane	I
Pentoxol	I
Propyl alcohols	I
Phenol	S
Propylene dichloride	I
Tetrachlorethane	S
Tetrahydrofuran	S
Toluene/ethyl alcohol (95%) (60:40 by weight)	S
Toluene	I
VM&P Naphtha	I
Xylene	I
Xylene–n–butyl alcohol (60:40 by weight)	I

[a]Key: S, Completely soluble; I, Insoluble or not completely soluble.
Source: Ref. 3.

TABLE 7 Properties of Butvar Dispersion BR

Property	Description/Value
Form	Milk-white aqueous dispersion of plasticized polyvinyl butyral
Total solids	50.0–52%
Viscosity	500–1500 mPa[a]
pH	8.0–10.5
Particle size	Most particles between 0.25 and 1.5 μ
Particle charge	Anionic
Plasticizer content	40 parts per 100 parts of resin (28.6% of solids)
Pounds per gallon at 25°C	8.4
Grams per liter	1008

[a]Determined on a Brookfield viscometer, LVF, no. 3 Spindle, 30 rpm, 25°C.

TABLE 8 Western Pine Association Knot Sealer, WP578: Brush Application

Material	% by weight
Butvar B–90 (Monsanto Chemical Co.)	3.3
Durite P–97 (Borden Chemical Co.)	40.0
SDA 35A, 95% ethanol	56.7
	100.0

Source: Ref. 3.

TABLE 9 Metal Coating 2009: Spray or Roller Aplication

Material	% by weight
Diacetone alcohol	17.4
n–Butanol	17.4
SDA–35A, 95% ethanol	7.7
Xylene	34.7
Methylon 75–108 (Bakelite Thermosets, Ltd.)	5.1
Epon 1007 (Shell Chemical Corp.)	13.0
Butvar B–90 (Monsanto Chemical Co.)	2.0
10% Phosphoric acid (diluted with a portion of the given solvent mixture)	2.7
	100.0

Cure cycle: 15 minute air dry + 30 minutes at 190°F and 20 minutes at 400°F

Source: Ref. 3.

REFERENCES

1. K. Toyoshima, in *Polyvinyl Alcohol* C. A. Finch, Ed. New York: Wiley, 1973, pp. 381–411.
2. T. P. Blomstrom, in *Encyclopedia of Polymer Science and Engineering*, 2nd ed., Vol. 16, J. I. Kroschwitz, Ed. New York: Wiley-Interscience, 1989.
3. *Butvar/Formvar Brochure*, publication No 6070E, Montanto Chemical Company, Plastics and Resins Division, 800 North Lindbergh Avenue, St. Louis, MO 63167.
4. *Butvar Dispersions BR Data Sheet*, Monsanto Chemical Company, Plastics and Resins Division, 800 North Lindbergh Avenue, St. Louis, MO 63167.
5. A. F. Fitzhugh and R. N. Crozier, *J. Polym. Sci.*, *8*, 225 (1952); *9*, 96 (1952).
6. K. Hoshi, S. Tasaka, and T. Yoshimi, Jpn. Kokai, 62/46, 953 (1987); Taiyo Yuden Co. Ltd., [*Chem. Abstr.*, *106*(22), 181, 481h.]
7. T. T. Boersma, *J. Imaging Technol.*, *13*(2), 75, (1987).
8. E. Lavin and J. A. Snelgrove, "Polyvinyl acetals, in *Encyclopedia of Chemical Technology*, 3rd ed., Vol. 23. New York: Wiley-Interscience, 1983, pp. 798–816.

47
Polyimides

B. H. Lee

Ciba-Geigy Corporation
Ardsley, New York

Polyimides are condensation polymers that contain the imide structure $-CO-\overset{|}{N}-CO-$ as a linear or heterocyclic unit along the polymer main chain. They exhibit exceptional thermal, thermooxidative, and chemical resistance, and good radiation resistance and dimensional stability, while maintaining an excellent balance of mechanical and electrical properties.

1.0 CHEMISTRY AND PROPERTIES

Aromatic polyimides are generally produced by a two-step polycondensation reaction of aromatic dianhydride with either aromatic diamine or aromatic diisocyanate in a suitable reaction medium. They have the following general structure:

$$\left[\begin{array}{c} \\ \end{array} \right]_n$$

The direct production of high molecular weight aromatic polyimides in a one-step polymerization could not be accomplished because the polyimides are usually insoluble and intractable. The polymer chains precipitate from the reaction media (whether solution or melt) before high molecular weights are obtained. Therefore, processing of the aromatic polyimides can by accomplished only with the first step intermediate amic acid varnish,

435

which is still soluble and fusible. Were it not for the processing difficulties associated with known polymers, polyimides would be enjoying success in many more new application areas. This processing problem was partially overcome by the development of copolymers. The two major commercial polyimide copolymers are an amide-imide known as Torlon, a product of Amoco, and an ether-imide produced by General Electric under the trade name Ultem. Another approach to this processing problem was by incorporating a soft aromatic segment and/or aliphatic moiety in the polymer main chain. Among the various efforts to overcome the processing difficulties, one commercial success was achieved by incorporation of totally asymmetric diamionophenylindane isomeric mixture into the polyimide backbone; it is marketed as Matrimid 5218 from Ciba-Geigy. This material is soluble in relatively nonpolar solvents and is characterized also by exceptionally high glass transition temperature and high thermooxidative stability.

Key properties of aromatic polyimides are their outstanding high temperature resistance, toughness, good dielectric properties, low flammability, and high resistance to radiation and structural deformation under load at elevated temperatures.

2.0 USES

Polyimides are primarily used in the aerospace, automotive, and electronic industries, where a need exists for materials with long-term high temperature capabilities. Molded polyimides are now widely used in loading bearing applications such as struts, chassis, and brackets in automotive and aircraft structures because of their high flexural modulus and compressive strength. Their excellent high temperature dimensional stability, chemical resistance, and natural lubricity make them ideal bearing materials in such applications as jet engines, appliances, and office equipment. Polyimide films are used as insulation for electric motors, aircraft parts, missile wire cable, magnetic wire, and flat flexible cable.

The electrical properties of polyimides make them ideally suitable for applications in the electrical and electronic markets. They are being used in place of glass and ceramics for high heat connectors, switches, housings, and controls. They are also used in the fabrication of injection-molded printed circuit boards. Polyimide coatings are increasingly being used in microelectronic applications. Major applications are in the following areas:

Interlayer dielectrics on silicon and gallium arsenide integrated circuits for multilevel devices.

Passivativation, thermal–mechanical buffer, and α-particle protection coatings on integrated circuits and other circuitry.

Masking for multilayered resist processing; for negative-profile liftoff processing; for harsh processing such as ion implantation or dry etching.

The principal features that make polyimides suitable for microelectronic applications are:

Polyamic acid varnishes and/or polyimide solutions can be easily spun on planarized layers, exposed, and etched with existing equipment.

Polyimides have good dielectric properties.

Polyimide coatings are tough and resilient, and have good thermal and chemical stability. They also have good radiation resistance.

3.0 COMMERCIAL INFORMATION

Several commercially available aromatic polyimides, such as Kapton (Du Pont), IP-2080 (Dow), Matrimid 5218 (Ciba-Geigy), Ultem (General Electric), and LARC-TPI (Mitsui-Toatsu Chemicals), are used in the form of films, moldings, adhesives, and composite matrices. Numerous lacquers or varnishes based on similar chemistries are also available from the basic resin suppliers. Several polyimide coating solutions also have been introduced commercially.

REFERENCES

1. K. L. Mittal, Ed., *Polyimides*, Vols. I and II. New York: Plenum Press, 1984.
2. J. W. Verbicky, Jr., in *Encyclopedia of Polymer Science and Engineering*, 2nd ed., Vol. 12. New York: Wiley, 1988, p. 364.
3. J. J. King and B. H. Lee, in *High Performance Polymers: Their Origin and Development*, R. B. Seymour and G. S. Kirshenbaum, Eds. New York: Elsevier, 1986, p. 317.
4. Y. K. Lee and J. D. Craig, in *Polymer Materials for Electronic Applications*, E. D. Feit and C. W. Wilkins, Jr., Eds. Washington D. C.: American Chemical Society, 1982, p. 107.

48
Parylene Coating

William F. Beach

Bridgeport, New Jersey

1.0 PROCESS

The parylene process [1,2] is a means of applying a pinhole-free coating with exceptional conformality and control of thickness. The coatings so produced have excellent dielectric as well as barrier properties. The parylene coating, composed solely of poly (*p*-xylylene) (PPX), a family of linear, high molecular weight organic polymers, is grown directly on a substrate by vapor deposition polymerization (VDP). The gaseous *p*-xylylene monomer (PX) is transformed into a solid polymer coating without passing through an intermediate liquid stage. Since surface forces have no opportunity to alter the cross-sectional profile, the result is a coating of extraordinary uniformity of thickness and continuity. No postdeposition cure is necessary to complete the coating chemistry. The parylene process affords exceptional control of coating thickness. While typically used in thicknesses of 1–10 μm, continuous parylene films have been demonstrated at thicknesses under 500 (0.05 μm). In principle there is no upper limit to the thickness to which a parylene film might be grown, but practical constraints of time and cost place an upper bound in the vicinity of 100 μm.

The parylene process is further distinguished by the fact that it is conducted at room temperature. Parylene growth rates actually decrease at high temperatures. There is an advantage in operating the process at subambient temperatures, if such operation is feasible. Another distinguishing feature of the parylene process is that it proceeds without the assistance of a catalyst. Thus the coating is of remarkable chemical purity with respect to catalyst residues, which in other coating systems can be ionic or ionogenic, or leachable.

The monomer is exceptionally reactive. It cannot be stored. It can be handled only as a rarefied, low pressure gas. It is therefore necessary to generate monomer as it is required by the coating process. Monomer is conveniently generated by the pyrolytic cleavage of its dimer, di-*p*-xylylene (DPX), a [2.2] paracyclophane. Monomer generation from dimer proceeds in quantitative yield with no by-products. Since the temperatures for monomer generation and consumption are so different, monomer transport from one site to the other during deposition is a practical necessity. Such transport is done most efficiently when all other gases are absent. For this reason the commercial process is conducted within a vacuum system.

The composition of the coating can be modified to some extent by attaching substituents to the ring carbons of the DPX molecule. Although many versions of parylene process feedstock DPX are known, those that are commercially available at this time include DPXN, the base hydrocarbon; DPXC, with an average of one chlorine atom per aromatic ring, and DPXD, averaging two clorine atoms per aromatic ring. The coatings prepared starting with these dimers are called Parylene N, Parylene C, and Parylene D, respectively. Substituents reduce the volatility of the monomer, causing the processing characteristics of the modified versions to be somewhat different.

The discovery of the base member of the family, poly(p-xylylene), was first reported in 1947. However, it was not until 1965, when the Gorham process starting with DPX was announced,[3] that the polymer family became industrially viable. Figure 1 is a schematic of the parylene equipment, with typical operating conditions indicated.

2.0 PROPERTIES

The glass transition temperatures of the parylenes are in the vicinity of room temperature. However, the parylenes are crystalline polymers, and as such retain substantial physical strength and solvent resistance at temperatures approaching their crystalline melting points (for Parylenes N, C, and D: 420, 290, and 380°C, respectively). Long-term contact with solvents results in a mere few percent swelling. Equilibrium moisture absorption is very low. Significantly, no mode of moisture-induced degradation is chemically feasible. The dielectric constants and loss factors of the parylenes are low and invariant over a wide range of frequencies. Although permeable, the parylenes at a given thickness are superior as barriers to other organic polymers that can conveniently be prepared as coatings.

The parylenes are vulnerable to attack by oxygen, particularly at elevated temperatures and in the presence of ultraviolet radiation. In air, 10-year use temperatures for Parylenes N, C, and D are projected to be 60, 80, and 100°C, respectively. In oxygen-free environments, the thermal endurance of each is substantially better.

While it is tempting to categorize the parylene vapor deposition polymerization process with the superficially similar processes of evaporation, sputtering, or chemical vapor deposition of metals or inorganics, there are important distinctions to be made. In the latter processes, the growth action is confined to the outer substrate surfaces, while the parylene polymerization chemistry actually occurs under the surface of the growing coating. As a result, the parylenes deposit in a condition of compressive stress and adhere to organic substrates, which are permeable to monomer by an interpenetration mechanism. Adhesion to impermeable metallic or inorganic substrates such as aluminum or silica can be achieved by pretreatment with an organosilane primer such as γ-methacryloxypropyl trimethoxysilane (A–174).

3.0 APPLICATIONS

The parylenes first found use in electronics construction. Because of its exceptionally low and frequency-independent dielectric constant, Parylene N continues to be used as the functional dielectric in high quality miniature film capacitors. Very early on, the parylenes were qualified under MIL–I–46058, the specification for coating printed circuit assemblies. Parylene C continues to enjoy a reputation as a high performance coating for military cir-

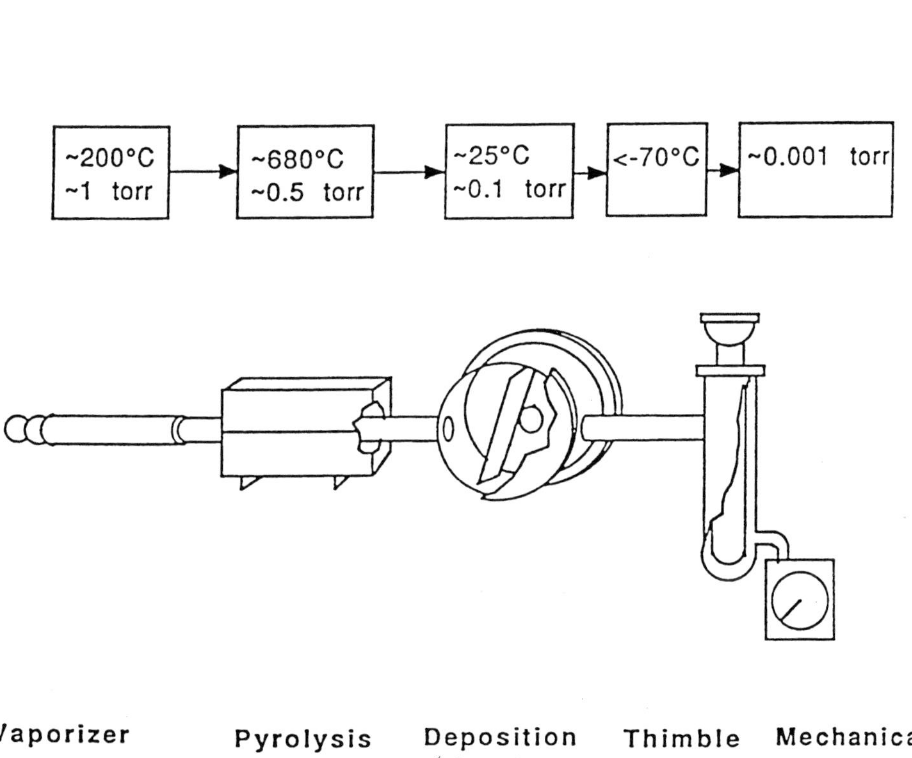

Figure 1 Schematic of parylene deposition apparatus.

cuitry, particularly in avionics. Parylene C is also an Underwriters Laboratories (UL) approved conformal coating. The use of the parylenes as circuit coatings has extended to hybrids. The exceptional conformality of the parylenes has served them well in these applications, but so far has impeded their use in multilevel interconnection schemes, where a planarizing dielectric is sought. In the manufacture of miniature electric motors, such as those used in wristwatches, parylene is used as an insulating coating on the armature. Parylene's exceptional thickness control permits the winding of a maximum number of turns, and therefore superior motor performance. Furthermore, a parylene process variation in which the parts are tumbled during deposition permits economies through the coating of thousands of parts at a time.

Parylene's use today has broadened to such diverse missions as the immobilization of loose particles that otherwise would result in early device failure (Winchester drives and hybrids), the modification of surface abrasiveness on ferrite toroids, and the reinforcement and preservation of embrittled paper in old books and museum artifacts.

REFERENCES

1. W. F. Beach, C. Lee, D. R. Bassett, T. M. Austin, and R. Olson, "Xylylene polymers," in *Encyclopedia of Polymer Science and Engineering,* 2nd ed., New York: Wiley-Interscience, 1988.

2. (a)   S. M. Lee, in *Kirk–Othmer Encyclopedia of Chemical Technology,* 3rd ed., Vol. 24. New York: Wiley-Interscience, 1983, pp. 744–771. (b)   M. Szwarc, *Polym. Eng. SCE.,* *16*(7), 473–479 (1976). (c)   W. F. Gorham and W. D. Niegisch, in *Encyclopedia of Polymer Science and Technology,* Vol, 15. 1971, pp. 98–124. (d)   L. A. Errede and M. Szwarc *Q. Rev.* 12, 301–320 (1958)

3. W. F. Gorham U.S. Patent 3,342,754 (Sept. 19, 1967); Union Carbide Corp.

49
Expanding Monomers

Claire Bluestein and Murray S. Cohen

Epolin, Inc.
Newark, New Jersey

1.0 INTRODUCTION

Monomers that cure with zero shrinkage or even expand instead of shrinking on polymerization produce films that can better adhere to difficult substrates. The cured coatings provide protection against corrosion and abrasion, as well as chemical and moisture resistance. In other areas of application, strengthened, impact-resistant, fiber-reinforced composites made from expanding monomer compositions are also reported.[1]

Expanding monomer technology is of interest for many applications, including matching of optical assemblies with exact precision, joining of optical fibers without loss of transmission; corrosion-resistant coatings on metals; clear, protective, weather-resistant coatings for aircraft windows, headlamps, and canopies; longer lasting adhesives; and better fitting medical and dental restoratives. The coatings can range from hard and tough to soft and flexible.

2.0 FUNDAMENTALS

The chemistry of expanding monomers is based on ring opening polymerization of bicyclospiranes. The three fundamental types of these monomers are the spiroorthocarbonates (*I*), the spiroorthoesters (*II*), and the bicyclic orthoesters (*III*). The original concept was advanced in the early 1970's by William J. Bailey of the University of Maryland. He continued experimental development, published a series of papers [2,3] and obtained a U.S. patent (4,387,215,) in June 1983. Epolin Inc. was founded in 1984 to commercialize, manufacture, and market systems based on polymerization of these ring opening monomers, and was granted exclusive license to Prof. Bailey's patent.

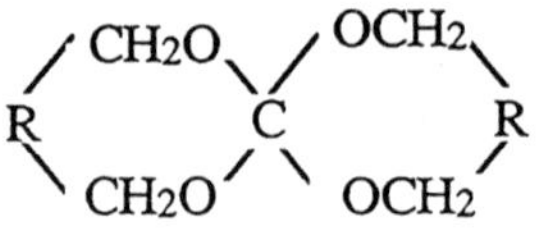

I

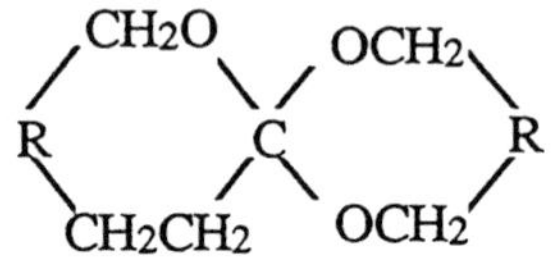

II

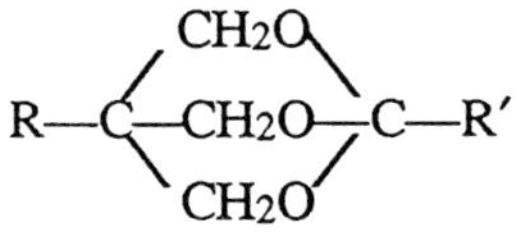

III

In the majority of the polymerizations studied to date, the fundamental bicyclospirane ring opening reaction is catalyzed cationically, although unsaturated bicyclospiranes have been prepared that are initiated via free radical catalysis. The general polymerization reaction of a spiroorthocarbonate is shown below.[4]

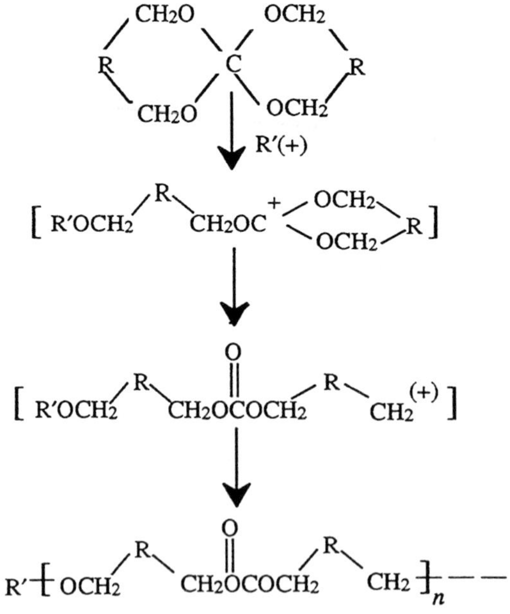

Prof. Bailey reasoned that the bicyclic ring structure monomers would form as solids, tightly packed into a crystal lattice; and indeed most of the symmetrical spiroorthocarbonates are high melting, highly crystalline solids. Rotational and transitional freedom is highly restricted, as noted in the lack of solubility of these monomers in most common solvents at ambient temperature. During polymerization both rings open nearly simultaneously, yielding a linear polymer having many degrees of freedom. Furthermore, the symmetry is destroyed. This basically explains why the density is lowered (expansion occurs), and mobility increases (lowering melting point and increasing solubility).

The syntheses require a number of unit operations to prepare the monomers, resulting in a high product cost. Thus the finished coatings containing a monomer are priced beyond what most industrial coatings applications will bear. This economic factor together with the necessity of working with high melting, relatively insoluble monomers has, for many years, kept the expanding monomers in the realm of academic investigation.

2.1 Putting Chemistry Into Practice

As a practical matter, it is not desirable to attempt homopolymerization of expanding monomers, since the polymerization would entrap insoluble crystalline monomer and terminate itself long before completion.

A marketable coating in today's world of technological development must meet the many needs of the end user, a manufacturer. The system must be designed to work with existing engineered application machinery, which operates with fluid liquids free of agglomerates. Systems also must be nonpolluting: putting no hazardous solvent into the air and no undesirables into the water supply. The greatest demand is for curing (or drying) with low energy usage from the utility sources. All these considerations made it reasonable to seek a liquid system in which the expanding monomer blends with available monomers and oligomers containing no solvent and one that has a low energy cure.

During work on a grant from the U.S. Air Force, Epolin discovered a new method for formulating coatings based on the same principles, but without the requirement of presynthesis of the expanding monomers. The research funds from the Air Force have fostered work on coatings in two major applications areas—coatings on metals, particularly aluminum, and coatings on clear plastics. Ring opening polymerization with expansion is ideally suited to the hard surfaces of metals and plastics, and it also works well on glass. More recently Epolin was granted U.S. Patent 4,738,999 (April 1988), which covers the early part of this work[4]. A new coating system that meets all the described criteria was developed.

The schematic representation in Figure 1 shows that as a liquid conventional coating cures to a gel point, shrinkage occurs. The expanding monomer coating in Figure 2 does the same. This initial shrinkage is not a problem, nor is it really part of true shrinkage, since the system is still mobile and has not set. Beyond the gel point, however, the critical shrinkage occurs in a conventional coating. Even after it is set or cured, it will continue to cross-link and shrink further upon aging, thus eventually delaminating from its substrate (Fig. 1). The true shrinkage is that measured from the gel point on.[5,6]

However, the coating containing expanding monomer (Fig. 2) will expand in volume slightly or remain the same as it cures beyond the gel point—depending on formulation. Furthermore, the slower reacting cationically initiated spiro monomer will typically continue to polymerize and further expand after the set point has been noted. We have observed increases in hardness, adhesion, and scratch and chemical resistance after 3–4 weeks on some of our coatings. This is the basis for the longevity and superior behavior of coatings containing expanding monomers.

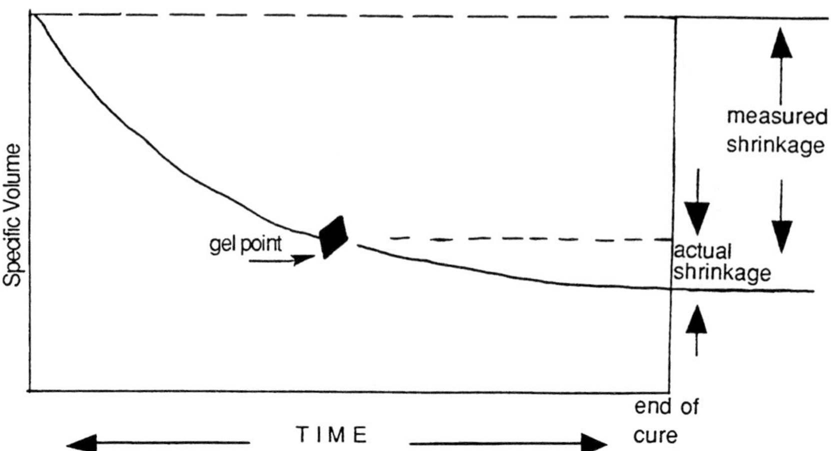

Figure 1   Change of specific volume during cure of a conventional coating.

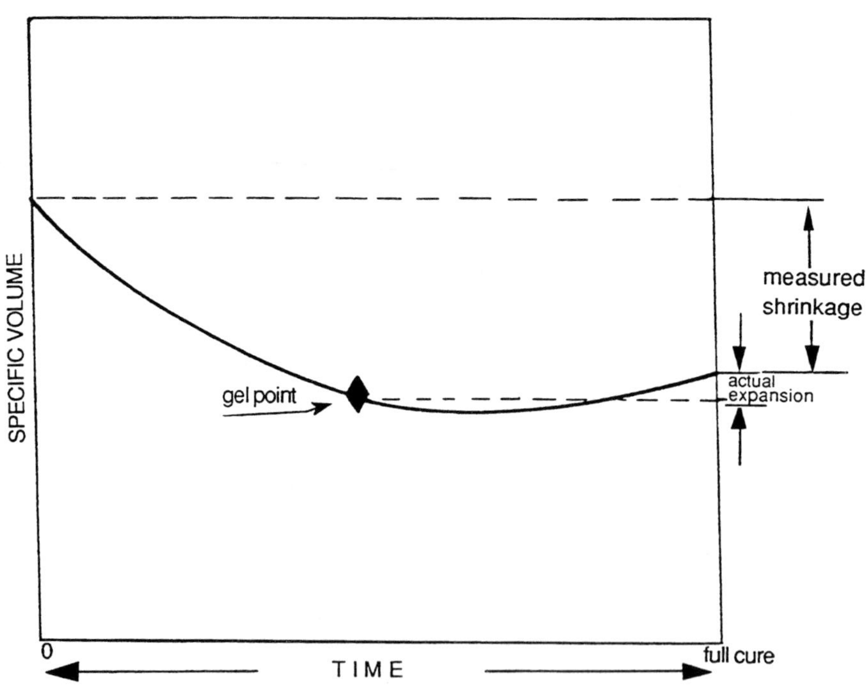

Figure 2   Change of specific volume during cure of an expanding monomer coating.

3.0 PROPERTIES

3.1 Ultraviolet-cured Expanding Monomer Based Coatings

The new systems that are available are based on ultraviolet light initiated cure of 100% reactive liquid systems. These coatings contain spiroorthoesters, liquids that react more readily than the spiroorthocarbonates. The expanding monomer blends into a matrix resin system based on various commercial epoxy resins or mixed epoxy and acrylic resins. There is a range of products suited for each application technique and for adhesion to the various metals, plastics, or glass products to be coated. At the same time, coatings can be modified to be clear, high gloss, low gloss, or pigmented. These coatings, being relatively low in viscosity for 100% solids compositions, can readily be adapted to most application equipment. They can be applied by spray, spin coat, dip coat, roller coat, curtain coater, screen print, etc. The range of liquid properties is as follows:

99.5% Nonvolatiles.
Clear and colorless or pigmented
Viscosity range, 50 to >9000 cP.
Density, 1.08 or 9 lb/gal.
Odor is mild.
Shelf stability in dark container > 6 mos.

Cure of these coatings involves the commercially available cationic sulfonium salt photoinitiators. UV lamps having sufficient power (at least 200 W/in.) in the 300 nm wavelength range are required, or electron beam radiation may be used. As with other cationic cure coatings, these cure conditions do not require absence of air or oxygen, but they are sensitive to presence of moisture. Table 2 shows the range of properties that can be obtained after cure of the present line of coatings.

Extensive testing according to the methods of the American Society for Testing and Materials (ASTM) for chemical and moisture resistance has been performed, particularly on aluminum substrates. The clad aluminum panels have held up for more than 1000 hours in the ASTM B–117 salt spray method without the usual chromated base coats. Both clear and gray pigmented panels have held up under these conditions. Additional testing may involve

Table 2 Typical Properties of Epolin UV-Cured Spiroorthoester Coatings and Adhesives

Property	Value
Cured coating thickness	0.2–0.8 mil
Pencil hardness	B to 6H
Adhesion as cured, ASTM D–3359	5A
Adhesion after immersion tests, ASTM D–3359	5A
Refractive index	1.55–1.59
Glass transition temperature	47–70°C
Salt spray testing, ASTM B–117	> 1000 hours

placing a coated and cured panel into boiling water for several hours. The best coatings do not haze up or delaminate under these conditions.[7]

3.2 Thermally Cured Expanding Monomer Based Coatings

In an earlier development, corrosion-resistant coatings were prepared for steel utilizing a dispersion of norbornene spiroorthocarbonate (NSOC) (IV) in an epoxy resin cured with a commercial amine hardener (preferably Ciba-Geigy XU264) together with a Lewis acid thermal catalyst. This work was performed under an SBIR contract for NASA. The coating was supposed to be able to resist attack from aqueous hydrochloric acid. This coating composition required cure for 14 hours at 110°C. The steel panels so coated withstood 5 cycles between 10% HCl for 2 hours and a 175°C oven for 30 minutes without signs of corrosion. Furthermore, coated and cured panels were tested via the ASTM B–117 salt spray method and the best samples passed 500 hours.[8] This system has not generated commercial interest due to the high cost of NSOC and the need for extended high temperature cure time.

$$IV$$

REFERENCES

1. M. R. Piggott, *Controlled Matrix Contraction Composites*. Patented: Canada, 1,176,812 (1984); Australia, 548,698 (1986); Europe, 0041843 (1987).
2. W. J. Bailey et al., in *Ring Opening Polymerization*, "Ring opening polymerization with expansion in volume." ACS Symposium Series No. 59. Washington, DC: American Chemical Society, 1977, pp. 38–58.
3. W. J. Bailey, M. J. Amone, B. Issari, et al., "Polymerization with expansion in volume for improved matrices and adhesives," *Proc.ACS Div. Polym. Mater.*, *54*, 23 (1986).
4. M. S. Cohen, C. Bluestein, and M. Dunkel, in *Proceedings, RadCure '84* (September 1984). Dearborn, MI: Society of Manufacturing Engineers.
5. M. S. Cohen, C. Bluestein, and M. Dunkel, Paper FC85–425, in *Proceedings, RadCure Europe '85* (May 1985). Dearborn, MI: Society of Manufacturing Engineers.
6. C. Bluestein, and M. S. Cohen, in *Proceedings of the Sixth International Meeting on Radiation Processing*, Ottowa, Canada, June 1987.
7. C. Bluestein, M. S. Cohen, and R. Mehta, in *Proceedings, RadTech '88*, Vol. I. Northbrook, IL: RadTech International, 1988, p. 389.
8. M. S. Cohen, C. Bluestein, R. Kozic, and D. E. Fergesen, *Proc. ACS Div. Polym. Mater. Sci. Eng.*, *54*, 12–17 (April 1986).

50
Nitrocellulose

Daniel M. Zavisza

Hercules Incorporated
Wilmington, Delaware

When most people think of nitrocellulose, they think of guncotton, a material that was developed for explosives or gun propellant. But they are only partially correct. Nitrocellulose is one of the oldest and most widely used film formers adaptable to a number of uses. It is derived from cellulose, a material from plants, and therefore a renewable source. Soluble nitrocellulose possesses a unique combination of properties such as toughness, durability, solubility, gloss, and rapid solvent release. As the film former in lacquer systems, it affords protective and decorative coatings for wood and metal. In addition, it finds use in flexible coatings for paper, foil and plastic film, printing inks, and adhesives. This chapter briefly covers the properties, uses, and handling procedures for nitrocellulose and the formulations made from it.

1.0 PREPARATION

Nitrocellulose is the common name for the nitration product of cellulose. Other names include cellulose (tri)nitrate and guncotton. The commercial product is made by reacting cellulose with nitric acid. Cellulose is composed of a large number of β-anhydroglucose units, which are joined together into a chain. The anhydroglucose units are six-membered rings having three hydroxyl (—OH) groups attached to them. The number of anhydroglucose units in the typical cellulose chain ranges from 500 to 2500 in chemically purified cellulose.

1.1 Degree of Substitution

Nitric acid can react with these three hydroxyl groups of the anhydroglucose units to form the nitrate ester. Fully nitrated cellulose would then be a trinitrate—that is a nitrate having a degree of substitution of 3. The calculated nitrogen content of such a fully nitrated cellulose is 14.14%.

In practice, however, the maximum nitrogen level that can be achieved is 13.8%. This corresponds to a degree of substitution of 2.9. At this level nitrocellulose does not possess properties that are useful for coatings use. Film forming properties are better at degrees of substitution between 1.8 and 2.3.

1.2 Degree of Polymerization

The degree of polymerization of nitrocellulose is the number of anhydroglucose units that are linked together to form one molecule. The more units are linked, the higher the viscosity of the nitrocellulose in solution at a given concentration. Commercial nitrocellulose is categorized into grade by the viscosity of a 12.2% solids solution at 25°C in terms of centipoise or the time it takes, in seconds, for a metal ball of a specified size to fall a measured distance through the solution. The solvent system used is usually 55 parts by weight toluene, 25 parts denatured ethanol, and 20 parts ethyl acetate.

1.3 Types and Grades

Nitrocellulose is divided into types according to the nitrogen content of the product, which reflects higher or lower degrees of substitution. The lowest commercially useful nitrogen content is 11% (10.0–11.2%). This corresponds to a degree of substitution of 1.8–2.1. This type is further classified into viscosity grades ranging from 30–35 cp[1] to 40–60 seconds. These grades are useful in lacquers for paper and foil, low odor lacquers, sealers, fillers, printing inks, and plastics. At this level of nitrogen, nitrocellulose is more tolerant of alcohol than higher nitrogen types.

A second type has an average nitrogen content of 11.5% (11.3–11.7%). Viscosity grades of 0.5 and 5–6 seconds are available for use in lacquers for coating cellophane, paper, and textiles.

The next type contains an average of 12% nitrogen (11.8– 12.2%) and is available in a large number of viscosity grades, from 18–25 cp to 2000 seconds. This type is more tolerant of aromatic hydrocarbon solvents, such as toluene, and less tolerant of aliphatic hydrocarbons. This grade is compatible with many resins and has many uses. It is used in coatings for wood, metal, paper, textiles, and foil; for lacquer emulsions for wood and metal; and for architectural finishes, adhesives, cements, and inks. Higher nitrogen levels (>13%) find use in the manufacture of gun propellants and explosives.

Solubility and viscosity in solution and compatibility of nitrocellulose with a variety of modifiers, such as plasticizers, resins, and pigments, determine its usefulness in preparing lacquers and coatings. The chapter appendix is a table of the typical properties of nitrocellulose containing 12% nitrogen, the type most commonly used for lacquers and coatings.

2.0 SOLUBILITY

2.1 Solvents and Diluents

The generally used method of formulating nitrocellulose coating systems features volatile solvents that dissolve the nitrocellulose and its modifiers to form a homogeneous system (with the exception of pigments and fillers).

---

[1]The viscosity of very low molecular weight nitrocellulose is measured and described in centipoise. The falling ball method is used for times of a quarter-second and higher. This results in a change of units used to describe the viscosity grades.

The resulting formulation can then be applied to the substrate by one of a number of methods such as brushing, spraying, or curtain coating. The solvent evaporates to leave a solid film on the substrate.

True solvents are liquids that will dissolve nitrocellulose completely. For 12% nitrogen nitrocellulose, these are ketones, esters, amides, and nitroparaffins. Some solvents, such as ethyl or isopropyl alcohol, will not dissolve nitrocellulose on their own. They may be added to true solvents without precipitating the nitrocellulose. These are termed "cosolvents." Aliphatic and aromatic hydrocarbons are nonsolvents. Termed "diluents," they may be added to nitrocellulose solutions in limited amounts without precipitation, to lower cost and improve solubility of resin modifiers. Aromatics may usually be added to a greater extent than aliphatics. At the lower nitrogen level of 11%, more hydrocarbon can be added to the nitrocellulose solution without precipitation.

Dilution ratio is the ratio of the volume of diluent that can be added to a given volume of nitrocellulose solution in a true solvent before separation of the nitrocellulose takes place. For example, starting with 8 g of nitrocellulose in 100 m. of solvent, butyl acetate, the dilution ratio for toluene is 2.75 compared with 1.5 for MV&P naphtha (an aliphatic hydrocarbon mixture). Different solvent–diluent combinations will have different dilution ratios and must be measured separately. When ethanol, a cosolvent, is combined with a true solvent, the dilution ratio for a given diluent is higher than with the true solvent alone. Raising the temperature will usually lower the dilution ratio for a diluent–solvent mixture.

2.2 Viscosity Effects

The viscosity of a polymer solution will normally increase with the solvating power of the solvent at a given concentration, because this allows the polymer to stretch out further and to become entangled with other polymer molecules more easily. A poorer solvent forces the polymer to become more compact and to occupy a smaller volume, resulting in a lower solution viscosity.

Toluene is a poorer solvent for nitrocellulose than esters or ketones. Up to a point, however, it may be added to solutions of nitrocellulose in true solvents, such as esters on ketones, without precipitating the nitrocellulose. The effect is that the viscosity of the resulting solution is lower for solutions containing toluene than without it at the same solids level. This technique allows a formulator to maximize the solids level for a given application viscosity. The most common diluents used in this way are toluene and xylene.

Solutions prepared of nitrocellulose exhibit a viscosity drop on standing. A solution of half-second grade at 20 wt% solids in an ethanol–toluene–ethyl acetate solvent blend exhibits, for example, a viscosity of 3.8 seconds by a falling ball method 1 hour after addition of the solvent but 3.5 seconds after 24 hours. This viscosity loss will stop after a finite time.

2.3 Blushing

Formulating lacquers that contain too great a percentage of rapidly evaporating solvents may result in a white chalky appearance upon drying. The cause of this is cooling of the film surface below the atmospheric dew point by the evaporating solvent, condensing water onto the film. If sufficient moisture is condensed, the nitrocellulose can precipitate. This blushing problem is worsened during humid weather.

The remedy for blushing is to add a slowly evaporating solvent to the solvent blend. This has the effect of retarding the evaporation rate and lowering the cooling effect. Examples of such solvents are diisobutyl ketone, methyl amyl acetate, amyl acetate, methyl amyl ketone, and 2-butoxyethanol.

Blushing or haze formation can also occur when a poor solvent, which has a slow evaporation rate is used with a fast, good solvent. The good solvent will evaporate first, causing precipitation of the nitrocellulose by the poor solvent. Too much high boiling hydrocarbon diluent will do this.

2.4 Solution Preparation

The type of mixing apparatus is usually the preparer's choice. A vertical tank with propeller or disc agitators works well. Laboratory samples are easily dissolved by tumbling or rolling jars containing the nitrocellulose and solvent. If a solvent blend is to be used, it is best to disperse or wet the nitrocellulose with the cosolvent or diluent first, then add the active solvent portion. This procedure will reduce the time required to effect solution. Solvent blends containing higher ratios of diluent to active solvent will take longer to dissolve the nitrocellulose than those with lower diluent ratios.

3.0 FILM PROPERTIES

Film properties of nitrocellulose are affected by the solvents used, the casting technique, the drying conditions, as well as other parameters. Different types of modifier are also used to alter nitrocellulose dry film properties. Nitrocellulose is only one ingredient in lacquers an coatings. The following description of modifiers is intended to give a flavor of how nitrocellulose systems may be modified.

3.1 Plasticizers

Plasticizers are nonvolatile materials added to control flexibility and elongation of a film. Plasticizers should be nonvolatile, colorless, odorless, and tasteless. They should be nontoxic and should provide a maximum flexibility with minimum loss in film strength and toughness. They should not destabilize the film chemically.

Plasticizers fall into two types: solvent and nonsolvent. Solvent-type plasticizers are those that exhibit complete miscibility with nitrocellulose in all proportions. Examples are dibutyl phthalate and diisononyl phthalate.

Nonsolvent-type plasticizers neither dissolve nor cause formation of colloidal nitrocellulose at room temperature. However, they are compatible with nitrocellulose in solution and in the dry film. Examples of nonsolvent-type plasticizers are castor oil and polymeric or polyester-type plasticizers. The polymeric plasticizers improve flexibility and have very low volatility, are nonmigrating and nonspewing, and do not leave the film at elevated temperatures.

3.2 Resins

Resins are used in nitrocellulose coating compositions to improve the degree of film build by increasing the solids content at a given viscosity (i.e., spray viscosity). Depth, gloss, and adhesion can also be promoted by added resin.

Natural resins, such as shellac, dammar, elemi, and mastic, were some of the first resins used to modify nitrocellulose. Most of these resins are not film formers in themselves and do not improve tensile strength, flexibility, or elasticity. The use of natural resins is limited.

Some synthetic resins now available are designed to have specific properties that make them generally more adaptable for use in lacquers than natural resins. Alkyd resins are widely used. They are prepared by reacting a polyhydric alcohol, such as pentaerythritol, with a polybasic acid, such as phthalic acid. Alkyd resins modified with drying oils or unsaturated fatty acids are good film formers. Addition of nitrocellulose to these resins accelerates the drying time and often eliminates the need for baking.

Phenol–formaldehyde resins have excellent resistance to alcohols, acids, and alkalies. When used with nitrocellulose, however, there may be compatibility problems and poorer color stability. Other useful resins are acrylic resins, vinyl resins such as polyvinyl butyral and polyvinyl acetate, certain polyamides, epoxies, and low molecular weight polyesters.

Pigments are added for producing opaque, colored finishes. Because nitrocellulose tends to be degraded in sunlight, some pigments extend the service life of films exposed to sunlight. Certain pigments should be avoided because they show alkaline reactions, which cause nitrocellulose degradation.

3.3 Cross-Linkable Coating Systems

Soluble nitrocellulose contains residual hydroxyl groups that may be utilized to prepare coatings that are cross-linked to other vehicle materials through these hydroxyls. Some groups that react with these hydroxyls are methylolamino and alkylated methylolamino groups, isocyanates, and epoxides.

Catalyzed nitrocellulose–alkyd–aminoplast systems are systems that will cure by acid catalysis. The nitrocellulose and the alkyd provide the hydroxyl groups. The aminoplast can be an alkylated urea–or melamine–formaldehyde resin. These systems are catalyzed to cross-link with a strong acid, such as p-toluenesulfonic acid or phenyl acid phosphate. The coatings produced are tough and solvent resistant and find application on kitchen cabinets. Once catalyzed, the system must be used promptly because the solution will gel with time.

Nitrocellulose–urethane systems are prepared from the reaction of a polyisocyanate with nitrocellulose. The coatings may be toughened by the addition of a polyol such as a polyester, acrylic, vinyl, or alkyd polyol. Each of these polyols will give different properties. conventional urethane catalysts, such as zinc octoate, may be used to speed up the cross-linking.

Epoxy-containing resins may be used; but since the available hydroxyls in nitrocellulose are less reactive than the usual polyols in epoxy systems, they are not used extensively for coatings systems.

3.4 Safety Considerations

Some basic safety precautions must be observed because nitrocellulose is a very flammable material. It is sold wetted with some material such as ethanol or isopropanol, which lowers its flammability. It should never be allowed to dry, because then it is extremely flammable. No spark-producing sources, flames, or heat or static electricity sources should come into proximity of nitrocellulose. Nitrocellulose containers should be kept tightly closed. Smoking must be prohibited when handling nitrocellulose or its solvents, diluents, or solutions. Safety procedures are contained in National Fire Protection Association (NFPA) Standard 35, Manufacturing Chemists Association Chemical Safety Data Sheet SD–96, the Hazardous Materials Regulations of the U.S. Department of Transportation (009) and the U.S. Occupational Safety and Health Standards, Part 1910.

APPENDIX

Typical Properties of RS Nitrocellulose

General Properties

Odor of material	None
Taste of material	None
Color of film	Water-white
Clarity of film	Excellent

Physical Properties of Solid or Film

Bulking value in solution	0.0704 gal/lb
Specific volume in solution	16.26 in. ^{3}/lb
Specific gravity of cast film	1.58–1.65
Refractive index, principal	1.51
Light transmission, lower limit of substantially complete	3130

Electrical properties of unplasticized film

Dielectric constant at 25–30 °C	
60 Hz	7–7.5
1 kHz	7
1 MHz	6
Power factor at 25–30 °C	
60 Hz	3–5%
1 kHz	3–6%
Electric charge on rubbing with silk	Negative

Mechanical Properties of Unplasticized Film

Tensile strength at 23°C, 50% relative humidity	9000–16,000 psi
Elongation at 23°C and 50% relative humidity	13–14
Flexibility of 3-to 4-mil film, MIT double folds under 200-g tension	30–500
Hardness, Sward, of glass	90%
Softening point range (Parr)	155–220°C

Solubility and Compatibility Characteristics

Solvents, principal types	Ester, ketones, ether-alcohols
Resins, compatible types	Almost all
Plasticizers, compatible types	Almost all including many vegetable oils
Waxes and tars, compatible types	None
Compatible cellulose derivatives	Ethyl cellulose, cellulose acetate, ethylhydroxyethyl cellulose

Chemical and Physical Properties of Unplasticized Clear Film

Moisture absorption at 21°C in 24 hours in 80% relative humidity	1.0%
Water vapor permeability at 21°C	2.8 g/cm^2/cm/h x 10^6
Sunlight	
effect on discoloration	Moderate
effect on embrittlement	Moderate
Aging, effect of	Slight
Water	
effect of cold	Nil
effect of hot	Nil
General resistance:	
acids, weak	Fair
acids, strong	Poor
alkalies, weak	Poor
alkalies, strong	Poor
alcohols	Partly soluble
ketones	Soluble
esters	Soluble
hydrocarbons	
aromatic	Good
aliphatic	Excellent
oils	
mineral	Excellent
animal	Good
vegetable	Fair to good

Source: Nitrocellulose (technical brochure), Hercules Incorporated, Wilmington, DE.

51

Soybean, Blood, and Casein Glues

Alan Lambuth

Boise Cascade, Boise, Idaho

1.0 SOYBEAN GLUES[1-8]

1.1 Preparation

Of all the available agricultural "seed meals," only soybean meal has ever developed any significant use as a raw material for wood glues. The principal reason is that its protein content, the active constituent for adhesives, is the highest available among commercial seeds and legumes. The protein assay of oil-free soybean meal ranges from about 35 to 55% on a dry basis and averages 44–50% in commercially blended soybean meals and flours. The remaining dry meal content consists of about 30% carbohydrates, 3% fiber, and 6% ash.

In addition to protein, soybeans contain a very high percentage of triglyceride oil, which is useful in many ways. As a result of this unusually high content of both edible protein and unsaturated oil, soybeans have been a very important agricultural crop to mankind for about 5000 years. In view of this long history as a foodstuff, the use of soybeans in adhesives for wood is a very recent development, dating only from about 1920.

In practice, the oil is removed from coarse-ground soybean meal by high pressure extrusion or solvent extraction. The two products are then sold separately into their respective markets. The extracted soybeans, as meal or flour, are widely used for human nutrition around the world. For these food applications, soybean meal is deliberately heated or "toasted" during processing at temperatures above 160°F to enhance the digestibility of its proteins and carbohydrates. When soybean meal is intended for adhesives uses, the processing temperature is carefully maintained below 160°F, to preserve the alkaline solubility of its protein content. To further prepare it for adhesive applications, oil-free soybean meal is ground to an extremely fine flour, mostly through a 325-mesh screen. At this fineness, measurements for process control must be made by a standard test of fine powders for specific surface, calibrated in square centimeters per gram. The normal range of specific surface for "adhesive grade" soybean flour is 3000–6000 cm2/g.

457

1.2 Wood Glues

To become an adhesive for wood or paper, soybean flour must first be wetted with plain water and then reacted with a strong alkali such as sodium hydroxide or trisodium phosphate. (The alkali should not be present during initial wetting, or permanent lumps of dry flour will form.) The chemical action of the alkali on wetted soybean flour particles is to unfold or "disperse" their complex protein structure, making all functional sites available for reaction with wood functional groups. The carbohydrate content of soybean flour becomes similarly dispersed. While this alkaline dispersion step is essential for converting soybean proteins into useful adhesives, it also starts a slow process by which the same proteins gradually lose viscosity and adhesive efficiency through hydrolytic destruction. As a result, soybean glues have a definite working life, usually 6–8 hours at inside temperatures. Fortunately, the acids in wood neutralize most of the dispersing alkali shortly after glue application, so the dry adhesive bonds of soybean glues to wood surfaces are preserved indefinitely.

As the strong alkali in soybean glue reacts with wood, it causes a distinctive brown discoloration called alkali stain. This prominent glue line discoloration has limited the use of soybean adhesives in fine furniture and paper products, for example. As a result, soybean adhesives have found their widest application in structural and paint-grade wood products such as plywood, millwork, flush doors, and prefabricated assemblies. It is possible to prepare nonstaining soybean glues for wood and paper by using much milder alkaline dispersing agents such as calcium or ammonium hydroxide. However, the full adhesive potential of soybean protein is not developed by these moderate alkaline treatments. The resulting glues yield significantly lower bond strengths, adequate for paper and chipboard but not really satisfactory for structural wood joints.

Typical high and low alkali adhesive formulations are given in Tables 1 and 2. The high alkali glue mix was used successfully from about 1930 to 1960 to bond interior-grade softwood structural plywood.

The addition of hydrated lime and sodium silicate in the high alkali formulation accomplishes two purposes: it helps maintain a level glue viscosity for a longer working life, and it improves the water resistance of the cured adhesive bond by forming water-soluble proteinates. Orthophenyl phenol imparts long-term mold and bacteria resistance to soybean adhesive bonds when glued wood products are intended for use in highly humid locations. It is currently approved by the U.S. government for this purpose.

Since the alkaline Ph of calcium hydroxide is so moderate, lime and soybean flour can also be blended into a single dry package, requiring only the addition of water (in two steps) to prepare the adhesive. This formulation has been used extensively as a briquetting binder for charcoal and other powdered materials and for paper laminating to wood.

One of the big advantages of soybean glues over many synthetic resin adhesives is their capability to be cured either hot or cold. Hot curing is accomplished in a conventional steam or oil-heat press at temperatures between 230 and about 270°F. For plywood, a pressure of 175 psi is generally employed. Curing time is about 1.5 minutes per quarter inch of panel thickness, the higher press temperatures being used for the thicker panels. On dry wood, the cold curing of soybean glues is accomplished in an unheated press at about 150–175 psi in 15 minutes. During this clamping cycle, the soybean glue film develops a sufficient gel strength by dehydration into dry wood to hold the plies tightly in contact when pressure is removed. Complete adhesive cure develops at room temperature over the next several days, but machining can begin in about 6 hours. This widely used and patented procedure was

Table 1 Typical High Alkali Formulation for Soybean Glue

Component	Parts by weight
Water at 60–70°F	175
Adhesive-grade soybean flour	97
Pine oil or diesel oil defoamer	3[a]
Mix 3 minutes or until smooth	
Water at 60–70°F	145
Mix 2 minutes or until smooth	
Fresh hydrated lime (as a slurry in)	12
Water at 60–70°F	24
Mix 1 minute	
50% sodium hydroxide solution	14
Mix 1 minute	
Sodium silicate solution	25[b]
Mix 1 minute	
Orthophenyl phenol	5
Mix 10 minutes	

[a]Normally blended with the soybean flour for dust control.
[b]8.90% Na_2O, 28.7% SiO_2, 41° Baumé.

called the Noclamp Process. It is unique to soybean-based adhesives and certain low soluble blood glues.

1.3 Blends

Another unique feature of soybean proteins as adhesives is their compatibility with other protein adhesive materials to yield "blend glues" of enhanced performance properties. The two most widely used combinations have been with dried soluble animal blood and with

Table 2 Typical Low Alkali Formulation for Soybean Glue

Component	Parts by weight
Water at 60–70°F	225
Adhesive-grade soybean flour	97
Pine oil or diesel oil defoamer	3[a]
Mix until smooth	
Water at 60–70°F	150
Mix until smooth	
Fresh hydrated lime (as a slurry in)	30
Water at 60–70°F	50
Mix 5 minute	

[a]Normally blended with the soybean flour for dust control.

Table 3 Extracted Soybean Protein Formulation

Component	Parts by weight
Water at 65–75°F	385
Extracted soybean protein	100[a]
China clay	100[a]
Powdered Sodium sulfite	1[a]
Pine oil or diesel oil defoamer	1[a]
Mix 15 minutes	
50% sodium hydroxide solution	6
Mix 15 minute	
Hexamethylenetetramine in	3
Water at 65–75°F	6
Mix 5 minute	

[a]Optionally blended together.

milk casein. In each case, the somewhat granular dispersed consistency of the soybean glue adds useful working and curing properties to those of the two animal-derived proteins. The normal combining limits are between about 20:80 and 80:20 soybean to animal protein, depending on the performance properties desired. Examples of these blended protein adhesives may be found in Sections 2.0 and 3.0.

For applications requiring extremely smooth and fluid soybean glues, more in the nature of size coatings, soybean proteins are available as extracted and dried powders at somewhat higher cost. Typical uses would be as paper overlay adhesives, boxboard adhesives, water-base paint binders, and paper sizes. For structural wood bonding, they are not as cost effective as soybean flour adhesives and they also tend to flow excessively in wood joints under pressure. Table 3 describes a typical paper overlay glue.

This glue is extremely smooth and stable. Its formulation illustrates the use of a formaldehyde donor, in this case hexamethylenetetramine, to partially cross-link the dispersed protein for longer working life and improved water resistance. Other formaldehyde sources and addition compounds have been widely used with soybean glues for these purposes. The additions are always small, about 1%, and are usually made at the end of the mix.

For sizes and coatings, other moderately alkaline salts are frequently used to disperse fine soybean flour or extracted soybean protein. These include the sodium or potassium carbonates, phosphates, and borates, as well as ammonium hydroxide. The borates offer some advantage with soybean flour, since they also complex the carbohydrate constituents to yield increased viscosity and improved tack. None of these compositions can be considered to be adhesives in the structural wood bonding sense, however.

2.0 BLOOD GLUES[4,6,9–13]

2.1 Preparation

Animal blood has a very long history of use as an adhesive in lime mortars, cements, and wood glues. From the Far East to the Baltic to the early Americas, there have been legends and reports of independently discovered adhesives applications for blood. In all cases, the raw material was fresh whole blood, which was subject to rapid spoilage. This undoubtedly limited broader historical uses for blood protein glues.

Blood adhesive technology as we know it today began in about 1900 with the development of a commercial method for drying fresh whole blood without causing it to lose water solubility. Once this had been accomplished, blood proteins could be dried and stored indefinitely for use on demand. The only preparatory step needed was the removal from freshly collected animal blood of the clotting substance fibrin, to make the blood stable for processing.

The most commonly available dried animal bloods are beef and hog, with a lesser quantity from sheep. On a worldwide basis, their primary uses at this time are as protein-rich feed supplements and edible binders for domestic animal and pet foods. In certain cultures these bloods are used extensively for human nutrition. Poultry blood, which has a very high lysine content (and too low an intrinsic viscosity for adhesives), is used exclusively as a feed supplement.

For most food applications, whole blood is quickly coagulated to total insolubility with dry heat or steam. For adhesive applications, it is very carefully dried in vacuum pan ovens or spray dryers to yield controllable levels of cold water solubility. Dried bloods in the range of 80 to about 93% solubility (marketed as high soluble bloods) dissolve almost completely in cold tap water. On the addition of alkali to the water, they become extremely smooth, livery gels. Dried blood particles in the range of 25–40% solubility (low soluble bloods) are really wetted only in cold water. They require the addition of a fairly strong alkali to the water to become completely dissolved and dispersed into useful adhesive form. Their dispersed consistency is always quite grainy, a characteristic that is particularly useful for certain types of adhesive, described later. Dried bloods in the 40–70% solubility range have intermediate dispersed consistencies, tending toward smoothness as the solubility level rises.

2.2 Formulating

The adhesive constituents of dried animal blood include serum albumin and globulin and also red cell hemoglobin. Collectively, they provide nearly 100% adhesive-functional proteins in dried blood solids. Due to differences in composition and proportions of these proteins in the blood from various animals, there is wide variation in alkaline dispersed viscosity levels among soluble bloods, hog yielding the lowest and beef the highest. There are also significant viscosity variations within blood samples from a single species due to differences in age, activity, and nutrition. As a result, commercial lots of dried blood for adhesive uses are always blended in large quantities to help maintain uniformity of glue performance.

In formulating adhesives from soluble animal bloods, they dry powder must be initially wetted and redissolved in plain water. The wetted blood is then subjected to one or more alkaline dispersing steps to unfold the protein molecules and render them fully adhesive. (If the initial water is alkaline, permanent lumps will form.) Unlike the vegetable proteins in legumes such as soybeans, high soluble bloods can be adequately dispersed with moderately alkaline compounds including hydrated lime and ammonia to become useful wood adhesives. Simple dispersions of this kind represented the earliest class of blood-based adhesives discovered and used around the world. Much later it was learned that the addition to simple blood dispersions of chemical cross-linkers such as aldehydes greatly increased their water resistance. Glues of this type were successfully used during and after World War I to bond aircraft propeller laminations and other structural components. In this form, blood

Table 4   Typical Commercial Formulation for Blood Glue

Component	Parts by weight
90% soluble dried animal blood	100
Water at 60–70°F	80
Mix 3 minutes or until smooth	
Water at 60–70°F	60–120[a]
Mix until smooth	
Ammonium hydroxide, specific gravity 0.90	6
Mix 3 minutes	
Powdered paraformaldehyde (sift in slowly while mixing)	15
Allow to stand 30 minutes	
Mix briefly until glue is fluid and smooth	

[a]Variable for viscosity control.

glues represented the most water-resistant wood bonding adhesives available until the advent of phenol–formaldehyde resins about 1932. Table 4 gives a commercial example.

Ammonium hydroxide yields sufficient dispersion at moderate alkalinity to expose most of the adhesive-functional polar groups on the protein structure. The paraformaldehyde actually gels the blood protein briefly, but it thins out again to yield a workable adhesive viscosity. The useful life is about 8 hours. This glue delivers adequate bonds when cold pressed, but it provides the most water-resistant bonds if hot pressed to a minimum attained temperature of 160°F.

Table 5   Typical Commercial Formulation for High Alkaline Blood Glues

Component	Parts by weight
Water at 60–70°F	300
20% Soluble dried animal blood	75
200-mesh wood flour	22[a]
Diesel oil (defoamer)	3[a]
Mix 3 minutes or until smooth	
Water at 60–70°F	330
Mix 2 minutes or until smooth	
Fresh hydrated lime (as a slurry in)	10
Water at 60–70°F	20
Mix 1 minute	
50% sodium hydroxide solution	16
Mix 10 minutes	
Sodium silicate solution	35
Mix 5 minutes	

[a]Normally dry blended with the dried blood for easier handling and dust control.
[b]8.90% Na_2O, 28.70% SiO_2, 41° Baumé.

During World War II and for 20 years thereafter, blood protein glues were used extensively in the manufacture of interior and intermediate grades of structural plywood (water-resistant but not waterproof). By this time it was found that the highly alkaline multistep dispersion techniques employed with soybean glues yielded excellent consistency and adhesion when applied to dried animal bloods in the lower range of solubility. The resulting glues were used in large quantities to bond structural plywood until about 1960. (See Table 5).

Since prior heat treatment of the blood itself has already reacted or "denatured" a significant portion of its protein content toward insolubility, glues of this type bond strongly and develop significant water resistance when cured without heat. In addition, the lime and silicate dispersion steps yield protein derivatives, which further insolubilize these blood glues up on cold curing. For optimum bond strength and weather resistance, however, hot pressing to a minimum glue line temperature of 160°F is still recommended. Whether cured hot or cold, blood glues of this type normally meet the published mold resistance requirements for plywood without the addition of preservatives.

Another form of blood glue employed extensively from World War II until about 1960 (and again during the petrochemical shortage of the 1970s) involves blends of high soluble blood with "adhesive-grade" soybean flour. In this combination, blood protein provides the water-resistant bonds and rapid hot cure, while the soybean flour provides the desired granular consistency for machine application and assembly time tolerance and, in addition, reduced cost. A commercial example of these so-called blend glues is given in Table 6.

Unlike straight, low soluble blood adhesives, soybean-blood blend glues must be hot pressed to ensure adequate water resistance and must also contain a preservative. As an alternative to orthophenyl phenol, final additions of alkaline phenol-formaldehyde liquid resins (resoles) have proved effective for imparting mold resistance to these blend glues. About 15 lbs of a 43% solids phenolic plywood adhesive resin would serve well in the formulation of Table 6.

Up to this point, the adhesives discussed have relied on alkaline dispersed animal or vegetable proteins as the active ingredients. Another type of protein-containing wood glue is commercially significant, namely, alkaline phenol-formaldehyde (PF) resin adhesives fortified with soluble dried blood. In this case, the PF resin solids are the principal adhesive constituents, yielding completely waterproof bonds. The blood proteins are also fully durable when cured in a matrix of phenolic solids. In addition, however, they provide the benefit of considerably shorter hot press curing times compared with straight resin adhesives. The blood proteins in this mix, fully dispersed by resin alkalinity, quickly develop a very adequate initial heat-cured bond. The phenolic resin polymers cure more slowly but completely by means of latent heating as the bonded product (plywood) is not-stacked after a relatively short pressing cycle. For an example of this combined adhesive, see Table 7.

The presence of blood protein in this alkaline adhesive mix allows the normal phenolic resin hot press curing times to be reduced about 30%. The blood solids are considered to be additive to the phenolic resin solids in terms of contributing to exterior durability.

Currently, a blood-phenolic resin glue similar to this one is being used as a foamable adhesive that permits glue application by extrusion onto dry veneer. After veneer assembly, plywood is produced by hot pressing in the normal manner. The only formulation change needed is a substitution of surfactant for defoamer to promote air entrainment by the dispersed blood protein. (A special intensive mixing device is employed.) This "air extension" of the phenolic adhesive permits a 25% reduction in glue application weights with no apparent loss of adhesion or durability.

Table 6 Typical Commercial Formulation of a Blood-Soybean "Blend Glue"

Component	Parts by weight
Water at 60–70°F	180
Adhesive-grade soybean flour	62
93% soluble dried animal blood	25
200-mesh wood flour	10[a]
Pine oil or diesel oil defoamer	3[a]
Mix 3 minutes or until smooth	
Water at 60–70°F	260
Mix 2 minutes or until smooth	
Fresh hydrated lime (as a slurry in)	8
Water at 60–70°F	16
Mix 1 minute	
Sodium silicate solution[b]	40
Mix 1 minute	
50% Sodium hydroxide solution	10
Mix 5 minute	
Orthophenyl phenol	5
Mix 5 minutes	

[a]Normally blended with the soybean flour and dried blood for dust control.
[b]8.9% Na_2O, 28.70% SiO_2, 41° Baumé.

Table 7 Typical Commercial Formulation of PF Resin Adhesive Fortified with Dried Blood

Component	Parts by weight
Water at 70–80°F	291
200-mesh nutshell flour, or	100
equivalent cellulosic extender	
Industrial wheat flour	30
Dried 90% soluble animal blood	30
Mix 2 minutes or until smooth	
50% sodium hydroxide solution	30
Mix 20 minutes	
Diesel oil (defoamer)	6
41% solids alkaline phenol	1500
formaldehyde plywood adhesive	
resin, added slowly	
Mix 5 minutes or until smooth	

A few additional observations may assist in understanding the proper applications for blood protein adhesives.

Virtually all blood glues are dark reddish-brown. As a result, they are generally excluded from applications in which low color glue lines are a requirement. (e.g., furniture, fine millwork, paper bonding).

Alkaline-dispersed blood glues are unusually sensitive to curing with heat. Thus, their use is well worth considering when rapid hot pressing is essential.

When cured cold, alkaline low soluble blood glues perform as well as soybean glues in utilizing a patented short clamping cycle of 15–20 minutes for plywood and flush doors. The gel strength developed in this length of time is adequate to maintain the bond between the plies after pressure is released while full cure is still developing (U.S. Patent 2,402,492).

The working life of alkaline blood glues is about 8 hours at room temperature. By this time, the dispersing alkali itself has begun to degrade the adhesive protein structure through hydrolysis. The addition of complexing chemicals such as formaldehyde donors helps to control this reaction and lengthen useful adhesive life.

Blood proteins have some use in paper bonding if the colored hemoglobin is removed from the remaining serum constituents. The resulting protein mixture is similar to casein in cost and performance and its utilized in much the same way.

3.0 CASEIN GLUES[9,14–19]

3.1 Preparation

According to archaeological evidence, the adhesive qualities of casein curd from milk were recognized by civilizations as early as that of the Egyptians. The record of the actual use of casein as a glue is more detailed from medieval European history, however. By that time, aqueous mixtures of casein with lime or other alkaline materials were being used for furniture gluing, paint pigment binding, and canvas sizing, to name just a few applications. In the area of furniture and musical instrument assembly, casein glues competed with gelatin glues extracted from animal bones and hides. The gelatin glues were applied hot and yielded quick bonds on cooling. However, they remained permanently sensitive to heat and moisture. Alkaline casein glues stayed fluid and sticky at room temperature for a considerable length of time, permitting the complex assembly of parts, but they did require clamping until cured. The hardened casein glue bonds had a substantial degree of heat and moisture resistance. It is these attributes of casein glue that have characterized its performance from early history down to the present; namely, long working or assembly time tolerance, cold-clamping until cured, and significant heat and moisture resistance.

Casein protein is recovered from skim milk by acid precipitation at a pH of about 4.5. Mineral acids can be used to promote precipitation, or the milk can be cultured with a *Lactobacillus* that converts milk sugar to lactic acid, which in turn precipitates the casein. In either case, the precipitated casein curd is washed free of acid, dried, and ground. Casein is frequently identified for sale by the method of its precipitation (i.e., "lactic acid casein").

Because of the widespread use of milk and its by-products as foods, the price of industrial casein tends to rise and fall with the balance of production versus demand for dairy foods on world markets. Currently, the price is $2.50 per pound for edible-grade ground casein, which causes casein-based glues to be fairly expensive adhesive systems. At other times the price per pound may be a dollar or less. Regardless of current price, the unique properties of casein glues will generally justify their continued use.

In preparing casein for use as a wood glue, the ground product is screened to provide a particle size range of about 30 to 60 mesh. Coarse particles are recycled. Finer particles become the raw material for a variety of paper sizes and adhesives, ranging in application from foils to labels to cigarette paper gluing and bookbinding. The reason for the limited particle size range in casein woodworking glues is related to their formulation and storage as one-package dry compositions requiring only the addition of water to yield working adhesives. These dry compositions contain a substantial proportion of ground casein plus one or more alkaline salts in granular form that provide dispersing alkalinity as soon as water is added. The entire dry composition is oiled with a petroleum product, typically diesel oil, to prevent the absorption of atmospheric humidity during storage and to slow the rate of solubility of all constituents when water is added. This permits all the casein particles to become uniformly wetted in essentially neutral water. Since some alkalinity is present in the mixing water almost immediately, any finely ground casein particles present in the dry composition will quickly form permanent encapsulated lumps. Thus, casein fines simply cannot be present.

3.2 Formulation

In commercial practice, at least two alkaline agents are used in the formulation of modern casein woodworking glues. A carefully controlled proportion of hydrated lime is present to promote a gradual reaction with the dispersed casein protein to form calcium caseinate, which contributes significantly to water resistance. Strongly active alkali metal salts such as trisodium phosphate, sodium carbonate, and sodium fluoride (which forms sodium hydroxide by double decomposition with lime) are also included to accomplish complete protein dispersion. Together, they convert the casein granules into a slick, viscous consistency, which is an excellent cold or hot press-curing adhesive for wood. Some cellulosic filler may also be included to provide glue application control and gap-filling properties. Table 8 gives a typical casein lumber laminating composition. The dry ingredients are intensively blended in an appropriate mixer while defoamer is sprayed in slowly to provide uniform distribution. The dimethylol urea, a very effective protein cross-linker and denaturant, is

Table 8 Typical Casein Lumber Laminating Composition

Dry glue composition	Parts by weight
30–60-mesh lactic acid casein	30
30–60-mesh sulfuric acid casein	30
200-mesh wood flour	10
Fresh hydrated lime	13
Granular trisodium phosphate	8
Granular sodium fluoride	4
Powdered dimethylol urea	0.1
Diesel oil (defoamer)	2.9
Sodium orthophenylphenate	2
	100

added in small variable amounts for glue viscosity control. The casein materials described in Table 8 would be mixed as follows. 200 lb of water at 60–70°F and 100 lbs of dry glue (as in Table 8) for

1. Mix 2 minutes or until smooth and thickening begins.
2. Let stand 15 minutes or until thinning has occurred.
3. Mix 2 minutes or until smooth.

The early thickened stage through which these adhesives pass requires strong agitation, preferably with counterrotating paddles and a sidewall scraper blade.

Casein adhesives of this type were widely used for lumber laminating, specialty gluing, and millwork assembly from 1930 to 1970, more or less, before being supplanted by the newer synthetic resin adhesives. Their long assembly time tolerance and cold cure made the casein adhesives ideal for manufacturing very large curved laminated beams, for example. Similarly, their gap-filling properties, moisture resistance, and deep adhesion to almost any wood surface contributed to their reputation as one of the finest general-purpose wood adhesives ever developed from a natural source.

One current application for casein glues that has not yet been preempted by synthetic resin adhesives is its use in panel and flush door assembly. For this application, casein is used in combination with adhesive-grade soybean flour to yield alkaline-dispersed protein glues of composite performance. Casein contributes deep bonds, tacky consistency, and heat resistance (particularly important for fire doors), while soybean flour contributes the granular consistency that permits quick water loss and a short clamping cycle cold cure. The fire door application (Table 9) is essentially unique among protein glues. As before, this is a single-package adhesive requiring only the addition of water.

The dry ingredients are intensively blended in an appropriate mixer while the defoamer is sprayed in slowly to provide uniform distribution. From the standpoint of glue consistency, the proportion of casein shown is about the maximum at which short cycle clamping of doors and millwork is still possible. Mixing directions are as follows. 200 lb of water at 60–70°F and 100 lb of dry glue (as in Table 9) for

Table 9   Single-Package Casein Glue for Use in Fire Doors

Dry glue composition	Parts by weight
"Adhesive-grade" soybean flour	58
60-mesh lactic acid casein	19
Fresh hydrated lime	7
200-mesh wood flour	5
Granular sodium carbonate	5
Granular sodium fluoride	2
Granular trisodium phosphate	1
Diesel oil (defoamer)	3
	100

1. Mix 2 minutes or until smooth and thickening begins.
2. Let stand 15 minutes or until thinning has occurred.
3. Mix 2 minutes or until smooth.
4. Add 50 lb of water at 60–70°F, and mix 2 minutes or until smooth and fluid.

The second water addition may be varied to obtain the desired final viscosity. In general, casein glues perform best on wood in a viscosity range of about 4000–8000 cp at room temperature. The application rate to wood surfaces is about 70–90 pounds per 1000 square feet of joint area.

Casein-containing adhesives for all forms of paper gluing are distinctly different in chemistry. For example, calcium and magnesium salts should *not* be present. Since these glues must be only mildly alkaline to avoid staining paper surfaces, a different approach to protein dispersion is employed. Typically, this involves preparing a stable casein solution containing 15–20% casein solids by using a mild alkali such as ammonium hydroxide or borax and then heating the solution to about 170°F for 10–15 minutes to complete the dispersion. Rapid cooling should follow to prevent degradation. Small amounts of preservatives, protein cross-linkers, and defoamers are frequently added to yield a storable product of the desired viscosity and running properties.

Casein solutions can be used alone, usually with further cross-linking and thickening, as label pastes, bag adhesives, and cigarette paper seam glues, and for a wide variety of other precision paper bonding applications. For these purposes, the same performance characteristics important to wood bonding recommend their use, namely, a sticky, wipe-resistant consistency, strong bonds to cellulose fibers, and significant water resistance.

Casein solutions are often combined with synthetic latex such as neoprene acrylic or styrene-butadiene to yield strongly adhering yet resilient paper coatings and adhesives. These combinations are specifically used for laminating metal foil to paper. The casein "glue" imparts a degree of rigidity for paper handling which the latex themselves tend to lack.

REFERENCES

1. G. H. Brother, A. K. Smith, and S. J. Aricle, "Soybean protein." Washington, D.C., U.S. Department of Agriculture, Bureau of Agricultural Chemistry, 1940.
2. R. S. Burnett, *Soybeans and Soybean Products*, New York, Wiley-Interscience, 1951.
3. A. L. Lambuth, "Soybean adhesives," in *Handbook of Adhesives*, New York, Reinhold, 1962.
4. T. D. Perry, *Modern Wood Adhesives*, New York, Pitman, 1944.
5. Douglas Fir Plywood Association [now American Plywood Association],"Mold resistance of plywood made with protein adhesives, technical report, Tacoma, WA, DFPA, 1952.
6. U. S. Department of Agriculture Forest Products Laboratory, "Bonding wood and wood products," in *Wood Handbook: Wood as an Engineering Material*, Agriculture Handbook 72, Madison, WI, 1987, USDA Forest Products.
7. J. Bjorksten, "Cross linkages in protein chemistry," in *Advances in Protein Chemistry*, Vol. 6. New York, Academic Press, 1951.
8. Technical Association of the Pulp and Paper Industry, "Protein and synthetic adhesives for paper coating," Monograph No. 9. Atlanta, 1952, TAPPI.
9. R. H. Bogue, *The Chemistry and Technology of Gelatin and Glue*, New York, McGraw-Hill, 1922.
10. U.S. Department of Agriculture Forest Products Laboratory, "Blood albumin glues: Their manufacture, preparation and application," Laboratory Report No. 281–282, Madison, WI, USDA Forest Products Laboratory, 1938 (rev. 1955).
11. A. L. Lambuth, "Blood glues," in *Handbook of Adhesives*. New York, Reinhold, 1962.

12. L. B. Lane and J. J. Frendries, "How blood adhesives are used." *Adhes. Age*, May, 1961.
13. Technical Association of the Pulp and Paper Industry, "Protein and synthetic adhesives for paper coating," Monograph No. 22, Atlanta, TAPPI, 1961.
14. E. Sutermeister and F. L. Browne, *Casein and Its Industrial Application*, New York, Reinhold, 1939.
15. U.S. Department of Agriculture Forest Products Laboratory, "Casein glues: Their manufacture, preparation and application," Research Note FPL-0158, Madison, Wi, USDA Forest Products Laboratory, 1967.
16. U. S. General Services Administration, "Adhesives, casein-type, water and mold resistant," Federal Spec. MMM A–125, Washington, D.C., GSA, 1955.
17. D. M. Weggemans, "Adhesives application charts," *Adhes. Age*, 16, October, 1973.
18. H. K. Salzberg, "Casein glues and adhesives," in *Handbook of Adhesives*, New York, Reinhold, 1962.
19. Technical Association of the Pulp and Paper Industry, "Protein and synthetic adhesives for paper coating," Monograph No. 9, Atlanta, TAPPI, 1952.

52
Fish Gelatin and Fish Glue

Robert E. Norland

Norland Products, Inc., North Brunswick, New Jersey

1.0 INTRODUCTION

Fish gelatin, or fish glue, is a proteinaceous material extracted from the skins of deep coldwater fish such as cod, haddock, and pollock. The skin and bone of all animals and fish contain collagen, which can be hydrolyzed in hot water and dilute acid to form soluble gelatin.[1-3]

The difference between gelatin and glue is in the extent of processing. A gelatin processed for photographic use would be exceptionally pure, of high molecular weight, and devoid of specific chemical impurities. An edible gelatin would be free of heavy metals, microorganisms, and any impurity that would make it unsuitable for human consumption. Inedible gelating or glue does not require as much refining and is suitable for industrial or adhesive applications, where less stringent requirements are found. Fish gelatin coatings are used in all these applications.

2.0 PROPERTIES

Fish gelatin is a long chain protein molecule containing 20 different amino acids. It is amphoteric; that is, it can react as a base or acid. End groups include hydroxy, carboxy, and amino. Reactivity of each group will depend on the pH of the gelatin solution. For example, the reaction of phthalic anhydride with gelatin on the alkaline side will produce a gelatin that will precipitate at pH 4.

Fish gelatin is insolubilized by the addition of salts of polyvalent cations such as ferric sulfate or chrome alum.[4] This principle finds use in the light sensitivity of fish gelatin coatings containing bichromates. The action of ultraviolet light on the bichromated gelatin results in the formation of polyvalent chromium ions, which insolubilize the light-struck area, leaving the unexposed areas water soluble.

Aldehydes such as formaldehyde, glutaraldehyde, and glyoxal will insolubilize gelatin. The aldehydes react with the amino end group, and this occurs more quickly on the alkaline side.

Fish gelatin solutions are liquid at room temperature, hence make good coating vehicles. Many of the reactions above can be tailored so that water solutions can be coated and dried, and the dried coating then becomes insoluble.

Dried fish gelatin films are insoluble in organic solvents, but a water solution will tolerate water-miscible solvents. The following are tolerance levels of various solvents in 100 parts of 45% fish gelatin.[5]

Acetone	25 parts
Ethyl alcohol	50 parts
Methyl Cellusolve	95 parts
Dimethylformamide	110 parts

Dried fish gelatin coating will not soften at elevated temperatures and can withstand temperatures up to 260°C. Baking at high temperatures ($\leq 275°C$) will insolubilize gelatin coatings, but they can be stripped with dilute caustic and by proteinaceous enzymes.

The following are typical properties of fish gelatin and glue.[5]

Color	Clear, light amber
Solids	45% in water
Viscosity at 20 °C	6000–8000 cp
Average molecular weight	60,000 d
Gel point	5–10 °C
Ash	0.1%
pH	4.6–5.4

3.0 APPLICATIONS

3.1 Remoistenable Coatings: Gummed Tape

Fish glue coatings are excellent for gummed tape, labels, and tags. The dried coating wets very easily and develops a tackiness that allows it to be bonded quickly to another surface. It is compatible with animal glue and can be added to increase the latter's tackiness.

The addition of a humectant will reduce curl of the coated paper. Glycerine, sorbitol, or glycols are suitable. Use 5–10% based on dry weight.

3.2 Photoresists for Photochemical Machining

Photochemical machining involves the manufacture of thin metal parts using a photoresist to image an acid-resistant stencil on the metal sheet.[6,7] Photoengraving glue, a clarified fish gelatin, is used as the base for a water-soluble photoresist.[8] The advantage over using a solvent-type organic photoresist is the ease of processing, as no organic solvents are required. Water solutions are used in the entire process. One suggested formula consists of 100 parts by weight,photoengraving glue (45%) 10 parts by weight ammonium bichromate, and 125–200 parts by weight water.

The amount of water to be added can be varied to give a specific coating viscosity. A coating thickness of 0.15–0.25 mil (4–6 μm) is sufficient for etching cold rolled steel, stainless steel, and copper up to 10 mils thickness. This coating is sensitive to ultraviolet light (300–400 nm) and will have a shelf life of 2–5 days, depending on solids and storage temperatures.[9]

Care must be taken in the use of ammonium bichromate. It is listed as a hazardous chemical, and users must conform to environmental laws. Untreated wastewater containing bichromate should not be discharged into the sanitary sewer or any other water supply.

3.3 Dyed Patterns on Glass

Bichromated photoengraving glue images also have good adhesion to glass. These images can be dyed using anionic water-soluble dyes. The same formulation given above can be used, but viscosities must be reduced to obtain a thickness of 0.10 mil (2.5 µm). Images of 10 µm wide or less are possible, and these can be reproduced very accurately.

For very fine reticles on glass, silver wash-off emulsions can be made using fish gelatin. Silver chloride is precipitated in diluted fish gelatin by the dropwise addition of silver nitrate and sodium chloride. Upon completion, the emulsion is precipitated by the addition of 10% polystyrene sulfonate based on dry gelatin weight and adjustment to a pH 4. This precipitates the emulsion, and the remaining liquid containing dissolved salts is poured off. The emulsion is redissolved by neutralizing and sensitized by the addition of ammonium bichromate or an organic sensitizer such as 4.4'-diazidostilbene-2,2'-disulfonic acid. Spin coating on glass will result in a 0.1 mil (2.5 µm) coating. After exposure to ultraviolet light through a negative and washing an image will appear on the glass. This image can be made black by using a photographic developer such as Eastman Kodak Dectol.

3.4 Ceramic Stencils

Colored pigments can be ball milled into fish glue. These inks can be stamped or stenciled onto ceramic parts and the resultant image fired to fix the pigment to the ceramic part.

3.5 Electrical Insulators

Fish gelatin, purified to have low ash, can be coated and then baked to reduce the water sensitivity. The coating will have electrical insulating properties.

3.6 Temporary Protective Coatings

Because of their water solubility, fish gelatin coatings can be used as temporary protective coatings on metal parts during mechanical processing and handling. When complete, the parts are washed with a dilute alkaline cleaner to provide a chemically clean and sterile surface.

3.7 Plating Release Agents

Fish gelatin is a liquid protein solution of high molecular weight that has a good attraction to clean metal surfaces. When a clean metal plating master is dipped into a dilute solution, a monomolecular layer attaches itself to the metal surface, which acts as a releasing agent after plating. After plating on the master, the negative can be separated easily by starting at one edge and peeling the two pieces apart. The master is then recleaned, dipped into the dilute gelatin solution, and replated.

3.8 Fish Gelatin in Photographic Coatings

Fish gelatin can be used in the silver halide emulsion system. Because fish gelatin solution are liquid at room temperature (down to 10°C), silver halides can be precipitated at tempera-

Table 1 Effect on Gelling Temperature of
Varying Amounts of Animal and Fish Gel

Animal gel (%)	Fish gel (%)	Gelling temperature (°C)
100		32
75	25	30
50	50	26
25	75	24
15	75	20
10	90	19
5	95	17
	100	8

tures less then required for animal gelatin. The resultant emulsion can be washed of soluble salts by using any of the common precipitating agents such as polystyrene sulfonate.[3]

Photographic film usually requires a sub layer between the emulsion and the film, to increase the adhesion of the emulsion to the film. Photographic film also requires a back coating of gelatin to balance the curling stresses of the emulsion coating. Fish gelatin can be used in both these applications.

3.9 Gelatin Capsules

Fish gelatin added to animal gelatin will reduce the softening point of animal gelatin capsules, and the resultant capsules will dissolve more readily. Table 1 gives an example of the effect on gelling temperature of the addition of fish gelatin to animal gelatin.

4.0 CONCLUSION

Fish gelatin and fish glue have the following features:

Supplied as a pourable liquid.
Completely water soluble.
Excellent adhesion to metal, rubber, glass, leather, cork, wood, and paper.
Act as protective colloids to suspend small particles or monomers in solution.
Coatings can be made water resistant and insoluble in water.
Insoluble coatings will accept water-soluble dyes.
Insoluble in organic solvents.
Dry to a hard, smooth finish.

REFERENCES

1. R. A. Berg, "Fish gelatin in coating applications," in *Image Technology 1985* (38th Annual SPSE Conference), Atlantic City, NJ, pp. 106–109.
2. K. H. Gustavson, *The Chemistry and Reactivity of Collagen*, New York, Academic Press, 1956.

3. R. E. Norland, "Fish gelatin, technical aspects and applications," in *Proceedings of the Fifth Royal Photographic Society Symposium on Gelatin*, Wadham College, Oxford, 1985, pp. 266–271.

4. *Technical Data Photoengraving Glue Bulletin 1*, Norland Products Inc., New Brunswick, NJ, 08902.

5. *Technical Data HiPure Liquid Gelatin*, Norland Products Inc., New Brunswick, NJ, 08902.

6. "Chemically milling precision parts," Am. Mach., Feb. 8, 1971.

7. J. F. Holhan, *Electronic World*, December 1965.

8. J. Kosar, *Light-Sensitive Systems*, New York, Wiley, 1965.

9. M. Sasaki and K. Honda, *Photosensitivity of Gelatin Films Containing Hexavalent Chromium Oxysalts*, Graphic Arts of Japan, 1972, pp. 25–33.

53

Waxes

J. David Bower

Hoechst Celanese Corporation, Somerville, New Jersey

1.0   INTRODUCTION

A discussion on the use of waxes for coatings must begin with an explanation of what constitutes a wax. Because of the diverse nature of products included in the class of waxes, this is not a simple objective.

There are few references that discuss waxes and their applications. In 1956 A. H. Warth published a book that covered the types and applications of waxes available[1]. More recent general references on wax are a two-volume work entitled *Industrial Waxes* by H. Bennett[2] and *Wax: An Introduction* by R. Sayers[3]. The Chemical Specialties Manufacturers Association (CSMA) has also published a booklet entitled *Technology of Waxes*, which contains several articles dealing with the chemistry and physical properties of waxes of various types[4].

These are the major sources for information on waxes. Numerous articles are available, which may discuss one specific type of wax or cover an application area, often only casually mentioning wax usage.

2.0   DEFINITION

In the medieval days (and earlier) the definition of waxes was very simple. Wax was the substance produced by bees that prevented the honey from running out the bottom of the hive. Beeswax was the only wax known. In 1848 Justus Liebig characterized beeswax as consisting predominantly of long chain fatty acids, esters, and alcohols. Several other natural waxlike products from plant and animal sources had a similar chemical makeup. The definition of a wax was still fairly vague. If the chain length of a substance was longer than the C_{16}–C_{18} associated with fatty acids, in a chemical sense it was a wax.

In the twentieth century, a large number of new products were introduced. Some of these were waxlike but did not conform to the chemical definition. People wanting wax defined by physical properties won out over those supporting a chemical definition. Therefore,

Table 1 Definition of Wax from a German Technical Society

1. 20°C (68°F) kneadable solid to brittle hard.
2. Macro to microcrystalline, translucent to opaque but not glasslike.
3. Melting above 40°C (104°F) without decomposition.
4. Relatively low viscosity slightly above the metling point.
5. Consistency and solubility are strongly dependent on temperature.
6. Capable of being polished (i.e., buffable) under slight pressure.

Source: Einheitsmethoden, Abteilung M—Wachse, proposed in 1954 and
revised in 1975 by the Deutsche Gesellschaft für Fettswissenschaft.
The term "wax" is a collective name for a number of natural or synthetic
substances. If, in borderline cases, more than one of these properties cannot
be met by a substance, it does not represent a wax in the actual meaning of
this definition.

products that have waxlike properties but are not based on long chain esters (e.g., paraffins
and low molecular weight polyethylenes) are referred to today as waxes.

In 1954 the Deutsche Gesellschaft für Fettwissenschaft (DGF) established a wax definition based on physical properties and abandoned the efforts to maintain the chemical definition. A revised version was published in "DGF—Einheitsmethoden Abteilung M—
Wachse" in 1975 (Table 1). The failure to put specific limits on most of the listed properties
(e.g., no upper limit on melting property and no specific viscosity range) demonstrates the
diverse properties of waxes. Only five of the six DGF properties must be met for a product to
fall under the definition of wax.

3.0 TYPES

One generally attempts to classify waxes into one of the three basic categories of animal,
vegetable, or mineral. But even this approach encounters difficulties. Mineral waxes are in
reality primordial vegetation deposits subjected to pressure and millions of years of aging to
form oil and coal (i.e., aged vegetable wax). A fourth classification is also needed for the
numerous synthetic waxes that are derived from the three main sources. The question also
arises as to how much refining and alteration of a wax can be made before it is not longer
considered a natural wax but a synthetic wax.

3.1 Animal Waxes

Animal waxes are derived from two separate sources. One source is by-products produced
by the animal (beeswax). The second source is from the animal itself (fatty acid derivatives).

Lesser known waxes include Chinese and shellac waxes obtained from secretions of
various insects and wool wax refined from lanolin (extracted from sheep's wool). Commercially, they are found only in a few specialized areas.

3.1.1 Fatty Acids

Fatty acids and their derivatives come from processing tallow and lard. These products are
produced in very large volume. The production is often considered to be an industry in and
of itself, and these materials are not always thought of as wax products. Chemically they

consist of C_{16} and C_{18} esters of glycerine. Typical products are stearic acid, glycerol tristearates, and glycerol monostearates.

3.1.2 Beeswax

Beeswax is the oldest known wax. It consists of long chain fatty acids and long chain monovalent alcohols. The exact composition of beeswax is extremely complex, and even today an exact duplicate based on synthetic materials has not been accomplished.

3.2 Vegetable Sources

Vegetable waxes are obtained from the leaves, stems or berries of plants. These waxes consist mainly of alkyl esters. Chain length and type vary considerably depending on the surface.

3.2.1 Leaf Waxes

Leaf waxes are obtained by extracting wax from the leaves. They are also referred to as palm waxes.

The leaf wax of greatest commercial significance, carnauba wax consists mainly of esters of alkyl alcohols (C_{24}–C_{34}) and alkyl acids (C_{18}–C_{30}). There are also substantial amounts of diesters and hydroxy esters. These hydroxy esters are believed to provide some of the unique paste forming properties and the ability to form microemulsions associated with carnauba wax.

Carnauba wax is extracted from the leaves of the carnauba palm (*Copernica cerifera*), which grows in the north and northeast regions of Brazil. The tree also grown in other regions (Sri Lanka, West Central Africa), but only in Brazil are weather conditions suitable (torrential rains followed by a hot, dry windy period) to form significant wax deposits on the leaves.

3.2.2 Stem Waxes

Stem waxes include flax wax, sugarcane wax, and bark wax (Douglas fir). Efforts to develop these waxes commercially have met with limited success.

3.2.3 Berry or Fruit Waxes

A large number of waxlike products are derived from the fruit of various plants. Some waxes exist in the natural state as an outer coating on the fruit (bayberry and Japan waxes). Another source is refined and modified vegetable oils (i.e., hydrogenation). The cocoa fatty acid groups (C_{12}–C_{14}) represent the largest segment, but these are marginally included in the category of waxes because most derivatives are soft and low melting.

Castor wax is derived from castor oil, which is obtained from the seeds of *Ricinus communis*, a plant that grows in most tropical regions. Castor oil is predominantly (90%) the triglyceride of 12-hydroxystearic acid, with a double bond in the 9–10 position. Hydrogenation of the double bond forms the castor wax.

3.3 Mineral Waxes

Mineral waxes are obtained from oil and coal deposits. The name "mineral wax" has nothing to do with the composition but derives from the source (which is beneath the earth's surface). They are sometimes referred to as "earth waxes."

3.3.1 Paraffins

Paraffins are extracted form oil. The amount and type of wax present depends on the source of the oil. The paraffin is normally separated from the oil, distilled into several fractions, and sold as commodity items by the oil refineries.

3.3.2 Microcrystalline Waxes

Microcrystalline waxes are normally the higher molecular weight fractions from the paraffin extractions. They vary in hardness and melt properties depending on molecular weight distribution and branching. The name "microcrystalline" evolved from the discovery that these higher melting paraffins were not amorphous but consisted, to a high degree, of microcrystalline structures.

3.3.3 Montan Waxes

Montan waxes are extracted from soft coal deposits located predominantly in Germany, where the wax content in certain coal deposits is 10–15%. The coal is crushed and extracted with solvents to yield a dark brown to black crude montan wax with a wax content of 50–60%.

3.4 Synthetic Waxes

3.4.1 Polyethylene and Polypropylene Waxes

Polyethylene waxes are made by polymerization of ethylene to form ethylene chains similar to polyethylene resin but much lower in molecular weight (2000–10,000). The waxes can be oxidized by a variety of methods to produce waxes that are emulsifiable.

Polypropylene waxes are formed by controlled polymerization of propylene.

3.4.2 Modified Montan Waxes

Modified montan waxes are derived form coal. The crude montan waxes are subjected to a series of solvent extractions and oxidized with chromium trioxide (CrO_3) generically referred to as the Gersthofen process. The CrO_3 selectively attacks tertiary carbon structures, leaving behind only the linear montanic esters and acids. After oxidation has been completed, a light-colored product consisting mainly of montanic acid is formed. The montanic acids can be reesterified with various alcohols and polyols and partially saponified to give a series of montan ester wax products having various properties.

3.4.3 Miscellaneous Synthetic Waxes

Fischer–Tropsch waxes are made by reacting carbon monoxide (from the coal gasification process) with hydrogen gas to form a hydrocarbon wax.

Polyglycol waxes are high molecular weight chains of ethylene oxide that form a wax-like product.

Oxidized hydrocarbons (paraffins, microcrystalline, and polyethelyene waxes) are formed by oxidation (generally air blown) of hydrocarbon waxes to form a large variety of emulsifiable waxes.

4.0 COATING APPLICATIONS

Various waxes are used in different application areas for coatings to enhance certain physical properties.

4.1 Mechanism

Waxes generally function by one of two basic mechanisms.

The "ball bearing" mechanism is a dispersion of discrete wax particles throughout the coating matrix. The particles that protrude above the film surface prevent the abrading media from contacting the coating surface. Instead, the other surface glides over the protruding wax particles, causing little damage to the surface.

The "migration" mechanism involves migration of the wax to the film surface, where it aids in formation of a smooth glossy film. In some applications, the plastic nature of the wax aids in filling voids formed as the solvent evaporates or the specific volume changes during curing of the resinous components.

4.2 Property Enhancement

The major properties affected by waxes are surface related.

4.2.1 Gloss

A high gloss is obtained from waxes predominatly by the migration mechanism. The waxes tend to fill the microvoids formed as the coating dries to provide a smooth continuous film. Often only small amounts of waxes are needed (0.5–2% of solids) to obtain a gloss improvement. In fact, too much wax can lead to formation of wax solid on the surface, resulting in a dull haze. A high gloss can be obtained by buffing the surface (property 6 in Table 1), but this is not practical in most situations. If the excess max were to continue to migrate, the haze could return after several days (or weeks).

4.2.2 Matting

A mat (flat) surface is the opposite of high gloss, but both effects can be obtained using waxes. To obtain a matting effect, the type of wax and method of incorporation must be chosen to maximize the "ball bearing" mechanism. The wax particles that protrude above the surface diffuse the light to reduce glare. Particle size is very critical in obtaining a smooth, silky mat finish tat does not feel rough or look dull or hazy. Dispersibility in the paint or varnish vehicle is important to obtain a uniform surface.

Silicas are often used as matting agents, but their high density can lead to problems with settling out. Combinations of waxes and silica can lead to optimum effects. Using agents with differences in refractive indexes can further enhance matting effects without forming a haze.

4.2.3 Slip

Waxes can reduce the coefficient of friction by aiding in the formation of a smooth surface, as discussed in Section 4.2.1, or by reducing the number of contact points by gliding over the wax particles (ball bearing mechanism) discussed in Section 4.1.

Soft oily-type agents (soft, microcrystalline waxes or silicones) can also be used to form a slippery lubricating film on the surface. Mar resistance and blocking generally are very poor with this approach.

4.2.4 Abrasion Resistance

The ability of waxes to improve abrasion resistance is generally related to their ability to increase slip.

4.2.5 Antiblocking

Blocking can be reduced by incorporating hard wax particles (ball bearing mechanism) to reduce surface contact. A compromise in gloss must often be accepted. Some hard waxes can reduce blocking by absorbing oils and reducing the migration to the surface of other soft tacky components present in the coating formulation.

4.2.6   Antisettling/Antisagging Agents

When dissolved and cooled in solvent-based systems, many waxes form a thixotropic dispersion. The formation of a gel, which will liquify when stirred, will reduce the tendency of denser materials (pigments, fillers, etc.) to settle out.

4.3   Incorporation Methods

How a wax is incorporated into a coating formulation can affect the end performance. Particle size and uniformity of dispersion are critical factors. The best method depends on the type of wax and the desired end effect. Four basic methods are used.

4.3.1   Wax Compounds

Solvent-dispersed pastes are formed by heating the wax in a suitable solvent (ideally the same as the vehicle used in the coating formulation) and quickly cooling to room temperature under high shear conditions. The rapid cooling and high shear rates are required to prevent the formation of large seed crystals, which would appear as blemishes in the coating when applied.

Minor variations in the production of wax compounds can cause changes in the particle size distribution. Because of the sensitivity related to production conditions of wax compounds, most users rely on wax compound producers instead of preparing such waxes in-house.

4.3.2   Micronized Powders

Air milling and classifying the wax powder gives the optimum control of particle size. Micronized powders are based mainly on very head, high melting waxes such as polyethylene waxes, amides, and polytetrafluoroethylene (PTFE). According to the wax definition, PTFE is not a wax, but because it is used in the same area associated with waxes, it is often included in discussions on wax applications. Attempts to prepare these waxes in a compound form is difficult. Their crystallinity invariably leads to formation of larger particles if the procedure is not properly done.

The dry powder reduces the handling of flammable and hazardous solvents often associated with waxy compounds. Both solvent-and water-dispersible grades are available. On the negative side, improper handling can lead to dusty conditions and explosion hazards. Micronized powders are also more expensive on a per-pound basis (however, greater efficiency can lead to lower use levels).

4.3.3   Milling

The wax can be obtained as a coarse powder and milled in the same manner as pigments, or along with the pigments. This approach provides the end user with wax at the lowest cost. However, the time involved and the cost of the equipment rarely justify the savings in raw materials. Those who currently use this approach do so because they already have the equipment, which would be idle otherwise.

4.3.4   Emulsification

This approach is limited to water-based systems. The emulsion can be prepared in-house, provided the equipment is available, or it can be supplied by outside producers. The wax must be an emulsifiable type. Most commonly used for emulsification are oxidized polyethylene waxes and the refined montan waxes.

REFERENCES

1. A. H. Warth, *Chemistry and Technology of Waxes*, 2nd ed., New York, Reinhold, 1956.
2. H. Bennett, *Industrial Waxes: Vol. I, Natural and Synthetic Waxes; Vol. II, Compounded Waxes and Technology*, New York, Chemical Publishing Co., 1975.
3. R. Sayers, *Wax: An Introduction*, London, European Wax Federation and Gentry Books Limited, 1983.
4. "Wax technology," extracted from *Proc. Chem. Spec. Manuf. Assoc. 61st Mid-year*, pp. 86–111, 1975.

54
Carboxymethylcellulose

Richey M. Davis

Hercules Incorporated
Wilmington, Delaware

1.0 INTRODUCTION

Carboxymethylcellulose (CMC) is a charged, water-soluble polysaccharide derived from cellulose found in plants. It is used extensively in aqueous paper coatings for controlling coating viscosity. Aqueous paper coatings are colloidal suspensions consisting primarily of pigment, such as kaolin clay, and binders such as starch or latex. The solid concentrations typically range from 55 to 70%. The properties of a coating are determined by colloidal forces acting between particles and by the properties of the suspending aqueous phase. CMC affects the colloidal forces by adsorbing on clay pigment and by increasing the viscosity of the aqueous phase. These effects will be discussed after a brief description of the coating process.

2.0 COATING APPLICATION

For the best performance in a paper blade coating machine, a coating should have a shear thinning viscosity that decreases as the shear rate increases. Typical viscosities range from about 3000 mPa · s at shear rates below 1000 sec^{-1} when the coating is mixed and pumped to less than 100 mPa · s at shear rates above 1,000,000 sec^{-1} when the coating is applied to the paper. The linear velocity of paper in a coating machine can vary from less than 450 m/min to greater than 1400 m/min.

Once the fluid coating has been applied to the moving paper web, the paper passes through a section of the machine in which the excess coating is removed, often by a doctor blade. As the coating passes under the doctor blade, the paper adsorbs water from the coating. The paper then goes to a dryer section, where most of the water is removed to form a solid coating layer on the paper. If the coating's water phase viscosity is too low or if the

This work was sponsored by Aqualon Company.

coating particles form large aggregates with a porous structure, too much water is lost prematurely from the coating under the blade. Solid aggregates form under the blade and scratch the coated paper. Excessive water loss also raises the coating solids until the coating becomes shear thickening (the viscosity increases with the shear rate). This causes undesirable streaking on the paper. Shear thickening occurs as a result of increased collisions between the colloidal particles in the coating. These collisions form aggregates, which raise the viscosity. For good coating performance, it is important to viscosity the water phase and to suppress the tendency of the solid particles to aggregate upon collision. The viscosity of the water phase increases as the concentration of CMC increases.[1] Particle aggregation is a direct consequence of the colloidal forces acting between the particles. These forces and the role of CMC in affecting them are discussed next.

3.0 COATING FORMULATION

The colloidal chemistry of paper coatings is governed by the attractive and repulsive forces between the clay, binder, and polymer. Kaolin clay comes in the form of thin platelets that have negative charges on the edges and faces at pH values exceeding 7.[2] Paper coatings are used typically at a pH of 8–9. Clay particles are attracted to each other by van der Waals forces, which are particularly strong when platelets approach to within 1 nm of each other.[3]

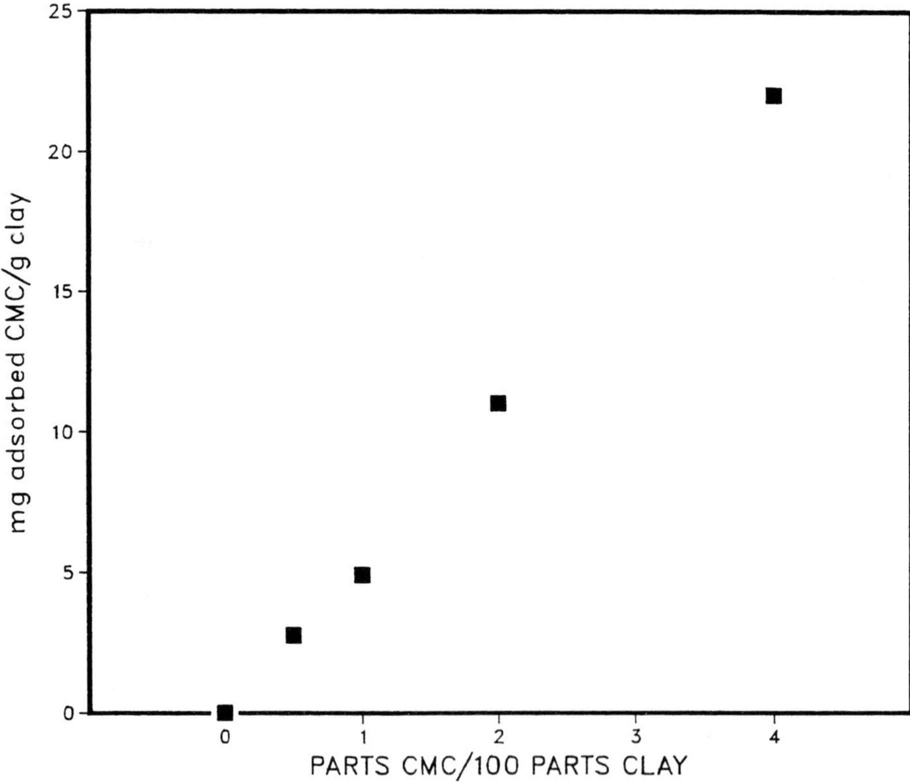

Figure 1 Adsorption isotherm for kaolin clay and CMC at pH = 7 and 25°C. Adapted from Ref. 4 with permission.

Most clays are treated with an anionic dispersant such as a low molecular weight polyacrylate, which adsorbs on the clay surface. This increases the net negative charge on the platelets and causes them to repel their neighbors, thus reducing the tendency for particle aggregation. A polymeric dispersant also imparts stability through steric stabilization, in which adsorbed polymer chains on one clay platelet repel polymer chains on neighboring platelets as a result of osmotic pressure.[3]

Negatively charged CMC adsorbs onto adsorbs onto kaolin clay and acts as a stabilizer. Figure 1 shows the adsorption isotherm for CMC with molecular weight of 110 kg/mol onto clay.[4] The amount of adsorbed polymer per mass of clay increases as the added polymer concentration increases. This polymer adsorption increases the amount of negative charge and hence the zeta potential on the clay, as shown in Figure 2. The zeta potential, which is the electrostatic potential near the surface of a clay aggregate, ranged from 20 to 30 mV, which is typical for suspensions that are moderately stabilized by electrostatic repulsion.

The effect of CMC on coating viscosity at high shear rates is illustrated by the first curve in Figure 3, where a concentration kaolin clay suspension without CMC begins to shear thicken at a shear rate of 18,000 sec^{-1}.[4] Adding 0.10 part of CMC per 100 parts clay increases the viscosity and also slightly reduces the shear thickening tendency. As the CMC concentration increases to 0.21 part, the suspension becomes shear thinning. Increasing the

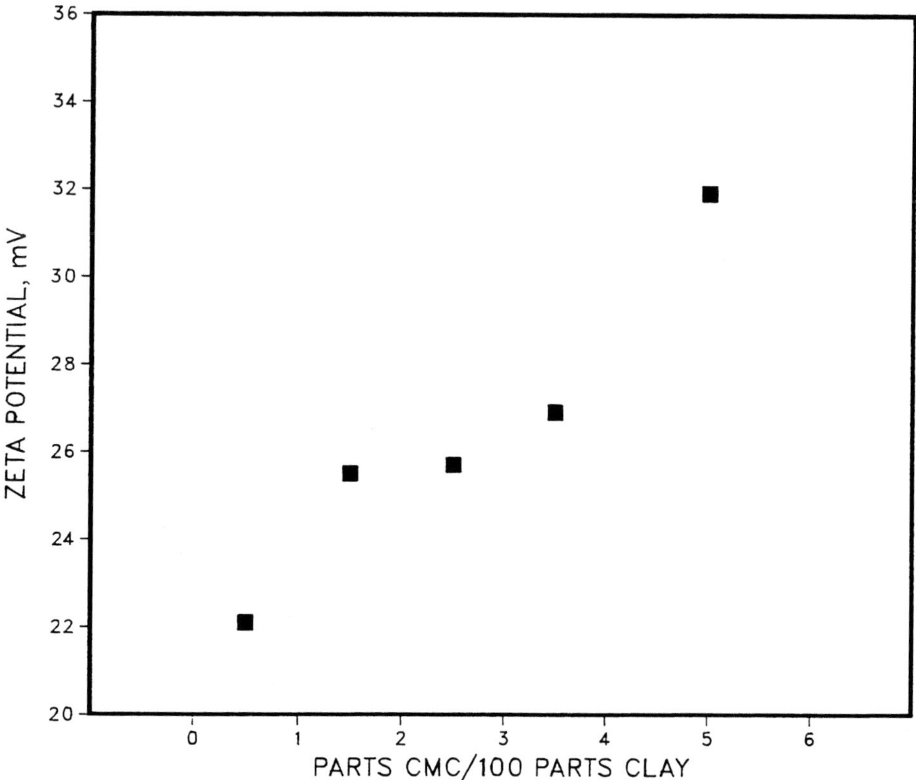

Figure 2   Zeta potential of kaolin clay with adsorbed CMC at pH = 7 and 25°C Adapted from Ref. 4 with permission.

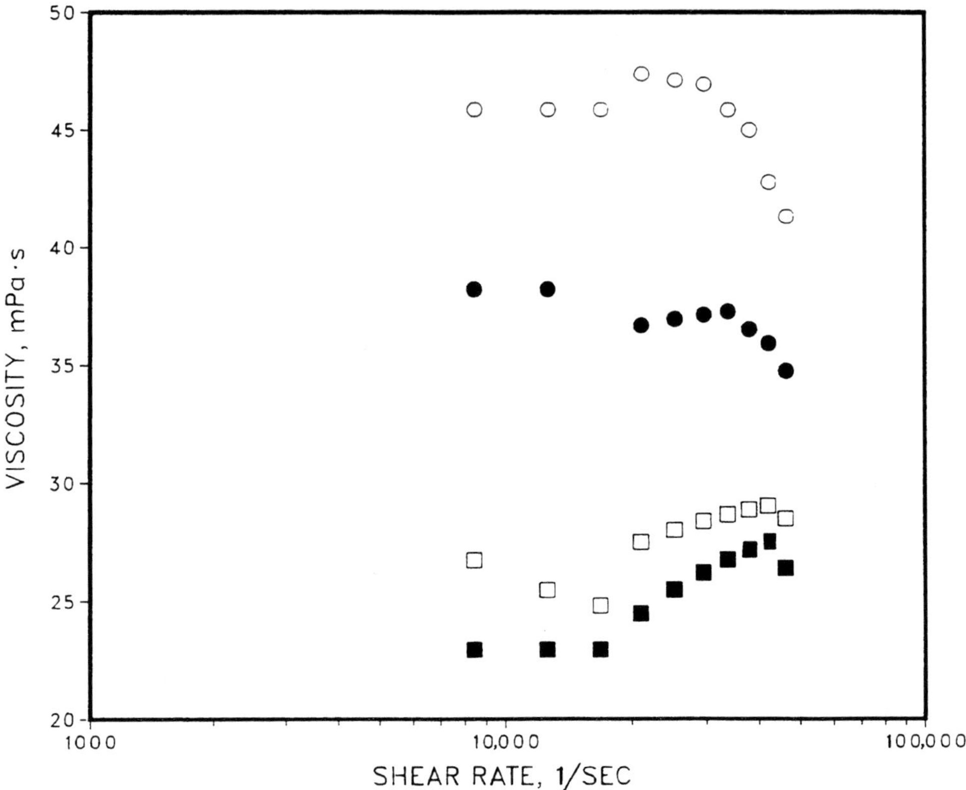

Figure 3 High shear viscosity for kaolin clay and CMC; ■ clay volume fraction = 0.39. Distilled water control: □ 0.10 part CMC/100 parts clay; ● 0.21 part CMC/100 parts clay; ○ 0.31 part CMC/100 parts clay (Adapted from Ref. 4 with permission.)

CMC level to 0.31 part makes the suspension even more shear thinning. The adsorbed CMC increases the degree of electrostatic and steric stabilization, thus inhibiting collision-induced flocculation.

There are no quantitative models that predict accurately the onset of shear thickening viscosity in paper coatings at very high shear rates. There are, however, numerous theories for the viscosity of concentrated suspensions at low shear rates. There were reviewed by Metzner.[5] A useful, general model was developed by Krieger.[6] The suspension viscosity η is correlated with the water phase viscosity η_s and the solid volume fraction Φ as

$$\eta = \eta_s \left[\frac{1 - \Phi}{\Phi_m} \right]^{-B\Phi_m} \tag{1}$$

where Φ is the ratio of the solids volume and the total volume; Φ_m and B are dimensionless parameters determined by fitting viscosity data to Equation 1.

An analysis of viscosity data using Equation 1 suggests that paper coatings, like other concentrated suspensions, flow in ordered layers along streamlines rather like cards in a deck sliding over each other.[4] At low shear rates, this ordered structure results in a low viscosity. Above a critical shear rate, this ordered structure breaks down and the particles in the suspension collide far more frequently, resulting in flow-induced agglomeration and the resulting shear thickening viscosity are suppressed by treating the coating with CMC.

REFERENCES

1. Aqualon Cellulose Gum (Sodium Carboxymethylcellulose), (product bulletin), Aqualon Company, Wilmington, DE 19850–5417.
2. A. E. James, and D. J. A. William, *Adv. Colloid Interface Sci.*, *17*: 219 (1982).
3. H. Van Olphen, *An Introduction to Clay Colloid Chemistry*, 2nd ed. New York: Wiley-Interscience, 1977, p. 48.
4. R. M. Davis, *TAPPI J. 70*(5), 99 (May 1987).
5. A. B. Metzner, *J. Rheol., 29* 739 (1985).
6. J. W. Goodwin, in *Science and Technology of Polymer Colloids*, G. W. Poehlein, Ed. 1982, p. 552.

55

Hydroxyethylcellulose

Lisa A. Burmeister

Aqualon Company
Wilmington, Delaware

1.0 INTRODUCTION

Hydroxyethylcellulose, (HEC) is a nonionic, water-soluble polymer derived from cellulose. It is used in water-borne paints for its action as a thickener, stabilizer, and/or suspending agent. HEC not only imparts mechanical and chemical stability to the paint system, it plays a significant role in controlling the rheology of the paint during manufacture, storage, and application. HEC is the most commonly used thickener in latex paints. In fact, it is often referred to as the industry standard for thickening.

Physically, hydroxyethylcellulose is a white, free-flowing powder that dissolves readily in hot or cold water. Available in a variety of types and grades, it can be used to produce aqueous solutions with a wide range of viscosities and a pseudoplastic nature. Upon drying, solutions of HEC form clear, gel-free films.

The nonionic nature of HEC is an advantage in latex paints. Nonionics can be used over a wide pH range, and, unlike polyelectrolytes, are compatible with salts and charged species such as surfactants, latex particles, and colorants.

More recently, a hydrophobically modified hydroxyethylcellulose (HMHEC) was introduced to the latex paint industry. HMHEC is an associative cellulosic polymer designed specifically for use in latex paints. It is a nonionic water-soluble polymer that contains both hydroxyethyl and long chain alkyl groups. This unique dual substitution differentiates the new-generation rheological modifier from traditional HEC. Like HEC, hydrophobically modified hydroxyethylcellulose thickens the aqueous phase of the paint, but substantial viscosity is built through association of the polymer's hydroprobes with the paint components. HMHEC combines the desirable properties of HEC with the enhanced rheology control (particularly with respect to spatter resistance and film build) required for today's coatings. A detailed discussion of HMHEC's chemical composition and rheological performance in latex paints is provided by Shaw and Leipold.

491

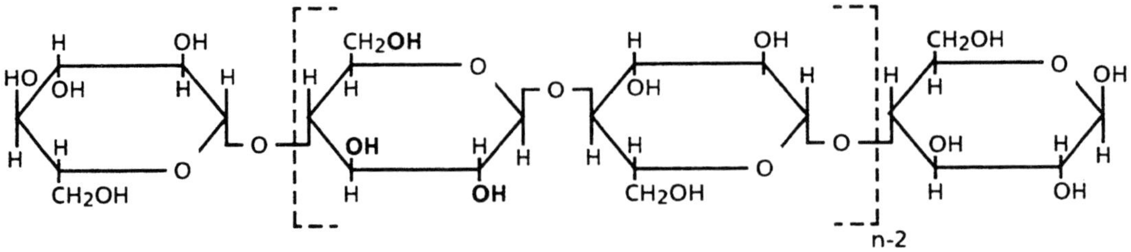

Figure 1 Structure of cellulose.

2.0 CHEMICAL COMPOSITION OF HYDROXYETHYLCELLULOSE

The cellulose molecule is a polymeric chain composed of repeating anhydroglucose units (Fig. 1). There is considerable intrachain and interchain hydrogen bonding, which results in a highly ordered, highly crystalline structure. It is for this reason that cellulose is not soluble in   water.

   The reactions used to make derivations of cellulose are straightforward. Cellulose is reacted with alkali, such as sodium hydroxide, to form alkali cellulose. The alkali treatment is necessary to disrupt the cellulose crystallinity. The swollen chain is then ready for the addition of appropriate substituents. Substitution on the cellulose chain causes disorder and forces the chains apart so that water may enter and solvate the chain. Each anhydroglucose

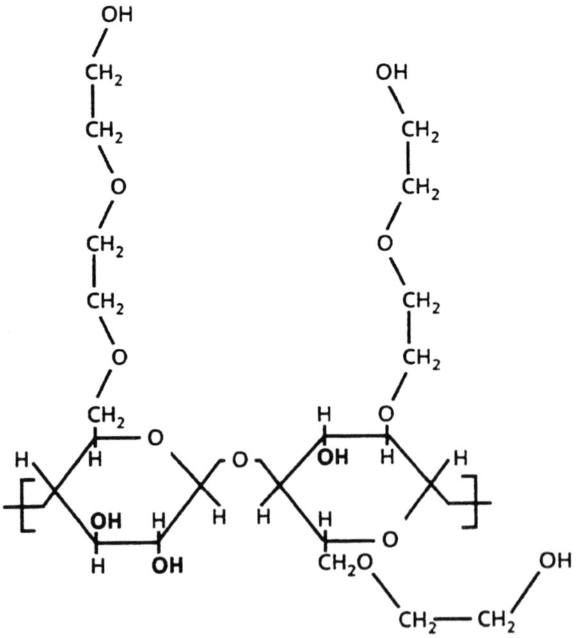

Figure 2 Idealized structure of hydroxyethylcellulose

unit in the cellulose molecule has three reactive hydroxyl groups. The number of hydroxyl groups substituted in any reaction is known as the degree of substitution (DS). Technically, all three hydroxyls can be substituted. The product from such a reaction would have a DS of 3.

Hydroxyethyl groups can be introduced into the cellulose molecule in two ways. First, ethylene oxide reacts with the hydroxyls in the cellulose chain. Second, ethylene oxide, reacting with previously substituted hydroxyls, can polymerize to form a side chain.

The average number of moles of ethylene oxide that attach to each anhydroglucose unit in cellulose in the two ways described is called "moles of substituent combine," or (MS).

Solubility in water is achieved as the degree of substitution is increased. By selecting appropriate reaction conditions and moles of substituent, complete and quick solubility in water is obtained. HEC with an MS of 2.5 is most frequently used in latex paints because of its optimal water solubility. An idealized structure of HEC is shown in Figure 2.

3.0   TYPES AND GRADES OF HYDROXYETHYLCELLULOSE

By controlling the chain length of the cellulose backbone, HEC can be produced in a wide range of viscosity types. However, economics indicate that medium (2% Brookfield viscosity = 4500–6500 cPs (mPa · s at 25°C) or higher viscosity grades are used to prevent sagging, and they improve can stability of some latex paint formulations. Most paints require 0.1–1% by weight of HEC to be thickened to the desired viscosity.

Hydroxyethylcellulose is also produced in a grade having superior biostability. Water is the continuous phase in latex paints. In addition, many of the components of a latex paint are manufactured in an aqueous environment. Whatever water is present, microbial growth may occur. The growth of these microorganisms is accompanied by the production of enzymes, including the cellulytic types. Cellulase enzymes break the bonds between adjacent unsubstituted anhydroglucose units in the cellulose backbone. This degradation reduces the molecular weight of the polymer, causing it to no longer be an effective thickening agent.

The process by which biostable HEC is manufactured results in a product in which the hydroxyethyl groups are distributed more uniformily along the cellulose backbone, as opposed to forming longer side chains. The more uniform distribution protects significantly more bonds from enzyme attack.

4.0   CHEMICAL AND PHYSICAL PROPERTIES OF HYDROXYETHYLCELLULOSE

Most water-soluble polymers have a tendency to lump or agglomerate during the dissolving process. When the polymer is added to water, the outer layer of the agglomerate swells to a gel very quickly, making it difficult to hydrate the inner portions. The time required for the polymer to dissolve completely is governed by the degree to which it agglomerates. To avoid agglomeration, HEC particles are surface treated with a substance, typically glyoxal, which cross-links the surface and produces a product that is temporarily insoluble in water. This temporary insolubility permits thorough dispersion of discrete particles of HEC. Once the glyoxal cross-links have been hydrolyzed by the water, the HEC will begin to dissolve without agglomerating.

Hydrolysis of the glyoxal cross-links is influenced both by temperature and pH. Figure 3 shows that increasing the temperature and increasing the pH both shorten the hydration time. (Hydration time is defined as the time between addition of HEC to water and the time

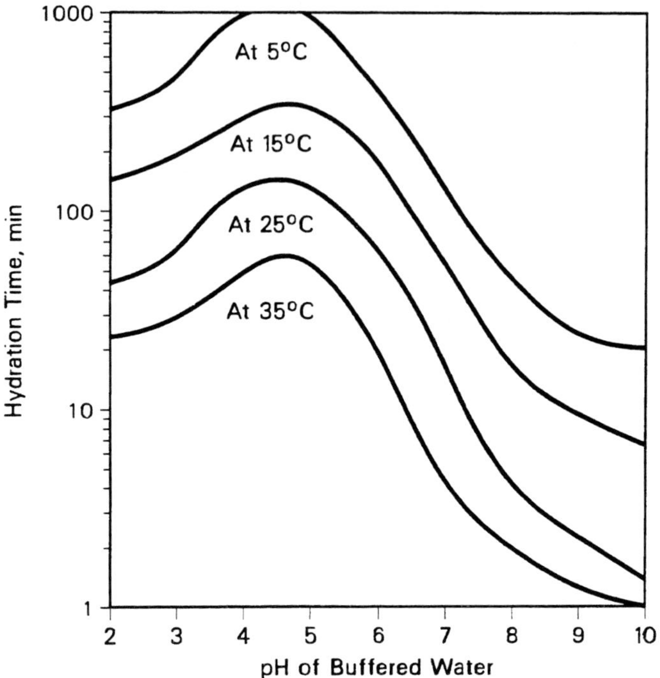

Figure 3   Effect of pH and temperature on the hydration time of HEC.

when viscosity starts to develop.) In fact, raising the water pH to levels of 8 and above decreases the hydration time so much that the surface treatment is no longer an effective means of improving dispersion.

5.0   INCORPORATING HYDROXYETHYLCELLULOSE INTO LATEX PAINTS

An important benefit of hydroxyethylcellulose for latex paint manufacturers is that it can be added to the batch at several different points. This gives the formulator process flexibility. Most often, at least some of the HEC required to thicken a batch of paint is added to the pigment grind. This increases the viscosity of the grind and makes it more efficient dispersing medium. However, some manufactures find it possible to achieve a good pigment dispersion in a low viscosity grind, hence add HEC later. HEC can be used in the powder form (as is), dissolved in water, or slurried in a nonsolvent before addition to ensure that a gel-free paint is manufactured.

6.0 APPLICATION PROPERTIES OF LATEX PAINTS THICKENED WITH HYDROXYETHYLCELLULOSE

The primary purpose of a thickener in a latex paint system is to control rheology, or flow properties. Hydroxyethylcellulose thickeners influence the rheology primarily by thickening the aqueous phase. Fairly strong hydrogen bonds are formed between the polymer chain and water molecules. The resulting chain association forms a concentrated network. Because HEC thickens the water only, its efficiency is independent of the latex, surfactant, or dispersant type. Formulation insensitivity is the greatest benefit of HEC. Also, very little viscosity change or phase separation seen in shelf stability, heat stability, or freeze–thaw stability testing is attributable due to the aqueous phase thickening mechanism. Sections 6.1–6.3 describe the effect of molecular weight or viscosity type of HEC on the important properties of leveling, sag resistance, film build, and spatter resistance.

6.1 Leveling and Sag Resistance

Very low shear rates are present in the paint during the leveling process (disappearance or reduction of brush marks in the paint immediately after application). In general, high molecular weight grades of HEC should be used if improved leveling is desired. Paint viscosity curves (Fig. 4) show that paints with higher molecular weight thickeners have the lowest viscosities in the low shear rate region. Low shear rate viscosity data have been found to be a good predictor of leveling when the only variable in a series of paints is the molecular weight of the HEC thickener. When comparing one formula with another, it is important to examine also the rate of viscosity recovery after shearing, to be able to predict leveling.

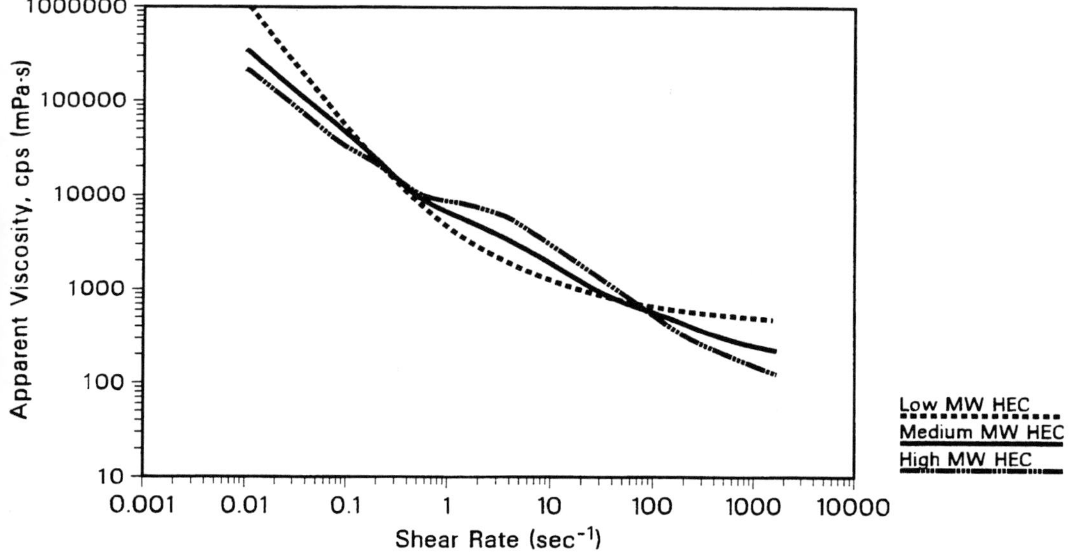

Figure 4 Influence on paint viscosities of molecular weight of HEC and shear rate.

A second cause of poor leveling is wicking. When the water is adsorbed quickly into the substrate, the brush marks are frozen. To alleviate this condition, a greater amount (lower molecular weight) of HEC should be used.

6.2 Film Build

Good film build, a prerequisite for one-coat hiding, has been shown to be directly related to viscosity during the high shear processes of brush and roller application. This viscosity is controlled by thickener concentration. Therefore, to increase the film build of a formula while maintaining the Stormer viscosity, a greater amount of a low molecular weight HEC should be used.

6.3 Spatter Resistance

Spatter resistance, also a high shear application property, does not depend on thickener viscosity, but on the elasticity of the aqueous phase. Like film build, spatter resistance can be predicted from the polymer solution rheology. Higher molecular weight thickeners are more elastic, and consequently impart elasticity to the paint. Paint made from lower molecular weight grades of HEC is more resistant to spatter during roller application.

During roller application, threads of paint are pulled off the nip of the roller. When the thickener has some degree of elasticity, these threads are stabilized and can be stretched farther before breaking. More spatter is created when these threads are stabilized and stretched than if they break close to the nip of the roller.

REFERENCE

1. K. G. Shaw and D. P. Leipold, *J. Coatings Technol.*, *57* (727), 63–72.

56
Antistatic and Conductive Additives

B. Davis

ABM Chemicals Limited,
Stockport, Cheshire, England

1.0 CONDUCTIVITY ADDITIVES FOR SPRAY PAINTS

In the electrostatic spray technique, the atomized paint particles receive a high voltage in the "gun." Subsequently the paint particles are attracted to the grounded workpiece, resulting in highly efficient paint utilization with good coverage of any awkward areas (Fig.1).

Paints formulated largely from nonpolar solvents have an electrical resistance that is far too high for satisfactory application in this way. It is necessary to reduce the paint resistance below 1 MΩ, so very large proportions of polar solvent would be necessary; therefore a search for more efficient chemical additive revealed that some quaternary ammonium compounds (Fig. 2) are particularly effective alternatives.

Choosing the best quaternary ammonium compound involved the investigation of various chemical structures, molecular weights, and anions. Using commercially available blends of raw materials and avoiding potentially corrosive ions, it has been possible to develop a particularly effective product, namely Catafor CA. Bearing in mind the customer preference for pourable liquids, an 80% solution in butanol is normally used to give very low resistivities with small additions, as illustrated with xylene in Figure 3.

When using Catafor CA in paints, it is advantageous to add a modest amount of polar solvent because a synergistic effect is usually observed (Table 1).

The advantages of using quaternary ammonium salts as a conductivity additive can be summarized as follows:

1. Catafor CA, containing quaternary ammonium salts, is compatible with most types of paint thus avoiding reformulation.
2. The fluid 80% solution is quickly and easily dissolved.
3. Only low dosage is needed to produce a large drop in resistivity (0.5–3% in practice).
4. The same low dosage will reduce surface tension to a suitably low value.

497

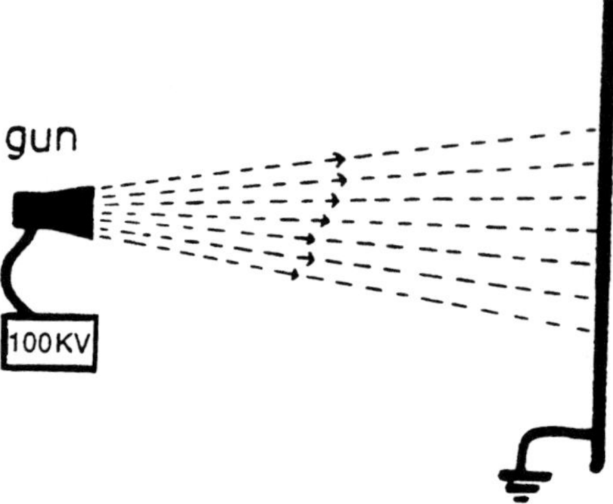

Figure 1   Schematic diagram of a electrostatically charged spray.

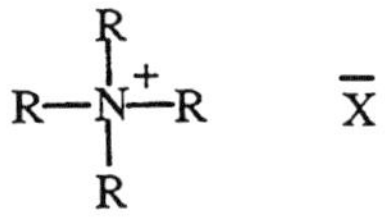

$$R-\overset{\displaystyle R}{\underset{\displaystyle R}{\overset{|}{\underset{|}{N}}}}{}^{+}-R \qquad \bar{X}$$

Figure 2   Structure of quaternary ammonium compounds

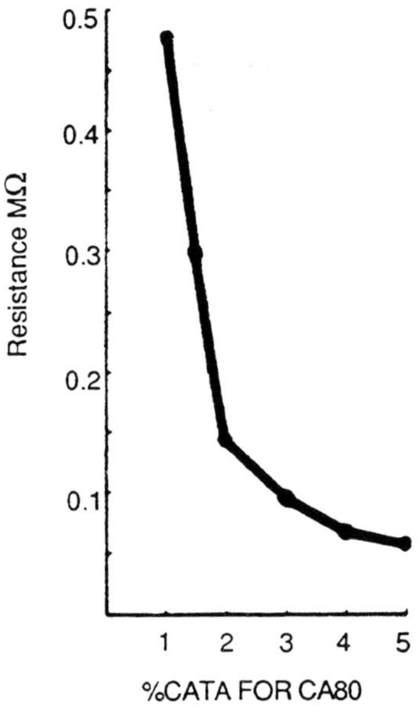

Figure 3   Effect of quaternary ammonium salt on electrical resistance in paint.

Table 1 Effect of Catafor CA in a Long Oil Alkyd Paint

Paint–solvent combination	Resistance (MΩ)
Paint + hydrocarbon solvent	20
Paint + hydrocarbon solvent + 1% Catafor CA80	1.8
Paint + hydrocarbon solvent + 10% methyl ethyl ketone + 1% Catafor CA80	0.29

5. Yellowing effects are not usually observed, even in sensitive white paints.
6. Effects on the paint film properties (evenness, general appearance, hardness, corrosion resistance, etc.) are minimal.

2.0 ANTISTATIC ADDITIVES FOR PLASTIC MATERIALS

The many adverse effects of static electricity on insulating plastic materials are well recorded.[1,2] These include textile processing difficulties, poor sound reproduction from records, excessive dust and dirt collection on surfaces, and dust or solvent explosion hazards. One of the most important hazards associated with modern technology is electrostatic discharge, which can cause critical damage to microchip circuits, so essential in electronics.

Various types of antistatic agent are reviewed in the literature,[1] and it appears that surfactant molecules are particularly effective. Nonionic, anionic, cationic, and even ampholytic types have been used as internal antistatic agents, depending on the type of polymer. For example, nonionics are recommended for polyolefins, but cationics are particularly effective in the more polar polymers,[2] such as polyvinyl chloride, polystyrenes, and polyurethanes.

The large influence of relative humidity on the effectiveness of antistatic agents suggests that a film of water is held on the plastic surface by the presence of the ionized or polar groups of the surfactant molecular projecting from the surface, with the hydrophobic tail anchored in the polymer. Generally recognized levels of surface resistivity[3] for various types of polyurethane are shown in Figure 4.

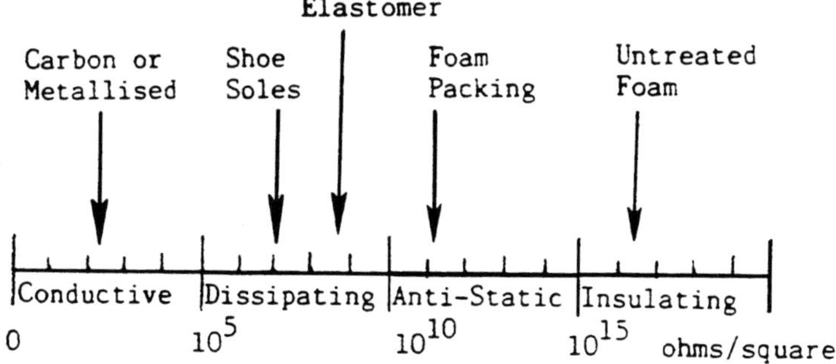

Figure 4 Surface resistivity spectrum.

Very conductive plastics are usually made by filling with large quantities (10–30%) of carbon or powdered metal to produce a continuous conductive path through the polymer bulk, and it is unlikely that organic additives, relying on moisture adsorption, can ever produce such low resistivities. However quaternary ammonium compounds have proved to be very effective at producing a useful increase in conductivity, and the active ingredient is sold in a variety of solvents (e.g., butanediol, butyl oxitol, ethylene glycol, trichloropropyl phosphate) to suit various needs.

The use of quaternary ammonium antistatic agents can be restricted by the high temperatures reached in making or molding plastic materials. However, when high temperatures exist for short periods only, catafor CA can be used in plastics (polyurethane, PVC, polystyrene, rubber, etc.). External application is also useful, when the effect is more temporary. It can be used on fibers and in the cleaning of surfaces, when it is sufficiently substantive to give some antistatic effect.

REFERENCES

1. *Kirk-Othmer Encyclopedia of Chemical Technology* Vol. 3, 3rd ed. New York: Wiley Interscience, 1983, p. 149.
2. R. Gachter and H. Muller, in *Plastics Additives Handbook*, 2nd ed. Munich: Hanser, 1984, Chapter 12, p. 565.
3. A. Lerner. "A new additive for electrostatic discharge control in foams and elastomers," *Proceedings of the Society of the Plastics Industry*, Texas, 1984, pp. 331–334.
4. B. Davis, "Industrial applications of surfactants," Royal Society of Chemistry Special Publication 59, 1987, pp. 307, 317.

57

Silane Adhesion Promoters

Edwin P. Plueddemann

Dow Corning Corporation, Midland, Michigan

Silane adhesion promoters are organofunctional silicon compounds that promote adhesion of coatings to substrates, especially improving their resistance to debonding under humid conditions. It must be emphasized that these silanes are not related to polydimethylsiloxane fluids, which cause cratering and poor repaintability in coatings. Silane adhesion promoters may help overcome such problems of poor wetting and intercoat adhesion in coatings.

Silane adhesion promoters are generally formulated into primers or added at relatively low levels to coatings, in contrast to silicone resins, which are typically used at about 30% level in silicone-alkyd or silicone-acrylic formulations. These silanes are ambifunctional compounds of the general structure $(CH_3O)_3$ Si—R—X, where X is an organofunctional group chosen for compatibility with the organic coating and the alkoxysilane portion provides bonding to mineral substrates[1].

The alkoxysilane may be prehydrolyzed to silanols that are reactive with hydrated metal oxide surfaces and contribute siloxane cross-links. Methoxysilanes also react directly with metal hydroxides and then cross-link in the presence of atmospheric moisture.

Silane adhesion promoters are often effective in improving the adhesion of coatings to plastic surfaces or to oily metal surfaces. As additives in paints, they may be useful in improving intercoat adhesion.

A recent trend is to formulate adhesion promoters by mixing silane monomers, partially prehydrolyzing the silanes, or by mixing silanes with polymer precursors such as epoxies and melamines. Silane adhesion promoters are offered by silane manufacturers Dow Corning Corporation, Union Carbide Corporation, Petrarch Chemicals Division of Dynamit Nobel, and Peninsular Chemical Research, as well as by proprietary formulators such as Hughson Chemical Division of Lord Corporation. Adhesion promoters supplied by Dow Corning and recommended for coatings are described in Table 1.

Walker[2] reviewed a study of silane primers and additives for adhesion of a two-part epoxy paint with polyamide cure, and a two-part aliphatic isocyanate adduct cured polyester paint on mild steel, aluminum, cadmium, copper, and zinc surfaces. Silanes were tested as primers (essentially monolayer coverage on metal), and as 0.1–0.4% additive based on total

501

Table 1 Dow Corning Adhesion Promoters for Coatings

Product	Type
21 Additive	Amino (50% in butanol)
25 Additive	Formulated silane-melamine
Z6030	Methacrylate
Z6032	Vinyl benzylamine
Z6040	Epoxy
X1–6100	Mixed silanes
1200 Primer	Silicate
1205 Primer	Formulated silane-epoxy

paint. Results, summarized in Table 2 on page 503 for iron and aluminum, indicate that the diamine-functional silane gave uniformly improved retention of adhesion under wet conditions and gave greater recovery of adhesion when the panels were dried. Methacrylate-, epoxy-, and mercaptofunctional silanes were also useful as adhesion promoters. Results were similar on cadmium, copper, and zinc. Paints with some silane additives showed little deterioration in performance during 9 months of storage.

Some general recommendations on use of silane adhesion promoters of Table 1 for coating on glass, aluminum, and steel are summarized in Table 3 on page 504.[3]

Silane adhesion promoters are also effective in bonding coatings to plastic surfaces[4]. Primers 1200, 1205, and 25A are generally useful on engineering thermoplastics and thermosetting resin surfaces, while 1205P and Z6032 are most useful on polyolefins and vulcanized rubber surfaces.

REFERENCES

1. E. P. Plueddemann, *Silane Coupling Agents*, New York, Plenum Press, 1982.
2. P. Walker, *J. Coatings Technol.*, *52*(668), 33 (1980).
3. P. G. Pape and W. A. Finzel, Dow Corning Corp., private communication.
4. E. P. Plueddemann, in *Adhesion Aspects of Polymeric Coatings*, K. L. Mittal, Ed., New York, Plenum Press, 1983, p. 363.

Table 2 Wet[a] and Recovered[b] Adhesion of Paints on Metals (modified with diamine-functional silane)

| | | Mild steel | | | | Aluminum | | | |
| | | Wet | | Recovered | | Wet | | Recovered | |
Paint	Metal treatment	MPa %	Detached	MPa %	Detached	MPa %	Detached	MPa %	Detached
Polyurethane	Degreased	5.6	100	6.8	100	3.8	100	9.9	100
	Degreased, primed	7.4	90	12.5	90	11.8	30	22.8	30
	Degreased, 0.2% additive	9.8	60	20.2	100	9.5	100	14.5	100
	Sandblasted	11.8	95	20.8	60	8.5	100	13.6	80
	Sandblasted, primed	22.7	30	29.14	0	14.9	20	22.0	30
	Sandblasted, 0.4% additive	21.3	20	29.8	0	15.6	70	25.2	30
Epoxide	Degreased	7.2	100	10.9	100	5.6	100	11.2	100
	Degreased, primed	28.7	100	29.2	10	11.4	30	20.5	40
	Degreased, 0.2% additive	17.7	10	26.6	20	25.3	0	26.9	0
	Sandblasted	9.2	100	20.9	100	8.5	100	13.7	80
	Sandblasted, primed	26.3	50	27.8	40	12.9	40	24.9	30
	Sandblasted, 0.2% additive	23.8	0	25.3	20	28.2	0	28.7	0

[a]After 1500 hours of immersion in water.
[b]Dried in air for 48 hours.
Source: Adapted from Ref. 20.

Table 3 Recommendations for Adhesion of Coatings to Metals

Coating	Primer	Additive
Acrylic solutions		
thermoplastic	1200P, X1–6100, Z6040	21A, Z6032
thermosetting	1200P, 25A, Z6040	25A, Z6040, Z1–6100
Acrylic emulsions		
thermoplastic	21A, Z6040, Z1–6100	25A, Z6032, Z6040, Z1–6100
thermosetting	21A, 25A, Z6040	25A, Z6030, Z6040, Z1–6100
Alkyd	1200P	None
Polyester	1200P, 25A, Z6040	21A, Z6032, Z6040, X1–6100
Two-package urethane		
polyester	1200P, 25A, Z6032, Z6040	21A, 25A, Z6030, Z6040
acrylic	1200P, 1205P, 25A, Z6040	21A, 25A, Z6032, X1–6100
Epoxy		
baking	1205P, 21A, 25A, Z6040	21A, 25A, Z6032, X1–6100
air dry	1200P, 1205P, 25A, Z6040	21A, 25A, Z6032, Z6040
PVC plastisols	21A, Z6032	—[a]

[a]A partially hydrolyzed 21A devolatilized in higher alcohols has been offered as Q1–6012.
Source: Dow Corning Corporation.

58
Chromium Complexes

J. Rufford Harrison

E. I. du Pont de Nemours & Company
Wilmington, Delaware

1.0 INTRODUCTION

Chromium complexes are widely used for changing the characteristics of paper, glass, and other hydroxylic surfaces, and the surfaces of materials such as polyethylene, which can be made functional by corona discharge and similar methods. These binuclear compounds have the approximate formula shown in Figure 1. The chromium end of the molecule attaches itself to the substrate, whose—OH groups are in effect converted to —R groups. The radical R is of two general types: long chain saturated hydrocarbon, incorporated, incorporated by the use of a fatty acid, and unsaturated hydrocarbon, incorporated as methacrylic acid. The fatty acids are used to change the physical properties of the surface, whereas the function of the methacrylic acid is to change its chemical properties.

The chromium in these complexes is exclusively trivalent. As is discussed below, the health hazard associated with hexavalent chromium compounds is not present in these products, which have the approval of the U.S. Food and Drug Administrative (FDA).

2.0 MANUFACTURE

The starting point for these materials is a solution of a basic chromium chloride, which is converted to the complex by reaction with the appropriate acid. The products are green solutions, with such depth of color that they appear black. In addition to being classed as flammable liquids, they are corrosive, since some of the chlorine (see Fig. 1) is hydrolyzed and therefore exists in the form of HCl. in use, both hazards are removed by dilution and neutralization.

3.0 METHACRYLIC ACID TYPES

The Du Pont Company makes methacrylatochromium complexes under the name Volan bonding agent. Although the methacrylic acid types are used to coat substrates, the coating

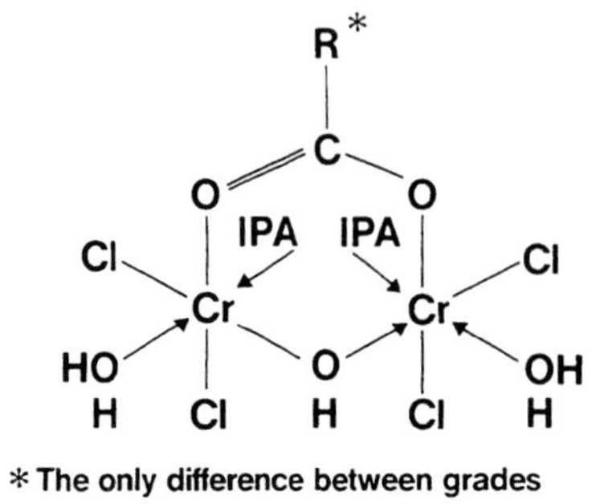

Figure 1   Typical chromium complex (IPA = isopropyl alcohol).

is always reacted further before the end use is reached. They are therefore not normally considered to be coatings per se, and are discussed only briefly.

A methacrylate group can polymerize with any other molecule undergoing vinyl polymerization. If Volan-treated glass fiber, for instance, is impregnated with an unsaturated polyester, then when the latter is polymerized, the chrome complex polymerizes with it, thereby forming a strong, hydrolysis-resistance bond between the resin and the glass. The methacrylatochromium complex will bond almost any inclusion to an unsaturated polyester, or to any other thermosetting vinyl resin such as a styrenic or an acrylic; fillers, reinforcing agents, pigments, or other solid additives can all be treated in this manner.

A property of particular interest in these methacrylatochromium coatings is that they disperse static electricity, so that treated powders, for instance, remain free flowing; some bonding agents give rise to so much static that treated products become very difficult to use.

4.0   END USES

Almost any carboxy group will form a complex with chromium, provided the molecule does not include a group that complexes more strongly. The common fatty acids all form these complexes, but until the number of carbon atoms reaches about 14, the products have no properties of interest. In the range from C_{14} to C_{18} the acids form complexes that give rise to properties like those of wax, namely water resistance and release. Du Pont markets these materials under the name Quilon. They do not contribute per se grease resistance, but they do make excellent cross-linking agents for polyvinyl alcohol (PVA), which has very good resistance, a disadvantage that is eliminated by cross-linking with a chromium complex.

Like the methacrylato complex, those based on fatty acids also act as antistatic agents.

4.1   Release

The main use for these complexes is in the manufacture of release paper. Using the "Scotch tape test," the strength of release on untreated paper may be about 1200 g/in., which falls to perhaps 40 g/in. if the paper is treated with silicone, the best-known and most-used release agent. With a chromium complex, the release is likely to be in the range of 500–600 g/in. What the user of chromium is seeking is a composite structure—the release paper and the material that is adhered to it—that has enough adhesive force to stay together under a range of conditions; yet the release must be clean, without fiber tear, when required.

The three main uses in this category are as follows.

4.1.1 Bakery Liners

As mentioned in Section 1.0, these complexes are not toxic, and they do have FDA approval. They are used on papers for baking a variety of products, and for packing a variety of products that are not baked. The substrate is usually, but not exclusively, glassine or parchment; more porous papers can be used if some provision is made to retain the treatment solution at the surface. Although release is usually needed on only one side, both sides are coated; bakery liners are often reused—often several times on each side—and the extra coating increases the lifetime of the paper.

4.1.2 Decorative Laminates

High pressure laminates made by impregnating several layers of paper and other materials with a thermosetting resins, and heating the composite between steel cauls. Adhesion to the steel is prevented by an intermediate layer of release paper, coated sometimes on both sides, sometimes on only one. Since sheets of decorative laminate are made in pairs, back to back, another layer of release paper is used between them.

4.1.3 Vinyl Casting

Textured soft vinyl is often made by casting a vinyl plastisol onto a textured release paper. The latter is a quite heavy grade.

4.2 Water Resistance

Chromium-treated paper will transmit water vapor but is resistant to liquid water. Until paper succumbed to competition from polyethylene, chromium complexes were used in the manufacture of ice bags. Some chromium complex is still used for water resistance, but the use is now much smaller.

4.3 Grease Resistance

A layer of polyvinyl alcohol is often used on paper to provide grease resistance, but if liquid water or high humidity are major factors, PVA cannot be used without being cross-linked. Chrome complexes are often used for this purpose. The PVA is first dissolved in water, and the resulting solution is used to dilute the Quilon. The product has enough stability for use in a paper mill. Interleaving paper for bacon slices is an example of a use requiring this product.

5.0 APPLICATION METHODS

As indicated above, the chromium complexes are provided as concentrated, acidic, alcoholic solutions. In use they are diluted, often as much as one one hundred-fold, with cold water (or with isopropanol, if the substrate is not readily wetted by water). They are then usually neutralized to a pH of about 3.5. The common alkalies such as caustic or ammonia cannot be used without specialized equipment; they may produce localized regions of high pH, and chromium compounds precipitate. Instead a buffer solution (formate–urea) is used, or a material that produces ammonia upon hydrolysis—urea or hexamethylenetetramine (HMT). The best of these is HMT, but it does not have FDA approval. Manufacturers of bakery liners therefore use formulations based on urea. A typical formulation is as follows:

1 part chromium complex solution (commodity, as supplied)
50 parts water at 20–30°C (say 70–90°F)
0.02–0.03 part urea

Table 1   Some Typical Application Equipment

On-machine	Off-machine (converting plants)
Size press	Doctor roll or knife
Spray	Meyer rod
	Gravure
	Offset
	Corrugator

On very porous papers it may be necessary to apply a base coat-alginates, ethylated starches, and polyvinyl alcohol are all used-to retain the chrome complex at the surface. If it is not feasible to apply this second coating, the efficiency of the chrome complex solution can be increased by incorporating PVA in it to reduce penetration.

Although the complexes as shipped are almost indefinitely stable, once neutralized they begin to hydrolyze, and will eventually form a sludge. They do, however, have stabilities of a few days, particularly if an effort is made to maintain the temperature lower than 30°C, and if they can be run readily in a paper of textile mill.

Another precaution is to avoid contaminating the treatment bath with components of the paper. The latter is often alkaline, and too much contamination can result in precipitation of chromium in the feed tank. Perhaps the easiest way to avoid this problem is to use a small recycle tank, fed on demand from the larger mix tank that is never contaminated by recycle.

The treatment solution can be applied to the paper by almost any method that is logical for the available equipment; see Table 1. On a paper machine, for instance, it is often applied on a size press. But this may not be feasible if the machine has only one press that is used for applying a different coating, such as starch or PVA. In such a case the chromium solution can be sprayed on to the paper. On converting equipment there are more options. The solution can be sprayed or rolled onto the paper, it can be offset, it can be doctored—almost any equipment capable of applying a liquid to paper can be used.

Coating weights range from 1 to 5 ponds of the original complex commodity—the solution as shipped—per 10,000 square feet of surface. One cannot be more precise than this, since much depends on the paper itself, particularly on its porosity. The first-time user is advised to use the higher end of the range initially; he can then gradually reduce the add-on until he discovers his minimum.

6.0   GOVERNMENTAL AND OTHER REGULATIONS

Since none of the chromium is in the hexavalent state, these materials have FDA approval for use on paper in contact with dry, aqueous, and fatty foods. Corresponding approvals for use on films and foil have not been sought, but presumably could be obtained. Grades of these complexes based on the C_{14} acid, which is obtained from coconut oil, conform to Jewish dietary laws.

Regulations covering the disposal of chromium-containing materials are generally local and should be checked before use. The commodity itself can be fed slowly to a biotreatment pond; the chromium will precipitate and go out with the sludge, and the fatty acid will be biodegraded. Approval of the appropriate authority should be sought first, however, to ensure that the compound does not interact with something else that is being fed at the same time.

Alternatively, the commodity can be pyrolyzed (i.e., incinerated under nonoxidizing conditions). The residue will include trivalent chromium,which can usually be deposited in a landfill after consultation with the appropriate authority. Some HCl will be formed during pyrolysis and may be permitted to escape through the smokestack subject to local approval; alternatively, it may be scubbed out. Since the HCl is corrosive, these materials should not be disposed of routinely in this way unless the incinerator is corrosion-resistant.

Incineration under oxidizing conditions may be safe, depending on other products being handled at the same time. There is always the possibility that some of the chromium will oxidize, in which form it may appear in either the ash or the volatiles. This possibility should first be checked on a small scale; if any hexavalent chromium is produced, the method should not be selected. Pyrolysis would avoid this problem. Biotreatment may be used also for treatment of solutions.

Treated paper may be pyrolyzed or incinerated, with the same consideration as described above, or it may be deposited in a landfill. Alternatively it may be repulped, in which case most of the chromium will go out with the broke.

Some landfilling operations place limits on heavy metals. The composition of any proposed waste should be discussed with the operators.

59
Nonmetallic Fatty Chemicals as Internal Mold Release Agents in Polymers

Kim S. Percell, Harry H. Tomlinson, and Leonard E. Walp

*Witco Corporation,
Memphis, Tennessee*

1.0 INTRODUCTION

A major problem in injection molding is removal of the part from the mold. If the part has a large surface area or a certain type of surface, it may be almost impossible to remove the piece without damage. The force of the ejector can be increased to some extent, but this increases the possibility of damage to the piece. The use of some type of mold treatment is well known in the industry. The most common treatment is spraying a silicon-or fluorocarbon-type mold release agent directly on the mold. This procedure makes it considerably easier to remove the piece from the mold, but unfortunately it causes other problems. The spray-on mold release is typically good for 8–10 moldings and then must be resprayed. Continued respraying eventually causes a buildup on the mold, which must be removed. Both the respraying and the cleaning cause an increase in cycle time.

The use of an internal mold release agent can eliminate or lessen the problems of spray-on mold release. It would not be necessary to continually spray the mold, and there would not be as much buildup on the mold when internal mold release agents are used.

One theory about how mold release additives work is that the additive exudes to the surface in the time between injection and ejection and serves to lubricate the boundary area during ejection. The mold release agent may also serve to reduce electrostatic attraction during ejection in some cases. If these assumptions are true, then a number of things can effect the usefulness of an additive: solubility in the resin, rate of migration in the resin, lubricity of the additive, melting point of the additive, and ability of the additive to reduce electrostatic attraction, to name some. Since most of these interrelationships are not well known, it follows that a large part of mold release is done on a trial-and-error basis. Eventually what will and will not work can be predicted for a given resin, but these predictions are made only on the basis of how similar compounds behave.

In addition to the absence of a proven theory of how mold release agents work, there is no standard test for mold release effectiveness. It is necessary, of course, that there be tight adherence of the piece to the mold, but many mold configurations can accomplish that. The relative rankings of mold release agents should be the same, however. The method we have devised is an effective test procedure and does appear to produce useful information.

2.0 TEST PROCEDURE

The key element in our mold release testing is a quartz piezoelectric device attached to the ejector rod of the injection molding machine. It is known that a quartz crystal generates an electric current when compressed and, furthermore, the amount of current is proportional to the amount of compression. By amplifying and recording the current, the amount of force needed to eject a piece from the mold can be measured. If an additive improves mold release, a lower current is generated and the difference can be measured.

The mold is designed to produce a cylindrical piece about 5 cm in diameter, 4.5 cm long, and 0.4 cm thick. The piece is designed such that there is a maximum of contact between the resin and the mold. As the injected resin cools, it contracts around the mold, making it adhere tightly to the inner mandrel of the mold. Because of this tight adherence, the effectiveness of mold release agents can easily be seen. After the mold opens, a steel plate begins to push the piece off the mandrel. The greatest amount of force is that which is needed to break the piece free from the mandrel and start it forward. Progressively less force is needed to push the piece as it goes further toward the end of the mandrel. The mold release force is that force which is needed initially to start the piece moving.

This force is sensed by the piezoelectric device, which sends a signal to an amplifier and subsequently to a recorder. The signal is recorded on a strip chart and the amount of force is calculated. At least 10 injections and measurements are made and averaged for each number reported.

Results are reported as a percent decrease of mold release force rather than as absolute numbers. There are two reasons for this. First, relative, rather than absolute, numbers are being measured. Resin with additive is being compared to resin without additive, instead of absolute numbers being measured. Second, the baseline drifts slightly from day to day. If results are reported as absolute numbers, a different control value would need to be reported for each additive, adding considerable confusion. When percent differences are calculated from the data, there is no day-to-day drift on repeated tests, indicating that both the baseline and the experimental values drift in the same direction, by the same amount. The reproducibility of test values is good, giving rise to an error value of ±1.0 unit in the reported percentage numbers.

3.0 EXPERIMENTAL

All injection moldings are done on an Arburg 221E/150R injection molder. The molder has 25 ton clamp capacity and 1.5 ounce shot capacity and is equipped with a hydraulic ejector. The compression sensor is a 201A04 force ring transducer made by PCB Piezotronics, wired to a model 484 B line power unit (PCB Piezotronics). The power unit is coupled to a Gould 2200S recorder.

Additives are masterbatched into resins on a Brabender extruder. The additive is mixed with the resin (usually at a level of 1.0%) as thoroughly as possible. Extruder barrel temperatures and screw speed are determined by a combination of manufacturers' recommendations and in-house experimentation. The mixture of additive and resin is passed through the extruder into a water bath and then into a pelletizer. The pelletized resin is passed through the extruder two more times to ensure complete mixing. Resins that are sensitive to hydrolysis are dried in between extrudings.

Conditions for injection molding, also, are determined by a combination of manufacturers' recommendations and in-house experimentation. When the molder is at operating temperature, about 15–20 injections of resin without additive are made to allow it to stabilize. Another 10 injections of resin without additive are made, and measurements of mold release force are taken and averaged, to establish a baseline to compare to resin with additive. After the baseline has been established, molding using resin with additive (diluted to the desired level) is started. Six or seven injections are made before any measurements are made. Another series of 10 injections is made in which mold release forces are measured and averaged. Settings on the injection molder are not changed during testing, and the semiautomatic mode allows for reproducible molding. The mold temperature is held constant by a circulating water bath. The time between moldings is also kept constant.

4.0 RESULTS

4.1 Results

4.1.1 Acrylobutyl Nitrile
In acrylobutyl nitrite (ABS) it was found that the materials that had the greatest ability to improve mold release were esters of long chain saturated fatty acids and long chain saturated fatty alcohols. Of such compounds, cetyl esters of palmitic or stearic acid make the best mold release agents (see Table 1). Similar esters of shorter or longer chain length, or esters made from unsaturated materials, are not quite as good. Esters of other types (methyl esters, ethylene glycol esters, etc.) are even less useful as mold release agents.

None of the nonester materials tested showed the effectiveness of the wax esters. Neither fatty acids nor fatty amines are good mold release agents. Fatty amides also are not good

Table 1 Effectiveness of Mold Release Agents in ABS

Release agent	Level (ppm)	Reduction of ejection force (%)
Cetyl palmitate	10,000	11.2
Cetyl stearate	10,000	11.4
Cetyl palmitate–stearate	10,000	10.9
Cetyl myristate	10,000	7.6
Myristyl palmitate	10,000	8.7
Stearyl stearate	10,000	7.4
Benzenyl benzenate	10,000	6.9
N,N–Ethylene bisstearamide	10,000	2.3
Cetyl stearate	7,000	10.5
Cetyl stearate	6,000	8.9
Cetyl stearate	5,000	3.5
Fluorocarbon spray-on	—	11.1

Table 2 Mechanical Properties of ABS With Cetyl Stearate Present

	Cetyl stearate (ppm)		
Property	0	5000	10,000
Tensile strength at yield, psi	7172	6561	6221
Elongation at break, %	3.95	3.38	2.83
Izod impact, ft–lb/in.	3.99	4.14	4.47

release agents. Ethylene bisstearamide is sometimes added to ABS, probably as a flow improver, but it does not significantly improve mold release.

The optimum level of ester additive has been determined to be 6000–8000 ppm. At levels above this amount there is little increase in mold release effectiveness, and at levels below there is not enough mold release enhancement.

Only slight changes are observed in tensile strength and Izod impact resistance when tested at room temperature at levels up to 10,000 ppm (see Table 2). The percent elongation at break also shows little change when tested at room temperature.

4.1.2 Acetal

The most effective mold release agents in acetal are fatty amides in general and fatty bisamides amides in particular. Ethylene bisoleamide shows 25.1% reduction of mold release force at 5000 ppm but causes a visible darkening of the resin during processing. Ethylene bisstearamide, a saturated amide, is nearly as good, showing 23.3% reduction at a level of 5000 ppm; it does not cause any color problems. Other secondary amides, such as stearyl stearamide and stearyl erucamide, give mold release improvement nearly as good as the bisamides (see Table 3). Primary amides such as erucamide, oleamide, and stearamide also show good mold release enhancement in acetal.

Table 3 Effectiveness of Mold Release Agents in Acetal

Release agent	Level (ppm)	Reduction of ejection force (%)
N, N'–Ethylene bisstearamide	7500	26.0
N, N'–Ethylene bisoleamide[a]	5000	25.1
N, N'–Ethylene bisstearamide	5000	23.3
Stearyl stearamide	5000	21.1
Stearamide	5000	20.4
Erucamide	5000	21.8
N, N'–Ethylene bisstearamide	2500	15.2
N, N'–Ethylene bisstearamide	1000	5.3
Fluorocarbon spray–on	—	22.1

[a]Causes resin to darken during processing.

Table 4 Mechanical Properties of Acetal With N, N–Ethylene Bisstearamide Present

| | N,N'–Ethylene bisstearamide (ppm) | | |
Property	0	2500	5,000
Tensile strength at yield, psi	9965	9883	9818
Elongation at break, %	36	39	41
Izod impact, ft–lb/in.	1.30	1.32	1.35

None of the nonamide materials examined has the mold release effectiveness of the amides. The best ester mold release agent is cetyl palmitate, which exhibited a 16.5% reduction in mold release force at a 5000 ppm treatment level.

The optimum amount of ethylene bisstearamide is 5000 ppm. When used above that level, there is little increase in effectiveness; below that amount, the maximum effectiveness is not reached.

The use of fatty amides as mold release agents has negligible effect on mechanical properties (see Table 4).

4.1.3 Polybutylene Terephthalate

Fatty bisamides are the best mold release agents in polybutylene terephthalate (PBT). Both saturated and unsaturated bisamides show about 10% reduction of ejection force when used at a level of 5000 ppm. The bisoleamide, however, causes some darkening of the resin during processing (see Table 5).

Other mold release agents, such as fatty esters, amines, and acids, are not as effective as the amides.

The optimum amount of ethylene bisstearamide in PBT is between 2500 and 5000 ppm (see Table 5). Amounts less than 2500 ppm show a steep decline in effectiveness, while amounts above 5000 ppm do not show any more effectiveness than 5000 ppm.

The use of fatty amides as mold release agents in PBT has negligible effect on mechanical properties when tested at room temperature (see Table 6).

Table 5 Effectiveness of Mold Release Agents in PBT

Release agent	Level (ppm)	Reduction of ejection force (%)
N, N'–Ethylene bisoleamide[a]	5000	9.8
N, N'–Ethylene bisstearamide	5000	9.4
Stearamide	5000	6.7
Erucamide	5000	6.2
N, N'–Ethylene bisstearamide	2500	7.7
N, N'–Ethylene bisstearamide	1000	2.7
Fluorocarbon spray–on	—	8.8

[a]May cause resin to darken during processing.

Table 6 Mechanical Properties of PBT With N, N–Ethylene Bisstearamide

	N, N'–Ethylene bisstearamide (ppm)		
Property	0	2500	5,000
Tensile strength at yield, psi	9640	8664	8840
Elongation at break, %	299	249	227
Izod impact, ft–lb/in.	1.00	1.05	1.030

Table 7 Effectiveness of Mold Release Agents in Polypropylene

Release agent	Level (ppm)	Reduction of ejection force (%)
Glyceryl monostearate		
45% α-monoester	2500	23.9
60% α-monoester	2500	23.8
90% α-monoester	2500	22.4
Glyceryl distearate	2500	15.9
12% α-monoester		
Erucamide	5000	25.8
Fluorocarbon spray-on	—	23.1

Table 8 Mechanical Properties of Polypropylene with and Without Glyceryl Monostearate

	Glyceryl monostearate, 45% α-ester (ppm)	
Property	0	2500
Tensile strength at yield, psi	4848	5035
Elongation at break, %	295	60
Izod impact, ft–lb/in.	0.65	0.72

4.2 Polyolefins

4.2.1 Polypropylene

Glyceryl monostearate (GMS) has been found to be the best mold release agent in polypropylene. At a level of 2500 ppm, GMS shows about 24% reduction of mold release force. The reduction in mold release force is the same regardless of whether the GMS has 45, 60, or 90% α-monostearate (see Table 7). The remainder of the monostearate is β–monostearate, distearate, and small amounts of tristearate. When the amount of α-monostearate is less than 45%, there is a decrease in mold release enhancement.

Other glyceryl monoesters have also been tested, and only glyceryl monolaurate is a good a mold release agent as GMS. In addition to glycerol esters, ethylene glycol distearate and monostearate, cetyl palmitate, and methyl stearate have been examined. None is as good as GMS.

The results of these experiments indicate that the greater the amount of free hydroxyl in the ester, the more effective the mold release agent, up to a certain amount. Perhaps the hydroxyl groups make the additive less soluble in the resin, thus making it exude more to the surface. After a certain amount of hydroxyl has been reached, the migration to the surface reaches a maximum and further increases do not further enhance migration.

Fatty amides also show some utility as mold release agents in polypropylene. Erucamide shows 25.8% reduction of mold release force when used at a level of 5000 ppm. This amount of reduction is comparable to GMS, although the GMS is used at a lower level.

The use of glyceryl monostearate or erucamide as mold release agent in polypropylene has negligible effect on the mechanical properties (see Table 8).

4.2.2 High Density Polyethylene

The best mold release agent in high density polyethylene (HDPE) is erucamide. At a level of 2500 ppm it shows a mold release force reduction of about 22%. Other primary amides are also fairly effective as mold release agents, although not as good as erucamide (see Table 9). Secondary amides and bisamides are not as good mold release agents as the primary amides.

In addition to primary fatty amides, ethoxylated fatty amines are useful mold release agents in HDPE. At a level of 5000 ppm, ethoxylated oleyl amine and ethoxylated tallow amine show nearly a 20% decrease in mold release force (see Table 9). The use level is higher than the primary amide usage level, but the ethoxylated amines are also known as antistatic agents and therefore could solve two problems with one additive.

Table 9 Effectiveness of Mold Release Agents in HDPE

Release agent	Level (ppm)	Reduction of ejection force (%)
Erucamide	5000	30.8
Erucamide	2500	20.2
Stearamide	5000	26.7
Ethoxylated oleyl amine	5000	18.9
Ethxylated tallow amine	5000	19.8
Glyceryl monostearate	5000	14.8
Flurocarbon spray-on	—	20.7

Table 10   Mechanical Properties of HDPE with Various Additives

Property	Level (ppm)	Tensile strength at yield (psi)	Elongation at break (%)
No additive	—	3810	551
Erucamide	2500	3825	410
Erucamide	5000	3840	513
Ethoxylated oleyl amine	5000	3813	402
Ethoxylated tallow amine	5000	3877	425

Fatty esters have also been tested for mold release, but with the exception of glyceryl monostearate, they do not have much usefulness as mold release agents. GMS, used at a level of 5000 ppm, shows a 14.8% reduction of ejection force, but this is not as good as the primary amides or the ethoxylated amines.

The use of primary amides or ethoxylated amines as mold release agents in HDPE has negligible effect on mechanical properties when tested at room temperature (see Table 10).

4.2.3   Linear Low Density Polyethylene

The results of testing for mold release in linear low density polyethylenes (LLDPEs) are quite similar to those in HDPE. Erucamide is the most effective mold release agent (see Table 11).

The primary amides can be used at lower concentrations, because they are more efficient in LLDPE. Erucamide shows 30.3% reduction of mold release force when used at a level of 1000 ppm, while in HDPE it must be used at a level of 5000 ppm to achieve the same effect.

The ethoxylated amines are also useful mold release agents and exhibit the same increased efficiency reported for the primary amides (see Table 11). As in HDPE, the ethoxylated amines are also useful as antistatic agents.

Table 11   Effectiveness of Mold Release Agents in LLDPE

Release agent	Level (ppm)	Reduction of ejection force (%)
Erucamide	5000	49.8
Erucamide	2500	43.1
Erucamide	1000	30.2
Ethoxylated tallow amine	2000	30.9
Ethoxylated oleyl amine	2000	29.6
Glyceryl monostearate	2000	26.0
Flurocarbon	—	11.1

5.0 CONCLUSIONS

It has been shown that it is possible to measure qualitatively the effectiveness of internal mold release agents in injection molding. Numerous fatty chemicals were tested in polyolefins and engineering resins. The chemical type that is the most effective mold release agent in a particular resin varies widely with resin type. The required mold release pressure can be reduced for each of these resins without a significant change in the mechanical properties of the resin. One or more preferred mold release agent has been suggested for each resin.

ACKNOWLEDGMENT

Some of the information in this chapter is covered under an existing patent and a pending patent.

60
Organic Peroxides

Peter A. Callais

Pennwalt Corporation, Buffalo, New York

1.0   INTRODUCTION

Organic peroxides are derivatives of hydrogen peroxide, HOOH, wherein one or both hydrogens are replaced by an organic group (i.e., ROOH or ROOR)[1-5]. They are thermally sensitive and decompose by homolytic cleavage of the labile oxygen-oxygen bond to produce two free radicals:

$$ROOR' \xrightarrow{\Delta} RO\cdot + \cdot OR' \tag{1}$$

The temperature activity of organic peroxides varies from below room temperature to above 100°C, depending on the nature of the R groups. In addition to thermal decomposition, certain organic peroxides can be decomposed by activators or promoters at temperatures well below the normal decomposition temperature.

A major application of these compounds is as free radical initiators in the polymerization of vinyl and diene monomers in the plastics and coatings industries. They are also used as cross-linking and modifying agents for polyolefins, as vulcanizing agents for elastomers, and as curing agents for polyester resins.

2.0   TYPES AND PROPERTIES

Peroxide manufacturers now offer over 50 different organic peroxides in more than 100 formulations including dilutions in solvents, pastes, and filler-extended grades. In most cases, these formulations are designed for specific applications and to allow shipping and handling with a reasonable degree of safety.

The major classes of commercial organic peroxides are shown in Table 1. Decomposition rates of peroxides are commonly reported in terms of half-life ($t_{1/2}$) temperature, that is, the time at which 50% of the peroxide has decomposed at a specified temperature. Table 1

Table 1   Commercial Peroxide Classification

Peroxide type	Structure	10 Hour $t^{1/2}$ Range °C
Diacyl peroxides	 $\overset{\text{O}\quad\text{O}}{\overset{\|\quad\|}{\text{RCOOCR}}}$	20–75
Acetyl alkysulfonyl peroxides	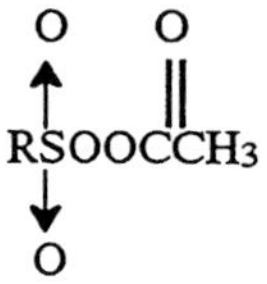 RSOOCCH_3 (O↑ ... ↓O)	32–42
Dialkyl peroxydicarbonates	 $\overset{\text{O}\quad\text{O}}{\overset{\|\quad\|}{\text{ROCOOCOR}}}$	49–51
tert-Alkyl peroxyesters	$\overset{\text{O}}{\overset{\|}{\text{R' COOR}}}$	49–107
oo-tert-Alkyl o-alkyl monoperoxycarbonates	$\overset{\text{O}}{\overset{\|}{\text{ROOCOR'}}}$	90–100
Di-*tert*-alkyl peroxyketals	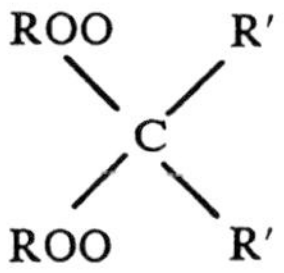 ROO, R' / C \ ROO, R'	92–115
Di-*tert*-alkyl peroxides	ROOR'	117–133
tert-Alkyl hydroperoxides	ROOH	133–172
Ketone peroxides	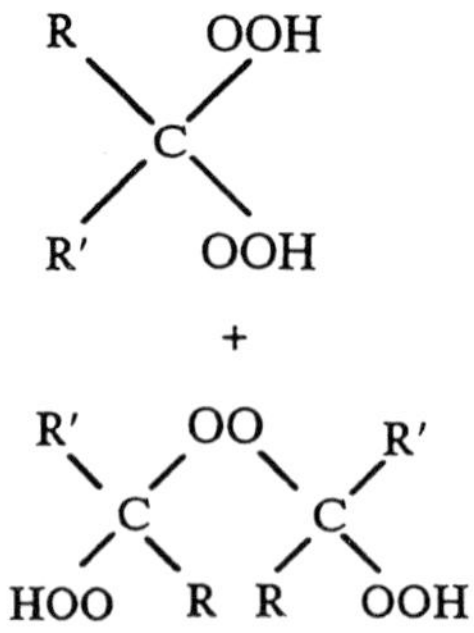 R, OOH / C \ R', OOH + R', OO, R' / C–C \ HOO, R, R, OOH	

list the 10-hour $t_{1/2}$ temperature ranges for the major organic peroxide types. Peroxides of certain types, such as hydroperoxides and ketone peroxides, are primarily used in combination with promoters and are employed at temperatures much lower than their measured 10-hour $t_{1/2}$ temperature.

2.1 Peroxide Selection

To a large extent, the half-life range at which an organic peroxide decomposes determines the application and controls overall process efficiency and product quality. In their product bulletins, organic peroxide producers often provide half-life data over a wide range of temperatures. For optimum efficiency, however, a peroxide is usually chosen so that the half-life under the reaction conditions is in the range of 10–20 minutes. This ensures the steady generation of radicals at such a rate that the heat of reaction can be safely contained and a high conversion of monomer to polymer results.

It is important to note, however, that peroxide half-life data are usually determined in select inert solvents and low peroxide concentration. Decomposition rates can be affected by solvent polarity, radical-induced decomposition, and peroxide concentration[5].

The temperature activity is not the sole consideration in selecting an organic peroxide for a particular application. Other factors to be taken into account include cost, solubility, safety, efficiency and type of radicals produced, necessity for refrigerated storage and shipment, compatibility with production equipment, effect on the final product, and the ability to be activated.

2.2 Radical Types

Although organic peroxides initially cleave at the oxygen-oxygen bond, other bond cleavages can and do occur, either simultaneously with or sequentially to the oxygen-oxygen bond dissociation. The relative stability of the R radical determines whether a peroxide undergoes single-or multiple-bond homolysis. The relative stability of the R radicals, in turn, can be correlated with the carbon-hydrogen or oxygen-hydrogen bond dissociation energies of the parent compound, as shown in Table 2[6–10].

The higher the bond dissociation energy, the less stable (more reactive) is the corresponding radical that is formed by removing the hydrogen atom. Thus, the phenyl radical is significantly more reactive than the alkyl radicals. The relative stability of alkyl radicals is in the order of *tert*-alkyl > *sec*-alkyl > *n*-alkyl > methyl. The methyl radical is about as reactive as an alkoxy radical.

One reaction that follows the peroxide linkage dissociation is the β-scission reaction:·

$$R-\overset{\overset{\textstyle R'}{|}}{\underset{\underset{\textstyle R''}{|}}{C}}-O\cdot \quad\xrightarrow{\ \beta\text{–scission}\ }\quad R'-\overset{\overset{\textstyle O}{\|}}{C}-R'' \ + \ R\cdot \qquad (2)$$

One of the R groups splits off to form a new alkyl radical and a ketone. The group that splits off forms the most stable radical. If the radical that splits off is more stable than the alkoxy radical, the β-scission reaction will be fast and the predominant initiating species will be the alkyl radical rather than the alkoxy radical. The rate of β-scission also depends on tempera-

Table 2 Bond-dissociation Energy of Typical Compounds

Bond in parent compound	Dissociation energy [kcal/mol (kJ/mole)]
$(R)_3C$—H	91 (381)
$(R)_2CH$—H	95 (397)
RCH_2—H	98 (410)
CH_3—H	104 (435)
RO—H	105 (439)
$\overset{\overset{\text{O}}{\|}}{RC}$O—M	112 (469)
 ⬡—H	112 (469)
HO—H	119 (498)
ROO—H	90 (377)

ture and pressure; that is, more scission occurs if the alkoxy radical is generated at higher temperatures and pressures.

For example, Figure 1 illustrates the β-scission reaction for two alkoxy radicals: one formed by the decomposition of a t-butylperoxy compound to the t-butoxy radical, the other by the decomposition of a t-amylperoxy compound to the t-amyloxy radical[11]. In this illustration, the initiating species for the t-butyl peroxide is either the t-butoxy radical or the methyl radical. The β-scission reaction is slow, since both radicals are highly reactive based

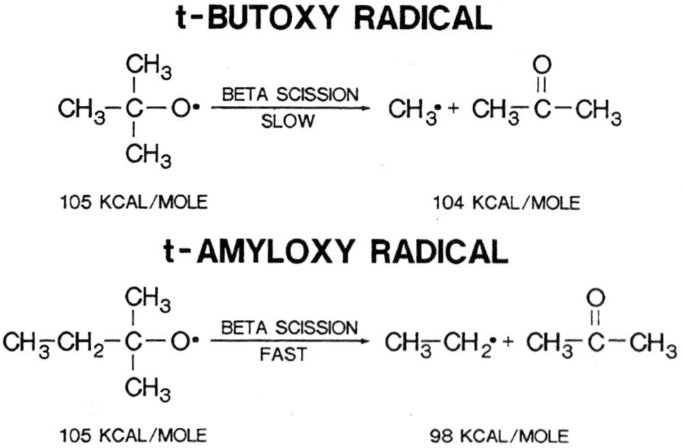

t-BUTOXY RADICAL

$$CH_3-\overset{\overset{\displaystyle CH_3}{|}}{\underset{\underset{\displaystyle CH_3}{|}}{C}}-O\bullet \xrightarrow[\text{SLOW}]{\text{BETA SCISSION}} CH_3\bullet + CH_3-\overset{\overset{\displaystyle O}{\|}}{C}-CH_3$$

105 KCAL/MOLE   104 KCAL/MOLE

t-AMYLOXY RADICAL

$$CH_3-CH_2-\overset{\overset{\displaystyle CH_3}{|}}{\underset{\underset{\displaystyle CH_3}{|}}{C}}-O\bullet \xrightarrow[\text{FAST}]{\text{BETA SCISSION}} CH_3-CH_2\bullet + CH_3-\overset{\overset{\displaystyle O}{\|}}{C}-CH_3$$

105 KCAL/MOLE   98 KCAL/MOLE

Figure 1 The β–scission reactions.

on the bond dissociation energies in Table 2. On the other hand, the t-amyloxy radical undergoes β-scission prodominantly to yield acetone and a lower energy ethyl radical.

The radical type can be either beneficial or detrimental, depending on the application. Phenyl, alkoxycarbonyloxy, alkoxy, and methyl radicals are capable of hydrogen abstraction, which can lead to cross-linking, chain branching, and polymer degradation. *tert*-Alkyl, *sec*-alkyl, *n*-alkyl, and *tert*-cumyl radicals are less energetic, more selective radicals. These radical types contribute to greater chain linearity and a narrower molecular weight distribution.

3.0 APPLICATIONS IN COATINGS

Organic peroxides are commonly used in the preparation of solution acrylic thermoplastic and thermosetting resins and the graft modification of alkyd resins. Some organic peroxides that are suitable as initiators for acrylic polymers and copolymers are listed in Table 3[12–15].

Efforts in recent years have been directed toward the production of hydroxy-functional acrylic oligomers for use in high solids coatings applications[16–19]. To achieve high solids coatings, acrylic polyol resins of low molecular weight and a narrow molecular weight distribution must be used to ensure an acceptable solution viscosity in the coatings formulation. In the synthesis of these resins, azonitrile and peroxide initiators, alone or in combination with high concentrations of a chain transfer agent and high temperatures, are employed[20–26]. However, certain peroxide initiators that abstract hydrogens cause chain branching, a broadening of the molecular weight distribution, and a high solution viscosity. In addition, the use of mercaptan chain transfer agents is known to produce objectionable odors, as well as color and light instability in the coating[26].

In a recent report, the use of t-amyl peroxides has been shown to produce acrylic oligomers with low molecular weight, narrow molecular weight distribution, low color, and low solution viscosity without added chain transfer agent[27,28]. The t-amyl peroxide initiated resins also gave better gloss and gloss retention in accelerated weathering.

Table 3 Organic Peroxide Initiators for the Polymerization of Acrylics

| | Half-life temperature (°C) | | |
Peroxide	10 h	1 h	Ref.
Dibenzoyl peroxide	73	91	12
t–Butylperoxy–2–ethylhexanoate	77	95	13
t–Butylperoxybenzoate	102	120	13
Dicumyl peroxide	116	136	14
Di–t–butyl peroxide	126	149	14
t–Amylperoxy–2–ethylhexanoate	75	92	15
t–Amylperoxybenzoate	100	121	15
t–Amylperoxyacetate	100	120	15
1,1–Di–(t–amylperoxy)cyclohexane	93	112	15
2,2–Di–(t–amylperoxy)propane	108	128	15
Ethyl 3,3–di–(t–amylperoxy)butyrate	112	125	15

In the graft modification of alkyd resins, the proportion of the graft copolymer formed has been shown to be dependent on the nature of the radical initiator used[29,30]. *t*-Butyl peroxides have been shown to be preferable to azonitriles because of the formation of a more reactive radical favoring graft copolymer formation.

4.0 SAFETY FACTORS AND PRODUCERS

Organic peroxides are useful initiators because of their thermal instability. Since peroxide initiators encompass a wide temperature activity range, a safe temperature for one peroxide may be unsafe for another. Manufacturers' recommendations should be carefully followed for handling, storage, and disposal. In addition, compounds such as transition metals, amines, strong acids and bases, and reducing agents can accelerate peroxide decomposition. Consequently, peroxides should be kept free of contamination. Only peroxides and peroxide formulations that can be shipped and utilized with a reasonable degree of safety are produced commercially.

5.0 FUTURE TRENDS

The current trend in organic peroxide research is to commercialize new multifunctional initiators, that is, peroxides with functionality other than simply being free radical sources. For example, organic peroxides that contain a hydroxyl group have recently been introduced commercially[31].

Polymeric peroxides that contain the peroxide linkage in the main chain of the polymer, on the end of a polymer, and pendent to the polymer chain have been investigated[32–37]. Organic peroxides containing a ultraviolet light absorbing[38] or a hindered amine light stabilizer[39,40] group have been used to synthesize polymers with the stabilizer group chemically bound to the polymer.

REFERENCES

1. D. Swern, *Organic Peroxides*, Vol. 1, New York, Wiley-Interscience, 1970.
2. D. Swern, *Organic Peroxides*, Vol. 2, New York, Wiley-Interscience, 1971.
3. C. S. Sheppard and O. L. Mageli, in *Encyclopedia of Chemical Technology*, Vol. 17, 3rd ed., H. F. Mark, D. F. Othmer, C. G. Overberger, and G. T. Seaborg, Eds., New York, Wiley-Interscience, 1982, p. 27.
4. R. L. Pastorino and R. N. Lewis, in *Modern Plastics Encyclopedia*, Vol. 64, 10A, R. Juran, Ed., New York, McGraw-Hill, 1988, p. 165.
5. C. S. Sheppard, in *Encyclopedia of Polymer Science and Technology*, Vol. 11, 2nd ed., H. F. Mark, N. H. Bikales, C. G. Overberger, and G. Menged, Eds., New York, Wiley-Interscience, 1988, p. 1.
6. D. M. Golden and S. W. Benson, *Chem. Rev.*, *69*, 125 (1969).
7. P. Gray, A. A. Herod, and A. Jones, *Chem. Rev.*, *71*, 289 (1971).
8. A. A. Zavitsas, *J. Am. Chem. Soc.*, *94*, 2779, 2788 (1972).
9. R. T. Sanderson, *J. Am. Chem. Soc.*, *97*, 1367 (1975).
10. R. S. Nangla and S. W. Benson, *J. Phys. Chem.*, *83*, 1138 (1979).
11. R. J. Kirchgessner, V. R. Kamath, C. S. Sheppard, and S. E. Stromberg, *Mod. Plast.*, *61*(11), 66 (1984).
12. *Diacyl Peroxides* (product bulletin), Lucidol Division, Pennwalt Corporation, Buffalo, NY, 1986.

13. *Peroxyesters* (product bulletin), Lucidol Division, Pennwalt Corporation, Buffalo, NY, 1986.
14. *Dialkyl Peroxides* (product bulletin), Lucidol Division, Pennwalt Corporation, Buffalo, NY, 1985.
15. *t-Amyl Peroxides* (product bulletin), Lucidol Division, Pennwalt Corporation, Buffalo, NY, 1985.
16. M. Takahashi, *Polym. Plast. Technol. Eng., 15,* 1 (1980).
17. L. W. Hill and Z. W. Wicks, Jr., *Prog. Org. Coatings, 10,* 55 (1982).
18. R. H. Kuhn, N. Roman, and J. D. Whitman, *Mod. Paint Coatings, 71*(5), 50 (1981).
19. R. F. Storey, in *Surface Coatings*, A. L. Wilson, J. W. Nicholson, and H. J. Prosser, Eds., London, Elsevier Applied Science, 1987, p. 69.
20. C. J. Bouboulis, U.S. Patent 4,739,006 (1988).
21. D. Rhum and P. F. Aluotto, U.S. Patent 4,075,242 (1978).
22. Y. Eguchi and A. Yamada, U.S. Patent 4,687,882 (1987).
23. W. R. Berghoff, U.S. Patent 4,716,200 (1987).
24. R. A. Gray, *J. Technol., 57*(728), 83 (1985).
25. R. Buter, *J. Technol., 59*(749), 37 (1987).
26. D. Rhum and P. F. Aluotto, *J. Technol., 55*(703), 75 (1983).
27. V. R. Kamath and J. D. Sargent, Jr., *J. Coat. Technol., 59*(746), 51 (1987).
28. V. R. Kamath, U.S. Patent pending.
29. F. M. Merrett, *Trans. Faraday Soc., 50,* 759 (1954).
30. D. H. Solomon, *J. Oil Colour. Chem. Assoc.,45* 88 (1962).
31. J. Sanchez, U.S. Patent 4,525,308 (1985).
32. A. J. D'Angelo and O. L. Mageli, U.S. Patents 4,304,882 (1981), 3,952,041 (1976), 3,991,109 (1976), 3,706,818 (1972), and 3,839,390 (1974).
33. A. J. D'Angelo, U.S. Patent 3,671,651 (1972).
34. R. A. Bafford, U.S. Patent 3,800,007 (1974).
35. R. A. Bafford, E. R. Kamens, and O. L. Mageli, U.S. Patent 3,763,112 (1973).
36. O. L. Mageli, R. E. Light, Jr., and R. B. Gallagher, U.S. Patent 3,536,676 (1970).
37. H. Ohmura and M. Nakayama, U.S. Patent 4,659,769 (1987).
38. C. S. Sheppard and R. E. MacLeay, U.S. Patents 4,042,773 (1977) and 4,045,427 (1977).
39. T. N. Myers, European Patent Appl., 233,476 (1987).
40. P. A. Callais, V. R. Kamath, and J. D. Sargent, *Proc. Water-Borne Higher Solids Coatings Symp., 15,* 104 (1988).

61
Pigment Dispersion

Theodore G. Vernardakis

Sun Chemical Corporation
Cincinnati, Ohio

1.0 INTRODUCTION

The dispersion of pigments in fluid media is of great technological importance to the coatings manufacturers who deal with pigmented systems. The basic aim is to change the physical state of pigments to achieve desired effects in specific application systems. The dispersion process involves the breaking down and separation of the aggregated and agglomerated particles that are present in all pigments in their normal form after their manufacture. Dispersion is not considered to be a process of pulverization, but rather a process of particle separation, homogeneous distribution of the particles in a medium, and stabilization of the resultant system to prevent reaggregation, reagglomeration, flocculation, and settling. The process of dispersion must be done efficiently and in the shortest time possible to draw out of the pigment its maximum color properties at the least cost.

The topic of pigment dispersion in fluid media has been covered extensively in the literature[1-6]. The theoretical aspects of pigment dispersion apply equally well to inorganic and organic pigments. In this chapter, the practical examples of surface treatments apply primarily to organic pigments, but similar treatments can be carried out on inorganic pigments as well.

2.0 A BRIEF INTRODUCTION TO PIGMENTS

2.1 Pigment Definition

Materials are colored by the use of pigments or dyes. Pigments are colored, black, white, or fluorescent particulate organic or inorganic solids, usually they are insoluble in, and essentially physically and chemically unaffected by, the vehicle or substrate in which they are incorporated. They alter appearance by selective adsorption and/or by scattering of light[7].

Pigments usually are dispersed in vehicles or substrates for application (e.g., in inks, paints, plastics, or other polymeric material). Pigments retain a crystal or particulate structure throughout the coloration process.

529

As a result of the physical and chemical characteristics of pigments, pigments and dyes differ in their application: when a dye is applied, it penetrates the substrate in soluble form, after which it may or may not become insoluble. When a pigment is used to color or opacify a substrate, the finely divided, insoluble solid remains throughout the coloration process.

Organic pigments are highly colored, inert synthetic compounds that are usually brighter, purer, and richer in color than inorganic pigments. Generally, however, they are less resistant to sunlight (some fade badly on exposure to light), to chemicals (greater tendency to bleed in solvents), and to high processing temperatures (lower heat stability); quite often too, they are more expensive than inorganic pigments. Pigments are classified by the "Colour Index" according to specific pigment name and constitution number. For example, phthalocyanine blue is known by the C.I. name Pigment Blue 15 and, its C.I. number is 74160, while titanium dioxide is C.I. Pigment White 6, C.I. 77891. The great number and variety of organic and inorganic pigments make it impossible to treat them all in this chapter. References should be consulted for information on pigment types, chemical and physical properties, methods of preparation, grades, specifications, and applications. See, for example, References 8–11.

2.2 Pigment Particles

Pigments are normally produced in a wet presscake form, which upon drying and grinding or spray drying assumes the form of a fine dry powder. Presscakes, either at their normal pigment content (20–40%) or as "high solids" (50–60%), are used by the manufacturers of aqueous pigment dispersions for paint, textile, and ink applications, as well as by those who produce flushed colors for oil ink or coatings applications. Dry pigment powders are used in a host of other systems such as solvent inks, coatings, and plastics. Pigments in the presscake or dry powder form are composed of fine particles, normally in the submicrometer size range. Their color properties are generally influenced by particle size and particle size distribution; therefore, an assessment on the degree of dispersion must, above all, be considered in terms of these critical measurements [12]. In general, color properties, such as strength, transparency, gloss, rheology, and lightfastness of all pigmented systems, are affected to a greater or lesser extent by the size and distribution of the pigment particles in the dispersion. For example, phthalocyanine blue is first prepared commercially in a "crude" pigment form having a large particle size, up to 25 μm. As such, it has little color value and must therefore be reduced to smaller, finer particles to enhance its coloristic properties. After particle size reduction (down to 0.03–0.15 μm), an excellent pigment is obtained, which exhibits a high degree of tinctorial strength, transparency, and gloss. Typical electron micrographs of these two materials, showing particle size, are reproduced in Figure 1.

Pigment particles normally exist in the form of primary particles, aggregates, agglomerates, and flocculates. Primary particles are individual crystals and associated crystals as they are formed during the manufacturing process. (Fig. 2). They may vary in size depending on the conditions of precipitation and growth, which are controlled by the pigment manufacturer. The scanning electron photomicrograph of Figure 2 for micronized sodium chloride (although this is not a pigment), is used only to illustrate the individual and associated crystals that make up the primary particles of a compound.

Aggregates are collections of primary particles that are attached to each other at their surfaces or crystal faces and show a tightly packed structure. Agglomerates consist of primary particles and aggregates joined at the corners and edges in a looser type of arrangement. Aggregates are formed during the manufacturing process in the course of the ripening period of the precipitates. Agglomerates, most often, are formed during the drying of the

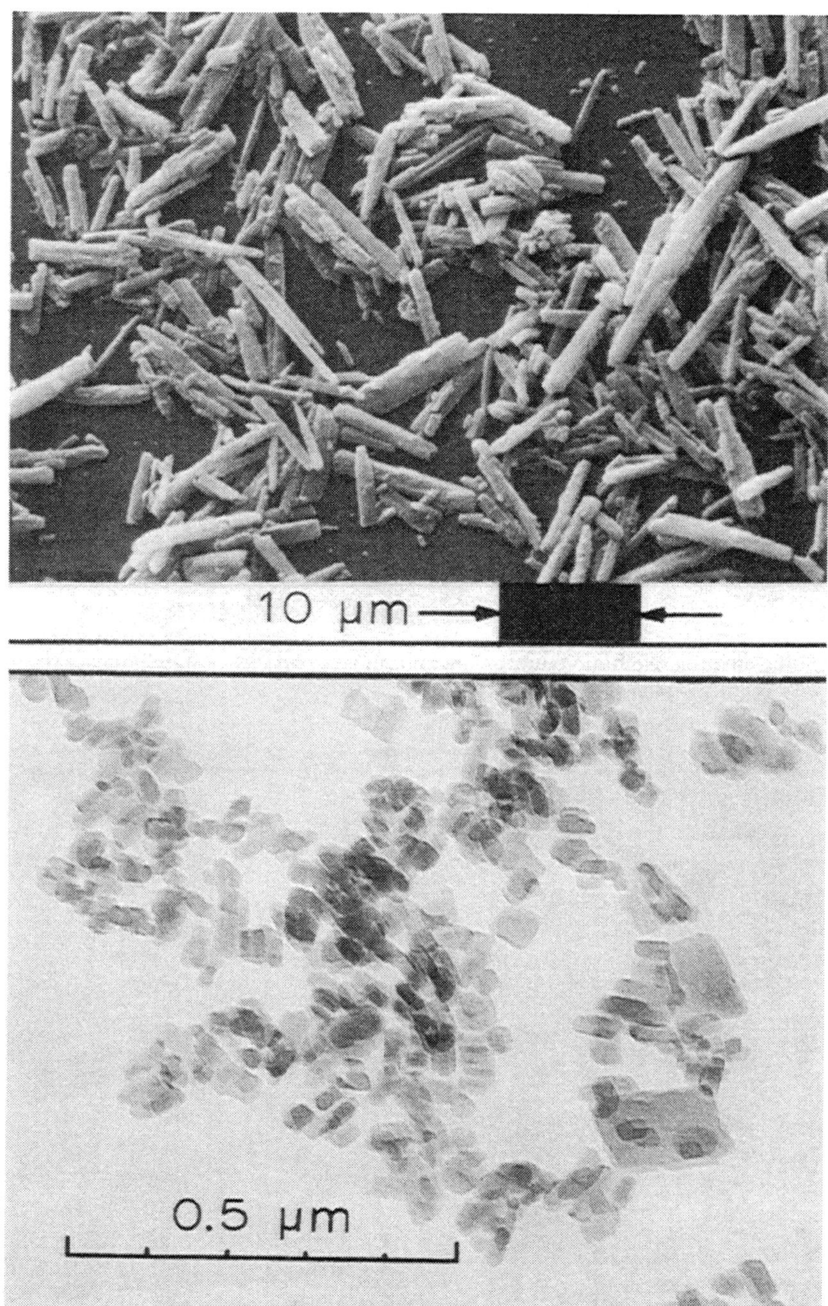

Figure 1 Scanning electron photomicrograph of copper phthalocyanine blue crude (top) and transmission electron photomicrograph of copper phthalocyanine blue pigment (bottom) showing particle size differences; Pigment Blue 15.

Figure 2   Scanning electron photomicrograph showing primary particles: (a) individual crystals and (b) associated crystals of micronized NaCl.

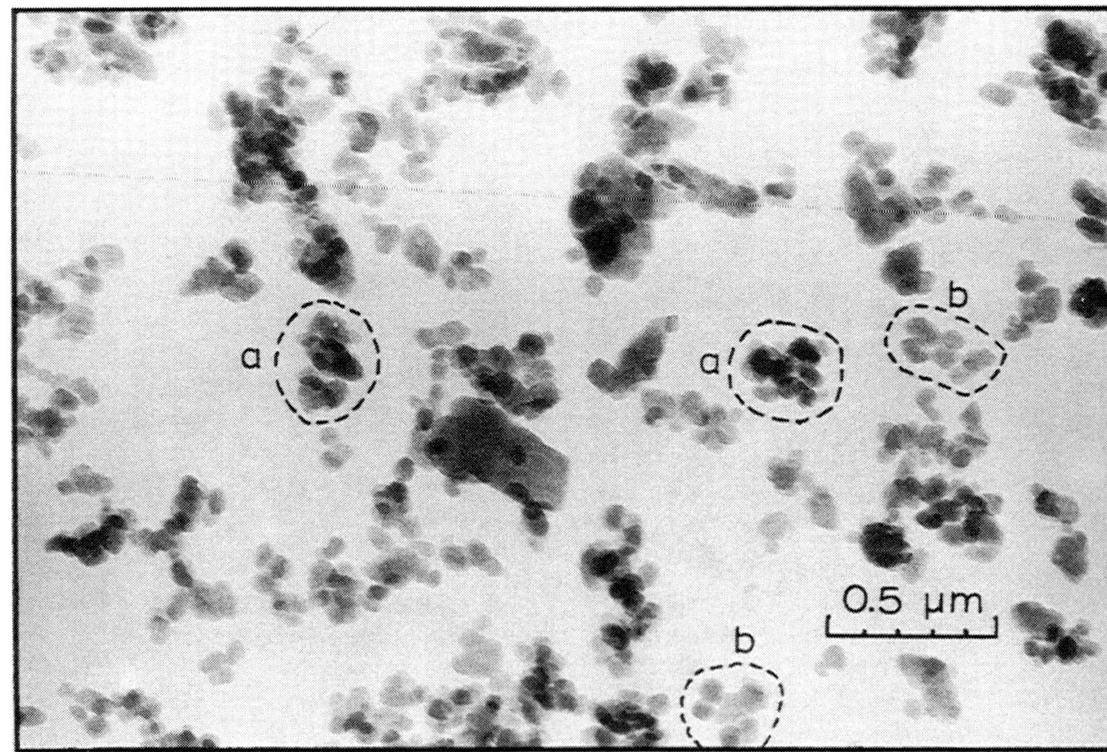

Figure 3   Transmission electron photomicrograph showing (a) aggregated and (b) agglomerated pigment particles. D&C Red No. 30, Vat Red 1.

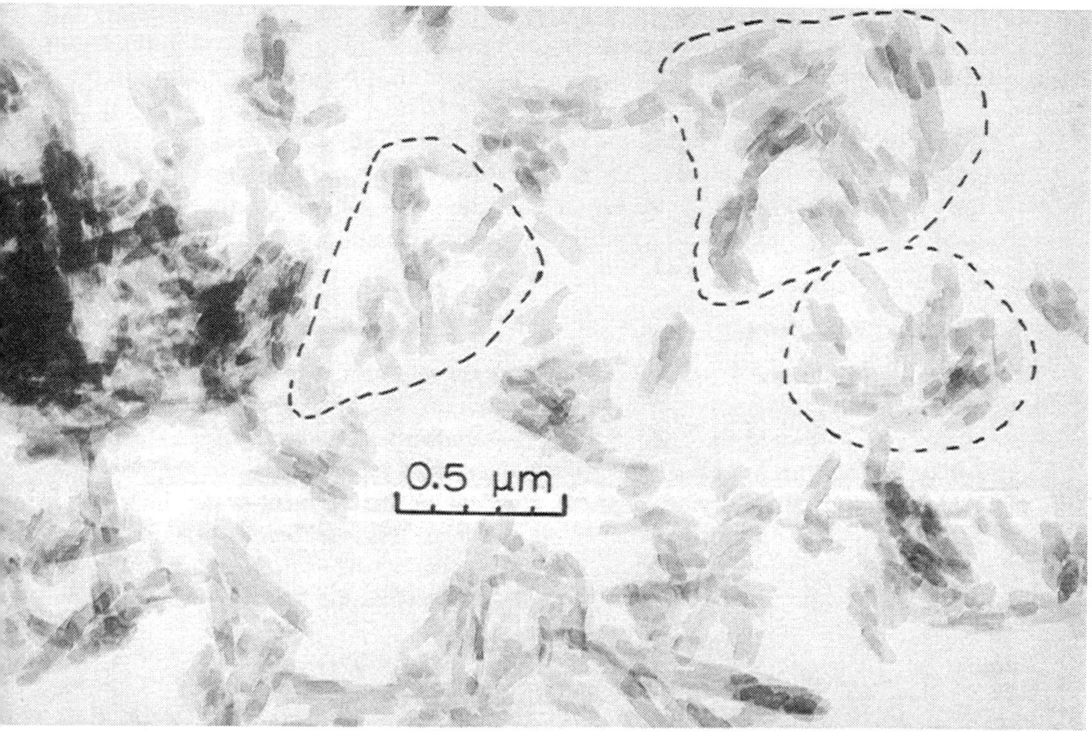

Figure 4 Transmission electron photomicrograph showing flocculated pigment particles. Dimethylquinacridone magenta, Pigment Red 122.

presscakes and the subsequent dry milling of the pigment lumps. Figure 3 shows typical arrangements of aggregated and agglomerated pigment particles.

Flocculates consist of primary particles, aggregates, and agglomerates, generally arranged in a fairly open structure, as shown in Figure 4. Flocculates may be broken down easily under shear, but they will form again when such shear forces are removed and the dispersion is allowed to stand undisturbed.

3.0 THE DISPERSION PROCESS

The primary purpose of dispersion is to break down pigment aggregates and agglomerates to their optimum pigmentary particulate size (down to individual single particles, if possible) and to distribute these pigment particles evenly throughout a medium (i.e., the carrier). Usually the carrier is a liquid or a solid polymeric material that is deformable at high temperatures during processing. To achieve the optimum benefits of a pigment, both visual and economic, it is necessary to obtain as full a reduction as possible to the primary particle size. After all, the color strength of a pigment depends on its exposed surface area: the smaller the particle size, the higher the surface area, and thus, the stronger the color. Furthermore, the pigment is generally the most expensive constituent of any pigmented system; therefore the user normally wants to obtain optimum performance with the smallest possible amount of pigment. Ideally, a good pigment dispersion consists chiefly of primary particles, with only a minimum of loose aggregates and agglomerates. In practice, reduction to the primary particle size is largely determined by the nature of the pigment (i.e., its dispersibility), by the dispersion system and processing equipment, and by the end–user requirements of the product.

Dispersion should not be confused with pulverization. The latter is simply a comminution process whereby large pigment lumps are broken down to smaller units, which constitute the powder form. Pulverization does not break down the aggregated, agglomerated, and flocculated particles into primary particles. Dispersion, however, accomplishes this effectively.

3.1 Pigment Wetting

It is generally recognized that the dispersion process consists of three distinct stages: wetting, deaggregation–deagglomeration, and stabilization. The wetting stage involves the removal from the surface of the pigment particles of adsorbed molecules of gas, liquid, and other materials and their replacement with molecules of the vehicle. In other words, the pigment–air interface in dry pigment powders or the pigment–water interface in presscakes is replaced by the pigment–vehicle interface. This is accomplished through preferential adsorption. The efficiency of wetting depends primarily on the comparative surface tension properties of the pigment and the vehicle, as well as the viscosity of the resultant mix.

3.2 Particle Deaggregation and Deagglomeration

After the initial wetting stage, it is necessary to deaggregate and deagglomerate the pigment particles. This is usually accomplished by mechanical action with devices such as ball mills, bead mills, and two–roll mills. As the pigment powder is broken down to the individual particles, higher surface areas become exposed to the vehicle and larger amounts of it are required to wet out newly formed surfaces. During this stage of deaggregation, the amount of free vehicle in the bulk diminishes; therefore, the viscosity of the dispersion increases. At higher viscosities, shear forces are greater and the breaking down and separation of particles become more efficient. It is this process of mechanical breakdown of the aggregates and agglomerates that demands a high energy input and can become quite costly. Some easily dispersible pigments have been developed lately to aid in the reduction of energy requirements. Such pigments are produced by surface treatment of the pigment during manufacture, with the purpose of reducing or inhibiting agglomeration–aggregation formation. In many cases, such treatments are highly specific to a single ink, paint, coating, or plastic medium.

3.3 Dispersion Stabilization

The third stage of great importance in the dispersion process is the stabilization of the pigment dispersion. This ensures that complete wetting and separation of the particles has been reached, and also that the pigment particles are homogeneously distributed in the medium. If the dispersion has not been stabilized, flocculation may occur as a result of clumping together of the pigment particles. Flocculation is generally a reversible process. Flocculaes typically break down when shear is applied and will form again when the shear is removed. Where a pigment dispersion is not stabilized by the action of resin molecules in the vehicle, the use of surfactants or polymeric dispersants can be considered. Such additives may be used directly during pigment manufacture, or they may be incorporated in the vehicle.

4.0   THE ROLE OF SURFACE ENERGY

It is well known that molecular forces at the surface of a liquid are in a state of imbalance. The same is true of the surface of a solid, where the molecules or ions on the surface are subject to unbalanced forces of attraction normal to the surface plane. Such atoms do not have all their forces satisfied by union with other atoms. As a result, there is a net force, which tends to pull the surface molecules into the bulk. The opposing force, which resists this inwardly pulling force, is known as the surface tension or surface energy. All solids and liquids have surface energies to a greater or lesser degree. To satisfy these surface forces, liquids and solids tend to attract and retain on their surfaces dissolved substances in the solution or gasses from the surrounding atmosphere. These forces are short–ranged attractive forces, known as van der Waals or London forces, and they play a very important role in particle aggregation, wetting, and dispersion stabilization.

4.1   Surface Energy and Surface Area

Pigments having a very small particle size exhibit high surface area and consequently high surface energies. As large pigment particles are broken down into several smaller particles, new surfaces are constantly created, contributing to a higher surface area and thus a higher surface energy.

Let us assume that a pigment powder has a surface area S of 60 m²/g and a density ρ of 1.0 g/cm³. Its basic particle diameter D from

$$D = \frac{6}{\rho S}$$

will be 0.1 μm. If these particles are cubic in structure, and if, for the sake of simplicity, we assume that a 1 cm³ of pigment is broken down into particles 0.1 μm in size, then 1×10^{15} particles will be produced. We assume also that the particles are in perfect cubic packing. To get an idea of the area created by the new surface, we need only compare the surface area of 6 cm² for the 1 cm cubic particle to the surface area of 600,000 cm² (60 m²) for the 1×10^{15} cubic particles that are 0.1 μm in size. The increase in surface area is 100,000–fold. The new surfaces produced are tremendously large. The surface energies associated with these new surfaces are also quite large. These van der Waals surface energies create the attraction between the submicrometer particles that come together to form the aggregates and agglomerates.

4.2   Surface Energy and Pigment Wetting

Surface energies play an important role in the wetting and stabilization of pigment dispersions. For wetting to be effective, the wetting energies of the pigment–vehicle interface must be greater than the sum of the adsorption energy (this is because of substances on the pigment surface) and the attractive energy that holds the pigment particles together. Generally, lower energy (low surface tension) liquids, such as aliphatic and aromatic hydrocarbons, will spread over, or wet, higher energy surfaces. Quite often, it happens that a liquid does not spread over a pigment surface completely. This occurs when a high energy liquid

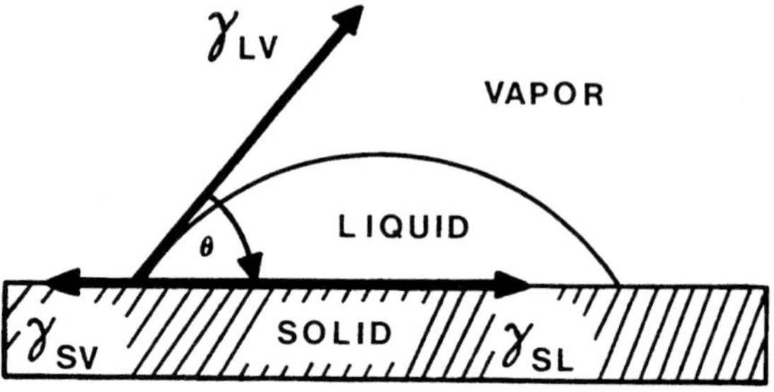

Figure 5   Partial wetting of a solid surface by a liquid in accordance with the Young–Dupré equation.

(high surface tension), such as water, will not entirely wet out a high energy surface. In this case, the wetting energy is equal to or less than the sum of the adsorption and interparticle attraction energies, and wetting may be either partial or nonexistent. The liquid will not spread entirely over the surface, as shown in Figure 5. The relationship that describes such a system is given by the Young–Dupré equation as follows:

$$\gamma_{SV} = \gamma_{SL} + \gamma_{LV} \cdot \cos\theta$$

where γ_{SV}, γ_{SL}, and γ_{LV} are the interfacial energies at the solid–vapor, solid–liquid, and liquid–vapor interfaces, respectively, and θ is the contact angle. For complete wetting, the contact angle is zero ($\cos\theta$ becomes unity), and the liquid spreads entirely over the solid surface, For $\theta > 0$, wetting either is incomplete or does not occur at all.

4.3   Surface Energy and Destabilization of the Dispersion

Surface energies play an important role in the destabilization of the dispersion. Particles dispersed in liquid media are in constant motion (thermal or Brownian movement). As they move through the medium, they collide with other pigment particles. The frequency of these collisions depends on the size of the particles and on the viscosity of the medium. During such collisions, the particles will attract and may join with other particles because of the powerful short–range London–van der Waals attractive forces, which in effect are surface energies. These forces are electrical in nature (type) and are due to the interaction of the dipoles that are present in the particles, as permanent dipoles (polar particles) or induced dipoles (nonpolar but polarized particles).

Once the particles have come together, they may reaggregate or form flocculates if their surface is not protected, and they will settle to the bottom of the container. This is an undesirable effect for the ink, paint, or coatings manufacturer. Therefore, to prevent reaggregation or flocculation, such dispersions must be stabilized.

4.4   Surface Energy and the Acid–Base Concept

The idea of surface energy in pigments has been closely related to the acid–base concept, advanced by Sorensen,[13] who has used it to describe the interaction between pigments, binders, and solvents, to obtain optimally stable pigment dispersions having the best application properties in fluid ink systems. Such interrelation between the surface energies of these three components in a dispersion can be characterized by their acid–base properties. Pigments can be classified as acidic (electron acceptors), basic (electron donors), amphoteric (electron acceptors and donors), or neutral. Binders and solvents can be similarly characterized.

Acidic pigments should be used with basic resins (polyamide, melamine, alkyd), while basic pigments should be used with acidic resins (vinyl, acrylic, maleic). Amphoterics can be used with both resins. Neutral pigments should be surface treated to improve their dispersion characteristics. The solvent must have the same acid–base character as the pigment, whereby the interaction between the solvent and the pigment surface is minimized and at the same time the interaction between the resin binder and the pigment surface is maximized. In other words, there should be no competition between solvent and binder for the pigment surface; to obtain maximum dispersion and stability, only the binder should adsorb.

5.0 MECHANISMS FOR THE STABILIZATION OF DISPERSION

5.1 Charge Stabilization

Dispersions may become stable through two generally accepted mechanisms: charge stabilization and steric or entropic stabilization. Charge stabilization is due to electrical repulsion forces, which are the result of a charged electrical double layer surrounding the particles as shown in Figure 6. The charged electrical double layer developed around the particles extends well into the liquid medium, and since all the particles are surrounded by the same charge (positive or negative), they repel each other when they come into close proximity.

5.2 Steric or Entropic Stabilization

Steric stabilization is due to steric hindrance resulting from the adsorbed dispersing agent, the chains of which become solvated in the liquid medium, thus creating an effective steric barrier that prevents the other particles from approaching too close. This phenomenon is also called entropic stabilization because, as the coated particles approach each other, the solvated chains of the adsorbed dispersant lose some of their degrees of freedom, resulting in a decrease in entropy. Such lowering in entropy gives rise to repulsive forces, which keep the particles away from each other. This type of steric or entropic stabilization is also represented in Figure 6.

6.0 SURFACE TREATMENT

6.1 Surfactants

Surface active agents or, simply, surfactants are substances that are used to lower the interfacial tension between a liquid and a solid. Such is the case for pigments in fluid media, with the expressed purpose of improving pigment dispersibility by improving pigment wetting characteristics, preventing reaggregation, and increasing the stability of the dispersion. A surfactant molecule typically contains two groups of opposite polarity and solubility. The hydrophilic group is the polar, water–loving part, while the lipophilic group is the nonpolar, oil–loving part of the molecule.

Surfactants are characterized by their HLB value (hydrophile–lipophile balance), which is a ratio of the hydrophilic to lipophilic groups on the molecule and gives an indication of

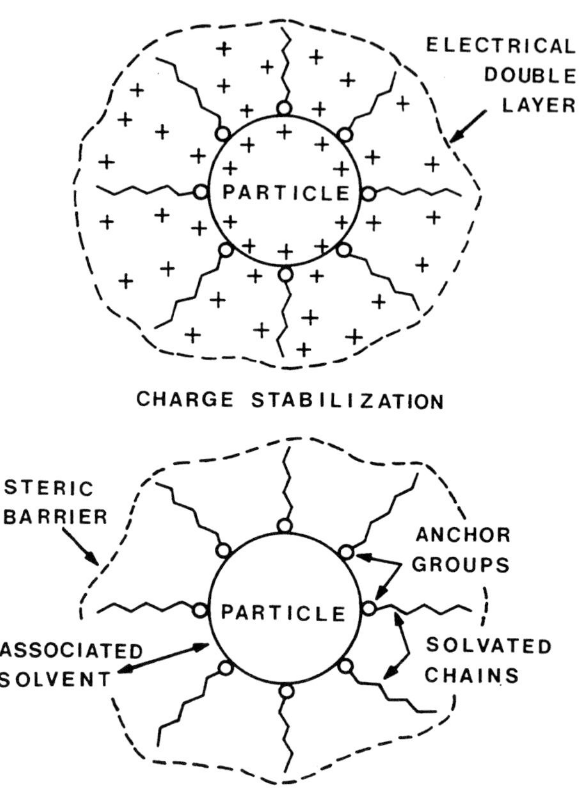

Figure 6   Charge and steric or entropic stabilizations.

their solubility in water or oil–solvent systems. High HLB values mean that the surfactant is soluble in water (an abundance of hydrophilic groups). Low HLB values, on the other hand, mean that the surfactant is soluble in oil or solvents (an abundance of lipophilic groups).

Surfactants attach themselves to the pigment particles via preferential adsorption, as shown in Figure 7 for aqueous and nonaqueous systems. In aqueous systems, the lipophilic (or hydrophobic) groups are adsorbed on the particle surface, and the hydrophilic (or lipophobic) groups extend into the bulk of the aqueous phase to form an effective protective barrier around the particle. In the case of non–aqueous solvent systems, the hydrophilic groups of the surfactant are attached to the particle surface and the lipophilic groups (tails) extend into and are solubilized by the solvent.

larized functional groups, which can serve as adsorption sites for the hydrophilic or lipophilic groups of the surfactants. For instance, organic pigments typically contain groups

such as nitro ($-NO_2$), hydroxyl ($-OH$), carbonyl ($-\overset{|}{C}=O$), amide ($-NH-\overset{|}{C}=O$), methoxy ($-O-CH_3$), chlorine ($-Cl$), bromine ($-Br$), sulfonate ($-SO_3$), carboxylate ($-COO^-$), and metal ions such as Ba^{+2}, Ca^{+2}, Mn^{+2}, and Cu^{+2}, which can function as the anchoring sites for the hydrophilic or lipophilic groups of the surfactants.

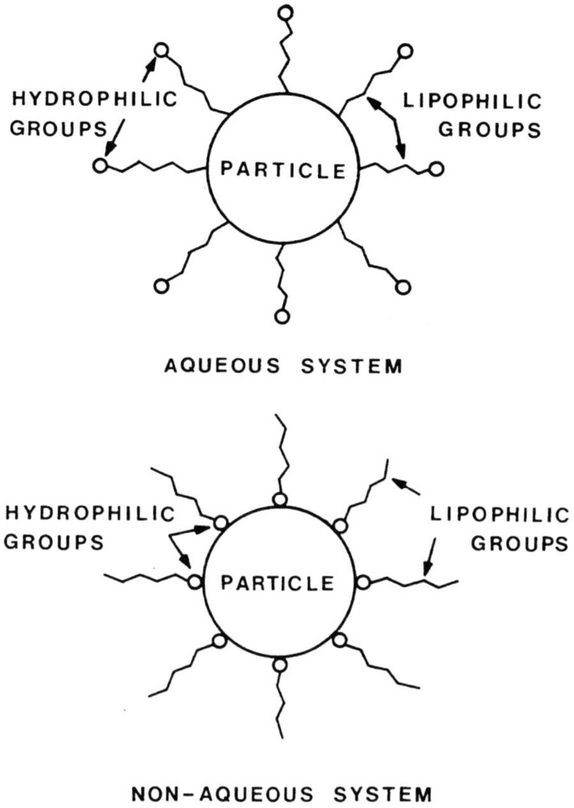

Figure 7   Surfactant attachment on pigment particles in aqueous and nonaqueous dispersions.

It is well known, however, that classical surfactants do not always improve the dispersion characteristics of pigments, especially when pigment surfaces are low in polarity or nonpolar and are dispersed in nonpolar vehicles. With such pigments and vehicles, dispersants and surface modifying agents of other types must be used to improve wetting and dispersibility and to prevent flocculation of pigment particles.

6.2 Polymeric Dispersants

Polymeric dispersants or "hyperdispersants" [14] are claimed to be more effective dispersion stabilizers for nonaqueous systems. These substances have a two–part structure, one consisting of an anchoring functioning group (or groups), and the other consisting of a polymeric solvatable chain to which the functional group is attached. They are in effect polymeric surfactants or dispersants but were developed for use in specific nonaqueous systems, where classical surfactants have limitations. When they are used as dispersants for organic pigments, it is preferable that they have multiple anchoring groups on one polymeric chain, because organic particles are not as strongly polar as inorganic particles. Such dispersants may be of a fatty polyester type, containing a carboxy group at the end [e.g., poly(12–hydroxystearic acid)] with the carboxy group functioning as the anchor and the polyester group as the solvated chain. Others with multiple anchor groups, are fatty polyureas and polyurethanes, which may even contain polymeric solvatable groups instead of the long fatty chains.

6.3 Surface Modifying Agents

Surface modifying agents are another group of additives that can be used to aid the dispersion of organic pigments in organic media. These agents are often pigment derivatives (e.g., large flat dye molecules), which provide improved resistance to flocculation and greater stability to the dispersion. The pigment derivative is adsorbed onto the pigment surface via the van der Waals attractive forces, which act over a large area, because such large planar dye molecules lie flat on the pigment surface. They may be used either alone or in conjunction with a polymeric dispersant. When used alone, they introduce or increase on the surface of nonpolar or low polarity pigments, the number of polar sites, which are necessary to interact with the resin in the vehicles, to stabilize the dispersion. When used together with the polymeric dispersant, they provide anchoring sites on which the anchor groups of the dispersant will become attached. In this context, they can be used synergistically with dispersing agents, at which time they are called colored synergists.

7.0 SURFACE TREATMENT DURING PIGMENT MANUFACTURE

Generally, surface–treated pigments are more easily dispersible, produce more stable dispersions in fluid media with improved flow, and impart higher strength and gloss to the printed films, when compared with untreated pigments. Surface treatments can be carried out at different stages of pigment manufacture. Some pigments are prepared directly as finished products, while others are in the form of a crude pigment that must be conditioned into the pigmentary state.

Use of surfactants is typically made at the initial stage of pigment manufacture. During the precipitation of the intermediate (e.g., diazo in the preparation of azo pigments), surfactants are used to wet out and control the fineness of the precipitate, and they may also act as promoters to accelerate the azo coupling reaction. At the second stage, during the precipitation of the pigment (e.g., in the azo coupling reaction), surfactants may be used in the dispersion of the pigment particles as they are being formed—for example, in azo yellows, which are precipitated in the pigmentary state, or in the dispersion of the precursor (dyestuff), as in the case of metallized azo reds (which are first formed as sodium salts), to control the salt formation (barium, calcium, etc.), and thus produce the final pigment. At the third stage, during the conditioning of the pigment, surface treatments are used to pigment particle dispersion, for coating the pigment surface to prevent aggregation, and for controlling the growth of crystal particles. If particles are too difficult to filter, use of a specific additive (flocculant) sometimes induces controlled flocculation and facilitates filtration. Complex formation with additives may also be carried out during this conditioning stage to stabilize the particles and increase dispersibility, as is the case with diarylide yellows, which may be surface treated with fatty amines to produce Schiff base stabilization and result in easily dispersible pigments.

8.0 SURFACE TREATMENT OF PIGMENTS: APPLICATION

8.1 Organic Pigments

The published literature dealing with surface treatments of organic pigments patented or otherwise, is so extensive that no attempt is made to review it, although it may be referred to, occasionally. Readers are urged however, to consult the review by Hayes,[15] which covers the role of classical surfactants, polymeric dispersants, and pigment derivatives in surface treatments. Other reviews of interest are those by Topham,[16] Merkle and Schafer,[17] and Hampton and McMillan,[14] the latter dealing specifically with polymeric dispersants. Further examples will be presented here.

It is well known to the pigment manufacturer that rosination is perhaps the oldest surface treatment known, especially for azo pigments, where rosin (abietic acid) is precipitated onto the pigment surface as the barium or calcium salt. It can also be used to treat other pigments, such as copper phthalocyanine blue,[18] and for a host of similar applications, in a polymerized form. Along the same lines, long chain carboxylic acids (fatty acids) have also been used to treat pigment surfaces[19]. A likely arrangement of these molecules adsorbed on the surface is shown in Figure 8. The hydrophilic anchor groups are attached to the surface, with the lipophilic groups projecting outward. The use of rosin has been mentioned because of its historical significance and because it is still widely used today, since it is one of the least expensive surface treating agents.

In the course of a study by the author for the development of a diarylide yellow AAOT (Colour Index Pigment Yellow 14) for flexographic ink applications, it was found that a desirable product was one prepared in the presence of an amine–type ethoxylated guanidine weakly cationic surfactant, in combination with a polar tetramethyl decynediol solvent.[20] It appears that these two surface active agents worked synergistically to produce a strong, transparent and nonflocculating pigment, as opposed to products in which the surfactant or the solvent or both were absent from the preparation. Transmission electron micrographs

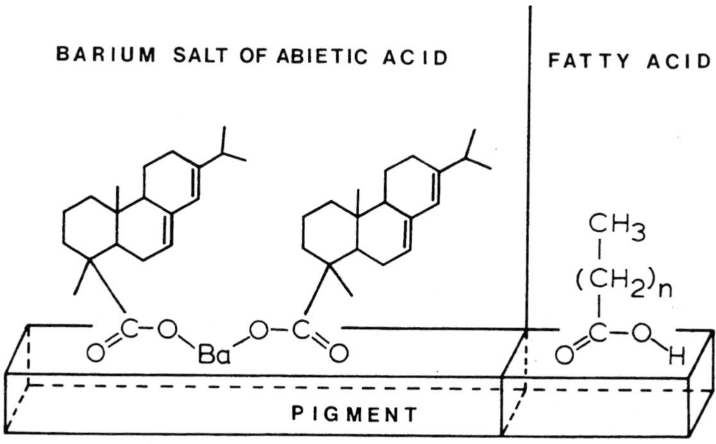

Figure 8 Treatment of pigment surfaces with rosins and fatty acids.

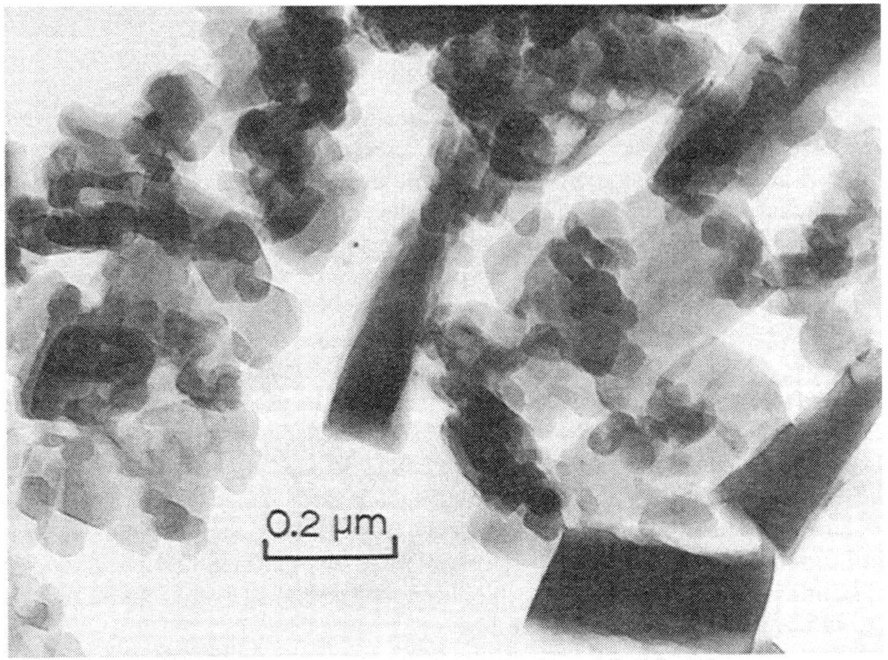

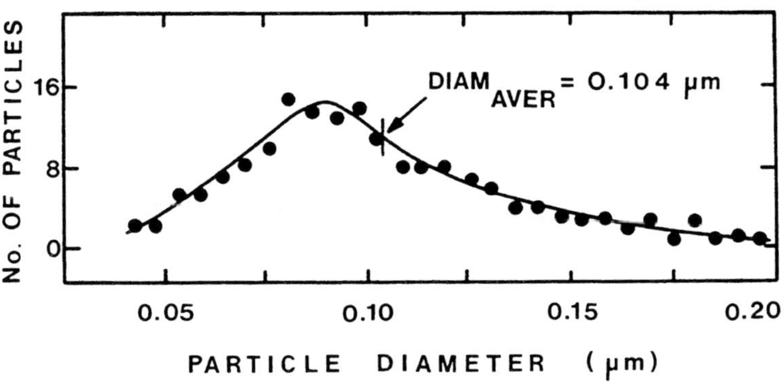

Figure 9   Transmission electron photomicrograph and particle size distribution of an untreated diarylide yellow AAOT, Pigment Yellow 14.

and particle size distributions for such pigments are shown in Figures 9 and 10. It is apparent that the particle size of the treated sample is smaller (average particle diameter$=0.073$ μm) and much more uniform (narrow distribution) than that of the untreated sample (average particle diameter $= 0.104$ μm and wider distribution). To show pigment flocculation in flexographic inks, optical photomicrographs were obtained. Figure 11, a micrograph of the liquid ink on a glass slide with cover for the untreated pigment, exhibits flocculation. Figure

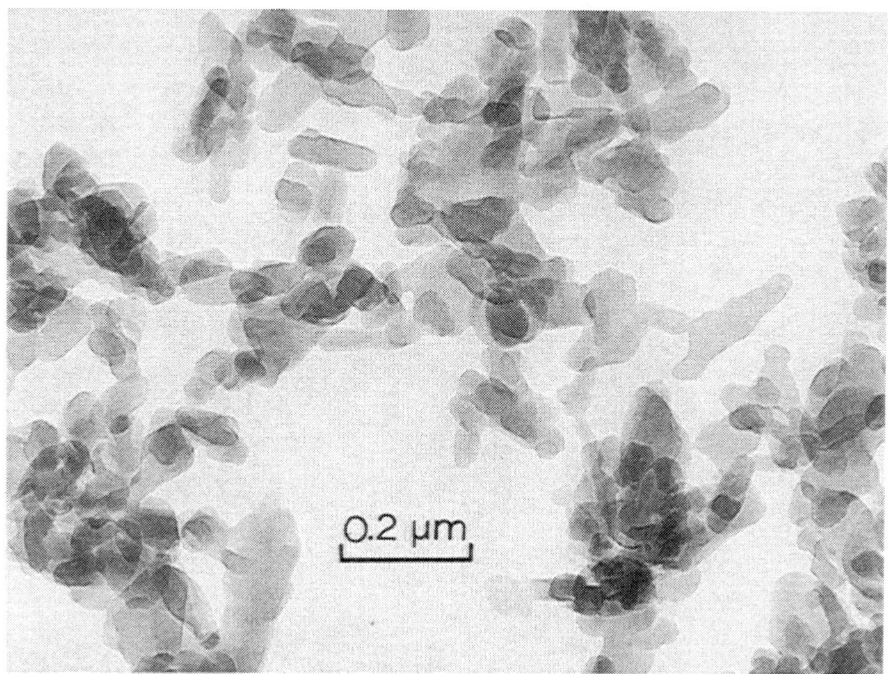

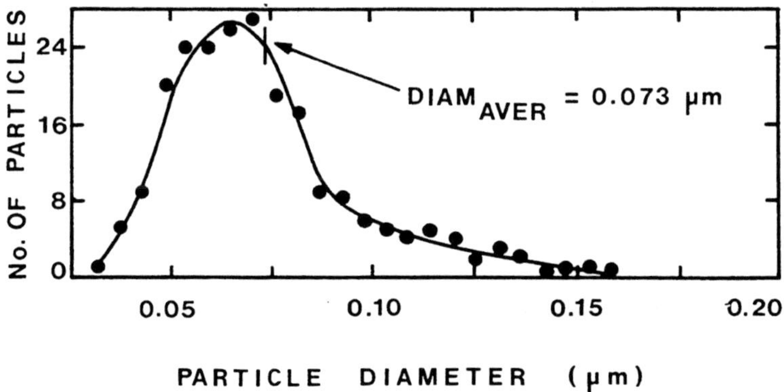

Figure 10   Transmission electron photomicrograph and particle size distribution of a surface-treated diarylide yellow AAOT, Pigment Yellow 14.

12, on the other hand, represents the liquid ink prepared with the surface–treated pigment and shows that this is a non-flocculating pigment.

Polymeric dispersants[21] such as poly(12–hydroxystearic acid) are reportedly used both as free acid and as a salt with a variety of organic toners; these agents show more effectiveness when reacted with a primary amine (3–dimethylaminopropylamine, 3–octadecylaminopropylamine, etc.). The latter types can be used with pigment derivatives to

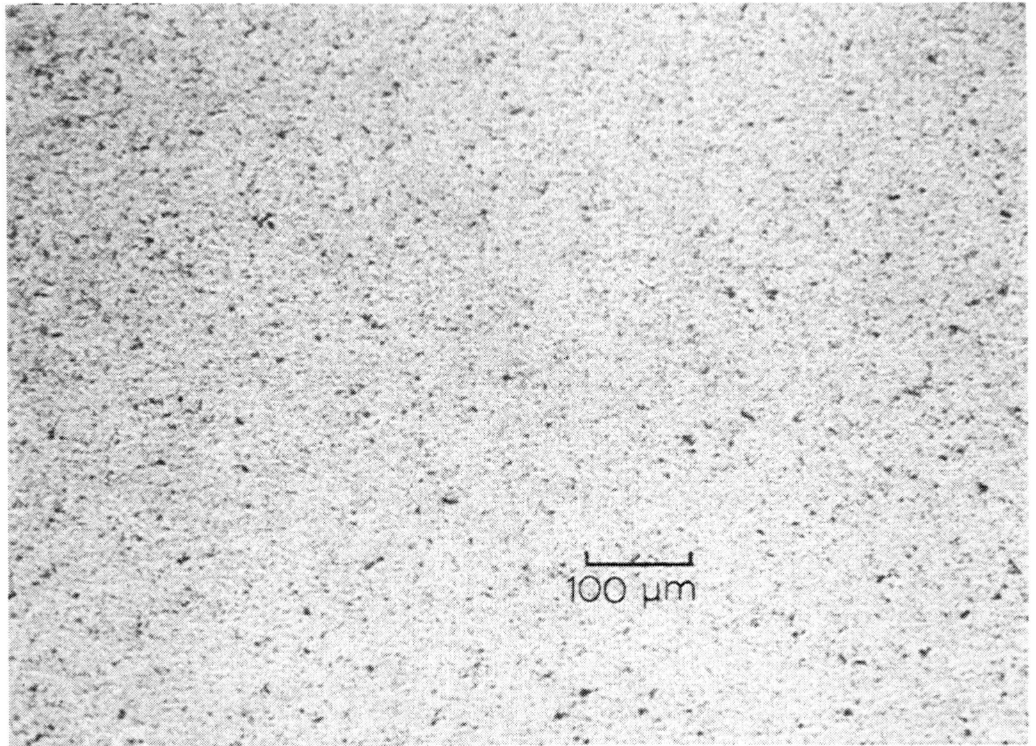

Figure 11 Optical photomicrograph of liquid ink prepared with untreated diarylide yellow AAOT. Shows flocculation of pigment particles. Same pigment as that of Figure 9.

Figure 12 Optical photomicrograph of liquid ink prepared with surface–treated diarylide yellow AAOT. Does not show flocculation of pigment particles. Same pigment as that of Figure 10.

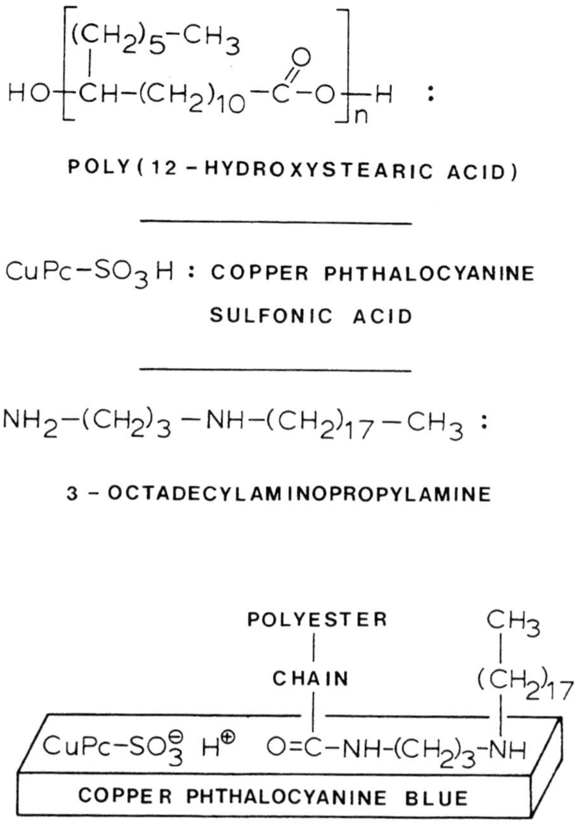

$$HO\left[CH-(CH_2)_{10}-\overset{\displaystyle O}{\overset{\|}{C}}-O\right]_n H \quad : \quad \underset{\displaystyle (CH_2)_5-CH_3}{\overset{\displaystyle \mid}{}}$$

POLY (12-HYDROXYSTEARIC ACID)

$CuPc-SO_3H$: **COPPER PHTHALOCYANINE SULFONIC ACID**

$NH_2-(CH_2)_3-NH-(CH_2)_{17}-CH_3$:

3-OCTADECYLAMINOPROPYLAMINE

$$\begin{array}{cc} \text{POLYESTER} & CH_3 \\ | & | \\ \text{CHAIN} & (CH_2)_{17} \\ | & | \\ CuPc-SO_3^{\ominus}\ H^{\oplus} & O=C-NH-(CH_2)_3-NH \end{array}$$

COPPER PHTHALOCYANINE BLUE

Figure 13 Surface treatment of copper phthalocyanine blue, Pigment Blue 15, showing the synergistic effect between sulfonated copper phthalocyanine and a polymeric disperant on the pigment surface.

produce a synergistic effect on the pigment surface for improved dispersion. An example is copper phthalocyanine sulfonic acid[22]. The mechanism of synergism is illustrated in Figure 13 for the surface treatment of copper phthalocyanine blue. A great number of other phthalocyanine derivatives have also been prepared and used as pigment stabilizers for phthalocyanine blue[23].

Phthalocyanine pigments may be conditioned from the crude state to the pigmentary form, for example, by milling the "crude" with a phthalocyanine derivative[24] such as a sulfonated phthalimidomethyl phthalocyanine[25] in the absence of any milling of grinding aid[26]. These large planar molecules appear to lie flat on the copper phthalocyanine surface, as shown in Figure 14, and they impart stability to the dispersions when used in printing inks, paints, and coatings, without any additional conditioning of the milled product.

Pigment derivatives are by no means limited to phthalocyanines. Quinacridone pigments have been surface treated with sulfonated quinacridone derivatives[27] either as the sulfonic acid form or as the metal sulfonate salt, with a wide range of metals possible. As in the

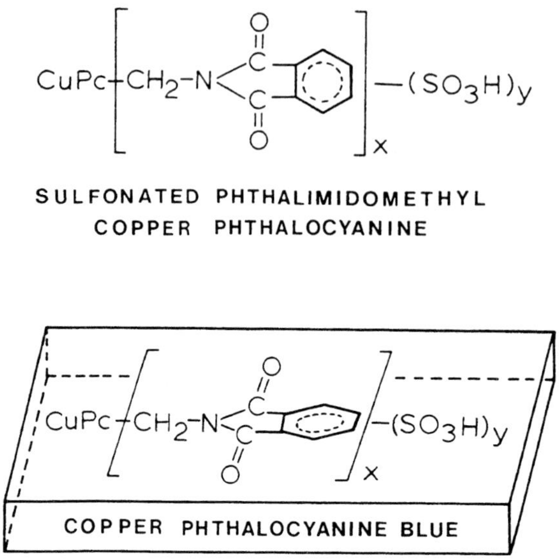

$$\text{CuPc}\left[\text{CH}_2\text{-N}\underset{\underset{O}{\overset{\overset{O}{\parallel}}{C}}}{\overset{\overset{O}{\overset{\parallel}{C}}}{<}}\right]_x\text{-(SO}_3\text{H)}_y$$

SULFONATED PHTHALIMIDOMETHYL
COPPER PHTHALOCYANINE

COPPER PHTHALOCYANINE BLUE

Figure 14 Surface treatment of copper phthalocyanine blue, Pigment Blue 15, with a sulfonated copper phthalocyanine additive, surface modifying agent.

preceding cases, the planar sulfonated quinacridone molecules appear to lie flat on the quinacridone pigment surface and thus improve considerably the dispersion properties of the pigment, especially when used in coating applications. Figure 15 represents the arrangement of sulfonated quinacridone derivative on the pigment surface.

Pigment derivatives of azo reds, [28] oranges, and yellows [29] have also been used for surface treating the corresponding pigments. With azo yellows, treatments can be carried out in situ with fatty amines to produce easily dispersible products through a Schiff base reaction between the $\overset{|}{-}C=O$ (carbonyl) groups of the pigment and the —NH$_2$ groups of primary amines, to form $\overset{|}{-}C=N-$ Schiff bases.[29-32] Derivatives of monoarylide and diarylide yellow pigments can also be prepared by reacting the pigment with a primary diamine and a glycidyl ether[33] to produce a Schiff base. The structure of one of these derivatives is shown in Figure 16 for Pigment Yellow 12, AAA yellow. Again, the planar pigment molecule appears to lie on the pigment surface, and the long chains project outward into the vehicle to produce stabilization of the dispersion.

8.2 Inorganic Pigments

Titanium dioxide, in the two naturally occurring crystal forms, anatase and rutile, is the most important white pigment, which provides maximum opacifying power. Normally, TiO$_2$ pigments are not used in their pure form because of their poor dispersibility in a variety of resins and solvents. Generally, they are surface coated with small amounts of alumina, silica, or both (up to 3% total, on TiO$_2$) to increase the functionality of the surface (active adsorption sites for the resin molecules) and to improve dispersibility and impart stability to the dispersion, especially in alkyd resin paint systems. For alumina–coated titanium diox-

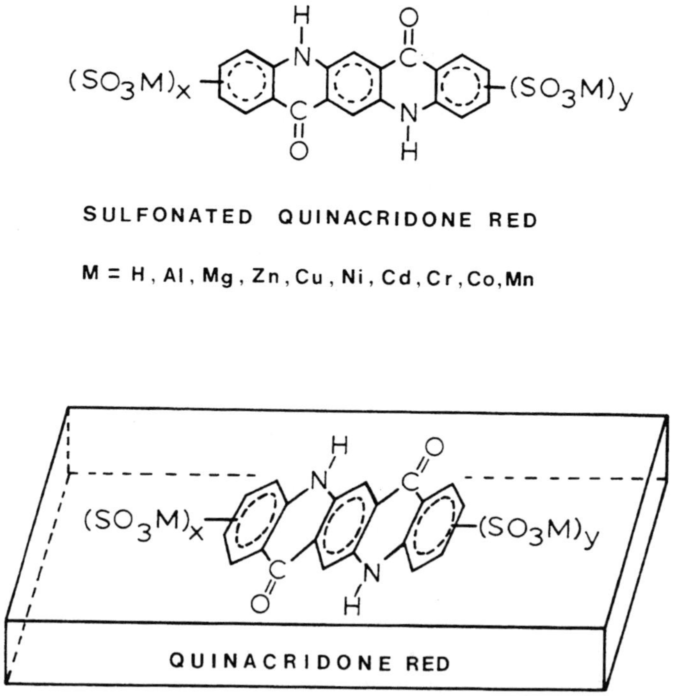

SULFONATED QUINACRIDONE RED

M = H, Al, Mg, Zn, Cu, Ni, Cd, Cr, Co, Mn

QUINACRIDONE RED

Figure 15 Surface treatment of quinacridone red, Pigment Violet 19, with a sulfonated quinacridone additive, surface modifying agent.

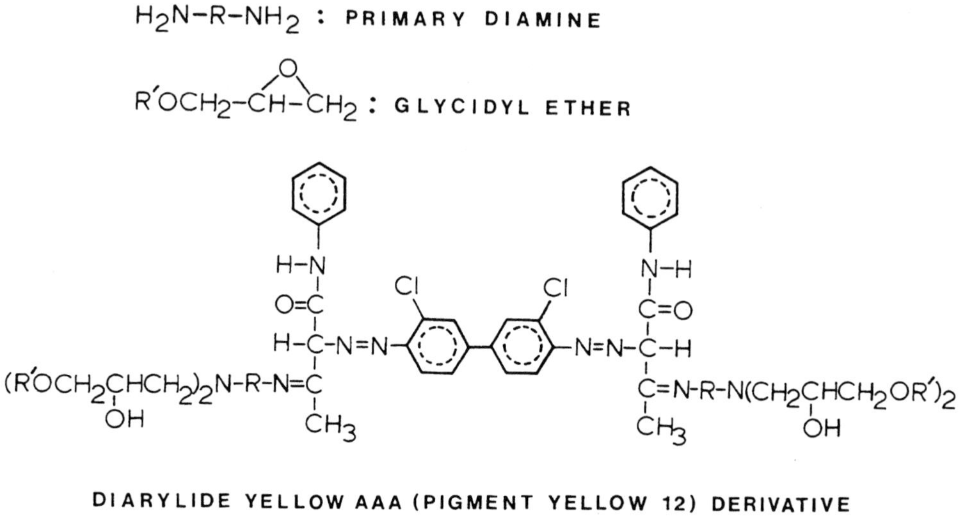

$H_2N-R-NH_2$: PRIMARY DIAMINE

$R'OCH_2-CH-CH_2$: GLYCIDYL ETHER

DIARYLIDE YELLOW AAA (PIGMENT YELLOW 12) DERIVATIVE

Figure 16 Diarylide yellow AAA, Pigment Yellow 12, derivative; Schiff base.

ide,[34] the highly basic sites on the alumina surface, which are much more basic than the sites on the TiO_2 surface, cause specific adsorption of the acidic functional groups of the alkyd resin molecules. The remaining parts of the resin molecules (long chains) extend away from the surface, creating a considerable amount of steric hindrance around each pigment particle, thus resulting in steric stabilization of the dispersion.

Alumina–coated titanium dioxide, iron oxide red, and other inorganic pigments and fillers can be surface treated with alkanolamines (aminoalkanols), having the general formulas:

$$R_1\text{---}CH\text{---}CH_2\text{---}NH_2, \quad R_2\text{---}CH\text{---}CH\text{---}R_3, \text{etc.}$$
$$\underset{OH}{|} \qquad\qquad \underset{OH}{|} \quad \underset{NH_2}{|}$$

where R_1, R_2, and R_3 are alkyl groups containing from 1 to 22 carbon atoms in the chain.[35] The dispersibility of these pigments is increased considerably when used in paint formulations containing air drying alkyd resin vehicles. The stability of the dispersion is similarly improved because of the steric stabilization imparted to the pigment particles by the R_1, R_2, and R_3 long chain alkyl groups.

Organic isocyanate adducts[36] are used as effective dispersing agents for several classes of inorganic pigments, including zinc oxide, iron oxides, Prussian Blue, cadmium sulfide, ultramarine, vermillion, and chrome pigments (zinc, barium, and calcium chromates). These agents improve the dispersion characteristics and the flocculation resistance of the above–listed pigments when incorporated into conventional alkyd paint vehicles with organic solvents, where these systems also contain a substantial amount of titanium dioxide.

9.0 THE CHARACTERIZATION AND ASSESSMENT OF DISPERSION

The extent to which a pigment is dispersed in the medium or the degree of dispersion is normally assessed in terms of color strength, gloss, brightness, and transparency, and it also has an effect on the rheological properties of the system[37–39]. Since all these properties are governed by the size and distribution of the pigment particles in the dispersion, one can, today, measure these properties using any of the latest particle size analyzers based on the light scattering principle of the dispersed particles[12]. With these instruments, a very dilute suspension is required, and it is necessary to know the refractive index and viscosity of the suspending medium. The average particle diameters and the particle size distributions obtained are those of individual particles, aggregates, agglomerates, and flocculates in the dispersion. The advantages of these instruments are that they are quite easy to operate, they give results rapidly, and they allow the dispersion process to be followed at different times and at different stages.

One such instrument is the Coulter model N4 Submicron Particle Analyzer. Figure 17 represents the particle size results for a green–shade phthalocyanine blue, C.I. Pigment Blue 15:3, in an aqueous dispersion. The distribution is quite narrow, and the mean particle diameter is 0.117 μm. These results are very similar to those obtained from inspection of the transmission electron micrographs of Figure 1 for the same phthalocyanine blue pigment in the dry powder form, showing that very little aggregation exists in the dispersion.

Such particle size analyzers, based on light scattering, can be used very effectively to study particle size changes that occur during the dispersion of pigments in fluid systems. Furthermore, time studies may be carried out on the flocculation of pigments by determin-

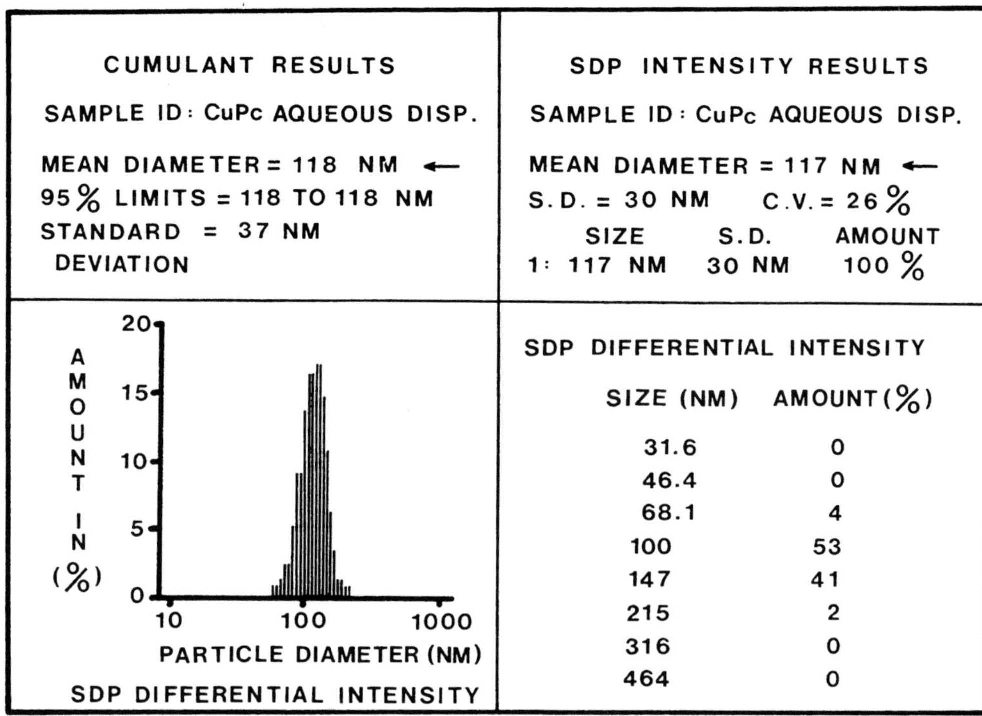

Figure 17 Particle size results of a copper phthalocyanine blue, Pigment Blue 15:3, aqueous dispersion by the Coulter Model N^4 Submicron Particle Size Analyzer, used to assess the degree of dispersion (SDP = Size Distribution Program).

ing particle size immediately after dispersion and then later, after the dispersions have been allowed to stand for certain periods. This gives a measure of the stability of the dispersion.

10.0 CONCLUSION

There is no question as to the desirability and effectiveness of a fully dispersed and stabilized pigmented system. Such a dispersion brings out the optimum color properties of the pigment in terms of color strength, gloss, transparency, and rheology. When a pigment is completely dispersed, it contains a larger number of primary particles; therefore, a smaller amount is required to produce the necessary coverage and color strength than would be necessary for a pigment that was not as well dispersed and contained a larger number of aggregates, agglomerates, and flocculates.

The trend today is toward production of more and more easily dispersible pigments, as counterparts to the easily dispersible azo yellows, which are already used widely in certain printing ink systems. Pigment manufactures are always improving pigment dispersibility, through the use of surface treatments, in terms of surfactants, polymeric dispersants, and pigment derivatives. The end result is the achievement of complete dispersion easily and quickly. And since this is an energy intensive process, as far as the dispersion equipment utilized, less energy is required, resulting in greater economic benefits for the pigment user.

REFERENCES

1. G. D. Parfitt, Ed., *Dispersion of Powders in Liquids* 2nd ed. New York: Wiley, 1973.
2. T. C. Patton, *Paint Flow and Pigment Dispersion*, 2nd ed., New York: Wiley, 1979.
3. V. T. Crowl, *J. Oil Colour Chem. Assoc.*, 55, 388 (1972).
4. O. Hafner, *J. Oil Colour Chem. Assoc.*, 57, 268 (1974).
5. W. Carr, *J. Oil Colour Chem. Assoc.*, 61, 397 (1978).
6. D. M. Varley and H. H. Bower, *J. Oil Colour Chem, Assoc.*, 62, 401 (1979).
7. H. M. Smith, *Polym. Paint Color J.*, 175, 660 (1985).
8. P. A. Lewis, Ed., *Pigment Handbook*, Vol. 1, 2nd ed. New York: Wiley, 1988.
9. G. D. Parfitt and K. S. W. Sing, Eds., *Characterization of Powder Surfaces*. London: Academic Press, 1976.
10. H. P. Preuss, *Pigments in Paint*, Park Ridge, NJ: Noyes, 1974.
11. W. M. Morgans, *Outlines of Paint Technology*, Vols. 1 and 2, 2nd ed. London: Charles Griffin & Co., 1982.
12. T. G. Vernardakis, *Am. Ink Maker*, 62(2), 24 1984.
13. P. Sorensen, *J. Paint Technol.*, 47, 31 (1975).
14. J. S. Hamptom and J. F. MacMillan, *Am. Ink Maker*, 63(1), 16 (1985).
15. B. G. Hays, *Am. Ink Maker*, 62(6), 28 (1984).
16. A Topham, *Prog. Org. Coatings*, 5, 237 (1977).
17. K. Merkle and H. Schafer, in *Pigment Handbook*, Vol. III, T.C. Patton, Ed., New York: Wiley 1973 pp. 157–167.
18. A. E. Ambler and R. W. Tomlinson, U.S. Patent 3,296,001 (Jan. 3, 1967); ICI.
19. T. C. Rees and R. J. Flores, U.S. Patent 4,032,357 (June 28, 1977); Sherwin–Williams.
20. T. G. Vernardakis, *Dyes Pigments*, 2, 175 (1981).
21. J. F. Stansfield and A. Topham, U.S. Patent 3,996,059 (Dec. 7, 1976); ICI.
22. P. K. Davies, L. R Rogers, J. F. Stansfield, and A. Topham, U.S. Patent 4,057,436 (Nov. 8, 1977; ICI.
23. Anon., British Patent 1,544,839 (Apr. 25, 1979); BASF.
24. W. H. McKellin, H. T. Lacey, and V. A. Giambalvo, U.S. Patent 2,855,403 (Oct. 7, 1958); American Cyanamid.
25. V. A. Giambalvo and W. L. Berry, U.S. Patent 3,589,924 (Jun. 29, 1971); American Cyanamid.
26. S. L. Johnson, G. McLaren and G. H. Robertson, U.S. Patent 4,448,607 (May 15, 1984); Sun Chemical.
27. E. E. Jaffe and W. J. Marshall, U.S. Patent 3,386,843 (June 4, 1968); DuPont.
28. J. Mitchell and A. Topham, U.S. Patent 3,446,641 (May 27, 1969); ICI.
29. J. Mitchell and A. Topham, British Patent 1,139,294 (Jan. 8, 1969); ICI.
30. Anon., British Patent 1,080,115 (Aug. 23, 1967): KVK.
31. F. Dawson, J. Mitchell, L. R. Rogers, W. Todd, and A. Topham, British Patent 1,096,362 (Dec. 29, 1967); ICI.
32. G. H. Robertson, U.S. Patent 4,220, 473 (Sep. 2, 1980); Sun Chemical.
33. R. J. Schwartz and T. Sulzberg, U.S. Patent 4,468,255 (Aug. 28, 1984); Sun Chemical.
34. M. J. B. Franklin, K. Goldsbrough, G. D. Parfitt, and J. Peacock, *J. Paint Technol.*, 42, 740 (1970).
35. H. Linden, H. Rutzen, and B. Wegemund, U.S. Patent 4,167,421 (Sept. 11, 1979); Henkel.
36. F. Hauxwell, J. F. Stansfield, and A. Topham, U.S. Patent 4,042,413 (Aug. 16, 1977); ICI.
37. W. Carr, *J. Oil Colour Chem. Assoc.*, 65, 373 (1982).
38. K. Tsutsui and S. Ikeda, *Prog. Org Coatings*, 10, 235 (1982).
39. R. Polke, *Am. Ink Maker*, 61(6), 15 (1983).

62

Colored Inorganic Pigments

Peter A. Lewis

Sun Chemical Corporation
Cincinnati, Ohio

This chapter describes the chemistry, manufacture, and properties associated with the major classes of colored inorganic pigments as used in the coatings industry. Thus pigmentary inorganic whites such as titanium dioxide are not covered.

1.0 THE COLOUR INDEX SYSTEM

The Colour Index System is a coding system developed under the joint sponsorship of the Society of Dyers and Colourists (SDC) in the United Kingdom and the Association of Textile Chemists and Colorists (AATCC) in the United States. The system is referred to as the "Colour Index." In referring to pigments using the Colour Index System, we may describe, for example, Molybdate Orange as Pigment Red 104, Colour Index number 77605.

The Colour Index names for pigments are abbreviated as follows:

PB = Pigment Blue	PBk = Pigment Black
PBr = Pigment Brown	PG = Pigment Green
PM = Pigment Metal	PO = Pigment Orange
PV = Pigment Violet	PR = Pigment Red
PW = Pigment White	PY = Pigment Yellow

2.0 PIGMENT SELECTION

Once a formulator has decided on the shade required for a particular application, the next most important criteria of any pigment are its fastness properties. It is pointless to formulate a coating with a pigment that will not withstand the exposure specifications of the coatings end use. Such specifications can extend to requiring as much as 5 years of outdoor exposure in states such as Florida and Arizona.

551

Due attention must be paid to the appropriate manufacturers' literature to ensure that a pigment has been chosen that will satisfy the end-use criteria for fastness to light, solvents, heat, and chemicals, as well as flocculation and crystallization. Once a pigment class has been selected that will perform adequately in the end-use application, the formulator can consider such other factors as economy.

The formulator should be aware, however, that the fastness properties of a pigment will be affected by the medium into which it is incorporated. Thus, even though intrinsically the pigment may feature the required properties, it is still necessary for the final pigmented coating to be tested in the end-use application. Use of "base coat–clear coat" formulations in the automotive and industrial marketplace in recent years has had a major effect on pigment choice in this area, since incorporation of UV absorbers in the clear coat has done much to extend the fastness properties of the pigments used in the base coat.

3.0 INORGANIC BLUES

3.1 Iron Blue

The most common and economical inorganic blue is iron blue, Pigment Blue 27, a complex ferriferrocyanide having the formula $FeNH_4Fe(CN)_6 \cdot xH_2O$. Originally called Prussian Blue, the pigment was discovered by Diesbach in 1704 and was characterized as having a very jet masstone, a red tint, and a very hard texture. Development of a pigment that had a softer texture and improved dispersion by the French chemist Milori led to the introduction to the marketplace of a product chemically identical to Prussian blue but known as Milori Blue.

$$Na_4Fe(CN)_6 + FeSO_4 + (NH_4)_2SO_4 \rightarrow Fe(NH_4)_2Fe(CN)_6 + 2Na_2SO_4 \tag{1}$$

$$6Fe(NH_4)_2Fe(CN)_6 + 3H_2SO_4 + NaClO_3 \rightarrow 6FeNH_4Fe(CN)_6 + NaCl + 3(NH_4)_2SO_4 + 3H_2O \tag{2}$$

Iron blue is manufactured by reacting ferrous sulfate with sodium ferrocyanide in the presence of ammonium sulfite to yield a leukcoferricyanide known as Berlin White (Eq. 1). This intermediate stage is then dissolved in sulfuric acid and oxidized with sodium chlorate to give iron blue (Eq. 2).
Various specialty grades of iron blue exist that differ in masstone, tint strength, dispersion, and oil absorption.

Chinese Blue is a fine grade that exhibits a green undertone as compared to the original Prussian Blue. Bronze blue is a variety that exhibits a surface bronze that depends on the angle at which the surface is viewed.

Iron blue exhibits good resistance to weak acids but very poor resistance to even mildly alkaline systems. Iron blues also show marked tendency to lose their blueness if stored in a paint system that contains oxidizable vehicles such as linseed oil. An extension of this phenomenon occurs when the paint is applied to a surface, since the color will return as the film dries and atmospheric oxygen reverses the reaction.

Although iron blues feature good lightfastness properties when used at masstone or deep tint levels, this feature rapidly becomes unacceptable if the pigment is used highly extended with white. Use of iron blues in the coatings industry has been greatly reduced because of the availability of the more economical and durable copper phthalocyanine blue.

3.2 Cobalt Blue

Cobalt blue is a mixed metal oxide or solid solution with a typical composition of $CoAl_2O_4$ that is a crystalline matrix of Al_2O_3 (65–70%) and CoO (30–35%). The pigment is Pigment Blue 28, and it is made by calcining cobalt oxide and aluminum oxide at temperatures approximating 2400°F. The calcined pigment is reduced to a fine powder using physical means such as hammers or air milling.

Cobalt blue is chemically inert, extremely heat stable, and lightfast; it is hard and abrasive and has a high hiding power due to its relatively high opacity. The material is not as strong or as clean as copper phthalocyanine blue and as such is relatively more expensive. The major markets the product serves are the coatings and plastics market, with a particular niche in the vinyl sidings industry.

3.3 Ultramarine Blue

Ultramarine blue is catalogued as Pigment Blue 29 and is also known under more common names such as Laundry Blue, Dolly Blue, and lapis lazuli. Chemically the pigment is a sodium aluminum sulfosilicate of inexact composition but closely related to the zeolites and having a suggested formula of $Na_6Al_6Si_6O_{24}S_4$. Two chemically distinct types of ultramarine blue, a green shade and a red shade, can be made by varying the calcination process by which the product is manufactured. Although their basic characteristics are the same, the green shade contains more aluminum and less sulfur than does the red shade.

The major use for ultramarine blue, as one of its common names implies, is as a laundry aid in soap and detergents. Less than 5% of the production is used in the coatings industry. Its uses are in the area of interior gloss and emulsion paints that require excellent alkali resistance and lightfastness.

3.4 Cobalt Chromate

Cobalt chromate, also called Cerulean Blue, is related to cobalt blue in that it too is a mixed metal oxide. Cobalt chromate, however, contains chromium in the matrix in addition to cobalt and aluminum. The empirical formula is $Co(Al,Cr)_2O_4$, and the compound is catalogued as Pigment Blue 36. The material is a crystalline matrix of cobalt oxide (30–35%), chromium oxide (30–60%), and aluminum oxide (10–40%). The pigment is made by calcining an appropriate mixture of these metal oxides at a temperature of 2400°F.

Cobalt chromate is used for premium coatings, where outstanding long-term durability and weatherfastness are a prerequisite of the desired end use.

4.0 INORGANIC BROWNS

The major inorganic browns used in today's coatings applications are those based on oxides of iron; they are available as the naturally mined product or as the synthetically produced material. The coatings market is the largest consumer of iron oxides, irrespective of their hue. Until 1920 all the major iron oxide pigments were wholly from naturally occurring deposits with little changes other than that of physical separation from impurities found alongside the natural ores. Natural iron oxides suffer from the obvious disadvantage of non-uniformity, since their properties, particularly particle size, vary with their source. For this reason natural oxides have traditionally carried names such as Raw Sienna, Burnt Sienna, Turkish Umber, Raw Umber, and Burnt Umber.

Oxides produced synthetically offer the user a more uniform product, with minimal batch–to–batch variation and controlled physical properties.

4.1 Natural Iron Oxides

By definition, these iron oxides are obtained from naturally occurring deposits of ore. The source mines for these pigments can be classified as either "iron ore mines" or "pigment mines." Mines of the former type operate primarily as suppliers of ore as feedstock for blast furnaces with only a small offtake for the pigment industry. Nevertheless this relatively small output represents the major source of red oxides. The pigment mines are in existence solely to supply crude ore for pigments. The earth's crust contains massive deposits of both hematite (red iron oxide) and magnetite, the mythical lodestone (black iron oxide). Deposits of the Yellow Ochers and Brown Siennas and Umbers are relatively sparse.

A typical mining operation may involve dumping crude ore on an incline followed by slurring this ore into an aqueous suspension using high pressure water jets. The slurry is next directed through a series of crude washing stages to remove coarse sand and rock particles. This removal of suspended impurities continues as the slurry passes into a separator tank and then through a Dorr bowl rake. The ore leaving the rake overflows to a separating tank, from which it is withdrawn under controlled conditions. The slurry is dried as a thin aqueous layer using a rotary drum drier.

For a particle to be described as pigmentary, it is generally accepted that the particle size should be smaller than 25 μm. For natural iron oxides to possess this physical property, the dried ore must be pulverized and classified to eliminate agglomerates. After classification, the particle size is defined by reference to the percentage of particles that will pass through a 325-mesh sieve—that is, a screen with a mesh size of 44 μm.

In the drying process steam is used as the heat source, to permit the hydrated oxides to dry without loss of water of crystallization and consequent color shift. The process of calcining, where temperatures in excess of 1200°F are reached, involves the treatment of the hydrated oxides at temperatures that will result in the complete loss of water to form the corresponding Burnt Siennas and Burnt Umbers.

In today's industry, fluid energy mills and micronizing or jet mills are used to grind and deaggregate the dried agglomerates. With many commercial pigmentary oxides, the final stage in processing involves the blending of batches showing variations outside the accepted standard specification to produce a standardized product.

4.2 Iron Oxide Browns

Pigment Brown 7 is an iron oxide brown that is available in shades ranging from light red to deep purple brown. The empirical formula is Fe_2O_3 and as a natural product it contains varying amounts of clay. Offered as Metallic Brown, made from calcined hematite (PR 102) and Burnt Sienna, made from calcined limonite (PY 43), the composition of the pigment varies depending on both the source of the ore and the conditions of calcination. Major deposits are worked in Cyprus and in Italy; in the United States, there are deposits in Georgia and Virginia.

Pigment Brown 7:x is a ferrosoferric oxide derived from ores containing 25% manganese dioxide having a very indistinct composition that may be described as $Fe_2O_3 \cdot xMnO$ with varying proportions of clay. Classical names include Raw Umber, Burnt Umber, and Turkish Umber. Principal producers of the "umbers" continue to be the United States, Cyprus, and Turkey.

Synthetic brown oxide is Pigment Brown 6. Also known as brown magnetic iron oxide, the pigment is produced by controlled oxidation of Pigment Black 11. Chemically the material is Fe_2O_3 with a proportion of FeO and $Fe_2O_3 \cdot H_2O$. Pigment Brown 11 is magnesium ferrite made by calcining a blend of ferric oxide and magnesium oxide. Chemically the pigment is $MgO \cdot Fe_2O_3$. The volume of brown iron oxides used in coatings is generally low, since most users achieve browns by formulating blends of yellow, red, and black pigments.

Brown iron oxides have good chemical resistance and high tint strength and are permanent pigments. They are ideal pigments for use in wood stains and furniture finishes.

4.3 Mixed Metal Oxide Browns

The other browns used in the coatings industry are Pigment Browns 24, 33, and 35. Each is a mixed metal oxide of variable composition.

Pigment Brown 24 is chrome antimony titanium buff rutile made by calcining oxides of titanium, chromium, and antimony. The pigment is a yellowish brown with excellent fastness to heat, light, and weather.

Pigment Brown 33 is zinc iron chromite brown formed by calcining a mixture of the oxides of zinc, chromium, and ferrous and ferric iron. The product is a dark brown with outstanding durability, chemical resistance, and baking fastness.

Pigment Brown 35 is iron chromite brown made by calcining chromium oxide with a mixture of ferric and ferrous oxides. Again, typical of this class of pigments, the fastness properties are excellent.

5.0 INORGANIC GREENS

5.1 Chrome Green

Chrome greens, Pigment Green 15, are blends or mixtures of a green-shade chrome yellow (PY 34) and iron blue (PB 27). Varying the ratio of these two substituents gives greens with a light yellow shade to a deep, dark shade. An empirical formula is $PbCrO_4 \cdot xPbSO_4 \cdot yFeNH_4Fe(CN)_6$.

The pigments provide good hiding, high tint strength, and moderate chemical resistance at a relatively good price. Bake temperatures up to 300°F can be tolerated. Use of chrome greens is restricted to exterior and industrial coatings applications because of the lead content of the pigments.

5.2 Chromium Oxide Green

Chromium oxide green, Pigment Green 17, finds a special application in camouflage paints because of its ability to reflect infrared light. The pigment is pure, calcined chromium oxide, Cr_2O_3, manufactured by reduction of sodium bichromate with sulfur or carbon.

$$Na_2Cr_2O_7 + 2C \rightarrow Cr_2O_3 + CO\uparrow + Na_2Co_3 \tag{3}$$

The pigment is used where its price can be justified by the resultant excellent light and chemical fastness properties the pigment gives to the coating.

5.3 Hydrated Chromium Oxide Green

Also known as viridian green or Guignet's green, this material is Pigment Green 18 and is chemically a hydrated chromic oxide of formula $Cr_2O_3 \cdot 2H_2O$ made by hydrolizing the product of the calcining of sodium bichromate with boric acid. The pigment is a bright, blue-shade green with exceptional chroma and brilliance that has outstanding fastness properties in both masstone and deep tints.

5.4 Mixed Metal Oxide Greens

Cobalt chromite green, Pigment Green 26, is a mixed metal oxide of variable composition made by calcining cobalt and chromium oxide to give $CoCr_2O_4$. The pigment is a blue-shade green with excellent light, weather, and heat fastness.

Cobalt titanate green, Pigment Green 50, is another mixture made by calcining cobalt and titanium oxides to produce Co_2TiO_4. The pigment offers high infrared reflectivity in camouflage paints and outstanding light, weather, chemical, and heat fastness.

6.0 INORGANIC ORANGES

6.1 Cadmium Orange

Also known as cadmium sulfoselenide orange, this pigment, Pigment Orange 20, is a solid solution produced by calcining cadmium sulfide with cadmium selenide at a temperature approximating 1000°C. Change in the stoichiometry of the cadmium sulfide–cadmium selenide mixture results in a range of pigmentary products with a shade range from bright yellow (PY 35) to a bright red (PR 108). Barium sulfate added or formed in the process extends the pigment and gives the lithopone version of the orange designated by a suffix as Pigment Orange 20:1.

Cadmium orange is used in the coatings industry in industrial color coding applications, where chemical and heat resistance are prime requisites of the applied coating.

6.2 Chrome Orange

Chrome orange, Pigment Orange 21, is a basic lead chromate formed under alkaline conditions to give a product of empirical formula $PbCrO_4 \cdot xPbO$. Shades from a yellow-shade to a red-shade orange can be produced, depending on the alkalinity of the reaction mass.

Because of the pigment's low cost and moderate lightfastness, the major coatings use of chrome orange is in the production of protective coatings. As with all lead-containing pigments, the color will rapidly darken on exposure to atmospheric sulfur. Restrictions on the use of lead-containing products has resulted in a marked reduction in consumption of chrome orange in recent years.

A variation of chrome orange is Pigment Orange 21:1, a basic lead silicochromate of formula $PbSiO_3 \cdot 3PbO$. This pigment's major outlet is in road traffic paints.

6.3 Cadmium Mercury Orange

Known by the Ciba-Geigy trade name of Mercadium Orange, Pigment Orange 23 is a solid solution of mercury sulfide in cadmium sulfide. Careful control of the formation of the mixed crystal enables a product of controlled hue to be produced. The pigment is manufactured by precipitation of the sulfides of mercury and cadmium from a solution containing their soluble salts followed by calcining this precipitate in an inert atmosphere.

$$x\text{CdSO}_4 + (1-x)\text{HgSO}_4 + \text{Na}_2\text{S} \rightarrow x\text{CdS} + (1-x)\text{HgS} + \text{Na}_2\text{SO}_4 + 2\text{O}_2 \qquad (4)$$

An extended or lithopone-type pigment, Pigment Orange 23:1, can be produced by using barium sulfate as the precipitant, thus producing insoluble barium sulfate as a by-product of the reaction sequence which is retained with the pigment.

$$x\text{CdSO}_4 + (1-x)\text{HgSO}_4 + \text{BaS} \rightarrow x\text{CdS} + (1-x)\text{HgS} + \text{BaSO}_4 + 2\text{O}_2 \qquad (5)$$

Cadmium mercury oranges are extremely heat stable, withstanding temperatures up to 700°F, and have excellent chemical resistance and outstanding weatherability and solvent fastness.

7.0 INORGANIC REDS

7.1 Iron Oxide Reds

As with the iron oxide browns, we again see pigments that are available as both natural and synthetic products. Iron oxide red, Pigment Red 101, is available under such names as Hematite, Persian Gulf oxide, Mars red, Turkey red, Indian red, Ferrite red, and Rouge. The synthetic material is chemically Fe_2O_3, whereas the naturally occurring product is Fe_2O_3 with varying amounts of FeO and clay.

Synthetic red iron oxides represent the largest class of synthetic iron oxides manufactured. Four principal manufacturing routes are employed.

1. Copper as red oxide is formed by calcining iron sulfates.

$$12\text{FeSO}_4 \cdot \text{H}_2\text{O} + 3\text{O}_2 \rightarrow 2\text{Fe}_2\text{O}_3 + 4\text{Fe(SO}_4)_3 + 12\text{H}_2\text{O} \qquad (6)$$

2. Synthetic red oxide is produced from synthetic black oxide by calcining the material in a controlled atmosphere containing oxygen.

$$4\text{FeO} \cdot \text{Fe}_2\text{O}_3 + \text{O}_2 \rightarrow 6\text{Fe}_2\text{O}_3 \qquad (7)$$

3. Precipitated red oxide is prepared in an aqueous medium by growing seed nuclei in the presence of a ferrous salt and scrap steel. Control of the pigments particle size during the process results in control of the darkness/lightness of the oxide.

4. Synthetic red oxide may be produced by calcining synthetic yellow iron oxide ($\text{Fe}_2\text{O}_3 \cdot \text{H}_2\text{O}$) to give the dehydrated product, which is Fe_2O_3. The wide range of shades available in the red oxide class, coupled with their acid and alkali resistance, heat stability, and the economics associated with using these pigments, accounts for the large tonnage of these pigments used annually in the coatings industry.

7.2 Molybdate Orange

Although given the common name classification of an orange, this pigment is in fact Pigment Red 104 with empirical formula $\text{PbCrO}_4 \cdot x\text{PbMoO}_4 \cdot y\text{PbSO}_4$. Other common names for this pigment include Chrome red, Molybdate red, and Chrome vermilion. The pigment is in fact a solid solution of lead chromate, lead molybdate, and lead sulfate. The material is produced by adding a solution of sodium chromate, sodium molybdate, and sodium sulfate under controlled conditions into a solution of lead nitrate at a temperature between zero and 40°C to precipitate the mixed crystal. Control of the particle size distribution and crystalline shape determines the actual hue of the manufactured pigment from red-shade yellow to red-shade orange.

Molybdate orange is an opaque pigment with high solvent fastness, moderate heat fastness, and good economy. The pigment has poor alkali and acid resistance, and the non-treated grades will exhibit considerable darkening on exposure to atmospheric pollutants, particularly in heavily industrialized areas. The pigment is also sensitive to overgrinding when, if care is not exercised, the crystal will disproportionate to give a yellower, weaker looking product.

The major outlet for Molybdate orange is in the coatings industry, with an estimated 60% of the pigment produced going into coatings. In decorative and industrial paints Molybdate orange is used as the basis for many red and orange formulations, particularly in combinations with organic pigments such as quinacridone, when used for orignial equipment manufacturers (OEM) finishes in the automotive industry, in countries that still allow lead to be used in OEM coatings formulations.

7.3 Cadmium Red

Pigment Red 108 is cadmium sulfoselenide red, a solid solution of cadmium sulfide and cadmium selenide produced by calcining a coprecipitate of cadmium sulfide and cadmium sulfoselenide. The amount of cadmium selenide incorporated into the solid solution dictates the final hue of the pigment as does also, to a lesser extent, the temperature of calcination. The empirical formula is $CdS \cdot xCdSe$.

A lithopone version, $CdS \cdot xCdSe \cdot yBaSO_4$, Pigment Red 108:1, is also available as an extended product coprecipitated with barium sulfate.

The pigment exhibits excellent stability to heat, alkali, organic solvents, and light when used at masstone and high tint levels. It offers clean bright shades of high chroma. Lighter shades will show poorer resistance to light and moisture than will the more intense shades.

7.4 Mercury Cadmium Red

Another inorganic pigment, known by the Ciba-Geigy trade name of Mercadium Red, is Pigment Red 113. This calcined coprecipitate of mercuric sulfide with cadmium sulfide has the empirical formula $CdS \cdot xHgS$.

An extended lithopone type also exists, Pigment Orange 23:1, which can be produced by using barium sulfate as the precipitant, thus producing insoluble barium sulfate as a by-product of the reaction sequence, which is retained with the pigment.

Not as heat resistant as cadmium reds but offering economy, good hiding, good solvent resistance, and excellent brightness, mercury cadmium reds do show a marked sensitivity to acids, which precludes their use in coatings that are likely to come into contact with such substances.

8.0 INORGANIC VIOLETS

8.1 Ultramarine Violet

This violet is Pigment Violet 15, a sodium sulfosilicate prepared by oxidation of Pigment Blue 29, Ultramarine blue. The redness of hue of the violet is determined by controlling the degree of oxidation. Chemically the pigment is $Na_4H_2Al_6Si_6O_{24}S_2$.

The pigment possesses a brilliant hue and offers good lightfastness and heat stability. It will react with metals to form sulfides and finds use in cosmetic applications and in nontoxic acrylic poster paints and artists' colors.

8.2 Manganese Violet

First introduced at the turn of the century, this pigment, Pigment Violet 16, is manufactured by slurrying together a mixture of manganese dioxide and diammonium phosphate in phosphoric acid at high temperatures. The mixture is dehydrated during the reaction sequence to produce the pigment as $MnNH_4P_2O_7$. The manufacture is unusual in that unlike many of the inorganic reactions described in this chapter, it does not proceed via a calcination process.

The pigment does not have a bright hue; it has only moderate opacity and poor tint strength. It has excellent acid resistance and poor alkali resistance. On the positive side, the pigment has high bleed resistance and lightfastness. The major outlet for the pigment is in the cosmetics industry and as a toner in the plastics industry.

8.3 Mixed Metal Oxide Violets

For completeness, the mixed metal oxide violets can be listed as follows

1. Cobalt violet phosphate, Pigment Violet 14, is a crystalline pigment produced by calcining a mixture of cobalt and phosphorus oxides. The pigment is a blue shade violet of formula $Co_3(PO_4)_2$. Used in artists' colors and plastics, this pigment has poor tint strength and a dirty undertone but outstanding chemical, weather, and lightfastness.

2. Cobalt lithium violet phosphate, Pigment Violet 47, is produced by calcining cobalt, lithium, and phosphorus oxide to give $CoLiPO_4$, a red-shade violet. Again this pigment has poor tint strength, lacks brightness, and has a dirty tint tone. Although the pigment has excellent chemical, light, weather, and heat stability, it finds little use in coatings and is primarily used in plastics.

3. Cobalt magnesium borate, Pigment Violet 48, is a crystalline borate manufactured by calcining a mixture of the oxides of cobalt, magnesium, and boron to give a pigment of formula $(Co,Mg)_2B_2O_5$. Another red-shade violet, this pigment features the same properties as cobalt lithium violet phosphate and finds its major use in engineering resins.

4. Cobalt ammonium phosphate, Pigment Violet 49, is produced by a coprecipitation process to give a crystalline pigment of formula $CoNH_4PO_4 \cdot H_2O$. Used in artists' colors and as a toner in whites, this pigment exhibits excellent solvent and lightfastness. The pigment has poor acid and alkali resistance, poor bake stability, and poor tint strength.

9.0 INORGANIC YELLOWS

9.1 Strontium Yellow

This yellow is Pigment Yellow 32 and is chemically strontium chromate, $SrCrO_4$, prepared by precipitating a soluble chromate with a solution of an appropriate strontium salt. The pigment is used primarily in corrosion-inhibiting coatings. Its poor tint strength, low hiding power, and unsatisfactory acid and alkali resistance limit the pigment's use in the coatings industry.

9.2 Primrose Chrome Yellow

Primrose chrome yellow, Pigment Yellow 34, is a coprecipitate of lead chromate and lead sulfate of empirical formula $PbCrO_4 \cdot xPbSO_4$.

$$2Pb(NO_3)_2 + H_2O + Na_2Cr_2O_7 \rightarrow 2PbCrO_4 + 2NaNO_3 + 2HNO_3 \tag{8}$$

Primrose chrome is made by substitution a soluble sulfate for part of the sodium bichromate in Equation 8; 23–30% lead sulfate is contained in typical solid solutions, which are sold as Primrose chrome.

Primrose chrome is stabilized by the use of proprietary additives during the formation of the pigment crystals, such that the orthorhombic crystal form is produced exclusively. The pigment must not be precipitated as the monoclinic form; otherwise instability will result.

Many different grades of lead chromate based pigments are available to offer such improved properties as better chemical resistance, decreased tendency to darken on exposure to the atmosphere, and silica encapsulation to minimize the solubility of the lead in the pigment.

Primrose chrome offers a green shade pigment of good lightfastness, high opacity, low rheology, and low cost. The coatings industry is the largest consumer of Primrose chrome yellow, followed closely by the printing ink and plastics industries.

9.3 Cadmium Zinc Yellow

This pigment is yet another coprecipitate or solid solution. For this pigment, Pigment Yellow 35, cadmium sulfide is coprecipitated with zinc sulfide followed by calcination to give pure cadmium zinc yellow, $CdS \cdot xZnS$. The hue of the pigment can be varied by altering the amount of zinc sulfide used in the solid solution. A primrose hue is achieved with levels of 14–21% of zinc sulfide, whereas 1–7% will give a redder shade or "golden" hue.

A lithopone version, Pigment Yellow 35:1, is available. As with all "lithopones," this pigment is merely manufactured in the presence of barium sulfate such that an intimate precipitate is produced containing barium sulfate as part of the solid solution with formula $CdS \cdot xZnS \cdot yBaSO_4$.

The major outlet for cadmium zinc yellows is in the plastics industry. The pigments offer bright, clean, high hiding colors with outstanding resistance to heat, light, and solvents. The pigments, however, have poor fastness to mineral acids and will fade markedly when used at low tint levels or in the presence of moisture.

9.4 Zinc Chromate

Zinc chromate or zinc yellow is Pigment Yellow 36. A lithopone version, extended with barium sulfate, also exists as Pigment Yellow 36:1. The pigments are bright green-shade yellows made by precipitating hydrated zinc potassium chromate from the reaction of sodium bichromate with zinc oxide and potassium chloride.

Zinc chromate is used principally in corrosion-inhibiting coatings. It has poor tinting strength and poor resistance to mineral acids and alkalies, thus limiting its use within such systems.

9.5 Cadmium Sulfide Yellow

Simply calcined calcium sulfide, CdS, this pigment, Pigment Yellow 37, can be produced in hues ranging from a green shade to a red shade by varying the conditions during calcination. The pigment offers excellent stability to heat, light, mineral acids, and alkali. A major drawback is the tendency to fade in the presence of moisture.

The lithopone version, extended with barium sulfate, $CdS \cdot xBaSO_4$, is Pigment Yellow 37:1.

9.6　Iron Oxide Yellows

As with most of the iron oxides discussed, this pigment can be obtained as the natural oxide (Pigment Yellow 43) or the synthetic version (Pigment Yellow 42).

The natural yellow iron oxide is $FeO \cdot xH_2O$, containing various amounts of clay and other minor minerals. The pigment is also known as limonite and is mined in Africa (African ocher), India (Indian ocher), France (Ocher), and Italy (Sienna). The French material is reported to be of the highest quality.

The synthetic material can be produced in a variety of ways. A direct precipitation process using an alkali such as ammonium hydroxide and ferrous sulfate. The Penniman–Zoph process uses scrap steel and a ferrous salt to grow seed particles, and the aniline process reacts nitrobenzene with metallic iron to produce iron oxide and aniline, a primary aromatic amine, as a by-product. Irrespective of the manufacturing process, the product has the empirical formula $Fe_2O_3 \cdot xH_2O$.

Iron oxide yellows are economical pigments with excellent lightfastness, durability, opacity, and rheology. On the negative side, they have a dull masstone, which results in dirty tints and only fair tinting strength and baking stability. Nevertheless they find widespread use within the coatings industry because of their economics of use.

9.7　Mixed Metal Oxide Yellows

For completeness, the mixed metal oxide yellows can be listed as follows

1.　Nickel antimony yellow rutile, Pigment Yellow 53, is a product from the calcination of the oxides of nickel, titanium, and antimony, $(Ti,Ni,Sb)O_2$. A green-shade yellow with excellent lightfastness, chemical resistance, and heatfastness, this pigment offers only poor gloss and low tint strength.

2.　Zinc ferrite brown spinel, Pigment Yellow 119, is made by calcining the oxides of zinc and ferrous and ferric iron, $(Zn,Fe)Fe_2O_4$. A brown-shade yellow or buff with outstanding durability, chemical fastness, and heat stability, this pigment has moderate tinting ability compared to Pigment Yellow 53 but lacks brightness and has a dirty tint tone.

3.　Nickel niobium titanium yellow rutile, Pigment Yellow 161, is a crystalline solid solution prepared by calcining the oxides of titanium, niobium, and nickel, $(Ti,Ni,Nb)O_2$. This pigment is a green-shade yellow with low tint strength and dirty tint tones. Typical of the mixed metal oxides, the pigment offers high heat resistance and outstanding chemical stability and outdoor durability.

4.　Chrome niobium titanium buff rutile, Pigment Yellow 162, is a solid solution obtained from the calcination of titanium, chromium, and niobium oxides, $(Ti,Cr,Nb)O_2$. It gives a red-shade yellow with outstanding chemical, heat, and lightfastness, coupled with excellent durability. Again its tinting strength is poor, and the pigment lacks brightness.

5.　Manganese antimony titanium buff rutile, Pigment Yellow 164, is a pigment prepared by calcining the oxides of manganese, titanium, and antimony, $(Ti,Mn,Sb)O_2$. Typical of oxides are manufactured by calcination. Although it has low tint strength and lacks brightness and cleanliness, the pigment does offer excellent heatfastness and exterior durability, coupled with outstanding chemical resistance.

9.8　Bismuth Vanadate/Molybdate Yellow

The most novel inorganic yellow to appear in the pigment marketplace within the past decade is bismuth vanadat/molybdate, Pigment Yellow 184, which was introduced by BASF in 1985. The pigment is manufactured by dissolving bismuth nitrate, sodium vanadate, and sodium molybdate in nitric acid, precipitating a complex oxide of the metals, followed by calcining the complex coprecipitate to give the pigment as a polycrystalline product. The accepted formula is $4BiVO_4 \cdot 3Bi_2MoO_6$.

The pigment is a green-shade yellow used primarily for brilliant solid shades in both automotive and industrial coatings. The pigment has excellent weatherfastness, high opacity, and good gloss retention.

63
Organic Pigments

Peter A. Lewis

Sun Chemical Corporation
Cincinnati, Ohio

1.0 INTRODUCTION

A definition of a pigment, as distinct from a dyestuff, has been prepared by the Dry Color Manufacturers' Association (DCMA) in response to a request from the Toxic Substances Interagency Testing Committee. This definition should clarify the terms "pigment" and "dyestuff," which are often erroneously used interchangeably:

> Pigments are colored, black, white, or fluorescent particulate organic and inorganic solids which usually are insoluble in, and essentially physically and chemically unaffected by, the vehicle or substrate in which they are incorporated. They alter appearance by selective absorption and/or by scattering of light
> Pigments are usually dispersed in vehicles or substrates for application, as for instance in inks, paints, plastics, or other polymeric materials. Pigments retain a crystal or particulate structure throughout the coloration process.
> As a result of the physical and chemical characteristics of pigments, pigments and dyes differ in their application; when a dye is applied, it penetrates the substrate in a soluble form after which it may or may not become insoluble. When a pigment is used to color or opacify a substrate, the finely divided, insoluble solid remains throughout the coloration process.

2.0 THE COLOUR INDEX SYSTEM

The Colour Index System is a coding system as developed under the joint sponsorship of the Society of Dyers and Colourists (SDC) in the United Kingdom and the Association of Textile Chemists and Colorists (AATCC) in the United States. The system is referred to as the "Colour Index." (This system is registered as a trade name and the use of the "u" in "Colour" must be retained.) By giving a compound a "Colour Index Name" and a "Colour Index Number," a colored compound can be readily placed into a classification according to its

563

Table 1   Range of Colour Index Constitution Numbers

Chemical Class	Colour Index Constitution Number
Insoluble Azo	
1. Acetoacetyl	11640–11790
2. Heterocyclic hydroxy	12600–12825
3. Disazo	20000–29999
4. 2-Naphthol	12050–12211
5. 3-Hydroxy–2–naphthanilide	12300–12520
Precipitated Azo	
1. 2–Naphthol (sulfonic) acid	15500–16815
Precipitated Nonazo	
1. Xanthene	45000–45999
2. Triphenylmethane	42000–44999
Insoluble Nonazo	
1. Phthalocyanine	74000–74999
2. Anthraquinine	58000–72999
3. Quinacridone	73900–73999

chemical constitution and color. This description is recognized by many government bodies as adequate information for inclusion in hazard data sheets or material safety data sheets to fully identify the pigment in question and to provide accurate reference when including a pigment in any inventory listing. Thus, for example, phthalocyanine blue has a Colour Index name Pigment Blue 15 and the Colour Index number 74160.

The Colour Index names for pigments is abbreviated as follows:

PB = Pigment Blue	PBk = Pigment Black
PBr = Pigment Brown	PG = Pigment Green
PM = Pigment Metal	PO = Pigment Orange
PV = Pigment Violet	PR = Pigment Red
PW = Pigment White	PY = Pigment Yellow

Additionally, the Colour Index constitution number conveys information regarding the structure of the compound as shown in Table 1.

3.0   PIGMENT SELECTION

Once a formulator has decided upon the hue that is required for a particular coating, the next most important criteria of any pigment are its fastness properties. It is useless to formulate any coating with a pigment that will not withstand the exposure specifications of the coating's end use. Such specifications can extend to requiring as much as 5 years' outdoor exposure in states such as Florida and Arizona.

Due attention must be paid to the respective manufacturers' literature to ensure that a pigment has been chosen that will satisfy the end-use criteria for stability to flocculation and crystallization, as well as fastness to light, solvents, heat, and chemicals. Once a pigment

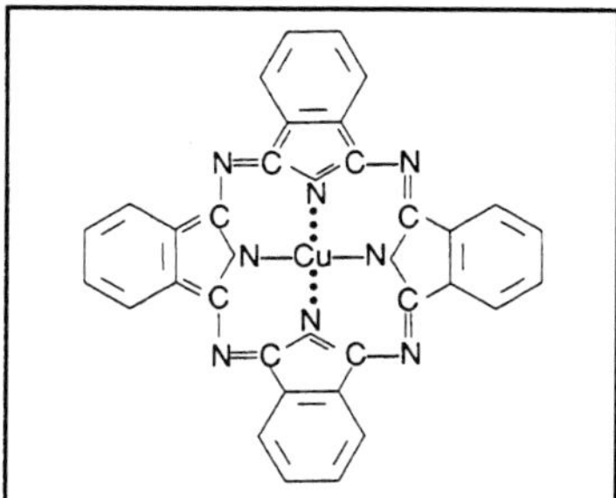

Figure 1   Structure of copper phthalocyanine blue (pigment blue 15).

class has been selected that will perform adequately in the end-use application, the formulator can consider such other factors as economy. The formulator should be aware, however, that the fastness properties of a pigment will be affected by the medium into which it is incorporated. Thus, even though intrinsically the pigment may feature the required properties, it is still necessary for the final pigmented coating to be tested in the end-use application.

3.1   Organic Blues

3.11   Copper Phthalocyanine Blue

The major blue used within the coatings industry is copper phthalocyanine blue (PB 15), with its usage far outweighing other blues such as Indathrone blue (PB 60).

Phthalocyanines are planar molecules with a tetrabenzotetraazoporphin structure as shown in Figure 1. Manufacture is comparatively easy despite the superficial complexity of the phtalocyanine molecule. Reaction of a phthalic acid derivative at temperatures approximating 190° C with a source of nitrogen such as urea and a metal or metal salt is usually all that is required to produce the appropriate metal phthalocyanine. Molybdate, vanadates, and certain compounds of titanium have been found to be useful catalysts for this condensation reaction.

Figure 2 illustrates the chemistry behind the production of copper phthalocyanine blue. This condensation reaction results in the formation of copper phthalocyanine in a crude, nonpigmentary form. The product has thus to be finished or conditioned to give the pigment grade of choice. Typically crude phthalocyanine blue is characterized by a crystal size of the order of 50 μm, a purity in excess of 92%, and a poor pigmentary strength.

Metal-free phthalocyanine blue (PB 16) is normally manufactured via the sodium salt of phthalonitrile. Acid pasting is used to condition the crude and give the pigment.

Copper phthalocyanine is commercially available in two crystal forms known as the α and β. The α form is described by the designations Pigment Blue 15, 15:1, and 15:2 and is a bright red-shade blue pigment. The β form is described as Pigment Blue 15:3 and 15:4 and is a bright green or peacock shade. The α form is meta-stable and requires special treatment to stabilize the crystal against its tendency to revert to the more stable, green-shade β crystal. If either of the unstable α crystal forms (PB 15 or 15:1) is used with strong solvents, conversion to the β form will occur upon storage of the system. Conversion from the α to the β form is usually accompanied by an increase in crystal size with subsequent loss of strength and shift to a greener hue.

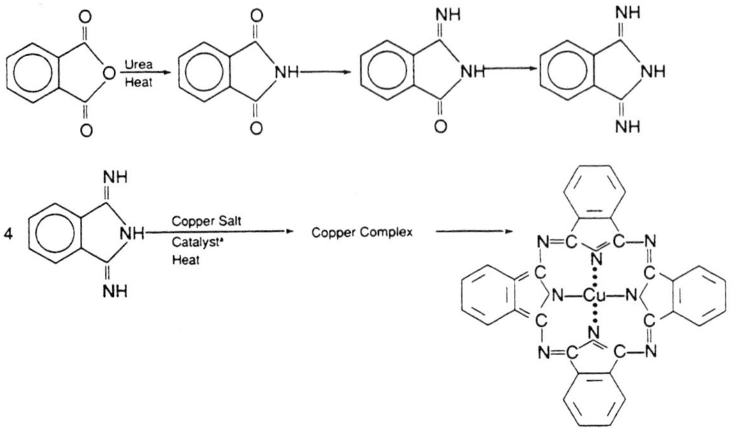

(a) Molybdate or vanadate.

Figure 2   Chemistry of copper phthalocyanine.

As stated earlier, copper phthalocyanine gives excellent service in most coatings applications, but there is considerable variation between both the chemical and crystal types available.

Pigment Blue 15 is an α crystal with the reddest shade of the types commonly available. It is the least stable of the family and as such is often referred to as crystallizing red-shade (CRS) blue. This crystal form cannot be used in any solvent containing systems.

Pigment Blue 15:1 is also an α crystal, but chemical modifications have been made to stabilize the structure against crystallization. Most commonly the molecule is chlorinated to the extent of introducing one chlorine molecule to give "monochlor" blue. Another technique involves the use of a substituted phthalocyanine, added to the pigment at levels approaching 10–15%, that confers crystal stability to the system. The monochlorinated grade is, as a consequence of the introduced chlorine atom, greener than the additive-stabilized crystal.

Pigment Blue 15:2, described as "noncrystallizing nonflocculating" red-shade blue is widely used within the coatings industry. The product is an α crystal that is stabilized against both crystallization and flocculation using additive technology.

Pigment Blue 15:3 represents the green-shade, β crystal phthalocyanine blue and, since it exists in the stable crystal form, it is less susceptible to crystallization. Most commercial grades of Pigment Blue 15:3, however, contain from 4 to 8% of the α crystal, which will be adversely affected by strong solvent systems. A 100% β blue is too dull, opaque, and weak to be commercially attractive; hence a proportion of the α crystal is left in the system, contributing considerably to the attractiveness of the system.

Pigment Blue 15:4 represents a β blue that has been modified with phthalocyanine-based additives to give a green-shade blue that is resistant to flocculation and can be used in strong solvent systems.

Copper phthalocyanine approximates the ideal pigment. It offers strength, brightness, economy, and all-around excellent fastness properties. Perhaps the pigment's only disadvantages are its tendencies to change to a coarse, crystalline, nonpigmentary form in strong solvents and to flocculate or separate from white pigments when used in paints and lacquers.

3.13 Miscellaneous Blues

Although the organic blues used in the coatings industry are primarily copper phthalocyanines, brief mention must be made of other blue pigments that find use in the coatings marketplace.

Indanthrone blue, Pigment Blue 60, belongs to the class of pigments described as "vat pigments." This pigment is expensive relative to copper phthalocycanine, and thus economic considerations are a limitation to its widespread use. Idanthrone blue is a very red-shade pigment with outstanding fastness properties.

Carabazole violet, Pigment Violet 23, is a complex polynuclear pigment that is a very valuable red-shade blue of high tinctorial strength. The pigment possesses excellent fastness properties, and only its relatively high cost and its hard nature limit its more widespread use. From an economic standpoint it costs approximately three times as much as

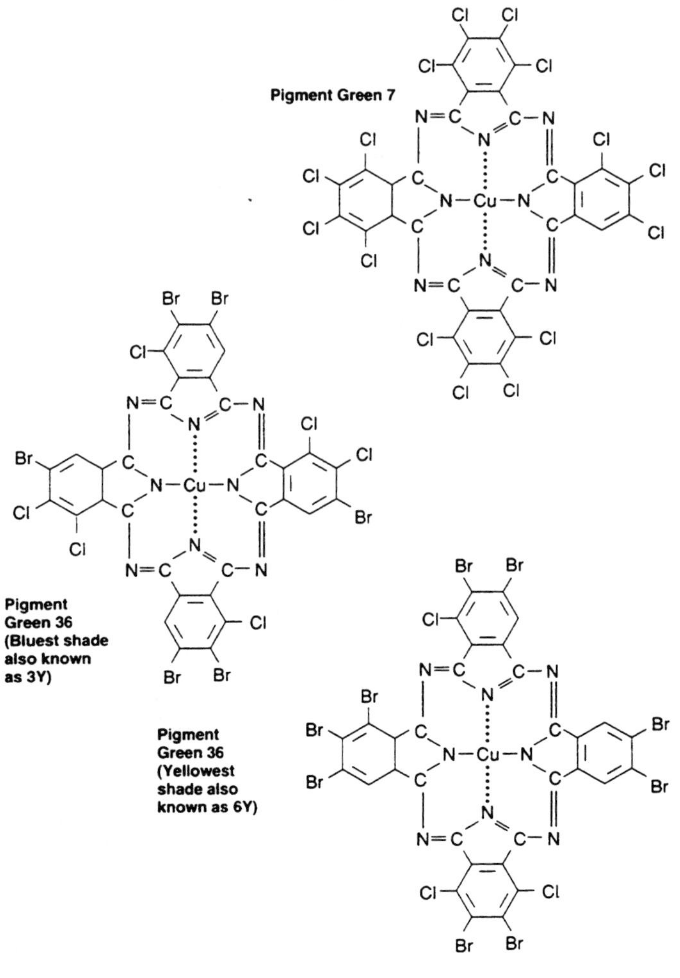

Figure 3   Structure of copper phthalocyanine greens.

phthalocyanine blue.

The pigment is used as a shading component in high performance coatings that call for particularly red-shade blue.

3.2 Organic Greens

3.2.1 Copper Phthalocyanine Green

The major green pigment used as a self shade in the coatings industry is based on halogenated copper phthalocyanine and, as such, is termed phthalocyanine green. The Colour Index names are Pigment Green 7 and Pigment Green 36.

Pigment Green 7, the blue-shade green, is based on chlorinated copper phthalocyanine with a chlorine content that varies from between 13 to 15 atoms per molecule.

Pigment Green 36, the yellower shade, is based on a structure that involves the progressive replacement of chlorine on the phthalocyanine structure with bromine. The composition of Pigment Green 36 varies with respect to the total halogen content, chlorine plus bromine, and in the ratio of bromine to chlorine. Figure 3 illustrates the proposed structures of the phthalocyanine greens. In practice, no single pigment consists of a specific molecular species; rather, each pigment is a complex mixture of closely related isomeric compounds.

These pigments are ideal, since their tinctorial and fastness properties allow their use in the most severe application situations. They possess outstanding fastness to solvents, heat, light, and outdoor exposure. They can be used in masstone shades and tints down to the very palest of depths.

Phthalocyanine greens are manufactured by a three-step process: crude phthalocyanine blue is first manufactured, then halogenated to give a crude copper phthalocyanine green, and finally conditioned to give the pigmentary product.

3.22 Miscellaneous Greens

Table 2 gives a summary of the properties of some other commercially available organic greens that may find some minor application in the coatings industry.

Table 2 Summary of Miscellaneous Green Properties

Colour Index Name	Chemical Name	Comments
PG 1	Brilliant Green (Triphenylmethane PTMA)	Brilliant, blue shade. Poor alkali and soap resistance, solvent bleed, and lightfastness. May be used in interior finishes.
PG 2	Permanent Green (Triphenylmethane PTMA)	Blend of Pigment Green 1 and Pigment Yellow 18. Bright yellow shade. Poor fastness overall.
PG 4	Malachite Green (Triphenylmethane PTMA)	Bright, blue shade. Poor fastness properties overall.
PG 8	Pigment Green B (Nitroso)	Yellow shade. Dull hue. Poor fastness properties. May be used in interior emulsions.
PG 10	Green Gold	Yellow shade. Loses metal in strong acid or alkali. Good lightfastness. Moderate solvent fastness. Used in automotive and exterior paints.

Table 3 Organic Orange Pigments

Colour Index Name	Colour Index Number[a]	Chemical type
PO 2	12060	Azo
PO 5	12075	Azo
PO 13	21110	Bisazo
PO 16	21160	Bisazo
PO 31	n/a	Bisazo condensation
PO 34	21115	Bisazo
PO 36	11780	Benzimidazalone (azo)
PO 38	12367	Azo
PO 43	71105	Perinone
PO 46	15602	Azo
PO 48	n/a	Quinacridone
PO 49	n/a	Quinacridone
PO 51	n/a	Pyranthrone
PO 52	n/a	Pyranthrone
PO 60	n/a	Benzimidazolone (azo)
PO 61	n/a	Tetrachloroisoindolinone
PO 62	n/a	Benzimidazolone (azo)
PO 64	n/a	Heterocyclic hydroxy
PO 67	n/a	Pyrazoloquinazolone

[a]n/a = Not assigned.

3.3 Organic Oranges

Table 3 lists the orange pigments that are available in today's marketplace. A broad subdivision can be made based on such chemical features of the molecule as azo based, benzimidazolone based, and miscellaneous oranges.

3.3.2 Azo-Based Oranges

The structure of the seven orange pigments that can be placed into the "azo" category are shown in Figure 4.

Pigment Orange 2, Orthonitraniline Orange, is prepared by the classical diazotization and coupling technique in which diazotized orthonitroaniline is couples to β-napthol. The pigment's major market is in the field of printing inks. Its use in coatings is not recommended because the pigment's solvent fastness is poor and its lightfastness inadequate.

Pigment Orange 5, Dinitroaniline Orange, is manufactured by a diazotization and coupling sequence in which diazotized dinitroaniline is coupled onto β-napthol. This pigment offers good lightfastness in full tone and moderate solvent fastness. As such, the pigment finds widespread use in latex-based paints and, with the exception of high bake enamels, in both architectural and industrial coatings.

Pigment Orange 13, Pyrazolone Orange, is synthesized by coupling tetrazotized 3, 3'-dichlorobenzidine onto 3-methyl-1-phenyl–pyrazol-5–one. The pigment is a bright, clean yellow-shade product that is tinctorially stronger than Pigment Orange 5. It may be recommended for interior coatings, particularly as a replacement for lead-based oranges.

PO 2

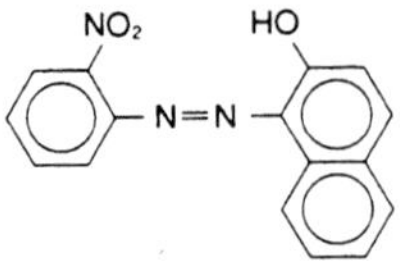

Orthonitroaniline Orange

PO 5

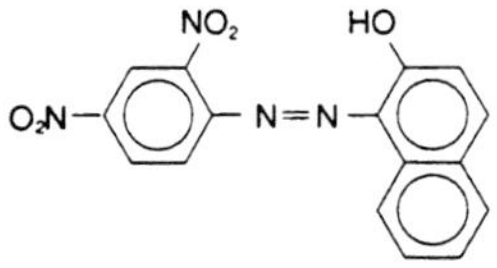

Dinitroaniline Orange

PO 13

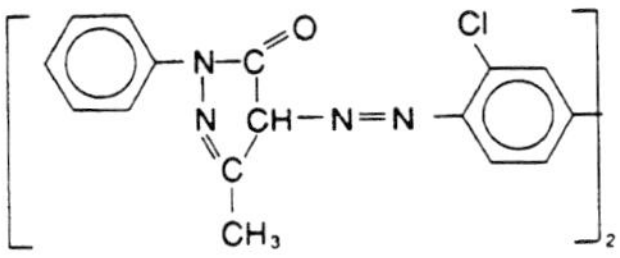

Pyrazolone Orange

PO 16

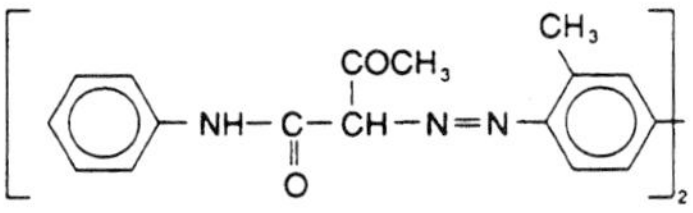

Dianisidine Orange

PO 34

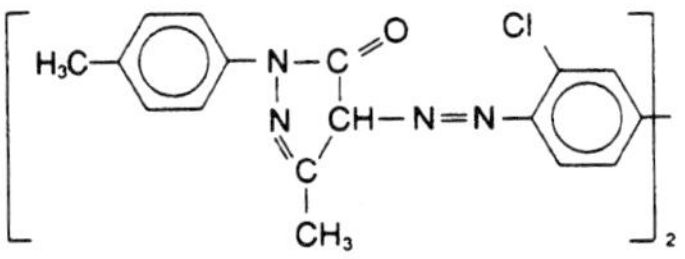

Tolyl Orange

PO 38

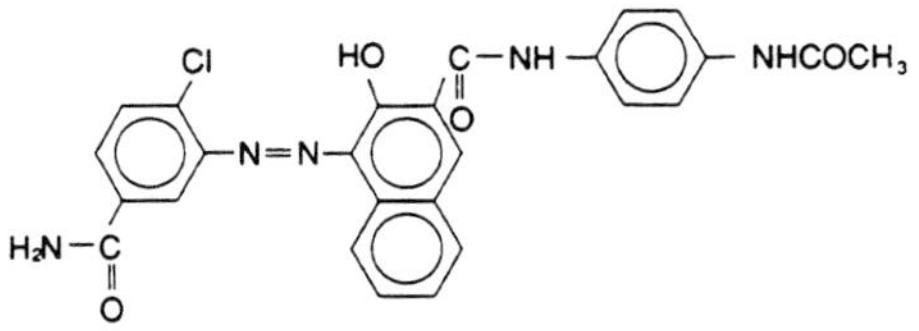

Naphthol Orange

PO 46

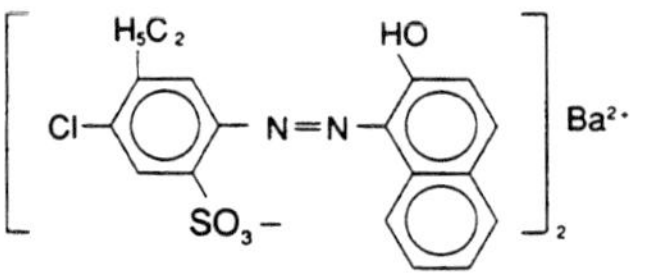

Clarion Red

Figure 4 Structure of azo-based oranges.

Pigment Orange 16, Dianisidine Orange, is a diarylide orange that is prepared by coupling tetrazotized 3,3′-dimethoxybenzidine onto acetoacetanilide. The pigment finds use in baking enamels, since its heat-fastness is superior to that of other orange pigments with similar economics. Usage in interior coatings at full tone levels is also recommended.

Pigment Orange 34, Tolyl Orange, is produced by coupling tetrazotized 3,3′-dichlorobenzidine onto 3-methyl-1-(4′-methyl-phenyl)-pyrazol-5-one. The pigment is a bright, reddish orange that offers moderate lightfastness and good alkali resistance, but poor solvent fastness. As such, the material is used in interior coatings applications, particularly where a lead-free formula is specified.

Pigment Orange 38, Naphthol Orange, is manufactured by coupling diazotized 3-amino-4-chlorobenzamide onto 4′-acetamido-3-hydroxy-2-napthanilide. The pigment is a bright reddish orange that exhibits excellent alkali and acid fastness, moderate solvent fastness, and acceptable lightfastness when used at full tint. As such, the pigment finds use in baking enamels, latex, and masonry paints.

Pigment Orange 46, Clarion Red, is a metallized azo pigment manufactured by coupling diazotized 2-amino-5-chloro-4-ethylbenzene-sulfonic acid onto β-napthol and metallizing the product with barium to yield the pigment. The pigment has poor lightfastness, inferior alkali resistance, and inadequate solvent fastness, hence is not recommended for use in coatings.

3.3.2 Benzamidazolone–Derived Oranges

The three benzimidazolone-derived oranges contain the azo chromophore and are all based on the 5-acetoacetylaminobenzimidazolone as the coupling component.

Pigment Orange 36 is the product of coupling diazotized 4-chloro-2-nitroaniline to the benzimidazolone. Pigment Orange 60 is the product of the coupling of 4-nitroaniline to the benzimidazolone. Because of the proprietary nature of this product, the structure of Pigment Orange 62 has not been fully declared (Fig. 5 illustrates two typical structures).

Pigment Orange 36 is a bright red-shade orange of very high tint strength. The opacified form of this pigment offers excellent fastness properties to both heat and solvents and a hue similar to the lead containing pigment, Molybdate Orange (PO 104). As such, Pigment Orange finds major use in lead-free automotive and industrial high performance coatings.

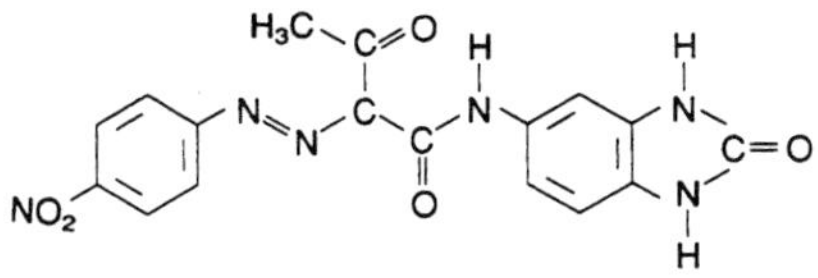

Figure 5 Structure of the benzimidazolone oranges.

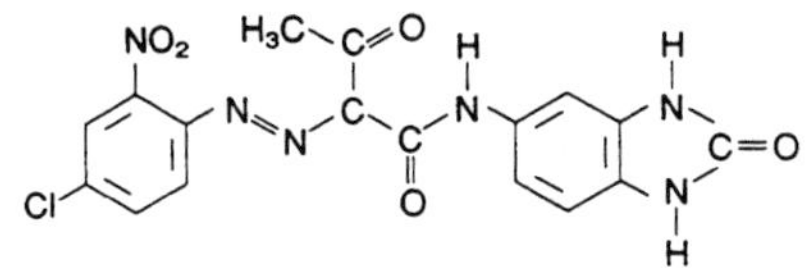

Pigment 60 is a transparent, yellow-shade orange that also exhibits excellent heat and solvent fastness with an exterior durability that allows the pigment to be used in high quality industrial and automotive finishes.

Pigment Orange 62 is also a yellow-shade orange that shares the lightfastness properties of the other two oranges. Currently it is used in oil-based inks and artists' colors. Its use in the coatings industry has yet to be fully explored.

3.3.3 Miscellaneous Oranges

The structures of Pigment Orange 53, a pyranthrone, Pigment Orange 64, a heterocyclic hydroxy, and Pigment Orange 67, a pyrazoloquinazolone, have not been fully declared. Table 4 summarizes the properties of this class of pigments, which represents a series of oranges that are finding increased application in the coating industry.

Table 4 Summary of the Properties of the Miscellaneous Oranges

Colour Index Name	Common Name/ Description	Properties
PO 43	Perinone	Red shade. Strong, clean, vat pigment with excellent fastness properties. Used in metallized finishes and high grade paints. Shows slight solvent bleed.
PO 48	Quinacridone Gold	Yellow shade. Excellent lightfastness. Lacks brightness in masstone. Used in metallic finishes.
PO 49	Quinacridone Deep Gold	Red shade. Dull masstone. Excellent durability. Used in metallics.
PO 51	Pyranthrone Orange	Medium shade. Excellent fastness to solvent, light, and heat. Dull in tint. Exhibits slight solvent bleed. Used in air dry and bake enamels.
PO 52	Pyranthrone Orange, red shade	Vat pigment with excellent fastness to solvent, light, and heat. Dull in tints. Slight solvent bleed. Used in air dry and bake enamels.
PO 61	Tetrachloroisoindolinone orange	Medium shade. Exhibits some solvent bleed. Used in metallic automotive finishes.
PO 64	Bright shade red	Excellent solvent and lightfastness. Used in industrial coatings.
PO 67	Yellow shade	Excellent brilliance in full shade. Good gloss retention. Very good weather- and lightfastness in full shade. Used in industrial and automotive coatings.

3.4 Reds

Metallized Azo Reds

Many of the reds used in the coatings industry fall into the chemical category of azo pigments because the azo chromophore — N = N — is a feature of the molecule.

A further subdivision may be made into acid, monazo metallized pigments such as Manganese Red 2B (PR 48:4) and Calcium Lithol (PR49:2) and nonmetallized azo reds such as the Naphthols (e.g. PR 17 and PR 23) and Toluidine Red (PR 3). Typically, each of the acid, monoazometallized pigments contains an anionic grouping such as a carboxylic (—COOH) or sulfonic acid (—SO3H) group, which will ionize and react with a metal cation such as calcium or manganese to form an insoluble, metallized pigment.

Nonmetallized pigments do not contain an anionic group in their structure and, as such, will not complex with a metal cation.

All azo reds contain one or more azo groups and are produced by similar reaction sequences. The initial reaction sequence, described as diazotization, involves reacting an aromatic primary amine with nitrous acid, formed in situ by reacting sodium nitrite with hydrochloric acid at low temperatures to yield a diazonium salt. Invariably the diazonium salt that is formed by this process is unstable and should be keep cold to avoid any decomposition.

The diazonium salt is reacted quickly with the second half of the pigment, which is called the coupler. The coupling reaction takes place rapidly in the cold to yield the sodium salt of the pigment. This sodium salt is all but useless as a pigment for the coatings industry because of its marked tendency to bleed even in the weakest of solvent systems. The pigment is, therefore, metallized to confer improved properties on the product. The pigment suspension is then filtered and washed to remove any residual inorganics derived from the reaction.

Figure 6 illustrates the structures of the different metallized azo reds that are readily available.

The *Lithol Reds* are primarily Barium Lithol (PR 49:1) and Calcium Lithol (PR 49:2). Although limited in their application in the coating industry, these pigments do find some use at masstone levels—that is, the pigment is not tinted with a white tint base—where their fastness properties are acceptable.

The pigments are bright reds with high tint strength and good dispersion characteristics. The barium salt is lighter and yellower in hue compared to the calcium salt.

Permanent Red 2B is the generic name that encompasses Barium Red 2B (PR 48:1), Calcium Red 2B (PR 48:2), and Manganese Red 2B (PR 48:4). The major use of the calcium and barium salts is in baked industrial enamels, which are not required to be fast to outdoor exposure. Use in alkaline systems is severely restricted because of the poor alkaline fastness of these salts.

The barium salt is characterized by a clean, yellow hue and, although slightly poorer than the the bluer calcium salt in lightfastness and tinting strength, it does possess a slight advantage in bake stability.

Manganese Red 2B has a sufficiently improved lightfastness to be used in implement finishes. The manganese salt is slightly bluer, dirtier, and less intense when compared to the calcium salt.

Rubine Red also known as Calcium Lithol Rubine (PR 57:1) is a clean, blue-shade red pigment exhibiting the high tinting ability typical of the azo reds of this class. Its major use in coatings is in interior systems that call for an inexpensive red with good solvent and heat resistance. Again, to maintain maximum fastness properties, use of the pigment at near to masstone levels is recommended.

PR 57

PR 48

PR 52

PR 200

PO 46

PR 53

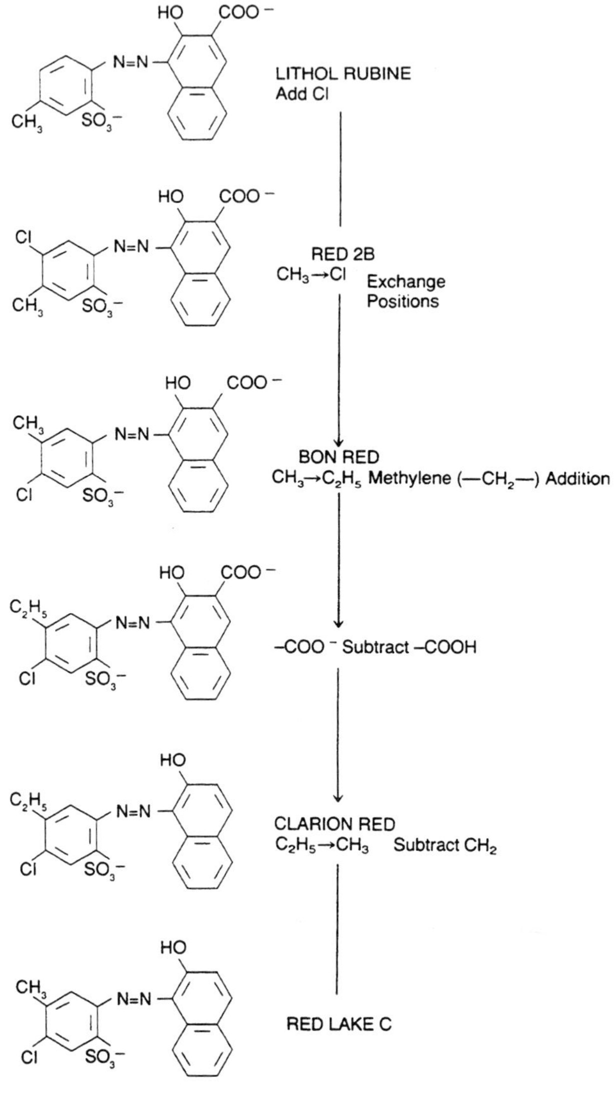

Figure 6 Metallized azo reds.

BON red, used as calcium BON Red (PR 52:1) or Manganese BON Red (PR 52:2), is characterized by outstanding cleanliness, brightness, and purity of color. The manganese salt offers a very blue-shade red with improved lightfastness compared to the calcium salt. As such, this salt is suitable for exterior coatings applications.

BON Maroon, (PR 63:1) is illustrated in Figure 7; the manganese salt of BON Maroon is of considerably more importance than either the calcium or barium salts. Its lightfastness is such that the pigment can be used at masstone levels for implement finishes.

3.4.2 Non-Metallized Azo Reds

As implied by their classification, the nonmetallized azo reds do not contain a precipitating metal cation and, as such, offer increased stability against hydrolysis in strongly acidic or alkaline environments.

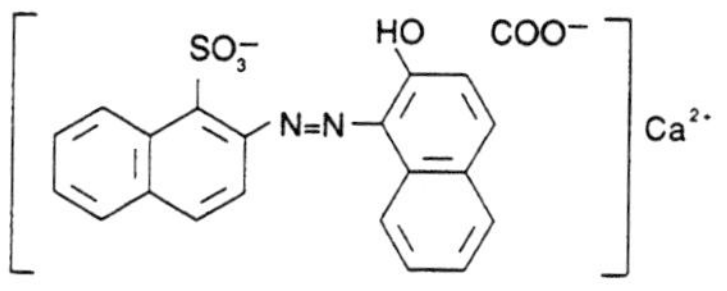

2-naphthylamine-1-sulfonic acid ⎯⎯→ 3-hydroxy-2-naphthoic acid (BON)

Figure 7 BON maroon.

Synthesis of this class of pigment follows the previously described classical method of diazotization of a primary aromatic amine followed by coupling of the resultant diazonium salt. No anionic groups capable of accepting a metal cation are present in the molecule; thus metallization is not a factor in their synthesis. Typical nonmetallized reds are Toluidine Red (PR 3) and the wide range of Napthol Reds as represented by Pigment Reds 17, 22, and 23.

Toluidine Red is used in full shade in such coatings applications as farm implements, lawn and garden equipment, and bulletin paints, where a bright, economical red of adequate lightfastness is required. Because of the pigment's poor durability in tint shades, it is rarely used at any level other than a full shade.

The individual properties of the Napthol Reds depend on the specific composition of the product as well as the conditioning steps used during pigment manufacture. As a class, they are a group of pigments that exhibit good tinctorial properties combined with moderate fastness to heat, light, and solvents.

Unlike the metallized azo reds, the Napthol Reds are extremely resistant to acid, alkali, and soap. These properties lead to their use in latex emulsion systems and masonry paint.

In terms of performance and economic characteristics, the Napthols form a link between the toluidine reds at the lower end of the scale and the perylene and quinacridone reds at the higher end.

3.4.3 High Performance Reds

Pigments for the exacting standards of today's automotive coatings are required to show satisfactory durability to outdoor exposure in such states as Arizona and Florida for 2 and possibly 5 years before being approved for use in automotive finishes. Similar requirements are placed on pigments chosen for use in automotive repair systems and marine coatings.

High performance reds fall into four basic classes; quinacridone reds and violets, reds based on vat dyestuffs known to include the perylene reds, reds derived from the benzimidazolone diazonium salts, and the disazo condensation reds.

Quinacridone Reds Quinacridones may be described as heterocyclic pigments in that their structure comprises a fused ring structure in which the ring atoms are dissimilar. In the case of quinacridones, the ring atoms are carbon and nitrogen (Fig. 8). Addition of differing

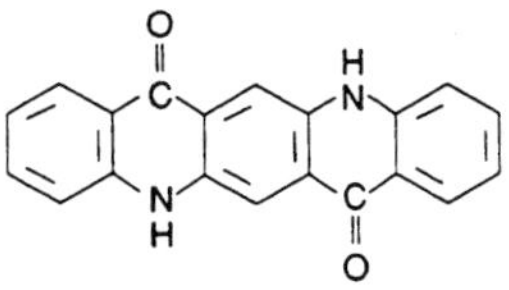

Figure 8 Structure of translinear quinacridone.

auxochromic groups such as methyl (—CH$_3$) and chlorine (—Cl) gives Pigment Red 122 and Pigment Red 202, respectively.

The two most commercialized routes in the synthesis of quinacridone (PV 19) involve either the oxidation of dihydroquinacridone or the cyclization of 2,5-diarylaminotereph-thalic acid as outlines in Figure 9. Subsequent conditioning leads to the desired crystal modification. Use of 2,5-diarylaminoterephthalic acid at the cyclization stage yields the un-substituted trans linear quinacridone. Use of 2,5-ditoluidinoterephthalic acid yields the 2,9-dimethyl quinacridone, Pigment Red 122.

As a group of high performance pigments,quinacridones find their primary uses in auto-motive finishes, both metallic and solid shades, industrial finishes, and exterior finishes.

The pigments combine excellent tinctorial properties with outstanding durability, sol-vent fastness, lightfastness, heatfastness, and chemical resistance. Table 5 lists the shades currently available.

Vat Reds   The vat red pigments, based on anthraquinine, include such structures as Anthraquinone Red (PR 177), Perinone Red (PR 194), Brominated Pyranthone Red (PR 216), and Pyranthrone Red (PR 226), as illustrated in Figure 10.

These anthraquinone-derived pigments are manufactured via a series of complex reac-tions to include such processing as sulfonation, nitration, halogenation, condensation, and substitution.

Perylene Reds   Perylene reds provide pure, transparent shades and novel styling effects when used in metallic finishes. These pigments offer improved flow characteristics when used in highly pigmented coatings formulations such as those required for high solids base coat/clear coat systems, as well as high tranparency and good bleed resistance.

The perylenes possess high color strength, good thermal stability, excellent light- and weather fastness and, with the exception of Pigment Red 224, excellent chemical resistance.

Perylenes may also be described as vat pigments and in fact are the only class of vat pigments to be specifically developed as pigments rather than as dyestuffs. Almost all the

Table 5   Types of Quinacridone

Colour Index Name	Hue	Comments
PO 48	Gold	Quinacridone quinone
PO 49	Deep Gold	Quinacridone quinone
PR 122	Magenta-yellow	2,9-Dimethyl quinacridone
PR 192	Red-yellow	Unsymmetrical monomethyl quinacridone
PR 202	Magenta-blue	2,9-dichloroquinacridone
PR 206	Maroon	Mixed solid solution
PR 207	Scarlet	4,11-dichloroquinacridone
PR 209	Yellow-shade red	3,10-dichloroquinacridone
PV 19	Violet-blue	β-quinacridone
	Red-yellow	γ-quinacridone
PV 42	Maroon	Mixed solid solution

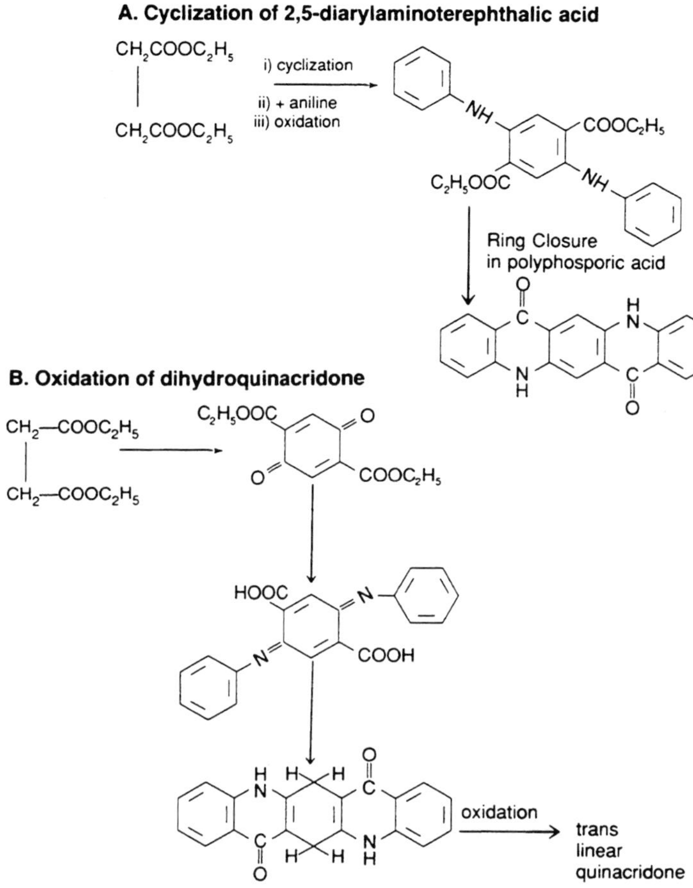

A. Cyclization of 2,5-diarylaminoterephthalic acid

B. Oxidation of dihydroquinacridone

Figure 9   Routes to quinacridone.

perylene pigments have a structure as shown by the generic formula illustrated in Figure 11; that is; they are based on N,N'-substituted perylene-3,4,9,10-tetracarboxylic diimide. A notable exception is Pigment Red 224 (Fig. 12), which is actually derived from the perylene tetracarboxylic dianhydride.

Acenaphthene is oxidized to 1,8-naphthalic anhydride followed by ammonation to yield the naphthalimide. The napthalimide is condensed in a fused caustic medium to yield the perylene-3,4,9,10-tetracarboxylic acid diimide. The diimide can then be conditioned to convert the crude into pigment and achieve Pigment Violet 26, or methylated to give Pigment Red 179.

The diimide may be hydrolyzed to produce the dianhydride, Pigment red 224, or condensed with various aromatic or aliphatic amines to give such pigments as those featured in Figure 13. As with many of the pigments already described, the perylenes have to be conditioned to obtain the compounds in a pigmentary form.

Thioindigo Reds   The thioindiogoid chromophore serves as a nucleus for a wide range of red to violet pigments. The thioindigo reds include Pigment Reds 86, 87, 88, 181, and

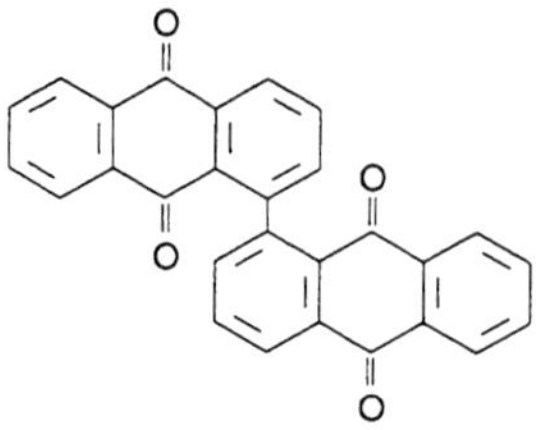

Anthraquinone Red (PR 177)

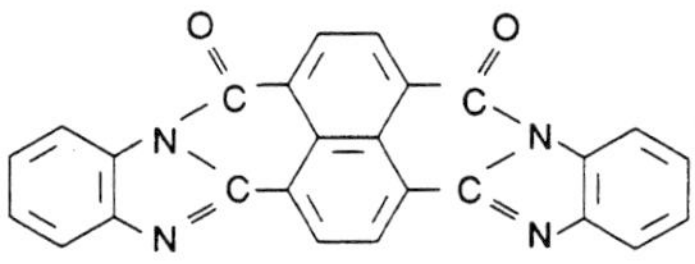

Perinone Red (PR 194)

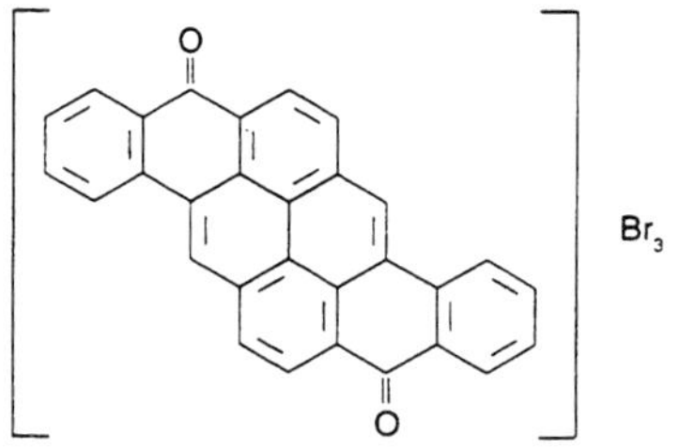

Brominated Pyranthrone Red (PR 216)

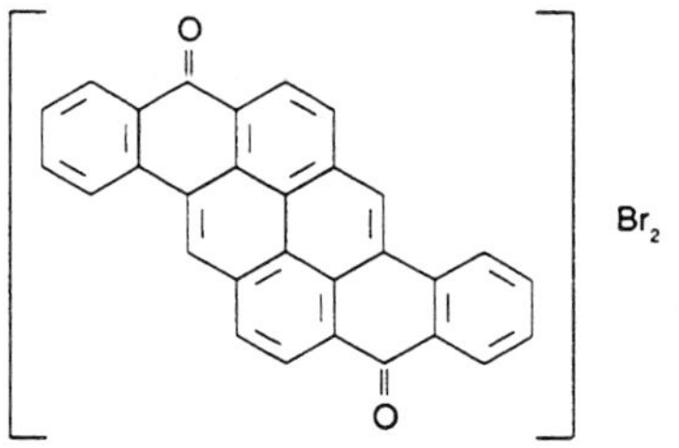

Pyranthrone Red (PR 226)

Figure 10 Structure of translinear quinacridone.

198. These pigments are noted for their brightness of shade and all-around good fastness properties. These properties have resulted in the use of the thioindigoes in quality coatings, Pigment Red 88 is the prime pigment, followed by Pigment Red 198.

Benzimidazolone-Based   The benzimidazolone pigments feature Pigment Reds 171, 175, 176, 185, and 208. These pigments are azo-based compounds that contain the benzimidazolone structure as part of their coupling component.

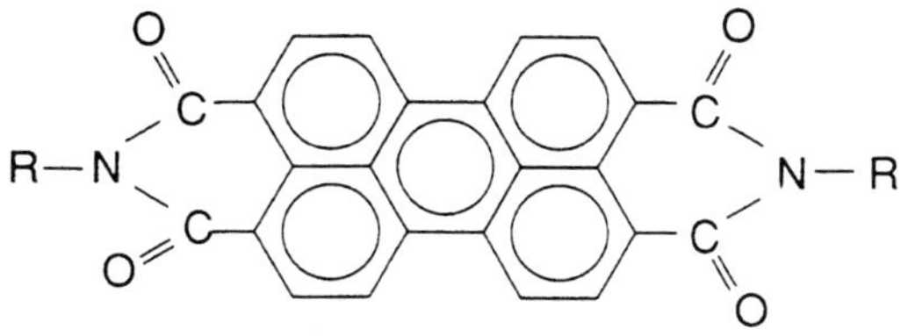

R = $-C_6H_3OC_2H_5$ PR 123   Vermillion
R = $-C_6H_3(CH_3)_2$ PR 149   Scarlet
R = $-CH_3$ PR 179   Maroon
R = $-C_6H_5OCH_3$ PR 190   Red
R = $-H$ PBr 26   Bordeaux
R = $-C_6H_5Cl$ PR 189   Yellow Shade Red

R = ⟨⟩—N=N⟨⟩   PR 178   Red

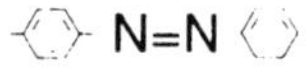

Figure 11   Perylene structures.

Although used primarily for the coloring of plastics because of their outstanding thermal stability, this class of pigments does find application in the coatings industry. Benzimidazolones show excellent fastness to light, good weatherability, and excellent fastness to overspraying at elevated temperatures. The benzimidazolone pigments are prepared using the diazotization and coupling techniques described earlier.

Generally, after the coupling process, the pigment is treated to obtain a uniform, controlled crystal growth and particle size distribution.

Disazo Condensation Reds   These pigments feature such properties as high tinctorial strength and fastness to solvents and heat. Disazo condensation reds have found considerable use as replacement pigments for lead containing pigments.Their outstanding fastness properties have resulted in their use in high quality industrial finishes.

Figure 14 illustrates three typical structures of the disazo condensation reds. Colour Index names for these three pigments have not been assigned. The synthesis sequence generally is similar for each of the disazo condensation pigments. The azo components are initially coupled to yield monoazo dyestuff carboxylic acids, which are converted to acid chlorides before final conversion to the disazo by condensation with the arylide component to yield the pigment in question.

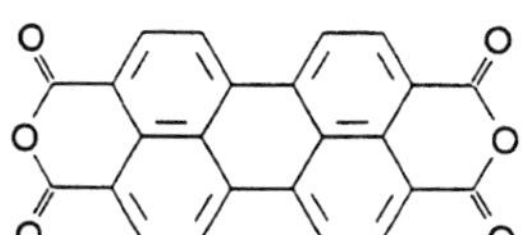

Figure 12   Pigment red 224.

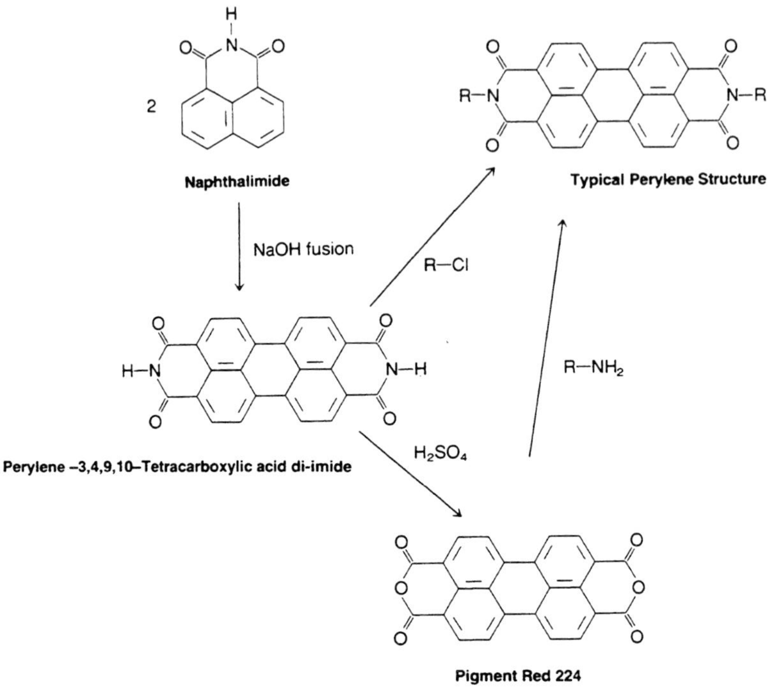

Figure 13 Perylene synthesis.

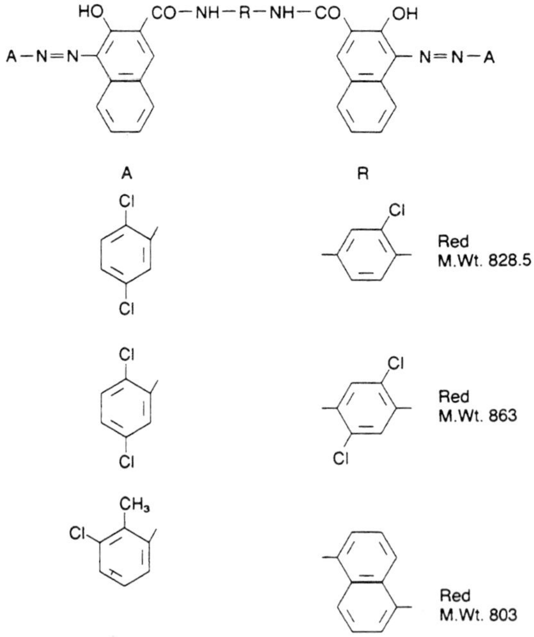

Figure 14 Structures of the disazo condensation reds.

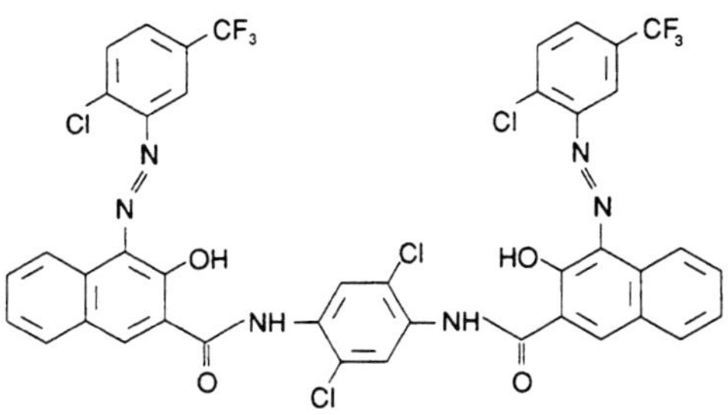

Figure 15   Structure of pigment red 242.

Miscellaneous High Performance Reds   In recent years a number of novel high performance reds have been commercialized by such companies as Sandoz and BASF. New from Sandoz are Pigment Reds 242, 214, and 257.

Pigment Red 242 is shown in Figure 15. The product is a yellow-shade red with a clean bright shade and very good all-around fastness properties. It is recommended for lead-free coatings formulations for the production of high quality finishes and bright red shades.

Pigment Red 214 (Fig. 16) is a bluish red with properties similar to those of Pigment Red 242.

Pigment Red 257, (Fig. 17) is a nickel complex pigment with a red violet masstone and a magenta undertone; its fastness properties are similar to those of the quinacridone pigments.

3.5 Organic Yellows

Yellow pigments can be subdivided into four broad classification based on their chemical constitution. These classifications are comprised of monoarylide yellows, diarylide yellows, benzimidazolone yellows, and heterocyclic yellows.

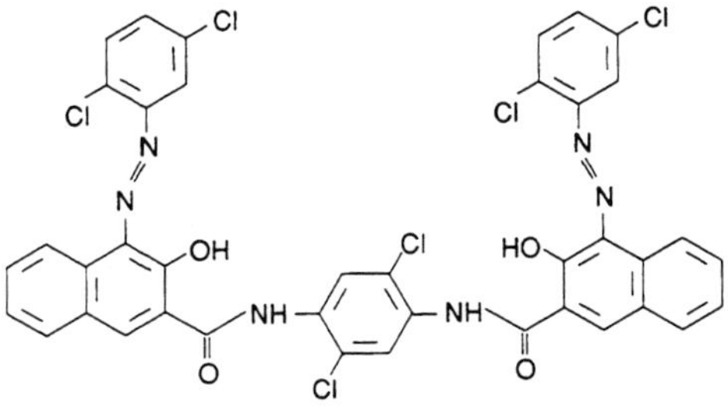

Figure 16   Structure of pigment red 214.

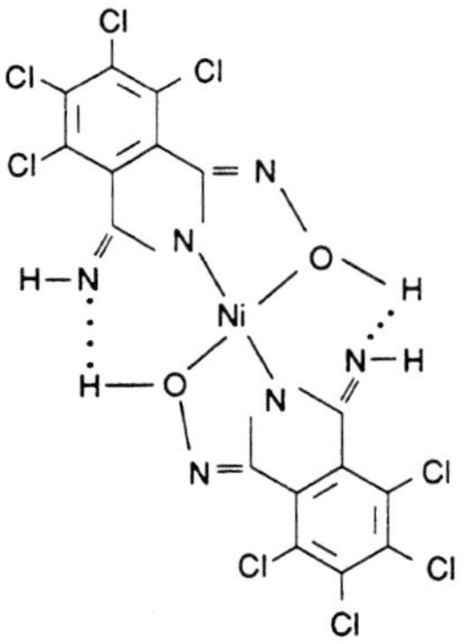

Figure 17   Structure of pigment red 257.

3.5.1 Monoarylide Yellows

Monoarylide yellows are all azo pigments; their manufacture is based on the diazotization procedure, followed by coupling. The structures of the major momoarylide yellows are represented in Figure 18.

Pigment Yellow 1 is often referred to as Hansa Yellow G for historical reasons. This pigment is a bright yellow that finds outlets in trade sales, emulsion, and masonry paints. The pigment's major drawbacks are its poor bleed resistance, poor lightfastness in tint shades, and markedly inferior bake.

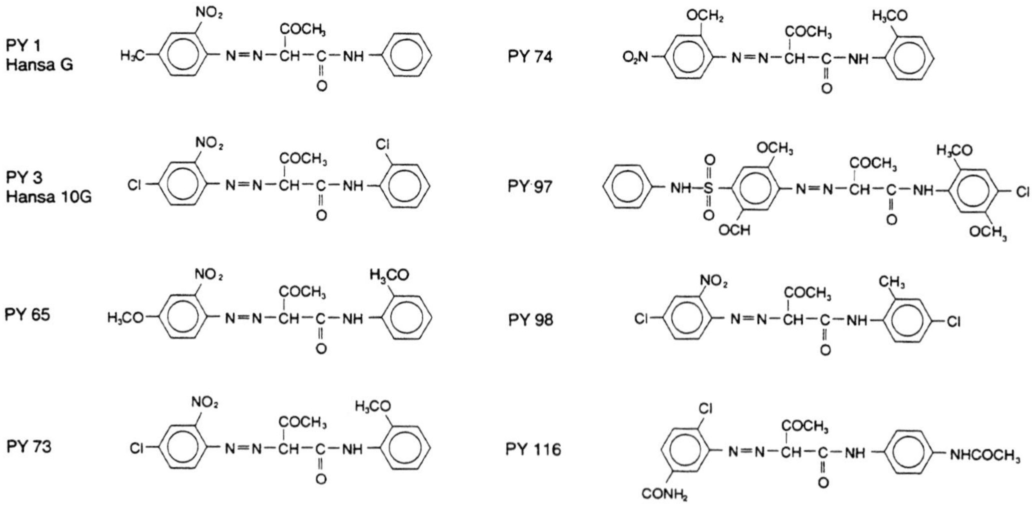

Figure 18   Monoarylide yellow structures.

Pigment Yellow 3, referred to as Hansa Yellow 10G, is a pigment that is considerably greener and cleaner than Pigment Yellow 1. The pigment finds a market in trade sales, water-based emulsions, and masonry paints. The pigment suffers from the same deficiencies of poor bleed resistance and poor tint lightfastness exhibited by Pigment Yellow 1. Pigment yellow 3 is, however, suitable for exterior use at high tint strength.

Pigment Yellow 65 is a monarylide yellow that finds use in trade sales, latex, and masonry paints. The pigment offers a lightfastness in full shade of 7 and 6–7 in tint, a considerable improvement over Pigment Yellow 1.

Pigment Yellow 73 is a pigment close in shade to Pigment Yellow 1 which again finds use in trade sales, latex, and masonry paints. It is not, however, considered to be durable enough for exterior applications. Pigment Yellow 73 plays an important role in interior, intermix systems because of its stability against recrystallization in the presence of glycols and wetting agents as used in aqueous systems.

Pigment Yellow 74 offers the user a pigment that is suitable for outdoor applications and considerably stronger and somewhat greener than Pigment Yellow 1. Again usage in trade sales, latex, and masonry paints is widespread.

Pigment Yellow 97 is a newer pigment that shows the advantages associated with a smaller particle size, less agglomeration, and an improved particle size distribution when compared to the other monoarylide yellows. Pigment Yellow (7 gained significant use in trade sales applications primarily because it was available when demands for lead-free formulations were increasing. Additionally, the pigment finds use in high quality decorative paint.

Pigment Yellow 98 is a pigment that has met with only limited commercial success to date. The pigment is similar in shade to Pigment Yellow 3, but considerably stronger and more heat stable; it finds use in trade sales, masonry, and latex-based paints.

Pigment Yellow 116 is a pigment similar in shade to light chrome yellow (PY34). Pigment Yellow 116 exhibits improved chemical resistance compared to the other monoarylide yellows, showing improved fastness to solvent, heat, and light. Its major area of use is currently in the formulation of lead-free coatings.

In summary, the monoarylide yellows constitute the most important class of organic yellows as consumed by the coatings industry. Because of the absence of any lake forming groups (e.g.—COOH or —SO₃H) in their structure, they possess excellent alkali fastness, which enables them to be used in all major aqueous paint formulations. Additionally, when compared to their inorganic lead containing counterparts, the chrome yellows, the monoarylides offer considerably higher tinting strength, bright yellow hue of high chrome, and less tendency to darken on exposure to atmospheric pollutants.

3.5.2 Diarylide Yellows

Figure 19 illustrates the structure of the diarylide yellows, and Table 6 presents a summary of the properties of this class of yellows.

Basically, an inspection of the structures indicates, this class of yellows has a backbone structure that hinges on 3,3'-dichlorobenzidine with properties that differ depending on the nature of the coupling component. Properties that are common features of this group of yellows are low cost, reasonable heat stability, and moderate chemical resistance.

The major market for the diarylide yellows is the printing ink industry. The diarylide yellows are approximately twice as strong as the monoarylide yellows; additionally they offer improved bleed resistance and fastness to heat. However, none of the diarylide yellows has durability adequate for the pigment to be considered in exterior coatings applications; thus the diarylide yellows should not be used in any outdoor situations.

Table 6 Summary of Diarylide Yellow Properties

Colour Index Name	Common Name[a]	Comments
PY 12	AAA Yellow	Poor lightfastness. Poor bleed resistance. Major use in printing inks.
PY 13	MX Yellow	Redder shade than PY12. Improved heat stability and solvent fastness. Major use in printing inks.
PY 14	OT Yellow (274–1744)	Green shade. Some use in interior finishes. Poor tint lighfastness.
PY 16	Yellow NCG	Bright green shade. Improved heat and solvent fastness. Used in full shade for coatings.
PY 17	OA Yellow (275–0562)	Green shade. Some use in interior finishes. Poor lightfastness.
PY 55	PT Yellow	Red shade. Some use in interior finishes. Poor lightfastness. Isomer of PY 14.
PY 81	Yellow H10G	Bright, green shade. Same shade but much stronger than PY 3.
PY 83	Yellow HR (275–0570) (275–0050)	Very red shade. Improved transparency, heat stability, and lightfastness over PY 12. Some use in interior finishes.
PY 106	Yellow GGR	Green shade. Poor tint lightfastness. Major use in packaging inks.
PY 113	Yellow H10GL	Very green shade. More transparent than PY 12 and offering better heat and solvent fastness. Some interior finish use.
PY 114	Yellow G3R	Red shade. Improved solvent and lightfastness over PY 12. Major use in oil-based inks.
PY 126	Yellow DGR	Similar shade to PY 12 but offering improved heat and solvent fastness. Major use in printing inks.
PY 127	Yellow GRL	Bright, red shade. Poor lightfastness. Major use in offset inks.
PY 152	Yellow YR	Very red, opaque product. Poor lightfastness. Some use in interior finishes as a lead chrome replacement.

[a]Numbers in parentheses are codes used by Sun Chemical Corporation.

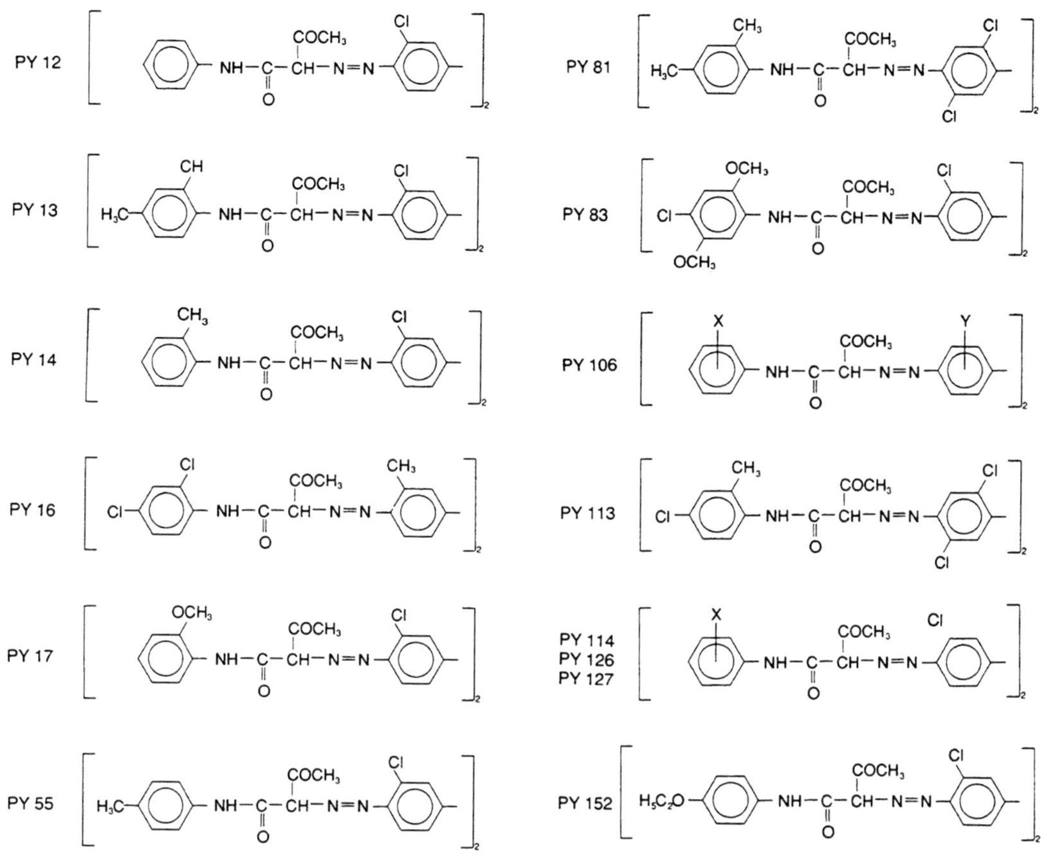

Figure 19 Diarylide yellow structures.

Table 7 Summary of Benzimidazole Yellow Properties

Colour Index name	Common name	Comments
PY 120	Yellow H2G	Medium shade. Good solvent fastness. Excellent lightfastness. Used in industrial finishes.
PY 151	Yellow H4G	Greener shade. Good solvent fastness. Excellent lightfastness. Industrial and refinish applications.
PY 154	Yellow H3G	Green shade but redder than PY 151. Good solvent fastness. Excellent lightfastness. Industrial and automotive refinish applications.
PY 156	Yellow HLR	Redder shade. Transparent. Good exterior durability in full shade and tint. All exterior coatings and refinish applications.
PY 175	Yellow H6G	Very green shade. Good solvent fastness. Excellent lightfastness. All exterior coatings and refinish applications.

3.5.5 Benzimidazolone Yellows

Figure 20 illustrates the structure of the organic yellows that fall into the classification of benzimidazolone yellows because each is an azo pigment derived from 5-acetoacetyl-aminobenzimidazolone.

The exceptional fastness to heat and the excellent weatherfastness of this class of pigments are attributed to the structural presence of the benzimidazolone group.

Used initially for coloring plastics, these pigments are finding increased use in coatings systems, where their excellent weatherfastness, heat stability, and fastness to overstriping are required (e.g. when formulating high quality industrial finished). Table 7 gives a summary of the properties of the benzimidazolone yellows.

P. Y. 120

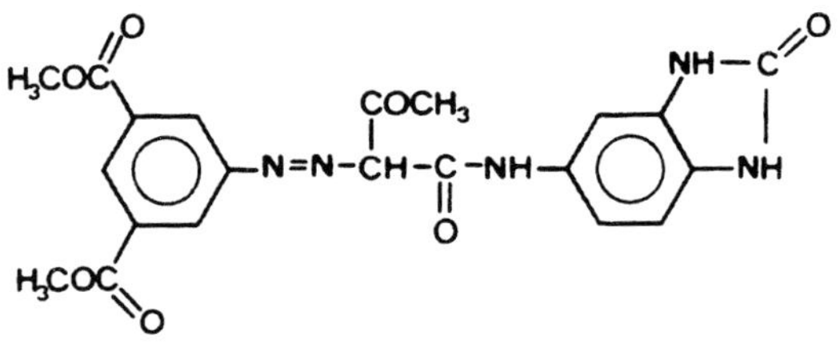

P. Y. 151

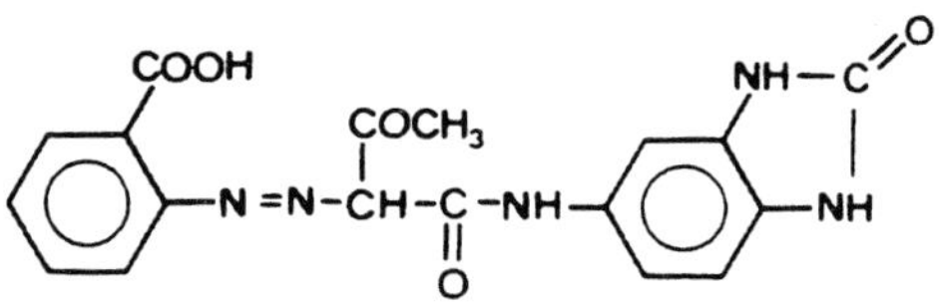

P. Y. 154

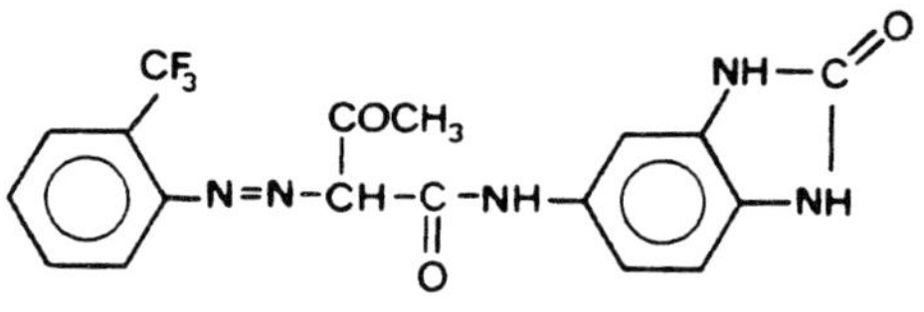

P. Y. 156

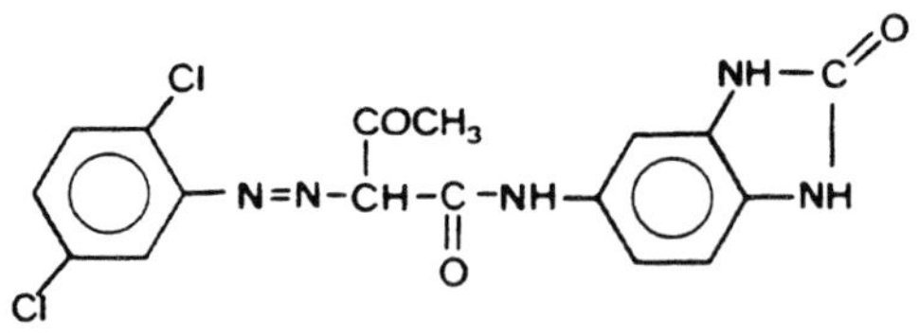

P. Y. 175

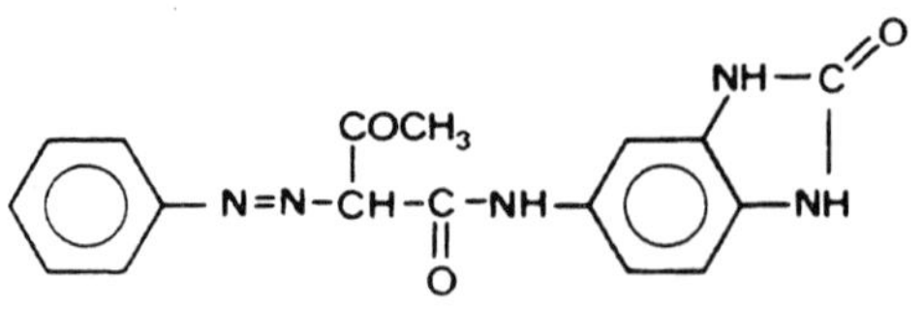

Figure 20 Benzimidazolone yellow structures.

Figure 21 Structure of heterocyclic yellows.

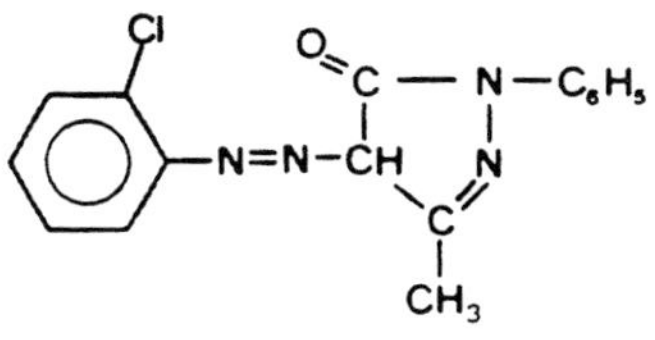

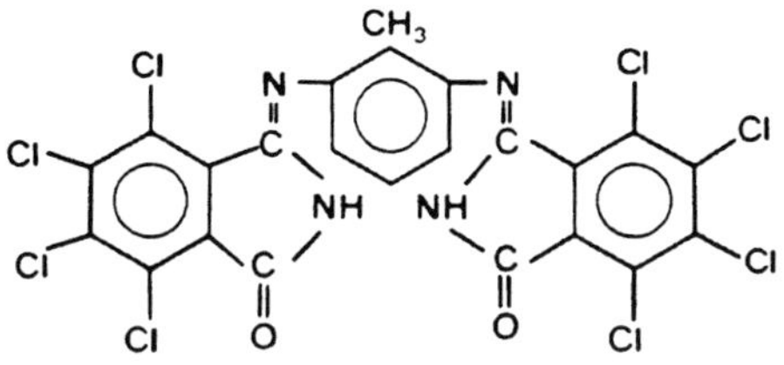

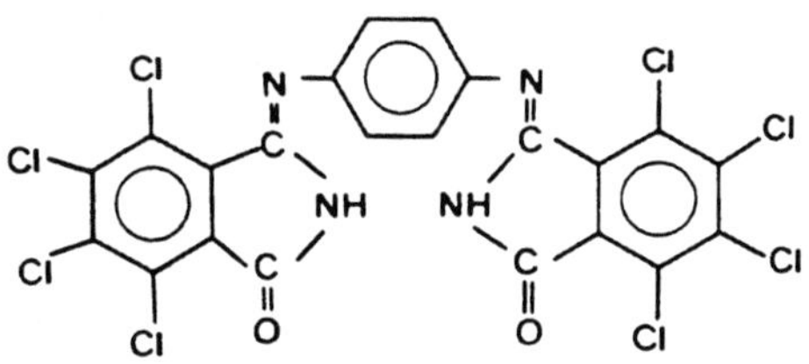

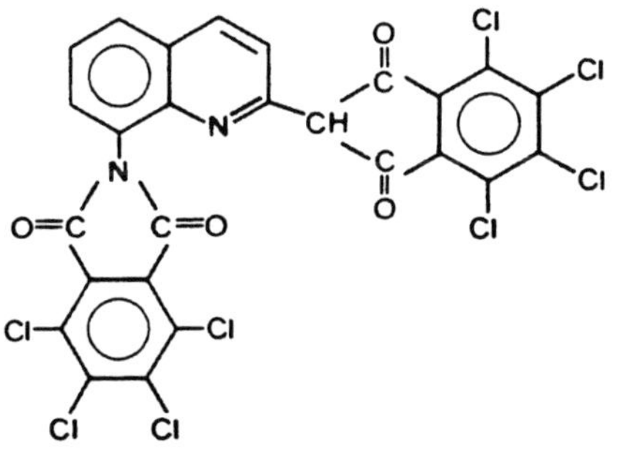

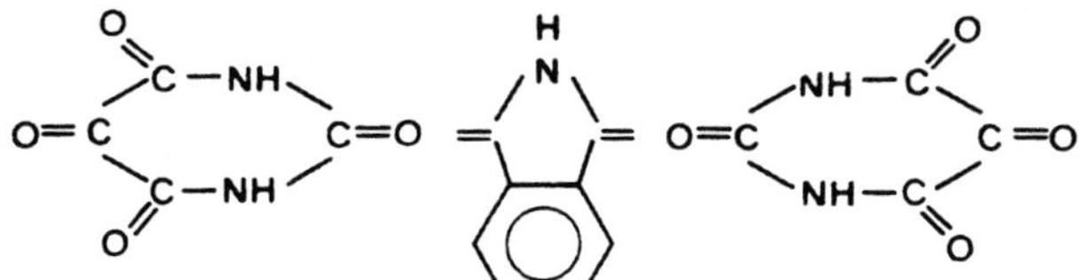

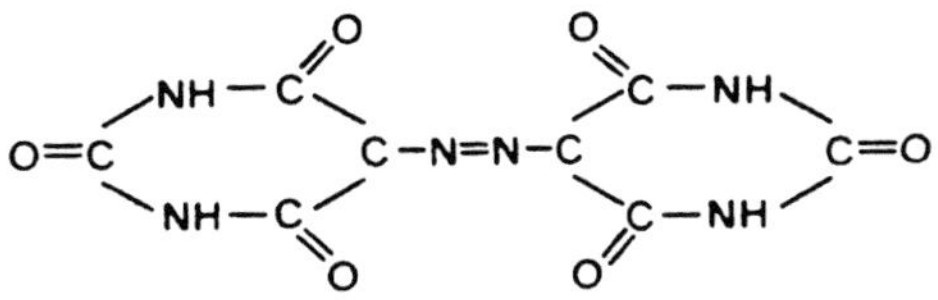

Table 8 Summary of the Properties of the Heterocyclic Yellows

Colour Index name	Common name /description	Comments
PY 60	Arylide yellow	Very red shade. Moderate light and solvent fastness. Trade sales, latex and masonry paints. Good acid and alkali fastness.
PY 101	Methine Yellow	Bright yellow. Highly transparent and exceptionally brilliant. Industrial finishes and specialty coatings. Only moderate bleed and alkali fastness.
PY 109	Tetrachchloroisoindoline	Green shade. Excellent brightness strength and durability. Automotive finishes.
PY 117	Greenish yellow complex of an azomethine	Excellent chemical, light, and heat fastness. Specialty finishes,
PY 129	Azomethine Yellow	Very green shade. Excellent chemical, light, and heat fastness. Industrial and specialty coatings.
PY 138	Green shade quinophthalone	Clean hue and excellent overall fastness properties. High quality industrial and automotive finishes.
PY 139	Red-shade isoindoline	Similar in masstone to medium chrome (PY 34). Excellent light and solvent fastness. Industrial and automotive coatings.
PY 150	Pyrimidine Yellow	Very green shade. Good heat and lightfastness. Industrial coatings.
PY 153	Red-shade Nickel Dioxine	Excellent fastness properties. Specialty coatings and baking enamels. Poor acid resistance.
PY 155	Azo Condensation Yellow	Green shade. Excellent overall fastness properties in full shade. Industrial and specialty coatings.
PY 173	Isoindolone Yellow	Very green shade. Excellent fastness properties. Industrial and specialty finishes.
PY 182	Triazinyl Yellow	Medium shade. Excellent fastness properties at masstone levels. Industrial finishes.

3.5.4 *Heterocyclic Yellows*

This class of pigments contains an assortment of yellows that all contain a heterocyclic molecule in their structure as presented in Figure 21. In spite of the apparent degree of complexity in the synthesis of such structures, these new high performance pigments continue to be introduced into the marketplace to fulfill the exacting demands of the consuming industries. Compounds such as Isoindoline Yellow (PY 139) and Quinophthalone Yellow (PY 138) are particular examples of such complex, novel pigments.

All these yellows offer the user additional high performance pigments that find applications in high quality coatings, where the end use can justify the economics of purchasing. Table 8 summarizes the properties of these heterocyclic pigments.

Part 4
Surface Coatings

64

Flexographic Inks

Sam Gilbert

Sun Chemical Corporation
Northlake, Illinois

1.0 INTRODUCTION

Flexographic inks, like most other inks, are based on colorants (pigment or dye), various combinations of resins, and various combinations of solvents (organic and water). The particular combination of materials that is necessary for a given ink is determined by the flexographic process,—its physical and mechanical attributes (e.g. press applications and plates), and by the end-use applications, which incorporate considerations of substrates and finished product requirements.

2.0 PROCESS

For printing any given package we must first look at the process. If, in fact, that package will be printed by flexography, then certain ink attributes are predetermined.

The flexographic inking system is based on either a three-roll system or a system that uses enclosed doctor blades. The three-roll system uses a rubber pickup roll that picks up the ink and transfers it to an anilox roll (engraved metal or ceramic roll). The ink fills the engraved cells and is then transferred to plates (either rubber or plastic), where the image is raised in relief from the background, nonimage area. A final transfer then occurs from the plate to the substrate.

In the enclosed doctor blade system, ink is applied directly to the anilox roll, with the doctor blade being used to remove excess ink. Ink is then transferred through the balance of the process as described above.

In both cases the amount of color (ink) to be applied to the substrate is determined by the viscosity of the ink. The amount of the ink carried by an anilox roll is dependent on the volume of the engraved cells in that roll. Higher viscosity with any given anilox will generally provide more color.

To provide easy ink transfer through the inking system, the actual viscosity of most inks is very low (0.1 —1.0 poise). The solvents used to make the inks and to adjust the viscosity

593

of the finished ink at the press are very volatile. Flexo presses can run at speeds from 50 to 2000 ft/min. The solvent used will vary in volatility from very fast (ethyl acetate) to very slow (glycol ethers), depending on press speed and ink formulation. The major solvents used are alcohols (ethyl, N-propyl), acetates (ethyl, N-propyl), and aliphatic hydrocarbons (heptane). Water containing inks constitute a large percentage of the inks used.

Another press consideration is the printing plate type that will be used. Rubber plates demand certain solvents, and photopolymer plates demand different solvent combinations. The solvent system used in the ink must be one that will not attack the plates but will provide good ink transfer from the plate to the substrate.

3.0 SUBSTRATE

The next consideration is the substrate to be printed: paper, polyethylene, polypropylene, foil, or some other material.

The attributes desirable in the package will determine the substrate. What is to be packaged? Is it a solid? A liquid? Is it perishable, corrosive, moisture sensitive, or heat sensitive? Does it need protection from light, oxygen, or grease? Will the product be processed in the package? Must it withstand cooking temperatures or storage as a frozen food? Will the Environmental Protection Agency (EPA), the Food and Drug Administration (FDA) or other government body be involved? How will the package be formed, filled, and sealed?

These conditions determine the package type and the substrates to be used. The chosen substrate will partly define the proper composition of ink vehicles.

4.0 COLORANTS

The choice of colors will complete the process of putting together an ink. There are two kinds of colorants available for use in printing inks—dyes and pigments. For the most part dyes do not offer the light and product resistance that pigments offer, and their use in flexographic inks has declined substantially.

The pigments used in manufacturing most flexo inks are classed as either organic or inorganic. They are colored, largely insoluble compounds. Each compound as found in nature or as manufactured produces its own unique color.

5.0 VEHICLE

The varnish portion of the ink, more properly referred to as the vehicle, is a composite of resins, various additives, and solvents.

Basically, the resin may be thought of as a virtually transparent component that suspends and supports the pigment on the printed surface. It is the vehicle that provides adhesion to the substrate and provides most of the properties required. For the proper formulation of inks, the resins must be soluble in the desired solvents, offer good transfer, and have good, fast solvent release.

When the ink is being manufactured, sufficient pigment is added to the resin system to obtain the desired color strength. The pigment must be dispersed so that agglomerates are eliminated and all pigment surfaces are wetted with the vehicle; otherwise adhesion of the ink will be poor and ink properties will suffer.

To obtain the correct color strength on the substrate, the viscosity of the ink must be controlled. High viscosity produces high color strength but poor ink properties. Low viscosity produces low color strength. The solvent or water used to adjust viscosity is only the carrier. It is added to the mixture of pigment and resin to make the combination flow. After the ink has been applied to the surface to be printed, the solvent must be removed. This is done by evaporation, normally in hot air ovens. The air volume and velocity are more important than temperature level.

6.0 FORMULATIONS

A typical formula for use on polyethylene film might be:

Pigment	15.0%
Nitrocellulose resin	5.0%
Polyamide resin	20.0%
Wax additive	1.0%
Ethyl alcohol	35.0%
N-Propyl acetate	9.0%
Heptane	15.0%

This formula is, of course, solvent based. Formulas largely based on water also are quite common for use on paper. A typical formula might be:

Pigment	15.0%
Acrylic resin	4.0%
Acrylic emulsion	60.0%
Solvent	5.0%
Wax	1.0%
Antifoaming agent	0.5%
Water	14.5%

Water systems for use on many different substrates are now being developed and will be quite common in the future.

65

Multicolor Coatings

Robert D. Athey, Jr.

Athey Technologies
El Cerrito, California

1.0 INTRODUCTION

A standard paint is a single shade attained by a mixture of many finely ground pigments so that one perceives a homogeneous shade. But natural products are seldom homogeneous in color. They are striped and spotted with dark and light regions of varying size. To reproduce this appearance in a painted surface, multicolor coatings are used, which yield a multicolor appearance in one coat. Zola[1] discusses multicolor coatings of a natural appearance, with striations from brush application and mottling from roller or spray application. Multicolor paints are used as interior architectural coatings for public buildings and have been used in automotive trunk interiors.

2.0 PRACTICE OF THE ART

Little is published on the practice of formulation or making the multicolor coating. Zola patents show multicolor formulations.[2–8] The major sources of information are the patents and publications,[9–23] and the commercial brochures describing the products used and specific formulation technology.

2.1 Continuous Phase

The continuous phase in a colloidal mixture is called the dispersion medium (e.g., water in milk, air supporting the droplets of water in a fog). The continuous phase must be a fluid that has a sufficiently low viscosity at high shear rates to allow easy application of the paint. If it is too stiff, the drag on the brush will tire the painter, or the fluid will not be properly atomized in a spraying application. However, at low shear rates the viscosity must be sufficiently high to slow the settling out or the agglomeration and coalescence of particles. Colloidal additives are used to protect such dispersions. Protective colloids are added to increase storage stability. Methyl cellulose, polyvinyl alcohol, and various nonionic surfac-

597

tants are the most often used protective additives. These materials adsorb onto the surface of the suspended materials to form a jellylike layer that inhibits coagulation. Another approach is to induce an electrical charge on the globule surface, so that the globules will repel each other. Ionic surfactants, salts, or charged clay particles are commonly added to induce surface charges.

The stabilization of suspended globules is absolutely necessary to prevent the coalescence and mixing of different colors. This ability must extend during the coating drying as well: the individual spots must have distinct boundaries with the next spot of a different color. This requires that the paint dry quickly.

2.2 Dispersed Phase

The suspended materials should not coalesce within the globule. The viscosity of the paint globules must be sufficiently high to maintain the pigment dispersed and to inhibit coalescence of adjacent globules. The pigment dispersion medium and the polymeric binder used must be insoluble in the continuous phase liquid. Pigment binder must be chosen to yield a sprayable system that does not cobweb during spraying. Usually the formulation requires a slowly evaporating solvent or a low molecular weight binder, or both. The polymeric binder should be curable to yield the solvent resistance and scrubbability needed in a wall coating.

2.3 Combining Dispersed and Continuous Phases

Formulations for dispersed and continuous phases are made separately. The phases should remain separate after combining, and that requires incompatibility between the components of separate phases. An ingredient in one phase should not be extracted by the other phase. Both phases should be of nearly equal density to minimize settling or creaming of the dispersed phase, which might encourage coalesence of the dispersed phase globules.

REFERENCES

1. J. C. Zola, *Mod. Paint Coatings*, p. 17, August 1975.
2. S. Grasko and S.S. Kane, U.S. Patent 3,370,024.
3. J. C. Zola, U.S. Patent 3,458,328.
4. J. C. Zola, U.S. Patent 3,573,237.
5. J. C. Zola, U.S. Patent 3,811,904.
6. J. C. Zola, German Patent 1,261,974; *Chem. Abstr.* 68,88261y.
7. J. C. Zola, U.S. Patent 2,591,904.
8. J. C. Zola, U.S. Patent 3,725,089.
9. J. L. Petty, U.S. Patent 3,058,932.
10. R. L. Sears, U.S. Patent 3,138,568.
11. D. K. Lesser, U.S. Patent 3,809,119.
12. F. F. Gourion, French Addn. 88224; *Chem. Abstr.* 67,65475q
13. C. H. Coney, U.S. Patent 3,356,312.
14. R. Rutkowski, German Patent 1,723,730; *Chem. Abstr.*, 80, 60081m.
15. Y. Shiota, et al., Japanese Patent 79, 87738; *Chem. Abstr.*, 92, 24442w.
16. Nippon Carbide Industrial Co., Ltd. Japanese Patent 7i6i, 2487; *Chem. Abstri.*, 85, 79787j.
17. D. A. Hilliard, U.S. Patent 3,038,869.
18. P. Talley, *Paint Ind.* 74(5), 10 (May 1959).
19. D. A. Hilliard, *Off. Dig.*, p. 4, June 1960.
20. E. W. Fllick, *Solvent Based Paint Formulations*, Park Ridge, NJ, Noyes Data Corp., 1977, p.415.

21. M. Schwefsky, U.S. Patent 2,658,002.
22. K. Nelson, U.S. Patent 3,024,124.
23. H. Hondo and Y. Nakamuro, U.S. Patent 3,360,487.

66

Paintings Conservation Varnish

Christopher W. McGlinchey

The Metropolitan Museum of Art
New York, New York

1.0 INTRODUCTION

To understand the properties of a varnish that satisfy the requisites of paintings conservation, it is necessary to introduce the rudiments of conservation and the artist's original reasons for varnishing. It is also important to mention that no varnish currently exists that satisfies all the needs of a paintings conservator, though great progress has recently been made.[1-3]

2.0 EFFECT OF VARNISHES

Picture varnishes used for the restoration of paintings are intended to create the same optical effect on the painting that was achieved by the artist but subsequently lost as a result of chemical and physical changes of the original varnish layer. An artist, for instance, relies on varnish to create the perception of the third dimension and the saturation of colors as well as to achieve some degree of glossiness in the process. The deleterious ramifications of a visibly aged varnish are not confined to the obvious yellowing of the colors; in addition, the lightening of dark passages and the darkening of lighter areas may result in a reduction in the range of color one visualizes. How these changes affect the painting depend on the style and subject matter of the painting as well as the materials artist used. It is however safe to say that in any circumstance these changes divert the purpose of the picture.

It is most often changes in the varnish layer that can obscure or confuse an artist's original intentions. Unfortunately, badly aged varnish is not the only factor that can upset the balance of a painting: some pigments are fugitive and thus fade; binding media can yellow; and furthermore, the various components of the painting as a whole respond differently to climatic changes in the environment, causing more brittle areas to crack under stress. Fortunately, the sensitive, well-trained conservator, through scholarly and scientific assistance, best understands how these inevitable acts of time affect the balance of a painting and how the work was originally intended to read. With the greatest caution, using fully reversible

601

materials and techniques, conservators can restore much of this balance, depending on the extent of change that has overcome the painting.

The effect of varnish on a painting cannot be ignored; in fact, there have been some movements in art history that preferred the mat appearance of an unvarnished picture, most notable the Impressionists and Cubists.[4-6] (The responsible conservator will honor the artists's choice, and refrain from ever varnishing such a picture.) On the other hand, an artist who chooses to varnish a painting has most certainly decided to do so from the inception of the painting process; this is known because varnish affects a painting so profoundly that when present, it enters the artist's thoughts as an integral component of the formulation for the desired affect.

Neglecting the optical effect of varnishes on paintings while attempting to use them for other reasons often produces calamitous results. For instance, conservators who apply varnish as "protective" coatings would cause drastic changes in tonal values that have been carefully achieved by the artist; consequently these values become far removed from what the artist originally intended. In this circumstance one would prefer an inert coating that does not affect the substrate optically: unfortunately, and for obvious reasons, no such coating exists. The protection of a painting is much more successfully managed through control of light levels, filtering out ultraviolet radiation, and minimizing large temperature and relative humidity fluctuations in the ambient climate.

3.0 TYPES OF VARNISHES

Briefly, artist's natural varnishes used for centuries have ranged from the very early use of egg whites as a surface coating to recipes of oil varnishes and to the more recent use of natural resins that are soluble in mineral spirit. (For a more complete treatment of artists' varnish recipes, see References 7 and 8.) The most common mineral spirit soluble natural resins are mastic and dammar, which are predominantly tetracyclic triterpenoids derived, respectively, from the tree families Dipterocarpaceae and Anarcardiacae.[9,10] Dammar has been well liked because of its ease of solvation and relative nonyellowing properties, which are attributed to its low acid value compared to other natural resins. Mastic has highly admired brushing and handling properties, which enable subtle optical effects, but it tends to yellow somewhat more than dammar. Both these resins are rather brittle, with poor mechanical strength compared to the durability of modern industrial coatings. However there is little need for this relative inflexibility to be a serious concern, since the museum environment is mild compared to the rigors of outdoor tests that most industrial finished must not fail.[3,9]

Mineral spirit soluble resins in theory are acceptable as conservation varnishes because they dry only by solvent loss (i.e., by a lacquer–type drying process) and not any subsequent polymerization: this property permits these resins to be removed at a later date, without affecting the paint layer. Solubility happens to be a criterion of great importance to the paintings conservator because any additions by the conservator's hand must be reversible and distinct from the original, though not necessarily in appearance. In practice these mineral spirit soluble resins oxidize, thus requiring more polar solvents for their solvation, thereby introducing a greater threat to leaching some components of the painting's binding media.[11,12] Even though chemical changes cause natural varnish resins to yellow, bloom, and embrittle, few synthetic resins can achieve the same optical qualities under many of the frequently encountered conservation circumstances; hence these natural varnishes are still in use by conservators today. (Recently, de la Rie along with this author demonstrated that in a UV–free environment with the use of a hindered amine light stabilizer and a benzotriazole, stabilization of dammar is possible.[13]

Thus far, the three most important criteria for conservation have been introduced. The criterion of utmost importance is respect for the artist's thoughts and achievements; second, any conservation must be reversible and not permanently bonded to the artist's original work; and finally, the material's stability must be considered. Stability is the single most distinguishing factor above the capabilities of an artist's varnish; whereas the artist is usually more concerned with optical effect than with stability, the conservator is accountable for both properties. With the scientific developments of polymeric materials of the twentieth century, a variety of synthetic resins have come into use by artists as sell as conservators; these materials possess remarkable stability or remarkable handling properties but unfortunately not both.

Synthetic varnish resins that dry only by solvent loss, not subsequent polymerization, range from low molecular weight oligomeric ketone resins to higher molecular weight polyvinyl acetates, and copolymers of alkyl methacrylates. Though the high molecular weight synthetic resins possess remarkable stability and durability, their optical properties are seriously compromised when compared to the former low molecular weight resins. On the other hand, the synthetic low molecular weight (e.g., ketone) resins often suffer from much of the same degradative problems as the natural resins do. Work has been carried out by de la Rie and Shedrinsky[14] to reduce the carbonyl groups of ketone resins to an alcohol functionality to prevent the homolytic bond cleavage associated with carbonyls (Norrish type I and II reactions) and furthermore to esterify a fraction of the —OH groups to reduce the brittleness of the resin.

It has been shown by de la Rie[4] that films cast from high molecular weight varnish resins have a low distinctness of image value when applied over a rough surface. Lee[15] has illustrated leveling properties of such coatings on rough substrates using cross–sectional photographic methods, achieving the same results. These experiments indicate that high molecular weight thermoplastic solution varnishes reproduce the contours of the substrate underneath and, if smooth, have the capability of being glossy; if rough, however, they will produce a mat effect as elucidated by the authors. Furthermore, de Witt[16] has shown from visual tests involving a group of conservators that varnishes possessing both a high refractive index and a low viscosity grade create an ideal optical surface, taking into consideration that a resin possessing these qualities should be of low molecular weight, and should have an oligocyclic structure, (the latter contributing to the material's high refractive index).

Since conservators are faced with the problem of having to create a specific saturation and gloss on a painting whose substrate may not be entirely uniform, it is strongly desired to use a varnish that would appear optically consistent over a rough or smooth surface. As discussed earlier, this is best achieved with the use of a low molecular weight resin. Recently, de la Rie and this author recommended some new materials to conservators for their experimentation, to determine whether the handling properties of these resins permit the desired optical effect. Ultimately, the appropriate optical effect is one of the essential qualities for understanding and appreciating paintings.

REFERENCES

1. E. R. de la Rie, "Stable varnishes for Old Master painting," Ph.D. dissertation thesis, University of Amsterdam, (1988).

2. R. L. Feller, N. Stolow, and E. H. Jones, *On Picture Varnishes and Their Solvents*, National Gallery of Art, Washington DC, 1985.

3. G. Thompson, "New picture varnishes," in *Recent Advances in Conservation*, G. Thomson, Ed., London: Butterworths, 1963.

4. E. R. de la Rie, "The influences of varnishes on the appearances of paintings," *Stud. Conserv.*, 32, 1–13 (1987).

5. John Richardson, "Crimes against the Cubists," *New York Review of Books*, June and October 1983.

6. R. Ruurs, "Matte or glossy? Varnish for oil paintings of the 17th century," *Intermuseum Conserv. Bull.*, 16, 1 (November 1985).

7. A. del Serra, "A conversation on painting technique," *Burlington Mag.*, 127(982), (January 1985).

8. R. Mayer, *The Artist's Handbook of Materials and Techniques*, New York: Viking, 1982.

9. C. Mantell, *The Technology of Natural Resins*, New York: Wiley, 1942.

10. J. Mills and R. White, "Natural resins: Their art and archaeology, their sources, chemistry and identification," *Stud. Conserv.*, 22, 12–31, (1977).

11. E. R. de la Rie, "Photochemical and thermal degradation of films of dammar resin," *Stud. Conserv.*, 33 (2), 53–70 (1988).

12. N. Stolow, "Some investigations of the actions of solvents on drying oil films," *J. Oil Chem. Colour Assoc.*, 40, 337–402, 438– 499 (1957).

13. E. R. de la Rie, and C. McGlinchey, in preparation; to be published in *Studies in Conservation*.

14. E. R. de la Rie, and A. Shedrinsky, "The chemistry of ketone resins and the synthesis of derivative with increased stability and flexibility," *Stud. Conserv.*, in press.

15. S. Lee, "An investigation into the contouring properties of some picture varnishes," Master thesis, Hamilton Kerr Institute, University of Cambridge.

16. E. de Witte et. al., "Synthesis of an acrylic varnish with a high refractive index," Ottawa: *ICOM Committee for Conservation*, 16 (4), 1–7, (1981).

67

Peelable Medical Coatings

Donald A. Reinke

Oliver Products Company
Grand Rapids, Michigan

1.0 INTRODUCTION

Peelable coatings used on packaging materials for disposable medical devices first became practical in the late 1960s with the introduction of a spunbonded polyethylene paperlike material (Tyrek) by E. I. DuPont Company. Previously, disposable or single-use medical devices were packaged in tear-open packages. Medical device manufacturers were only beginning to be concerned about the particulates generated when such packages were torn open. The industry was not regulated at this time, and the concern for product liability was not to come for several years.

A major change took place in the way medical devices were packaged when DuPont began marketing Tyvek as a bacterial barrier material for packaging. Tyvek is a unique material that meets all the critical requirements for medical device packaging. It is a good bacterial filter, that is water resistant, as well as puncture and tear resistant. Being made from polyethylene, it is stable during both gas and radiation sterilization. It is also nontoxic, hence could meet the same U.S. Pharmacopoeia requirements listed for medical devices.

With the availability of an acceptable packaging material, peelable coatings could be developed to seal this material to plastic films to form pouches or to thermoformed plastic trays as a lid.

2.0 COLD-SEAL COATINGS

Peelable coatings for medical devices generally fall into two groups. First, cold-seal or cohesive coatings, made from latex, are sealed by pressure alone and will seal to themselves only. This type of coating has been used for self-sealing envelopes, for candy wrappers, and in a variety of industrial packages. Peelability with cohesive coatings is achieved by coating a release layer between the adhesive and the substrate. This allows the adhesive to peel away from the substrate without leaving a residue or tearing fibers.

605

3.0 HEAT-SEAL COATINGS

The second type of coating, heat-seal adhesives, are more complex in their requirements than the cold-seal adhesives. The latter (and) generally are applied only in the seal area, seal only to themselves. The heat-seal adhesives are normally coated over the entire surface of the web and have to seal to a wide range of plastics. Heat and pressure are required to form a seal.

Many different types of heat-seal coating are used for medical packaging, but all must meet the same requirements. The most important requirement is that they be stable after sealing, so that a sterile package is assured for several years. Disposable devices are sterilized after packaging and must remain sterile until used. They must also have a porosity of under 200 seconds Gurley, to prevent the buildup of excessive internal pressure in the package during a normal gas sterilization cycle. The coating must be nontoxic and must meet the same requirements listed in the U.S. Pharmacopoeia for the plastics used in the device being packaged. Other desired attributes are good hot tack, clean color, enough strength to pass shipping tests, and good transfer of the adhesive onto the seal surface when the package is peeled open.

The coatings used on heat-sealable Tyvek and papers come in many forms. They can be solvent-type lacquers applied in a grid pattern to maintain porosity, water-based dispersions dried very carefully so that they do not melt, or hot melt adhesives applied in small dots.

The lacquer-type adhesives are applied by the gravure process at a rate of 3-6 pounds per 3000 square feet. The adhesive is applied in lines that cross at an angle, leaving an uncoated area between them resulting in good porosity. Also the lines break up the peel forces to give better peelability.

The water-based dispersions are based on ethylene-vinyl acetate copolymer dispersions blended with waxes and tackifiers to form a blend of particles. This blend is coated on the web by air knife coating or by gravure. The coating must be dried at a low temperature to keep the coating from melting and forming a film. The porosity is achieved by having large particles in the blend and by mud cracking of the coating as it dries. The dried coating weight is 5-8 pounds per 3000 square feet.

Hot melt coatings are a blend of an ethylene-vinyl acetate copolymer, terpene resins, and wax. These are melted together and applied by a gravure roller engraved with 50-70 cells per inch. The small dots of adhesive that result from this have a small uncoated area between them that makes the coating porous. The coating is applied at a rate of 9-12 pounds per 3000 square feet.

The optimum amount of coating will vary for each type of adhesive. The controlling factor is the melt viscosity of the adhesive at the sealing temperature. If the viscosity is low, a portion of the adhesive will be pushed into the substrate, reducing the amount of adhesive available to absorb stresses and to transfer during peeling. Good adhesion to a smooth surface with coated paper or Tyvek can be achieved with 5-12 pounds of adhesive per 3000 square feet. An accepted rule of thumb in the coating industry is that 15 pounds per 3000 square feet is equal to 0.001 inch of thickness.

While there are no regulations in this country requiring nontearing packages, the medical device industry places a high priority on this feature. Another desirable feature of the coating, which goes along with peelability, is the ability of the coating to transfer to the plastic surface to which it is sealed. The transfer pattern must be unbroken when the package is opened. This is interpreted to mean that the seal integrity has been maintained. A clear track through the seal area will cause the package to be rejected because there is a path for bacteria to enter the package.

Formulating a coating that is nontoxic and strong, with good hot tack and good adhesion to a variety of plastic surfaces, and is able to peel with a controlled peel strength, is a difficult task.

Several methods have been used to achieve peelability. The use of primer coats between the adhesive and the substrate is the one most often chosen; this creates as a parting layer between the adhesive and the substrate. Release coatings on the substrate surfaces are also used where paper instead of Tyvek is used.

Most people in the industry feel that the best method is to have an adhesive with a controlled cohesive strength. When the package is peeled open, the adhesive will split with a controlled force that is below the delamination strength of the substrate. By using this method, the surface fibers of the substrate are not raised or broken.

Cohesive failure of the adhesive can be achieved by formulating an adhesive that has two phases, one a strong adhesive to hold the package together and the other a weak friable material that breaks up the structure of the first phase. By varying the percentages of the two phases, the cohesive strength of the coating can be controlled within narrow limits.

The nature of peelable medical packaging materials is currently undergoing a change. There is a drive to reduce the cost of medical devices, which is causing a shift to paper and away from Tyvek. Also environmental concerns are pressuring medical device companies to use some method other than ethylene oxide gas sterilization. The requirement for a peelable package however still remains strong.

Further information on medical device packaging can be found in the proceedings of the Technical Association of the Pulp and Paper Industry's Coatings and Laminations conferences during the years from 1980-1987.

68

Conductive Coatings

Raimond Liepins

Los Alamos National Laboratory
Los Alamos, New Mexico

1.0 INTRODUCTION

In 1986 sales in the coatings industry exceeded $10 billion, and production approached a billion gallons.[1] The breakdown of sales was $4.1 billion for architectural coatings, $3.5 billion for industrial coatings, and $2.4 billion for specialty coatings. Conductive coatings—a minuscule part of these trade sales—have been used both as industrial coatings and as specialty coatings. Regulations of the Federal Communications Commission (FCC), in Docket No. 20780, which regulates electromagnetic emissions from computing devices, have provided a strong impetus for the commercial development of conductive polymeric materials (including coatings and paints). Since October 1, 1983, it has been necessary for any computing device that generated signals or pulses in excess of 10 kHz to comply with the emission standards set forth in the docket. Although conductive polymeric coatings have made inroads in areas where metallic coatings previously were used, progress has been slow.

A product related to conductive coatings is metallized plastic. The most important commercial processes for metallizing plastics are electroless plating, metal spraying, sputtering, and vacuum metallizing. The first commercial plating of plastics was recorded in 1905.[2] Metallizing of plastics occurred during World War II, and large-scale production started in the early 1960s. All these processes are now multimillion-dollar industries. Large quantities of plastics are metallized each year, with automotive items making up more than 60% of the market on a plated area basis.[3]

There are various reasons for metallizing plastics. In the automotive industry, metallized plastic combines the consumer appeal of metal with light weight. Electroless copper metallization is an indispensable part of the modern electronics industry. Printed circuit boards use electroless copper to coat nonconductive plastic surfaces to define the circuit patterns. Zinc arc and flame-spray techniques provide electromagnetic interference shielding on many plastics. The plastics that account for most of the substrates metallized are acrylonitrile–butadiene–styrene (ABS), polypropylene, polyphenylene oxide, epoxies, phenolics, polyimides, and polyesters. The commercial process for metallization of plastics merits separate discussion and is not further considered in this chapter.

Polymers (coatings) with conductivities greater than $1(\Omega cm)^{-1}$ are defined as conductive polymers (also metallically conducting plastics, synmetals).[4] Unfortunately, the literature is not clear-cut and often materials that are semiconductors with conductivities less than $1(\Omega cm)^{-1}$ are also called "conducting." This chapter discusses only coatings (polymers) and their applications requiring a conductivity of at least $1(\Omega cm)^{-1}$.

2.0 TYPES OF CONDUCTIVE COATINGS

We discuss, metallic, filled polymeric, polymeric, and organometallic conductive coatings.

2.1 Metallic

Substrates are coated with conductive coatings mainly for functional and rarely for decorative purposes. The most important commercial metallic coatings are nickel, copper, chromium, gold, palladium, platinum, silver, zinc, cadmium, iron, cobalt, tin, and lead.[3] The important commercial processes for metallizing plastics have already been mentioned and recent developments in this area are summarized in Section 5.0, "New Developments."

Some of the special techniques for metallizing plastics are summarized below.

Details of electroless metallizing technique of composites and/or thermoplastics that allows the repairing of damages areas in the field have been published.[5]

Surface metallization of molded liquid-crystal polymer parts has been accomplished.[6]

To increase the adhesion of a metal coating to an organic substrate, the substrate is heated to 0.6-0.8 T_{cure} of the substrate and held at that temperature while the metallic coating is deposited. This procedure provides for optimum intermixing between the metal atoms and the substrate, thus producing maximum adhesion.[7]

A simply technique for plating ABS plastics consists of mixing a mixed catalyst solution for simultaneous sensitization and activation of the surface.[8]

Adhesion between polyethylene and/or polypropylene and aluminum is improved if the plastic surface is pretreated with an oxygen ammonia, or sulfur dioxide low pressure plasma.[9,10]

An electroless metal plating of plastics involves a plastic reducible by one metal salt followed by a redox reaction between the reduced plastic and the main group metal salt to produce the plated plastic.[11]

A process for metallizing shaped articles of filled plastics with cuprous oxide has been described. The surface of the filled articles is subjected to a reducing agent such as a borohydride to convert the cuprous oxide to a conductive free-metal surface.[12,13]

2.2 Filled Polymeric

Plastic coatings and paints can be made conductive by adding metallic fillers. Typical metal powders that have been used in coatings, paints, and lacquers are nickel, copper, gold, aluminum, iron, cobalt, palladium, zinc, and platinum, as well as various of their alloys. A step beyond the use of powdered, more or less spherical, fillers is the incorporation of conductive fillers with high aspect ratio: length-to-diameter ratio of more than 100. This family of materials includes such fillers as metallized glass fibers, graphite fibers, metallized carbon fibers, quick-quenched aluminum flakes or fibrils, and fine diameter (6–8 μm) steel fibers.

The most common metallic conductor, copper, is not used because it oxidizes within the plastic and impairs its physical properties.

Depending on particle form (sphere, flake, fiber), size and orientation, there is a certain critical concentration at which conductivity increases drastically (many orders of magnitude). To achieve significant conductivity with more or less spherical fillers (carbon black, metal powders), volume percent loadings above 40 are required. A striking difference in conductivity is observes when a high aspect ratio particle of the same material is used. Thus, with metal or metallized glass fibers of aspect ratio of 750 as fillers, metallic conductivities can be achieved at as low as 3 vol %.[14]

At critical concentration, the filler can form a continuous phase through the matrix in the form of microscopic conductive channels (percolation). The use of high aspect ratio conductive fillers, when they are properly oriented, will always produce a superior conductivity at a lower volume percent concentration than comparable spherical fillers. However, for these materials to have high efficiencies, they also must survive the preparation and application processes. For example, the initial high aspect ratio of a very brittle material may quickly reduce to powder in a high shear mixing environment, if proper techniques are not developed for incorporating such brittle materials.

Some specific examples of uses for filled polymeric coatings usage are summarized below.

Organic solvent-based coatings of acrylic or urethane resins containing either nickel, silver, copper, or graphite powders are used in the electromagnetic and radio frequency interference applications. For cost reasons the nickel-filled acrylic coatings are most often used. Lacquers based on silver give the highest conductivities and, therefore, the best screening attenuation, but they are the most expensive. Copper lacquers are cheaper and the attenuation is similar to that of silver; the oxidation problem of copper lacquers has been solved. Lacquers containing graphite are the cheapest, but their performance is also the least effective.[15,16]

An initiation layer for the electroless deposition of copper, nickel, Ni-P, Ni-B alloy coatings has been described. The layer was formed by applying a polyester or its polymerization material solution containing such metal powders as nickel, copper, silver, aluminum, cobalt, palladium, zinc, or platinum.[17]

Large differences in the specific conductivity of filled conductive plastics (coatings were demonstrated between materials filled with 15 vol % aluminum alloy fiber, $\sigma = 0.25$ $(\Omega \cdot cm)^{-1}$, and 15 vol % brass fiber, $\sigma = 170$ $(\Omega \cdot cm)^{-1}$.[18]

Techniques have been developed for generating and incorporating in situ metallic filler in organic coatings or paints. For example, the use of $AgNO_3$ and erythorbic acid, formaldehyde, paraformaldehyde, or terephthalic aldehyde will result in the generation of metallic silver in a chosen organic coating or paint system.[19,20]

2.3 Polymeric

Although the unfilled, molecularly conductive polymeric materials have attracted much attention recently, the successful developments in surface-metallized plastics make it unlikely that inherently conductive polymers will be a significant factor in the marketplace in the near future. Also, the commercially entrenched, filled conductive polymeric materials seem to become more and more efficient as time goes by.

One projection of worldwide markets for filled and inherently conductive polymers is as follows:[21]

	1982	1987	1992
Filled conductive, lb	1×10^6	1×10^7	1.5×10^7
Inherently conductive, lb	0	1×10^5	1.5×10^6

This projection, made in the mid-1900s, is clearly very optimistic about the prospects of inherently conductive polymers.

The major problems of molecularly conductive polymeric materials (coatings) are still their instability under environmental conditions and difficult or impossible processing from solution or melt. Consequently, they have to be synthesized and treated in their final form of application. In the undoped state, the basic polymers are either semiconductors or insulators. It is only after they are doped (oxidized or reduced) chemically or electrochemically that they become metallically conductive.[4] Doping can be done by means of electron-accepting (oxidizing) agents, such as AsF_5, iodine, bromine, SbF_5, $NOPF_6$, $Al(ClO_4)_3$, or $HClO_4$; or electron-donating (reducing) agents, such as lithium, sodium, potassium, or sodium naphthalide. Unfortunately, most of the doped materials generated are extremely reactive because of the carbonium ions and carbanions formed upon doping and thus are not stable under ambient conditions.

Recently, some progress has been made in this area. The first environmentally stable, conducting polymer was polypyrrole. Coatings or films of it were prepared electrochemically from aqueous solutions of sodium n-alkylsulfate, sodium n-alkylsulfonates, and disodium 1,10-decane disulfonate or toluenesulfonic acid. The products possessed conductivities as high as 230 $(\Omega\cdot cm)^{-1}$. The films possessed good mechanical properties and environmental stability in excess of 5 years (thus far tested).[22,23]

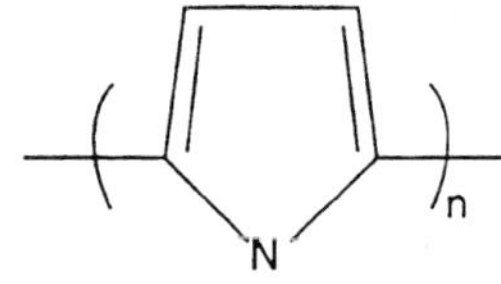

Another heterocyclic, conducting, environmentally stable polymer is polythiophene. Polythiophene and its derivatives can be synthesized and doped chemically or electrochemically to conductivities as high as 100 $(\Omega\cdot cm)^{-1}$.[24,25]

Environmentally stable polycarbazole was synthesized and I_2 doped to a conductivity of 5 $(\Omega\cdot cm)^{-1}$.[26]

A polymer that is attracting more attention recently is polyaniline. It can be prepared both chemically and electrochemically and when doped with various acids in aqueous solutions, it achieves conductivities as high as 10 $(\Omega \cdot cm)^{-1}$.[27–29]

A recent and exciting development is the preparation of aqueous polypyrrole and polyaniline latex.[30–33] Spherical, submicrometer polypyrrole latexes have been prepared using poly(4-vinyl pyridine)-based steric stabilizer. The use of such stabilizers allowed the polypyrrole particles to be controllably and reversibly aggregated or stabilized, depending on the pH of the dispersion medium. The solid state conductivity of films can be as high as 2 $(\Omega \cdot cm)^{-1}$.

Except for the electrochemically prepared polypyrrole in a continuous sheet form, obtainable in developmental quantities from Badische Anilin- und Soda-Fabrik (BASF), Ludwigshafen, West Germany, none of the other materials have passed the research stage.

2.4 Organometallic

Although most of the current interest has been in unfilled molecularly conducting polymer systems, conductivity in metal containing polymers has also been a topic of interest. Almost all the molecularly conducting polymers are doped systems and, thus, may contain metals from the doping agent. Such systems, however, are air and moisture sensitive and difficult to handle. In general, organometallic polymers do not have these instability problems, but their processibility is still a problem.

A new class of low-dimensional materials—for example, polymeric metallophthalocyanines with such metals as aluminum, chromium, tin, and gallium at the center of the ring—has been synthesized and made conducting [~1 $(\Omega \cdot cm)^{-1}$] by halogen doping (partial oxidation).[34,35]

Various transition metal ions have been coordinated with conjugated ligands to synthesize polymers with metal ions within the main chain.[36] Ligands such as tetrathiosquarate, tetrathionaphthalene, tetrathiafulvalene, and tetrathiooxalate have been used. The tetrathiooxalate complexed with nickel ions gave linear polynickel tetrathiooxalate oligomers with conductivities as high as 20 $(\Omega \cdot cm)^{-1}$.

A completely different approach to depositing conducting organometallic coatings involves the use of a low pressure plasma (LPP) environment. The LPP environment may be used to deposit a polymeric organometallic coating (or powder) from organometallic monomers, [37–42] or it may be used to convert a deposited organometallic coating into a metallic one.[20,43,44] The metals introduced in the organometallic coatings by LPP/organometallic monomers were iron, tin, mercury, tantalum, lead bismuth, and metal coatings by the LPP posttreatment were gold, platinum, palladium, silver, and lead. The generation of metal surface coatings from certain organometallic coatings can be also achieved by thermal means (controlled pyrolysis).[20] The advantage of the LPP process is that it permits a metallic coating to be formed on a heat-sensitive substrate without the use of elevated temperatures. The process also permits formation of adhering gold and platinum coatings otherwise difficult to deposit on plastic substrates.

None of the conducting organometallic coatings or their deposition processes have gone beyond the research stage. However, the conducting organometallic coatings effort is very new compared with the other types mentioned before.

3.0 COMMERCIALLY AVAILABLE CONDUCTIVE COATINGS

Whereas the metallized plastics effort is a multimillion-dollar industry, the commercial application of conductive polymer coatings as a paint, lacquer, ink, adhesive, or a solution of some kind forms a very small industry indeed. Below are summarized several typical products available commercially.

A proprietary aluminum containing paint, AG 9680, manufactured by A.I. Technology, Inc., Princeton, New Jersey, is claimed to approach the shielding effectiveness of 70–75 dB, an effectiveness similar to that of pure silver.

A silver-filled silica matrix elastomer, Aremco-Shield 615, has been developed by Aremco Products, Inc., Ossining, New York. This material has been formulated into a conductive paint that can be applied by either brush or spray; it cures at room temperature and bonds to metals, glass, and plastics.

A silver lacquer (Eccocoat CC-2) and an elastomeric, silver-filled, conductive coating (Eccocoat CC-40) have been developed by Emerson and Cuming, Canton, Massachusetts. Both the lacquers and the elastomer coatings may be applied by dipping, spraying, silk screening, roll coating, or brushing. In most cases a simply spray coat is adequate to produce a highly conductive surface [up to 20 $(\Omega \cdot cm)^{-1}$] using the air-dry method. Oven curing will give improved conductivity.

A rather extensive series of conductive coatings under the trade name Evershield has been developed by the E/M Corp. of West Lafayette, Indiana. The series of products consists of a graphite-filled acrylic resin system, EC-G-102, intended mainly for applications for electrostatic charge dissipation; a high performance nonoxidizing copper-filled acrylic resin system, EC-C-301, is suitable for spray gun application. The coating has an attenuation performance of 50–70 dB at 10–1000 MHz. A popular nickel-filled acrylic resin system, EC-N-501, easily paintable and with superior adhesion characteristics for a wide variety of plastic substrates, is also available. It has an attenuation performance of 50–60 dB at 30–1000 MHz.

4.0 APPLICATIONS

4.1 Shielding from Electromagnetic Interference

The advent of the FCC Docket 20780, which regulates electromagnetic emissions from computing and communication devices used in industrial and residential locations, has really provided a stimulus for the industry to come up with cost-effective methods for limiting the level of electromagnetic interference (EMI). To meet the set standards, the manufacturers have adopted a variety of methods for controlling EMI. These methods have ranged from redesigned basic circuitry to incorporation of conductive shielding materials in the devices. The incorporation of conductive shielding may take different routes[45]:

Use of metal enclosures.
Metallic foil tapes.
Metal coatings on plastic enclosures.
Conductive paints on plastic enclosures.
Conductive plastic enclosures.
Flexible laminates with metal foil.

The shielding effectiveness of a homogeneous medium, such as a conductive coating, is related to the propagation of the electromagnetic field through the coating. The shielding effectiveness is directly related to the electronic and magnetic properties of the coating; therefore, for best shielding effectiveness, materials with both high relative magnetic permeability and high electrical conductivity are necessary. Thus, it has been found that the various metals and alloys form the following "series" in decreasing order of effectiveness:

Ag > Cu > Au > Al > Zn > brass > Ni > Sn > steel > stainless steel

Figure 1   Stealth fighter.

Currently, the most cost-effective and the most problem-free materials for shielding are claimed to be nickel-filled acrylic or polyurethane conductive paints.[46]

4.2 The Stealth

One of the more glamorous applications of conducting coatings has been in the "stealth" technology. Until the Pentagon revealed the top-secret Stealth (Fig. 1) fighter on November 10, 1988, little could be said about its materials technology. The so-called stealth materials can provide a minimal radar profile for military aircraft and naval vessels. This profile is achieved primarily by a combination of geometric design and materials properties. The active components consist of several classes of materials: carbon fiber composites, radar-absorbing coating, ferrite layers, and interference layers in the form of certain pigment-filled polymer coatings.[21] From other EMI work, it had been known for years that incorporating short, electrically conducting fibers (e.g., stainless steel) into thermoplastics in as low as 0.5 vol % fraction gave sufficiently high levels of dc conductivity for electromagnetic shielding purposes.[47] One of the radar-absorbing coatings has been iron fibril-filled elastomeric epoxy.[48] Presumably, such coatings are most effective against short wavelength radars, and the longer the wavelength, the thicker the required coating.

4.3 Miscellaneous

A novel application of conducting coatings is to use them as ambient-responsive elements for remotely readable indicator devices.[49]

Spacecraft charging problems have required efforts in the separation, shielding, and filtering to decouple discharges from susceptible circuits, and this decoupling involves the use of conductive surface coatings.[50]

Conductive coatings are also used on waveguides, radio frequency reflectors, and radar dishes, and as radiofrequency shielding on capacitor plates.[51]

Electrochemically generated, conducting polyheterocycle coatings are good candidates for electrochromic displays.[52,53]

Aluminum is being used on video disks and in large-area solar cells.[54]

Conductive paints are used to enhance the electromagnetic field attenuation of equipment rooms or entire buildings.[55]

Conducting coatings are used to coat rusty areas and bolts to suppress background noise in high frequency transmitters and receivers on ships and aircraft.[19]

Conductive coatings are candidates for passivating layers against photocorrosion of photoelectrodes and modifiers of electrodes for electrochemical reactions.[56]

5.0 NEW DEVELOPMENTS

The field of laser direct-write metallization of organometallic films has been expanding rapidly as the potential for accomplishing one-step pattern definition and metallization is realized.[57,58]

Recently, ion implantation has been conducted on a variety of polymers with a variety of ion species. The result has been significant increases in electrical conductivity. Thus, for example, Kapton H film implanted with 1 MeV $^{15}N_2^+$ at several doses reached a conductivity of 1 $(\Omega\cdot cm)^{-1}$. It was suggested that the ion implantation led to graphite particle formation, probably along the penetrating ion tracks.[59]

Conducting silver lines have also been generated by focusing a laser beam on a polyimide film containing dispersed silver nitrate.[60,61]

A novel chemical method was used to generate an Ag–Hg alloy on the surface of a polyacrylamide film. The technique involved holding the polyacrylamide aqueous solution with $AgNO_3$ in a mercury-saturated atmosphere.[62]

A recent symposium on laser and particle-beam chemical processing for microelectronics described laser direct writing of aluminum, copper, palladium, and carbon.[63–67]

REFERENCES

1. T. J. Miranda, in *Applications of Polymers*, R. B. Seymour and H. F. Mark, Eds. New York: Plenum Press, 1988.
2. H. Narcus, *Trans. Electrochem. Soc.*, **88**, 371 (1945)
3. J. I. Kroschwitz, Ed.-in-Chief, *Encyclopedia of Polymer Science and Engineering*. Vol. 9. New York: Wiley, 1985.
4. C. C. Ku and R. Liepins, *Electrical Properties of Polymers, Chemical Principles*. New York: Hanser, 1987.
5. G. J. Shawham and B. R. Chuba, in "Materials—Pathway to the future," *SAMPE*, **33**, 1617 (1988).
6. Y. Ikenaga, T. Kanoe, T. Okada, and Y. Suzuki, *Ausz. Eur. Patentanmeld. I*, 3(12), 485 (1987).
7. P. S. C. Ho, P. O. Hahn, H. Lefakis, and G. W. Rubloff, *Ausz. Eur. Patentanmeld. I*, 2(29), 1367 (1986).
8. S. John and N. V. Shanmugam, *Met. Finish.*, **84**(3), 51 (1986).
9. R. Liepins, Camille Dreyfus Laboratory Annual Report, Research Triangle Institute, Research Triangle Park, NC, Dec. 31, 1970, and unpublished results after 1970.
10. J. Fredrich, I. Loeschcke, and J. Gahde, *Acta Polym.*, **37**(11–12), 687 (1986).
11. L. J. Krause, "Electroless metal plating of plastics," U.S. Patent 4,600,656 (July 15, 1986).
12. R. Cassat, "Metallizing electrically insulating plastic articles," U.S. Patent 4,590,115 (May 20, 1986).
13. R. Cassat and M. Alliot-Lugaz, "Metallization of electrically insulating flexible polymeric films," U.S. Patent 4,564,424 (Jan. 14, 1986).
14. D. E. Davenport, in *Conductive Polymers*, R. B. Seymour, Ed. New York: Plenum Press, 1984, p. 39.

15. J. E. McCaskie, in *Modern Plastics Encyclopedia 1985–1986*, **62**, No. 10A, J. Aranoff, Ed. New York: McGraw-Hill, 1985, p. 381.
16. V. Krause, *Kunsts. Plast. (Munich)*, **78**(6), 17 (1988).
17. R. A. Baldwin, A. J. Gould, B. J. Green, and S. J. Wake, British Patent 2,169,925A (July 23, 1986).
18. A. Yanagisawa, H. Koyama, K. Suzuki, and T. Nakagawa, *Proceedings of the 31st International SAMPE Symposium*, April 7–10, 1986, p. 1583.
19. J. C. Cooper, Ru. Panayappan, and R. C. Steele, in *Proceedings of the 1984 IEEE National Symposium on Electromagnetic Compatibility*, April 24–26, 1984, San Antonio, TX, p. 233.
20. R. Liepins, B. Jorgensen, A. Nyitray, S. F. Wentworth, D. M. Sutherlin, S. E. Tunney, and J. K. Stille, *Synth. Met.*, **15**, 249 (1986).
21. A. Wirsen, "Electroactive polymer materials." National Technical Information Service Publication PB86-185444, January 1986.
22. W. Werner, M. Monkenbusch, and G. Wegner, *Makromol. Chem., Rapid Commun.*, **5**, 157 (1984).
23. H. Naarmann and N. Theophilous, *Synth. Met.*, **22**, 1 (1987).
24. T. Yamamoto, K. Sanechika, and A. Yamamoto, *J. Polym. Sci., Polym. Lett. Ed.*, **18**, 9 (1980).
25. S. Hotta, T. Hosaka, and W. Shimotsuma, *Synth. Met.*, **6**, 317 (1983).
26. S. T. Wellinghoff, Z. Deng, J. Reed, and J. Racchini, *Polym. Prepr.*, **25**(2), 238 (1984).
27. A. G. MacDiarmid, J. C. Chiang, M. Halpern, W. S. Huang, S. L. Mu, N. L. D. Somasiri, W. Wu, and S. I. Yaniger, *Mol. Cryst. Liquid Cryst.*, **121**, 173 (1985).
28. J. C. Chiang and A. G. MacDiarmid, *Synth. Met.*, **13**, 183 (1986).
29. W. S. Huang, A. G. MacDiarmid, and A. P. Epstein, *J. Chem. Soc. Chem. Commun.*, **1987**, 1784.
30. R. B. Bjorklund and B. Liedberg, *J. Chem. Soc. Chem. Commun.*, **1986**, 1293.
31. S. P. Armes and B. Vincent, *J. Chem. Soc. Chem. Commun.*, **1987**, 289.
32. S. P. Armes, J. F. Miller, and B. Vincent, *J. Colloid Interface Sci.*, **118**(2), 410 (1987).
33. S. P. Armes, M. Aldissi, and R. D. Taylor, "Non aqueous polypyrrole colloids," LA-UR-88-3855, submitted to *J. Chem. Soc. Chem. Commun.*
34. P. M. Kuznesof, R. S. Nohr, K. J. Wynne, and M. E. Kenney, *J. Macromol. Sci.-Chem.*, **A16**(1), 299 (1981).
35. T. J. Marks, *Science*, **227**, 881 (1985).
36. J. R. Reynolds, J. C. W. Chien, F. E. Karasz, and C. P. Lillya, *Polym. Prepr.*, **25**(2), 242 (1984).
37. R. Liepins and K. Sakaoku, *J. Appl. Polym. Sci.*, **16**, 2633 (1972).
38. E. Kny, L. L. Levenson, W. J. James, and R. A. Auerbach, *J. Phys. Chem.*, **84**, 1635 (1980).
39. G. Smolinsky and J. H. Heiss, *Org. Coat. Plast. Chem.*, **28**, 537 (1968).
40. R. K. Sadhir and W. J. James, in *Polymers in Electronics*, T. Davidson, Ed. ACS Symposium Series No. 242. Washington, DC: American Chemical Society, 1984.
41. R. Liepins, M. Campbell, J. S. Clements, J. Hammond, and R. J. Fries, *J. Vac. Sci. Technol.*, **18**(3), 1218 (1981).
42. E. Kny, L. L. Levenson, W. J. James, and R. A. Auerbach, *Thin Solid Films*, **85**, 23 (1981).
43. R. Liepins, "Method of forming graded polymeric coatings on films," U.S. Patent 4,390,567 (June 28, 1983).
44. R. Liepins, "Method of forming metallic coatings on polymer substrates," U.S. Patent 4,464,416 (Aug. 7, 1984).
45. R. W. Simpson, Jr., in *Proceedings of the 1984 IEEE National Symposium on electromagnetic Compatibility*, April 24–26, 1984, San Antonio, TX, 1984, p. 267.
46. D. Staggs, in *Proceedings of the 1984 IEEE National Symposium on electromagnetic Compatibility*, April 24–26, 1984, San Antonio, TX, 1984, p. 43.
47. B. Bridge, M. J. Folkes, and H. Jahankhani, in *Inst. Phys. Conf. Ser. No. 89*, Session 8, p. 307, 1987.
48. T. A. Hoppenheimer, *High Technology*, p. 58, December 1986.

49. R. H. Baughman, R. L. Elsenbaumer, Z. Igbal, G. G. Miller, and H. Eckhardt, in *Electronic Properties of Conjugated Polymers*, H. Kuzmany, M. Mehring, and S. Roth, Eds. New York: Springer-Verlag, 1987, p. 432.

50. K. J. DeGraffenreid, in *Proceedings of the 1985 IEEE International Symposium on Electromagnetic Compatibility*, April 24–26, 1984, San Antonio, TX, 1985, p. 273.

51. Emerson and Cuming, Technical Bulletin 4-2-14, Canton, MA.

52. M. Gazard, J. C. Dubois, M. Champagne, F. Garnier, and G. Tourillon, *J. Phys. Paris Colloq.*, C3, 537 (1983).

53. F. Garnier, G. Tourillon, M. Gazard, and J. C. Dubois, *J. Electroanal. Chem.*, **148**, 299 (1983).

54. W. J. Miller, in *Modern Plastics Encyclopedia 1985–1986*, **62**, No. 10A, J. Aranoff, Ed. New York: McGraw-Hill, 1985, p. 380.

55. H. E. Coonce and G. E. Marco, in *Proceedings of the 1985 IEEE International Symposium on Electromagnetic Compatibility*, April 24–26, 1984, San Antonio, TX, 1985, p. 257.

56. H. Munstedt, in *Electronic Properties of Polymers and Related Compounds*, H. Kuzmany, M. Mehring, and S. Roth, Eds. New York: Springer-Verlag, 1985, p. 8.

57. M. E. Gross, A. Appelbaum, and P. K. Gallagher, *J. Appl. Phys.*, **61**(4), 1628, (1987).

58. R. Liepins, B. S. Jorgensen, and L. Z. Liepins, "Process for introducing electrical conductivity into high-temperature polymeric materials" (submitted for patent).

59. T. Hioki, S. Noda, M. Kakeno, A. Itoh, K. Yamada, and J. Kawamoto, in *Proceedings of the International Ion Engineering Congress*, Sept. 12–16, 1983. Kyoto, Japan, 1984, p. 1779.

60. A. Auerbach, *Appl. Phys. Lett.*, **47**(7), 669 (1985).

61. A. Auerbach, *J. Electrochem. Soc.*, **132**(6), 1437 (1985).

62. J. Y. Lee, H. Tanaka, H. Takezoe, A. Fukuda, and E. Kuze, *J. Appl. Polym. Sci.*, **29**, 795 (1984).

63. T. Cacouris, G. Scelsi, R. Scarmozzino, R. M. Osgood, Jr., and R. R. Krchnavek, *Meter. Res. Soc. Proc.*, **101**, 43 (1988).

64. J. E. Bouree and J. E. Flicstein, *Mater. Res. Soc. Proc.*, **101**, 55 (1988).

65. A. Gupta and R. Jagannathan, *Mater. Res. Soc. Proc.*, **101**, 95 (1988).

66. L. Baufay and M. E. Gross, *Mater. Res. Soc. Proc.*, **101**, 89 (1988).

67. A. M. Lyons, C. W. Wilkins, Jr., and F. T. Mendenhall, *Mater. Res. Soc. Proc.*, **101**, 67 (1988).

69

Silicone Release Coatings

Richard P. Eckberg

General Electric Company
Schenectady, New York

1.0 INTRODUCTION

Silicone release coatings are vitally important to the tag and label industry, which could not exist in its present form without reliable release agents. Silicones possess unique physical and chemical properties that make this class of substances ideal for the purpose of releasing pressure-sensitive adhesives. Silicone release agents worth \$130–\$150 million were sold worldwide in 1988, contributing to products whose total value exceeds \$3 billion.

The term "silicones" as commonly used refers to linear (two–dimensional) polydimethylsiloxanes, which may be structurally depicted as follows:

$$\left[\begin{array}{c} CH_3 \\ | \\ -SiO- \\ | \\ CH_3 \end{array}\right]_x$$

where x is an integer greater than 1. Silicon is tetrafunctional, so an infinite number of silicone polymers may be devised with different organic groups replacing methyl, or with three-dimensional resin structures wherein silicon atoms are incorporated in the polymer structure via three or four—Si—O—linkages. Since, however, the low surface tension, nonpolarity, chemical inertness, and low surface energy responsible for the outstanding release characteristics of silicone coatings all derive from the linear dimethylsilicone structure, this discussion focuses on linear polymers.

Silicone coatings that release pressure-sensitive adhesives have been in use for some 35 years. The chemistry and applications of silicone release coatings have undergone remarkable change during this time, with the pace of development accelerating in recent years. In the face of increasingly sophisticated and demanding requirements, silicones remain the

621

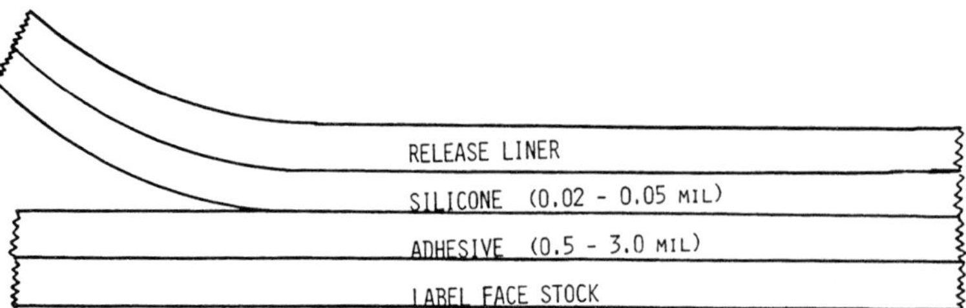

Figure 1   Pressure-sensitive label stock laminate (not drawn to scale).

only proven means of providing pressure-sensitive adhesive release for the tag and label industry.

The laminate structure normally used by the label industry is illustrated in Figure 1.

The liner most often used is paper, usually a machine-calendered (i.e., supercalendered kraft), clay-coated, or glassine paper designed to minimize penetration during coating and curing of silicone. Good holdout is essential for complete coverage at economically viable coat weights (< 1.2 g/m^2). Converters are also increasing their use of plastic films and film laminates to take advantage of plastics' dimensional stability and strength. Such substrates are often thermally sensitive, requiring suppliers to develop novel low temperature cure or radiation cure silicone systems to meet the new demands.

In use, the laminate construction herein depicted is split into two separate lamina: the release liner–silicone lamina is peeled away from the label–adhesive lamina. Silicone coatings therefore function as a vehicle for the adhesive, permitting it to be applied far from where it is ultimately used. The low surface energy of a cross-linked polydimethylsiloxane surface (usually ~ 22 dynes/cm) compared with the surface energy of organic pressure-sensitive adhesives (typically > 30 dynes/cm) prevents tight physical bonding of the adhesive to the silicone. That is, nonpolar silicone coatings are not compatible with highly polar adhesives, besides possessing minimal attractive forces toward other molecules. Chemically, the carbon–silicon and oxygen–silicon bonds of a dimethylsilicone molecule are unreactive, so that properly cured silicones are not prone to chemical interaction with pressure-sensitive adhesives with which they are in contact.

2.0   THERMAL-CURED SILICONE RELEASE AGENTS

Silicone release coatings widely used today can be considered to be thermosets: they are applied in liquid form (100% silicone fluids or dispersed in solvent or water) to a release liner, then irreversibly converted to cross-linked films via thermally accelerated chemical reactions between reactive silicone polymers and silicone cross-linkers. "Thermal cure" is a somewhat misleading term, because the chemical reactions that effect cross-linking into a cured state take place at ambient conditions (25°C), as well as at oven temperature (~ 150°C). Upon conversion to absolute (Kelvin) temperature scale (room temperature = 300 K; oven temperature = 425 K), it is clear that an oven need furnish only half again as much energy to a reactive chemical system as is already present in the coating pan environment to bring about cross-linking. Mixing reactive silicone polymers, catalysts, and cross-linkers together therefore initiates the curing process; maintaining a complete coating mix at a coatable viscosity while still permitting rapid oven cure mandates the use of dispersing media, inhibitors, or other modifications, as described in this section.

Heat-accelerated silicone cross-linking (curing) reactions are of two types: condensation cure and addition cure. Condensation cure systems for paper release applications were the earliest development of this technology. Condensation cure utilizes silanol-chain-stopped dimethylsilicone polymers, polymethyl–hydrogensiloxane cross-linkers, and condensation catalysts. The polymers are illllustrated below:

$$HO \left[\begin{array}{c} CH_3 \\ | \\ Si\ O \\ | \\ CH_3 \end{array} \right]_n H \qquad (CH_3)_3\ Si\ O \left[\begin{array}{c} H \\ | \\ Si\ O \\ | \\ CH_3 \end{array} \right]_m Si\ (CH_3)_3$$

SILANOL POLYMER   METHYLHYDROGEN SILICONE

where n can vary from about 50 to more than 4000, while m is much less than n; m normally is 10–50. SiH is a very reactive chemical species that readily condenses with silanol (SiOH) groups, forming extremely stable siloxane bonds and liberating hydrogen in the process:

$$\equiv SiOH + \equiv SiH \longrightarrow \equiv SiOSi \equiv + H_2$$

Many different catalysts accelerate or initiate this condensation reaction; metal soaps and driers such as dibutyltin acetate are the most efficient and economical, and are therefore in general use.

Condensation cure systems are applied as solutions in organic solvents (toluene or heptane, or mixtures thereof), or as oil-in-water emulsions, because in the absence of a dispersing medium, a catalyzed mixture of a silanol-stopped silicone plus polymethyl–hydrogensiloxane cross-linker sets up to an insoluble cross-linked gel in a few minutes at room temperature. There is no known means of retarding the condensation reaction sufficiently at room temperature to permit solvent-free coating without rendering the composition uncurable at oven temperatures. Solvent (or water in the case of emulsions) therefore acts as a bath life extender through the dilution effect, plus permitting easy, convenient coating of the silicone material.

Although use of solvents or water mandates high oven temperature and solvent recovery, and entails fire or explosion risk, such materials are readily coated via simple techniques such as direct gravure, reverse roll, metering rod, and doctor blade. Coating out of a solvent vehicle also gives the silicone supplier wide latitude in silanol molecular weight; such dispersion products as General Electric SS-4191 consists of approximately 30 wt % solutions of high molecular weight silanol gums ($\overline{MW}$ in aromatic solvents. Even at 70% solvent, these products as furnished have viscosities exceeding 10,000 cps, requiring further dilution to about 5 wt % silicone solids with more solvent to render them coatable. The cross-linker is normally packaged in the silanol solution; catalyst is added to the fully diluted bath at time of use.

Controlling silanol molecular weight is a proven means of controlling the release characteristics of the cured condensation-cross-linked coating. Long chains of polydimethylsiloxane between cross-linking sites provide a rubbery, elastomeric coating; shorted inter-cross-link intervals lead to higher cross-link density and a harder, more resinlike coating. The rubbery coatings provide tight (high) release, which displays a marked dependence on

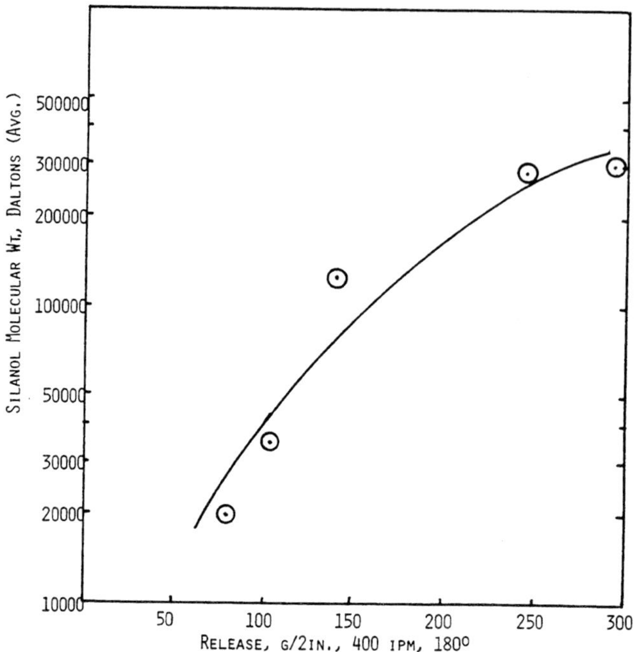

Figure 2   Release of an acrylic pressure-sensitive adhesive (Monsanto Gelva 263) as a function of silanol molecular weight.

delamination speed in comparison to the low (easy) release independent of stripping speed obtained from highly cross-linked silicone films. Accordingly, silicone suppliers offer several different molecular weight silanol-based dispersion products, permitting the end user to obtain a desired range of release. The relationship between silanol chain length and nominal release level is graphically shown in Figure 2.

Addition cure silicones resemble condensation cure silicones in some respects: both types of system rely on thermally accelerated cross-linking reactions between polymethylhydrogen siloxane cross-linker molecules and a separate reactive dimethylsiloxane polymer. Addition cure processes utilize catalyzed reaction of unsaturated organic groups attached to otherwise unreactive dimethylsilicones with SiH groups present on the cross-linker. Polymers in use are vinyl-functional silicones, the general structure of which may be represented as follows:

$$X + Y = 50 \text{ to } >4000; \ Y \geq 0$$

The curing reaction is an addition to the SiH group across the olefin double bond, also known as a hydrosilation process;

$$\equiv\!Si\!-\!CH\!=\!CH_2 + HSi\equiv \longrightarrow SiCH_2CH_2Si\!\equiv$$

With each polymer containing multiple reactive sites, it is clear that a very complex highly cross-linked thermoset structure is obtained upon cure.

Hydrosilation can be triggered by heat, peroxides, azonitriles, and high energy ultraviolet or gamma radiation. These particular reactions are free radical in nature and are therefore not well suited for normal coating equipment, as radicals so generated are quenched by atmospheric oxygen. Hydrosilation catalyzed by silicone-soluble Group VIII precious metal compounds (certain complexes of platinum, rhodium, and the like) is the best means of curing solventless silicone coatings.[1-7] These catalysts effectively promote SiH + Si-vinyl cross-linking at concentrations in the part-per-million range, making their use economical despite their cost. The addition cure reaction is an example of homogeneous catalysis. The precise mechanism is not firmly established,[8,9] but it is thought to be a concerted series of events in which the catalyst metal atom acts as a site to which reactive Si—H and olefin groups simultaneously bond in the proper proximity and stereochemical configuration for addition to occur.

This description to addition cure silicone chemistry is still incomplete. Whatever the system chosen, useful pot life of catalyzed mixtures of vinyl silicone and methylhydrogen silicone polymers seldom exceeds a few minutes at room temperature, like the condensation cure silicone system described previously. However, use of an important ingredient—the inhibitor—permits solvent-free coating of addition cure silicone release agents. Inhibitors function by tying up catalyst atoms at low temperature while still allowing rapid cure to occur at elevated temperature. An ideal 100% solids silicone coating composition combines instant cure at oven temperature with infinite pot life in the coating pan. Actual commercial systems depart from the ideal behavior, as shown in Figure 3.

A clue to inhibitor function is provided by the hydrosilation mechanism. During the cross-linking process, SiH and Si-vinyl groups must at some time be coordinated to the platinum metal center. An inhibitor molecule that prevents access to catalyst atoms at low temperature must not block coordination sites at oven temperature. Inhibitors are therefore often selected on the basis of colatility, or because complexes formed with platinum are unstable and break down at high temperature. Many of these proprietary additives are described in the patent literature, ranging from classical complexing or chelating agents to electron-deactivated olefins and acetylenes.[10-15]

Addition cure silicone release agents are available as solvent-free, low viscosity vinyl silicone fluids, as solvent-dispersed vinyl silicone gums (analogous to most silanol-based condensation cure products), and as emulsions. In each case, sufficient vinyl functionality is built into the linear silicone molecules to promote formation of highly cross-linked resinous cured coatings. These products therefore provide uniformly low (premium) release from most pressure-sensitive adhesives. Controlled release is not obtained by varying molecular weight of these vinyl silicone polymers (unlike the silanol case), and thus a different approach has been taken by silicone suppliers to combine the advantages of addition cure chemistry (particularly solventless pasckages) with a controllable range of release.

Studies of cured dimethylsilicone release coatings by electron spectroscopy for chemical analysis have confirmed that the surface is much more organic than would be predicted from the stoichiometry of the $(CH_3)_2SiO$ polymer unit.[16] An adhesive in a laminate construction is therefore largely in contact with unreactive, bulky, freely rotating methyl groups; more polar —Si—O—Si— polymer backbones concentrate beneath the coating surface. A depiction of this postulated structure is offered in Figure 4.

Since the surface orientation of Si—CH_3 groups governs the release characteristics of highly cross-linked nonelastomeric silicone coatings, it follows that breaking up this nonpolar, featureless "methyl landscape" by inclusion of materials that alter the polarity of the silicone should alter the release of the coating. This is in fact accomplished by adding vinyl-functional silicone resins to the basic linear vinyl silicone polymers.[17] "Resin" is here defined as nonlinear silicone structures bearing high concentrations of

$$O \atop | \atop O - Si - O \atop | \atop O$$

functionally. Solventless high release additives are therefore mixtures of vinyl silicone resins with vinyl silicone fluids. Since these resins are normally friable solids when isolated, their blends with vinyl silicone fluids are much more viscous than the unblended fluids, which limits the amount of resin that can be present in solventless high release additives. Nonetheless, a wide range of controlled release can be obtained by use of suitable blends of high and low release solventless vinyl silicone products, as is illustrated in Figure 5.

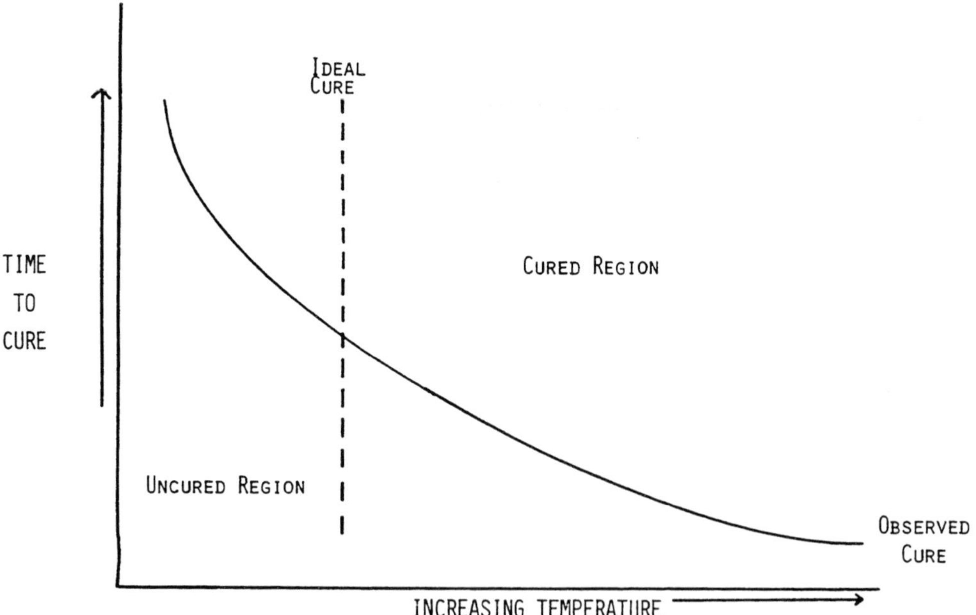

Figure 3　Cure time as a function of temperature.

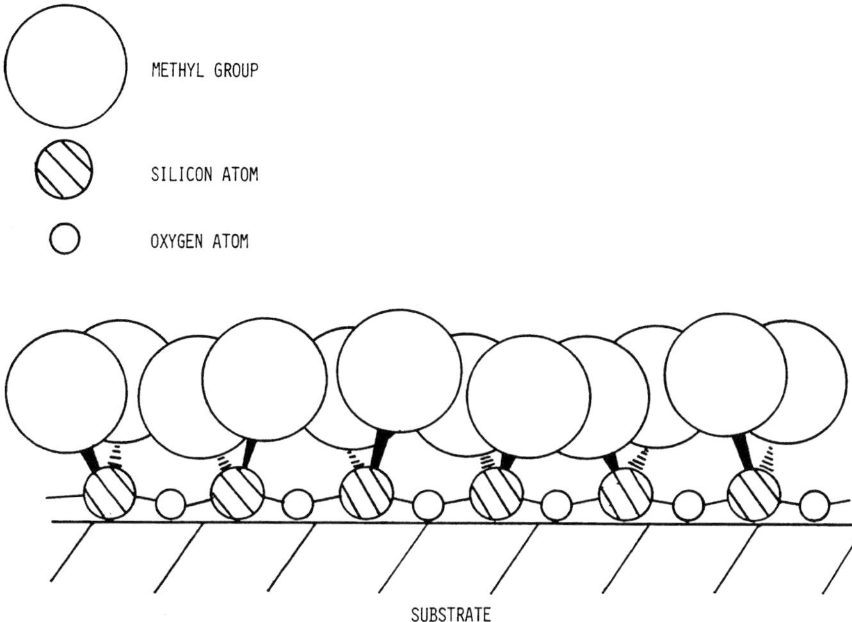

Figure 4 Representation of silicone release coating.

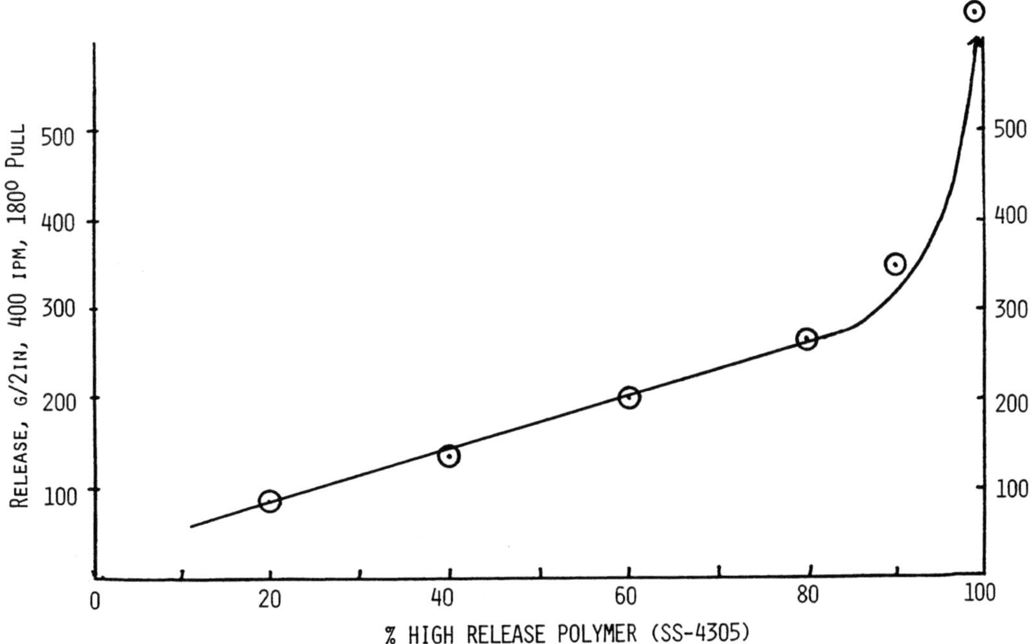

Figure 5 Solventless silicone (GE SS4300/SS4305 system) controlled release versus a pressure-sensitive adhesive based on styrene–butadiene rubber.

Two solventless silicone product designs are available. The simplest system consists of three products:

Low release polymer + catalyst + inhibitor.
High release polymer + catalyst + inhibitor.
Cross-linker.

It has recently been discovered that more efficient use of expensive hydrosilation catalysts results from maintaining catalyst and inhibitor apart until time of use.[18] Multicomponent systems taking advantage of this consist of:

Low release polymer.
Low release polymer + high catalyst concentration.
Inhibitor.
High release polymer.
Cross-linker.

Separation of the solventless release package into its individual components permits converters to custom formulate their coating baths to optimize performance for any given application; determining the catalyst and inhibitor levels that provide best combinations of cure and bath life permits end users to derive the most benefit per silicone dollar. While in principle the hydrosilation reaction is zero order in catalyst, increasing the catalyst concentration speeds cure at a given temperature and lowers temperature required for cure at a given oven dwell time. This effect is illustrated in Figure 6.

Another approach to lowering siliconizing cost is the use of silicone-compatible olefin extenders. In principle, a nonvolatile unsaturated hydrocarbon miscible with addition-curable dimethylsilicone fluids should cure in the coating with little effect on release performance.[19] In practice, replacement of silicone with less expensive organic olefins (or acetylenes) has not been firmly established as a means of reducing coating cost without performance penalty.

More than 30 domestic solventless silicone coating sites are currently in operation; new release coating capacity expected on stream in the near future will add to this number. These machines use offset gravure heads with multiple differential roll speeds to provide continuous coverage at low deposition. This coating technique works best with coating materials of less than 2000 cps viscosity. Silicone suppliers therefore furnish solventless silicones in viscosity ranges of 200–2000 cps (lower viscosity results in too much penetration of paper substrates), limiting polymer molecular weights to roughly, 5000–20000 daltons. Novel coating techniques have been reported that do not require gravure rolls and are claimed to work with higher viscosity materials.[20]

3.0 RADIATION-CURABLE SILICONE RELEASE AGENTS

Solventless silicone coating application extends to radiation-curable silicones, for there is no reason to furnish UV or electron beam (EB) processable materials in solvent carriers. Polydimethylsiloxanes are not cross-linked by UV or EB radiation at economically practical intensity or dose, respectively. Practical radiation cure of silicones requires silicone polymers incorporating radiation-sensitive organofunctional groups, as illustrated below:

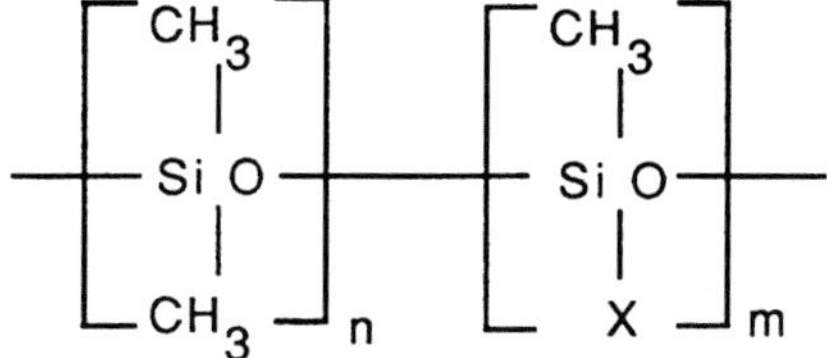

X = mercaptan, methacrylate, acrylate, epoxy, vinyl ether

In addition, UV cure normally requires high concentrations of photosensitizers or photoinitiators, such as benzophenone, benzoin ethers, and cationic-type "onium" salts. The presence of polar organofunctional moieties plus photocatalysts in radiation-cured release agents causes significant performance differences between radiation-cured and conventional thermally cured silicones. Nonetheless, industry demands for low (or "zero") temperature processing of silicone coatings to permit use of thermally sensitive films and to prevent demoisturization of papers have prompted considerable efforts by major silicone suppliers to develop, then improve, radiation-curable products. Radiation-curable silicones are now available from several sources.

Acrylated and methacrylated silicones were the focus of the earliest patented work in this area.[21] Acrylated silicones specifically developed for release applications were introduced

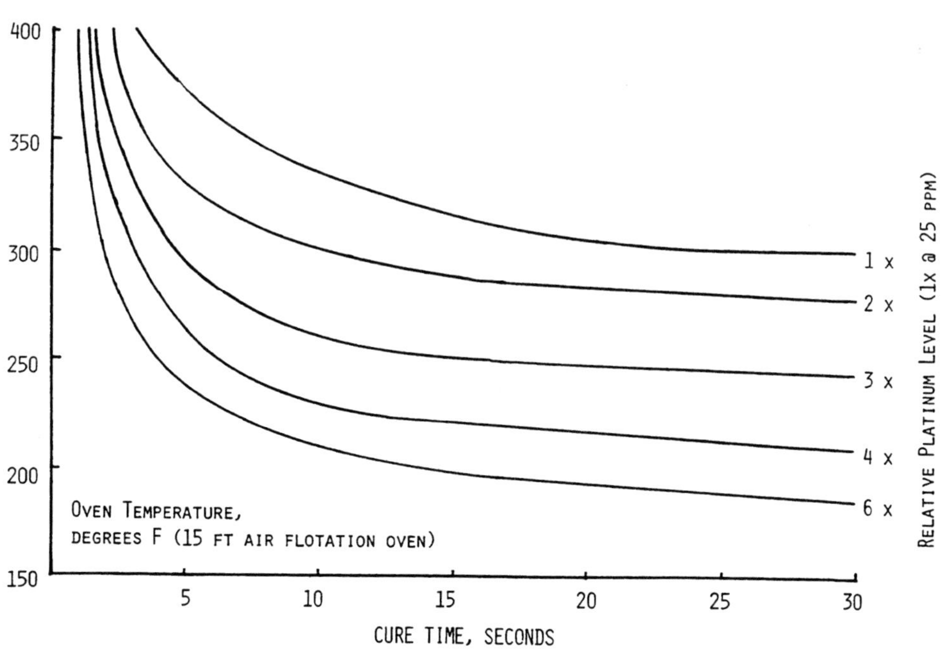

Figure 6 Cure as a function of catalyst concentration.

by Goldschmidt[22] and are in commercial use at a small number of coating facilities equipped to perform EB cure. These materials have an important performance drawback inherent in free radical acrylate cross-linking chemistry: since cure is subject to severe inhibition by atmospheric oxygen, efficient inerting (< 500 ppm O_2) of EB or UV cure chambers with nitrogen is essential for fast, complete cure to occur. While nitrogen blanketing is not impossible, inerting adds complexity and cost to the coating operation.

Cure chemistry pioneered by W. R. Grace & Company overcomes oxygen inhibition problems, which interfere with radiation cure of acrylates. Mercapto-olefin addition is initiated by UV light in the presence of suitable photosensitizers, or by EB radiation. The reaction is analogous to hydrosilation:

$$—RSH + CH_2 = CHR' \longrightarrow —RSCH_2CH_2R'$$

The chemistry has been extended to release coatings by development of mercaptoalkyl-functional silicone polymers.[23–26] While mercaptovinyl silicone systems can be UV cured in ambient atmosphere, market acceptance of products based on this technology has been slowed by their objectional odor (skunk fragrance is derived from mercaptans) and by the tendency of unreacted mercaptan residues in the cured coatings to chemically react and bond (via addition) to free acrylate usually present in cross-linkable acrylic pressure-sensitive adhesives. The same issues affect hybrid acrylic silicone–mercaptosilicone systems developed to take advantage of the oxygen insensitivity of mercapto-olefin addition.[27]

Certain "onium" (sulfonium and iodonium) salts are known to be capable of initiating photopolymerization of epoxides[28] and vinyl ethers.[29] Epoxy-functional silicones are readily prepared,[30] so application of cationic UV cure to silicones was an obvious extension of the technology.[31–33] A major performance advantage inherent in epoxy silicone–iodonium salt photocurable systems results from the non-free-radical nature of this cross-linking. This particular cure mechanism is not subject to oxygen inhibition, making UV-curable epoxy silicone based release agents particularly well suited to wide web converting operations, since nitrogen blanketing is not needed.

The epoxy silicone UV cure system has been shown to combine exceptionally fast ultraviolet cure response with premium, stable release versus cross-linkable acrylic, styrene–butadiene rubber, and hot melt adhesives.[16] As with other radiation-curable silicone release systems,[34] however, controlled release additives capable of providing a broad, predictable range of release for the UV epoxy silicone coatings have remained elusive. Another problem associated with these cationic cure silicone materials is substrate-dependent performance. Excellent cure, anchorage, and release are obtained when corona-treated films and film laminates (such as polyethylene, polyethylene kraft, polypropylene, polystyrene, and certain polyesters) are coated, but slower cure, higher release, and inconsistent anchorage often occur when porous paper substrates are utilized. In many respects this difficulty is analogous to problems encountered when solventless silicones were introduced and converters were first attempting to coat 100% solids silicones on paper release liners. More highly densified (or glassine) papers, perhaps treated with acidic coatings or sizing agents, ought to prove compatible with the UV-curable epoxy silicone release agents.

4.0 THE FUTURE

Solvent-dispersed thermal cure release coatings (condensation and addition cure) will be an important segment of the silicone coating market for years to come, particularly in Far East and Third World countries, where environmental and regulatory pressure to reduce the use of solvents is not a great factor. New coating capacity in North America and Europe will be predominantly solventless, with radiation cure capability built into new coating lines or retrofitted to existing 100% silicone solids equipment as acceptance of new silicone curing technology becomes widespread.

Silicone suppliers will accelerate development and commercialization of new and improved radiation cure and "zero temperature" cure products. An interesting area of research is room temperature cure silicones, consisting of one-part packages whose cure is triggered by moisture rapidly permeated through thin release coatings after they are applied.[35] Addition cure approaches to lower temperature cure 100% solids silicones have focused on more active catalysts and more volatile inhibitors; combining thermal and UV addition cure via ultraviolet-light-activated platinum hydrosilation catalysts[36] or UV-sensitive inhibitors[37] is another promising area of development. Few potential radiation cure silicone chemistry options have evaded development efforts, but vinyl ether-functional silicones (in concert with appropriate iodonium salt catalysts) seem to be capable of combining extremely fast noninerted UV cure[38] with EB curability, too.[39]

Interesting and exciting new silicone chemistry is undergoing rapid acceptance by the traditionally conservative converting industry. The result is that yet-unknown liner materials and applications of silicone release agents will expand pressure-sensitive tag and label markets. Research and development by silicone manufacturers will drive this growth.

REFERENCES

1. B. D. Karstedt, U.S. Patent 3,814,730.
2. B. A. Ashby, U.S. Patent 3,159,601.
3. H. F. Lameroux, U.S. Patent 3,220,972.
4. S. Nitzche et al., U.S. Patent 3,814,731.
5. G. Chandra et al., U.S. Patent 3,928,629.
6. A. Ossko et al., U.S. Patent 3,865,858.
7. M. Bargain, U.S. Patent 3,836,489.
8. J. F. Harrod et al., *J. Am. Chem. Soc.*, **86**, 1776 (1964); **87**, 16, 1133 (1965); **88**, 3491 (1966).
9. L. N. Lewis et al., *J. Am. Chem. Soc.*, **108**, 7228 (1986).
10. A. Berger et al., U.S. Patent 3,882,083.
11. G. V. Kookootsedes et al., U. S. Patent 3,445,420.
12. R. P. Eckberg, U.S. Patent 4,256,870.
13. R. P. Echberg, U. S. Patent 4,347,346.
14. R. P. Eckberg, U.S. Patent 4,476,166.
15. S. Shirihata et al., U.S. Patent 4,465,818.
16. R. P. Eckberg, in *Radtech '88 Conference Proceedings*, New Orleans, 1988.
17. P. Sandford, U.S. Patent 4,123,604.
18. M. E. Grenoble et al., U.S. Patent 4,448,815.
19. A. J. Dallavia, U.S. Patent 4,526,953.
20. F. S. McIntyre et al., in *Proceedings of the 1987 TAPPI Polymers, Laminations, and Coatings Conference*, San Francisco, 1988.
21. J. D. Nordstrom et al., U.S. Patents 3,577,256; 3,650,813.
22. G. Koerner et al., U.S. Patent 4,306,050.
23. R. Viventi, U.S. Patent 3,8166,282.

24. J. Bokerman et al., U.S. Patent 4,052,059.
25. J. A. Colquhoun, U.S. Patent 4,070,525.
26. R. P. Echberg et al., U.S. Patent 4,558,147.
27. F. Hockemeyer et al., U.S. Patent 4,571,349.
28. J. V. Crivello et al., *J. Polym. Sci.*, **17**, 977, 1047 (1979).
29. J. V. Crivello et al., in *Radcure IV Proceedings*, Chicago, 1982.
30. F. D. Mendecine, U.S. Patent 4,046,930.
31. R. P. Eckberg et al., U.S. Patent 4,279,717.
32. R. P. Echberg et al., U.S. Patent 4,421,904.
33. R. P. Echberg et al., U.S. Patent 4,547,431.
34. R. H. Bickford, in *Radtech '88 Conference Proceedings*, New Orleans, 1988.
35. G. R. Homan et al., U.S. Patent 4,525,566.
36. T. J. Drahnak, U.S. Patent 4,510,094.
37. R. P. Eckberg, U.S. Patent 4,670,531.
38. J. V. Crivello et al., U.S. Patent 4,617,238.
39. S. C. Lapin, in *Radcure '88 Conference Proceedings*, Baltimore, 1988.

70

Silicone Hard Coatings

Edward A. Bernheim

Exxene Corporation, Corpus Christi, Texas

1.0 INTRODUCTION

Silicone (polysiloxane) hard coatings are finishes of superior abrasion resistance and inertness to hostile chemical and environmental conditions. They consist of several monomers and other ingredients, and the makeup of the formulations varies from manufacturer to manufacturer. Among the highly varied components are monomeric silanes, dimerized silanes, silane hydrozylates, silaceous materials, leveling agents, flow control agents, cross–linking agents, and catalysts of various types. Silicone coatings are solvent–borne coatings. Some of the possible solvents are alcohols and glycol ethers. This includes such alcohols as isopropanol, propanol, ethanol, *n*–butanol, isobutanol, and methanol.

Polysiloxane coatings are applicable to many substrates, but the majority of applications are on nonmetallic surfaces, especially plastics. Silicone coatings can be dyed or pigmented, but for the most part these coatings are used as clear top coatings. They have excellent light transmission and actually improve the optical properties of the material that is coated.

Some of the plastics that are used with polysiloxane coatings are polycarbonate, acrylic, polyarylate, polysulfone, vinyls, nylons, polyester, cellulose acetate, cellulose acetate–butyrate, and polyolefins, just to name a few. Despite this plethora of materials, acrylics and polycarbonates are generally the plastic of choice.

Silicone hard coatings do not have automatic adhesion to all plastic surfaces. Materials such as polycarbonate and acrylics will, in many cases, have tape adhesion without need of surface treatment, etch, or primer coat. Many of the other substrates cannot be coated successfully without pretreatment. Many of the materials require either a corona discharge or a plasma treatment. For long–term outdoor stability, polycarbonate requires a primer coat to be applied first.

Plastic, to be coated, comes in many forms. Silicone coatings are used on cast sheet stock, extruded sheet stock, molded parts, lenses, windows, films, etc. The application methods are also quite varied. Some of the methods are spraying, flow coating, spin coating,

633

dip coating with various withdrawal speeds and, with some formula modifications, pad coating and roller coating. The coatings are not suitable for screen or gravure application.

Hard silicone coatings are useful as antifog, antistatic, photochromic, color–dyed, pigmented, UV absorbing, UV stabilized for exterior exposure, chemical resistance, 5–minute curing, and tinted coatings. Many of these properties can be combined in one coating. The coatings are used in such diverse areas as the automotive, electronic, computer hardware, architectural and architectural glazing, recreation, sporting goods, protective eyewear, safety, and optical industries.

The coatings cure thermally at baking schedules that vary with the heat stability of the substrate. The baking schedule can be influenced by such process variables as air–flow, nonair media, and uniformity of heating. There are many types of silicone coating and because of this most data given will have large variances. The cure time of these coatings vary from 5 minutes to 3 hours. Since these are solvent–borne coatings, if they are placed in heat immediately after coating, the higher area of the coating strata will cure first, leaving solvent entrapped in the film. Solvent entrapment in the coating entails such possible problems as lower abrasion resistance, weaker adherence to the surface of the substrate, chemical resistance, and loss of some durability.

The parts must be free of dust, dirt, grease, and mold release. When not in use, this type of coating should be stored at a temperature of 40° F. Storage of the coatings at low a temperature increases the life of the coating by three or four times that of the same coating under ambient conditions. The storage life for polysiloxane coatings is anywhere from 2 weeks to about 17 months at room temperature. With this wide range of stabilities, the stability of the coating chosen should be confirmed by querying the manufacturer. In many cases, it is wise to have the coating shipped by refrigerated truck or by air freight.

The price of the coatings may vary from $70 to $200 per gallon. The coverage can vary from 300 square feet per gallon to as much as 2000 square feet per gallon, depending on drip reusability, dilution, and coating viscosity.

2.0 SUBSTRATES

The data in Table 1 suggest how silicone hard coatings are used. The coatings are applicable for many other substrates. Solvent addition will change coating properties and give additional adhesion in some cases. With treatment, the use of these coatings on metals is quite possible. The solvent flash–off time will also have an effect on the ultimate cure schedule. The more complete the removal of solvent, the quicker the time of cure. Hardness is increased by longer cure times.

3.0 USES

Table 2 gives a number of current uses for polysiloxane coatings. These very versatile coatings are alcohol based, which makes them quite mild even on sensitive substrates. The biggest problem in determining a new use is the heat stability of the material to be coated. The length of the cure time will also help to determine whether the process is feasible.

4.0 APPLICATION OF THE COATING

A typical application procedure is summarized as follows.

Table 1 Cure schedules and pretreatment of some substrates

Plastic substrate	Typical cure temperature ($^\circ$F)	Typical cure time (minutes) (min)	Prime coat, or surface treatment (recommendation)
Polycarbonate	250	5–60	Primer (exterior and some interior use
Acrylic	190	20–120	None
Polyarylate	190–325	20–60	Primer or solvent addition
Vinyl	140	60–180	Corona or plasma discharge
Cellulose acetate	150	30–60	Alkali etch
Cellulose acetate –butyrate	145–200	30–60	Alkali etch
Polyester	180–375	4–60	Corona or plasma discharge

Table 2 Some uses of polysiloxane hard coatings

Train windows	Plane windows
Glazing for windows	Car headlight
Car tail lights	Auto parts
Computer screens (low gloss)	Electronic components
Gauge covers	Instrument covers
Electronic sensors (antifog properties help continuous readability)	
Computer housings	Instrument housings
Protected printed parts	Racquetball glasses
Ski goggles	Swimming goggles
Sunglasses	Safety glasses
Prescription glasses (may require tintable coating)	
Chemical splash glasses	Fireman's face shields
Lenses for gas masks	Full face shields
Scuba masks	Plano lenses
Shields for self-contained breathing apparatus	
Bus windows	Truck windows
Motorcycle shields	Watch crystals
Surgical shields	Outdoor lighting
Operating theater lights	Dental lights

1. Remove all protective masking and wrapping material.
2. Check the parts for optical quality.
3. Clean the parts. Some of the methods are solvent cleaning with alcohols and other mild solvents, ultrasonic cleaning with aqueous cleaning solution or trichlorotrifluoroethane, vapor degreasing, and roll cleaning machines that remove particulate with a high tack, nondeposit surface.
4. Before coating, treat the parts to prevent any possible static problems. This can be accomplished in a number of ways.
5. Perform annealing, which may be necessary in some parts, depending on substrate, coating, and the stress in the ultimate part.
6. Consider such other factors in the coating operation as temperature at which the material must be maintained, temperature of the room, maximum and minimum humidity, material of construction for coating vessels, and need of filtration.
7. Be sure that the parts have their solvent flash off under very clean conditions. Baking should be performed for the required time period and temperature.

The following coating processes may be used:

Dip Coating. The part is immersed in the bath and withdrawn by hand or by a mechanized method. Slow withdrawal produces even coating thickness from top to bottom and no drip buildup at the bottom. Faster withdrawal methods produce parts faster, but with less coating uniformity and a greater buildup at the bottom of the part.

Flow Coating. This consists of running a filtered stream of coating across the top of a part and letting the excess material drip off at the bottom. This method is good for coating one side of a part.

Spray Coating. This easily automated method is suitable for coating one or both sides of a part. The costing might have to be diluted to be sprayed.

5.0 Conclusions

Hard polysiloxane coatings have become important in many industries. In the overall framework of modern coating technology, these coatings are in their infancy. Their tremendous hardness and excellent properties in very hostile environments make them candidates for expansion into other areas of use.

71

Pressure-Sensitive Adhesives and Self-Adhesive Products

Alexander Zettl

BASF AG, Ludwigshafen/Rhein, Germany

1.0 INTRODUCTION

Adhesives are substances that join surfaces together by nonmechanical means. Without them, modern bonding techniques would be impossible. Yet from historical records, we know that gluing is one of the oldest means of bonding materials known to man. The ancient Egyptians, for instance, used glues to join wooden articles together.[1]

Nowadays, as 3500 years ago, basically three stages are involved in using adhesives.

The liquid adhesive is applied.
The adherents are joined together.
The adhesive film is set by drying, cooling, or chemical reaction.

In 1845 the Americans Shecut and Day, and in 1882, the German Beiersdorf paved the way for a new bonding technique.[2,3] They patented an adhesive for surgical plasters that obviated the need for the processing stages of the conventional method. These plasters were the first application of the principle of pressure-sensitive adhesion.

The essential advantages of pressure-sensitive bonding over conventional bonding are as follows.

No need to apply adhesive to one of the adherends, which is always a flexible strip of material.
No need for the user to set the adhesive coating, because this stage is taken care of during manufacture.
The desired adhesion is obtained by lightly pressing the adhesive layer against the second adherend.

These unusual features are made possible by the physical composition of pressure-sensitive adhesives. In simple terms, such adhesives are extremely viscous, viscoelastic liquids,

637

which, owing to certain flow behavior under slight pressure, form bonds to almost all surfaces.[4]

Pressure-sensitive adhesives are coated onto flexible carrier materials in the form of broad, jumbo-sized rolls. They are subsequently converted into so-called self-adhesive products by being split into narrower rolls or printed and punched. These products can be categorized as tapes, labels, and miscellaneous articles.

Self-adhesive products are so easy to handle that they speed up operations in industry, commerce, offices, and households. For this reason, they are primarily used in countries with high labor costs and high standards of living.

2.0 COMPOSITION

The most important component of a pressure-sensitive adhesive is the elastomer. It forms the supporting structure and imparts the internal strength or cohesion. The type of polymer employed is frequently used to denote the type of adhesive e.g., rubber adhesive, acrylic adhesive.

Many of the elastomers and polymers used to manufacture pressure-sensitive adhesives are not inherently tacky. Permanent tackiness is obtained by mixing them with resins. Nonadhesive elastomers include natural rubber, styrene-butadiene rubber, and block copolymers. Polyacrylics are much more diverse in their composition. Depending on the monomers used and the molecular weight, they can range from being nontacky and hard (or elastomeric), through to tacky, soft to highly cohesive pressure-sensitive adhesives.

The composition of a pressure-sensitive adhesive derived from a nontacky elastomer is governed by the desired properties. The mass fractions usually lie in the following ranges.

Elastomers	30-50%
Resins	20-50%
Plasticizers	0-40%
Fillers	0-40%
Stabilizers	1-2%

The elastomers and resins, and sometimes the plasticizers, are usually balanced combinations of two or three different types.

In principle, polyacrylics are ready for processing to pressure-sensitive adhesives as soon as they leave the reactor and indeed are often used in this way. However, to take advantage of large-scale production, acrylic dispersions in particular are frequently manufactured to feature just the basic properties. In subsequent processing stages, dispersions, resins, etc.[5] are mixed to fine-tune the properties:

Acrylic dispersions	50-98%
Resins	0-50%
Plasticizers	0-20%
Fillers	0-1%
Adjuvants	0-2%

For example, 98 parts of one or more acrylic dispersions could be mixed with two parts of such adjuvants as wetting agents, defoamers, and thickeners. In this particular modification, the processing properties of the adhesive dispersion have been adapted to suit the coating conditions without any significant alteration having been made to the acrylic polymers.

A mixture of 50 parts acrylic dispersion and 50 parts resin dispersion would lie at the other end of the wide range of possible modifications. This type of modification basically emulates the properties of nontacky elastomers. The adhesion/cohesion of the acrylic polymer is increased through the addition of the resin.

3.0 PROCESSING

Adhesive must be fluid for application to the carrier. This can be effected by carrying out coating from solution, from a dispersion of polymers in water, from a hot melt, or as a reactive substance.

Coating is followed by conversion to the usable state by removal of the solvent or water in a drier, by cooling the melt or, in the case of reactive substances, initiating a chemical reaction. Examples of these techniques are rubber solutions, polyacrylic solutions, polyacrylic dispersions, and hot-melt adhesives. Reactive pressure-sensitive adhesives are infrequently used.

Coating from solution has been been employed continually since the inception of pressure-sensitive adhesion and is still the most common technique for processing pressure-sensitive adhesives today. However, legislation designed to protect the environment, and the costs of solvents and drying techniques, are allowing slow but steady inroads to be made by alternative, solventless technologies.

Aqueous dispersions of pressure-sensitive adhesives were first described in a patent applied for by IG Farben Werk, Ludwigshafen (today the main works of BASF AG) in 1943.[5] The polyacrylic dispersion described in this patent was marketed at the start of the 1950s and is still used to manufacture specialties.

BASF AG soon launched numerous follow-up products, and competitors sprang up quickly. However, the use of these dispersions was limited for a long time to special products, such as a polyvinyl chloride film, insulating tape, and sound insulation materials. Only at the start of the 1970s were dispersions, mostly those based on polyacrylics, able to gain entry into the huge market for self-adhesive labels and, to a lesser extent, that of self-adhesive tapes. Improvements in quality, apart from reducing environmental pollution, accelerated the rapidly increasing use of polyacrylics.

Consideration was first given to solventless coating from the melt as far back as 1953.[7] However, it was not until the arrival much later of a new class of product (e.g., block polymers styrene-butadiene-styrene and styrene-isoprene-styrene) that hot melt adhesives gained practical importance.[8] By the start of the 1970s, hot melt adhesives based on these new materials were used for a large number of products. At the same time, hot melt adhesives required less complicated processing equipment than solvent-borne and dispersion adhesives.

The oil crisis in 1973, combined with incipient concern for the environment (solvents at that time were simply discharged into the atmosphere), helped hot melt adhesives to gain enormous popularity within a short period of time. Toward the end of the 1970s, however, it was realized that problems that had originally been originally classified as unimportant had become insurmountable, and hot melt adhesives soon reached the limits of their potential application. One of the main difficulties was that the flow properties required for hot melt processing could be achieved only at the expense of cohesion, and this trade off seriously impaired the properties of the finished article on application.

The next logical development of hot melt adhesives would be a new generation of reaction substances, which ideally would flow readily at room temperature and from which finished self-adhesive mix would be formed during an exothermic chemical reaction taking place after coating. Varying the composition of the monomer for cross linking and, if necessary, the amount of energy input, should make it possible to modify the properties of the self-adhesive mix. Such reactive pressure-sensitive adhesives are not yet in widespread use. Just how intensely this interesting area is being researched is attested to, however, by the proliferating patent literature.[9]

4.0 SELF-ADHESIVE PRODUCTS

From their early beginnings in the medical sector, self-adhesive products had soon developed into industrial self-adhesive tapes by the turn of the century.[10] the step toward large-area and stackable self-adhesive articles was taken in 1932 when the adhesive coat was faced with a second strip of material.[11] This idea was commercially exploited by Stanton Avery, who began manufacturing and marketing the first self-adhesive price labels in 1935.[12,13] Following a period of stagnation in the 1940s on account of World War II, industry started applying the principles of pressure-sensitive adhesion on a large scale in the 1950s.

A new and steadily growing industrial sector coated the most diverse materials with various pressure-sensitive adhesives and created a range of self-adhesive products that embraced thousands of articles.

Broadly speaking, self-adhesive products can be divided into three categories: tapes, labels, and miscellaneous products. Adhesive tapes are substrates that are coated with a pressure-sensitive adhesive, which is sometimes covered by a release material. They are cut into different widths and packaged in the form or rolls. Self-adhesive products are called labels when they are separated by a knife cut and faced with a release material.

Adhesive tapes fulfill many functions, including those of sealing, joining, fastening, and protecting. By contrast, labels have mainly only one role and that is to act as information carriers.

Although not as important in terms of quantity, the group of miscellaneous articles is all the more impressive by virtue of its sheer diversity. It includes everything that cannot be classified as tape or label. Examples are medicinal articles, decorative film, film for protecting maps, and adhesive memo pads.

The examples that follow are intended to illustrate the otherwise largely unknown range of applications of self-adhesive products.

Let us start with adhesive tape. Most people are familiar with the standard household and office tapes, 10-25 mm in width and 30 m in length. Only specialists will know that there are similar, industrial packaging tapes in rolls of 50-75 mm in width and up to 2000 m in length. They are used in hand dispensers and automatic machines for sealing cartons.

Sensitive surfaces such as polished metal and varnished wood are protected during further processing, transportation, and storage by self-adhesive paper or film that is removed before the articles are used without leaving any residue.

Double-sided adhesive tapes are superseding conventional bonding methods. They bond so reliably that even mirrors can be safely attached to tiles and cupboard doors. And in the printing industry, for instance, flexography could not enjoy its present popularity if it had not been for the efficiency of double-sided adhesive tapes (Fig.1).

Figure 1

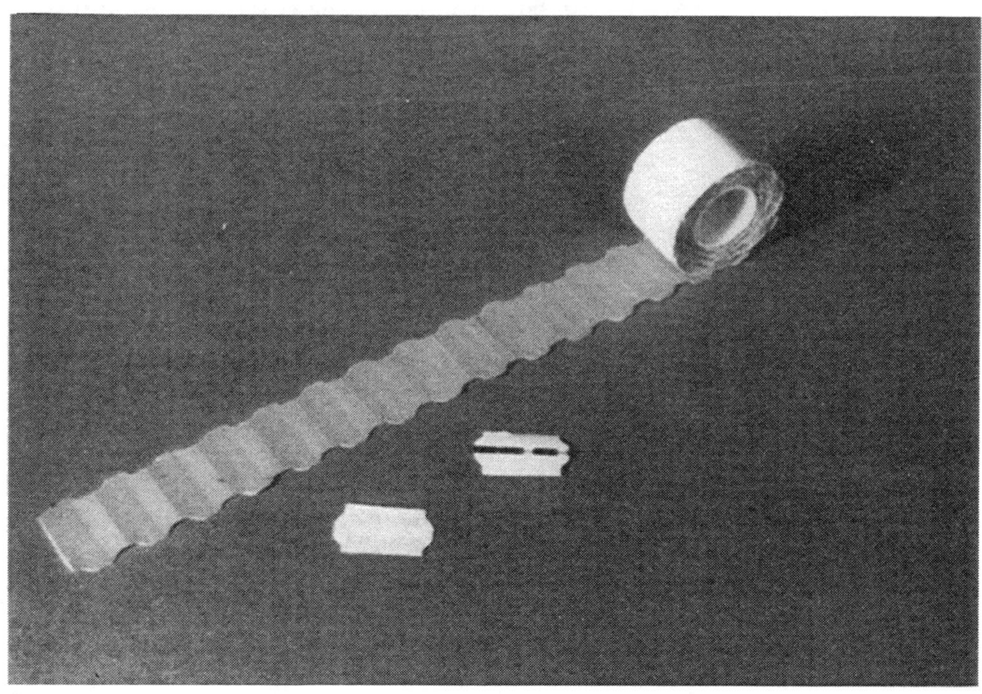

Figure 2

The retail sector could not dispense with price labels. However, net income is often seriously diminished by the theft rate. A very promising means of reducing inventory shrinkage is afforded by new antitheft labels that allow articles to be priced and rendered theft-proof in one stage. A strip embedded in the adhesive layer of the label triggers an alarm if it is not deactivated by being moved past a weak magnetic field at the point of sale (Fig. 2).

Self-adhesive name tags are fashionable nowadays. A further development is self-adhesive visitors' tags. These resemble labels and consist of two or three sheets of duplicating paper. When they have been filled out, the original is retained by the visitor and the copies are given to the doorman or are used for other purposes.

Price and visitors' labels are examples of articles that are discarded after one use. At the other end of the spectrum of applications of pressure-sensitive adhesives lies symbolically the construction design of an audio cassette. It depicts the tolerance limits of the cassette housing and the positions of the upper archiving and lower identification labels. In this case, the high quality, printed self-adhesive label serves to identify the product and brand and is thus a functional part of a technical consumer product (Figs. 3 and 4).

Long-life dry batteries are also an application area of pressure-sensitive adhesive labels. Like antitheft labels, they are one of a few cases in which the label not only acts as an information carrier but also fulfills a further function, namely that of an insulating film. The use of a composite film adhesive label instead of a steel jacket and insulating film has increased the capacity of alkaline-manganese cry batteries to the extent that new generation of long-life batteries last 30% longer than the older ones.

Advertising and decorative display areas can be creasted just as quickly with a roll of self-adhesive film as with a spray gun and paint. And they are just as durable. The self-adhesive technique offers much greater scope for design and drastically reduces the off-road time of vehicles.

The latest product to originate from this sector is a totally new innovation. Adhesive memo pads combine the concepts of pressure-sensitive adhesion and the memo pad. In just a few years, they have scored unparalleled success in all areas of life ranging from the office to the household.

5.0 PROPERTIES

The bonding behavior of a pressure-sensitive adhesive is governed by the mechanical or viscoelastic behavior of the adhesive and the surface and interfacial properties of the adhesive and substrate.

In the first stage of pressure-sensitive adhesion, the substrate is wetted by the adhesive. The lower the viscosity and the modulus of elasticity, the more easily are the points of contact formed. When the surface of the adhesive is close enough to the substrate, the second stage comes into effect and intermolecular forces followed by surface and interfacial forces take over. Under comparable conditions, a state of equilibrium is attained after a certain contact time characteristic of every pressure-sensitive adhesive, and the bond is formed.

It has not yet proved possible to use these physically clearly defined material parameters to uniquely describe pressure-sensitive adhesives. For this reason, investigators are resorting to test methods that have been developed as the pressure-sensitive adhesive products have been used. However, the values for tack, adhesion, and cohesion provided by these methods provide sufficient information about the bonding behavior of a pressure-sensitive adhesive.

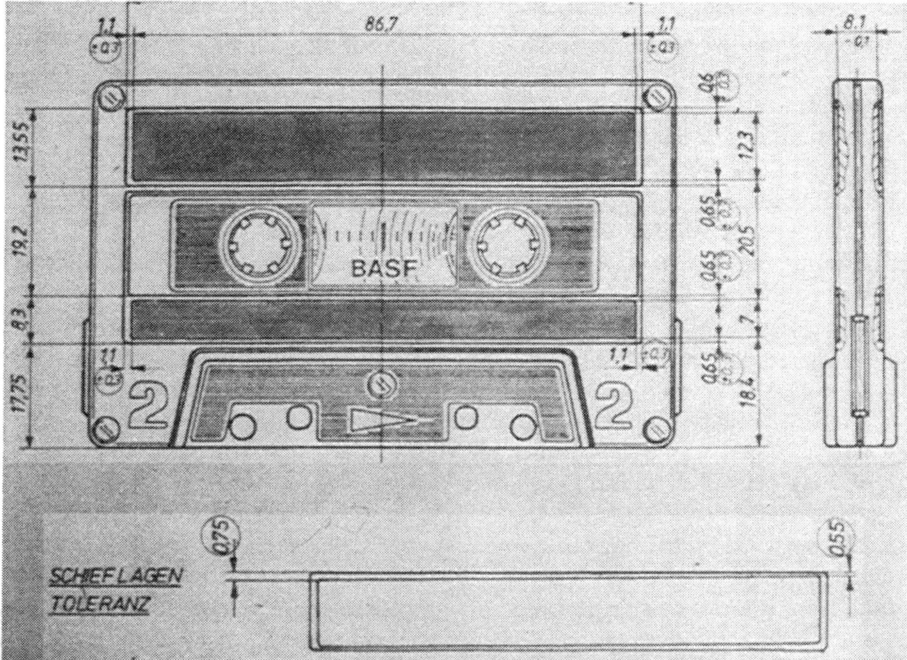

Figure 3

Figure 4

Adhesion or peel strength is the force required to induce separation from the adherend's surface. It is a measure of the strength with which the adhesive adheres to the substrate.[14]

Cohesion or shear strength is the internal strength of the pressure-sensitive adhesive. It is a measure of the ability to resist static load.

Tack is the ability of a pressure-sensitive adhesive to adhere to the surface when a slight pressure is applied. A pressure-sensitive adhesive with good tack properties forms the points of contact required for full adhesion after only fleeting contact with the substrate.[15]

The effect exerted on bonding behavior by such various stress factors as aging, UV radiation, and exposure to plasticizers and moisture is determined by simulating factors that will occur in application and noting the change in adhesion values.

The ability of a pressure-sensitive adhesive to adhere to the substrate is impaired by contamination, pick up of dirt, moisture, etc. on the surface of the adhesive prevents intermolecular forces from becoming effective at the adherend's surface and so adhesion cannot occur. The layer of adhesive must therefore remain covered until the moment of use.

With adhesive tapes, this requirement is accomplished in a simple manner by rolling the tape with the layer of adhesive facing inwards (i.e., the adhesive is covered by the reverse side of the carrier film).

Large-area materials, such as decorative films and labels, are covered with specially finished release material.

Protecting or covering the coating of pressure-sensitive adhesive is complicated by two factors. First, the coating must be readily and completely removable from the protective layer (e.g., release material or the reverse side of the tape). Second, in some cases, a high degree of adhesion to the adherend's surface is also required. Both these demands can be be met only if the adhesion of the pressure-sensitive adhesive film to the backing (A_B) is greater than cohesion (C), which in turn must be greater than the adhesion to the covering release material or reverse side of the tape (T_R):

$$A_B > C > T_R$$

Tack and adhesion on the one hand and cohesion on the other are mutually dependent but totally opposing properties. The quality of a self-adhesive article therefore depends considerably on the extent to which these properties (i.e., the inner strength and the adhesive strength) are balanced to suit the intended application.

Thus protective film must have a low adhesion but high cohesion, (i.e., it must not leave a residue when peeled off), and permanent adhesive labels must have very high adhesion and low to moderate cohesion. The demands imposed on the cohesion of labels are lower because, since they are usually permanent, there is no requirement for clean removal. In this case, the cohesion must simply be high enough to compensate for the resilience of the carrier on adhesion to curved surfaces.

Decorative films must fulfill only moderate adhesion and cohesion demands. However, the plasticizer in the carrier material (plasticized PVC film) must affect the adhesion characteristics of the pressure sensitive adhesive as little as possible.

The foregoing example demostrates that cohesion and adhesion constitute only some of the performance specifications. As applications grow in number and become more specialized, an increasing number of extra requirements, in some cases mutually exclusive extra requirements, re being placed on pressure-sensitive adhesives. They include:

Good resistance to plasticizers/ready removability with solvents.
High initial tack/good removability.
Low initial tack/good final adhesion.
High shear strength above room temperature/good tack below room temperature.

This list is only a very small selection of the demands imposed on pressure sensitive adhesives. Naturally, a pressure sensitive adhesive need not and cannot fulfull all of them.

6.0 TESTING

Various manufacturers' associations have devised their own methods for determing adhesion, cohesion, and tack. The best-known test methods are published by the following organization.

AFERA (Association des Fabricants Européens de Rubans Autoadhesifs.)
ATSM (American Society for Testing and Materials)
FINAT (Féderation Internationale des Fabricants et Transformateurs d'Adhesifs et Thermocollants sur Papier et autres Supports)
PSTC (Pressure Sensitive Tape Council)

All associations use the same basic tests to measure adhesion, cohesion, and tack. However, the results obtained by any one method cannot be compared directly with those of another, because differences exist in the test procedures used. The principles underlying the test methods are discussed next (Fig. 5).

6.1 Peel Strength

The peel strength, which is a measure of the adhesion during true adhesive failure, is determined by pulling the strip back over itself at an angle of 180°. The force required for separation is recorded and the mean value calculated. The bonding time is crucial here because the values determined increase with increase in contact time.

6.2 Shear Strength

In the static shear strength test, a specified area of the teststrip is bonded and subjected to a specified load. Shear strength is a measure of cohesion, but only when true cohesive failure occurs. Often, the weight is not allowed to act perpendiccularly. Instead, it acts as at 2° to the point of adhesion, such that an angle of 178° is formed. The purpose of this is to avoid premature collapse through vibration-induced peeling. The time required for the weight to fall is recorded. The test is usually performed at 23° C, but higher temperatures are also employed.

6.3 Tack Test

This test is surrounded by mush uncertainty, as attestes by the large number of different methods employed. Each test has more or less pronounced advantages but has at least one major disadvantage. In all cases, the surface contact is not measured directly. Instead, the force required for separation is measured; this usually means that strong peeling effects are more or less included as well.

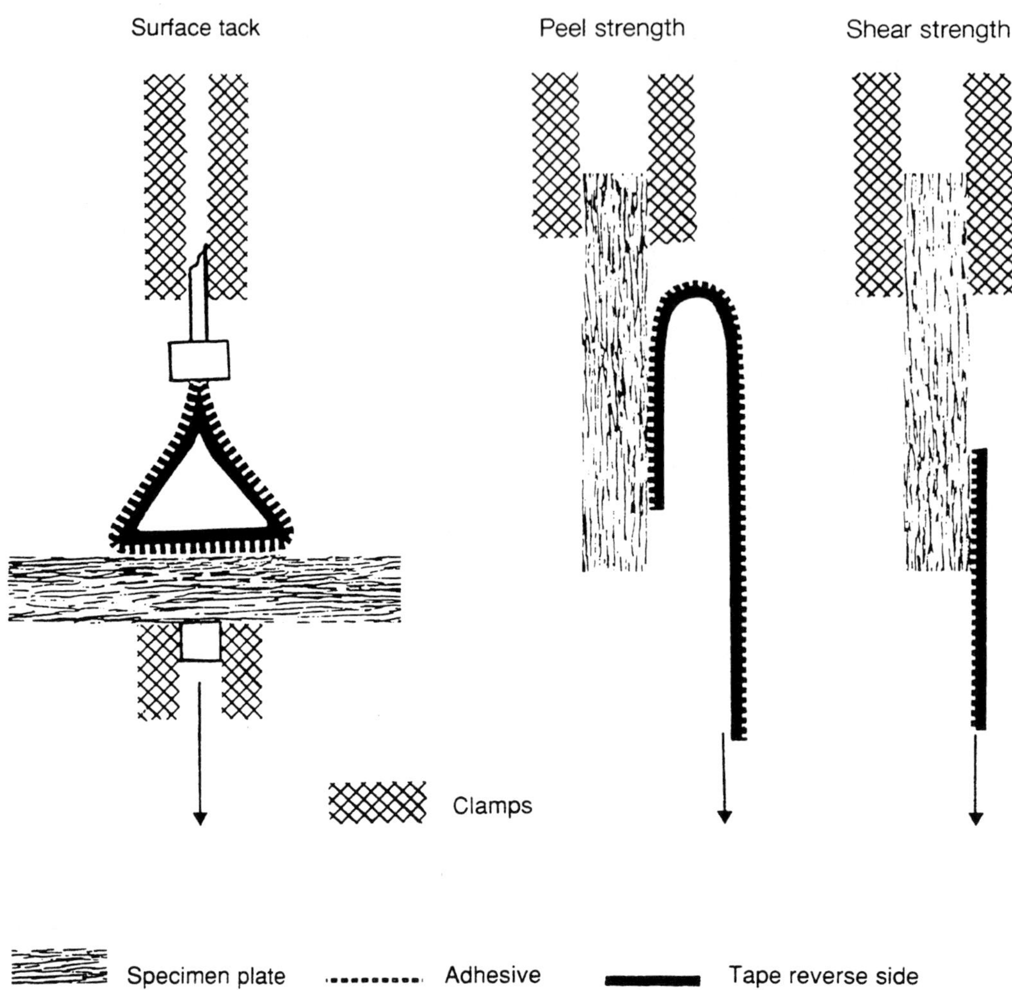

Figure 5

6.3.1   Surface Tack or Quick Stick by the Loop Test

The adhesive strip is formed into a loop and moved by the testing device toward the test surface at constant velocity. The contact area is allowed to form naturally without pressure being exerted, then the device moves in the opposite direction and records the force required for moving the strip.

6.3.2   Rolling Ball

A ball of specified diameter and weight rolls down an inclined plane onto a layer of adhesive, which arrests its movement. The distance travelled by the ball before it stops is measured. The shorter this distance (the lower the value), the greater is the tack.

6.3.3   Polyken Probe Tester and Zosel Tack Tester

A probe is allowed to impinge on the adhesive layer, and the force required to remove it is measured.[4,16] The velocity and contact time are variable but must always be the same for

Table 1 Factors Influencing Adhesion

Property	Adhesive layer	Carrier	Release	Substrate	Test Condition
			Factors[a]		
Chemical composition	•	•	•	•	
Mechanical (viscoelastic) properties	•	•		•	
Surface tension	•	•		•	
Surface structure	•		•	•	
Thickness	X	•	•	•	
Stiffness		•			
Drying conditions	•	•	•		•
Contact time					X
Rate of separation					X
Angle of peel					•
Contact pressure					X
Temperature					•

a• Influential but very complex, X adhesion increase to a certain extent.

comparable measurements. The Zosel tack tester records the change in force during the contact and separating periods, whereas the Polyken specimen probe tester records only maximum values.

6.3.4 Rotating Drum and Druschke Toothed-Wheel Tack Tests

Since the adhesive tape rotates on a cylinder, the contact times are very short and yield a large number of different results. This increases the statistical reliability of the value determined.[15,17]

6.3.5 Finger Tack Test

A very subjective yet widely used method for testing tack consists in evaluating the adhesion of a self-adhesive article to the finger or thumb. This method is primarily used as a preliminary evaluation by experienced technicians.

Unlike the determination of physical characteristic data, the testing of adhesive values can be influenced by many parameters that affect the test results.

Despite the large number of influences, more or less reproducible results can be obtained provided sample preparation and test conditions are specifed exactly and adhered to.

Factors influencing adhesion are summarized in Table 1.

REFERENCES

1. Heinz Lucke, "Kleb– und Dickstoffe: Herkunft, Aufbau, Wirkung," *Adhesion*, 6 (1985).
2. William A. Shecut, and Horace H. Day, U.S. Patent 3965 (1845).
3. P. Beiersdorf, German Patent 20,057 (1882).
4. A. Zosel, "Adhesion and tack of polymers: Influence of mechanical properties and surface tensions," *Colloid Polym. Sci.*, 263, 541 (1985).
5. A. Zettl, "Modification of Acrylic Dispersions, In *Handbook of Pressure–Sensitive Adhesives Technology*, (D. Satas, ed.) 2. Ed., New York: Van Nostrand Reinhold, 1989.
6. H. Fikentscher and C. Schuster, German Patent 862 957 (1953); BASF AG.
7. F. Salditt, Swiss Patent 327 731 (1953); Lohman KG.
8. J. T. Harlan, U.S. Patent 3,239,478 (1966) Shell Oil Company.
9. K. C. Stueben, "Radiation curing of pressure–sensitive adhesives: A literature review, *Adhesive Age*, June 1977.
10. Beiersdorf AG, *100 Jahre Beiersdorf: 1882–1982*, 1982.
11. Harry E. Carpenter, U.S. Patent 2,030,135 (1932).
12. Fasson, *From Pioneers to Market Leaders* Netherlands, 1986.
13. Ray Stanton Avery, U.S. Patent 2,220,071 (1937).
14. BASF, *Technische Information Haftklebstoffe*, Teil 1.
15. W. Druschke, *Adhesion und Tack von Haftklebstoffen.* Vortrag Afera–Tagung, 1986.
16. 64/1957 / F. Wetzel, ASTM Bulletin 221, Society for Testing and Materials, Philadelphia: American 1957, p. 64; F. H. Hammond, ASTM Special Technical Publication 360, Philadelphia: ASTM, 1963, p. 123.
17. R. F. Ball et al., "Measuring of tack using a rotating drum method," *Adhes. Age*, May 20, 1968.

72

Self–Seal Adhesives

Larry S. Timm

Findley Adhesives, Inc.
Wauwatosa, Wisconsin

1.0 INTRODUCTION

Cohesives, often referred to as "self–seal adhesives," are coatings which in a dry state exhibit the unique ability to selectively stick to themselves when brought together under pressure.

1.1 Adhesion and Cohesion

Natural rubber is the ingredient common to all self–seal adhesives. It is this raw material, generally utilized in the latex form, which confers or facilitates the ability to cohere.

Natural latex consists of long chain, high molecular weight molecules, the ends of which are thought to contain an extra double bond protected with a weak hydrogen–bonded protein molecule. It is believed that when such molecules are positioned in intimate contact and mechanical or thermal energy is introduced, the following reaction takes place. First, the weakly hydrogen–bonded protein molecule breaks from the long chain and second, stored elastic energy from within the molecule is released, causing molecular vibration. The subsequent entanglement of the chains and joining at the now–available bonding sites produces a molecular inseparable condition referred to as a cohesive seal (Figs. 1–3).

In application, the greater the number of bonding sites utilized to impart adhesion, the smaller the number remaining to facilitate subsequent adhesion.

Laboratory evaluation on the functionality of self–seal formulations confirms the significance of achieving an appropriate adhesive/cohesive balance relative to a specific substrate.

Through product selection or formula modification, it is possible to enhance or accentuate the adhesive or the cohesive properties, but not, however, without inversely diminishing the other set. A product exhibiting the appropriate adhesive/cohesive balance after a consideration of the seal functionality requirements mandated by the application and other pertinent performance parameters. In addition, product selection can be made only after the influence of the specific substrate on the adhesive/cohesive balance has been acknowledged.

649

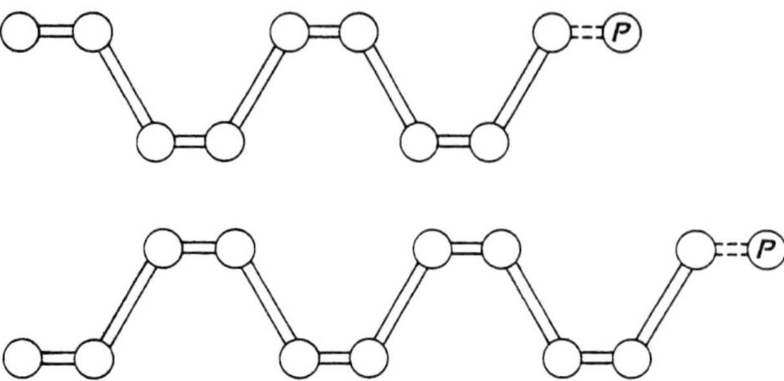

Figure 1 Long chain natural latex molecules.

Coating weight also affects seal functionality. Approximately 3.5 g/m² is requires to ensure proper film formation, generating optimum adhesive functionality; 5.0 g/m² is generally suggested as the coating weight offering the most reliable performance. Consequently, the adhesive/cohesive balance is typically evaluated at this coating weight, whereas 7.0 g/m² represents perhaps the highest commercially viable coating weight for one–pass applications on high speed converting equipment. This coating weight is employed only in applications that require enhancement of the cohesive functionality over that which is available through formulation.

Drying or, more significantly, insufficient drying, can influence the performance of the product as well. Excessive retained moisture is the equivalent of a plasticizer addition, minimizing adhesion while accentuating cohesion. Additionally, retained moisture can create a roll blocking problem. Polyamide–based overprint lacquers typically employed for release can themselves become softened in the presence of excessive moisture.

Cohesives are dried at temperatures ranging from 50 to 90°C; however, drying is best accomplished at a web temperature of 60–70°C and through the use of a scrubbing zoned drying, or otherwise minimizing the amount of moisture–laden recirculated air.

1.2 Seal Performance

Eminently important in evaluating, testing, and anticipating seal performance is an understanding of the significance of the mode of failure. Simply stated, observe and record whether the seal strength obtained represents an adhesive, cohesive, or substrate failure. In these respective cases, the self–seal is not adhering to the substrate, the self–seal is not adequately sticking to itself, or delamination or fiber tear/film destruct, is occurring.

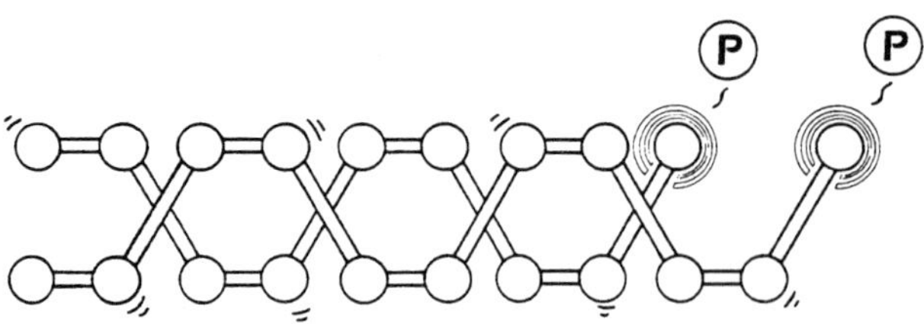

Figure 2 The effect of energy in seal formation: molecular excitation; severing of the "protein" molecule.

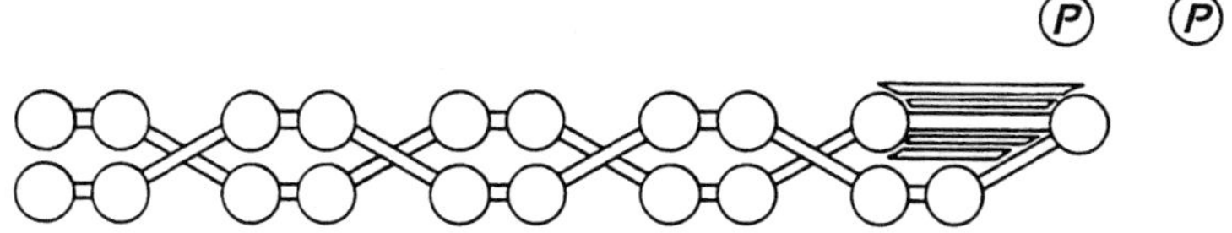

Figure 3 Cohesive seal consisting of both chemical bonds.

Some further descriptive commentary on the mode of failure also is appropriate:

Webby: adhesive failure (en masse release from one or the other side).
Leggy: adhesive failure (tenacious, elongating, rubber–like bands).
Stringy: cohesive failure (short or fine elastic strands, splitting seal).
100% cohesive failure (visually appears undisturbed, as before sealing).

Thus, significant in evaluating seal performance is the recording of the procedures utilized in testing. Test methods used should relate as closely as possible to the actual conditions employed in commercial seal formation.

Self–seals are pressure–respondent. The greater the pressure, the more immediate the cohesion response. However, the introduction of mechanical energy can help to form a bond and generate a suitable response in a mature seal. A seal is considered to be mature after 1 hour, but some minimal increase in strength may continue to develop for up to 24 hours before a fully stabilized condition is achieved. Typically associated with an increase in seal strength is a change in the mode of failure. For instance, a 100% cohesive failure at 300 g/25 mm could yield a substrate destruct seal at 800 g/25 mm with increased pressure or aging. No such similar change or improvement would be realized from a seal exhibiting an adhesive failure. It is only the developing cohesive response that is subject to this improvement.

As already mentioned, it is an accepted practice to utilize polyamide–based overprint lacquers to inhibit roll blocking. Careful consideration must, in fact, be given to selecting an ink–overprint system that is specific to the substrate and suitable for use in conjunction with self–seals. This not only assures proper release and eliminates ink pick, but also minimizes the potential for offset contamination. Offset contamination is represented by an accumulation of migrating slip or release modifiers from the printed surface to the surface of the self–seal, diminishing or inhibiting the subsequent cohesive response.

2.0 APPLICATION TECHNIQUES

In relaying techniques and procedures developed to facilitate cohesive applications via gravure, it is appropriate to define some of the inherent product limitations and physical characteristics that affect their use.

2.1 Handling

Cohesives are perishable. They have a suggested shelf life, and proper rotation of inventory is required.

Cohesives must be protected from freezing; ideally they are stored at 10–30°C in tightly closed containers.

Although as a general rule cohesives are not subject to settling, they do, in fact, stratify, as a result of their low viscosity and variable specific gravity, and must therefore be well stirred before use.

Cohesives are susceptible to contamination. Transfer of material should be made only to clean, water–rinsed containers.

Particulate matter and/or solvents, including alcohol, will reduce mechanical stability, and cohesive formulations should not be adulterated. Any such modifications can interfere with product performance.

It is advisable to discard used cohesive rather than returning it to a storage container; this eliminates the possibility of cross–contaminating fresh material.

Cleanup is best accomplished with cool water. Ammoniated water or alkali detergent may also be used. Completely dried material may be stripped or peeled from equipment. The use of solvent is not recommended.

Cohesives are preserved with ammonia, which is highly corrosive to copper and/or copper alloys. Additionally, the incorporation of copper ions into a cohesive will cause accelerated degradation. Therefore, any such contact must be avoided.

As copper is typically employed as the base for gravure cylinders, it is imperative that a hard, impervious chrome plating be utilized over 100% of the cylinder. Sometimes a nickel flash is employed immediately under the chrome for added protection.

Cohesives stabilized for machinability may be shear sensitive, and excessive shear must be avoided.

Cohesives should not be pumped by means of gear pumps routinely used for inks. A diaphram pump can be utilized. However, a systaltic pump employing alkali–resistant flexible tubing is preferred.

2.2 Procedures

Experience has shown that shear generated by blade doctoring can be minimized through maintaining as steep an angle as possible (i.e., 90° tangent to the cylinder). It is suggested that a well–dressed, rigid blade assembly be utilized; typically blue steel 150–200 μm doctor, 350–500 μm backer, recessed 1.5–3.5 mm (Fig. 4).

To further reduce shear generated at the doctor blade in pattern applications, scavenger bars on overall etching of cylinder ends should be considered. Oscillating the blade through the widest parameters and at its fastest speed will also aid in the elimination of particles formed in the land areas. Also suggested is the under cutting of the impression roll to eliminate inadvertent backside coating, should shear–generated particles cause the ends of the blade to lift (Figs. 5,6).

Ammonia, used as a volatile preservative, functions as the secondary stabilizer as well. During intermittent or low usage runs, aeration may cause a reduction in the ammonia level sufficient to create machining difficulties. In such situations, nondetergent 26° Be (29.4%) ammonium hydroxide should be added to a pH of 9.5–10.5.

Cohesives are quick drying, facilitating their utilization on high speed converting equipment. However, unlike solvent–borne inks, once dry they will not redisperse in themselves. Therefore to minimize drying in the application station, it is suggested that all other "make–ready" be accomplished before cohesive is added to the pan. Once added, the gravure cylinder must be kept rotating, ideally flooding the cylinder when not actually coating. Before it is dropped or reseated, the blade should be wiped with a rag moistened with water, remove any dried material or other contaminants.

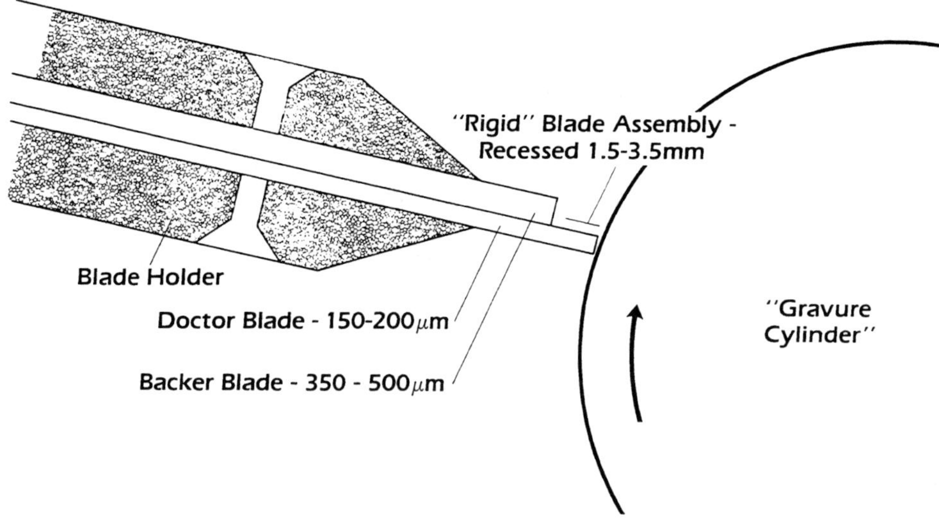

Figure 4 Schematic of a rigid blade assembly.

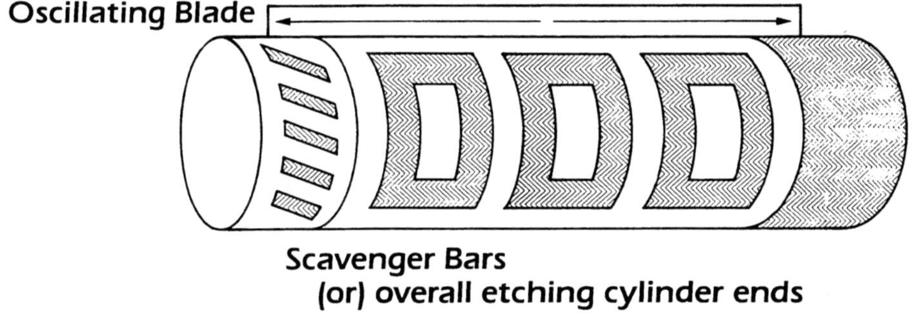

Figure 5 Schematic of a gravure cylinder with scavenger bars, 100% etching of a cylinder end, and oscillating doctor blade.

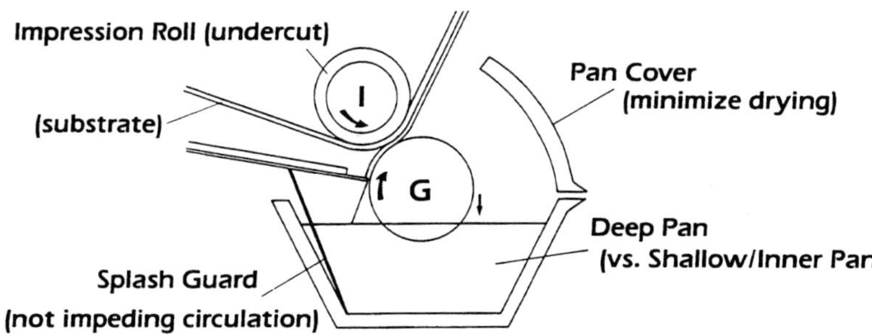

Figure 6 Schematic of a gravure station.

The pan ideally should be deep, thereby increasing volume, and covered if possible, with the flow not restricted by inappropriate positioning of the splash guard.

Additionally, thought should be given to minimizing hot drafts from the driers onto the application station.

Rewind tension should be carefully controlled to avoid blocking from too tight a roll. Cohesives exhibit antitelescoping frictions, and low rewind tensions are indeed functional in this respect. Cohesive–coated wrap typically produces noise through the rewind and slitting operations because of the inherent tendency to cling.

Converted rolls should be cradled or stored on end, rather than on their sides. This is to reduce intermittent blocking, which could be caused by the weight of the roll.

73

Sol–Gel Coatings

Lisa C. Klein

Rutgers University
New Brunswick, New Jersey

1.0 Introduction and Basic Processing

The sol–gel process has been suggested as an alternative to conventional methods such as sputtering, chemical vapor deposition (CVD), and plasma spray for applying thin ceramic coatings. Sol–gel thin films are in a few notable cases technically sound alternatives to these methods as well as commercially viable alternatives.

The technology of sol–gel thin films has been around for more than 20 years. The process is surprisingly simple. A solution containing the desired alkoxide is prepared with a solvent and water. It is applied to a substrate by spinning, dipping, or draining. A tacky film forms that is typically 1 μm thick, uniform over large areas, and adherent to the substrate. The sol–gel process is able to apply a coating simultaneously to the inside and outside of complex shapes. The film dries quickly to a microporous coating of oxide.

The chemistry of the sol–gel process can be simplified to:

$$\text{Condensation} - \text{Polymerization: } 2M(OR)_{x-1}OH - H_2O + (OR)_{x-1}M - O - M(OR)_{x-1}$$

$$\text{Hydrolysis: } M(OR)_x + H_2O - M(OR)_{x-1}OH + ROH$$

where M is a metal species (Si, Al, Ti, Zr, etc.) and R is an organic group (methyl, ethyl isopropyl, etc.). These reactions generate an oxide skeleton in the solution. Upon contact with the substrate and exposure to atmosphere, the solution gels, that is, becomes rigid. Remaining solvent can be evaporated readily. These coatings can be heat treated to densify them. Densification means that the microporosity closes and eventually is eliminated. The limitation to this step is the temperature to which the substrate and the film can be subjected. Hermetic films can be achieved in some but not all applications.

Among sol–gel processed materials, thin films are the true leader, and some examples are given for abrasion–resistant, corrosion–resistant, optical, and electrical thin films.

655

2.0 ABRASION–RESISTANT COATINGS

Alumina coatings have been developed because they provide higher hardness on the surfaces of many substrates, in particular stainless steel. In porous and dense conditions, a simple application of alumina on a surface can provide scuff resistance. This means that ceramic coatings have less tendency to show scratches, because ceramics in general have higher hardness than metals and plastics. Some technological problems in the application and adherence of sol–gel alumina coatings come up from thermal expansion mismatch on a large scale.

Oxynitride films have been prepared by nitriding silica gel films. These films are harder than silica alone because they incorporate nitrogen. Gel–prepared oxynitride films incorporate nitrogen in larger proportions than thermally nitrided conventional films. These films offer both mechanical durability and chemical durability in hermetic packages for microelectronic devices, where there is a large demand for dielectric films.

3.0 CORROSION–RESISTANT COATINGS

Early on, phosphate coatings were applied to silicate glass panels to improve their chemical resistance to attack by water. This improvement was accomplished in 0.2 μm. This process made it possible to make a cheaper window, which showed the chemical durability of a more expensive window with the use of a small amount of materials. This points out one of the primary advantages of the sol–gel process in the form of films. The material use is efficient. Any excess material is recovered and can be used again, resulting in little waste. Even relatively costly components such as zirconia can be incorporated in the coating. The added durability of the coating protects the bulk of the material without requiring incorporation of the costly component throughout the material.

4.0 OPTICAL THIN FILMS

There are a number of optical thin films. First, there are rear–view mirrors for cars. These coatings consist of titania–silica–titania interference filters that give the effect of total reflectance. The coatings are applied successively with a backing on in between. Second, there are solar reflecting films for windows. These coatings consist of palladium–containing titania films that show selective absorption (e.g., Irox by Schott Glaswerk). Third, there are antireflective coatings of several types. Both borosilicates and titania silicates have been developed for this application. The index of refraction in this case can be graded by changes in the composition, changes in the microstructure, and changes in the porosity. In comparison to Vycor, the process developed at Sandia National Laboratories, in Albuquerque, New Mexico, shows better long–term behavior, structural stability, and resistance to abrasion in accomplishing antireflection on solar cells and flat plate or tubular solar collectors. Antireflection in the ultraviolet range can be achieved with thoria and hafnia films. Broadband antireflection properties have been used in laser optics applications. Other selective wavelength films have been developed with transition metal oxides.

5.0 ELECTRONIC AND MAGNETIC FILMS

Thin films have been applied from solution for numerous electronic and magnetic applications. Indium tin oxide (ITO) has been developed for electrically conducting transparent films for displays. Cadmium stannate is another tranparent conductor. Opto–electronics such as lead–lanthanum zirconate titanate (PLZT) have been prepared. Iron oxide layers have been used for magnetic films. Vanadia layers have been used for antistatic films. Barium titanate has been prepared in thin layers for multilayer capacitors. Potassium tantalate has been use for capacitors as well. The advantage of thin films from gels for a dielectric is that the thinner the layer, the higher the capacitance or the higher the density of devices possible on a chip.

Tungstate films have been prepared for electrochromic displays. Titania films have been made for photoelectrodes. The area of transition metal oxide films from solutions grows as the filed of microelectronics comes up with more and more devices.

Recently, the interest in supercomputers has spilled over into the sol–gel processing community. The rare earth–transition metal oxide compounds, which show superconductivity at liquid nitrogen temperatures, are difficult to process in practical geometries with the exception of thin films. The sol–gel process, which allows for molecular–scale mixing of the components in these compounds, is especially promising for depositing thin films.

6.0 ADVANTAGES OF SOL–GEL COATINGS

Acceptance of the sol–gel process for thin films is mostly a question of educating the user of thin films and coatings on how to apply these films. The number of commercial products available to the general public in this area is small. There are the spin–on coatings for doping silicon wafers (Accuspin from Allied–Signal Electronic Chemicals and Liquicoat from— Chemicals). There are the liquid oxides for various applications (Atolon from Nippon Soda). Many industries may have internal uses for what can be classified as sol–gel coatings or sputtered coatings.

To summarize sol–gel coatings, there are limitations. The main limitation is that only films around 1 μm thick can be applied with ease by a sol–gel process. Thinner films can be applied, but thicker films cannot. Thick films suffer from the same problems of drying, shrinkage, and cracking that plague bulk samples. However, for thin films there are the applications cited above and probably others that have been overlooked.

An important feature to emphasize is the microporous nature of the films in the early stages. It appears that microporosity has not been exploited for applications in which a high surface area coating is desired. Some applications to look for in sol–gel coatings are catalytic surfaces, membrane properties, and self–lubricating surfaces.

BIBLIOGRAPHY

Brinker, C. J. and M. S. Harrington, *Solar Energy Mater.*, 5, 159 (1981).
Dislich, H. and P. Hinz, *J. Non–Cryst. Solids*, 48, 11 (1982).
Gruninger, M., J. B. Wachtman, and R. Haber, *MRS Symp. Proc.*, 54, 823 (1986).
Klein, L. C., *Annu. Rev. Mater. Sci.*, 15, 227 (1985).
Klein, L. C., *Ceram. Eng. Sci. Proc.*, 5, 379 (1984).
Klein, L. C., Ed., *Sol–Gel Technology for Thin Films, Fibers, Preforms, Electronics and Specialty Shapes.* Park Ridge, NJ: Noyes Publications, 1988, 407 pp.
Livage, J., and J. Lemerle, *Annu. Rev. Mater. Sci.*, 12, 103 (1982).
Mukherjee, S. P., and W. H. Lowdermilk, *J. Non–Cryst. Solids*, 48, 177 (1982).
Pantano, C., and P. Glaser, *J. Non–Cryst. Solids*, 63, 201 (1984).

74

Radiation-Cured Coatings

Joseph V. Koleske

Charleston, West Virginia

1.0 INTRODUCTION

Curing coatings by means of radiation represents one of the new techniques that is replacing the use of conventional or low solids, solvent-borne coatings. Radiation-cured coatings offer a manufacturer several important features. These include:

High solids—usually 100% solids.
Low capital investment (with certain specific exceptions).
Low energy curing costs—low power requirements and elimination of solvent costs.
Rapid cure speeds.
Ability to cure a variety of substrates, including heat-sensitive substrates such as plastics
   and parts for the electronics industry.
Increased productivity.
Shorter curing lines and decreased floor space requirments for operating line and for liquid
   coating storage.
A variety of different chemistries from which to select, and thus broad formulating latitude
   from the wide variety of formulation ingredients available.

The main sources of actinic energy for curing coatings by radiation are electron beam and ultraviolet light.* In 1984, Pincus[1] indicated that there were four suppliers of electron beam (EB) equipment and more than 40 suppliers of ultraviolet light (UV) equipment. The ninth edition (1987) of the *Radiation Curing Buyer's Guide* lists the same number of EB suppliers and about 50 suppliers of UV equipment. In the United States, there were about 100 electron beam units and about 25,000 ultraviolet light units operational in 1983–1984.[1] These figures include laboratory, pilot, and production units. With the industry growing at about

---

*It is realized that other radiation processes such as microwave, infrared, and gamma rays can be used to cure coatings. However, this chapter is only concerned with electron beam and ultraviolet light radiation, which are the most important commercial processes,

659

10–15% per year,[2–4] it is very reasonable to expect that these numbers had increased by the end of the decade. Equipment for EB curing is significantly more expensive than that for UV curing, and it represents the exception listed above in the "low capital requirement" advantage of radiation curing. Even a laboratory-sized EB unit will run into six figures, whereas simple conveyorized UV systems can be obtained for a few thousand dollars. Furthermore, energy consumption is high with EB units and low UV systems.

In 1980 about 80 million pounds of formulated radiation-curable product were sold in the United States. This number increased to 127 million pounds in 1985 and was expected to increase to 216 million pounds in 1990.[4] About 88% of the 1985 total was cured by ultraviolet light and the remainder by electron beam. Thus, it is readily apparent that radiation curing is a specialty area, but it is one that is growing much faster than the overall coatings market. A recent study[5] indicates that the market will be 336 million pounds by 1995, which represents a 9.2% compounded growth rate from the prediction for 1990.

There are two basic technologies involved in the radiation-cured coatings market. The older and more well-known technology involves free radical chemistry, and the cure of compounds containing unsaturation (i.e. acrylates, styrene/unsaturated polyester, and the like). The other technology involves cation chemistry and the cure of cycloaliphatic epoxides and compounds that will copolymerize with them under the influence of Brønsted or Lewis acids.

2.0 EQUIPMENT

2.1 Electron Beam

With electron beam curing, energy transfer is caused by energetic or accelerated electrons.[6–9] Therefore, photoinitiators usually are not used in EB cure formulations and there are no photoinitiators fragments, which might have an odor or otherwise detract from properties in the cured coating. The electron source is a filament that is heated inside a vacuum tube. The electrons are accelerated to a high energy ($\sim 10^5$ V) by an impressed electrical field with the degree of acceleration increasing with increasing applied voltage. High voltages (150,000–300,000 V) are involved. The accelerated electrons pass through a metallic foil window and are sent on to the compound, which is capable of absorbing their energy and thus undergoing polymerization. The ability of the electrons to be absorbed by the compound (i.e., the radiation-curable formulation) depends on the material's density, and therefore depth of penetration is inversely proportional to density. The electron bombardment of the formulated coating abstracts hydrogen atoms from some of the molecules, thereby generating short-lived free radicals that can initiate the polymerization of acrylates and similar compounds that will interact and react with such species. Clear coatings of up to 20 mils (500 μm) and pigmented coatings of about 15 mils (400 μm) can be cured with EB equipment.

2.2 Ultraviolet Light

In general, ultraviolet light units operate with electromagnetic radiation that is in the optical region of 200–760 nm. They do produce infrared radiation of 760 nm to 1.0 mm, but this energy is thermal and acts either to anneal the cured coating and relieve internal stresses and strains or to enhance cure rate in cationic cure systems. Of course, in certain formulations this thermal energy can have a deleterious effect by causing volatilization of reactants, and it is minimized in some equipment. In other cases, such as cationic UV curing, the thermal energy can be highly beneficial by kinetically enhancing the cure rate of the compounds.

Basically, in the case of UV curing, compounds susceptible to rapid polymerization are contacted with initiating species obtained by photolysis of a photoinitiator. In the UV cure* of such compounds, a photoinitiator that is capable of photolyzing or degrading to an active species is added to the formulation. When light of the proper wavelength strikes the photoinitiator, the active species is generated and polymerization rapidly takes place. Certain photoinitiators generate free radicals, and these are used to cure acrylates. Other photinitiators generate cations, which are used to cure cycloaliphatic epoxide-based systems. It should be readily apparent that matching the output of the ultraviolet light source with the absorption spectra of the photinitiator is an important aspect of this technology.

There are four different ultraviolet light curing technologies. These are as follows.

Medium pressure mercury vapor lamps (in certain countries, low pressure and high pressure
    mercury vapor lamps are also used).
Electrodeless vapor lamps.
Pulsed xenon lamps.
Lasers.

Medium pressure mercury vapor (MPH) lamps have been used commercially for about 20 years. The bulb is an evacuated quartz tube that contains metallic mercury and has electrodes at each end. Electrical energy is supplied through the electrodes, and an arc is struck between them. This heats mercury in the tube to a plasma, which emits UV, visible, and infrared radiation. Output ranges from about 100 to 400 in., and a warm-up time of 10–15 minutes is often recommended for the bulb to be fully operational. The systems cam be doped to alter the emission spectrum, but usually only mercury is used.

Electrodeless lamps have been in use for about a decade and currently are very popular in the industry.[10,11] The vacuum tube UV bulb is manufactured from quartz, which is invisible to the UV radiation. The bulb contains either mercury or other proprietary metals and gases. The systems is activated by microwave or radiofrequency energy (i.e. it does not have electrodes). Because of the nature of this activation system, it essentially has instant on and off operation. There are available a variety of doped bulbs with various output spectra that can be "matched" to that of different photoinitiators. Bulb lifetime is usually 5–10 times longer than that of standard MPH lamps.

Xenon lamps are quartz tubes filled with doped xenon. The lamp is powered by pulsed electrical current.[10,12] These units offer very low heat ouput along with short time, extremely high peak intensity output (some types have as high as 8000–10,000 W output, although most operate at much lower intensities). The output spectrum is continuous with this source rather than that of the discrete line types as from the MPH lamps. However, in certain instances mercury is added to the tube contents to enhance curing in the region of mercury's spectral lines. The output has also been modifies with other metals such as iron and beryllium.

Argon ion and nitrogen lasers in combination with specific photinitiators that have a strong absorbance at the emission line of the laser have been used to cure multifunctional acrylates.[13,14] Although the studies are interesting and may hold promise for the future, at

---

* Ultraviolet light photons are of relatively low energy (~103 V) and are not capable of abstracting hydrogen atoms from molecules. Thus, a photoinitiator is needed when ultraviolet light curing technology is practiced.

present this is considered to be a research area of potential interest for the electronics industry,

3.0 CHEMISTRY

Most of the following discussion deals with ultraviolet light technology. The reason for this is that photinitiator breakdown is important to cure. In a general sense, the same compounds that will cure with the free radical photoinitiators will cure with electron beam.

3.1 Photoinitiators

3.1.1 Free Radical Type

Free radical generating photoinitiators are of two general functional types. [15,16] The first type involves a mechanism known as *homolytic fragmentation*, in which a compound such as a benzoin alkyl ether undergoes a photochemically induced fragmentation into highly active free radicals as described in Figure 1.

The second type of free radical generation functions through a mechanism that is termed *electron transfer*. This mechanism involves a photolytic excitation of the photinitiator from a fround state singlet to an excited triplet state. This is followed by electron transfer to a hydrogen atom doner, such as dimethylethanolamine (DMEA), and the formation of highly excited free radicals as described in Figure 2.

Typical commercial photoinitiators include compounds such as 2,2–diethyoxyacetophenone, 2,2–dimethoxy–2–phenyl acetophenone, hydroxycyclohexylphenyl ketone, benzophenone,–triethylamine, 2–methyl–1–[4–(methylthio)phenyl]–2–morpholino–propane–1, 1–phenyl–1,2–propane dione–2–(o–ethoxycarbonyl)oxime, and benzoin methyl, isopropyl, isobutyl, and other alkyl ethers.

Free radical generating photoinitiators of the foregoing types are inhibited or inactivated by oxygen as a result of a complex that forms between the light-activated photoinitiators and molecular oxygen. This effect can be overcome by inerting the coating with nitrogen during cure, by adding waxes to the system, or by use of excess photoinitiator. Air that has been dispersed in the coating system during formulation contains oxygen, and it acts as a stabilizer. However, formulations containing very active photinitiators of this type have a tendency to polymerize during storage if this oxygen is depleted over a period of time. Compounds that will help prevent such instability include phenothiazine and Mark 275 stabilizer.

Tertiary amines will act as photosynergists,[17,18] and they greatly enhance curing rate of compounds such as those described above. Ureas and amides also have been described as synergists for benzophenone.[19] Compounds that have been used to accelerate cure rate of

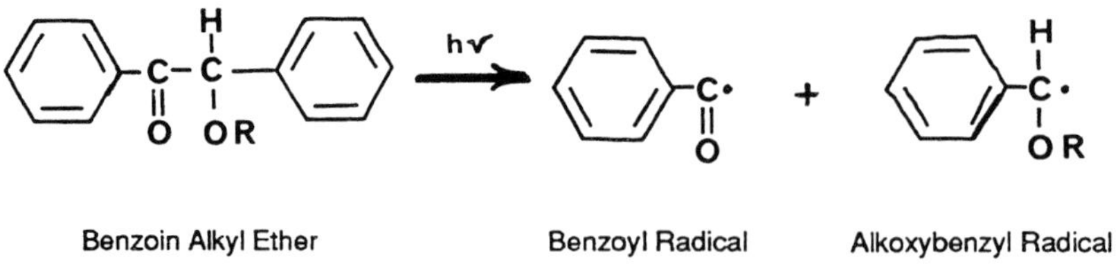

Benzoin Alkyl Ether Benzoyl Radical Alkoxybenzyl Radical

Figure 1 Homolytic fragmentation.

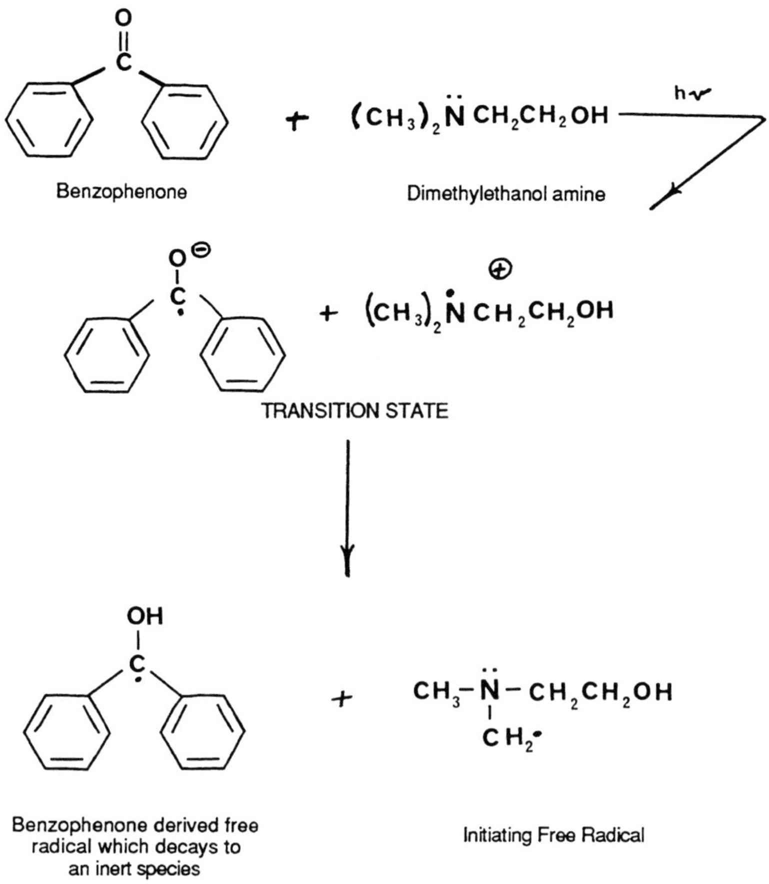

Figure 2 Electron transfers.

pigmented systems include isopropyl–thioxanthone, ethyl–4–dimethylaminobenzoate, and 2–chlorothio–xanthone.

3.1.2 Cationic Type

Although there are various types of photoinitiators that photolyze to yield a cationic species capable of polymerizing cycloaliphatic epoxides and active hydrogen compounds of the hydroxyl type or vinyl ethers, only the arylsulfonium salts are commercial at present. These types include aryldiazonium salts, aryliodonium salts, iron–arene complexes, aluminum complex–silanols, and the commercial arylsulfonium salts.[20–24]

Aryldiazonium hexafluorophosphates and tetrafluoroborates decompose under the action of UV light and yield Lewis acids such as BF_3 and PF_5, nitrogen, and other fragments.[25–27] These photoinitiators were used in the infancy of cationic UV cure of cycloaliphatic epoxides. Although they were quite active for first-generation products, the disadvantages of thermal instability, which led to short shelf life, and of nitrogen evolution, which led to pinholes and bubbles in films thicker than about 0.2 mil, inhibited commercial use and led to their replacement by the onium salts in the marketplace.

The polymerization of epoxides with aluminum complex–silanol photoinitiators has been described [28,29] The technology is not being practiced in the United States, but it may be in use in Japan. The iron–arene complexes represent a new type of cationic photoinitiator that was recently described.[30,31] When photolyzed, these compounds degrade to yield both Lewis acid type catalysts and free radicals. Since these compounds are relatively new, detailed information about them is not available.

Various investigators studied the onium salts of iodine or the Group VI elements.[32–37] Currently, the arylsulfonium salts are commercially used as photoinitiators. These compounds do not have the deficiencies of the diazonium salts because there is no nitrogen evolution on photolysis and, if protected from ultraviolet light, the systems can have ambient-condition shelf lives in excess of 2 years. When UV light interacts with the onium salts, an excited species is formed. This species undergoes homolytic bond cleavage to yield a radical cation, which extracts a hydrogen atom from a suitable donor and generates another free radical species. The new compound then gives up the proton for formation of a strong Brønsted acid. The Brønsted or protic acid that is the polymerization catalyst is of the form HMF_6 where M is a metal such as antimony, arsenic, or phosphorus. This catalyst is long-lived, and the cationic polymerization of the epoxide system can continue in the "dark" after initial exposure to UV light until the available epoxide is exhausted or the polymerization is terminated by some other mechanism. Thus, the onium salts generate both cationic species and free radicals and can be used in radiation-activated, dual-mechanism systems.

Note that the onium salt photoinitiator is a blocked or latent photochemical source of the strong Brønsted acid that acts as a catalyst/initiator for the formulated system. Because of the acidity of the UV-generated catalyst or inititator, it is necessary to keep the formulated system (substrate, coating equipment, etc.) free from basic compounds that would neutralize the acid and either negate or slow down cure rate. Even very weak basic compounds will react or interact with the strong acidic species.

3.1.3 Dual-Mechanism Curing

Since the cationic photoinitiators generate both free radicals and Brønsted acids when exposed to ultraviolet light, it is possible to combine acrylates that will cure with free radicals and epoxides that cure with the protic acids. Free radical generating photinitiators such as 2,2-diethoxyacetophenone can be added, if an additional source of free radicals is necessary. Experience has shown that this usually is not necessary. Of course, the benzophenone–amine systems describes earlier should not be used. Little can be found in the literature[38–40] about this interesting topic, but dual-mechanism curing should prove to be a useful technique in the future and merits further study.

Dual-mechanism systems that involve free radical chemistry coupled with thermal chemistry are also known. Dual-cure plastisols[41] and dual-cure pigmented[42] coatings have been reported. The combination of ultraviolet and infrared radiation for curing coatings, which can be beneficial for either of the dual-mechanism systems, has also been discussed.[43,44]

3.2 Formulation

The subject of formulation involves both free-radical-curable ingredients and how they are used, and cationic-curable ingredients and how they are used. If one takes an extremely simplistic approach, only a single species that will cure under the influence of radiation is needed for a coating. However, most formulations contain a variety of ingredients to achieve the property balance that is needed to meet a given set of performance criteria.

Most formulations contain a base oligomeric or multifunctional compound, a flexibilizer (which may be the base compound) a low viscosity, reactive diluent (which may be mono– or multifunctional), for viscosity control, a multifunctional acrylate for high cross-link density, a photoinitiator (unless electron beam cure is used), and usually a surfactant or flow and leveling agent. Combinations of more than one material of each type are usually used to optimize properties. The performance characteristics of specific end uses may also require use of a slip agent, nonreactive organic or inorganic fillers and/or pigments, adhesion promoters, flattening agents, an augmenting flow and leveling aid, or some other additive-type ingredient.

3.2.1 Free Radical Systems

In general, systems based on acrylates are used in free radical cure systems. [45–48] The base material is usually an epoxy acrylate or a urethane acrylate. Epoxy acrylates are the reaction products of acrylic acid with various diglycidyl ethers of bisphenol A. The epoxy acrylates contain acrylate and hydroxyl functionality, but they do not contain epoxide functionality and should not be confused with the epoxides to be discussed in the cationic cure systems section. The term "epoxy acrylate" is a widely used and accepted misnomer; as can be seen by the structural formula in Figure 3, "acrylated epoxy" would be a better term to use for these compounds.

Urethane acrylates are often prepared by end capping a polyether, polyester, or caprolactone polyol with a diisocyanate and then reacting this isocyanate prepolymer with an hydroxyalkyl acrylate.[49–51] The chemistry is more complex than this simple description, and in order of component addition is important to minimizing viscosity. Both epoxy acrylates and urethane acrylates have a high viscosity ($\geq 10^6$ cp) at ambient temperatures.[52] To facilitate manufacture, handling, and later formulation, the compounds are often made in a low viscosity mono- or multifunctional acrylate, which will later serve as a reactive diluent and/or cross-linking agent in a fromulated system.

In either the neat or the diluted form, these base compounds are diluted to application viscosity with more mono- or polyfunctional acrylate. Polyfunctional acrylates increase cross-link density and improve solvent resistance, increase hardness, and increase glass transition temperature. Care is always taken to avoid skin and eye contact and inhalation of the acrylates because of the potential for skin irriatation, sensitization, and toxilogical reaction. It is important to read the Material Safety Data Sheets supplied with these and all other compounds. The manufacturer's precautions should be carefully read and heeded when working with these and other chemicals. The ACS Monograph *Chemical Carcinogens* (No. 182, published in 1984) is a useful general reference.

Acrylates are available from a variety of sources. In addition to the urethane acrylates and epoxy acrylates described above, polyfunctional acrylates such as trimethylol propane triacrylate, pentaerythritol triacrylate, 1,6-hexane diol diacrylate, tripropylene glycol diacrylate, and tetraethylene glycol diacrylate, are available. There has been a trend to increase molecular weight by alkoxylation of compounds used to make multifunctional acry-

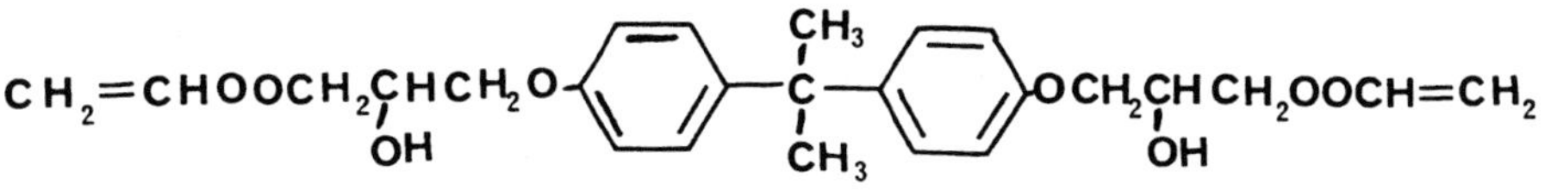

$$CH_2{=}CHOOCH_2\underset{\underset{\textstyle OH}{|}}{C}HCH_2O{-}{-}\underset{\underset{\textstyle CH_3}{|}}{\overset{\overset{\textstyle CH_3}{|}}{C}}{-}{-}OCH_2\underset{\underset{\textstyle OH}{|}}{C}HCH_2OOCH{=}CH_2$$

Figure 3 An "epoxy acrylate."

lates, to give them better handling and health characteristics. Monofunctional compounds useful as reactive diluents include *N*–vinyl–2–pyrrolidone, 2–ethylhexyl acrylate, dicyclopentadiene acrylate, hydroxyalkyl acrylates, hydroxylactone acrylates, ethoxyethoxyethyl acrylate.

Specific formulations are highly varied, and performance requirements guide or dictate ingredient levels. Many formulations can be found in the cited literature or other literature available from material manufactures.

3.2.2 Cationic or Epoxy Systems

The most important formulating ingredient in a cationic UV cure system is a cycloaliphatic epoxide of the 3,4–epoxy cyclohexylmethyl–3,4–epoxy cyclohexane carboxylate or bis(3,4–epoxy cyclohexylmethyl) adipate type.[53] Systems usually contain from 100% to about 30–49% cycloaliphatic epoxide. When this epoxide is used alone or at very high concentrations, strong, hard, and brittle coatings that are useful on rigid substrates result. These rigid coatings can be flexibilized and toughened in various ways. Although commercial, compounded flexibilizers/tougheners exist[20] for these systems, various polyols such as the propylene oxide[54] or caprolactone polyols[55] can be used. Polyester adipates can be used, but the relatively high acidity of these polyols can lead to shortened shelf life because the cycloaliphatic epoxides are well-known acid scavengers[56] and will readily react with any carboxylic or other acid groups in this system. This will either increase viscosity or cause gelation. Other flexibilizing agents include epoxidized soy bean and linseed oil epoxides and epoxidized polybutadiene. Care should be exercised when incorporating these compounds in the formulation because they can cause significant softening, along with flexibilization, and little or no increase in toughness.

Relatively small amounts (~1–20%) of the diglycidyl ethers of bisphenol A can be added to systems. However, the light absorbing characteristics of these compounds lead to a decrease in cure rate and in depth of cure. In addition, the compounds cause rapid increases in viscosity. Novolac epoxies appear to cure well in cationic systems, but their high viscosity is rapidly reflected in formulation viscosity.

Low molecular weight epoxides available under trade names[20] can be used as reactive diluents. Although somewhat slower in reactivity than many other cycloaliphatic epoxides, limonene mono– and diepoxide can be used as reactive diluents. Vinyl ethers can act as reactive diluents and cure rate enhancers in cationic cure, cycloaliphatic epoxide based sytems.[57,58] These compounds have not been fully investigated, but the available evidence suggests that they have formulating potential.

Since nonbasic, active hydrogen compounds react under cationic conditions with the oxirane oxygen of cycloaliphatic epoxides to form an ether linkage between the compound and the ring and a secondary hydroxyl group on the epoxide ring[54], low molecular weight alcohols, ethoxylated or propoxylated alcohols such as butoxyethanol, and similar compounds can be used as reactive diluents in cationic systems. However, since these compounds are monofunctional, they can act as chain stoppers—although they do generate the secondary, ring-attached hydroxyl group, which can further propagate polymerization or chain extension—and can be used only in limited amounts, about 1–10%, that are dependent on molecular weight. Low molecular weight glycols (diethylene glycol, 1,4–butanediol, etc.) can also be used. Such compounds may enhance cure rate by providing a source of active hydrogen; but, when used at permissible low levels, the glycols do not enhance toughness. In certain instances, inert solvents such as 1,1,1–trichloroethane are used to decrease viscosity and/or increase coverage from a given volume of coating. However, most end users prefer systems that only contain reactive components.

As mentioned above, the reaction mechanism of epoxides and hydroxyl groups[53,54] is such that a new hydroxyl group is generated for every hydroxyl group that is present. Thus, the initial hydroxyl content of a formulation is conserved after the reaction is complete. Although low levels of hydroxyl groups will often enhance adhesion, too many of these groups can detract from performance characteristics and cause adhesion loss under wet, moist, or high humidity conditions.

Specific formulations are highly varied, and performance requirements guide or dictate ingredient levels. Many formulations can be found in the cited literature or other literature available from material manufacturers.

4.0   END USES

Radiation-cured coatings, which are often taken to include inks, adhesives, and sealants, are used in a large number of ways for rigid and flexible metal, plastic, glass, paper, and wood substrates. Particular end-use arenas include the communication, construction, consumer products, electronics, graphic arts, medical/dental, packaging, and transportation markets. Specific end uses for radiation-cured compounds are numerous and include coatings for appliances, beer and beverage can bodies and ends, book covers, bottles and bottle caps, catalogs, closures, compact discs, cosmetic cartons, credit cards, decorative and functional foils and films, decorative mirrors, electronic components, flocked fabric, furniture, labels, magazines, magnetic tape, natural and simulated wood paneling, optical fibers, orthopedic casts and splints, photoresists, plastic cups and containers, printed circuit board assemblies (conformal coatings), record album jackets, solder masks, steel can ends for composite paper–metal cans, toys, transfer letters, and vinyl flooring.

REFERENCES

1.   A. H. Pincus, *Radiat. Curing*, 11 (4), 16 (November, 1984).
2.   Anon., "Revolution inf radiation curing," *Chem. Week*, p. 20, April 13, 1983.
3.   A. H. Pincus, "Radiation curing markets and marketing data," in *Proceedings of Radiation Curing VI*, Chicago, 1982.
4.   J. Weisman, "Radiation curing in the United States—An overview," in *Proceedings of Conference on Radiation Curing Asia*, Tokyo, 1986, p. 11.
5.   G. E. Ham, personal communication.
6.   R. Kardashian and S. V. Nablo, Adhes. Age, p. 27, December 1982.
7.   S. V. Nablo, *Radiat. Curing*, 10(2), 23 (May 1983).
8.   P. N. Cheremisinoff, O. G. Farah, and R. P. Ouellette, *Radio Frequency/Radiation and Plasma Processing*, Westport, CT. Technomic, p. 86.
9.   W. Karmann, "Radiation curing equipment," Paper FC 83–269, *Proceedings of RADCURE '83 Conference* Lausanne, Switzerland, 1983.
10. E. Finnegan, *J. Radiat. Curing* 9,4 (July 1982).
11. J. C. Pelton, *Radiat. Curing*, 10(2), 10 (May 1983).
12. W. R. Schaeffer, *"UV curing light sources—Equipment and applications,"* Paper FC85–768, presented at Finishing '85, Detroit, 1985.
13. C. Decker, *J. Polym. Sci. Polym. Chem. Ed.*, 21, 2451 (1983).
14. C. Decker, *J. Coatings Technol.* 56, 29 (1984).
15. G. Oster and N. L. Yang, *Chem Rev.*, 68,(2), 125 (1968).
16. H. G. Heine, H. J. Rosenkranz, and H. Rudolph, *Angew. Chem.*, 11, 974 (1972).
17. M. R. Sander, C. L. Osborn, and D. J. Trecker, *J. Polym. Sci.*, 10, 3173 (1972).
18. C. L. Osborn and D. J. Trecker, U. S. Patent 3,759,807.

19. A. F. Jacobine and C. J. V. *J. Radiat. Curing,* 1026 (July 1983).
20. Union Carbide Corp., Products for Ultraviolet Light-Cured Cycloaliphatic Epoxide Coatings, Publication F–60354, 1986.
21. 3M/Industrial Chemical Products Division, UV Activated Epoxy Curative FX–512, 1986.
22. General Electric Company, UVE–1014 Epoxy Curing Agent, Publication 1MF–212, 1981.
23. Asahi Denka KK, Opt Photoinitiators for Cationic Cure, 1986.
24. Degussa AG, "Degacure KI 85, 1986.
25. J. J. Licari and P. C. Crepeau, U. S. Patent 3,205,157 (1965).
26. S. I. Schlesinger, U. S. Patent 3,708,296 (1973).
27. S. I. Schlesinger, *Photogr. Sci. Eng.* 18, 387 (1975).
28. S. Hayase, T. Ito, S. Suzuki, and M. Wada, *J. Polym. Sci. Polym. Chem. Ed.,* 20, 1433 (1982).
29. S. Hayase and Y. Onishi, U. S. Patent 4,476,290 (1984).
30. K. Meier, "Photopolymerization of epoxides—A new class of photoinitators based on cationic iron–area complexes," Paper FC85–417, in *Proceedings of RADCURE Europe '85,* Basel, Switzerland, 1985.
31. K. Meier and H. Zweifel, *J. Radiat. Curing* 13 (4), 26 (October 1986).
32. J. V. Crivello and J. H. W. Lam, *Macromolecules,* 10, 1307 (1977).
33. J. V. Crivello, U. S. Patent 4,058,401 (1977); 4,138,255 (1979); 4,161,478 (1979).
34. G. H. Smith, U. S. Patent 4,173,476 (1979).
35. R. F. Zopf, *Radiat. Curing,* 9(4), 10 (1982).
36. R. S. Davidson and J. W. Goodwin, *Eur. Polym. J.,* 18, 589 (1982).
37. J. V. Crivello and J. L. Lee, *Polym. Photochem.,* 2, 219 (1982).
38. W. C. Perkins, *J. Radiat. Curing,* 8(1), 16 (1982).
39. F. A. Nagy, European Patent Application EP 82,603 (1983).
40. H. Baumann et al., East German Patent Application DD 158,281z (1983).
41. C. R. Morgan, "Dual UV/thermally curable plastisols." Paper FC83–249, in *Proceedings of RADCURE '83 Conference,* Lausanne, Switzerland, 1983.
42. A. Noomen, *J. Radiat, Curing,* 9(4), 16 (1982).
43. E. M. Barisonek, "Radiation curing hybrid systems," Paper FC83–254, in *Proceedings of RADCURE '83 Conference,* Lausanne, Switzerland, 1983.
44. J. L. Lambert, "Heating in the IR spectrum," Industrial Process Seminar, September 1975.
45. S. Saraiya and K. Hashimoto, *Mod. Paint Coatings,* 70(12), 37 (1980).
46. E. Levine, *Mod. Paint Coatings,* 73, 26 (1983).
47. C. B. Thanawalla and J. G. Victor, *J. Radiat. Curing,* 12, 2 (October 1985).
48. K. O'Hara, *Polym. Paint Colour J.* 175 (4141), 254 (1985).
49. L. E. Hodakowski and C. H. Carder, U. S. Patent 4,131,602 (1978).
50. M. S. Salim, *Polym. Paint Colour J.,* 177 (4203), 762 (1987).
51. B. Martin, *Radiat. Curing,* 13, 4 (August 1986).
52. G. Kühe, *Polym. Paint Colour J.,* 173, 526 (Aug. 10/24, 1983).
53. J. V. Koleske, O. K. Spurr, and N. J. McCarthy, "UV–cured cycloaliphatic epoxide coatings," in *14th National SAMPE Technical Conference,* Atlanta, 1982, p. 249.
54. J. V. Koleske, "Mechanical properties of cationic ultraviolet light-cured cycloaliphatic epoxide systems," in *Proceedings of RADCURE Europe '87,* Munich, West Germany, 1987.
55. J. V. Koleske, "Copolymerization and properties of cationic, UV–cured cycloaliphatic epoxide systems," in *Proceedings of RADTECH '88,* New Orleans, 1988.
56. Union Carbide Corp., "Cycloaliphatic Epoxide ERL–4221 Acid Scavenger-Stabilizer," publication F–5005, March 1984.
57. J. V. Crivello, J. L. Lee, and D. A. Conlon, "New monomers for cationic UV–curing," in *Proceedings of Radiation Curing VI,* Chicago, 1982.
58. GAF Corp., Triethylene Glycol Divinylether, 1987.

75
Nonwoven Fabric Binders

Albert G. Hoyle

Hoyle Associates
Lowell, Massachusetts

1.0 INTRODUCTION

A nonwoven fabric is precisely what the name implies, a fibrous structure or fabric that is made without weaving. In a woven or knit fabric, warp and/or filling yarns are made and intertwined in various patterns (weaving or knitting) to interlock them and to give the manufactured fabrics integrity, strength, and aesthetic value. By contrast, in manufacturing a nonwoven fabric, the yarn formation and yarn intertwining steps (weaving or knitting) are bypassed, and a web (fibrous structure) is formed using dry-lay or wet-lay formation techniques. This web is bonded together by mechanical entanglement or by the addition of a binder to create a nonwoven fabric.

This chapter describes the various binders available for nonwoven bonding with their applications, and provides a listing of resource contacts for latex, binder solutions, fiber, powder, netting, film, and hot melt binder suppliers.

2.0 BINDERS

The degree of bonding achieved, using any of several binders, is enhanced when the carrier fiber and binder are of the same polymeric family. Increasing the amount of binder in relation to the carrier fiber increases product tensile strength and also overall bonding. Binders used in nonwovens are of the following types: latex, fiber, powder, netting, film, hot melt, and solution.

At present, the binders most frequently used are latex, fiber, and powder, with fiber having the greatest growth potential for the future.

2.1 Latex

Latex binders are based mainly on acrylic, styrene-butadiene, vinyl acetate, ethylene-vinyl acetate, or vinyl/vinylidene chloride polymers and copolymers. Within any one series or group, very soft to very firm hands can be achieved by varying the glass transition temperature of the polymer. The lower the T_g, the softer the resultant nonwoven. These temperatures range from –42° to +100°C. in latex available today. Most latex are either anionic or non-ionic. Some have high salt tolerances, allowing for addition of salts to achieve flame retardancy. Some are self-cross-linkable, and others are cross-linkable by the addition of mela-mine– or urea–formaldehyde resins and catalysts to achieve greater wash resistance and high wet strength in cellulose-based nonwovens. Other latex by means of special surfactants can impart either hydrophobic or hydrophilic properties to the nonwoven. Still others, by means of certain surfactants, can be foamed easily to allow for foam application of the latex, resulting in greater economies in processing because of much lower energy demands for drying. Addition of thickeners to increase viscosity of the latex formulation makes for a much neater and cleaner application when print bonding is used. Addition of special heat-sensitive coagulants to the latex makes it possible to apply the latex to the fibrous web in a foamed state, and to gel or coagulate the foam in situ before it collapses while drying, result-ing in permanently foamed, open cell nonwoven structure that is very water-absorptive and water-retentive. By and large the acrylic-based latex are the most versatile ones in use, and they are also the most expensive to use.

2.2 FIBER

The main fiber binder types are morphologically classified as follows: amorphous homo-ploymer, amorphous copolymer, crystalline copolymer, and bicomponent fiber.

Classifying these by chemical origin, the ones most used today are polyester, polypro-pylene, polyethylene, polyamide, and vinyl chloride, vinyl acetate copolymer.

While latex binders are applied and processed by saturation, spray, foam, and print tech-niques, fiber binders are activated by thermal bonding methods. Thermal bonding, when first introduced, was used mainly in durable nonwovens and not used at all in disposables such as diaper top sheets and sanitary product covers—both exceedingly high volume items—for two reasons: (a) the higher prices of binder fibers when compared to latex, and (b) the properties needed in disposable nonwovens softness, strength, porosity, capability of transferring liquids to an inner absorptive medium, and no adverse reactions upon skin con-tact. As energy cost increased in the late 1960s and early 1970s, thermal bonding became a viable option for disposable as well as durable nonwovens.

2.21 *Amorphous Homopolymer*

The predominant amorphous homopolymer binder fibers were and still are polyester and polypropylene. The polyester binder fiber is blended with a polyester carrier fiber into a web by carding or air-lay methods and hot calendered to achieve a 100% bonded product. This gives a thin strong, low extensibility, papery product, which is being used successfully as a coating substrate and in electrical insulation. The 100% machine direction orientation from carding and relatively high binder content result in relatively low extensibility and high ten-sile strength, properties that are important , specifically, in coating substrates for tapes and for electrical insulation. Amorphous homopolymer polyester binder fiber is used exclu-sively in area bonding for industrial products.

Polypropylene carrier fiber can be self-bonded by making a blend of coarse and fine den-ier fibers and using the fine denier fiber as the binder for he coarse one. This procedure is practicable because polypropylene has a sharp bonding temperature and an almost simulta-neous melting point, in the vicinity of 165°C. when the blend is hot calendered, the fine denier fiber is melted while the coarse one is relatively untouched. A practical blend is one containing fibers of 6, 3, and 1.8 deniers per filament (d/f), and running the calender at

163–165°C. A highly compacted, fairly stiff nonwoven results, which is used in water filtration. A polypropylene fiber with a slightly lower melting point than regular polypropylene has been developed for use as a binder fiber. This fiber is capable of being used a binder for regular polypropylene fibers of any denier and, with the sharp melting range characteristic of polypropylene, is usable in point bonding of disposables containing fibers other than polyester, as well as in area-bonded products.

2.2.2 Amorphous Copolymer

With the advent of point bonding by hot calendering and the development of amorphous copolymers, mainly polyesters, work was reinitiated in adapting thermal bonding to disposable product needs. Overall area bonding is not acceptable in most disposables because open areas are needed for porosity and for the passing of liquids.

The development of amorphous copolymers with activation temperatures lower than amorphous homopolymers was an important development, allowing the making of a soft, absorbent, thin cover stock on a heated spot bonding calender at speeds high enough to make the process and product competitive.

Amorphous copolymer fibers are available with working temperatures in the calender from 227°C. down to 163°C., allowing a full range of light to medium-weight products, either 100% area bonded or else point bonded to be made at efficient processing speeds.

2.2.3 Crystalline Copolymer

Crystalline copolymer binder fibers are available in both polyamide and polyester types. These fibers are characterized by very sharp melting points, where the fiber is converted almost instantly to liquid when heat is applied. Microscopic studies show that the binder fiber, once molten, tends to congregate in droplets at carrier fiber interstices throughout the nonwoven.

Thermal bonding with this type of fiber can be effected in two ways: spot bonding with a heated spot bonding calender, and bonding with a through air oven with or without a consequent cold calendering step, using an overall bonding calender. Through air oven heating is the preferred method of bonding because greater control and consistency can be achieved by using it.

A crystalline copolymer binder fiber can be used to produce a spot-bonded product with open unbonded areas, a spot-bonded product with open bonded areas, an overall bonded bulky product, and an overall bonded compacted but not papery product.

2.2.4 Bicomponent Fiber

A bicomponent is a fiber formed by coextrusion of two different polymers. It is available commercially in polyamide, polyester, and polyolefin (polypropylene–polyethylene) combinations. The polyamides and polyesters are of the core sheath type, with the matrix fiber as the core and the binder fiber as the sheath. In the polyamide bicomponent, the core fiber is nylon 66 and the sheath fiber is nylon 6.

In the polyester bicomponent, both the carrier and the binder fibers generically are polyesters; the main difference is a lower melting or softening point in the binder fiber portion. The polyolefin bicomponent is of the side-to-side type. Polypropylene, the carrier fiber, is one side, and polyethylene, the binder fiber, is the other side. In all three instances— polyamide, polyester, and polyolefin—the melting or processing temperature of the binder fiber is sufficiently lower than that of the carrier fiber that the carrier fiber is not affected by the heat used in the bonding step.

Nonwovens containing bicomponent fibers are processed similarly to those containing crystalline copolymers, and the resulting nonwovens have properties similar to those made using crystalline copolymers. Overall economics tend to favor use of crystalline copolymers, but achieving the best quality product favors use of bicomponent fibers.

2.3 Powder

Because of refinements in powder application techniques and an increase in polymers available in powder form, use of powder as a thermal binder is now a viable process. Polyester, polyamide, polypropylene, and polyethylene are available in a variety of mesh sizes. The preferred method of bonding is radiant heat followed by cold calendering.

Products made with powder binders contain 10–20% binder. They have a soft hand and are porous. Possible markets are diaper top sheets, sanitary napkins, mattress pads, wipes, and medical–surgical products. Powder binders give characteristics and properties similar to crystalline copolymer fibers. Advantages of using powder over crystalline copolymer fibers are economies of powder versus fiber and versatility of binder selection. Disadvantages are less efficient application methods of powder versus fiber and permanence of location of powder once applied.

2.4 Netting

The netting type of binder is applied to a lightly prebonded web using a hot calender. The effect achieved is similar to that of spot or pattern bonding, and is one-sided. Low or high amounts of binder can be deposited at the patterned bond points, depending on the netting pattern. The main advantage of this process is that it can be off machine; it is relatively simple, and requires no large outlay for processing equipment.

2.5 Film

The film is combined with a web containing a carrier fiber and an amorphous binder fiber in a hot calendering operation. The end-product nonwoven is quite tough, very strong, very smooth on one side, and usually impermeable to vapor or liquid.

2.6 Hot melt

Hot melt thermal bonding is accomplished by applying a hot melt to a lightly prebonded web. The hot melt can be applied to the web directly from a patterned hot melt applicator roll or else cast on release paper and applied in a separate step in transfer printing or bonding equipment. Characteristics and properties of a nonwoven made in this manner resemble those of one made using a crystalline coploymer binder fiber or plastic netting as the binder.

Advantages of using this binder are the capability of applying special binder formulations in unique binding patterns.

2.7 Solution

Solution binders and coating adhesives are available in both aqueous and organic solvent mediums. In nonwoven bonding the only solutions used are aqueous based, and these are selected only for very specific applications.

Polyvinyl alcohol and some acrylics are used as prebinders or temporary binders for cellulose-based or fiberglass-based nonwovens when final bonding cannot or should not be preformed at that particular stage of processing or when regular binders are needed but only in very small amounts. Water resistance can be imparted by adding a cross-linking agent such as melamine– or urea–formaldehyde resin to the polyvinyl alcohol formulation or using a cross-linking acrylic followed by curing after the product is dried.

BIBLIOGRAPHY

Hoyle, A. G., "Properties and characteristics of thermally bonded nonwovens," in *TAPPI Nonwoven Division Workshop on Synthetic Fibers for Wet Systems and Thermal Bonding Applications*, Technical Association of the Pulp and Paper Industry, 1986, pp. 51–54.

Nonwovens Industry. Rodman Publications Inc., 26 Lake Street, P. O. Box 555, Ramsey, NJ 07446; phone (201) 825–2552.

Nonwovens World. 2700 Cumberland Parkway N.W., Suite 530, Atlanta, GA 30339; phone (404) 432-3186.

Textile World. McGraw-Hill Publications Company, 1221 Avenue of the Americas, New York, NY 10020; phone (212) 391–4570.

76

Fire-Retardant/Fire-Resistive Coatings

Joseph Green

FMC Corporation
Princeton, New Jersey

Joseph Green

FMC Corporation
Princeton, New Jersey

Paint-type coatings can be divided into three general classes: conventional paints, varnishes, and enamels; fire-retardant coatings formulated with halogen compounds with or without special fillers; and intumescent coatings designed to foam upon application of heat or flame for development of an adherent fire-resistive cellular char.

1.0   CONVENTIONAL PAINTS

Non-flame-retardant coatings usually give a low flame spread rating over asbestos–cement board, steel, or cement block. When the coatings are tested over wood and other flammable materials, flame spread ratings similar to those of the substrate are obtained.[1]

The fire-retardant effectiveness of paints is highly dependent on the spreading rate or thickness of the coating as well as the composition. When conventional paints are applied at the heavy rate common for fire-retardant coatings, they give flame spread indices comparable to those of fire-retardant paints. For example, coating of latex and flat alkyd paints applied to tempered hardboard at an effective spreading rate of 250 ft²/gal reduced the flame spread index of the uncoated substrate by factors of 3 and 5, respectively.[2]

2.0   FIRE-RETARDANT PAINTS

Fire-retardant coatings are particularly useful in marine applications. Ships are painted repeatedly to maintain maximum corrosion protection. As the layers of paint built, they pose a fire hazard even though the substrate is steel. In the event of fire, the paint may catch fire, melt, drip, and cause severe injury and damage to the vessel. Coatings are therefore formulated that do not sustain combustion; they should not spread the flame by rapid combustion nor contribute a significant amount of fuel to the fire.

Polyvinyl chloride containing 57% by weight chlorine is self-extinguishing. However, it is not a good vehicle for a flame-retardant coating because of its high melting point. This can

675

Table 1 Formulation for a Fire-Retardant Latex Paint

Ingredient	Parts by weight
Polyvinyl acetate copolymer emulsion (55% solids)	20
Chlorinated paraffin	5
Antomony oxide	11
Titanium dioxide	22
Mica	9
Thickeners, etc.	3
Water	30

be lowered substantially by copolymerization with other vinyl monomers such as vinyl acetate. To make these copolymers useful, addition of plasticizers and coalescing solvents is often necessary to give suitable application and performance properties. These additions dilute the overall concentration of chlorine thereby reducing the flame retardancy.[3]

Fire-retardant coatings are based primarily on chlorinated alkyds, alumina trihydrate, or a combination of chlorinated paraffins and antimony trioxide. Tables 1 and 2 give typical formulations. Flame spread test results depend both on the substrate and the thickness of the film.[4]

3.0 FIRE–RETARDATION MECHANISM

The combustion of gaseous fuel is believed to proceed by a free radical mechanism:

$$CH_4 + O_2 \longrightarrow CH_3 + H + O_2$$

$$H + O_2 \rightleftharpoons OH + O$$

$$CO + OH \longrightarrow CO_2 + H$$

Table 2 Formulation for an Alkyd-Based paint

Ingredient	Parts by weight
Long oil alkyd	22
Chlorinated paraffin	7
Antimony oxide	6
Titanium dioxide	29
Micronized talc	7
Whiting	11
White spirit drier, etc.	18

The H, OH, and O radicals are chain carriers and take part in a number of reactions in the flame zone.

The function of halogen containing compounds as flame retardants has been explained by the radical trap theory and takes place in the gas phase. In the foregoing reactions, liberated HCl or HBr competes for the radical species that are critical for flame propagation:

$$CH_4 + X \longrightarrow HX + CH_3$$

$$H + HX \longrightarrow H_2 + X$$

$$OH + HX \longrightarrow H_2O + X$$

$$O + HX \longrightarrow OH + X$$

The active chain carriers are replaced with the much less active halogen radical, slowing the rate of energy production and helping flame extinguishment.

Antimony oxide is known as a flame-retardant synergist when used in combination with halogen compounds. Volatile antimony oxyhalide (SbOX) and/or antimony trihalide (SbX$_3$) form in the condensed phase and transport the halogen into the gas phase.

Phosphorus compounds are also used as primary flame retardants. The flame-retardant mechanism for phosphorus compounds varies with the type of compound, the polymer, and the combustion conditions.[5] For example, some phosphorus compounds decompose to phosphoric acids and polyphosphates. A viscous surface glass forms and shields the substrate from the flame. If the phosphoric acid reacts with the polymer (e.g., to form a phosphate ester), subsequent decomposition results in a dense surface char. The coatings that form serve as a physical barrier to heat transfer from the flame to the substrate and to diffusion of gases; in other words, the substrate is isolated from heat, flame, and oxygen. This is the mechanism for fire-resistive intumescent coatings discussed below.

Triaryl phosphate esters are thermally stable, high boiling (>350°C) materials. They can volatilize without significant decomposition into the flame zone, where they decompose. Flame inhibition reactions, similar to the halogen radical trap theory, have been proposed.[6]

$$H_3PO_4 \longrightarrow HPO_2 + PO + etc.$$

$$H + PO \longrightarrow HPO$$

$$H + HPO \longrightarrow H_2 + PO$$

$$OH + PO \longrightarrow HPO + O$$

Alumina trihydrate (ATH) or magnesium hydroxide inhibits ignition by absorption of heat due to decomposition, releasing large volumes of water of hydration (> 30%).

4.0 FIRE–RESISTIVE INTUMESCENT COATINGS

Intumescent paints and mastics swell and char when exposed to heat and flame, giving a carbonaceous foam that insulates the substrate from heat, air, and fire. This may delay the onset of combustion of a wood or plastic substrate or delay the heat buildup and tensile loss of structural steel. In the latter case, thick coatings or mastics are used. These coatings are considered to be fire resistive and significantly more effective than fire-retardant coatings.

Table 3 Formulation for a Typical Intumescent
Emulsion Paint

Ingredient	Parts by weight
Polyvinyl acetate latex	18
Ammonium polyphosphate	22
Pentaerythritol	12
Starch	3
Dicyandiamide	16
Titanium dioxide	4
Water	25

The former offer protection to the substrate. Intumescent coatings require three basic ingredients: a carbonific or carbon producer such as pentaerythritol or other polyol, an acid releasing agent such as a phosphate, and s spumific or gas producer such as melamine. Polyols such as starch and dipentaerythritol (less water sensitive) are carbonifics. Other acid producers and their temperatures of decomposition are ammonium polyphosphate (215°C), monoammonium phosphate (417°C), and melamine phosphate (300°C); spumifics include urea (130°C), dicyandiamide (210°C), and melamine (300°C).[7]

When the coating or mastic is heated, the decomposing phosphate forms phosphoric acid. This esterifies the hydroxyl groups of the polyol, which subsequently decomposes, forming water and a carbonaceous char, regenerating the phosphoric acid. As the char forms, the paint binder softens and the spumific decomposes, liberating nonflammable gases. These gases expand the softened binder into a foam. A rigid carbonaceous foam is formed as charring of both carbonific and binder resin is completed.

Tables 3 and 4 give examples of intumescent systems.[4]

When used as a thick mastic coating on structural steel, such materials can prevent the steel from reaching the failure point for 3 hours, as indicated in Table 5.

A unique intumescent coating containing no water-sensitive polyols is composed of p,p'-oxybis(benzenesulfonamide) in a vinyl chloride–vinylidene chloride copolymer

Table 4 Formulation for a Typical Intumescent Solvent–
Thinner Paint

Ingredient	Parts by weight
Vinyl toluene–butadiene copolymer	7
Ammonium polyphosphate	28
Dipentaerythritol	8
Melamine	9
Chlorinated paraffin	9
Titanium dioxide	6
Mineral spirits	33

Table 5 Ratings on Steel Column

Coating thickness (in)	Rating (h)
0.31	1
0.39	1.5
0.58	2
0.78	2.5
0.97	3

Source: Ref. 8.

binder. The addition of melamine pyrophosphate improves the homogeneity of the system.[9] Dipentaerythriol can also be added to improve the homogeneity.[10]

Pitt-Char (PPG Industries) coating is a two-component epoxy–polyamide produce designed for structural steel. This intumescent epoxy coating requires no reinforcement and provides up to 3 hours of fire protection.[11]

5.0 MISCELLANEOUS COATINGS

Magnesium oxychloride mastics will protect steel by releasing large quantities of water upon heating and by relatively high temperature calcining (500°C). This material can be painted or sprayed and can be used in exterior applications.[12]

Polystyrene foam block can be protected against fire for specified time periods. Building plastics containing gypsum and perlite and an expanded vermiculite, portland cement, and limestone coating are particularly effective in protecting polystyrene foam block against the heat of a simulated mine fire.[13]

For an insight into potential methods of flame retarding polyurethane and epoxy coatings, an analogy with plastic flame retardants can be made. The most common way to achieve flame-retardant polyurethanes is by the addition of halogen, phosphorus, chlorophosphate, or chlorophosphonate compounds. Reactive halogen and phosphorus diols have also been used. Epoxy resins are generally flame retarded using the reactive tetrabromobisphenol A. Use of phosphorus–halogen compounds has been reported.

REFERENCES

1. C. A. Hafer, Final Report, Project No. 3-294-26, Southwest Research Institute, San Antonio, TX, 1970.
2. D. Gross and J. J. Loftus, *Fire Res. Abstr. Rev.*, 3(3), 151–158 (1961).
3. W. S. Chang, R. L. Scriven, and R. B. Ross, in *Flame Retardant Polymeric Materials*, Vol. 1, M. Lewin, S. M. Atlas, and E. M. Pearce, Eds. New York: Plenum Press, 1975, pp. 399–447.
4. A. G. Walker, *Proceedings of the Fire Retardant Chemical Association Meeting*, Pinehurst, NC, Oct. 27–30, 1985, pp. 51–61.
5. J. Green, *Plast. Compd.*, 10(3), 47 (May/June 1987).
6. J. W. Hastie, *J. Res. U.S. Natl. Bur. Stand.*, 77A(6), 733 (1973).
7. J. A. Shelton and M. G. Bivok, *Mater. Design*, 5, 41–42 (February/March 1984).
8. J. A. Seiner and T. A. Ward, *Poly. Paint Colour J.*, 178(4207), 75–78 (Feb. 10, 1988).

9. S. H. Roth, J. Green, and J. J. Seipel, U.S. Patent 3,714,081 (Jan. 30, 1973).
10. S. H. Roth, J. Green, and J. J. Seipel, U.S. Patent 3,714,082 (Jan. 30, 1973).
11. *Modern Paint Coatings*, pp. 188–190, October 1984.
12. S. I. Kawaller, *Oil Gas J.*, **71**(4), 78–80 (1973).
13. S. J. Luzik, *Fire Technol.*, **22**, 311–328 (1986).

77
Leather Coatings

Valentinas Rajeckas

Kaunas Polytechnic Institute
Kaunas, Lithuania

1.0 INTRODUCTION

Tanned leather is usually coated with a thin pigmented or lacquer coatings. One of the purposes of such a coating is decorative. The coating may also change some physical properties of leather: it may decrease water and air permeability, increase its rigidity, etc. Such changes depend on the coating type, especially on the polymer used as a film former. The properties also depend on coating formation technology: the coating may penetrate deeply into leather, or it may remain only on the surface. The coating technology chosen depends on the leather structure and the degree of its surface damage.

Tanned leather is the midlayer of an animal's hide—the derma, which is processed chemically and mechanically. During processing, leather becomes resistant to bacterial and fungal attack; its thermal resistance and its resistance to water increase. The derma consists basically of collagen protein having a fibrous structure. Collagen in the derma is in the form of a fibrous mat, and the fibers extend at varying angles with respect to the leather surface. The fiber diameter is 100–300 μm. Other proteins (albumin, globulin) and mucosaccharides are located between fibers and bond the proteinaceous materials into multifiber ropy structures. Such a multicomponent leather structure determines its capability to deform—its elasticity and plasticity.

Leather is used for many applications: footwear, gloves, clothing, purses, furniture upholstery, saddles, and a variety of other uses. Leather is processed differently for each application: different chemicals are used; their quantity and processing conditions may also be different. Thus leathers of different physical–mechanical properties are obtained: very soft, thin, and extensible for gloves and clothing, more rigid for footwear, and hard and stiff for soles. Often leather is dyed during processing. Dyeing may take place by the immersion of leather into a dye solution bath (usually in a rotating drum), or by covering the dry leather surface with a colored liquid coating. The latter technique confers a protective leather coating.

There is also another, but rarely used, method to form a surface coating: lamination of a polymeric film to the leather surface. In such cases, the surface is covered by a film, which is caused to adhere to the surface by pressing with a hot plate.

In general, there are several combinations of finished leather: undyed leather, dyed in a bath without a coating (aniline leather), surface dyed by applying a coating, and both bath dyed and surface coated. If the leather surface has many defects, these may be removed by grinding. In such cases, the coating is thicker and forms an artificial grain.

Coating of leather and coloring it by coating are accomplished by applying several coating layers. Initial grounding layers penetrate the surface of the derma and help to bond the leather. The midlayer is the most important; it has a protective and a decorative function. The top layer determines the gloss level, brings out the aesthetic qualities of the main layer, and often protects it from the effects of moisture.

The number of separate coating layers required to produce the desired leather surface quality depends on their composition and on the condition of the leather surface, which may be natural or polished. The formation of a leather surface coating generally consists of the following steps.

1. Application of a single nonpigmented seal coating. In the case of leather with a natural top grain, this coating is applied by spraying; in the case of a polished surface leather, by flow coating. This coating has a good penetration and it helps to bond collagen fibers and to form the ground for the formation of a new leather surface.

2. After some drying and light pressing (pressure 0.6–0.9 mPa and temperatures 55–60°C), the leather is coated with a pigmented ground coating. This coating may be applied by brushing.

3. After drying and sometimes light pressing, the main pigmented coating is applied. Depending on the color, it may consist of two or three separate applications; in the case of light colors, even four applications may be required.

4. The top coating is applied by spraying in a single or double application. Sometimes drying and pressure may be applied between the coats.

All these steps are not necessary in all cases. Depending on the leather's surface condition, sometimes the first ground coat may be omitted; quite often the second pigmented ground coating is omitted and its function is performed by the main coat.

2.0 CHARACTERISTICS OF LEATHER COATINGS

2.1 Main Coating

Leather coatings may be water- or solvent-borne. The main advantage of aqueous dispersions is the absence of flammable, volatile, and polluting organic solvents. Therefore, aqueous dispersions are used for leather coatings more often than solvent-borne coatings. That is especially true for footwear leather. Many different coating compositions are used.

The main coating ingredient is the film former or binder. Pigments are usually predispersed and added as a paste. Plasticizers are added to increase the flexibility, softness, and elasticity. Surface active agents improve wettability and therefore improve the adhesion. Waterborne coatings may employ thickeners to increase the viscosity. Cross-linking agents are used sometimes to improve temperature and chemical resistance. In the formulations for ground coatings, penetrating agents may be used; they usually are mixtures of surface active agents, organic solvents, and water. For solvent-borne coatings, organic solvents and diluents are used.

Various synthetic and natural materials are used as film formers. They are mainly aqueous emulsions of acrylic, and also of vinyl polymers and copolymers; polyurethanes and nitrocellulose, available as solvent solutions and aqueous dispersions; casein and its derivatives; and waxes and other materials capable of film forming and exhibiting a good adhesion to leather. The desired appearance of such coatings is also important in determining the choice of a binder.

Coatings are classified according to the binder used: thus we have emulsion, nitrocellulose, polyurethane, and casein coatings. Casein coatings, having the milk protein casein as the binder, are used less and less frequently. Casein as well as nitrocellulose is being replaced by various synthetic polymers.

2.2 Unpigmented Ground Coatings

The dermal structure consists of two main layers: the upper or gland layer and the lower or net layer. The gland layer is less dense and has large pores, and the fibers are bound less strongly together. The gland layer is further weakened and damaged when the leather surface is ground. Therefore, before applying a pigmented coating over a ground leather surface (and sometimes one that is not ground), an unpigmented ground coat is applied. This coating penetrates the gland layer of derma, bonds its structural elements, increases the bond between the gland and net layers, and in this manner, eliminates leather breakage in the finished product. In addition, the unpigmented ground coating provides improved adhesion of the main coating.

Unpigmented ground coatings are prepared from various latexes, which are diluted with water and contain penetrating agents, which increase the penetration of the ground coating into leather. Penetrating agents consist of organic water-miscible solvents, such as alcohol, nonionic or anionic surfactants, and water. These components are mixed in various ratios: solvent 25–40%; surfactants 10–15%; the remainder, water. Polymer concentration in the ground coating is 7–10%. Depth and rate of penetration depend on the coating's capability of wetting leather, the density of the leather upper layer, its viscosity, and the the chemical nature of the polymer. Penetrating agents improve the wetting capability. The penetrating capability of ground coatings can be characterized by the time required for a ground drop (0.1 ml) to be absorbed by leather. Absorption time of 3–8 seconds is acceptable. A ground coating penetrates to one-third of the leather thickness. The composition of ground coatings is selected empirically, because the relationship between the coating's reinforcing properties, its adhesion, rate of penetration, wetting capability, and emulsion particle size is not easily defined.

A ground coating recommended by Reichhold Chemie is of the following composition, in parts by weight:

Durlin-binder 5318	200
Durlin-binder 5534	70
Durlin-penetrator RCH-30	130
Water	600

Stall recommends the following ground coating:

Resin 6509	70
Resin 6582	30

| Penetrator 6507 | 60 |
| Water | 280 |

The consumption of ground coatings is 150–200 g/m².

After application of the ground coat, leather is allowed to relax and to dry. Drying conditions are very important. Leather is dried horizontally at a slow rate. The temperature should not exceed 30°C, and only a mild air circulation should be used. At such conditions, water evaporation from the surface and its migration into internal capillaries of the leather proceed slowly and do not cause the migration of dispersed polymer to the surface. The polymer remains uniformly distributed in the gland layer of the leather.

In addition to aqueous polymer emulsions, polyurethane solutions, capable of forming elastic films, are used. If polyurethane has free isocyanate groups, formation of chemical bonds between isocyanate and functional groups in collagen may develop. This increases the interaction between the ground coating and fibrous structure of the gland layer of the leather.

2.3 Aqueous Pigmented Coatings

Aqueous pigmented coatings are the most widely used coatings in leather manufacturing. In these coatings the film former is dispersed in water. They are prepared by blending pigment concentrates (paste consistency pigment dispersions, including a binder, which acts as a protective colloid with latex), various additives, and water to dilute to the required pigment concentration. Aqueous pigmented coatings are complicated colloidal systems; therefore their stability is especially important. They must not coagulate when diluted with water, when subjected to a mechanical force (mixing, pumping, applying by brush), and under the addition of some electrolytes. It is important that the pigments be protected from flocculation and that they do not precipitate. The protective colloid and emulsifier must be compatible with the latex film former and must form a uniform structure throughout the coating volume. The protective colloid function in various pigment pastes is performed by ammonium or sodium caseinates, methyl cellulose, carboxymethyl cellulose, or acrylic carboxylated copolymers. Film formers are acrylic and diene copolymer emulsions.

When formulating coatings, it is important to select components to ensure that the coating is elastic and resistant to aging, and that the pigments are uniformly distributed.

2.3.1 Acrylic and Diene Latexes

A latex may be blended by employing polymers that form soft and tacky coatings with polymers that form stronger and harder coatings. The elasticity temperature range may be expanded into lower temperatures by blending acrylic copolymers with diene latexes. However, diene copolymer latex films are less resistant to light. Therefore acrylic latexes are more suitable for white coatings.

Diene copolymer latexes are prepared by copolymerizing various diene monomers with acrylic or methacrylic acid esters. Such useful copolymers are methyl methacrylate–chloroprene (30:70), methyl methacrylate–butadiene–acrylic acid (35:65:1.5), piperylene–acrylonitrile–methacrylic acid (68:30:2), and many other copolymers. Films from these copolymers retain their elasticity at least down to −20°C and are useful for blending with acrylic latexes to extend their low temperature flexibility.

2.3.2 Casein

Casein is a protein prepared from milk. It is soluble in dilute alkalies. It is used as a binder in the preparation of pigment concentrates and also in casein and combined casein–emulsion coatings.

Modified casein is methylacrylate and ammonium caseinate emulsion polymerization product used as an additive in coating compositions with other, usually acrylic, latexes. Films of modified casein are elastic (elongation of 600–900%), strong (tensile modulus at failure 6–8 mPa), and soluble in water. However, they may be easily rendered hydrophobic by treatment with formaldehyde or solutions of polyvalent metal salts. Butadiene–ammonium caseinate copolymer latex has similar properties.

2.3.3 Wax Emulsions

Wax emulsions are water-dilutable dispersions at pH 7.5–8.5 and stabilized with nonionic surfactants. The basis is usually montan or carnauba wax. Wax emulsions are used as additives to pigment and top coatings.

2.3.4 Pigment Concentrates

To obtain well-colored leather, it is important that pigments be well dispersed in the binder and that a strong bond be formed between pigment particles and the binder. Direct pigment dispersion in the coating is difficult. Aqueous polymer emulsions used for leather coatings are not sufficiently viscous to maintain a uniform distribution of pigments. Furthermore, emulsions might not be stable enough to allow a direct addition of pigment. Therefore, the pigment is dispersed separately in the binder solution, yielding a stable dispersion, which can be safely blended with film forming emulsion and other additives to ensure the stability of the heterogeneous system. The pH should be similar in the two dispersions. Pigment concentrates, depending on the binder used, can be of varying composition: casein, where the binder is an aqueous alkaline casein solution, or a synthetic polymer base–mainly acrylic.

A pigment concentrate in casein may have the following composition (in parts by weight):

Pigments	14–60
Casein	3.8–8.6
Oil of alizarine	2.4–4.0
Emulsifiers	0.5–1.0
Antibacterial agent	0.5–0.9
Water	up to 100

To prepare such a composition, casein glue is made up first (18–20%); then antibacterial agent is added, followed by other additives. The pigment is dispersed employing suitable equipment until a stable dispersion is obtained. Casein binder is suitable for the dispersion of all pigment types.

In addition to casein-based pigment concentrates, pigment dispersions based on acrylic polymer are used. Pigments are dispersed in a thickened acrylic emulsion. If the film former in the coating is an acrylic latex, it mixes well with such pigment dispersions. Acrylic dispersions also are more effective in improving the coating elasticity, as compared to casein dispersions. The composition of pigment dispersions in acrylic latex may be as follows:

Pigments	11.2
Blend of two or three acrylic emulsions	85.3
Ammonium hydroxide (25%)	2.3
Oil of alizarine	0.7
Surfactants	0.5

Pigment concentrates based on acrylic emulsions are of a viscous paste consistency; they are easily dilutable with water, and they blend well with all aqueous emulsion film formers.

In addition to pigments, dyes may be used in coatings. In the preparation of pigment or pigment–dye blend coatings, attention must be paid to the pigment properties—their resistance to light, their opacity, and the elimination of such side effects as bronzing.

2.3.5 Nitrocellulose-Based Compositions

These products are used for nitrocellulose coatings, but most frequently, for top coatings over coatings of other types. Nitrocellulose solutions (lacquers) in organic solvents, or solution dispersions in water, are used.

Nitrocellulose lacquer is a solution of nitrocellulose in organic solvents and diluents compounded with plasticizers. Nitrocellulose is available in alcohol-soluble and -insoluble forms. The latter is used for leather coatings. Each type is available in several viscosity grades, depending on the molecular weight of the nitrocellulose. A compromise is usually made between the coating's physical properties, which improve with increasing molecular weight, and coating solids, which decrease with increasing molecular weight for a solution of required viscosity.

The solvents used are ethyl and butyl acetates, acetone, and methyl ethyl ketone. Alcohols (ethyl and isopropyl), while not solvents by themselves, enhance the solubility of nitrocellulose in other organic solvents. Diluents are miscible organic liquids that do not dissolve nitrocellulose, but decrease the solution viscosity. They are also less expensive than true solvents. Such diluents are toluene, xylene, and some aliphatic–aromatic hydrocarbon blends. The choice of solvents/diluents for nitrocellulose lacquer is determined by economics and by such properties as sufficiently low volatility, lack of water absorption, or capability to form azeotropic blends with water. For film formation it is important to have an optimum amount of alcohol, which has a relatively low volatility.

Nitrocellulose is brittle, and therefore plasticizers are used in compounding nitrocellulose coatings. Plasticizers used are alkyl phthalates, castor oil, camphor oil, and others.

Nitrocellulose lacquer is a clear, water-white, easily dilutable, viscous liquid containing 15–18% solids. The tensile strength of nitrocellulose film is 1.5–1.8 mPa; elongation at break is 50–60%.

Aqueous nitrocellulose dispersions also contain some organic solvents, which facilitate the coalescence of nitrocellulose lacquer particles. Film formation from nitrocellulose dispersions that do not contain any solvent is difficult. Both types of dispersion are used: oil in water and water in oil. Nitrocellulose coatings are used as top coatings over aqueous emulsion coatings. The mechanical properties of nitrocellulose films obtained from aqueous dispersions are poorer than those obtained from solutions.

Leather that does not require vapor and air permeability (e.g., leathers used for applications other than footwear or clothing) may be coated entirely with nitrocellulose, starting with the ground coat and ending with the top coat. For the ground coat, aqueous nitrocellulose coatings are mainly used; the main coat consists of a pigmented nitrocellulose enamel, and the top coat is a clear nitrocellulose coating.

2.3.6 Polyurethane Coating Compositions

Coatings described here are used for all polyurethane coatings and also as top coats for coating of other types. Polyurethane solutions in organic solvents and aqueous dispersions are used. Coatings of this type have a good adhesion to leather (because of the urethane groups). Polyurethane lacquers may be clear and colored (pigmented or dyed). Many different polyurethane structures are available: tridimensional, branched, linear. Thus coating properties may be obtained that exhibit variations in elasticity, resistance to solvents and abrasion, and thermal and light resistance. The main disadvantage of these coatings is that they mask the appearance of the leather surface and contribute to the artificial appearance of the product. This is mainly the case for coatings consisting entirely of polyurethane, but not for top coatings. Polyurethane coatings are also used to prepare varnished leather.

Polyurethane leather coatings are made from ester oligomers (molecular weight of about 2000) and isocyanates. Polyurethanes may vary widely in their properties. If oligomers have more than two hydroxyl groups, the resultant polyurethanes are three-dimensional structures and have a high chemical and temperature resistance. If ether oligomers or polyethers (polypropylene glycol, polytetramethylene glycol) are used, the resultant polyurethane film has a higher resistance to hydrolysis and better low temperature properties, retaining the elasticity at lower temperatures.

The synthesis of polyurethane is carried out in a solution of organic solvents, such as methyl ethyl ketone, dimethylformamide, butyl- and ethyl acetate, and their mixtures. Two main polyurethane polymerization processes are used: single stage and two stage. The two-stage process begins with the reaction between ester oligomers and isocyanate, which yields a low molecular weight prepolymer. In the second stage, under the effect of a chain extender (1,4-butanediol, diamines), the chains grow. In the manufacture of varnished leather, the second stage is usually carried out after coating application to the leather surface.

Polyurethane coatings can be applied by spraying or by flow coating. The coating is dried for about 16 hours at 30–40°C and low relative humidity (40–50%), with the leather sheets in the horizontal position.

Coatings with a metallic sheen may be prepared by the addition of metallic particles, prior to the application of a clear top coating.

A single-component polyurethane lacquer is a solution of thermoplastic, high molecular weight, linear polyurethane in organic solvents. Such lacquers are produced from linear polyesters and diisocyanates, and in dilute solution they are useful as top coats. Thermoplastic polyurethane films are elastic, their elongation at break reaches 1000%, and the elastic properties are retained over a wide temperature range (−30–120°C).

Aqueous polyurethane dispersions are obtained by the dispersion of polyurethane in water, employing emulsifiers. Aqueous dispersions are useful for pigmented coatings: ground coats, main color coat, and also for top coats.

Mixtures of polyurethane and nitrocellulose dispersions are also used as top coats. Such top coats dry faster, but their mechanical and aesthetic properties are poorer.

3.0 TECHNOLOGY OF DECORATION OF SKINS OF LARGE HOOFED ANIMALS WITH ARTIFICIAL GRAIN

The processing technology becomes more complex when it is required to form an artificial grain. The natural grain is ground down and a coating is applied in several applications. An artificial grain, including one imitating natural leather, is formed.

This process starts by grinding the grain surface two or three times. Grinding is performed by employing special grinding–polishing equipment and using #5-6 abrasive paper. After grinding, the leather surface is coated with an unpigmented ground coat of the following composition (in parts by weight):

Acrylic dispersion (40%) solids	100
Penetrator	40
Water	300

An unpigmented coating is applied by a curtain coater. Its consumption is about 0.2–0.3 kg/m^2 depending on the density of the leather's glandular layer. The leather is then stored for 8–24 hours, dried, stretched, pressed, and ground again. Pressing before grinding improves the grinding–polishing quality. The leather is dyed by running it twice through special equipment, which also involves drying, initially at 110°C, and then at 80°C at the end of the drying tunnel. The drying cycle lasts 1–2 minutes. In the next step the leather is pressed at a plate temperature of 50–70°C and a pressure of 14.5 mPa.

One or two additional color coatings are applied until the surface is sufficiently well covered. Light and intense colors may require three or four coating applications. The leather is dried at 65–75°C after each coating and pressed at a plate temperature of 65–75°C and a pressure of 14.7 mPa. After this smooth press, the grain is pressed in by a hydraulic press at a temperature of 70–80°C and 14.7 mPa pressure. The process is ended by spraying one or two layers of top coat. Nitrocellulose lacquer (or emulsified lacquer) or polyurethane lacquer is used for the top coat.

4.0 SOME NONSTANDARD COATING APPLICATIONS

Rohm & Haas Company suggest employing foamed color coatings. Such coatings consist of few components: latex film former, pigment dispersion paste, and foaming agent. These coatings contain 50% solids and are applied to a coating weight of 0.4–0.5 kg/m^2. The application technology is not much different from the standard application techniques: drying, pressing, and coating spraying are employed. The coating is elastic and resistant to flexing (endures 20,000–40,000 flex cycles). The method is more economical than traditional techniques.

Another method is casting on a silicone elastomer matrix with a negative leather grain. A polyurethane coating is applied (a prepolymer containing free hydroxyl groups and an isocyanate cross-linker). Coating thickness varies from 0.15 to 1 mm. A leather (which also could be split leather) is pressed against the uncured coating and after a contact time of 10–15 minutes, it is dried in a drier, where a well-adhered coating is formed. After such curing, the leather is removed from the matrix.

78
Metal Coatings

Robert D. Athey, Jr.

Athey Technologies,
El Cerrito, California

There are two facets to metal coatings—coatings on metal substrates, and metals as coatings on any substrates. The latter can be lumped together in a one-word category called "metallizing," which is done in many ways. The former, coatings on metal substrates, generally are thought of as paint-type materials but may include waxes, inks, and other coatings. The two topics will be dealt with separately, beginning with metallization , as those metal surfaces are often painted or coated for protection, as well.

1.0 METALLIZING

The objective of metallizing techniques is to place metal on the substrate for appearance or protection of some sort. The classes of metallization are many and complicated, but may be separated by their process details. Processes that apply metal to surfaces may use metal as individual atoms or ions, as the fluid molten metal, or as the solid metal. We deal with each separately.

1.1 Liquid Metal Processes

1.1.1 Galvanizing

The metal item (iron or steel) is dipped into a molten bath of zinc, then withdrawn, and the excess zinc allowed to drip off. The item is cooled and the zinc crystallizes on the surface (giving an appearance called "spangle"), with the cooling rate determining the size of the zinc crystals showing on the surfaces. This process can be made continuous for rod, wire, coiled sheet or pipe, and semi-continuous for reinforcing rod, pipe, and cut sheet.

The process isn't quite as simple as it sounds. There are iron/zinc compounds that form at the molten interfaces. The time of heating a molten zinc–iron interface governs the thickness of the interface containing these compounds, and the ratios of iron and zinc in the compounds at the interface. In addition, sal ammoniac (ammonium chloride) is used as a flux in the molten zinc, but it can appear as a blue coating or streak on the galvanized item, to the item's detriment. That "sali"-contaminated item should be recoated.

The zinc layer on the item acts as a physical barrier to corrosion. However, as soon as there is physical damage to the zinc layer, exposing the iron, the zinc acts as a sacrificial anode, giving the iron electrolytic corrosion protection, as well. Since zinc is soft and easily corroded, it will "wear" away, showing the white zinc oxide surface as the zinc layer becomes thinner. The white zinc oxide has substantial water solubility and washes away in rain. The zinc surface is often painted to give protection against these losses.

1.1.2   Flame/Plasma Spray

Although all the flame and plasma spray processes project liquid metal metal droplets through the air to the substrate, each starts with a solid metal wire or solid molten powder. The solid is taken into a device which heats it to the molten state, breaks it into microscopic droplets, and propels the droplets at the substrate. Longo edited a reference book on this topic[1].

The "flame" or "thermal" spray uses a modified oxyacetylene torch as the heat source. After the flame has been adjusted to its hottest, additional compressed air is blown into the flame. The wire or powder then fed in via a funnel, and the blast of liquid metal particles is pointed at the substrate. The "plasma" spray uses an electric arc as the heat source to melt the wire or powder feed into the compressed air stream.

The distance between the molten metal's origin and the substrate will determine the type of coating obtained. The close approach of the nozzle means that most particles will hit the substrate surface as a liquid. Many will stick, and transfer their heat to the substrate, but some particles will bounce off. If the nozzle is further away, the smaller liquid particles will solidify and bounce off the substrate, while the larger particles will still stick. If the nozzle is too far from the substrate, none of the particles will stick because they will have cooled too much. Losses from bounce-off or from premature cooling will be on the order of 25% of the weight of the metal sprayed.

The metals most commonly sprayed include copper, zinc, iron, and aluminum. Alloys and even metal oxides can also be sprayed, provided there is enough heat in the flame to make the particle soften. Since the sprayed metal is of lower density than the solid metal, there is occluded air, and even porosity. However, adhesion to most substrates is good, and is often assured by cleaning and roughening the surface (e. g., sand blasting). The inorganic coating may be built up enough to be machined, and may be strong enough to be a bearing surface.

1.1.3.   Other Liquid Metal Coatings

Any solid can be dipped into a molten metal.

1.2 Solid Metal Processes

1.2.1   Sherrodizing

An item of steel or iron is put in a drum with powdered zinc and steel balls, and the drum is rotated. The zinc powder is essentially hammered onto the surface of the steel/iron item as individual spots. The length of time in the rotating drum, the amount of zinc powder, the number of steel balls, and the number of items to be treated all govern the actual amount of zinc that ends up on the surface of the item to be treated. The zinc coating on the iron/steel acts as a corrosion (rust) inhibitive protection for the item.

1.2.2.   "Detaclad"   Process

The construction of metal laminates, such as the coinage products now used by the U.S. Treasury, is a simple process involving a layer of metal applied to another metal surface by the force of an explosion on one surface. If a silver is needed as an outer surface over copper, the explosive is on the side away from the copper in the set-up, as shown in Figure 1.

1.3 Vapor Processes

1.2.1 Vacuum Evaporation

The simplest of the vapor processes is vacuum evaporation. The items to be coated are put on racks that circle a central set of trays formed from electrical resistance heaters. A bell jar is lowered over the whole array, and sealed to be pumped down to 1 mm Hg of vacuum. The resistance heaters are fired off to evaporate the metal powder or slug in the tray, and the individual atoms of metal fly off in a straight line toward the bell jar in all directions. The items to be coated are hit by these atoms in a line of sight from the resistance heater tray. The atoms condense on all surfaces facing the heater trays, forming discrete globules at first, and covering the whole surface as the globules coalesce.

The items to be coated must be treated to make sure the metal wets and adheres. Such treatments might be simple washing or precoating with a polar lacquer, or as complicated as a plasma etch or a corona discharge surface treatment. Even nonpolar materials like poly(propylene) or acrylonite–butadiene–styrene resins can be successfully coated with good adhesion by vacuum evaporation. Common items so treated include toy cars and lipstick cases.

Good continuous film coatings can be formed on just about any surface, at thicknesses ranging from micrometers to millimeters. Indeed, the transparent gold coating put on aircraft wind shields for deicing/defogging is about 10 nm thick. This coating MUST be continuous, as any microhole in the coating would cause a blue "burn" spot as soon as current was applied.

Almost any metal may be vacuum evaporated, though some metals sublime, rather than undergoing the preferred heat/liquefy/boil process. The subliming metals don't do quite the line of sight depositions desired, and the vacuum fittings can become plugged. The most common are aluminum, copper, gold, and silver.

Since the metal deposited by this process is in the bare, unprotected, metal atom form, it may corrode quickly if it is so disposed in the electrochemical series. One must apply topcoat for corrosion protection immediately. Sorg[2] described vacuum metallized parts for

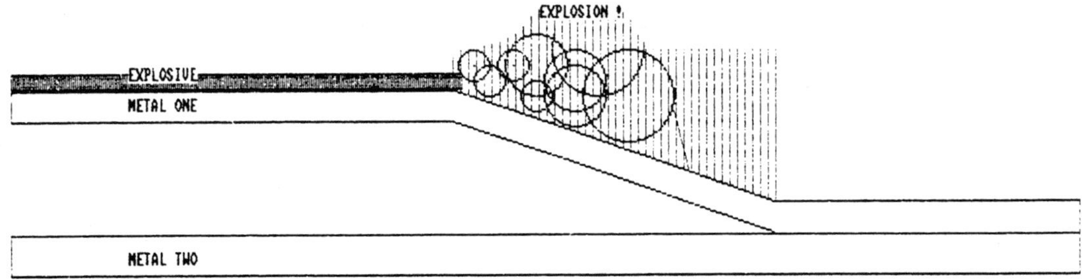

Figure 1 Explosive bonding of metal.

automobiles, while others compared a number of different approaches for automotive bright work[3]. Sorko Licensing Company has a patent on making mirrors from clear acrylic sheet[4].

1.3.2. Sputtering

The key elements in the sputtering process make it different from simple vacuum evaporation. The metal vaporized by an electron beam, which charges the metal atoms or droplets with a negative charge. The substrate is made the cathodic target (or is directly in front of a cathodic target). Thus, electromotive force is the driving force coating the substrate. It is used quite successfully to coat Mylar films for sunshade applications with copper, aluminum, and even stainless steel. Sputter plating onto actual polymers has been described from plumbing fixtures[5].

1.3.3 Chemical Vapor Deposition

A reference book on plasma deposition has been edited by Hochman and others[6]. The substrate is placed in a container that is then evacuated, and a gas or gases are drawn in. The container has a cylindrical electrode just within inner surface, and this is to react with each other or the surface of the substrate.

Monomers used in plasma deposition of polymer include methyl methacrylate[7], propylene[8], acrylic acid[9], acrylonitrile[10], norbornadiene[10], hexamethyldisiloxane[11], tetrafluoroethylene[12], ethylene[12], and styrene[12]. Nonmonomers (such as pentafluorobenzene[13], methane[14], or tetrafluromethane[15] also may deposit polymeric coatings. Colloidal metal may be included within the polymer film to generate a colored coating[15].

The advantage of plasma coating is the very thin film formed, with no pinholes and no contaminants (i.e., the additives needed in solvent or waterborne coatings).

1.4   Plating

1.4.1 Electroplating

A good place start investigation of plating processes is an annual handbook supplied by a trade magazine publisher[16]. Plating on plastics has been reviewed by Saubestre[17] and by Muller and associates[18]. Springer and associates[19] described details of polymer treatments[20,21] and morphology[22] as important in electroplating polymer surfaces. Safrenek[23] detailed the properties of electroplated metals and alloys. A zinc–nickel alloy electroplated onto steel is reputed to have better corrosion properties than galvanized steel[24].

1.4.2. Electroless Plating

The handbook cited earlier is a good reference for electroless plating as well[16]. The idea is simply to put a metal solution on a surface, and let it deposit the zero valent metal on the surface because of an added chemical. It turns out not to be quite so simple, because addition of the reducing agent to a solution does not guarantee that the metal will deposit on the surface, nor does it mean that the deposit will adhere if deposited. Hence, there are sequences of "washes" and "activators" that prepare the substrate to accept the final plating formulation. In the case of plating on plastics, several oxidative techniques (chromic acid solutions, plasma etching, nitric acid washes, . . .) that prepare a surface. One process uses three deposition steps, with clean water rinses between, to end up with copper on plastic. The three solutions are tin chloride, palladium chloride, and a formulated copper solution that has reducing agent along with rate modifier and surface modifier chemicals as well. Rigorous rinsing between the metal solutions is needed to assure that the last solution doesn't become a slurry of colloidal copper, because of "dragout" or contamination by preceding metal ions. The electroless plating deposits are continuous, conductive, and bright, but are thin—micrometers in thickness.

2.0   COATING ON METALS

Coatings for metals may be divided into two classes by main function: those that are decorative, and those that are protective. That is not to say that a protective coating can't be decorative, but that the coating's main function is protective, while it is chosen to be decorative as an additional option.

2.1   Decoration

The decorative aspect of coatings lies in several features that are dealt with elsewhere in this book. Among those aspects are color, gloss/flatness, and texture. Each entails specific approaches the formulator uses to attain the desired appearance, mainly involving choices of vehicle (the adhesive that sticks the pigment particles together and then to a substrate), pigments, the ratio of pigments to vehicle, and (occasionally) the additive that confers a certain property when pigment or vehicle cannot.

Yet another aspect of decoration is pattern. If and when there are multiple colors on a surface, the shape described at the color interfaces can be important. The camouflage paints strive to have what appear to be random splotches, because a regular geometric pattern (especially a straight line) attracts the eye to a potential target. The multicolor paints (described in Chapter 65) have the dots, splotches, smears, and lines simulating natural mineral or rock formation surfaces. And the pattern is a main point in signage or art work.

2.2   Protection

The metal substrate is generally thermodynamically unstable and is easily converted to the more thermodynamically stable oxide. However, the protection provided to the metal substrate is often designed to guard against sorts of attack other than chemical oxidation. Mechanical damage may aid the oxidation, by providing sites for the oxidant to work. Electrical exposure or damage can ease the chemical degradation. So, work on protecting a metal substrate must include consideration of the insults or attacks to be expected. In the instance of corrosion, the protection may aim to prohibit contact of the coating with oxygen or water—each a necessary element of the corrosion reaction. The aim may also be to prohibit contact with corrosion catalysts—salts, acids or alkalis. Schweitzer[25] reviewed corrosion and protective schemes, and one unit of the educational booklets from the Federation of Societies for Coating Technology that deals with corrosion protection[26a] while another covers corrosion and surface protection for painting[26b]. Wilmhurst[22] described aqueous maintenance paints for corrosion protection in Australia, while Campbell and Flynn[27] did so for the United States and Poluzzi[28] for Italy. A Golden Gate Society for Coatings Technology Technical Committee study[29] showed the waterborne systems have come to equal the solvent borne coatings for corrosion protection in aggressive environments (i.e., the Golden Gate Bridge).

At the National Coatings Center, we divide a "pitchfork" diagram (Fig. 2) to describe the three major corrosion protection schemes. The main emphasis was on improving ways to give corrosion protection through combinations of some of the corrosion protection schemes.

Examples of the barrier coatings are fairly straightforward. As noted earlier, the zinc in galvanizing is initially a barrier, and any impervious coating is a barrier, be it wax, asphalt, coal tar, polyethylene, or whatever. There are selective barriers that aim to protect against corrosion by blocking a specific corrosive element. The wax, polyethylene, coal tar, or asphalt is specifically aimed at prevention of water permeation to the metal surface, as water is

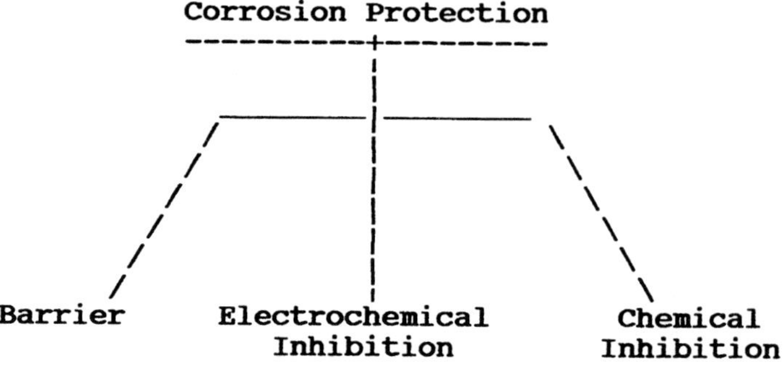

Figure 2 Corrosion protection schemes.

a specific corrosive substance. Other coatings (PVDC, Barex–type nitriles, etc.) aim to prohibit oxygen permeation. In both cases, the coatings are diffusion barriers, and the key is to have a coating that dissolves as little as possible of the permeant within the coating, and also inhibits permeation. A highly polar coating material is always more water permeable than the nonpolar material, because water dissolves so well in the polar material that the polar material acts as a pipeline rather than a barrier. It's a chuckle to hear of silicones or acrylates described as waterproofings, but you have to understand the implied statement that they protect against liquid water, while water vapor goes through them 100–1000 times faster than it would go through a hydrocarbon barrier. Munger[30] reviewed protective coatings for corrosion prevention.

The electrochemical protection schemes are bound up in converting the iron or steel into a cathode, since the corrosion reaction is an anodic oxidation of the zero-valent iron metal to ionic forms (usually ferrous). The easiest way is to simply contact the metal with something that is oxidized more easily (zinc, magnesium, aluminum, etc.) and let the iron be the cathode while the other metal is corroded as the sacrificial anode. Indeed, there is a substantial market in bars of magnesium or zinc that are attached to iron (pipelines, underground tankage) to act as sacrificial anodes. We've already noted that galvanized iron with the zinc surface damaged to penetrate to the iron is still electrochemically protecting the iron. The "zinc-rich" paints (having about 85% by weight of zinc metal powder and 15% binder) are also cathodic protectors of steel. BUT the sacrificial anode does send its corrosion products into the surroundings. In a tidal area, the sacrificial anode specified by some regulatory agencies feeds metal ions into the underground water while it is protecting the underground storage tank.

An alternative in cathodic protection is to impress a current onto the potentially corroding metal to make it cathodic. Here a battery, or a step down transformer is used with another power source. This technique is used on shipboard for iron hulled vessels, and the battery can be recharged when needed. Some pipelines are cathodically protected with impressed current. The advantage is that there are no ions lost to the surroundings, hence subsequent contamination of water.

The chemical inhibitors are many and varied items of commerce and may work in many ways. Most are sold because they've been shown to work, though no mechanism is proposed. Some of the surfactant type materials (oleyl sarcosine, for instance) may only add a physical barrier by adsorbing to the surface and blocking approach of oxygen, water, or catalytic ion. Other materials may adsorb on the metal and act as a pH modifier or buffer, as an amine would inhibit acid–catalyzed attack. Some inhibitors modify the electromotive potential at which corrosion occurs and are said to have "passivated" the surface. Most of the chemical inhibitors are low molecular weight compounds and can be washed away or otherwise rendered ineffective by chemical attacks or reactions. They are effective over short periods of time, and can be stabilized against erosions by formulation to some degree. For instance, there are oils into which corrosion inhibitors are formulated.

There are corrosion-inhibitive pigments. Seldom is there discussion of the mechanism by which these chemical inhibitive pigments work, and such characterization could extend their utility. It has been shown that nominally corrosion-inhibitive pigments don't work in some formulations (see SSPC literature and corrosion studies by the technical committees of the Northwest, New England, and Golden Gate Societies for Coatings Technology). Indeed, Nadim Ghanem (American University of Cairo) told of a basic lead carbonate formulation study in which the solvent-borne vinyl binder gave no corrosion protection, while a water-borne vinyl gave excellent protection. His hypothesis was that the coating needs to be permeable to water to get the lead ions to migrate to where they need to be to act as corrosion protectors. That may have also been the message in the Los Angeles Society for Coatings Technology work with aminosilane-treated talcs in latex formulations. Though many demonstrations of the effectiveness of the corrosion-inhibitive pigments exist, the mechanisms should be more thoroughly described to aid the formulator.

REFERENCES

1. F. N. Longo, Ed. *Thermal Spray Coatings: New Materials, Processes and Applications*, American Society for Metals Conference Book. Metals Park, OH: ASM, 1985.
2. R. T. Sorg, *Prod. Finish.* p. 72 July 1978.
3. Anon., *Ind. Finish.* p. 24 January 1978.
4. British Patent 1,369,056; *Chem. Abstr., 83*, 116382u.
5. M. Rogers, *Plast. Technol.*, p. 20, November 1982.
6. R. H. Hochman et al., Eds. *Ion Implantation and Plasma Assisted Processes*, American Society for Metals Conference Book. Metals Park, OH: ASM, 1988.
7. Fujitsu Ltd., Japanese Patent 8,089,833 Dec. 28, 1978; *Chem. Abstr. 5.* 39579m.
8. A. K. Sharma et al., *J. Appl. Polym. Sci. 26*, 2197 (1981).
9. R. Athey et al., *J. Coatings Technol., 57*(726), 71 (July 1985).
10. A. Morikana and Y. Asano, *J. Appl. Polym. Sci., 27*, 2139 (1982).
11. C. Arnold, Jr., et al., *J. Appl. Polym. Sci., 27*, 821 (1982).
12. M. R. Havens et al., *J. Appl. Polym. Sci. 22* (10), 2793 (1978).
13. D. T. Clark and M. Z. Abraham, *J. Polym. Sci., Polym. Chem. Ed. 19*, 2129 (1981).
14. N. Inagaki et al., *J. Polym. Sci., Polym. Lett. Ed., 19*, 335 (1981).
15. H. A. Beale, *Ind. Res. Dev.*, p. 135, July 1981.
16. M. Murphy, Ed., *Metal Finishing Handbook*, Hackensack, NJ.
17. E. B. Saubestre, in *Modern Electroplating*, 3rd ed., F. Lowenheim, Ed. New York; Wiley-Interscience, Chapter 28.
18. G. Muller et al., *Plating on Plastics*, Surrey, England; Portcullis Press.
19. J. Springer et al., *Angew. Makromol. Chem., 89*, 81 (1980).
20. J. Springer et al., *Angew. Makromol. Chem., 89* 81 (1980).
21. J. Springer et al., *Angew. Makromol. Chem., 89* 81 (1980).
22. A. G. Wilmhurst, *Aust. Oil Colour Chem. Assoc.*, p.12, November 1978.
23. W. H. Safranek, *The Properties of Electrodeposited Metals and Alloys,* American Electroplaters and Surface Finishers Society, 1986.
24. S. A. Watson, *Nickel. 4*(1), 8 (September 1988).

25. P. A. Schweitzer, Ed., *Corrosion and Corrosion Protection Handbook,* New York, Dekker, 1983.
26. (a) Unit 27, *Anticorrosive Barriers and Inhibitive Primers,* Philadelphia: Federation of Societies for Coating Technology, 1979. (b) Unit 26, *Corrosion and the Preparation of Metallic Surfaces for Painting,* Philadelphia: FSCT, 1979.
27. D. Campbell and R. W. Flynn, *Am. Paint Coatings J.,* p. 55, March 6, 1978.
28. A. Poluzzi, *14th Annual FATIPEC Congress Proceedings* 1978, p. 61.
29. R. Athey et al., *J. Coatings Technol.* 57(726), 71 (July 1985).
30. C. H. Munger, *Corrosion Prevention by Protective Coatings,* National Association Engineering, 1984.

79

Decorative Surface Protection Products

Jaykumar (Jay) J. Shah

Decora
Fort Edward, New York

1.0 INTRODUCTION

Decorative surface protection products, as the term suggests, are products that provide both decoration and protection to a wide variety of surfaces. These products are available to consumers in a variety of colors, decorative designs, and surface finishes. They are attractive, durable, washable, waterproof, and resistant to stains from food, beverages, and common household items. They must be conformable to a wide variety of surfaces, requiring no tools, water, or paste to be applied to any plain surface.

In most cases, they have informative and instructional carriers traditionally known as printed release liners, which are affixed by a flexible pressure-sensitive adhesive to a base sheet, also known as a primary substrate or a face stock. In some cases, the release liner has been eliminated by using an embossed primary substrate. Alternatively, the primary substrate may have been coated on one side with a release coating and on the other side with a low tack adhesive.[1] These products are known as self-wound adhesive coated decorative sheets. They are made in solid colors, decorative prints, and textured woodgrains.

2.0 HISTORY

Pressure-sensitive adhesive usage in making decorative surface protection products goes back to the early 1950s. Similar types of vinyl film were used by consumers on products such as printed and laminated table cloths and printed draperies. The idea of printed film led to the making of decorative surface protection products by a low tack, pressure-sensitive adhesive on printed film.

The construction of decorative products is shown in Figure 1.

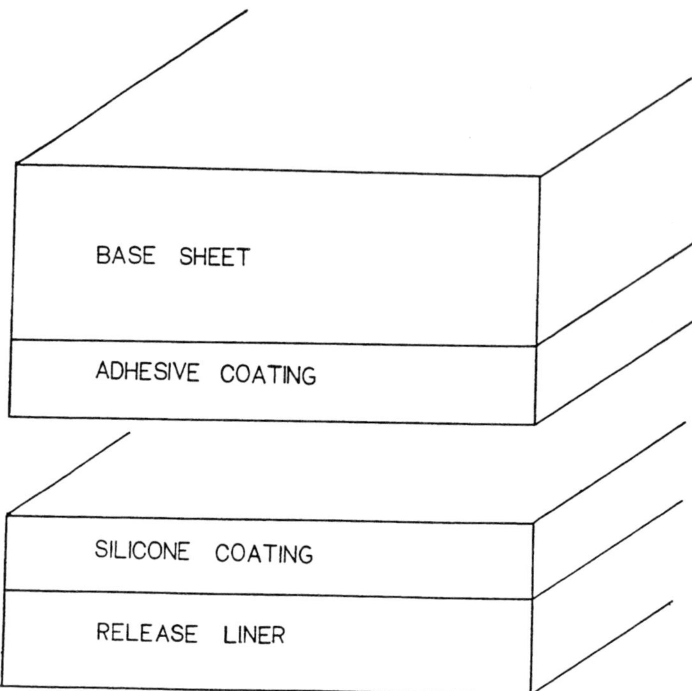

Figure 1   Construction of decorative products

3.0 PRODUCTS

The manufacturer planning to bring out a new product has many varieties to choose from. He should consider not only the physical, chemical, and thermal properties, but also the cost and ease of the processing method. The key to selecting the right material may be good communication between the processor and the material suppliers. The more rigid the specifications covering end use, the easier selection becomes.

3.1 Primary Substrate or Base Sheet

Typical substrates used are calendered plasticized vinyl films, extruded homo- or copolymers of polyolefin films, metallized polyolefins and polyester films, preimpregnated papers, treated cloth, and nonwovens.

Plasticized vinyl films have a dominant position as a decorative film product. These films are usually manufactured by the calendering process. The method of making such films is discussed in detail by Reip.[2] Abrasion and stain resistance coupled with unlimited coloring and design possibilities have made vinyl films one of the largest base sheets in the decorative coated product field.

Films extruded from homo- or copolymers of polyolefin are flexible but tough and durable, moisture proof, and grease resistant. The film industry, through recent coextrusion techniques, is producing single- and multiple-ply laminated polyolefin films that are printable, dimensionally stable, and opaque or clear. These materials are becoming more popular because of their lower cost.

Cotton or cotton–polyester fabrics were used earlier because they are more durable, tough, moistureproof, and washable, as well as more easily printable than films. However, they are not as flexible and conformable as films. Their higher cost have made them less popular. They have been replaced in some applications by supported or unsupported nonwovens made from natural or synthetic fibers.

Decorative products made with printed and metallized polyester or polypropylene films provide brightness and contemporary design. Their use is limited because of more difficult handling and higher cost.

Impregnated paper has been used more recently because of its low cost. Self-wound decorative products have been made using such paper. Because impregnated papers are printable and can be made self-wound, they have more applications. However, papers tear easily and are not as flexible as films, limiting their application.

The criteria to be considered to qualify a supplier for the base sheet and routine tests performed in accepting base sheets from qualified suppliers are described in Table 1.

3.2 Pigmented Coating for Decorative Printing

An organic coating is made up of two principal components, a printable vehicle and a pigment. A vehicle is a film forming ingredient that enables the coating to convert from a mobile liquid to a solid film. It acts as a carrier and suspending agent for the pigment. The pigments are the coloring agents.

Vehicles or extenders are composed of film forming materials, thinners (solvents), and water or water–solvent mixtures that control viscosity, flow, and coating thickness, as well

Table 1 Criteria to Select a Qualified Supplier for Base Sheet

Composition of film
Physical properties
  Tensile strength
  Elongation
  Modulus
  Tear strength
  Dimensional stability ASTM D-1204 methanol (both machine direction
    and cross direction)
  Opacity
  Volatiles content
  Surface finish (gloss)
  Hardness
  Color
Routine tests
  Opacity
  Color
  Gloss
  General observation to check hand, voids, pinholes, gel particles,
    dimensional stability, roll counter, roll profile
  Width and length of roll

Table 2 Characteristics for Qualifying a Printing Ink

Qualifying Tests	*Routine tests*
Blocking	Viscosity
Cracking	Solids content
Drying	pH (if waterborne)
Color in half-tone	Weight per gallon
Color in masstone	Opacity
Color strength	Gloss
Hiding strength	Color in masstone (proof press)
Gloss	Blocking
Viscosity	Adhesion (tape test)
Pigment content	Color match to standard
Cell clean-out	
Scratch resistance	
Orientability	
Weight per gallon	
Foaming	
pH (if waterborne)	

as driers, which facilitate application and improve drying qualities. Most common vehicles are made from acrylics and vinyl-based resins.

Pigments serve a decorative function. Closely associated with color is the hiding power function. There are white hiding pigments, colored pigments, and inert pigments. White pigments are used in making white bases for tinted and light shades. Colored pigments furnish the opacity and color of the finish. Earth-colored pigments are stable and are resistant to attack by heat, light, chemicals, and alkalies. The extender pigments used in the coating help control consistency, gloss, and smoothness, and leveling and filling qualities. They are chemically inactive.

Table 2 describes the characteristics that would be considered for qualifying a proper printing ink and routine tests performed on printing ink. The majority of base sheets and release papers can be printed by a direct gravure or rotogravure printing method. Flexographic printing and screen printing methods are also used in some cases.

3.3 Release Paper

Usually a bleached, semibleached, or unbleached high internal strength kraft paper is used. The most common types of paper used are sized machine-grade or machine finished. There are some products that use supercalendered-grade and polyethylyne-coated kraft paper. Some European and Asian manufacturers have used parchment or glassine-grade paper because of cost and ease of availability. The paper should exhibit exceptional smoothness and should have uniform thickness. Most paper used has a thickness of 2–3 mils. It should retain a minimum moisture of 3–4% for better runnability and to keep from tearing.

Characteristics to be considered for selecting a paper and routine physical tests are listed in Table 3.

Table 3 Criteria for Selecting a Release Paper

For better runnability
    Tensile strength (dry and wet), in machine and cross directions
    Moisture content
    Elmendorf tear
    Mullen burst strength
    Porosity
    Caliper
    Basis weight
For better coatability
    Sizing in paper for better silicone holdout
    Surface finish for better printability
    Heat resistance and shrinkage resistance for better dimensional
        stability while drying coatings
    Smoothness and porosity for better release coating
    Good formation to minimize cockling tendency
Routine physical tests on a qualified paper
    Caliper
    Basis weight
    Mullen burst strength
    Tensile strength in machine and cross directions
    Tear strength in machine and cross directions
    Surface smoothness
    Porosity
    Color

3.4 Release Coatings

Silicones are the most widely used materials for release coating applications. They give a low level of release and are useful for coating the release paper.

The basic chemistry and physical properties of silicone release coating is described in detail by Mary D. Fey and John E. Wilson.[3]

3.5 Pressure-Sensitive Adhesives

Although pressure-sensitive adhesive clearly takes a position of secondary importance in most products, careful selection of the adhesive should be an important function of a researcher.

The main function of the adhesive is to provide affinity to both the release liner and the base sheet used on these products. The adhesive should allow the base sheet enough time to move around the surface before bonding permanently. Another important function of the adhesive is to provide plasticizer migration resistance specifically when a plasticized vinyl film is used as a base sheet.

Another method of demonstrating improved pressure-sensitive properties is to compare the rate of buildup of peel adhesion to various surfaces with other candidates. The adhesive should provide good heat and humidity resistance and also better shrinkage resistance in the finished products.

Acrylics and vinyl–acrylic based pressure-sensitive adhesives appear to be the most promising compounds because of their UV and thermal stability and higher plasticizer migration resistance properties. These adhesives also provide better shrinkage resistance on the release papers and after the product is mounted on various surfaces. Basically these adhesives have the versatility to adhere to a variety of surfaces. These adhesives are available in the form of solutions and aqueous emulsions.

Several patents have been issued claiming the use of such adhesives on the decorative products.

Vinyl–acrylic adhesives for vinyl films have been suggested by R. L. Burke Jr.[4] The patent claims that adhesives of these types substantially eliminate migration of the plasticizers from vinyl films and thereby eliminate film shrinkage during aging.

Vinyl methyl ether based adhesives, as suggested by Helmut J. Mueller of BASF,[5] reduce the drop in adhesion strength caused by plasticizer migration. This phenomenon is due to the good compatibility of the vinyl methyl ether polymers with a great variety of solvents and plasticizers.

A new acrylic family is being developed specifically for low energy level surfaces such as polyolefin films.[6]

4.0 PROCESS OF MANUFACTURING

4.1 Application of Release Coatings

There are two basic types of converting operation for silicone release coatings: the off-line and the in-line coatings. In an off-line coating, only the silicone coating is applied as shown in Figure 2. The adhesive and the base sheet are married to the release-coated liner in a subsequent operation.

In the in-line operation there may be multiple coating heads and a laminating station. The entire product is made in one continuous operation. First, the silicone release coating is applied and cured. Then the adhesive is applied to the release-coated substrate and dried. Finally, the base sheet is laminated to the adhesive-coated release liner. This operation is shown in Figure 3.

The line speed for off-line production of silicone coatings is usually faster than the in-line silicone coating operation. Line speeds are usually slower for in-line operations and depend on the adhesive drying rate, the length of the oven, and the efficiency of the drying oven. Silicone release coating is applied by different coating methods, such as direct gravure, wire-wound rods, air knife, and the offset gravure. The majority of solvent silicone release coatings are applied by direct gravure or wire-wound rod methods. Most waterborne silicone release coatings are applied by the wire-wound rod or air knife methods.

Virtually all solventless silicone release coatings are applied by the three- or four-roll offset gravure or smooth roll coating method.

4.2 Adhesive Coating Application Process

Coating of solvent- or water-borne adhesives can be carried out by several different methods. The most commonly used methods of coating pressure-sensitive adhesives are direct gravure, wire-wound rod, reverse roll, direct roll, and knife over roll. Normally the reverse roll, direct roll, and knife over roll methods are used for high viscosity adhesive applications. The direct gravure and wire-wound rod methods are used for lower viscosity adhesives. The adhesive coating weights on these products may range from 0.2 to 1.0 oz/yard^2.

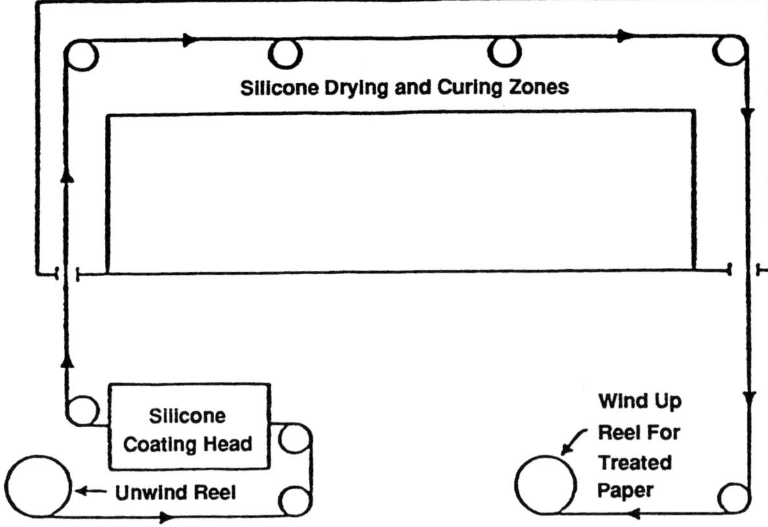

Figure 2 Off-line silicone coating.

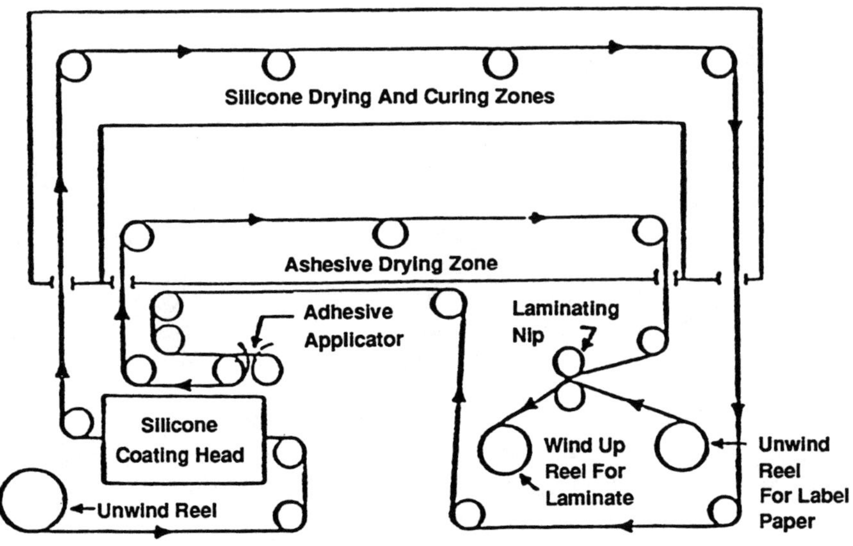

Figure 3 In-line silicone/adhesive coating and lamination.

Table 4 Routine Tests for Finished Products from Coated Rolls

Peel adhesion (0 min, 20 min, 24 h dwell) to varnished wood
  acrylic plastic sheet
  glass
  various other plastic surfaces
  painted wood
  stainless steel
  unpainted wallboard
Adhesive to adhesive anchorage
Release values from release liner
Adhesive coat weight
Total thickness
Free shrinkage at 158°F for 24 hours
Mounted shrinkage at 110°F for 7 days
Tensile strength
Elongation

The adhesive is usually transfer coated. Transfer coating eliminates the need for exposing heat-sensitive substrates such as vinyl and olefin films to higher heat drying ovens.

4.3 Finishing

The finished product from coated rolls is slit into several popular widths and lengths. All these products are slit either in-line or off-line. The most common slitting method is score knives, but the shear slit method is also used in some applications. The finished product is then tested. Table 4 describes the most common tests that may be applicable to the end-use requirements.

5.0 APPLICATIONS

The primary purpose of these products is to beautify surfaces and fulfill aesthetic objectives. Besides decorating, the products protect surfaces and keep them clean. They repair, refinish, and transform worn surfaces from floor to ceilings, as well as windows, doors and walls, furniture, jewelry boxes, books, and an unlimited number of other surfaces.

Some of the specialty decorative adhesive-coated products described here can solve specific household and office problems.[7]

"Frosty look" products are made from embossed translucent film that looks like frosted glass. The film lets light come through, but shuts out unattractive views. It can be applied to any window or clear surface, and offers privacy in applications from shower stall doors to office partitions.

Transparent film products are exceptionally useful because of their unique see-through properties. They are excellent for book covers, covering recipe cards, mending torn pages, sealing valuable documents, protecting pictures in photo albums, and similar applications. They are also good for windowsills, splash areas, and nursery areas, and for temporarily protecting glass, metals, and plastic while products made of these materials are handled in transit.

"Velvet look," known as "cushion all" products, with a velvety nylon surface protect, carpets from furniture dents and furniture from scratching. Other uses include lining jewelry drawers and boxes, tools, and sewing boxes, and covering surfaces for a decorative effect. These products are good for craft work, and for decorating walls, furniture, or the inside of picture frames.

"Stained glass look" and "leaded glass look" products are meant to provide a beautiful glowing radiance to any room. Light comes through the brilliant colors of these translucent and colorful patterns. The products are especially good for Christmas decorations or for problem windows year round.

"Quilt soft" products are attractive quilted materials providing protection from mildew and moths. They consist of two layers of film, softly padded with fiberfill and then quilted, and are used to line drawers, and vanities, closets, and pad shelfs. They provide a soft touch of luxury at low prices.

"Woodgrain vinyl" decorative products, available in a variety of patterns, add a new dimension of depth and beauty to furniture, walls, and shelving. They are woodlike patterns that are hard to distinguish from real wood. They are decorative and useful for splash areas.

"Burlap" products are available in various solid colors and add warmth of texture. Made from fabric combined with burlap, they are good for walls, corners, headboards, and bulletin boards.

Metallized or foil-based products can make any room in the house glitter with light. They can cover canisters, trays, boxes, desk accessories, lamp shades, books, table tops, wall panels, and furniture panels. Available in various designs and colors, these products reflect a contemporary mood.

"Wet look" products are designed to go slickly onto walls or lend a glossy sheen to furniture and accessories.

Some of the films and paper-backed products also have added fragrances, so the products not only decorate surfaces but also bring fresh flavor into the room.

This is not an all-inclusive list but an illustration of uses, provided for the reader's interest in hopes of furthering understanding of the versatility of such products in fulfilling market needs.

REFERENCES

1. M. C. Funk and E. A. Rugalska, Jr. U.S. Patent 3,575,788 (April 20, 1970).
2. H. Reip, *Adhes. Age, 15*(3), 17–23 (1972).
3. Mary D. Fey and John E. Wilson, Chapter 18 in *Handbook of Pressure-Sensitive Adhesive Technology*, D. Satas, Ed. New York, Van Nostrand-Reinhold, 1982.
4. R. L. Burke, Jr., U.S. Patent 3,547,882 (December 15, 1970); United Merchants and Manufacturers.
5. Helmut J. Mueller, in *Handbook of Pressure-Sensitive Adhesive Technology*, D. Satas, Ed. Chapter 14, p. 340. New York, Van Nostrand-Reinhold, 1982.
6. W. J. Sparks, "Advances in hot melt and waterborne acrylic PSAs," *Adhes. Age, 25*(3), 38 (1982).
7. Comark Plastic, Division of United Merchants and Manufacturers, "Ideas to Help You Sell More," 1967–1981, retail sales manuals.

80
Coated Fabrics for Protective Clothing

N. J. Abbott

Albany International Research Company
Dedham, Massachusetts

1.0 INTRODUCTION

Our need for protection from the environment in which we live or work is becoming ever more complex and demanding. We can travel easily to any part of the earth, even to the vacuum of space, and we expect to be able to live more or less normal and active lives no matter where we are or what we may wish to do. We have become more conscious of the potential dangers that surround us at home and at work, and we expect some protection from hazards that only a short time ago were thought of as risks that our way of life required us to accept without question. Consequently, we have recently seen advances in the design of a wide range of protective clothing of ever higher levels of sophistication, made possible by the continuing development of new materials capable of providing improved protection from many threats to our well-being.

Protective clothing is commonly designed to isolate the body from its surroundings through the use of an assembly that includes as one of its components a continuous film, either in the form of a coating or a lamination. This is essential whenever resistance to absorption or penetration of a liquid is required, and is often used also for gases, though these can sometimes be adsorbed or chemically modified in a layer that may not be totally impermeable. Often a structure designed to provide protection from a nonfluid threat—cold, for example—will function effectively only if it remains dry. Some degree of resistance to wetting may be provided by giving the fibers a water-repellent treatment. But totally effective resistance to the penetration of water into the structure requires that a film be added to the assembly, which provides improved wind resistance but otherwise plays only a secondary role in providing protection from cold.

Coating or laminating, then, is widely used in protective clothing, and the properties of the films, as well as how they are incorporated into the clothing assembly, are of prime importance to the production of an acceptable garment.

2.0 PROTECTION FROM WHAT?

Perhaps our most universal need is for protection from rain and cold. Other needs are more specialized, and generally apply only to that segment of the population whose livelihood exposes them to annoyances or dangers most of us seldom encounter. Firefighters, for example, and some industrial workers, confront sources of extreme heat which, without suitable protective clothing, could be life-threatening. Other workers run the risk of having molten metal splashed on them from a smelter or a welding operation. Many must handle toxic or corrosive liquids, or be exposed to toxic fumes in the course of their daily work. Farmers and other agricultural workers handle potentially dangerous chemicals in the form of fertilizers or pesticides. Those responsible for handling waste products from manufacturing operations run the risk that some of those products may be dangerous if handled improperly. Butchers and foresters need special protection against cutting because of the tools they must use. People working with viruses, bacteria, or other sources of infection, divers who must work underwater for extended periods of time, astronauts exposed to the vacuum of space, or people who run the risk of being exposed to X-ray or other electromagnetic or nuclear radiation may need virtually total isolation from their surroundings. Even football and hockey players must wear special protective pads to minimize the risk of injury.

The list of hazards from which we may want to be protected without serious impediment to our capability of performing necessary tasks is almost endless. Fortunately, the number of materials that can be used in protective clothing is also large, and the diversity of their characteristics is growing rapidly. Protection can now be provided against most threats, but not without some associated problems (see Chapter 81). To some extent, the protection can be obtained simply from the use of the right fiber in an appropriately designed garment. For the most part, however, fiber or fabric coatings are used to obtain the desired isolation of the body from a threat in the surrounding atmosphere.

3.0 COATING MATERIALS

3.1 Coating Types

Any compound that can be applied to the surface of a fabric is a candidate for use in protective clothing. Because each application has its own set of requirements, most coating materials will find a use for which they are suited. Consequently, the designer of protective clothing has a broad material base to work from. Unfortunately, however, the task of choosing the best material for a specific purpose is not as simple as looking up a table of properties in a handbook.

Fabric coatings are seldom composed of a single chemical compound. Rather, they are formulated as a mixture of many ingredients, each designed to provide or improve a particular characteristic. Additives may include viscosity modifiers to assist in application; reaction modifiers to accelerate or retard setting during processing; hardness modifiers; flame retardants; ultraviolet absorbers; antioxidants; surface friction modifiers; abrasion resistance enhancers; pigments to provide color or modify appearance; light and heat reflectors; inert fillers to provide bulk or opacity, or to reduce cost; compounds or chemical modifications to increase water vapor permeability; and even copolymers or polymer mixes as major property modifiers. Consequently, although the characteristics of the major coating types can and should be considered, it must be clearly understood that significant changes in these characteristics can be achieved through proper formulation.

Coating types that are used in protective clothing include acrylic, butyl rubber, cellulose acetate, fluorocarbon, fluorosilicone, chlorosulfonated polyethylene (Hypalon), chloroprene (neoprene), polyamide (nylon), polyester, polyolefin (polyethylene, polypropylene), polyurethane (polyester and polyether types), polyvinyl chloride, silicone, vinyl acetate, and various metals (particularly aluminum).

3.2 Method of Application

Generally coating materials are formulated for a specific method of application. Some may be best applied by blade coating; others are for calender coating; some must be used as films, either by a film lamination technique or in transfer coating; some are designed for extrusion coating; some for dip or saturation coating; some require solvent systems; some are water-based; some are 100% solids.

3.3 Properties

The following general information indicates typical properties of the base polymer in each case. As discussed above, these properties often can be significantly altered by chemical modification and the use of additives.

A general statement may be made about the range of temperatures within which these coatings are useful. The upper limit of this range is determined by the temperature at which the polymer softens, melts, or starts to decompose; it is unaffected by additives, though it may be altered by chemical modification of the polymer chain. The lower limit may be modified somewhat by formulation or chemical modification. In general, most of these polymers stiffen as the temperature is reduced, few remaining usefully flexible below about −40°C. Exceptions to this are the fluoropolymers and silicones.

3.3.1 Nonmetallic Coatings

Acrylic. Useful temperature range −30 to 175°C. Poor chemical resistance, fair abrasion resistance, good sunlight resistance. Ignites and burns without additive protection. Moderate cost.

Butyl Rubber. Useful temperature range −50 to 120°C. Poor chemical resistance to aromatic or chlorinated hydrocarbons, otherwise fair to good. Good abrasion and sunlight resistance, poor flammability resistance. Low to moderate cost.

Cellulose Acetate. Useful temperature range −20 to 120°C. Good chemical resistance to hydrocarbons, otherwise fair to poor. Fair abrasion and sunlight resistance, poor flammability resistance. Low cost.

Chlorinated polyethylene. Useful temperature range −40 to 160°C. Good chemical resistance except to chlorinated hydrocarbons and some other organics, excellent abrasion resistance, good sunlight resistance, burns but is self-extinguishing. Moderate cost.

Fluorocarbon. Useful temperature range −130 to 260°C. Excellent chemical resistance, fair abrasion resistance, excellent sunlight resistance; does not ignite or burn in air. High cost.

Fluorosilicone. Useful temperature range −70 to 200°C. Excellent chemical resistance, poor abrasion resistance, excellent sunlight resistance; does not ignite or burn in air. High cost.

Polyamide. Useful temperature range –40 to 120°C. Good chemical resistance except to acids and aromatic hydrocarbons. Excellent abrasion resistance, poor sunlight resistance; melts and burns. Moderate cost.

Polychloroprene. Useful temperature range –40 to 120°C. Good chemical resistance except to chlorinated hydrocarbons. Good abrasion and sunlight resistance; burns but is self-extinguishing. Low to moderate cost.

Polyester. Useful temperature range –40 to 120°C. Good chemical resistance except to chlorinated hydrocarbons, alkalies, and organic acids. Good abrasion and sunlight resistance; ignites and burns. Low to moderate cost.

Polyolefin. Useful temperature range –70 to 120°C. Poor chemical resistance to chlorinated hydrocarbons, otherwise fair to good. Good abrasion resistance, excellent sunlight resistance; ignites and burns. Low cost.

Polyurethane. Useful temperature range –50 to 120°C (depending on the type—this is a class of polymers covering a wide range of properties). Excellent chemical resistance to aliphatic hydrocarbons, otherwise fair, although polyether types have good resistance to organic chemicals. Excellent abrasion resistance, fair to good sunlight resistance; ignites and burns without additive protection. Moderate cost.

Polyvinyl acetate. Useful temperature range –20 to 100°C. Poor chemical resistance except to alkalies. Poor abrasion resistance, good sunlight resistance; ignites and burns. Low cost.

Polyvinyl chloride. Useful temperature range –20 to 100°C. Fair to poor chemical resistance except to aliphatic hydrocarbons, alkalies, and inorganic acids. Good abrasion and sunlight resistance; ignites and burns without additive protection. Low cost.

Silicone. Useful temperature range –115 to 320°C. Good chemical resistance to acids and alkalies, otherwise fair to poor. Poor abrasion resistance, excellent sunlight resistance; chars but does not burn. High cost.

3.3.2 Metallic Coatings

The commonest metal coating consists of aluminum that has been vacuum coated or otherwise deposited on a thermoplastic film, which is then thermally bonded to the fabric surface. Its function is to reflect infrared radiation. Consequently, its high reflectivity (95+%) must be retained through use. Its tendency to retain oily soils and to become brittle and flake off through flexing are problems that have never been satisfactorily overcome. Other metals, such as gold, silver, and nickel, are also used for special purposes.

4.0 MARKETS AND STANDARDS

The markets for protective clothing are small (by textile standards) and fragmented, because of the highly specialized nature of the individual needs addressed. Often the desired protection can be provided only by ingenious design, particularly of seams, closures, and means of ventilation. Consequently, protective clothing tends to be more expensive than conventional clothing, and consumer resistance to its use because of its awkwardness or lack of comfort further limits market size. However, a great deal of effort is going into the development of more satisfactory materials and innovative clothing design and manufacturing techniques, and we can confidently expect this to be a market of growing importance.

The need for standardization, particularly of testing procedures, is being answered by technical studies, as well as by the work of groups such as Committee F-23 on Protective Clothing of the American Association for Testing and Materials.

BIBLIOGRAPHY

Barker, R. L., and G. C. Coletta, Eds., *Performance of Protective Clothing*, symposium in Raleigh, NC, July 16–20, 1984. ASTM Committee F-23, ASTM Special Technical Publication 900, 1986.

U.S. Department of Commerce, *Protective Clothing: Arctic and Tropical Environments* (January 1970–July 1987). NTIS, PB87–854542/RPS, 1987.

U.S. Department of Commerce, *Protective Clothing: Fire and Radiation Environments* (January 1970–January 1988). NTIS, PB88–856950/RPS, 1988.

U.S. Department of Commerce, *Protective Clothing: Industrial Environments* (January 1985–January 1987). NTIS, PB87–854147/RPS, 1987.

U.S. Department of Commerce, *Protective Clothing: Survival, Aircraft and Combat Environments* (August 1981–December 1987). NTIS, PB8–854666/RPS, 1988.

81

Coated Fabrics for Apparel Use: The Problem of Comfort

N. J. Abbott

Albany International Research Company
Dedham, Massachusetts

Coatings or laminated films are commonly used in apparel fabrics to provide protection from wind, water, or other fluid (see also Chapter 80), or sometimes simply to answer the dictates of fashion. Almost any type of coating material may be used, particularly by the fashion designer, whose choice may be determined more by aesthetic considerations than by functionality or serviceability. But whenever a continuous coating is used in clothing, no matter what its type or purpose, there is one common problem: people are likely to find the clothing uncomfortable if worn for long periods of time, or during vigorous activity.

1.0 THE PHYSIOLOGY OF HEAT REGULATION

"Comfort" can be defined very simply and precisely. We feel comfortable when the temperature of our skin is within about 1° of 34°C and, of course, the physiology of our body controls that temperature very effectively under normal conditions. If there is a change in the ambient conditions or in any other factor, such as a reduction in physical exertion, which lowers the skin temperature, blood flow to that part of the body is increased. If, on the other hand, the skin temperature increases, there is an increase in the output of perspiration which, by its evaporation, produces a cooling effect. Activity levels that do not adversely affect our comfort, therefore, are those for which the body's mechanisms can control skin temperature without the accumulation of liquid sweat on the skin.

Moisture is carried away from the skin by two mechanisms: by evaporation at the skin and transmission of that vapor through the clothing (as vapor); and by absorption of liquid water into the clothing. While the second mechanism is useful in keeping the skin dry, it is of little value as a cooling device. Temperature control is best achieved through evaporation of perspiration from the surface of the skin.

The amount of heat (provided by the body) required to evaporate one gram of liquid perspiration from the skin at 34°C, and to dissipate it into the surrounding atmosphere, assumed for the purpose of this example to be at 27°C, 50% relative humidity, has been stated to be

713

$$E_T = L_{34} + E_C + E_V$$

where

L_{34} = latent heat of vaporization of water at 34°C = 578 calories

E_C = heat required to expand the vapor from 34°C, 100% RH, to 27°C, 100% RH = 9 calories

E_V = heat required to expand the air containing the evaporated water in order to reduce its relative humidity from 100% to the humidity of the Ambient air, 50% = 23 calories

Hence, the total cooling energy derived from evaporating one gram of liquid perspiration and dissipating it into the surrounding atmosphere is

$$E_T = 578 + 9 + 23 = 620 \text{ calories}$$

This is true only, of course, if the clothing is capable of transmitting the vapor to the ambient atmosphere without a change of phase or of temperature (other than that provided for in the calculation). It is clear, then, that the moisture vapor permeability of the clothing must be high enough to pass vaporized perspiration at the rate at which it is being produced at the skin.

2.0 VAPOR PERMEABILITY REQUIREMENTS

Any estimate of the vapor permeability requirements of clothing must be based on estimates of a number of factors, which depend on body size and physiology. Since these vary significantly from individual to individual, and even for one individual from one time to another, no precise values can be given. One of those factors consists of the metabolic rates corresponding to various levels of physical activity. Representative values of metabolic rates are shown in Table 1. Some of this metabolic energy is used to perform the work that is being done. Most of it, however, is turned into heat, which must be dissipated. Some of this is

Table 1 Metabolic Rates, Perspiration Production, and Vapor Permeability Requirements for Various Levels of Physical Activity

Activity	Metabolic rate (W)	Water evaporation rate (g/24 h)	Vapor permeability rate (g/m²/24 h)
Sleeping	60	600	200
Sitting	100	3,800	1,300
Gentle walking	200	7,600	2,500
Active walking	300	11,500	3,800
Active walking on the level, carrying a heavy pack	400	15,200	5,000
Active walking in the mountains, carrying a heavy pack	600–800	23,000–30,000	8,000–10,000
Very heavy work	1000+	38,000+	13,000+

expelled in respiration. The remainder must be dissipated through the evaporative cooling mechanisms discussed above. Depending on the design of the clothing, some portion of the vaporized perspiration may reach the surrounding atmosphere directly through vents in the clothing by means of a bellows-type action. The remainder must pass through the clothing fabric as vapor. Table 1 gives values for the required vapor permeability of the fabric, based on representative values of all these variables, as well as of the total surface area of the clothing.

It has been suggested that clothing that is to be worn during periods of physical activity have a moisture vapor permeability of 6000 grams per square meter per 24 hours. Even the densest uncoated sportswear fabric easily meets this requirement. The addition of an impermeable coating, however, reduces the permeability to a level often not more than 100 $g/m^2/24$ h. To ensure that clothing made from coated fabric is comfortable, this permeability must be raised by as much as 2 orders of magnitude.

2.1 Breathable Films

Three basic approaches have been taken to the manufacture of coatings that are permeable to water vapor, and at the same time, resistant to the passage of liquid water (so-called breathable films):

1. Puncture the film with needles, laser beams, or other means to produce an array of micrometer-sized holes.
2. Make the film from a material that can be broken up into fine, fiberlike strands, with micrometer-sized spaces between the fibers.
3. Create monolithic polymer membranes that contain no through-going pores, in which transmission of water vapor occurs through a process known as activated diffusion. In such membranes, the water vapor condenses and dissolves in the surface and then diffuses through to the other side of the film, where it desorbs and evaporates into the surrounding space.

The first of these approaches, involving the mechanical puncturing of a cast film, appears in some ways to be a simple and direct way of achieving the desired permeability. However, because the holes need to be extremely small if water resistance is to be maintained, and very closely spaced if the desired permeability level is to be attained, there is at present no totally successful, economically viable product available.

The production of microporous films by expanding and splitting a continuous film of appropriate morphology has been a more successful approach. Several such products are commercially available, the best known being based on a polytetrafluoroethylene film. Others based on polypropylene or polyurethane are also being produced.

The third approach, the production of a permeable monolithic film, is being pursued aggressively by many companies throughout the world. Most of these products are based on a modified polyurethane or polyester, and several are already in use, particularly in sportswear.

The best of these "breathable" materials have moisture vapor permeabilities as high as about 4000 $g/m^2/24$ h, which is high enough to keep a moderately active individual comfortable. It is not yet high enough to meet the needs imposed by vigorous activity, or the extreme requirements of, for example, the long-distance runner, hockey player, or firemen. But it is a significant improvement over the use of a regular continuous coating.

This is a rapidly changing area that is attracting a great deal of research and development effort. Recently, a new candidate material based on a modified amino acid [poly(R-methyl

L-glutamate)] has been announced. This coating was stated to have a moisture vapor permeability of 8,000–12,000 g/m²/24 h. We can confidently expect a proliferation of products to result and can look forward to a time when the long-term comfort of water-resistant clothing, even for the most active person, can be ensured.

BIBLIOGRAPHY

Fourt, L., and N. R. S. Hollies, *Clothing Comfort and Function*. New York: Dekker, 1970, pp. 21–30.

Greenwood, K., W. H. Rees, and J. Lord, "Problems of protection and comfort in modern apparel fabrics," in *Studies in Modern Fabrics*. Manchester, UK: The Textile Institute, 1970, pp. 197–218.

Hollies, N. R. S., and R. F. Goldman, *Clothing Comfort: Interaction of Thermal, Ventilation, Construction and Assessment Factors*. Ann Arbor, MI: Ann Arbor Science, 1977.

Keighley, J. H., "Breathable fabrics and comfort in clothing," *J. Coated Fabrics*, **15**, 89 (1985).

Lomax, G. R., "The design of waterproof, vapour-permeable fabrics," *J. Coated Fabrics*, **15**, 40 (1985).

Lomax, G. R., "Coated fabrics. Part I. Lightweight breathable fabrics," *J. Coated Fabrics*, **15**, 115 (1985).

Newburgh, L., *Physiology of Heat Regulation*. Philadelphia: Saunders, 1949, pp. 99–117.

Slater, K., "Comfort properties of textiles," *Textile Progress* (The Textile Institute, Manchester, UK), **9** (1977).

U.S. Department of Commerce, *Comfort Factors in Protective Clothing* (January 1978–April 1987). NTIS, PB87-857678, 1987.

82
Architectural Fabrics

Marcel Dery

Chemical Fabrics Corporation
Merrimack, New Hampshire

1.0 INTRODUCTION

In the late 1960's, the opportunity to economically encapsulate large, clear-spans dictated a light weight construction approach. The temporary nature of available fabrics was not objectionable, since many of the structures envisioned, such as halls for international expositions, required relatively short periods of actual use. This provided an impetus to reconsider the design implications for such structures and finally to a reconsideration of the materials of construction themselves.

To fully exploit the potential of the fabric option, there was little doubt that a new generation of structural fabric would be required: materials tough enough to withstand the rigors of handling by construction crews, virtually impervious to the ravages of weather, able to meet all applicable life safety codes including fire hazard, and sufficiently translucent to provide natural illumination in daylight hours.

2.0 PRODUCTS

While several available fabrics could meet some of these requirements, none would meet them all. Nevertheless, it seemed reasonable to believe that such properties could be engendered if the proper selection of materials were made. In retrospect, it now appears that the materials eventually selected, fiberglass and Teflon perfluoropolymer resins, may be unique in their ability to confer these properties in an efficient and cost-effective manner.

Glass in its fibrous form is an outstanding candidate for a woven reinforcement: it is pound for pound as strong as steel. It is incombustible, and it is compatible with the elevated temperatures required for processing in conjunction with the most incombustible resins.

Teflon perfluoropolymer resins are the most chemically inert plastics known and are particularly noted for their ability to withstand exposure to the ultraviolet radiation, moisture

Portions of this chapter were extracted from a presentation made by Dr. John A. Effenberger, Vice President and Technical Director at Chemical Fabrics Corporation.

and smogs associated with the outdoor environment. The flammability characteristics of these resins are equally outstanding: such materials will not support combustion in atmospheres containing less than 98% oxygen. Also, because of their lower heats of combustion, perfluoropolymers contribute substantially less fuel value than a comparable mass of hydrocarbon polymers. Finally, both the light transmission and flame-resistive properties can be expected to be maintained indefinitely, since these properties are inherent in the plastic and are not dependent on additives, which may bloom to the surface and oxidize, or be washed away or attacked by microorganisms. By working within these functional requirements, a family of permanent architectural fabrics was developed. Certain characteristics of the composite do present mechanical problems. The brittleness of fiberglass must be addressed without compromising its inherently high strength and modulus of elasticity. And its sensitivity to hydrolysis must be effectively counteracted. Additionally, the low abrasion resistance perfluoropolymers coatings had to be overcome. Last, a method for joining fabric panels into roofing elements with structural integrity equivalent to that of the fabric had to be developed.

The brittleness issue is addressed first by choosing the finest diameter filaments to assure maximum strand flexibility. The yarns are then plied and woven in a plain configuration with a high degree of openness to enhance elongation, tear strength, and translucency in a coated fabric. The woven fabric is subsequently heat set and treated with a finish to inhibit the penetration of moisture into the yarns during processing, to further enhance tear strength, and to control elongation. The effectiveness of this process is evidenced by the high initial tensile and tear strength and the retention of tensile strength upon folding or soaking in water. Typical mechanical specifications for Sheerfill architectural fabrics are shown in Table 1.

Perfluoroethylene, by nature, has a low abrasion resistance. Since architectural fabrics must withstand the rigors of weather, a method of enhancing this property had to be developed. A glass filler was added to the outermost coats, which greatly improved the abrasion resistance of the surface without affecting the self-cleaning properties. A self-cleaning property is inherent in perflouroethylene coated roofs. This leads to much lower maintenance cost over conventional roofs. The low coefficient of friction allows dirt, snow, and water to easily leave the roof.

This procedure for joining panels of fabric into a completed roof is as follows. Panels are lapped to provide a 3 in. seam area. A film of polyfluoroethylene resin is used as a hot melt adhesive. Because this joint must be as structurally sound as the fabric itself, it must be constructed to avoid creep of the adhesive under design load. These joints are normally as strong as the fabric itself and equally durable.

Aside from the purely mechanical aspects of architectural fabrics, critical considerations include weatherability, fire safety, acoustics, and solar–optical performance. Weatherability has been assessed both by accelerated Weather–o–Meter exposure and by continuing real-time exposure at various weather stations. Accelerated tests data indicate that it is realistic to expect the fabric to retain adequate structural properties for more than 20 years. The limited data available from real-time exposure tend to corroborate the expectation of exceptionally long life.

Permanent building codes in the United States have proven in the past to be most unyielding to fabric structures options. Sheerfill architectural fabric structures, however, have found acceptance under the most stringent of U. S. codes, and have been approved for every structure submitted, most of which involve high public occupancy.

Table 1 Typical specifications for architectural fabrics

| | Sheerfill | | | |
Property	I	II	III	Fabrosorb
Weight, oz/yd^2	44	38	37	14
Thickness, in.	0.036	0.030	0.030	0.014
Tensile strength, lb/in.				
warp	800	520	620	360
fill	700	430	480	280
Flexfold strength, lb/in.				
warp	700	440	500	305
fill	600	360	375	240
Tear strength, lb.				
warp	60	35	50	25
fill	80	38	55	20
Coating adhesion, lb/in.				
minimum average	15	13	13	4
minimum single	10	10	11	4
Solar transmission, %				
high transmission	11	13	15	23
low transmission	7	9	9	NA
Solar reflectance, % min	70	67	73	67
Fire resistance of				
roof coverings (class)	A	A	A	

Perhaps the most convincing test performed to substantiate the outstanding fire-resistive behavior of architectural fabrics is the ASTM–E–84 Tunnel Test. In such a test, an abestos–cement board receives a flame-spread rating of zero and red oak flooring is rated at 100. Materials rated below 25 are given Class A certification. The Teflon–fiberglass composites used in permanent structures all are rated Class A in this demanding test.

Fabrasorb Accoustical Fabric, manufactured by Chemfab, represents a fabric with high noise reduction capability over a broad frequency range. Fabrasorb is, like Sheerfill, a composite of Teflon and fiberglass. It, therefore, shares many of its outstanding properties: it is strong, resistant to moisture and mildew, and highly resistant to fire. However, it has a somewhat porous construction, which facilitates the attenuation of sound within the fabric. Thus, it has been found to offer highly significant advantages as a liner material for fabric structures, particularly where it may also serve as a plenum to channel warm air for snow melting along the inner surface of the outer fabric.

As a result of its more open construction, made possible largely by the reduced mechanical loading of the liner, the Fabrasorb liner has a relatively high solar transmission. Thus, in addition to its mechanical and acoustical functions, it is able along with the primary Sheerfill architectural fabric to provide an essentially double-glazed fabric roof with significant energy-conservant benefits to ordinary double-glazed windows.

Let us examine the solar–optical properties of architectural fabric in a general sense. The degree to which light may be transmitted through such fabrics is governed largely by the degree of openness in the woven fabric. A 400,000 square foot stadium roof has on the order of 20 billion point sources of light, each approximately 10–25 mils on edge. It is not difficult to understand why the transmitted light is of such a pleasing and diffuse quality.

The absolute level of solar transmission is on the order of 7– 16%, with the upper limit dictated by minimal tensile strength requirements and the lower limit dictated by minimal tear strength and coating adhesion requirements. Since the solar spectrum encompasses wavelengths beyond the visible range, the actual transmission of visible light is somewhat less than the solar transmission.

Energy savings may be realized with the use of a doubly glazed configuration by the reduced need for artificial lighting, which can account for up to 50% of the total energy demand in a department store setting, and a reduced refrigeration requirement that results from very low shading coefficients.

One of the most outstanding characteristics of these fabrics is their ability to reflect upwards of 70% of the incident solar energy. Such a superwhite external reflector in combination with a liner of Fabrasorb is capable of providing a doubly glazed roofing system with good light transmission (on the order if 4–8%) while providing summertime shading coefficients down to .08.

The calculated heat gains for architectural fabric glazings at comparable solar transmissions are substantially lower than those of relective glass glazings and suggest a real benefit to be derived from reduced refrigeration investment and reduced operating costs during the cooling season on a life-cycle cost basis. Such performance could make a fabric structure more attractive than a conventional structure with substantially lower initial costs when sited in an appropriate climate.

83
Gummed Tape

Milton C. Schmit

Plymouth Printing Company, Inc.
Cranford, New Jersey

1.0 INTRODUCTION

Paper tape with a coating of an adhesive that may easily activated by the application of water, and used to seal corrugated cartons, historically has been produced with a coating of animal glue. For the past 20–30 years, adhesive formulas based on thin boiling waxy maize starch have almost completely replaced animal glue in this product in the United States. In the European market, animal glue has been replaced by modified potato starch based formulas.

Over the years, in the United States, the demand for plain paper tapes for carton sealing has been decreasing and these tapes have been replaced by reinforced double–ply tapes. In other areas of the world, paper tapes command a larger share of the carton sealing market. A product line that is often included with reinforced carton sealing tape is manufacture's joint tape, a product used by manufacturers of corrugated cartons to form the tube of the carton.

2.0 PRODUCTS

Paper sealing tapes are described by the basis weight (24 in. × 36 in. minus 500) of the paper being coated. They are identified as light duty, medium duty, and heavy duty. The common base weight is 35 lb per ream for light duty, 60 lb for medium duty, and 90 lb for heavy duty tapes.

In the reinforced tape market there is little agreement among manufacturers as to what constitutes a difference in grades. The Gummed Industries Association (GIA)*, a trade association of manufacturers of gummed tape, has developed voluntary grading standards, but not all manufacturers apply them. The standards contain a formula method of grading glass reinforced tapes and define the grades as light duty, medium duty, and heavy duty.

---

*The Gummed Industries Association, Inc., P. O. Box 92, Greenlawn, NY 11740; phone (516) 261–0114.

Reinforced paper tapes, after a history of using randomly scattered sisal fibers, cotton yarn in a sine wave pattern, and rayon in various patterns as the reinforcement, now are nearly always made using glass fiber yarn. The most popular glass patterns have yarn in the machine direction along with a diamond pattern in the cross–machine direction. There still is some tape made with a scrim-type pattern of machine-and cross-directional glass fiber at right angles to each other.

Laminating adhesives are either hot melt or water borne adhesives. Hot melts include the traditional laminating asphalt, which has lost popularity over the years, and amorphous polypropylene. Any reinforced tape not using asphalt or a black laminating material is termed "nonasphaltic" by the industry. Water borne adhesives may based on polyvinyl acetate, polyvinyl alcohol, or other paper laminating polymers, often clay loaded for better surfacing and economy. The use of a water borne adhesives for the laminate will generally require a manufacturer to postgum the web after laminating, whereas a hot melt lamination allows a pregumming of the adhesive carrying ply. Postgumming is often slower and presents a more irregular surface to be coated with adhesive. The nonasphaltic tapes are also termed nonstaining, as opposed to the asphaltic tapes.

The paper stock used for reinforced tapes ranges from 23 to 40 lb per ream. The paper surface that is to be adhesive coated is preferable a machine glazed (MG) as opposed to a machine finished (MF) surface.

3.0 MANUFACTURING

Clean and dry warehousing is needed for the storage of base paper, as well as any bagged adhesive ingredients. Bulk dry powdered adhesive ingredients may be stored in silos, with unloading and in-plant transportation by pneumatic piping systems. The hot melt laminating materials may be received in block form and remelted, but most larger operations receive the material in hot tank trunks or in rail tankers, which can be reheated for unloading. Insulated bulk storage incorporating heat transfer oil heating is then used along with hot oil traced pumps and supply lines for in-plant transportation.

Adhesive is prepared most often in batch kettles, but continuous preparation is an alternative. Animal glue and starch formulations both need heat in their preparation and application. Kettles may be double walled for steam, hot water, or hot oil heat, or just insulated. Surface heating is too slow a method for raising the batch temperature, so steam injection is used. Agitation is important for the slurrying of the ingredients and for maintaining them in suspension during cooking. Regulating equipment is needed to give reproducible cooking cycles. Many formulas are sensitive to the rate of heating and especially to the length of exposure at high temperature. Adhesive may be fed directly to the coating equipment or sent to a tank farm for distribution to the coaters. Again, these holding units must be heated or insulated so that the adhesive will not gel or thicken too much.

In the United States most tape is coated by direct two-or three-roll coaters, while some of the European gummers are using slot die coaters, which can handle the higher viscosities of the potato starch formulas. Adhesive temperatures at the coating head may be anywhere between 150 and 190° F. In some cases the coating rolls themselves are internally heated so that adhesive stringiness may be overcome. In both two-and three-roll coaters, the adhesive is metered between two steel rolls, with the gap being controlled by a screw-driver wedge, or directly by a screw or cam. A three–roll coater has the third roll rubber covered to nip the paper against the metered adhesive on the upper steel roll. A two-roll coater uses a high tension device to hold the web in contact with the adhesive film on the upper steel roll. Most modern coaters are equipped with coating width sensing equipment that contains a closed loop arrangement for control of adhesive coating weight. Immediately after the coating head, the coated side of the web passes over small-diameter, reverse turning leveling rolls to remove any film splitting pattern from the adhesive surface. The web then passes through a conventional drying oven before being rewound for further processing. Several machines in the industry have the gumming head in line with the reinforcing and laminating equipment.

Laminating and reinforcement are carried out in one operation. The glass fiber yarn is formed into a pattern and fed between the two paper plies as they are nipped together. Most tape manufacturers have adopted the Scrim-masters*as the machine to form the cross-web glass fiber diamond pattern because of its ability to operate at a much higher speed than any other equipment. Older style weaving equipment had either shuttles that had to cycle over the full width of the web, or a wheel that carried the spindles of yarn from one edge to the other. Both these methods were slow operating at less than 200 linear feet per minute.

The Scrim-master uses a rocking motion shuttle or wand to weave the yarn between pins mounted on two chains. Metal tubes mounted in the wand pass around the chain pins. The chains travel in channels 6 in. apart during this weaving operation. As the chains leave the weaving wand section, the chain channels angle away from each other, forming a "Y" configuration that moves the chains out to just beyond the width of the web.While the chain is traveling in the expanding legs of the "Y", the glass fiber is running over the chain pins to form an expanded diamond matrix. At this point the channels again return to travel parallel with the web, carrying the expanded glass fiber matrix into a nip, where it is sandwiched between the paper piles.

The machine direction glass yarn is fed into this nip at the same time. the strands are held in place by a comb designed to give the desired spacing. The comb oscillates from 0.25 to 1 in. to ensure proper winding of the finished product—the yarn causes a protrusion in the lamination and will not allow the straight winding of a roll if the yarn is not oscillated.

Either of the two paper plies or both may be coated with the laminating adhesive before being brought together at the nip with the reinforcing fibers. Hot roll coaters or slot die coaters are used. The first nipping of the laminate sandwich is done in a soft or low pressure nip, followed by a heavily loaded nip. The paper side of the tape is to the rubber roll side of the nip the adhesive surface to the steel roll, thereby lessening the effect of any yarn protrusion on the gummed surface.

The adhesive coating shrinks considerably when it is dried, causing heavy side curl of the coated web. The glue surface is cracked to relieve some of the curl producing stress in the web. This is done by drawing the tensioned web over small radius steel bars. This is often done in line on the slitting equipment as the web is cut to width and length. Good breaking gives about 100 crack lines per inch and is not easily seen by the naked eye. Before slitting, a secondary machine may be used to pass the tensioned web over bars set at an angle to the web (30–45° is common), and then over a 90° bar. This gives a three-directional breaking pattern. To control the breaking and to prevent a pigtail curl from developing, the tension over each bar may need to be independently controlled and the radius of the bars may need to be different.

Plain paper tapes, reinforced tapes, and manufacturer's joint tapes are usually supplied in a nominal 3 in. width, with some grades being readily available also in 2.5 in. width. Roll length is highly variable, ranging from 375 to 800 linear feet for products used in the com-

---

* Nashua Industrial Machine, Pine Street Extension, Nashua, NH 03060.

mon bench type of tape wetting devices, while rolls for automatic taping equipment are run to lengths to give diameters of 20–24 in.

Most slitting of tape is done on single-shaft rewinders and with score cut knives. There is some shear cut slitting done as well. The smaller diameter rolls may be on cores or coreless, with inside diameters of 1.25–1.75 in. Larger diameter rolls are generally on cores with a wall thickness of at least 0.125 in and an inside diameter the same as the small rolls or a 3 in. inside diameter core. A bowed roller mounted between the slitter knives and the windup will cause the individual webs to separate slightly and help to separate the log of rolls when the shaft is removed.

Roll separation, inspection, and packing are done manually with no automation, except for carton sealing in some plants. The cartons are identified by a lot number so that any complaints can be traced through the manufacturing cycle. Finished product is palletized and stretch wrapped, or shrink wrapped for shipping or warehousing.

84

Transdermal Drug Delivery Systems

Gary W. Cleary

Cygnus Research Corporation
Redwood City, California

1.0 INTRODUCTION

A "transdermal approach or technology" can be defined as any method that allows a drug or biological marker to transit skin in either direction. A transdermal drug delivery system allows the drug or molecule to transit from the outside of skin, through its various layers, and finally into the circulatory system to exert a pharmacological action. There may be some exceptions to this definition—for example, a system may concentrate the drug in the skin layers or near the surface of the skin to exert a localized effect, such as wound dressings for antisepsis or improved healing processes.

The transdermal drug delivery approach offers new opportunities for old drugs and new avenues to medical therapy. Developing these new products is a complex task. Primarily, today's transdermal designs include not only drugs, pharmaceutical vehicles, and other excipients but also polymeric films, specialty coatings, pressure-sensitive adhesives, and release substrates. They involve many different scientific and manufacturing disciplines that are unfamiliar to both the pharmaceutical industry and the pressure-sensitive adhesive industry. New technology that can be brought to bear will expand the future opportunities.

In the 1870s, physicians and their associates were not only exploring new materials and adhesives for binding surgical wounds but including medicaments in these adhesive tapes to treat conditions that respond to drugs in the systemic circulation. Medicated plasters have been used to treat back pain, and iodine-impregnated in gauze pads were quite popular at one time, but these devices gained disfavor because of such problems as irritation, side effects, and changes in the drug regulatory environment.

Although, physicians have long been prescribing topical products for treatment of localized skin diseases, it was not until the 1950s in the United States that a drug was made commercially available for systemic circulation by means of a topical application. Topical delivery of nitroglycerin was achieved by means of an ointment that was rubbed into the skin and overwrapped with Saran film, which was secured to the skin by means of surgical tape. This was the forerunner of the current transdermal device that has pressure-sensitive adhesives as part of the product. It was not until 30 years later, the early 1980s, that a more sophisticated transdermal product appeared on the market.

Several transdermal delivery systems had reached the U.S. market as of 1988. They ranged in design from the amorphous ointments to solid state laminates. A review of the patent literature indicates a flurry of activity in developing many different designs. Scopolamine (1980), nitroglycerin (1981), clonidine (1985), and estradiol (1986) are drugs that have reached the U.S. transdermal market. Drugs that are sold outside the United States include isosorbide dinitrate and progesterone. The market to the pharmaceutical companies reached close to $500 million in 1987. The products have been accepted by physicians and by patients as a viable dosage form.

2.0 ATTRIBUTES OF A TRANSDERMAL SYSTEM

To begin a program of developing a transdermal, one should consider what it is that the product is to achieve; that is, one should develop a "product profile" that describes the essential attributes of the desired transdermal. Once the criteria have been determined, a list of attributes of the delivery system can be made to develop a product profile to provide direction to the formulator or "system designer". These attributes will describe the following:

Physical characteristics (system size and shape, thickness, construction, amount of drug, color, flexibility, etc.).
Functionality (necessary rate of release through skin, rate of release from the system, degree of adhesion to skin, length of time to adhere on skin site, method of applying system to skin, etc.).
Patient demography.
Degree of irritation tolerance by patient.
Cost to patient and to third-party payers.
Medical rationale and intended indications.
Required margins by pharmaceutical company.
Availability of raw materials.
Patentability.
Effect of regulatory environment.

For example, Key Pharmaceutical Company's first transdermal nitroglycerin product delivered 10 mg/day over a 20 cm² area (equal to 0.5 mg/cm² day). This transdermal which has been accepted by patients, was actually 90 cm² system because of the peripheral adhesive (see below, Type III design). Cosmetic appearance and size of the system will affect patient acceptance. Depending on the size that patients deem to be acceptable, the maximum total amount of drug that can be delivered may be as high as 50 mg/day. The size of the system will then depend on how much the skin will allow through, what blood level is needed to be achieved to elicit the desired pharmacological effect, and perhaps the patient's well-being. Skin irritation and sensitization are issues that must be addressed in the early stages of development. A skin reaction may be caused by the drug, or by any vehicles, enhancers, or polymer that may be present. The cost/benefit ratio of the transdermal compared to other dosage forms must be weighed, as well.

There are various ways to view the design of these transdermals. One way is to start by considering the design of the transdermal without regard to the mechanism of how the drug is released or controlled. Figure 1, a schematic side view of the laminate construction that is

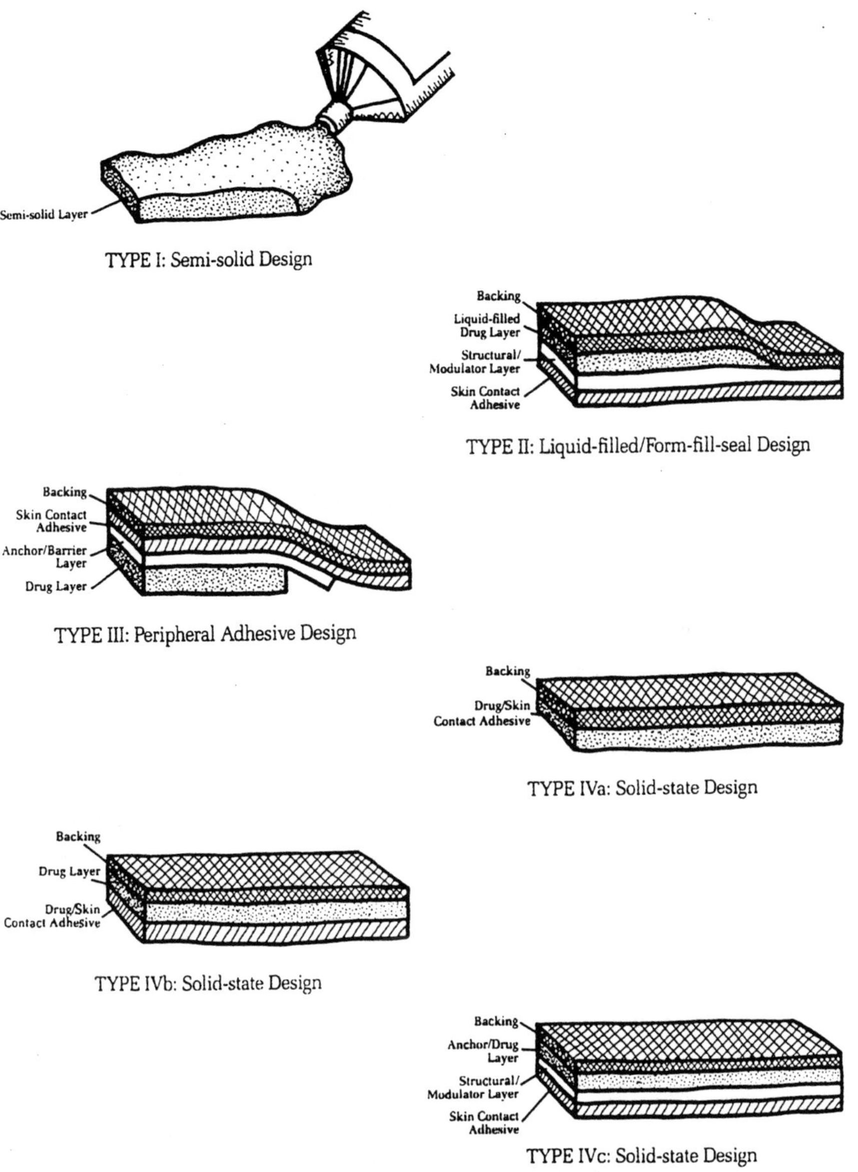

Figure 1 Schematic diagrams of four types of transdermal drug delivery system designs.

found in transdermal products, facilitates a visual conception of the actual product, its components, and how it might be formulated and fabricated. There are four types of designs on the market:

Type I: amorphous ointment, cream, lotion, or viscous dispersion that is applied directly to the skin—Nitrobid (Kremers-Urban).

Type II: amorphous liquid dispersion that is sealed between two laminate structures on all sides—Transderm Nitro and Estraderm (Ciba).

Type III: solid state polymer matrix attached to a peripheral adhesive tape—Nitro-Dur I (Key); Nitrodisc(Searle); Nitro TDS (Bolar).

Type IV: solid state polymer film structures laminated together —Transderm Scop V (Ciba); Transderm-clonidine(Boehringer-Ingelheim); Nitro-Dur II (Key); Deponit (Wyeth).

These four types are not all-inclusive, since new approaches and combinations of each are always possible. All these designs have the potential to deliver a drug to the skin so that the drug can migrate through the skin. Types II, III, and IV have the adhesive and film laminate structure built into the final product. By understanding these basic designs, their advantages and disadvantages, one can incorporate the most suitable diffusional mechanism, using the appropriate plasticizers or vehicles, polymers, films or membranes, and adhesives, and matching the diffusivity of the drug through skin to achieve the desired delivery rate and plasma profile of the drug. The types of material that have been used in transdermal products that have reached the marketplace include:

Pressure-sensitive adhesives: acrylates, silicone, and rubber-based adhesives.
Release liners: silicone and fluorocarbon coats on paper, polyester, or polycarbonate films.
Backings and membranes: ethylene–vinyl acetate, polypropylene, polyester, polyethylene, polyvinyl chloride, and aluminum films.
Specialty films: foams, nonwovens, microporous films, vapor-deposited aluminum films.

The selection of materials and the system design will affect many of the attributes listed in the product profile.

2.1 Developing a Transdermal

Transdermal technology is relatively new, and to develop such dosage forms is a complex procedure from concept to commercial manufacture. The development of a transdermal system calls for a background in several disciplines or fields of study. These disciplines are interrelated and need to be considered simultaneously in the design of a transdermal delivery system. The basic foundation includes an understanding in:

Skin properties.
Pharmaceutical formulation (pharmaceutics).
Polymer and material science.
Mass transport.
Adhesion theory and formulation.
Pharmacology (pharmacodynamics and pharmacokinetics).
Process engineering.
Fabrication, packaging, and manufacturing.

2.2 Selecting Drug Candidate

What criteria are used to select a drug as a good potential candidate for delivery through the skin? The selection criteria include several factors that must be weighed at the same time to proceed in the development of a transdermal delivery system:

Pharmacokinetic (what the body does to the drug) and pharmacodynamic (what the drug does to the body) information about the drug.
Ability of the drug molecule to permeate skin.
A targeted effective blood level.
Necessity for steady state delivery, or shape of blood level profile.
First-pass effect of the drug (high initial metabolism in the liver).
Amount of drug needed.
Estimated size of system.
Need for skin penetrant enhancers to increase the amount of drug throughout.
Skin toxicity (irritation, sensitization).
Side effects of drug.
Diffusional and solubility properties of drug.

3.0 FUTURE APPROACHES TO TRANSDERMAL DELIVERY SYSTEMS

Transdermal delivery systems have reached the marketplace and have achieved widespread acceptability in therapeutics today. However, to achieve success as a dosage form, the selection and development of a drug candidate for transdermal delivery must be justifiable economically for both the patient and the manufacturer and must have a sound therapeutic rationale. There may be opportunity to expand a particular drug's current market if that market is quite small. This was seen with nitroglycerin, which after a century of use had a market of less than $30 million/year. Five years after the first modern transdermal was introduced, however, the nitroglycerin market had exceeded $300 million/year.

Not all drugs penetrate the skin easily at suitable therapeutic rates. If ways were found to put them through skin at higher rates of diffusion, a new transdermal market could be made. Some ways to attain suitable permeation rates could be through optimizing the formulation, changing the lipid solubility or physical properties through synthesis of new derivatives (homologues, analogues, or prodrugs) to effect an improvement over the parent, or changing the skin permeability by means of chemical or electronic enhancers. Today, optimizing the formulation of a drug in a transdermal for passive diffusion is the predominant developmental approach. However, future approaches may include molecular modification of the drug, or using chemical or electronic means to increase skin permeation.

With the advent of post-World War II technology of electronics, medicine, and chemistry to produce a diversity of analytical instrumentation and sensitive drug assay methodology, combined with a better understanding of how the body affects drug metabolism, there have been advances in equipment and synthetic chemicals too produce new polymers, new film coatings, and new fabricating techniques. These innovations, as well as recent changes in the regulatory climate, and pressures to produce off-patent drug entities, are allowing the more venturesome to explore new ways to deliver drugs to the body. In this developmental climate comes a revitalization the technique of using the skin as the portal of entry for drugs by means of transdermal delivery systems.

85
Optical Fiber Coatings

Kenneth Lawson

DeSoto, Inc.
Des Plaines, Illinois

1.0 INTRODUCTION AND BACKGROUND

The concept of high speed data transmission on a beam of light was demonstrated by Alexander Graham Bell in 1880. His "photophone" used mirrors and sunlight to transmit low quality voice messages. Today's technology has developed to a point of passing extremely high data rates (> 2 Gbit/s) unrepeated through a high purity quartz fiber for distances exceeding 100 km.

Optical fibers offer many advantages over copper conductors and are gaining acceptance for the following reasons:

1. They carry more signals for greater distances without repeaters. A fiber with a 1 GHz bandwidth can transmit several thousand simultaneous telephone calls through use of state-of-the-art multiplexing techniques. Undersea cables have repeaters spaced as widely as 30–55 km, whereas copper cables require signal regeneration ever 2–5 km.
2. Electromagnetic interference does not occur.
3. Cross-talk is eliminated because the optical signal is maintained in fiber and "short-circuits" do not occur.
4. Weight and volume of fiber cables are greatly reduced (for the same signal carrying capacity).

These advantages have led to the development of numerous optical fiber telecommunication networks, with an increasing awareness that the initial growth of this technology for long-distance trunk lines will continue to accelerate as fiber replaces copper wire in the subscriber loop (residential and business installations).

1.1 Optical Waveguide Principles

In any transparent material, light travels more slowly than in a vacuum. The ratio of these speeds is the index of refraction (n). When light, traveling in a material of one refractive index, strikes a material having a lower refractive index, the light is bent back toward the

731

higher refractive index material. This phenomenon can be use to guide a light down a high purity glass rod, if the rod is clad with a material having a lower refractive index. Current telecommunication fibers employ a doped (germanium, phosphorus, etc.) silica core in combination with a lower refractive index ($n = 1.46$) silica cladding. The core will generally vary in thickness from 8 to 100 μm, whereas the outside diameter of the cladding is usually 125 μm. When made from pure silica, the resultant fibers have excellent strength (to 14,000 N/mm^2)[1], but they rapidly degrade as a result of the development of microscopic scratches, which are subject to growth, resulting in breakage when stressed (stress corrosion).

2.0 COATING

The application of a protective coating is required to preserve the optical fiber's strength and to protect it from lateral deformation, which can result in a reduced light signal (attenuation).

Attenuation is one of the major concerns of a fiber user. This is a measure of the light lost in passing through the fiber and is described in electronic terms as noise (i.e., decibels per kilometer: dB/km. This loss results from light scattering due to fluctuations of glass density, absorption due to impurities, and radiant signal loss due to microbending, the result of unequal forces or distortions of the glass fiber core.

Long-term durability of the fiber and its ability to transmit a signal of high fidelity is of paramount importance. Lifetimes of 25–40 years are generally expected, even though the anticipated environments vary from tropical climates, to arctic winters, to undersea water pressures. In addition, the fiber may be exposed to groundwater (having a pH of 2–12), solvents, steam, stress, and rapidly fluctuating temperatures. Under conditions of high stress and humidity, silica glass is known to undergo rapid stress corrosion through a process of hydrolysis, resulting in premature breaks of the communication link. To avoid this difficulty, manufacturers minimize flaws and conditions during fiber manufacture. To protect the glass, a protective buffer coating is applied within one or two seconds of drawing, or as soon as the glass has cooled to below 100°C. The buffer coating is designed to protect the glass from being scratched. However, it is mandatory that the coating be free of any particulate matter, which may cause microscratches on the glass surface, forming a site for crack propagation.

2.1 Handleability

To keep up with increasing demand, the coating or coating system must lend itself to high production speeds. During the commercial development period of the late 1970s, typical production speeds varied from 0.5 to about 1 m/sec. At these slow speeds several coatings were found to be acceptable. Two-component polysiloxanes were found to offer the best overall properties, namely good strength, and minimal temperature dependence (their flexibility and modulus remain largely unchanged over a temperature range of −55 to +85°C). Their durability is good, and their low modulus properties provide a soft coating, which protects the glass core from physical distortions and microbending. Major deficiencies of the silicones include relatively limited application and cure rates, high hydrogen generation, limited pot life (when catalyzed), and a high coefficient of surface friction, which requires overcoating. Other early coating systems included cellulose-based lacquers, Teflon extrusions, and polybutadiene rubbers.

2.2 Coating Requirements

In 1977 the demand for increase volume and productivity resulted in the development of ultraviolet light curable acrylate coatings, which met the needs for strength, longevity, and performance while also providing production speeds of 5 m/sec and faster.[2–4] By using combinations of acrylate-functional oligomers and monomers, it is possible to formulate coatings having a wide range of useful properties, including:

Modulus, 1–1000 MPa (125–150,000 psi)
Elongation, 10–250%
Hardness, 35 Shore A to 70 Shore D
Tensile strength, 0.5–40 MPa (70–6000 psi)
Low water absorption (< 3%)

It has long been theorized[5] that an optimal fiber coating system should consist of a very soft, low modulus, particulate-free coating against the glass, to provide for surface protection and microbending resistance. Microbending is the attenuation of the light signal caused by small bends of the core, resulting from nonuniform stresses on the fiber. The primary coating is followed by a harder, tough, secondary buffer to provide mechanical protection and water resistance, while offering a hard, slick surface that is compatible with subsequent cabling operations.

It is desirable that these coatings be applicable in-line, at maximum possible draw rates, and at several different outside diameters. Coating thicknesses may be from 3 µm to nearly 400 µm.

The properties usually desired and generally provided by multifunctional acrylate coatings include the following.[6]

1. Primary buffer coatings
    Good adhesion to the glass. Buffer removal or stripping can be accomplished with the use of solvents such as methylene chloride, or short lengths of coating can be removed mechanically. Another construction might have minimal bonding to the glass, to facilitate rapid, safe mechanical stripping of the buffer during installation.
    Low tensile modulus over a range of operating temperatures (– 55°C to +85°C).
    Long-stability, with excellent resistance to oxidation, hydrolysis, saponification, and ultraviolet light.
    A satisfactorily high index of refraction, so that the coating can be applied using laser forward-scattering for concentricity control, and an index of refraction sufficiently higher than the cladding that light launched into the glass cladding will be stripped from the fiber[7] (mode stripping) (n > 1.48).
    Low glass transition temperature (T_g) to allow the primary coating to function without going through a glass transition phase that will cause differential expansion and contraction, resulting in microbending.
    Low generation of hydrogen during the fiber's installed lifetime.
2. Secondary buffer coatings
    Low surface free energy (i.e., low coefficient of friction and tack) as to allow for easy handling in the cable plant.
    Low surface tack, to prevent binding or coupling of the coated fiber to the walls of a "loose tube" cable.

Good intercoat adhesion characteristics when overcoated with other components of a cable system (i.e., tape or extrusion coatings such as nylon, Hytrel Tefzel, etc.).

Excellent resistance to environmental factors such as water, acids, bases, solvents, and fungus growth.

High modulus, to resist external lateral forces that will deform the light path.

High-T_g, to avoid differential expansion and contraction during thermal cycling.

Low volatility, to resist outgassing during hot extrusion applications to tightly buffered cable processes and long-term changes in physical properties.

The ability to be colored or to accept coloring, which permits individual fiber identification.

Resistance to the chemical gels used to fill "loose tube" cables.

2.3 Composition of Fiber Coatings

More than 75% of the world's quartz optical fiber is coated with acrylate-based materials and cured with UV light. The classical reaction chemistry uses photoinitiators to absorb photons generated by medium pressure mercury, xenon, or doped mercury lamps. $1/cm^2$ of a wavelength of 300–390 n is generally sufficient for full cure in less that 1 second). Free radicals, which are produced by this reaction, cause rapid addition polymerization of the unsaturated acrylate groups. The proprietary chemistry is in the design and choice of the acrylate oligomers and reactive monomers. The oligomers are generally based on a range of polyesters, polyethers, and epoxy polymers, which have been esterified with acrylic acid (coreacted with a diisocyanate and a hydroxyfunctional acrylate). The reactive diluents are both mono-and multifunctional. Solvents are used only rarely for viscosity control. Stabilizers, inhibitors, flow aids, and a wide assortment of additives are used to gain required properties. Secondary buffers may include pigments or soluble dyes for color identification.

Two-component polymethylsiloxane coatings are used in decreasing quantities for quartz telecommunication fiber. A phenyl variation provides the high refractive index required for a primary buffer. but at a significant cost disadvantage. The usual low refractive index makes silicones the coatings of choice for most all-plastic fiber and high loss, "plastic clad silica" (PCS) fiber. Physical properties of the silicones are excellent, but line speeds are generally limited to speeds well below 2 m/sec. Silicones are also difficult to mechanically remove from glass fiber for splicing, because they leave a residue.

Fluorine-and silicone-modified acrylates are used for specialty applications, as are polyimides for high temperature resistance. A new class or inorganic, hermetic coatings is being commercialized to provide stronger, more durable fiber. These include various vapor-deposited coatings such as reactive silica and titanium nitrides and carbides, as well as molten aluminum. Gold and indium have also been used.

REFERENCES

1. C. L. Schlef, et al. *Radia. Curing*, p. 11, April 1982.
2. U. C. Paek, and C. M. Schroeder, *Appl. Opt.* V 1 20, (23/1), Dec. 1981 4028–4034 (December 1981).
3. K. R. Lawson, and O. R. Cutler, *J. Radiat. Curing*, pp. 4–10, April 1982.
4. *G. A. Perry*, "Performance of acrylate-coated single-mode fibers," – Proceedings of Interwire '87, *Atlanta, Oct. 27, 1987.*
5. J. A. Jefferies, and R. J. Klaiber, *Western Electric Eng.*, pp. 13–23, October 1980.

6. K. R. Lawson, *"Advances in UV cured coatings for optical fibers,"* in *Proceedings of Rad Cure Europe '83*, FC83–264, Society of Manufacturing Engineers.

7. K. R. Lawson, R. E. Ansel, and J. J. Stanton, *"UV cured coatings for optical fibers,"* in *Proceedings of Fifth Radiation Curing Conference*, Association of Finishing Processes – Society of Manufacturing Engineers.

86
Exterior Wood Finishes

William C. Feist

U.S. Department of Agriculture Forest Service
Madison, Wisconsin

1.0 INTRODUCTION

The primary functions of any wood finish (e.g., paint, varnish, wax, stain, oil) are to protect the wood surface, help maintain appearance, and provide cleanability. Unfinished wood can be use outdoors without protection. However, wood surfaces exposed to the weather without any finish are roughened by photodegradation and surface checking; in addition, they change color and slowly erode.

Wood and wood-based products in a variety of species, grain patterns, textures, and colors can be effectively finished by many different methods. Selection of the finish will depend on the appearance and degree of protection desired and on the substrates used. Also, different finishes give varying degrees of protection; therefore, the type, the quality, the quantity, and the application method of the finish must be considered when selecting and planning the finishing or refinishing of wood and wood products.

Satisfactory performance of wood finishes is achieved when thorough consideration is given to the many factors that affect these finishes. These factors include the properties of the wood substrate, characteristics of the finishing material, details of application, and severity of exposure. Some of these important considerations are reviewed in this chapter. Additional sources of detailed information are listed in References 1–7.

2.0 EXTERIOR SUBSTRATES

2.1 Wood Properties

Wood surfaces that shrink and swell the least are best for painting. For this reason, vertical- or edge-grained surfaces (Fig. 1) are far better than flat-grained surfaces of any species, especially for exterior, use where wide ranges of relative humidity and periodic wetting can

---

737

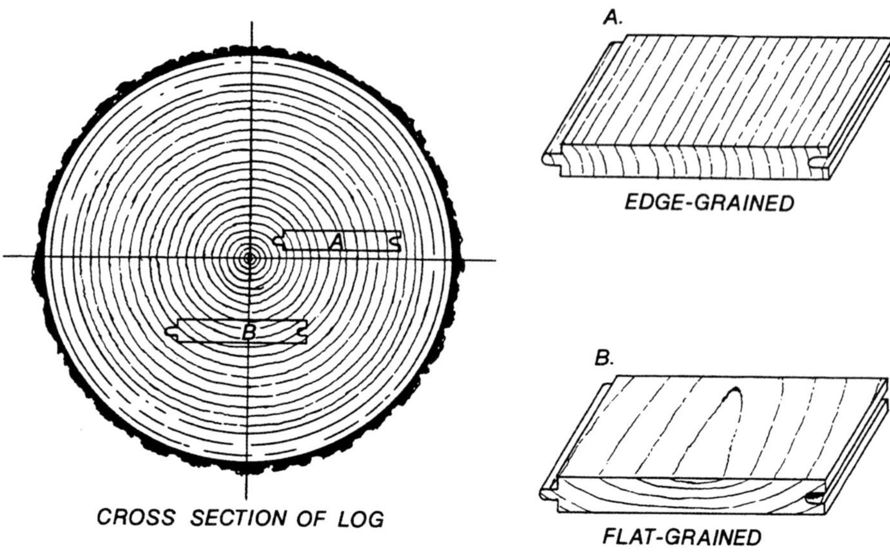

Figure 1 Edge-grained (or vertical-grained or quartersawed) board A, and flat-grained (or slash-grained or plainsawed) board B, cut from a lot.

Figure 2 Paint applied over edge-grained boards (top and bottom) performs better than that applied to flat-boards (middle).

produce equally wide ranges of swelling and shrinking. Table 1 lists the painting and weathering characteristics of softwoods and hardwoods.

Also, because the swelling of wood is directly proportional to density, low density species are preferred over high density species. However, even high swelling and dense wood surfaces with a flat grain can be stabilized with a resin-treated paper overlay (overlaid exterior plywood and lumber) to provide excellent surfaces for painting. Medium density, stabilized fiberboard products with a uniform, low density surface or paper overlay are also good substrates for exterior use. However, edge-grained heartwood of western red cedar and redwood are the species most widely used for exterior siding and trim when painting is desired. These species are classified in group I, the woods easiest to keep painted (Table 1). Edge-grained surfaces of all species are considered to be excellent for painting, but most species are available only as flat-grained lumber.

Species classified in groups II through V (Table 1) are normally cut as flat-grained lumber, are high in density and swelling, or have defects such as knots or pitch. The specific classification depends on the paint holding characteristics of each species. Many species in groups II through IV are commonly painted, particularly the pines, Douglas fir, and spruce, but these species usually require more care and attention than the edge-grained surfaces of group I. Exterior paint will be more durable on edge-grained boards than on flat-grained boards (Fig. 2) for any species with marked differences in density between earlywood and latewood, even if the species are rated in group I. Flat-grained boards that are to be painted should be installed in areas protected from rain and sun.

2.2   Wood and Wood-Based Products

Three general categories of wood products are commonly used in exterior construction: lumber, plywood, and reconstituted wood products, such as hardboard (a fiberboard) and particleboard). Each has characteristics that will affect the durability of any finish applied to it.

2.2.1   Lumber

Lumber is being used less and less as an exterior siding, but it was once the most common wood material used in construction. Many older homes have solid wood siding. The ability of lumber to absorb and retain a finish is affected by species, by ring direction with respect to the surface (or how the piece was sawn from the log), and by smoothness.

The weight of wood varies tremendously between species. Some common construction woods, such as southern pine, are dense and heavy compared with lighter weight woods, such as redwood and cedar. The weight of wood is important because heavy woods shrink and swell more than light ones. This dimensional change in lumber occurs as the wood gains or loses moisture. Excessive dimensional change in wood puts constant stress on the paint film and may result in early failure of the finish.

Ring direction also affects paint holding characteristics and is determined at the time lumber is cut from a log (Fig. 1). Most standard grades of lumber contain a high percentage of flat grain. Lumber used for board and batten siding, drop siding, or shiplap is frequently flat-grained. Bevel siding is commonly produced in several grades. In some cases, the highest grade of lumber is required to be edge-grained and all heartwood over most of the width for greater paint durability. Other grades may be flat-grained, edge-grained, or mixed-grain and without requirements as to heartwood.

Some species have wide bands of earlywood and latewood (Fig. 3). Wide, prominent bands of latewood are characteristic of southern pine and most Douglas fir, and paint will

Figure 1 Painting and Weathering Characteristics of Softwoods and Hardwoods[a]

	Ease of keeping well painted (I—easiest, V—most exacting)	Weathering		Appearance	
		Resistance to cupping (1—best 4–worst)	Conspicuousness of checking (1—least 2—most)	Color of heartwood (sapwood is always light)	Degree of figure on flat-grained surface
Softwoods					
Cedar					
Alaska	I	1	1	Yellow	Faint
California incense	I	—	—	Brown	Faint
Port-Orford	I	—	1	Cream	Faint
western red cedar	I	1	1	Brown	Distinct
white	I		—	Light brown	Distinct
Cypress	I	1	1	Light brown	Strong
Redwood	I	1	1	Dark brown	Distinct
Product overlaid with resin-treated paper[b]	I		1		
Pine					
Eastern white	II	2	2	Cream	Faint
Sugar	II	2	2	Cream	Faint
western white	II	2	2	Cream	Faint
Ponderosa	III	2	2	Cream	Distinct
Fir, commercial white	III	2	2	White	Faint
Hemlock	III	2	2	Pale brown	Faint
Spruce	III	2	2	White	Faint
Douglas fir (lumber and plywood)	IV	2	2	Pale red	Strong
Larch	IV	2	2	Brown	Strong

Pine					
Norway	IV	2	2	Light brown	Distinct
Southern					
(lumber and plywood)	IV	2	2	Light brown	Strong
Tamarack	IV	2	2	Brown	Strong
Hardwoods					
Alder	III			Pale brown	Faint
Aspen	III	2	1	Pale brown	Faint
Basswood	III	2	2	Cream	Faint
Cottonwood	III	4	2	White	Faint
Magnolia	III	2		Pale brown	Faint
Yellow-poplar	III	2	1	Pale brown	Faint
Beech	IV	4	2	Pale brown	Faint
Birch	IV	4	2	Light brown	Faint
Cherry	IV			Brown	Faint
Gum	IV	4	2	Brown	Faint
Lauan (plywood)	IV	2	2	Brown	Faint
Maple	IV	4	2	Light brown	Faint
Sycamore	IV			Pale brown	Faint
Ash	V or III	4	2	Light brown	Distinct
Butternut	V or III			Light brown	Faint
Chestnut	V or III	3	2	Light brown	Distinct
Walnut	V or III	3	2	Dark brown	Distinct
Elm	V or IV	4	2	Brown	Distinct
Hickory	V or IV	4	2	Light brown	Distinct
Oak, white	V or IV	4	2	Brown	Distinct
Oat, red	V or IV	4	2	Brown	Distinct

aWoods ranked in group V for ease of keeping well painted are hardwoods with large pores that need filling with wood filler for durable painting. When so filled before painting, the second classification recorded in the table applies.

bProducts are plywood lumber and fiberboard with overlay or low-density surface.

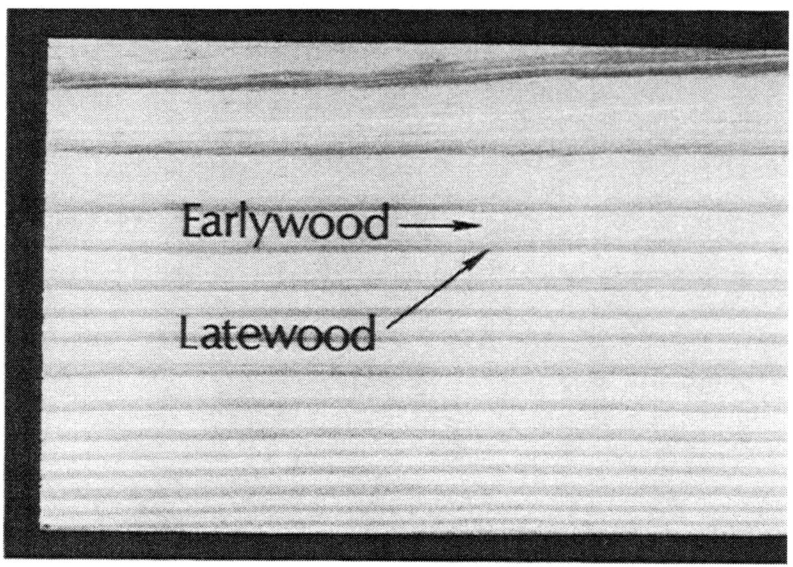

Figure 3 Earlywood and latewood bands in southern pine. These distinct bands often lead to early paint failure.

not hold well on these species (Table 1). In contrast, redwood and cedar do not have wide latewood bands, and these species are preferred when paint will be used.

2.2.2 Plywood

Exterior plywood with a rough-sawn surface is commonly used for siding. Smooth-sanded plywood is not recommended for siding, but it is often used in soffits. Both sanded and rough-sawn plywood will develop surface checks, especially when exposed to moisture and sunlight. These surface checks can lead to early paint failure when using oil or alkyd paints. Quality stain-blocking acrylic latex primers and top coat paints generally perform better. The flat-grained pattern present in nearly all plywood can also contribute to early paint failure. Therefore, if smooth or rough-sawn plywood is to be painted, special precautions should be exercised. Penetrating stains are often more appropriate for rough-sawn exterior plywood surfaces, but quality acrylic latex paints also perform very well.

2.2.3 Reconstituted Wood Products

Reconstituted wood products are made by forming small pieces of wood into large sheets, usually 4 x 8 feet or as required for a specialized use, such as beveled siding. These products may be classified as fiberboard or particleboard, depending on the nature of the basic wood component.

Fiberboards are produced from mechanical pulps. Hardboard is a relatively heavy type of fiberboard, and its tempered or treated form, designed for outdoor exposure, is used for exterior siding. It is often sold in 4 x 8 foot sheets as a substitute for solid wood beveled siding.

Particleboards are manufactured from whole wood in the form of splinters, chips, flakes, strands, or shavings. Waferboard, oriented strandboard, and flakeboard are three types of particleboard made from relatively large flakes or shavings.

Some fiberboards and particleboards are manufactured for exterior use. Film forming finishes, such as paints and solid color stains, will give the most protection to these reconstituted wood products. Some reconstituted wood products may be factory-primed with paint, and some may even have a factory-applied top coat. Also, some may be overlaid with a resin-treated cellulose fiber sheet to provide a superior surface for paint.

2.3 Water-Soluble Extractives

In some species of wood the heartwood contains water-soluble extractives, while sapwood does not. These extractives can occur in both hardwoods and softwoods. Western red cedar and redwood are two common softwood species used in construction that contain large quantities of extractives. The extractives give these species their attractive color, good stability, and natural decay resistance, but they can also discolor paint. Woods such as Douglas fir and southern yellow pine can also cause occasional extractive staining problems.

When extractives discolor paint, moisture is usually the cause. The extractives are dissolved and leached from the wood by water. The water then moves to the paint surface, evaporates, and leaves the extractives behind as a reddish-brown stain.

3.0 EXTERIOR FINISHES

Wood exposed outdoors undergoes a number of physical and chemical phenomena mostly caused by moisture influences, sunlight, and temperature. Being a product of nature, wood is also subject to biological attack by fungi and insects. Most of these stress factors interact and depend on numerous influences. Figure 4 illustrates stressing factors, influencing factors, and weathering effects on a finished wood surface.

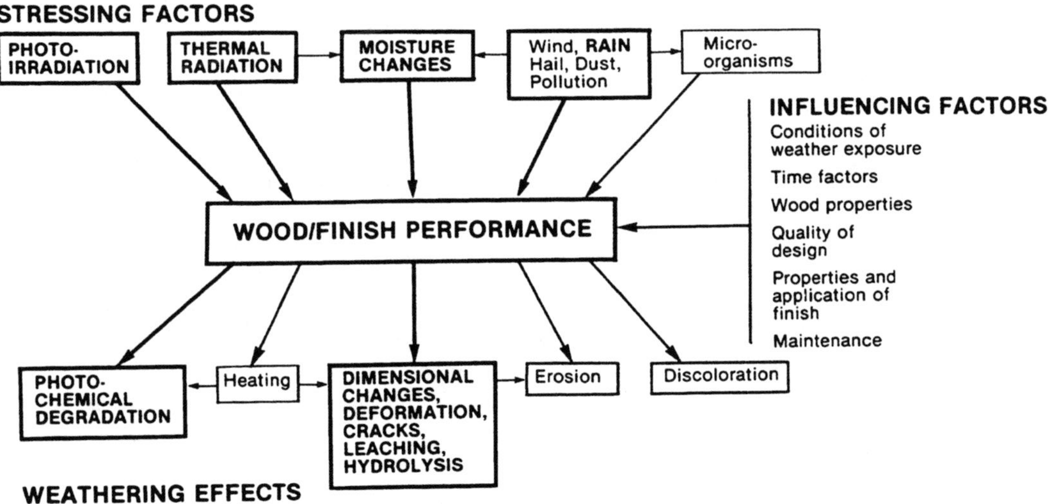

Figure 4 Schematic illustration of the weathering of finished wood.

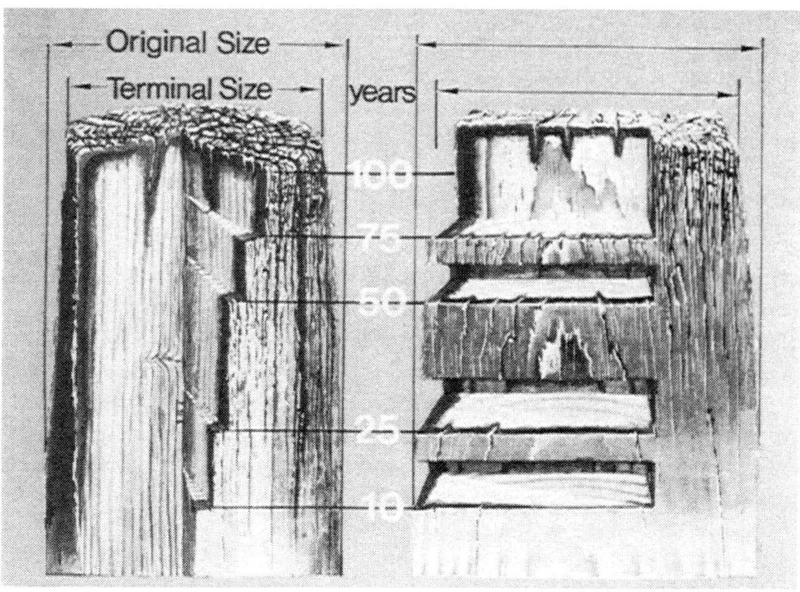

Figure 5 The weathering process of round and square timbers: cutaway shows that interior wood below the surface is relatively unchanged.

3.1 Natural Weathering

The simplest finish for wood is created by the natural weathering process. Without paint or treatment of any kind, wood surfaces gradually change in color and texture and then stay almost unaltered for a long time if the wood does not decay. Generally, the dark-colored woods become lighter and the light-colored woods become darker. As weathering continues, all woods become gray, accompanied by photodegradation and gradual loss of wood cells at the surface. As a result, exposed unfinished wood will slowly wear away in a process called erosion (Fig. 5).

The weathering process is a surface phenomenon and is so slow that most softwoods erode at an average rate of about 1/4 in. per century. Dense hardwoods erode at a rate of only 1/8 in. per century. Very low density softwoods, such as Western red cedar, may erode at a rate as high as 1/2 in. per century. In cold northern climates, erosion rates as low as 1/32 in. per century have been reported.

The physical loss of wood substance from the wood surface (erosion) during weathering depends not only on wood species and density, but also on growth rate, ring orientation amount or irradiation, rain action, wind, and degree of exposure. Erosion also occurs most rapidly in thin-walled fibers of earlywood in softwoods and, at a slower rate, in dense latewood. Accompanying this loss of wood substance are the swelling and shrinking stresses caused by fluctuations in moisture content. All this results in surface roughening, grain raising, differential swelling of earlywood and latewood bands, and the formation of many small parallel checks and cracks. Larger and deeper cracks may also develop, and warping frequently occurs (Fig. 5).

The weathering process is usually accompanied by the growth of dark-colored spores and mycelia of fungi or mildew on the surface; this gives the wood a dark gray, uneven, and unsightly appearance. In addition, highly colored wood extractives in such species as Western red cedar and redwood add to the variable color of weathered wood. The dark-brown color of extractives may persist for a long time in areas not exposed to the sun and where the extractives are not removed by the washing action of rain.

3.2 Applied Finishes

The outdoor finishes described in this section represent the range of finish types that are often used outdoors. Sources of information on the properties, treatment, and maintenance of exterior wood finishes are listed in the references, as is information on the suitability and expected life of the most commonly used finishes.

3.2.1 Paints

Paints are coatings commonly used on wood; they provide the most protection because they block the damaging ultraviolet light rays from the sun. They are available in a wide range of colors and may be either oil or latex based. Latex-based paints are waterborne, and oil or alkyd paints are borne by organic solvent. Paints are used for aesthetic purposes, to protect the wood surface from weathering, and to conceal certain defects.

Paints are applied to the wood surface and do not penetrate it deeply. The wood grain is completely obscured, and a surface film is formed. Paints perform best on smooth, edge-grained lumber of lightweight species. This surface film can blister or peel if the wood becomes wet or if inside water vapor moves through the house wall and into the wood siding because of the absence of a vapor retarding material. Latex paints are porous and, thus, will allow some moisture movement. Some oil-based paints are resistant to moisture movement, but they require organic solvents for cleanup.

Of all the finishes, paints provide the most protection for wood against surface erosion and offer the widest selection of colors. A nonporous paint film retards penetration of moisture and reduces the problem of discoloration by wood extractives, paint peeling, and checking and warping of the wood. It is important to note that paint is not a preservative. it will not prevent decay if conditions are favorable for fungal growth. Original and maintenance costs are often higher for a paint finish than for a water-repellent preservative or a penetrating stain finish.

3.2.2 Solid Color Stains

Solid color stains are opaque finishes (also called hiding or heavy-bodied). They are available in a wide range of colors and are made with a much higher concentration of pigment than the semitransparent penetrating stains. Solid color stains will totally obscure the natural color and grain of wood. Oil-based solid color stains form a film much like paint and as a result can peel loose from the substrate. Latex-based solid color stains are also available and form a film as do the oil-based solid color stains. Both these stains are similar to thinned paints and can usually be applied over old paint or stains if the old finish is securely bonded to the wood.

3.2.3 Semitransparent Penetrating Stains

Semitransparent penetrating stains are only moderately pigmented and, thus, do not totally hide the wood grain. These stains penetrate the wood surface, are porous to water vapor, and do not form a surface film like paints. As a result, they will not blister or peel even if moisture moves through the wood. Penetrating stains are alkyd or oil based, and some may contain a fungicide (preservative or mildewcide) as well as a water repellent. Moderately pigmented latex-based (waterborne) stains are also available, but they do not penetrate the wood surface as do the oil-based stains.

The semitransparent stains are most effective on rough lumber or rough-sawn plywood surfaces, but they also provide satisfactory performance on smooth surfaces. They are available in a variety of colors and are especially popular in the brown or red earth tones because they give a natural or rustic appearance. They are an excellent finish for weathered wood. Semitransparent penetrating, oil-based stains are not effective when applied over a solid color stain or over old paint. They are not recommended for use on hardboard.

An effective stain of the semitransparent penetrating type is the Forest Products Laboratory natural finish developed in the 1950s. This finish has a linseed oil vehicle, a fungicide to protect the oil and wood from mildew, and a water repellent that protects the wood from excessive penetration of water. It is available in durable red and brown iron oxide pigments that simulate the natural colors of redwood and cedar. A variety of other colors can also be achieved with this finish.

3.2.4 Water-Repellent Preservatives and Water Repellents

A water-repellent preservative may be used as a natural finish. It contains a fungicide or mildewcide (the preservative), a small amount of wax for water repellence, a resin or drying oil, and a solvent, such as turpentine or mineral spirits. Water-repellent preservatives do not contain coloring pigments. Therefore, the resulting finish will vary in color depending on the species of wood. The mildewcide also prevents wood from darkening (graying).

The initial application of the water-repellent preservative to smooth wood surfaces is usually short-lived. During the first few years, the finish may have to be applied every year. After the wood has gradually weathered to a uniform color, the treatments are more durable and the wood needs refinishing only when the surface starts to become unevenly colored by fungi.

Inorganic pigments can be added to the water-repellent preservative solutions to provide special color effects, and the mixture is then similar to the semitransparent pigmented penetrating stains. Colors that match the natural color of the wood and extractives are usually preferred. As with semitransparent stains, the addition of pigment to the finish helps stabilize the color and increase the durability of the finish.

Water-repellent preservatives may also be used as a treatment for bare wood before priming and painting or in areas where old paint has peeled, exposing bare wood, particularly around butt joints or in corners. This treatment keeps rain or dew from penetrating into the wood, especially at joints and end grain, and thus decreases the shrinking and swelling of wood. As a result, less stress is placed on the paint film, and its service life is extended. This stability is achieved by the small amount of wax present in water-repellent preservatives. The wax decreases the capillary movement or wicking of water up the backside of lap or drop siding. The fungicide inhibits surface decay. Water-repellent preservatives are also used as edge treatments for panel products like plywood.

Water repellent are also available. These are water-repellent preservatives with the fungicide, mildewcide, or preservative left out. Water repellents are not effective natural finishes by themselves but can be used as a stabilizing treatment before priming and painting.

3.2.5 Transparent Coatings

Clear coatings of conventional spar or marine varnishes, which are film forming finishes, are not generally recommended for exterior use on wood. Shellac or lacquers should never be used outdoors because they are very sensitive to water and are very brittle. Varnish coatings become brittle by exposure to sunlight and develop severe cracking and peeling, often in less than 2 years. Refinishing will often involve removing all the old varnish. Areas that are protected from direct sunlight by overhang or are on the north side of the structure can be finished with exterior-grade varnishes. Even in protected areas, a minimum of three coats of vanish is recommended, and the wood should be treated with a water-repellent preservative before finishing. Using compatible pigmented stains and sealers as undercoats will also contribute to a greater service life of the clear varnish finish. In marine exposers, six coats of varnish should be applied for best performance.

REFERENCES

1. F. L. Browne, *"Wood Properties and Paint Durability,"* Miscellaneous Publication 629, U.S. Department of Agriculture, Forest Service, Forest Products Laboratory, Madison, WI, 1962.
2. W. C. Feist and E. A. Mraz, *"Wood Finishing: Water Repellents and Water-Repellent Preservatives."* Research Note FPL-0124, U.S. Department of Agriculture, Forest Service, Forest Products Laboratory, Madison, WI.
3. J. M. Black, D. F. Laughnan, and E. A. Mraz, *"Forest Products Laboratory Natural Finish."* Research Note FPL-046, U.S. Department of Agriculture, Forest Service, Forest Products Laboratory, Madison, WI, 1979.
4. W. C. Feist, "Weathering of wood in structural uses, " in *Structural Use of Wood in Adverse Environments*, R. W. Meyer and R. M. Kellogg, Eds. New York: Van Nostrand Reinhold, 1982, pp. 156–178.
5. W. C. Feist and A. E. Oviatt, *Wood Siding   Installing, Finishing, Maintaining.* Home and Garden Bulletin 203, U.S. Department of Agriculture, Washington, DC, 1983.
6. W. C. Feist and D. N.–S. Hon, "Chemistry of weathering and protection," in *The Chemistry of Solid Wood*, R. M. Rowell, Ed., ACS Advances in Chemistry Series No. 207. American Chemical Society, Washington DC: 1984, pp. 401–454.
7. D. L. Cassens and W. C. Feist, *Finishing Wood Exteriors: Selection, Application, and Maintenance.* Agriculture Handbook No. 647, U.S. Department of Agriculture, Forest Service, Forest Products Laboratory, Madison, WI, (1986).

87

Pharmaceutical Tablet Coating

Joseph L. Johnson

Aqualon Company
Wilmington, Delaware

1.0 HISTORY

The coating of solid pharmaceutical dosage forms began in the ninth century B.C., with the Egyptians. At that time the primary solid dosage form was the pill, a hand-shaped spherical mass containing drug, sugar, and other diluents. A variety of materials were used to coat pills, such as talc, gelatin, and sugar. Gold and silver were also used. Many of these coatings proved to be impervious to chemical attack in the digestive tract; as a result, the pill never released its active ingredient and was thus ineffective.

The candy making industry was the first to develop and enhance the art of coating. It is most likely that the pharmaceutical industry adopted sugar coating technology for its own use. The first sugar-coated pills produced in the United States came out of Philadelphia in 1856. Coatings resistant to enteric or gastric fluids were developed in the 1880s. In 1953 the first compression-coated tablet was introduced, and in 1954 the first film-coated tablet was marketed.

2.0 REASONS FOR COATING TABLETS

There are many reasons for coating tablets; some aesthetic, some functional. One important reason is to enhance drug stability; that is, to protect the drug from oxygen, moisture, and light, the three key causes of drug degradation. Coating can also be used to separate reactive components in a tablet formula.

Another important reason for tablet coating is identification. Tablet coatings may take on a variety of colors. A coated tablet may also be imprinted with a symbol or word. In the case of the film-coated tablet, the tablet core may be embossed with a symbol or word that remains visible after the coating process. The definitive identification of a coated tablet has saved patients and health care professionals alike. Additionally, coating is used to uniquely identify a branded product.

Tablet coating is done for aesthetic reasons as well. Often the appearance of the tablet core is mottled or otherwise unattractive. Coating masks this. Many times, too, the drug itself has a bitter taste. Coating masks this as well.

Tablet coating can also be used to control the duration and site of drug release. Overall, tablet coating, through an additional step in the manufacturing process, is often vital.

3.0 TYPES OF COATING

There are two main types of tablet coating done today: sugar coating and film coating; film coating is the more popular. Coated tablets fall into three main subcategories depending on how the drug is released: immediate release, enteric release, and sustained release.

Immediate-release coating systems, as the name implies, allow immediate release of the drug compound to the body.

Enteric coatings are soluble only at a pH greater than 5 or 6. Thus the drug is not released in the stomach but in the small intestine. Enteric coatings are by far the most unreliable because of the wide and unpredictable variance in gastric pH profiles. Gastric pH varies considerable based on stomach content, age of the patient, and disease state.

Sustained-release coatings permit drug to dissolve slowly over a period of time. This helps to reduce dosing intervals and improves therapeutic reliability.

Film coating can be carried out using either an organic solvent system, such as ethanol or methylene chloride, or by using water as a solvent. The solvent film coating systems are fast disappearing because of cost, environmental, and safety concerns. Most film coating carried out today is done with aqueous systems.

3.1 The Sugar-Coated Tablet

The sugar-coated tablet is the most elegant solid dosage form produced today. Its glossy appearance, slippery feel, and sweet taste are unmatched by any other coated tablet. The sugar-coated tablet is also the most difficult and time-consuming to produce. The tablet consists of a core upon which layer after layer of coating material is slowly and carefully built up. In some cases this is done by hand in some cases automatically. In any event, there is still an art to sugar coating.

To successfully accept a sugar coating, the tablet cores must be robust. They are subjected to wetting and rolling in a coating pan with 50 kg or more of other cores. Generally the coating pan is spherical and has a solid exterior surface. Temperature-controlled air is introduced and removed from the pan via external ducts. The following procedure is used for the manual sugar coating of tablets.

The first step is to slightly waterproof the tablets by applying a coat of pharmaceutical-grade shellac. This prevents the cores from dissolving prematurely in the presence of the other coating liquids that are to be applied.

The second step is subcoating: a solution composed of acacia, gelatin, and sugar is applied to the tablets. The wetted cores are then dusted with dicalcium phosphate or calcium sulfate and allowed to dry. This step is repeated many times until a smooth rounded tablet form has been achieved.

The third step is the grossing coat. The cores are wetted with a sugar solution and dusted with titanium dioxide powder. This creates a very white base coat on which color may be applied.

The fourth step is the color coat. In this instance an insoluble opaque color solid is suspended in sugar syrup and applied to the tablet. No dusting of the cores takes place. The tablets are simply air dried.

The fifth step is the shutdown coat. In this step diluted sugar syrup is applied to the tablet and allowed to dry. This produces a very smooth finish in preparation for the last step.

The last step is polishing of the tablets. The tablets are placed in a canvas-lined drum. Beeswax or carnauba wax is dissolved in methylene chloride, and the solution is applied to the tablets, which are tumbled until the solvent evaporates and tablets achieve a very high shine.

In all, 40 or more separate layers are applied during the manual sugar coating process. The process takes between 5 and 8 eight-hour shifts to complete.

Automated sugar coating is generally faster. For example, the various syrups used in the coating process have the dusting powders suspended in them. The syrups are applied by spray. This process can be automated, to reduce the number of operators required. Perforated coating pans, which greatly enhance air throughput, are used almost exclusively. With greater air throughput, water evaporates more quickly, thus speeding the process. Using automated techniques, tablets can be sugar coated in about 16 hours.

3.2 The Film-Coated Tablet

The film-coated tablet consists of a core around which a thin, colored polymer film is deposited. Thus, a film-coated tablet gains about 3% of total tablet weight upon coating. The sugar-coated tablet undergoes a 100% weight gain. Overall, film coating is a much faster procedure, and much less prone to error.

The basic film coating formula consists of a film former, a pigment dispersion, a plasticizer, and a solvent. A variety of polymeric film formers can be used to coat tablets. By selecting the solubility properties of the polymer, one can produce an immediate-release, an enteric-release, or a sustained-release tablet.

The most popular immediate-release film formers are the water-soluble cellulose ether polymers. The two most common are hydroxypropylcellulose (HPC) and hydroxypropyl-methycellulose (HPMC). The low viscosity grades of these polymers are employed in the coating formula to maximize polymer solids concentration. Both these polymers are water soluble.

Water-insoluble film formers can also be used to prepare immediate-release coatings. These products fall into two categories: cellulose ethers and acrylate derivatives. The most common cellulose ether is ethylcellulose. This material is commercially available in two forms: as pure polymer and as an aqueous dispersion. The pure polymer is generally dissolved in an organic solvent; the dispersion is delivered out of an aqueous media. In both cases a certain amount of water-soluble component (up to 50% of the total polymer solids) is included in the coating formula, to provide immediate drug release.

The ethylcellulose and acrylate compounds are also used to formulate sustained-release products. Again, a water-soluble component is included in the coating formula. However, the level is very low: usually about 3% of total polymer solids. When the coated dosage form is exposed to water, the water-soluble component dissolves. This leaves a porous film surface through which drug diffuses.

The third class of coatings, the enterics, resist the attack of gastric fluids. As a result, drug is released only in the small intestine. Enteric coatings are prepared by using a polymer with pH-dependent solubility properties. Cellulose esters, substituted with phthalate groups, are the primary polymers used in this application, especially cellulose acetate phthalate. Polyvinyl acetate phthalate is also used. Acrylate derivatives are also capable of providing enteric release.

3.3 Compression Coating

Compression coating is a technique wherein a large tablet either completely or partially surrounds a smaller tablet. Essentially, a small tablet is compressed first and is then surrounded by powder, which undergoes compression. This type of coating technique requires the use of special tableting machinery and it is used to produce sustained-release tablets.

BIBLIOGRAPHY

The Theory and Practice of Industrial Pharmacy, Lachman, L., H. A. Leiberman, and J. L. Kanig, Eds. 1st – 3rd Ed., Philadelphia: Lea & Febiger.

Florence, A. T., Ed., *Critical Reports on Applied Chemistry*, Vol. 6, *Materials Used in Pharmaceutical Formulation*, London: Blackwell Scientific Publications.

Osol, Arthur, Ed., *Remington's Pharmaceutical Sciences*, 14th – 17th Eds. Easton, PA, Mack Publishing Company.

88
Textiles for Coating

Algirdas Matukonis

Kaunas Technical University
Kaunas, Lithuania

Woven or knitted fabrics, and various types of nonwoven product, may be used as coating substrates. The physical–mechanical properties and the end-use performance of the coated fabrics depend significantly on the type of coating polymer and the substrate characteristics. Textile structures used for the backing of coated fabrics are complex three-dimensional constructions. The properties of these textile structures are determined by particular properties of constituent fibers and the construction of yarns and fabrics, as well as finishing processes. Knowledge of the characteristics of backing and its mechanical behavior is essential for predicting and understanding the properties of various coated materials.

1.0 YARNS

Processing of coated fabrics involves a wide range of natural and man-made fibers. Cotton and other vegetable fibers are the most important natural fibers used for backing. The types of man-made fiber most widely used for backing are high wet modulus viscose, polyester, polyamide, acrylic, polypropylene, polyethylene, and aramid fibers. It is well known that fiber properties are determined by the nature of the chemical composition, by the molecular and fine structure of the constituent polymer, and by the external structure of fibers. The fibers mentioned are used in form of staple or yarns for the manufacture of woven, knitted, or nonwoven structures for backing.

There are two general classes of yarns: spun yarns made from natural and man-made staple fibers or their blends, continuous filament (multi- and monofilament) yarns.

Spun yarns are an assemblage of partly oriented and twisted staple fibers of relatively short definite length. The fibers in yarn are held together by twist, which causes the development of high radial forces and friction between fibers. Because of friction between fibers, the yarn obtains tensile strength and compactness. The fibers lie at varying angles to the axis of the yarn, with the fiber ends sticking out from the surface. The hairiness and the bulk of spun yarns play important roles with regard to absorbency and adhesion properties of back-

753

ing materials made from these yarns. The amount of twist also determines the mechanical properties, first of all the breaking force and extension of spun yarns.

Continuous filament yarns are made by extruding the fiber forming polymer (solution or molten mass) through the holes in a spinneret. Filaments obtained by this way are long continuous fiber strands of indefinite length. The number of filaments is determined by the number of holes in the spinneret. Continuous filament yarns are characterized by a smooth, compact surface formed by parallel packing of straight filaments with minimal air spaces between them. Yarn made from one continuous filament is called monofilament yarn. Continuous filament yarns may be twisted or intermingled, to obtain required degrees of compactness and structure.

For improving bulk, stretch, warmth of handle, or moisture absorbency of continuous filament yarns, the process of texturing or bulking is widely used. Texturing or bulking alters the shape of filaments of thermoplastic yarns by introducing crimp loops or crinkles by means of deformation and simultaneous heat setting. The air bulking of continuous filament yarns of all types is also a rapidly expanding process. Three classes of textured filament yarns are manufactured: bulk yarns, stretch (with high elongation) yarns, and modified stretch yarns.

Yarn fineness (size) or linear density is expressed in terms of mass of unit length or tex (the mass in grams of 1000 m of yarn).

2.0 FABRICS

2.1 Woven Fabrics

Woven fabric is a textile structure made by interlacing two sets of yarns at right angles to each other. The yarns running along the length of the cloth are called the warp; the yarns going cross-wise, the weft. The warp threads in the weaving loom are separated by raising and lowering the frames and warp threads to form a shed, through which the weft yarn is propelled. The manner in which the sets of yarns are interlaced is known as weave. The mean number of ends or picks required to produce the weave (i.e., one complete yarn interlacing pattern) is called the repeat.

The weave, together with yarn linear density and thread spacing, to a great extent determines the properties and the appearance of fabrics. Weaves are usually represented on design paper. The warp yarn is shown by a vertical row between two lines of paper; the weft, respectively, by a horizontal row. If the warp thread goes above the weft thread, the corresponding square is filled in; if under the square is left unfilled. In the binary notation used in computer programming, filled squares are represented by "1" and blank ones are indicated as "0."

The variety of weaves is extremely great. Only a very limited range of weaves, mainly the basic weaves are used for backing fabrics. There are three basic weaves: the plain weave, the twill weave, and the satin weave.

Plain weave (Fig. 1) is the simplest. Each warp yarn interlaces with each filling yarn alternately, on the one-up/one-down principle. Plain weave fabrics constitute the largest group of woven materials used for backing purposes.

Because the plain weave is characterized by the highest quantity of interlacings in comparison with other weaves, it increases the tensile strength, increases the tendency to wrinkle, and decreases absorbency more than in comparable fabrics made with weaves of other types.

Figure 1 Plain weave.

Twill weaves have yarn (warp or weft) floats on the surface of the fabric across two or more yarns of the opposite direction. The arrangement of warp and weft floats produces a diagonal pattern on the surface of fabric.

Twills are designated by a fraction, in which the numerator represents the number of warp threads that cover the weft thread and the denominator indicates the number of weft threads covering the warp yarns. A 1/3 filling faced twill is shown in Figure 2. It is clear that on opposite side of fabric, the 3/1 twill is arranged with the twill wale going in the reverse direction. Since the relative amount of interlacing in the twill weave is less than in a plain weave, yarns can be packed closer, producing a thicker cloth. On the other side, fewer interlacings diminish the interfiber friction, which contributes to a greater pliability, softness, and wrinkle recovery of fabrics, but makes for lower strength. For backing manufacturing, a 2/2 twill is widely used.

The satin weaves (Fig. 3) have long yarn floats (over four yarns minimum) with a progression of interlacing by definite number (over two yarns minimum). When warp yarns predominate on the face of the fabric, we have a warp-faced fabric—satin with a higher warp count. If the weft covers the surface, the fabric is called sateen. The filling count of sateen fabrics is higher than of warp ends. The few interlacings of satin weave fabrics increase the pliability and wrinkle recovery, but also increase yarn slippage and raveling tendency. Fabrics of this type have a smooth, lustrous appearance because of the long floats.

Manufacturing of heavy-weight coated materials requires corresponding woven backing with distinct thickness and mechanical properties. For this application, fabrics of two or

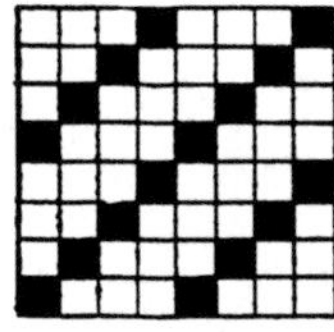

Figure 2 Filling faced 1/3 twill weave.

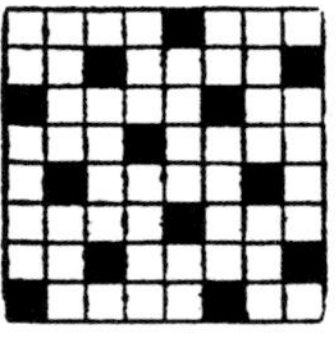

Figure 3   Satin weave: 1/4 yarn arrangement with the repeat of 5 × 5 yarns.

Figure 4   Four-layered fabric (derived from plain weave).

Figure 5   Weft-pile fabric.

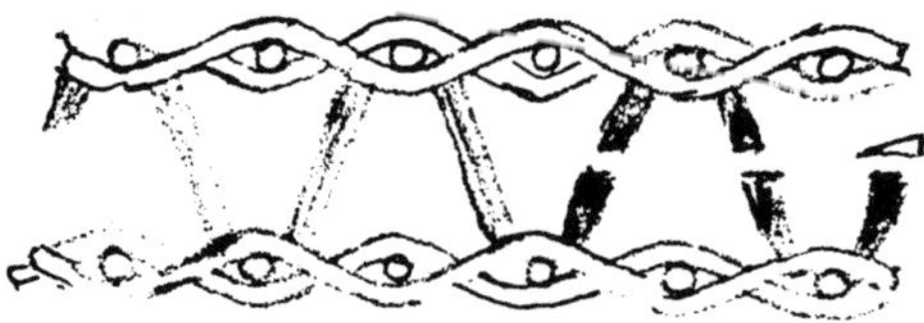

Figure 6   Warp-pile fabric (double-cloth method).

more layers are used. The construction of four-layered fabric based on plain weave is shown in Figure 4.

For improving adhesion and absorbency as well as the aesthetic properties, various fabrics can be napped during finishing on one or both sides, producing a layer of fiber ends on the surface of the cloth. The weave of fabric used for napping usually must be filling-faced because of the raising ability of the long weft floats.

The most dense and durable three-dimensional pile cover is produced by means of special techniques. There are two ways to manufacture woven pile fabrics: using weft-pile and warp-pile technologies. In the weft-pile fabric (velveteen and corduroy) an additional set of filling, usually staple yarns with floats, is used. After weaving, the surface floats are cut and brushed, producing a dense, stable pile cover (Fig. 5).

In warp-pile fabrics (velvets, plush, furlike fabrics), an extra set of warp staple or multifilament yarns is used. One very productive approach is the double-cloth method of warp-pile fabric manufacturing. Two parallel fabrics are woven in the special loom, face to face. The pile–warp interfacing connects both fabrics. As the pile is cut, two pile cloths are produced (Fig. 6). In weft-pile fabrics the tufts of pile are interlaced around ground warp yarn, and in the warp-pile fabrics they are interlaced around ground warp ends.

Some pile fabrics can be made very efficiently by tufting and punching extra yarns into woven base fabric by a series of needles, each carrying a pile yarn from a creel. The tufting pile can be cut or looped. The height of the pile depends on the type and end use of fabric. Velvet has a pile 1.5 mm high or shorter, velveteen not over 3 mm, plush usually 6 mm and longer, and furlike fabric 8–15 mm. Very important specific properties of all pile fabrics are the density of pile cover and resistance to shedding and pulling out. It must be noted that a coating polymer layer may be formed on the pile side of the fabric.

Other methods of producing pile fabrics, such as electrostatic flocking, using chenille yarn pile, etc., are also known.

The main structural characteristics of woven fabrics are linear density and count of constituent yarns, as mentioned previously, as well as weave, cover factor, and mass per unit area. Cover factor is expressed as follows:

$$ K = \frac{d_y}{a_y} \times 100 = d_y S_y \times 100 $$

where d_y is yarn diameter (mm), calculated on the base of linear density and apparent density of yarns, a_y is yarn spacing, and S_y is yarn count (number of threads per millimeter). Cover factor may be obtained for warp and weft yarns; it expresses the relative tightness of the fabric concerned. The magnitude of K in fabrics intended for coating varies in the ranges of 50–140% and 40–130% for warp and wefts, respectively. The mass per unit area (weight range) of fabrics depends on type and end use and varies from 40 to 400 g/m² or more.

Among the wide range of mechanical characteristics, there are several determining the field of use of coated woven fabrics. First of all, the fabric must have the required tensile strength and elongation. The tensile strength of fabric as well as of yarns is expressed in term of tenacity in specific units: centinewtons per tex (cN/tex). Tenacity is calculated on the basis of the breaking force of a 50 mm wide strip and the number of linear density of threads in the strained system.

For approximate calculations, it may be assumed that the breaking force of a loaded thread system is expressed as the sum of the loaded yarn's breaking force multiplied by a factor 0.8–1.2, depending on the weave, thread count, type of fibers and yarns, finishing

Table 1   Range of Breaking Characteristics of Woven Fabrics

Type of fabric	Breaking force (cN/tex)		Breaking extension (%)	
	Warp direction	Weft direction	Warp direction	Weft direction
Cotton	5–7	4–7	4–8	16–24
Linen	6–8	5–6	5–21	5–8
Wool	2–5	1–3	20–28	28–33
Viscose	7–8	4–5	16–20	15–23
Nylon	20–26	7–16	20–22	26–28

processes, and loading direction. In some cases the conditional value of tenacity is evaluated on the basis of breaking force and the whole mass of fabric strained (as in the case of nonwoven materials). The conditional values of breaking force and breaking extension of woven fabrics of various types are represented in Table 1. The strength of high tech fabrics made from high tenacity fibers (polypropylene, polyethylene, aramid, and others) may be much higher.

The tensile behavior under load of fabrics of different types—woven, knit, and nonwoven—is shown in Figure 7.

There are also other characteristics of fabrics that determine the usefulness of these materials for coating purposes. Important properties are tearing force, resistance to cyclic loading, bending stiffness, drape, porosity, resorbency, elasticity, and air permeability. Comparable information about the main properties of various fabric types is given in Table 2.

2.2 Knitted Fabrics

Knitted fabrics are also important backing materials. Knitted fabric is a three-dimensional textile structure made by means of needles from a series of interlocking loops from a single yarn or an assemblage of yarns. The basic construction element of knitted structures is the loop. The lengthwise rows of loops in knitted fabric are named wales, and those running across the fabric are known as courses.

Depending on the number of threads taking part in the arrangement of loops, the two fundamental classes of knit structures are known as weft-knitted and warp-knitted fabrics. If the fabric is constructed from a single yarn with the loops made horizontally across the fabric, the principle of structure forming is known as weft knitting. Warp-knitted fabric is a vertical loop construction made from one or more sets of warp by forming the loops.

The weft knits are produced on flat-bed or circular machines of various types. There are three basic structures of weft-knitted fabrics: plain, rib, and purl. Obviously a wide range of modifications may be derived. The plain weft-knitted fabric structure is shown in Figure 8a. The warp knits are made on two basic types of warp knitting machine—tricot and Raschel. The warp-knitted structures, as mentioned, require a set of threads fed from warp beams, each yarn producing a row of loops along the length of the material. The fabric is formed

Table 2 Comparison of Textile Properties

Fabric type	Thickness	Porosity	Specific volume	Roughness	Tenacity	Breaking extension	Tear force	Bending stiffness	Compres- sibility	Elastic modulus[b]	Drape co- efficient	Shear strain
						Properties[a]						
Woven	L/M	M	L	M	H	L/M	H	L/M	L	M	L/M	M
Knit	L	L/M	M	L	M	M	M/H	L	M/H	L	L	H
Stitchbonded (Malimo)	L	M	M	M	H	M	H	M	M	M	M	M/H
Adhesive- bonded	L/M	L/M	M	M	L/M	M	L	H	L/M	M	M/H	L
Stitchbonded (web)	M/H	H	H	H	L	M/H	L/M	M	H	M	H	L
Spunbonded	L	H	M	M	H	H	H	M/H	L	H	H	L

[a]L, low; M, medium; H, high.
[b]Modulus of elasticity (initial) expresses the ratio of stress to strain at the beginning of the stress–strain curve.

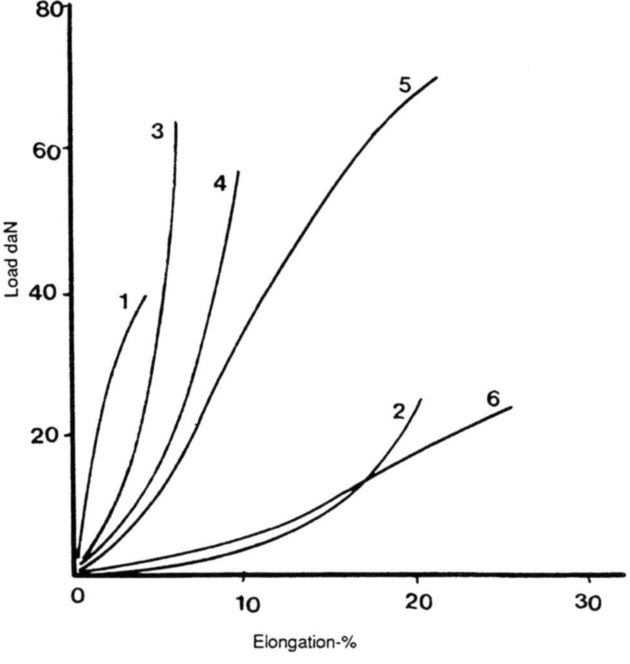

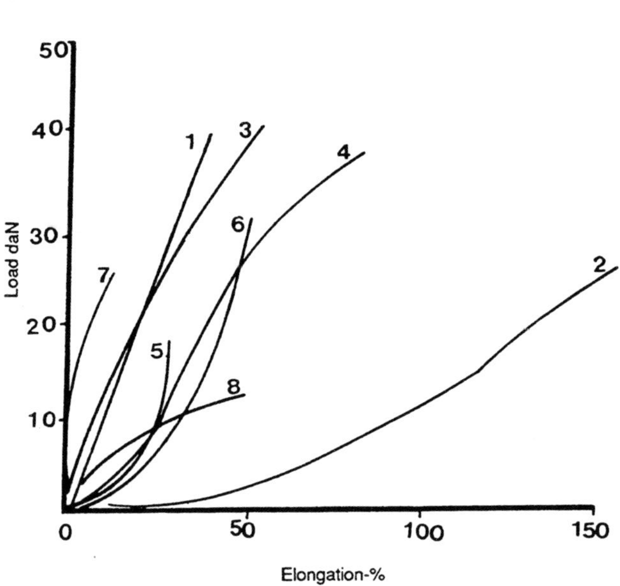

Figure 7　Load–elongation diagrams of different type of fabrics. (a) Woven fabrics: 1 and 2, cotton suiting; 3 and 4, linen; 5 and 6, nylon 6 taffetas (odd numbers—lengthwise extension; even numbers—crosswise extension). (b) Knits and nonwoven fabrics: 1 and 2, interlock double knit from viscose yarns; 3 and 4, tricot-charmà knit from nylon 6; 5 and 6, web stitchbonded cotton fabric; 7 and 8, adhesive-bonded nonwoven from viscose–nylon 6 blend (latex binding).

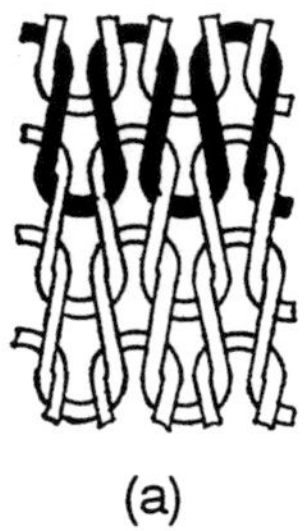

(a)

(b)

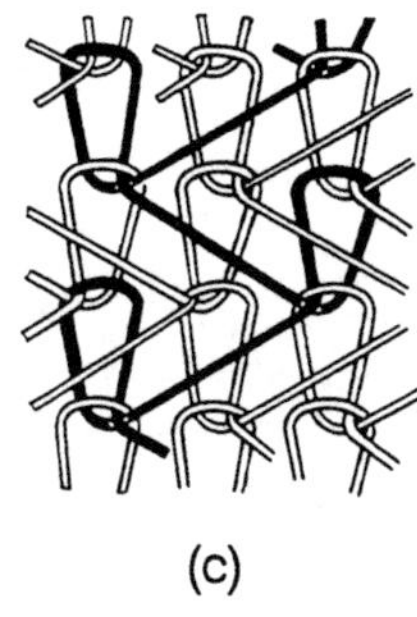

(c)

Figure 8   Knitted fabric structures: (a) weft-knitted fabric, (b) warp-knitted tricot structure, and (c) warp-knitted lockknit structure.

when the threads are moved from one needle to another on successive courses. Each warp set is controlled by yarn guides mounted in a guide bar. The number of guide bars depends on the number of warp sets. In different guide bars, yarns of different types may be fed. There are various structures of warp-knitted fabrics. For backing manufacture, tricot (plain jersey stitch), warp-knitted lockknit, and others are used. The warp-knitted tricot and warp-knitted lockknit structures are shown in Figures 8b and 8c.

A wide range of knitted fabrics can be made on the high speed multibar Raschel warp knitting machines. There are rigid and stretch fabrics, open-work fabrics, various tulles, nets, household fabrics, pile fabrics (velour, velvets), and industrial fabrics, including backing for coatings.

The main properties of knitted fabrics are determined by structure, which is characterized by the form and size of loops, the linear density of the constituent yarns, and the type of knitting structure. A very important parameter of knits is the cover factor, which expresses the relation of yarn diameter to loop length. Cover factor correlated with characteristics of stability and also with some mechanical properties.

The tricot fabrics have a high tensile and tear strength, as well as high resiliency, elasticity, and stability (see Table 2). Fabric of this type, if made from synthetic yarns, must be heat set to stabilize it against shrinkage. For tricot fabrics, a special finishing (e.g., antistatic, antisnap, flame-retardant) may be used.

The tenacity range of knitted fabric depending on type of knitting and type of fibers is 1–4 cN/tex (expressed by estimating the entire mass of fabric undergoing strain). The breaking extension is 50–250%.

Very important in achieving an increase in tensile strength has been the development of technology for manufacturing multiaxial warp-knit fabrics of various constructions, which are extremely suitable for backing uses. These fabrics can be produced on the tricot warp knitting and on Raschel machines. The basic principle is a sensible combination of the interlacing elements (loop and weft), which offers multiple fabric structure possibilities, especially dimensional stability. The simplest types of multiaxial warp-knit fabric are structures in which weft or warp or both systems are inserted in the knit structure. These additional straightened yarns are firmly bonded by the stitch construction and behave as load-bearing systems.

More complex constructions of multiaxial fabrics based on magazine weft insertion were recently developed. The number of load-bearing systems can be increased to three, four

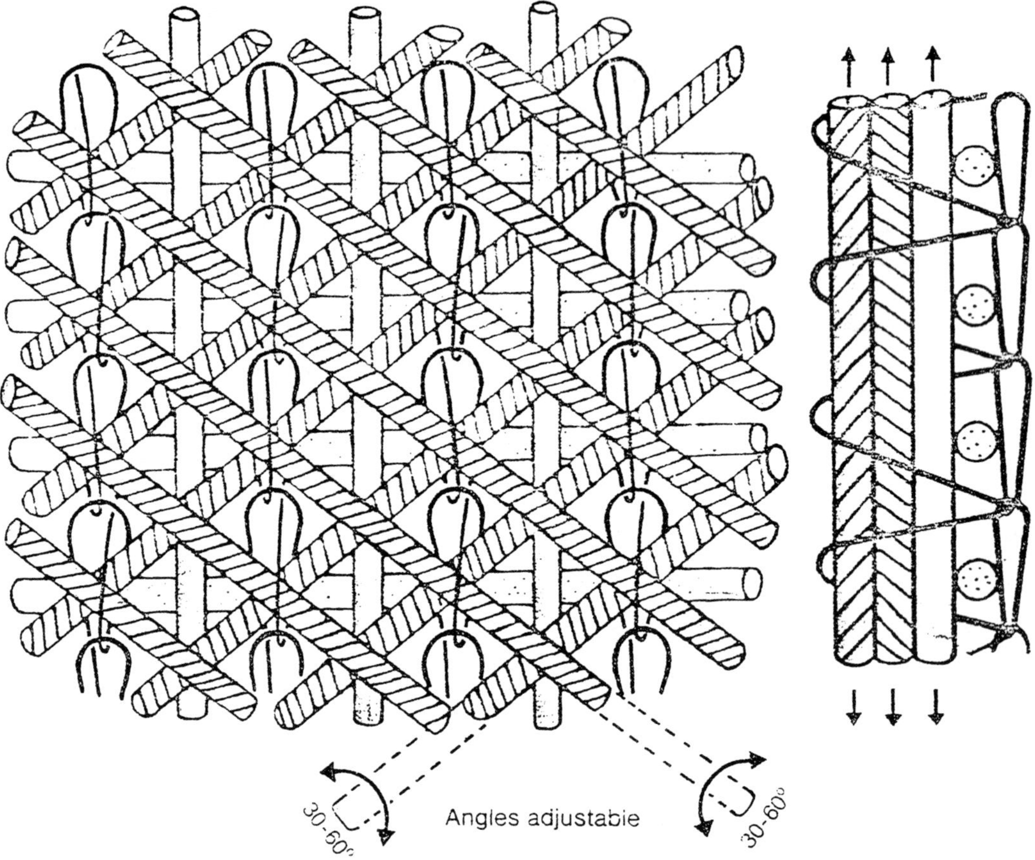

Figure 9 Multiaxial warp knit with four load-bearing systems, made on the Raschel machine (Karl Mayer, West Germany).

(Fig. 9), or five by adding a diagonal yarn arrangement with a varied angle of insertion. Fabric becomes resistant to stretch in all directions because of these reinforcing structural elements. Depending on the construction, any yarn type and linear density may be used for the reinforcing system, including aramid, glass fiber, carbon fiber, and all traditional kinds of fibers. It is also possible to use a prebonded web on one or both sides of the construction as protective layer of fabric (Fig. 10).

Multiaxial warp-knit fabric can be applied in all fields of household use, as well as for industrial applications.

The combined woven–knitted structure may be manufactured by means of the Metap machine (Czechoslovakia). The fabric consists of alternating lengthwise woven (85% of the whole surface) and knitted stripes. The properties of this fabric are analogous to those of woven structures. The fabric is suitable as a backing.

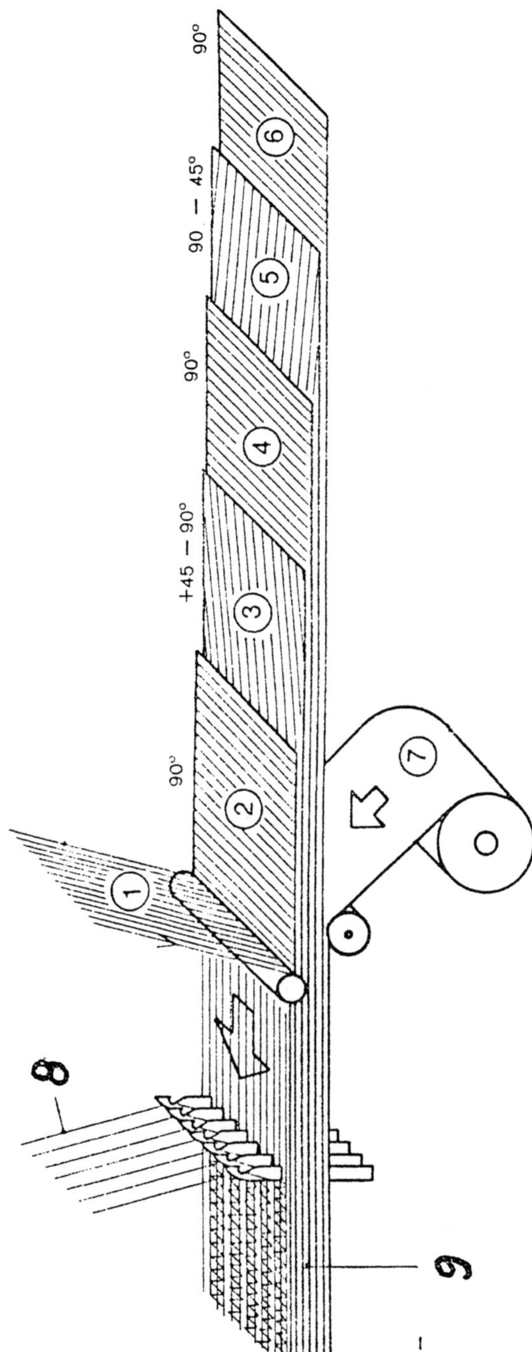

Figure 10 Multiaxial warp-knit fabric combined with prebonded web (made of Copcentra HS-ST machine, Liba, West Germany): 1, warp filler yarns; 2–6, weft yarn systems; 7, prebonded fiber web; 8, stitching warp system; 9, fabric formed.

2.3 Nonwovens

The term "nonwoven" is applied generically to textile fabrics made from fibers, using unconventional methods. There are three groups of nonwoven fabric technology: physical–chemical methods, mechanical technology, and combined technology. All techniques of nonwoven manufacture are characterized by a high operating speed and low fabric production costs in relation to conventional technologies.

2.3.1 Adhesive-Bonded Fabrics

Adhesive-bonded fabrics are made by the physical–chemical method, in which webs of fibers are strengthened by fiber-to-fiber adhesion. The web is prepared on special equipment based on carding or the aerodynamic principle; such machinery is capable of producing webs with random or oriented fiber distribution. The quality of the web determines to a large extent the quality of the nonwoven fabric. Adhesive-bonded fabric manufacturing uses a wide range of natural and man-made fibers or their blends.

Adhesion is achieved by means of the bonding agent, which may be an aqueous emulsion or a thermoplastic additive to the web. The adhesion of the bonding agent to the textile substrate must be good, and its cohesive strength must be adequate to withstand stresses during use. Because of the bonding action, the nonwoven fabric acquires strength and stiffness. The mechanical properties of fabrics (see Table 2) depend highly on binder content, fiber type and fiber orientation in the web. It is possible to produce a nonwoven of this type with the weight range of 15–500 g/m^2 of either high flexibility but low strength or low flexibility but high strength. For increasing the drape properties and flexibility, the print bonding method is used. The spacing formed in this way between bonded areas allows freedom of movement of the fibers, which increases the fabric's flexibility.

Thermoplastic materials used for bonding web fibers are powders, fibers, yarns, nets, and films. Widely used in thermobonding technology are polyester fibers (including the hollow type), polyethylene, polyamide, and special binder fibers. Most often, two-component fibers are used, formed from polymers with different melting temperatures. Thermoplastic materials in the form of bonding fibers are processed by a heat treatment (oven sintering or hot calendering). Thus, at a suitable temperature, the sheets of bicomponent fibers that are in contact in a web will remain bonded upon cooling. Nonwovens made in this way are characterized by a weight range of 15–80 g/m^2 and by good handle properties, porosity, and bulk. The portion of binder fibers to adhesive is 10–50%.

The tensile strength of nonwovens is usually expressed in term of tenacity, as in the case of woven fabrics:

$$\text{tenacity} = \frac{\text{breaking force}}{\text{mass per unit length}}$$

Breaking force is determined on the base of a strip 50 mm wide. Therefore, the mass per unit length is expressed as mass of fabric strip 0.05 m wide and 1000 m length (tex).

The tenacity of adhesive- and thermobonded fabrics is in range of 1–4 cN/tex.

2.3.2 Spunbonded Fabrics

The manufacture of spunbonded nonwoven fabrics consists of combining the preparation of webs with the production of man-made fibers. The whole sequence of operations, such as melt extruding and drawing of continuous filaments, arranging them on a moving collecting surface, forming a web, and bonding together by means of adhesive, thermobonding, or needlepunching, may be done in one process. Various man-made fibers may be used for the production of spunbonded nonwovens: viscose, polyester, nylon, polypropylene, polyethylene, and polyurethane fibers. The main spunbonded nonwoven properties depends on filament properties (linear density, tenacity, elongation, crimp, micromorphology). filament arrangement, and bonding parameters. These types of nonwoven fabrics are produced in weight range of 15–125 g/m². The use of randomly arranged continuous filaments contributes to a higher tear and tensile strength (5–8 cN/tex) in all directions, and also to good handle (see Table 2). By use of modified spunbonding techniques, it is possible to obtain fabrics with special properties, including those required for coating products (greater elasticity, elongation, and air permeability).

Trade names of spunbonded fabrics include Cerex (nylon 6 6), Reemay (polyester), Typar (polypropylene) and Tyvek (polyethylene).

Melt-blown fabrics also belong to the class of spunbonded nonwovens. The production process differs from the spunbonded method with respect to the principle of fiber production. The polymer here is melt-extruded through a row of die openings into a stream of hot high velocity air. The filaments formed are broken into fibers, which are entangled to form a web. The melt-blown fabric differs from the spunbonded one in that the web contains staple fibers rather than continuous filaments, with diameters in the range of 2–5 µm. This contributes to the softness, drapeability, opacity, and moderate strength of such fabrics. Strength may be increased by hot calendering. The weight range of melt-blown fabrics is 60–500 g/m².

2.3.3 Needlebonded (Needlepunched) Fabrics

Needlebonded fabrics are manufactured by a mechanical method. The principle of needlepunched fabric production is realized when the fibers move from one face of a web toward the other face as a result of the penetrating of the web by many barbed needles. Because of the transfer of fibers made by the needles of the punching machine, the fibers are interlocked, and the web obtains stability and strength. If required, the fabric may be finished by adhesive bonding or pressing, steaming, dyeing, and calendering. The weight range of fabric is 200–1500 g/m². The mechanical properties depend mainly on fiber characteristics, interfiber friction, web weight, and fabric finish treatment. Fibers of all types, and their blends, may be used for production of needlepunched fabrics. The strength of needlepunched fabrics varies in the range of 2–5 cN/tex. To increase fabric strength, additional backing in the form of a woven, knitted fabric or a film may be used. It is also possible to produce to patterned colored fabrics by means of colored layers and by needling fibers from the top layer through the surface layer, making loops on the face of fabric. Needlepunched fabrics with such special properties as flame retardancy, conductivity, and elasticity may be produced by using corresponding components.

2.3.4 Spunlaced Fabrics

Spunlaced fabrics are made by entanglement of the fibers in the web by means of streams of high pressure water jets. The web obtains the required bonding, which influences the strength, handle, drape, and air permeability of fabrics. the fluid fiber entangled fabrics may be made in the weight range of 20–70 g/m² and with a tenacity of 1.5–2.5 cN/tex. Polyester, polyamide, and other fibers may be used.

2.3.5   Stitchbonded Fabrics

There are several techniques for producing stitchbonded fabrics. The stitchbonding of fibrous web carried out on the Arachne (Czechoslovakia), Maliwatt (East Germany), and VP (USSR) is widely known. The web of natural or man-made fibers prepared by the carding process, with oriented or random arrangement of fibers, is stitched with yarns by means of warp knitting technology units. This technology is therefore sometimes known as "knit–sew." The type of stitch may be half-tricot, tricot, or others. The height of stitch varies from 1 to 6 mm. The gage of the VP machine is 2.5, 5, and 10; the gage of the Maliwatt varies from 3 to 22. Therefore the number of courses in the fabric made varies from 5 to 25 (on a 50 mm base). The range of fabric weight is 160–400 g/m² for the VP and the Arachne, and 100–1600 g/m² for the Maliwatt. The mass of stitching yarns contributes 10–30% of whole fabric mass because the stitching system forms a continuous net of warp knit, filled with fibers of web.

The mechanical properties of the fabrics produced by stitching a basic web depend on the type of fibers used as much as on the fabric stitching structure, causing friction between elements of fabric construction. The stitching yarns also contribute to fabric tenacity in the lengthwise direction. The tenacity of the fabrics considered is 1.6–3.5 cN/tex. The fabrics have good handle and tear resistance, especially in the cross direction (see Table 2). After stitching the fabric may be processed by dyeing, pressing, or steaming.

Because of their tensile strength, tear strength, and resistance to cyclic straining, web-stitched fabrics are often used as backings for artificial leather, industrial coated fabrics, and other purposes.

Stitchbonded fabrics made from yarns are produced using the Malimo (East Germany) technology. These fabrics are made with three sets of yarns. A warp system is fed from a warp beam. A set of weft threads is laid at an angle with the warp by means of a carrier. The warp and weft are stitched together using a third system—sewing yarns fed from a beam by means of compound needles. For stitching, a chain stitch as well as the tricot stitch can be used. It is possible to use all types of spun and continuous filament yarns for fabric manufacture. The range of fabric weight is 100–800 g/m². The fabrics usually are subjected to processing operations similar to those of woven fabrics.

Many types of fabric for apparel, household, and industrial end uses can be produced by the Malimo technology. The mechanical properties of yarn stitchbonded fabrics are similar to those of woven fabrics: good form stability, elasticity, drapeability, and thermal insulation characteristics. The tensile tenacity may reach 8–10 cN/tex lengthwise and 4–5 cN/tex crosswise, depending on the kind and density of yarns and stitch used.

There are several techniques for producing of loop-pile nonwoven fabrics, developed mainly on the basis of the Mali machines (East Germany) already mentioned. In the first of these technologies, known as Voltex, the fiber web is used to stitch pile loops on the existing basic fabric (conventionally woven, knitted, or nonwoven). The fabric produced has a loop-pile cover on one side. The range of basic fabric weight is 40–200 g/m². The weight of wool or man-made fiber web varies from 40 to 80 g/m². These fabrics are used for interlinings, artificial fur, and other applications.

The second method of manufacture of loop-pile nonwovens, known as Malipol (East Germany) is based on the pile stitching principle. The basic fabric, in the weight range of 150–180 g/m², is the same as that in the Voltex technology. The loops formed by the tricot knitting stitch cover one side of fabric. The pile yarns may be spun yarns of various types. The applications are the same as those for Voltex fabrics.

An analogous principle for the production of loop-pile stitchbonded nonwoven fabrics is also used in the Araloop (Czechoslovakia) and Kraftamatic (England) machines. The stitchbonded loop-pile fabric is produced from three sets of yarns on a Schusspol-type (East German) machine. The basic construction of the nonwoven is formed on the same principle as in Malimo technology. The loop piles are arranged by means of special mechanisms. The one-sided pile cover of man-made fiber yarns is stable and resistant to abrasion. The fabric serves for household uses (furnishing, carpets, etc.).

Pile fabrics also may be produced by flocking. This technique is carried out by the adhesive attachment of very short fibers (0.3–10 mm) in the erect position to the surface of a fabric. There are two basic methods for applying flock fibers: mechanical and electrostatic. After the flocking, the thermofixation process is carried out. The flocks are made mainly from man-made fibers. Pile height may be adjusted to suit the end use, which can be household goods or clothing, as well as some industrial applications, including backing for coated fabrics.

The mechanical properties of nonwovens in which a base fabric is used depend greatly on the characteristics of this basic construction element. The surface characteristics of a nonwoven obtained in this way will be determined by the pile system, its fiber composition, and the way in which the fibers or yarns are made up into a pile cover.

The main characteristics of the coated fabrics must be mentioned. They are fabric weight, thickness, tensile and tear strength, breaking extension, bending stiffness, resistance to cyclic bending, abrasion resistance, air permeability, transmission of water vapor, set after wetting and drying, and a number of particular properties influencing the performance of coated fabrics in actual use.

BIBLIOGRAPHY

Böttcher, Peter, Einsatz von Vliestofferzeugnissen und -verfahren in der Geotechnik, **69**, 557–561 (1988).

Collier, Ann M., *A Handbook of Textiles*. Exeter: Wheston, 1980.

Iljin, S. N., and M. Ch. Bernshtein, *Iskusstvennye Kozhi*. Moscow: Legkaya i pishtchevaya promyshlennost, 1983.

Martin E., "Material- und konstruktionsabhaengige Eigenschaften von Geotextilien," *Chemiefasern/Textilindustrie*, **36/88**, 114–115 (1986).

Matukonis, S., J. Palaima, and A. Vitakauskas, *Tekstiles medziagotyra*. Vilnius: Mokslas, 1976.

McConnell, R. L., M. F. Meyer, F. D. Petke, W. A. Haile, "Polyester adhesives in nonwovens and other textile applications," *J. Coated Fabrics*, **16**(1), 199–208 (1987).

"Vliesstoffe auf der Techtextil '86," *Chemiefasern/Textilindustrile*, **36/88**, 581–587 (1986).

89
Nonwovens as Coating and Laminating Substrates

Albert G. Hoyle

Hoyle Associates
Lowell, Massachusetts

1.0 INTRODUCTION

Before the advent of nonwovens, substrates available for coating and laminated products were woven and knit fabrics made out of spun yarns, mainly of cellulosic base. Properties of such fabrics were sharply defined and limited. They could be varied somewhat by changing the weave pattern, the yarn size, and the weight of the product. As a whole they were thick, had poor tear strength when coated, tended to lint, were uneven, had comparatively rough surfaces, and had holes or voids where the yarns intersected—poor properties when very thin and even coatings were needed.

Initial nonwovens of the carded and random air-lay type composed of synthetic fibers were an improvement in some respects but not all. Carded unidirectional webs were of good quality even at medium to low weights, but they were stiff and had too high an elongation for some end uses, as well as poor cross-directional strength and poor tear strength. Random air-lay fabrics had good isotropic strength and fair tear strength at low binder levels, but their quality was too poor for use as a coating substrate at anything lower than a weight of 85 g/m^2.

The introduction of finer deniers and continuous filament yarns in woven and knit fabrics used in coating substrates overcame some of the deficiencies of the spun yarn woven and knit fabrics, such as evenness of cover, roughness of surface, and minimum thickness. They also were an improvement over the initial nonwovens used, especially in regard to strength, strength/weight ratios, drape, and conformability for molding. However, these materials are much more expensive than either nonwovens or spun yarn wovens and knits.

Development of newer and finer fibers, binders, and processes for assembling them into nonwovens has resulted in a diversity of nonwoven substrates that are economically competitive and suitable functionally for use as substrates in the majority of coating and lamination applications.

Coated substrates are used in the following areas: home furnishings, construction, automotive, consumer products, filtration, and industrial applications.

"

2.0 SUBSTRATE STRUCTURES

2.1 Spunbonded Webs

Because of its diversity, ideal strength/weight ratio, and economics, the spunbonded web has the most potential for growth in the area of coating and laminating substrates. Thirty percent of the nonwovens made today in the United States are made by this process.

Spunbonded nonwovens are formed by extruding the proper formulation through a spinneret. The extruded continuous filaments are injected into an air stream, where they are drawn out and randomly distributed onto an open conveyor belt. Spunbonded filaments can be all of one composition or they can be coextruded filaments of two different compositions. Bonding of these materials can occur in a number of ways: thermobonding, either area of point bond, latex saturation, needlepunching, or hydrogen chloride gas bonding. The material can be made in a range of fiber fineness (denier/filament) to give an assortment of properties such as cover, quality, softness or stiffness, and strength. By changing the draft ratio as the material is deposited onto the conveyor from the extruder, one can achieve isotropic strength, unidirectional strength in the machine direction, or unidirectional strength in the cross-machine direction. In most instances balanced strength in both directions appears to be the most desirable.

2.2 Carded Unidirectional Webs

Carded unidirectional webs are made by blending together fibers of required polymeric composition and fineness, opening and mixing the fibrous blend, and feeding it into a card to form a web or mat having the fibers oriented predominantly in the machine direction. Bonding can occur in one of two ways: by incorporation of a binder fiber or powder in the fibrous mat, followed by the application of heat and pressure, or by application of a latex to the mat, followed by drying and possibly calendering. In both instances the web is thoroughly bonded together in length, width, and thickness. The fabric can vary considerably in stiffness, strength, and extensibility depending on the type and amount of binder that is applied. The web is stronger in the machine direction (MD) and more extensible in the cross-machine direction (CD) because of fiber orientation, but MD/CD strength ratios can range from 6 to 2 depending on fiber and binder selection and processing used.

2.3 Carded, Cross-Lapped, Needlepunched Webs

Carded, cross-lapped, needlepunched webs are made almost identically to the unidirectional ones except that they are processed through a cross-lapper after the carding step. This takes the fibers from a unidirectional machine direction orientation to an orientation in a direction diagonal to each other, and predominantly at 45° to the right or to the left of the (vertical) machine direction. This results in a fabric that has a balanced strength and extensibility in both machine and cross-machine directions. Bonding is performed by needling alone or combined with thermal bonding using a gapped calender. Coarse denier fibers (6 denier/filament and up) in the staple length range of 5–18 cm are used for this.

2.4 Air-Lay Needlepunched Webs

The air-lay needlepunch process uses the same type of opening and blending used for the carding process. The differences lie in the manner in which the fibrous web is formed and in the fiber orientation in the web. A carded web is mechanically formed and machine-direction oriented; the resultant nonwoven is stronger in the machine direction. An air-lay web is formed by opening the fiber, air transporting it to a rotating cylinder or a flat screen using a rapidly rotating tooth roll (licker-in roll) and a vacuum, releasing the vacuum, and transferring the web to a conveyor, where it proceeds to the bonding step. When the fiber is deposited onto the conveyor from the air stream by evacuating the air through the screen conveyor, it is deposited in a random manner, forming a web that is truly isotropic. Both long staple (2.5–7.5 cm) and short staple (0.3–2.1 cm) fibers can be processed into a nonwoven using this method, but different pieces of equipment are necessary (i.e., one being more suited for processing long staple fiber, the other for short staple fiber). The minimum weight at which a good quality web can be made, using this process and long staple fiber, is 75 g/m^2. Attempting to make lighter weights results in blotchiness and a generally poor quality in the webs. Coating substrate webs of this type are very rarely less than 85 g/m^2. After air-laying, the web is further processed by needlepunching.

Needlepunching depends on mechanical fiber manipulation to achieve the bonding effect. Manipulation consists of entangling fibers with each other by the use of barbed needles penetrating and receding from the web as the web is conveyed under a reciprocating bed of barbed needles. Needle configuration, needle length, barb shape, and barb spacing serve to give varying degrees of entanglement. The bonding formed in this process is more a series of three-dimensional floating anchor points rather than rigid bond points in a planar structure. The product formed is extensible, bulky, conformable, distortable, and extremely absorbable. Should bulkiness, absorbability, and decreased extensibility be required with additional strength and increased rigidity, this could be accomplished by adding thermal binder fiber to the fiber blend before making the air-lay web and passing the needlepunched material through a gapped hot calender to thermally bond it, in addition to bonding by needlepunching alone.

2.5 Poromerics

Poromerics or synthetic leathers are made using a nonwoven fabric as a base and selected polyurethane formulations as impregnants and coatings. The base nonwoven can be used as a substrate for other waterproof, air-permeable coated fabrics as well, using a vinyl formulation for coating. The nonwoven mat blend using fibers of a very fine denier, a portion of which is a high shrinkage fiber, is made either by carding and cross-lapping or by air-laying. A very light preneedling on one surface to give the web integrity for conveying without distortion is followed by intense needlepunching from both outer surfaces (400–1800 punctures/cm^2) to densify the web and give it leatherlike fiber orientation. To give the material density to act as a leatherlike substructure, further densifying is accomplished by passing the fabric through either wet or dry heat and shrinking it 30–50% in the machine direction. Then the material is impregnated and/or coated to form a synthetic leather.

2.6 Hydraulically Entangled Webs

Hydraulic entanglement is the process of entangling individual fibers with each other by using water jets with varying degrees of water pressure. The higher the water pressure, the greater the degree of entanglement. Depending on the backing substrate used during the entangling process, one can get a product that has a uniform surface or one that is apertured. Most coating substrate end uses require a uniformly surfaced product. In a sense one can consider hydraulic entangling to be a hydraulic needlepunching process. While hydraulic entanglement produces a material that is flatter and thinner than a needlepunched one, the bond formed has a lot in common with that produced by needlepunching. It is more like an anchor point than a rigid bond point, and the fabric made—like some needlepunched fabrics—tends to have low elastic recovery.

2.7 Wet-Lay Mats

The majority of the fibers used in wet-lay processing are in the length range of 0.3–0.6 cm. In papers, the fiber used is mostly wood pulp. In nonwovens, synthetic fibers, glass fibers, mineral fibers, and regenerated cellulose fibers are used as well as wood pulp. The ones most used are polyester, polyolefin, glass, and rayon; occasionally polyamide, aramid, vinyl chloride–vinyl acetate copolymer, and acrylic are also used. More and more, however, longer lengths are being introduced to increase the strength, cohesiveness, and toughness of wet-lay nonwovens. As in dry-lay processing, the fibers have to be opened, dispersed, and carried to the web forming stage. The difference is that in dry-lay the transportation medium is mechanical (carding or garnetting) or air, whereas in wet-lay the transportation medium is water. This process, wet-lay, resembles more the air-lay than the mechanical process. In both, the fiber is injected into, dispersed in, and carried by a fluid medium to a flat or a rotary screen, where the fluid is extracted and the fiber deposited onto the screen as a randomly oriented mat.

There are three main processes for wet-lay structuring: flat wire, inclined wire, and rotary wire cylinder. The one to use depends largely on the mat weight needed, the drainage properties of the fiber, and the fiber diameter. Wet-lay mats commonly used as coating substrates are bonded by using latex or by thermal bonding. In latex bonding the latex is either added to the fiber dispersion before web formation or flooded onto the wet-formed web in a second wet deposition step on the wire as the wet fiber mat goes by the second flooding station. These steps are normally followed by through-air drying and a series of dry cans to complete the process. Thermal bonding is accomplished by adding a thermal binder fiber to the initial fiber blend and following wet web deposition and drying with a hot calendering step to activate the binder fiber. Because of the multiplicity of fibers present in the mat due to the short length of most wet-lay fibers, the mats are of extremely good quality and are isotropic. Wet-lay mats tend to be very even, flat, and thin, and they can have MD and CD strength that is quite high, with a stiff or soft handle, depending on the type and amount of binder present.

2.8 Stitchbonded Materials

Stitchbonding is the process of bonding a web by using stitching yarns, filaments, fibers, or just the stitching needles themselves to do the bonding. In many respects, a stitchbonded fabric resembles a woven fabric in its properties, and that is because its construction is similar in many ways to a woven fabric. In some ways, it also resembles a needlepunched or hydraulically entangled fabric, especially in the lighter stitching constructions and when no yarn or thread is used in stitchbonding.

3.0 END-USE APPLICATIONS

3.1 Home Furnishings

3.1.1 Tablecloths

The nonwoven substrate for tablecloths could be a spunbond, a needled material, or a hydraulically entangled one. The substrate would be coated with a nonblocking vinyl formulation, dried, then color printed and/or embossed for aesthetic reasons. Properties needed in the substrate are softness, flexibility, isotropic strength, and puncture or snag resistance. The substrate could be a polyamide spunbonded with an acrylic latex binder, a needled polyester–rayon blend, or a hydraulically entangled polyester in the 45–55 g/m^2 range.

3.1.2 Upholstery

A typical nonwoven substrate for vinyl upholstery would have the following specifications: weight, 140 g/m^2; composition, 100% polyester fiber; process, needlepunching an air-lay or a carded cross-lapped material; tensile strength, 35–45 lb/in. MD and CD; and extensibility, MD 100%, CD 50%. Depending on end-use requirements, the substrate could also be composed of polypropylene, and it could be made using hydraulic entanglement or spunbonding as the bonding process.

3.1.3 Draperies

A suitable drapery composition would be a stitchbonded fabric coated with a pigmented opaque acrylic foam as an inner liner. Also acceptable would be a hydraulically entangled polyester used as a liner and reinforcement for a woven fabric (i.e., a flame-retardant drape made using a woven fiberglass fabric and a hydrolically entangled vinyl chloride base fiber web as a liner.

3.2 Construction Uses

3.2.1 Vinyl Floor Tile Reinforcement

A spunbonded, heat-bonded, copolyester fabric, weighing 50 g/m^2, with an MD and CD tensile strength of 35–40 lb/in., and an extensibility of 15%, MD and CD, is suitable as a floor tile base.

3.2.2 Roofing Membrane Substrate

A spunbonded polyester using an acrylic binder for heat resistance, instead of a copolymer thermal binder fiber system, in a weight range 25–100 g/m^2, is recommended as a roofing membrane substrate.

3.3 Automotive: Landau Tops, Interior Paneling, and Car Seats

A needlepunched, 100% polyester nonwoven laminated to a polyurethane foam on the bottom surface can be used as a coating base for a vinyl-coated fabric that is to be color printed and/or embossed. Substrates for tops are usually heavier, stronger, and less extensible than those used as substrates for car seats or door paneling. The needlepunched fabrics are made by carding followed by cross-lapping to bring the web up to the required weight, or they are made by the air-lay process. Substrates for landau tops are in the 225 g/m^2 weight range, while those for interior paneling and car seats are in the 140 g/m^2 range.

3.4 Consumer Products

3.4.1 Outer Garments, Handbags, and Luggage

Substrates for outer garments and luggage could be needlepunched fabrics with a high needled density to give good fiber entanglement and high strength, or highly hydraulically entangled materials with sufficient strength and weight for the particular end use. High tenacity polyester fiber would be preferred for use in these two instances because of the wear and abrasion associated with outer garments and luggage. Handbags can be made using hydraulically entangled materials with somewhat lower tensile strength, weight, and bonding, since physical property requirements for handbags are less stringent.

3.4.2 Clothing Labels and Wash-and-Wear Labels

Clothing labels are normally made with a medium to lightweight polyester spunbond that is impregnated and coated with a filled acrylic. The coating formulation and printing inks are selected for maximum printability, print durability, and fastness to washing and dry cleaning.

3.4.3 Footwear

A polyester-based poromeric material impregnated and coated with a suitable polyurethane formulation is an excellent leather substitute for shoes of all types.

3.5 Filtration: Microporous Membrane Substrates

Nonwoven fabrics per se are used in large volume as filtration media of all types. When used as coating substrates, their function is as a supportive reinforcement for a host of microporous membrane materials. Micromembranes, by the nature of their functionality, and for maximum filtering output, need to be as thin and as even as possible. To be ideal as a support and reinforcement, the substrate needs to have maximum isotropicity, evenness, and strength. Materials likely to be used as substrates for microporous membranes are (a) polyolefinic and polyester spunbonds, thermally bonded, made from fine denier fibers, with a high percentage of binder fiber and high calendering to achieve a very smooth surface for coating, and (b) wet-lay polyester or polyolefin mats with suitable latex or thermal binder systems, fine denier fibers, and calendering as the final finishing step to effect a smooth and even coating surface.

3.6 Industrial Applications

3.6.1 Tape Base

A 100% polyester unidirectional carded web containing a fiber binder and heat calendered to effect a high degree of bonding for strength and low extensibility makes an ideal tape base. In addition, it requires low MD and CD heat relaxation shrinkage (1–2%) to prevent curling of the coated product in the drying tower. Proper selection of the carrier fiber and the binder/fiber ratio are extremely important in these products.

3.6.2 Coated Papers

Lightweight polyolefinic and polyester spunbonds are suitable for use as coated papers when the coatings contain filled binders designed to give good wet strength, printability, and good abrasion resistance.

3.6.3 Flocked Fabric Substrate

Substrates being used for the flocked fabric market are lightweight spunbonded polyolefins and polyamides, latex-bonded, carded, calendered materials of various compositions, hydrolically entangled fabrics, and wet-lay mats.

3.6.4 Electrical Insulation

Unidirectional, carded, highly calendered, thermally bonded 100% polyester webs containing a high ratio of binder to carrier fiber are used extensively as electrical insulation. Impurities, which are electrically conductive, cannot be tolerated even in trace amounts in this

product. Evenness in thickness and weight, both crosswise and lengthwise, is very important for further processing (impregnation with a resin and/or lamination to a polyester film)
as well as for end uses in motor and transformer windings and slot insertions. Carrier fibers
and binder systems are constantly reassessed to improve higher temperature end use, which
is directly related to greater motor efficiency. Most products are in the weight range of
20–100 g/m^2.

BIBLIOGRAPHY

Hoyle, A. G. "Properties and characteristics of thermally bonded nonwovens," in *TAPPI Nonwoven Division Workshop on Synthetic Fibers for Wet Systems and Thermal Bonding Applications*, Technical Association of the Pulp and Paper Industry, 1986, pp. 51–54.

Hoyle, A. G. "Bonding as a nonwoven design tool," in *TAPPI Nonwovens Conference*, 1988, pp. 65–69.

Nonwovens Industry. Rodman Publications Inc., 26 Lake Street, P.O. Box 555, Ramsey, NJ 07446; phone (201) 825–2552.

Nonwovens World. 2700 Cumberland Parkway N.W., Suite 530, Atlanta, GA 30339; phone (404) 432–3186.

Textile World. McGraw-Hill Publications Company, 1221 Avenue of the Americas, New York, NY 10020; phone (212) 391–4570.

Index